# PRINCIPLES OF HEMATOLOGY

# PRINCIPLES OF HEMATOLOGY

*Peter J. Haen*

*Loyola Marymount University*

edited by
Linda Harris-Young

Book Team

Editor *Colin H. Wheatley*
Developmental Editor *Kristine Noel*
Production Editor *Carla D. Kipper*
Art Editor *Miriam J. Hoffman*
Photo Editor *Lori Hancock*
Permissions Coordinator *Vicki Krug*

**Wm. C. Brown Publishers**
A Division of Wm. C. Brown Communications, Inc.

Vice President and General Manager *Beverly Kolz*
Vice President, Publisher *Kevin Kane*
Vice President, Director of Sales and Marketing *Virginia S. Moffat*
Vice President, Director of Production *Colleen A. Yonda*
National Sales Manager *Douglas J. DiNardo*
Marketing Manager *Craig S. Marty*
Advertising Manager *Janelle Keeffer*
Production Editorial Manager *Renée Menne*
Publishing Services Manager *Karen J. Slaght*
Royalty/Permissions Manager *Connie Allendorf*

**Wm. C. Brown Communications, Inc.**

President and Chief Executive Officer *G. Franklin Lewis*
Senior Vice President, Operations *James H. Higby*
Corporate Senior Vice President, President of WCB Manufacturing *Roger Meyer*
Corporate Senior Vice President and Chief Financial Officer *Robert Chesterman*

Copyedited by Bea Sussman

Cover and part opening photo (scanning electron micrograph of normal red blood cells): © David M. Phillips/Visuals Unlimited

The credits section for this book begins on page 433 and is considered an extension of the copyright page.



A Times Mirror Company

Library of Congress Catalog Card Number: 94–70283

ISBN 0–697–13301–X



Printed in the United States of America by Wm. C. Brown Communications, Inc., 2460 Kerper Boulevard, Dubuque, IA 52001

10 9 8 7 6 5 4 3 2 1

*This text is dedicated to my wife, Annette, with love and appreciation*

# CONTENTS

## Chapter Five

*Structure of Hemoglobin 54*



## Chapter Six

*Functions of Red Blood Cells 68*



## Chapter Seven

*Metabolism and Fate of Red Blood Cells 78*



## Chapter Eight

*Introduction to Anemias 87*



## Chapter Nine

*Aplastic Anemias 97*



## Chapter Ten

*Megaloblastic Anemias 103*

## Chapter Fifteen



# SECTION THREE



## Chapter Sixteen



## Chapter Seventeen



## Chapter Eighteen



## Chapter Nineteen

## SECTION FOUR

# PREFACE

This book was primarily written as an upper division level undergraduate text for students who are planning careers in the health sciences. It is aimed especially for those individuals who want to become Medical Technologists and for students who plan to go to medical school.

I am aware that there are a good number of textbooks on hematology on the market. However, most of them contain too much material for a one semester introductory course in this subject area. Those textbooks tend to overwhelm the student with the tremendous amount of information found in them. Other hematology texts are written by a group of authors. Those books tend to have a lot of overlap and uneven language. Again, other books frequently lack a simple pattern of organization.

In writing this book I have tried to produce a text that does not overload the student with information. After working with a raw copy of this text for several years, I have come to the conclusion that the subject matter in it can be taught in a one semester course. Furthermore, I have tried to keep the language as simple as possible, so that the average student can read this text without the constant need of a dictionary. Finally, the organization of this book is also kept as simple as possible.

## ORGANIZATION

This textbook is divided into five sections. The first section contains introductory chapters. It starts with a discussion of the scope and history of the science of hematology. This is followed by an overview of the composition of blood, and it finishes with a description of the process of the formation of blood (hematopoiesis).

The second section starts with a discussion of the structure and functions of red blood cells. This is followed by an introduction to the concept of anemia. Finally there are a number of chapters that deal with the diseases of erythrocytes.

The third section has the same organization as section two. It starts with a description of the structure and functions of white blood cells and this is followed with a discussion of the major groups of disorders of leukocytes.

Section four starts with a description of the structure and functions of the components of hemostasis: blood vessels, platelets, and coagulation factors. This is followed by a discussion of the major disorders of hemostasis. The final section of this text is an appendix that contains a description of the most common procedures used in the hematology laboratory.


## ACKNOWLEDGMENTS

To begin with, I would like to thank a few people from the Wm. C. Brown Publishers Company who have been instrumental in getting this book into print. I would like to thank Mr. Colin Wheatley, Acquisitions Editor, who got me started on this project, for his faith in me and for his constant encouragement. My thanks also goes to Ms. Jane DeShaw, Developmental Editor, who followed the manuscript step-by-step and kept phoning me until the first and second drafts were finished. And finally, thanks to Ms. Kris Noel, Developmental Editor, who helped me with the final touches so that the book could get into production.

I would also like to thank all the reviewers of the manuscript: Kay Doyle, University of Massachusetts Lowell; Fred M. McCorkle, Central Michigan University; Keila Poulsen, Eastern Idaho Regional Medical Center; Helen M. Sowers, California State University-Hayward; Linda Harris-Young, Motlow State Community College. Their comments and criticisms have been very helpful in making this a better book. However, any errors and mistakes are mine only.

Finally, many thanks to my family, my Dean in the College of Science and Engineering, Dr. Gerald Jakubowski, and my colleagues in the Biology Department of LMU for their constant support and encouragement.



Peter J. Haen, Ph.D.
Loyola Marymount University
Los Angeles

## EDITOR'S NOTE

This book, *Principles of Hematology,* was dedicated by the author to his wife and companion, Annette. Since Peter passed away during the production of the book, it will serve as a fitting reminder to Annette and to their sons of Peter's love and devotion over many years of married life. I cannot and would not change Peter's apt dedication; however, this work will indirectly serve as a dedication to the memory of Dr. Peter J. Haen, a scholar, esteemed colleague and educator, and mentor and friend to many students at Loyola Marymount University.

Peter developed cancer near the completion of his work on the manuscript. As would be expected, he suffered with pain and discomfort; still he remained committed to finishing the book and pressed forward with the work when others would have perhaps given up. Peter was not only committed to finishing the book, but he was committed also to fighting the cancer. His was a courageous and determined fight. And, thus, this work on a subject that Peter dearly loved and taught for many years will serve as a monument to exemplary character and to the many outstanding qualities we all admired about him.

During production we were fortunate to enlist the assistance of Dr. Linda Harris-Young of Motlow State Community College, who oversaw the proofing, editing, and completion of the work in production. Before coming to Motlow State Community College, Dr. Young taught hematology, biology, medical microbiology, immunology, and environmental microbiology for many years at Jacksonville State College in Jacksonville, Alabama. She not only has a strong academic background, but she has had years of experience as a consulting microbiologist, researcher, and medical technologist. She is a member of the American Society for Microbiology, as well as other professional societies. Without her assistance we could not have continued, and we thank her for the dedication and contributions she brought to this work.

The Editor

# SECTION ONE

## *Background*

# Introduction

## *Definition and Scope of Hematology*

## DEFINITION OF HEMATOLOGY

Hematology may be defined as the scientific study of the structure and functions of blood in health and in disease. Although hematology is primarily a medical specialty, blood as a tissue is studied by many scientists, including a number that are not directly interested in its clinical aspects. For example, anthropologists, geneticists, biochemists, physiologists, immunologists, and taxonomists study the form and function of blood cells and plasma for a wide variety of objectives. The reasons why these scientists select blood rather than other body tissues for their investigations are many and varied, but there are four dominant reasons. First and foremost, it is easy to draw a sample of blood without interfering with the normal functioning of the organism. Second, it is easy to get more of the same blood at a later date, if needed. Third, and perhaps even more important, a blood sample provides us with a tissue that is still alive and capable of survival outside the body—without special arrangements for days or even weeks—with minimum treatment. Blood cells bring their own natural environment with them in the form of plasma, so the cells do not need to be maintained in expensive culture media at special temperatures. Fourth, blood consists of a wide spectrum of different cells—each group with its own structure and functions and unique genetic markers on its membranes and inside its cells. This contrasts with other tissues that consist of only two or three different cell types at most and that do not have their own fluid stroma.

## HEMATOLOGY AS A BIOMEDICAL SPECIALTY

Biomedical scientists are interested in blood mainly from a homeostatic point of view. Much of modern hematology is concerned with deductions based on changes in the amounts of cellular elements and variations in the concentrations of solutes in the circulating blood. Since blood forms an integral part of all other body tissues, any change in blood homeostasis will have an effect on the other tissues. For example, in iron deficiency anemia, the lack of iron results in the production of fewer and smaller red blood cells, and as a result less oxygen is carried to the tissues. This in turn has an effect on the patient, who will experience general weakness, shortness of breath, palpitations, and fatigue because other body tissues can no longer function optimally.

Further, since blood is in such close contact with all other body tissues, any change in these tissues will be reflected in the blood. For instance, people suffering from type I diabetes mellitus produce insufficient insulin due to nonfunctioning or malfunctioning endocrine glands in the pancreas, known as the islets of Langerhans. Insulin is a major hormone involved in regulating the level of blood sugar. Insufficient insulin results in a dramatic rise in the blood sugar level, which can be diagnosed by measuring the sugar content of the plasma or the amount of glycosylated hemoglobin.

Hematologists study the blood in health and in disease. They need to know what is normal before they can diagnose what is abnormal. Hematology therefore is a laboratory science in which we quantitatively and qualitatively observe the different components of blood in order to diagnose a great variety of diseases of humans and animals.

## SCOPE OF HEMATOLOGY

Hematologists study the normal and abnormal morphology and physiology of blood and blood-forming tissues. Anyone who wants to learn about hematology should become familiar with its different aspects, as follows:

**1. Origin and development of the various components of blood**

A thorough knowledge of the process of hematopoiesis (=blood formation), and hematopoietic tissues is essential before the causes of qualitative and quantitative variations of blood cells and many plasma proteins can be understood.

**2. Structure of the various components of blood**

A sound knowledge of the anatomy of the different blood cells and the histology of the various hematopoietic tissues is important. Normal shapes and sizes of the various cells and organs must be known before we can use abnormal size and shape as diagnostic parameters.

**3. Functions of the various components of blood**

A good understanding of the functions of the various blood cells and plasma proteins is necessary to correctly interpret the causes of anemias, leukemias, and other diseases of the blood.

**4. Regulation of levels of the various components of blood**

The modern-day science of hematology grew out of the concept of homeostasis, which holds that all elements of the blood are present in fairly exact quantities, and should remain that way, if a good state of health is to be continued. Maintenance of this internal fluid environment is regulated by a wide variety of signals; some are hormonal in origin, some are carried by nerve impulses, and other stimuli are produced in still different ways. The levels of some blood elements are monitored by the brain; other levels are monitored by the liver, the spleen, the kidneys, and other organs. The stimuli may affect the bone marrow, thymus gland, lymph nodes, and other hematopoietic tissues. In short, maintenance of blood homeostasis is an exceedingly complex process. One major aspect of the science of hematology is to study the regulation of the production of blood cells, platelets, and components of the plasma.

**5. Normal variations of blood components**

This aspect of hematology studies the limits of normal variations. It investigates the qualitative and quantitative differences in blood cells and plasma components between males and females, the old and the young, and well-fed and nutritionally deficient individuals. This aspect also observes changes that occur with pregnancy, variations that may result from environmental changes (e.g., changes in altitude), and so on. All these variations are studied to determine in each case what levels can be considered "normal." Only then can we determine what is "abnormal" and use these variations as diagnostic tools for the investigation of disease.

**6. Diseases associated with qualitative and quantitative variations of blood components**
One of the major interests of hematologists is to investigate the primary causes for abnormal blood patterns. They want to know whether the abnormalities are nutritional or hereditary in origin, are the result of some environmental factor such as chemicals, drugs, or radiation, or are due to some malfunction of the body, as would be the case with autoimmune hemolytic anemia. Once the causes are established (or even when the causes are not yet fully known), hematologists look for remedies and therapies to correct the abnormal blood pattern or at least to reduce its impact on the patient. Many clinical syndromes are due to qualitative and quantitative abnormalities of red blood cells. Diseases resulting from an abnormal shape or number of erythrocytes are collectively known as "anemias." Some are due to bone marrow failure (e.g., aplastic anemia), others are due to nutritional deficiencies (e.g., megaloblastic anemia), and still others are caused by hereditary factors (e.g., sickle cell anemia). Many diseases are also caused by the abnormal structure and function of white blood cells, resulting in a great variety of leukemias and other proliferative disorders, as well as a number of deficiency syndromes that result in impaired protection against invading pathogens.

**7. Understanding the hemostatic mechanism**
Hemostasis refers to the maintenance of the integrity of the blood circulation. Hemostatic mechanisms make sure that blood keeps moving properly to all parts of the body. The three major components of the hemostatic apparatus are (1) blood vessels, (2) platelets, and (3) plasma proteins. The lack of some plasma proteins may result in the movement of water from the blood vessels to the interstitial spaces of the tissues, resulting in a condition known as "edema." Other plasma proteins known as "coagulation factors" play a major role in plugging a hole in the blood vessels.

Maintaining the integrity of the circulation involves keeping the cells lining the blood vessels (endothelial cells) in place and functioning properly. This is mainly a function of the platelets. An intact endothelium prevents blood cells in the circulation from leaving at random and also prevents breaks in the blood vessels. In case a break does occur, due to chemical or mechanical injury, the hemostatic apparatus is mobilized to close the hole and to repair the damage within a short time. The hemostatic mechanism also prevents blood from clotting intravascularly. Abnormalities in one or more of the hemostatic forces will result either in bleeding disorders or in clotting problems (thrombosis). Hematologists study the mechanisms of hemostasis and the interactions of the various components of the clotting system in order to diagnose and treat abnormal bleeding or improper coagulation.

**8. Immunohematology and blood transfusions**
Since blood is such an essential component of life, many individuals may be saved when given blood donated by other people in cases of emergency, or during surgery, childbirth, or prolonged bleeding. However, we cannot transfuse just any blood. All blood cells have unique markers on their membranes that may prevent them from being accepted by certain recipients. Learning the blood type of both donor and recipient before a blood transfusion is attempted is important because a mismatch may have serious consequences for the recipient. The aspect of hematology that analyzes the presence of various markers on blood cells, and that investigates the effects of transfusion of blood, is known as immunohematology, and is another important branch of hematology.

## *Review Questions*

1. Define hematology. Give four different reasons why so many scientists study blood rather than other body tissues.
2. Explain why biomedical scientists are interested in blood mainly from the homeostatic point of view.
3. Provide two important reasons why the analysis of blood is such an outstanding diagnostic tool.
4. What major topics are studied in the science of hematology?
5. Define hemostasis. What are the three major components of the hemostatic mechanism?

# Chapter One

## *History of the Science of Hematology*

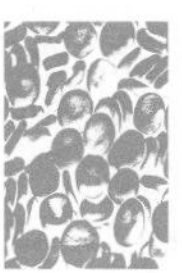

The scientific study of blood known as **hematology** started soon after the discovery of the microscope. However the interest in blood, and its relationship to health and disease, has been around since the beginning of humankind. From the earliest times onward, blood has been endowed with mysterious properties. Excessive blood loss was known to result in death, and therefore blood was suspected to be an essential part of life. Sublethal blood loss, however, gave one a chance to recover. Weakness was associated with loss of blood, but as the blood was replenished, so was the health and the strength of the person involved. No wonder that in ancient times, the ultimate sacrifice was the pouring of blood, symbolizing the giving of life. The drinking of blood was associated with strength and renewed life. These same symbols continue in some ceremonial aspects of Christianity. Even today the almost magical aspect of blood continues to influence our thinking. The Red Cross slogan "Give Blood, Give Life" is a good example.

## GREEK PERIOD

The Greek physician and philosopher **Hippocrates** is considered by many to be the father of the art and science of medicine (figure 1.1). One of the major contributions of Hippocrates was his analysis of blood, however faulty that analysis has been. Hippocrates wrote that the human body was made up of four elements: fire, air, water, and earth. To these elements corresponded the four "juices" of the body: blood, yellow bile, phlegm, and black bile (figure 1.2). Proof of the existence of these juices was found by Hippocrates in the condition of blood upon coagulation, when its component parts are separated (figure 1.3). In the lower, black part of the clot, he recognized black bile; in the upper part of the clot, which looked red, he saw the blood. Yellow bile was seen in the serum, and phlegm in the fibrin. The condition of the body was due to the existence and intermingling of these four elements and juices. If they existed in proper proportions, a person was in good health. However if the harmony between them was disturbed, sickness followed. Certain personality types were also associated with the dominance of one of these juices, and a person was classified as being either a sanguinicus (too much blood), a phlegmaticus (too much phlegm), a cholericus (too much yellow bile), or a melancholicus (too much black bile).

**Aristotle** (figure 1.4), generally considered to be the greatest of all biologists of antiquity, accepted the ideas of Hippocrates but added the concept of "coction," or cooking, as important in explaining the formation of blood. He theorized that foods were cooked in the stomach, and the vapors they produced ascended to the heart. This organ transformed the vapors into blood, and through the movement of blood, nutrients were transported to the body and assimilated. Further, he thought that the heart was the organ of the soul and intelligence, whereas the brain's purpose was to produce mucus and to prevent the blood from overheating by acting as a cooling system.

figure 1.1
Hippocrates, the Father of Medicine.

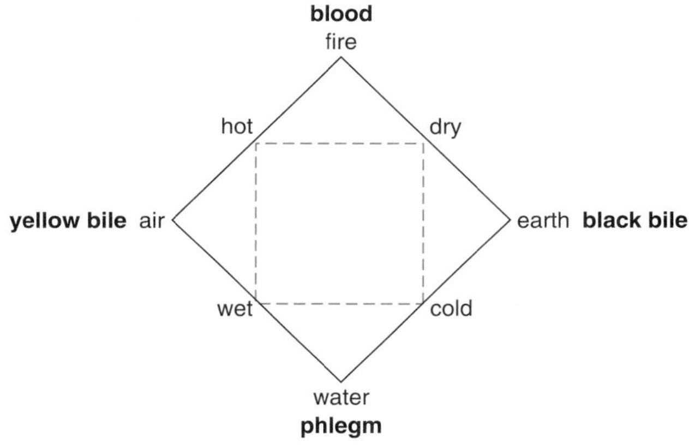


figure 1.2
Hippocrates' analysis: The basic elements and the corresponding juices of the body.

A contemporary of Aristotle, **Erasistratus** of Chios, theorized that the cause of all diseases was the condition of "plethora," which meant an excess of blood due to the presence of undigested food. According to Erasistratus this overproduction of blood could be cured only by starvation. He also studied the circulatory system in great detail and was the first to give us an accurate description of the heart and blood vessels. He traced the arteries and veins to the finest subdivisions visible to the naked eye, and he theorized that they were somehow connected by even smaller subdivisions. However he maintained with his contemporaries that the arteries contained only "pneuma" (air, spirit), and the blood was limited to the veins. This false concept is understandable

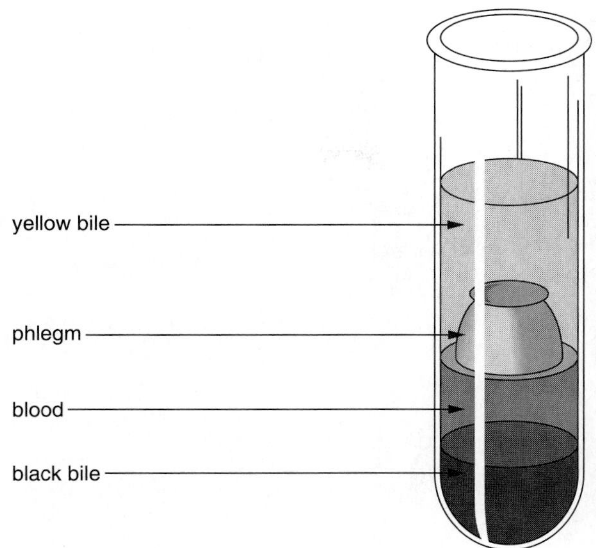


**figure 1.3**
Hippocrates' interpretation of the appearance of blood left standing until clotted. The dark mass on the bottom represents the black bile. The lighter colored mass above that represents the blood. The fibrin clot above the cell mass indicates the phlegm. The straw-colored fluid on top indicates the yellow bile.

figure 1.4
Aristotle (384–322 B.C.)

because when an animal is killed, blood spurts from the arteries—and by the time a dissection takes place, the thick-walled chambers of the heart and the major muscular arteries have pumped out all their blood and appear empty. Bleeding from arteries when injured was explained by the pneuma escaping through the hole in the artery and blood from the veins penetrating the arteries and then flowing from the wound.

## ROMAN ERA

Although the Romans conquered the Greeks, the Romans absorbed the Greek culture into their own civilization. Greek philosophy also had a great influence on Roman thought. The Romans were practical people and had the ability to apply the Greek discoveries in such areas as warfare, agriculture, administration, and governmental organization. However their original contributions to biology and medicine were small. Two contributors to the understanding of blood and the circulatory system were practicing physicians named Celsus and Galen.

**Aulus Cornelius Celsus,** a Roman nobleman who lived around the time of Christ, emphasized the importance of anatomy for medical practice. He knew that arteries contained blood under pressure, and not air, and described the danger of bloodletting by inexperienced people. He was very interested in infections and listed the four cardinal signs of inflammation: calor (heat), rubor (redness), tumor (swelling), and dolor (pain).

figure 1.5
Claudius Galen of Pergamum (129–199 A.D.)

**Claudius Galen** of Pergamum (figure 1.5) (A.D. 129–199) was the last great medical writer of antiquity. With his death, the creative period of Greek medicine came to an end, and his writings remained the unquestioned source of medical knowledge until the Renaissance. He was considered a man of great genius and produced hundreds of treatises on a variety of subjects. However he is most famous for his writings in the field of medicine. He is particularly famous for his description of the circulatory system and blood. He wrote that the products of digestion were transferred from the

figure 1.6
Leonardo da Vinci (1452–1519)

stomach to the liver by the blood vessels. In the liver these products were converted into blood. The useless parts of the food were absorbed by the spleen and converted into black bile, which was then excreted through the bowels. Excess water was removed from the blood via the kidneys. The concentrated blood was conveyed through the veins of the liver to the body and to the heart. Galen's work on the circulatory system was a strange mixture of physiological confusion and correct demonstration. Although he retired the faulty notion that the heart ventricles and the arteries contained only pneuma, his overall description of the circulatory system was less accurate than that of Erasistratus. But however confusing his ideas about blood and circulation were, they had a tremendous effect on the Western world, and were accepted without question by all physicians and biologists for about 10 centuries. Only with the arrival of the Renaissance spirit came any questioning of Galen's views. After much heated debate—and following the burning of several people at the stake for their correct beliefs regarding the origin and movement of blood—did the scientific world of that time finally admit that many of Galen's notions were faulty.

## RENAISSANCE PERIOD

After Galen's contributions, science and medicine entered the Dark Ages, which lasted until the fourteenth century, when the focus of thinking turned again from God to humans. The Renaissance was the period in which the human individual again became the central object of investigation. A new curiosity developed with respect to the human body and its functions. Among the pioneers of science and medicine were **Leonardo da Vinci** (1452–1519) (figure 1.6) and **Andreas Vesalius** (1514–1564). Around 1510 da Vinci produced a series of anatomical drawings, including some very detailed drawings of blood vessels. Sometime later Vesalius (figure 1.7) also produced an

figure 1.7
Andreas Vesalius (1514–1564)

figure 1.8
William Harvey (1578–1657)

anatomy atlas of great scientific accuracy that became the definitive textbook for many generations. Both men came very near to understanding the connection between arteries and veins, but they never grasped the concept of a closed system by proposing the existence of capillaries.

The first complete description of the circulatory system as a closed system was produced by the English scientist **William Harvey** (1578–1657) (figure 1.8). In 1628 he published his classic work, *De*

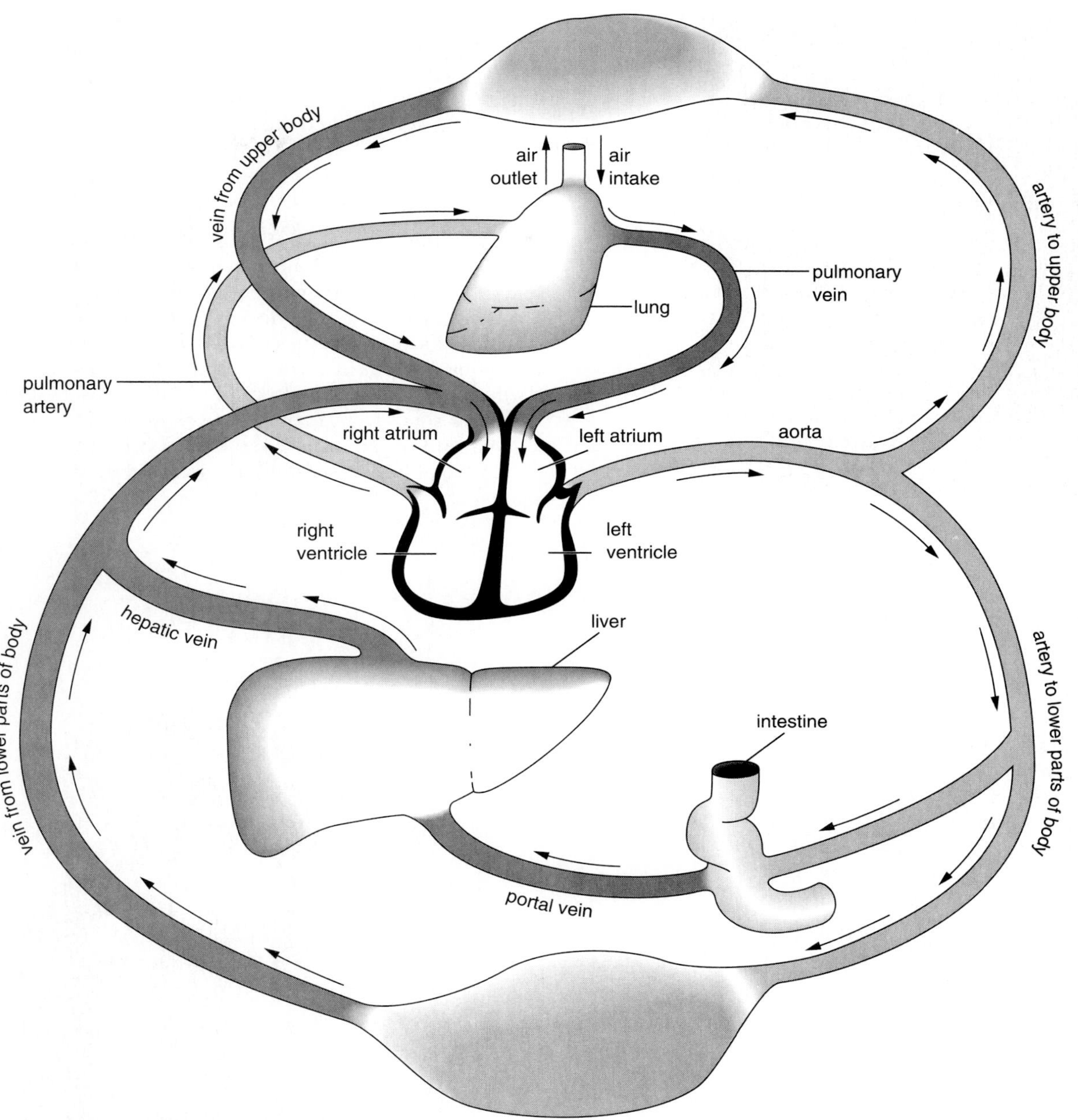


**figure 1.9**

The circulation of the blood according to William Harvey.

*Motu Cordis et Sanguinis* (On the Movement of the Heart and the Blood). In it he described the circulatory system as an arrangement where the blood flowed from the heart via the arteries to the capillaries in the tissues and back through the veins to the heart. This is illustrated in figure 1.9 and is essentially the same view we hold today. It must be noted, however, that Harvey never actually saw the blood moving through the capillaries. That observation had to wait until the development of the microscope sometime early in the seventeenth century.

The first complete description showing capillaries connecting the arteries with the veins was given by the Dutchman **Anton van Leeuwenhoek** (1632–1723) (figure 1.10). He looked through the transparent tail of a small eel with his microscope (figure 1.11) and wrote the following:

> "A sight presented itself more delightful than any mine eyes had ever beheld; for here I discovered more than fifty circulations of the blood in different places, while the animal lay quiet in the water, and I could bring it before my microscope to my wish. For I saw that not only in many places the blood conveyed through exceedingly minute vessels from the middle of the tail toward the edges, but that each had a curve, or turning and carried the blood back toward the middle of the tail, in order to be again conveyed to the heart. Hereby

figure 1.10
Anton van Leeuwenhoek (1632–1723)

it plainly appeared to me that the bloodvessels which I saw in the animal, and which bear the names of arteries and veins, are in fact, one and the same; that is to say, they are properly termed arteries so long as they convey the blood to the furtherest extremities of its vessels, and veins when they bring it back to the heart. And thus it appears that an artery and a vein are one and the same vessel, prolonged or extended."

The discovery of blood cells was first made by another Dutchman, **Jan Swammerdam** (1637–1680) (figure 1.12) in 1658. However this discovery is also attributed to van Leeuwenhoek, since his description of red blood corpuscles as elements of the blood predates the publication of Swammerdam's observation by nearly 50 years. Leeuwenhoek's descriptions are quite specific and comparative. He wrote in 1686 that mammals had round red blood cells, whereas those of frogs and various fishes were oval, and they also

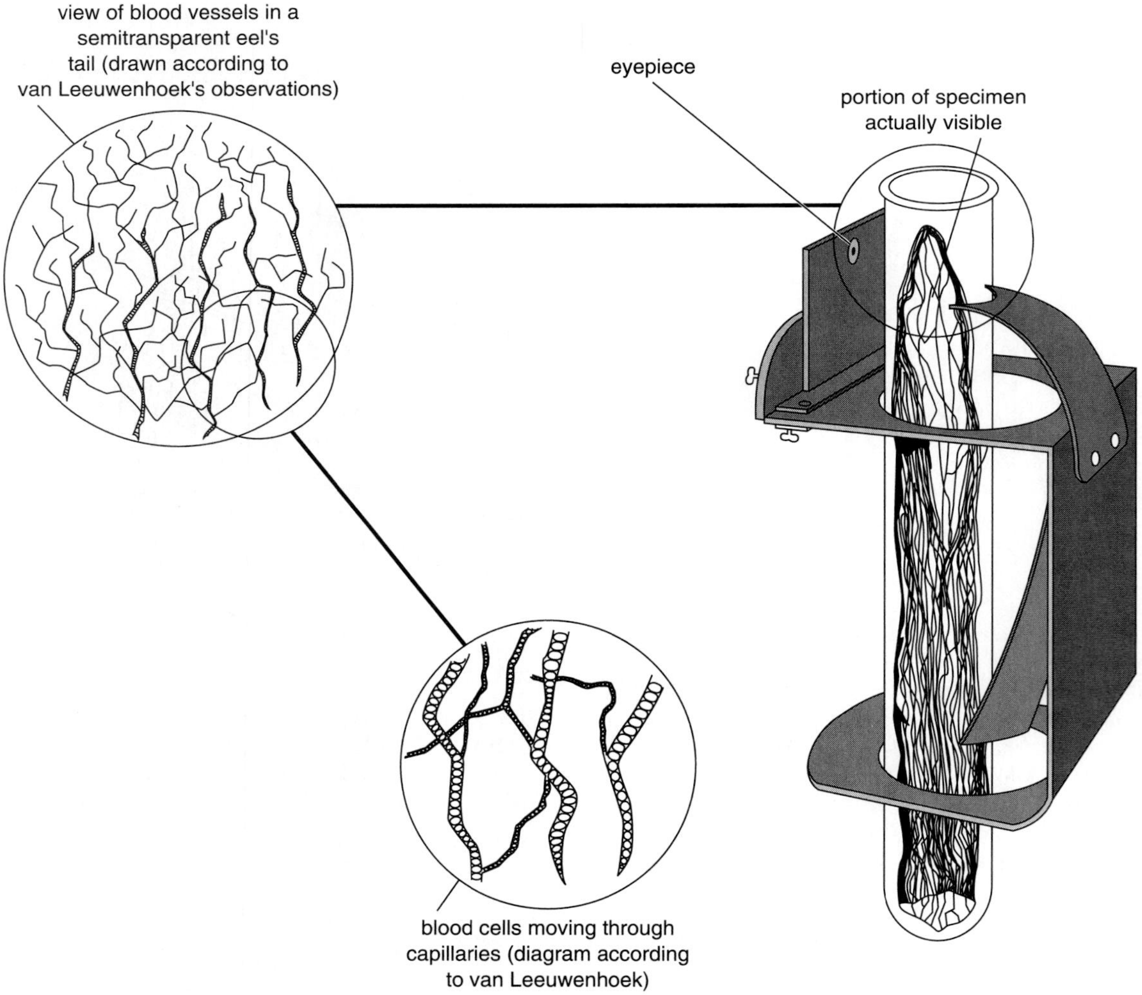

figure 1.11
van Leeuwenhoek's microscope, and what he saw when he viewed a tiny eel's tail through it. (c. 1670 setup with specimen in test tube)

figure 1.12
Jan Swammerdam (1637–1680)

had a "luminous spot in the middle" (a first description of a nucleus). Little more was discovered about blood and blood cells until the second half of the nineteenth century, when better microscopic studies were made possible with the introduction of fixatives and histological stains.

## NINETEENTH CENTURY

Although the existence of white blood cells was first noted by **William Hewson** (1739–1774), it was the German biologist **Paul Ehrlich** (1854–1915) (figure 1.13) who discovered that there were many different types of white blood cells. Ehrlich determined that the newly invented analine dyes had the capacity to brilliantly illuminate cells under the microscope. This discovery opened up a new era in histology and also laid the basis for modern hematology. By experimenting with various acidic and basic dyes, he described for the first time the different groups of white blood cells. Based on Ehrlich's work, other histologists started to experiment with various dyes and mixtures of dyes. In 1891 **Romanowsky** found that certain mixtures of acidic and basic dyes produced staining results superior to those obtained when these dyes were used sequentially. Many variations of the original Romanowsky stains have been developed since that time. The most useful and most widely used mixture is known as **Wright's stain.** This stain is a mixture of the acidic stain eosin and the basic dye methylene blue, with a solvent of methyl alcohol, which also serves as a fixative. These similar stains allowed scientists to distinguish between the various types of blood cells and to note abnormal inclusions in these cells, which are often associated with certain diseases. Since that time blood smears have been used extensively as an important tool in the diagnosis of diseases.

A real understanding of the functions of blood cells did not occur until the second half of the nineteenth century with the development of the science of physiology. It was **Karl Ludwig** (1816–1895) (figure 1.14) who first established that oxygen was absorbed during breathing and taken in by the red coloring matter (hemoglobin) in the blood corpuscles to the tissues. He noted that hemoglobin gave up part of its oxygen as the blood passed to the tissues and then returned to the lungs for a fresh charge. **E. F. Pflüger,** a student of Ludwig, showed that carbon dioxide was taken up by the blood from the tissues and discharged in the lungs.

figure 1.13
Paul Ehrlich (1854–1915)

figure 1.14
Karl Ludwig (1816–1895)

The discovery of hemoglobin as a respiratory pigment started a flurry of biochemical and biophysical investigations that have not abated to this day. Without a doubt, hemoglobin is the most investigated protein. Someone once said, What the rat has done for the understanding of physiology, hemoglobin has

figure 1.15
Johannes Müller (1801–1858)

figure 1.16
Claude Bernard (1813–1878)

done for biochemistry. We know more about its structure, functions, abnormalities, variations, and other characteristics than any other large biological molecule. The person who has contributed most to the understanding of the structure and function of the hemoglobin molecule is the English scientist **Max Perutz,** who received the Nobel Prize for his contributions.

The interest in the role of blood plasma also started with the rise of physiology. The great impetus to the study of physiology and consequently of blood, came from the German biologist **Johannes Müller** (1801–1858), (figure 1.15) who between 1833 and 1840 published his momentous work, *Handbuch der Physiologie des Menschen* (Handbook of Human Physiology). This became the definitive textbook for many generations and gave a great thrust to the growth and development of both biophysical and biochemical investigations of the body. He trained a number of gifted students who became famous scientists in their own right. Among them were **Theodor Schwann** (1810–1882), who made the discovery that all animal tissues are made up of cells; **Rudolf Virchow** (1821–1913), the father of the cell theory, which states that all cells are derived from previous cells; Karl Ludwig (previously mentioned), and **Hermann von Helmholtz** (1821–1896), who published important works of sight physiology. All these scientists did much to further the understanding of the structure and functions of blood.

The French physiologist **Claude Bernard** (1813–1878), who was a contemporary of Müller, also played a tremendous role in the advancement of physiology as a science (figure 1.16). Bernard made three great contributions to the science of physiology. First, he discovered the role of the nervous system in regulating the blood supply to the various parts of the body. Second, he promoted the idea of "internal secretion"—that is, that glands secrete substances directly into the blood. And last and most important, he produced the concept of homeostasis—that is, the stability of the internal fluid environment. He said, "La fixitée du milieu interieur est la condition de la vie libre," which freely translated means that in order for an organism to function optimally, its component cells must be surrounded by a medium of closely regulated composition. Since Bernard's death this principle of homeostasis has been overwhelmingly substantiated. It is now clear that the "interior milieu" of higher vertebrates is the subject of a multiplicity of complex regulatory mechanisms, and in consequence, its composition is controlled to within very fine limits. This fact is of central importance to the science of hematology, since much of the modern clinical diagnosis of diseases is based on the presence of qualitative and quantitative differences of the various components of blood and other body fluids.

The term "homeostasis" itself was first coined in 1929 by the American physiologist **Walter Cannon** (1871–1945) (figure 1.17), who established many of the normal limits pertaining to the composition of blood. What exactly is meant by homeostasis is perhaps best described by Cannon himself:

> "The constant conditions which are maintained in the body might be termed equilibria. That word, however, has come to have a fairly exact meaning as applied to relatively simple physiochemical states, enclosed systems where forces are balanced. The coordinated physiological processes which maintain most of the steady states in the organism, are so complex, and so peculiar to living beings—involving as they may, the brain and nerves, the heart, lungs, kidneys and spleen, all working cooperatively—that I have suggested a special designation for these states: Homeostasis. The word does not imply something set and immobile, a stagnation. It is a condition—a condition which may vary, but which is relatively constant."

figure 1.17

Walter Cannon (1871–1945)

## TWENTIETH CENTURY

Hematology came into its own as an important branch of the biomedical field in the twentieth century. Most of our knowledge of the structure and functions of blood has been elucidated for about the last 80 years. Understanding that many important diseases are the result of an abnormal structure of the various blood cells and plasma proteins has been properly articulated only in this century. With the acceptance of this concept came the idea of using variations in cell structure and blood plasma makeup as diagnostic tools for identifying a great number of clinical disorders. Today the analysis of blood is one of the most commonly used tools employed by clinicians to diagnose disease.

Scientific progress depends to a large extent on the introduction of new tools and techniques in the laboratory. This has also been true for the clinical laboratory. For about the last 50 years, many governments, private businesses, and universities have spent vast amounts of money on basic research, resulting in an explosion of scientific information. Further, the advances in pure research resulted in the development of many new techniques and instruments that could also be used by those in the applied sciences, including medicine. Development of microanalytic methods such as chromatography, electrophoresis, spectroscopy, X-ray crystallography, and microwave analysis have made it possible to isolate, purify, and identify many important compounds found in blood cells and plasma. The introduction of new microscopes including phase-contrast, fluorescent, and electron microscopes has allowed a close-up look at the various blood cells and blood vessels and helped us understand their functions and interactions. Improved tissue culture techniques allowed us to do many *in vitro* experiments that were impossible to perform *in vivo.*

New biochemical assays and immunoassays have made it possible to measure extremely small fluctuations in various body fluids. This in turn allowed us to better diagnose certain disorders. In short the explosion of scientific knowledge and technology in the twentieth century allowed hematology to develop into a mature science in its own right. The twentieth century also produced explanations for clotting and bleeding disorders. Although the existence of fibrinogen, and its conversion to fibrin clots, was known in the nineteenth century, it took until the last few decades to produce proper explanations for the process of coagulation. A number of mysteries still remain in this area; however most clotting factors and all the important coagulation pathways have been elucidated.

The subscience of immunohematology also started at the beginning of this century. The development of this branch of hematology is closely linked with the history of blood transfusion. The idea of blood transfusion has been around for a long time. With the almost magical qualities attributed to blood, it is not surprising that many people have tried in the past to perform blood transfusions. The first descriptions of blood-transfusion techniques appeared in the seventeenth century. **Christopher Wren,** better known as the architect of St. Paul's Cathedral in London, is said to have produced the first syringe. He used a small bird's quill as a cannula to which he attached a small bladder. This apparatus, somewhat analogous to a modern medicine dropper, was used to inject fluids intravenously. But because of problems of drawing blood in sufficient quantities without causing it to clot, few people were successful in transfusing blood. Those who did succeed frequently killed their patients in the process. As the result of a number of court cases and very little success, the enthusiasm for blood transfusions was never great until this century.

In 1886 **Jules Bordet** demonstrated that red blood cells could act as antigens and would produce antibodies when introduced in experimental animals. Around 1900 Paul Ehrlich showed that a similar response could be produced in individuals of the same species. The American scientist **Karl Landsteiner** (figure 1.18) proposed an explanation for this phenomenon. In a series of experiments performed around 1905, he discovered the existence of the ABO antigens on erythrocytes and proposed that all human beings could be divided into four major categories according to their blood type: A, B, AB, and O. In 1940 Landsteiner discovered the Rh system on red blood cells. For his work on the ABO antigens he was given the Nobel Prize in 1930. Discovery of the various blood groups, and the subsequent ability to determine to which blood group an individual belonged, made it possible to perform blood transfusions without serious consequences for the recipient. The first blood transfusions were performed at Chicago

figure 1.18

Karl Landsteiner (1868–1943)

General Hospital in 1937. The technique became quite popular during World War II, and it helped to save the lives of many severely wounded soldiers.

## FUTURE DIRECTIONS

Since there seems to be a constant shortage of usable blood for transfusions, a number of individuals have started a quest for a nonreactive, inert, artificial type of blood that would perform most or all of its functions without side-effects. Many prototypes of artificial blood have been tried in experimental animals. However, its use in humans has been limited because of the many side-effects. The first report on artificial blood used in a patient came in 1979, when the Japanese surgeon **Kenji Honda** gave one of his patients a liter of newly oxygenated perfluorocarbon emulsion because he could not obtain the rare O-negative blood. The patient survived the introduction of perfluorocarbons, which were removed sometime after the emergency operation. Since then perfluorochemicals have been used by a number of surgeons, but the quest for a perfect blood substitute is ongoing.

Finally, recent developments in immunology have opened up new avenues of treatment for such debilitating genetic diseases as beta-thalassemia major, certain forms of aplastic anemia and sickle-cell anemia, as well as certain types of leukemia. Many of these disorders could be cured if compatible bone marrow transplants could be given early in life. But finding compatible donors is frequently difficult. To surmount this problem, researchers have proposed transplanting stem cells—which give rise to all kinds of blood cells—into fetuses afflicted with these genetic blood diseases before they develop a functional immune system. Early animal trials have been highly successful, and human trials are currently under way. We can expect many more immunological breakthroughs in the treatment of hematological disorders in the future.

## *Further Readings*

Bendiner, E. The Man Who Did Not Invent the Microscope. *Hospital Practice,* Aug. 1984.

—. Andreas Vesalius: Man of Mystery in Life and Death. *Hospital Practice,* Feb. 1986.

—. Claude Bernard: Theater's Loss Was Physiology's Gain. *Hospital Practice,* Jan. 1979.

—. Ehrlich: Immunologist, Chemotherapist, Prophet. *Hospital Practice,* Nov. 1980.

—. Revolutionary Physician of Kings: William Harvey. *Hospital Practice,* Nov. 1978.

—. Renaissance Medicine: Alchemy, Astrology, Art, and Anatomy. *Hospital Practice,* June 1989.

Wintrobe, M. M. *Blood, Pure and Eloquent.* McGraw-Hill Book Company, New York, 1980.

# *Review Questions*

1. Why is blood associated with almost magical properties?
2. List the four basic juices of the body according to Hippocrates. How did he develop this classification?
3. According to Aristotle, where was blood formed? What was his concept of the role of the heart and the brain?
4. Why did the early Greeks think that the arteries contained pneuma?
5. Who first described the four cardinal signs of inflammation?
6. According to Galen, where was blood formed?
7. Who produced the first anatomical atlases describing in detail the position of the arteries and veins?
8. What was William Harvey's contribution to understanding the circulatory system?
9. Who first showed the existence of capillaries connecting the arterioles to the venules? How did he do it?
10. Who first described blood cells? In what century?
11. Who discovered the existence of different types of white blood cells? How was he able to distinguish between them?
12. What are Romanowsky dyes? What is the composition of Wright's stain?
13. Who discovered the function of hemoglobin as carriers of $O_2$ and $CO_2$?
14. Who received the Nobel Prize for his discoveries about the structure of hemoglobin?
15. Who first promoted the study of blood plasma?
16. What is the contribution of the French scientist Claude Bernard?
17. Who first coined the concept of homeostasis? What is it?
18. What was the great breakthrough concept, developed early in the twentieth century, that allowed hematology to develop into an important clinical science?
19. What are some important technical advances that allowed a rapid growth in understanding the structure and functions of the various blood components?
20. Who produced the first syringe? What did it look like? What was it used for?
21. Why did early attempts at blood transfusion usually result in severe side-effects and sometimes even in the death of the recipients?
22. Who discovered the ABO and Rh blood groups? When?
23. Why was it possible to perform successful blood transfusions only after 1940? Why did it become so popular a few years later?
24. Why do people keep looking for artificial blood? When was artificial blood first used on a human patient? What was the nature of that artificial blood?
25. What is a new way to treat young children and fetuses who have genetic hematological disorders such as homozygous sickle-cell anemia or beta-thalassemia major?

# Chapter Two

## *Composition of Blood*

Blood has been defined as a highly specialized tissue, which along with the circulatory system, is adapted to meet the needs of the body tissues and organ systems. These needs include such functions as exchange of gases, provision of nutrients, waste removal, and transport of chemical stimuli. Blood is composed of two phases: a liquid and a solid phase. The liquid phase is known as **plasma,** and the solid phase is made up of **cellular components** (table 2.1).

**Table 2.1 Components of the Blood**

Liquid phase: plasma
- Water
- Electrolytes (e.g., $Na^+$, $Ca^{2+}$, $HCO_3^-$, $Cl^-$)
- Proteins
- Miscellaneous: sugars, fats, vitamins, hormones

Solid phase: formed elements
- Erythrocytes
- Leukocytes
  - Granulocytes
    - Neutrophils
    - Eosinophils
    - Basophils
  - Lymphocytes
  - Monocytes
- Platelets

## PLASMA: THE LIQUID PHASE OF BLOOD

### Composition of Plasma

Plasma may be defined as a fairly complex mixture of proteins, electrolytes, and other chemical compounds, dissolved in water.

The most obvious component of plasma is *water,* constituting about 91% of the plasma. It functions not only as a solvent and a vehicle for the transport of blood cells and other components of the blood but is also extremely important in the temperature regulation of warm-blooded animals including humans. The metabolic processes carried out within the organism generate heat, and if this heat could not be disposed of properly, the temperature of the body would rise until it reached a stage no longer compatible with life. This excess body heat is lost in two ways: first, by direct radiation and second, by evaporation of water from the skin. The heat needed to change water from the liquid to the gaseous stage is provided by the body. This **latent heat of vaporization** keeps our body temperature constant. The only efficient way that water can be brought to the surface of the body is via the circulatory system.

Along with water, the **electrolyte** content of the plasma also plays an important role in the maintenance of the acid-base balance of the body. Most metabolic processes not only produce heat but also substantial amounts of acids, which must be appropriately neutralized and eliminated. Electrolytes and plasma proteins play a major role in buffering and transporting these substances to the organs of excretion. Some common electrolytes in the blood are shown in table 2.2.

About 7% of the plasma is made up of **proteins.** Knowledge of the number, structure, and functions of these plasma proteins has increased dramatically over the years. Currently several hundred different plasma proteins have been identified.

### Classification of Plasma Proteins

Plasma proteins may be classified in a number of ways. According to their function, they may be divided into such categories as enzymes, enzyme inhibitors, transport proteins, complement factors, immunoglobulins (or antibodies), coagulation factors, and fibrinolytic factors.

According to their structure they may be divided into two major categories: (1) simple proteins, made up of polypeptide material only and (2) complex proteins, which have additional materials added to their polypeptide core, such as a heme structure (in hemoglobins and cytochromes), or carbohydrate chains, producing compounds known as glycoproteins, or lipid material (lipoproteins).

Many years ago, when very little was known about the primary structure of proteins, they were classified according to their solubilities. At that time plasma proteins were classified into **albumins** (those soluble in water), and **globulins** (those not soluble in pure water but in dilute salt solutions). These terms have persisted to the present day in the clinical laboratory, where diagnosis of certain diseases is still based on variations in the albumin and globulin concentrations of the plasma.

Separation of proteins in a plasma sample is now commonly performed by a technique called **electrophoresis.** In this method a small amount of plasma is placed in a gelatinous medium, and an electrical current is passed through this medium. Since each protein has a different electric charge, each will move through the medium at its own characteristic speed. Most types of clinical electrophoresis use agar, polyacrylamide, or cellulose acetate media, and the typical separation time is less than 1 hour. During that time the plasma usually separates into five to six distinct fractions that can be stained with protein dyes. This staining allows each fraction to be measured quantitatively by a **densitometer.** A normal electrophoretic plasma protein pattern and scan is illustrated in figure 2.1. Note that there is usually only one albumin fraction and several globulin fractions. The albumin fraction is always the fastest-moving fraction, since it is made up of smaller proteins. Although there is

**Table 2.2 Electrolyte Composition of Blood (mEq/L)**

| Cations | Plasma | Anions | |
|---|---|---|---|
| $Na^+$ | 142 | $HCO_3^-$ | 27 |
| $K^+$ | 5 | $Cl^-$ | 103 |
| $Ca^{2+}$ | 5 | Protein | 16 |
| $Mg^{2+}$ | 3 | Miscellaneous | 9 |
| Total | 155 | | 155 |
| | **Erythrocytes** | | |
| $K^+$ | 125 | $HCO_3^-$ | 20 |
| $Na^+$ | 30 | $Cl^-$ | 74 |
| $Mg^{2+}$ | 12 | Miscellaneous | 73 |
| Total | 167 | | 167 |

usually only one albumin fraction, this fraction normally makes up about 60% of the total plasma protein content.

Globulins are usually divided into three groups according to the speed with which they move through the electrophoretic medium. These fractions are referred to as **alpha-, beta-,** and **gamma-globulins.** Most electrophoretograms show two alpha-globulin fractions, and occasionally they also show two beta-globulin fractions. However the gamma-globulin fraction, which is made up mainly of immunoglobulins (also known as antibodies), usually shows up as a single peak upon densitometry. As can be seen from table 2.3, each globulin fraction is made up of a number of plasma proteins, all with their own molecular weights and functions. More sophisticated forms of electrophoresis will allow each of these proteins under a common peak to be separated further. However for routine clinical analysis, the presence of five or six peaks, and especially the comparative height of each peak, is sufficient to produce a number of distinct electrophoretic patterns indicative of certain diseases (figure 2.2).

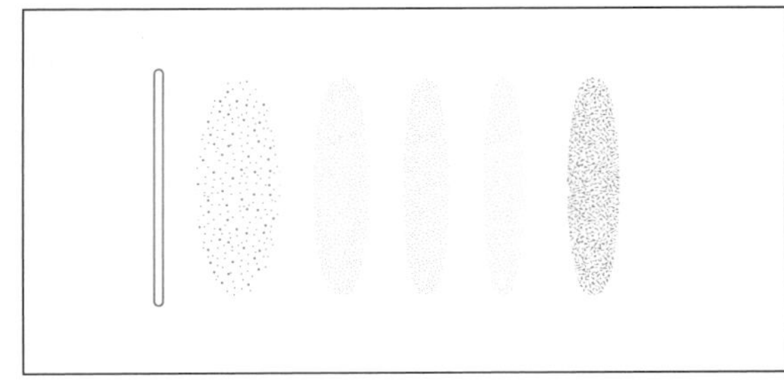


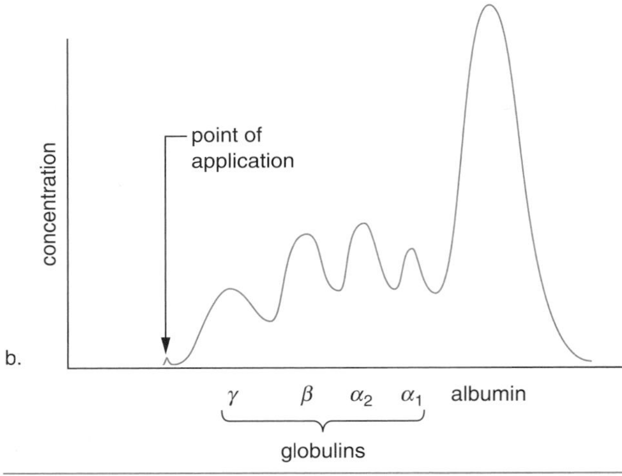


figure 2.1

(*a*) The electrophoretogram of human serum; (*b*) the densitometer pattern of the serum components.

## Functions of Plasma Proteins

Plasma proteins play important roles in the proper functioning and coordination of the body. As shown in table 2.3, plasma proteins have a large number of different functions, depending on the structure of the protein involved. Five of the major functions of plasma proteins are described as follows:

**1. Transport**

Many plasma proteins act as carriers for many simple and complex compounds involved in nutrition and hormonal integration. For many of these substances, the only way to reach their target organs is to become attached to these **carrier proteins.** This prevents them from becoming degraded before they reach their goal or before they have performed their function as a stimulus.

**2. Regulation of the movement of water between the intravascular and extravascular fluid compartments**

This is mainly a function of the **albumins.** In other words, albumins determine how much fluid stays within the blood vessels and how much moves into the interstitial spaces. The osmotic pressure of the albumins maintains the delicate balance between the two fluid compartments.

**3. Coagulation**

Certain plasma proteins play an essential role in the prevention of excessive loss of blood upon vascular injury by plugging any breaks in the circulatory system. This process is known as coagulation and is discussed in detail later in this book. The most important clotting factor is known as **fibrinogen** and can be found in the beta-globulin fraction of plasma. Fibrinogen is also found in large quantities in the blood: 200 to 400 mg/dL (=100 mL), whereas all other clotting factors are present in only minute amounts. When a clot occurs, the fibrinogen in the plasma is used up. When plasma is allowed to clot, the fibrinogen is changed to fibrin polymers, and the straw-colored, watery fluid left over is known as serum.

Closely linked to the coagulation factors are a series of plasma proteins known as fibrinolytic factors, which will prevent formation of spontaneous clots in the blood vessels. These factors also make sure that the coagulation process is limited to the area of the break in the circulation and does

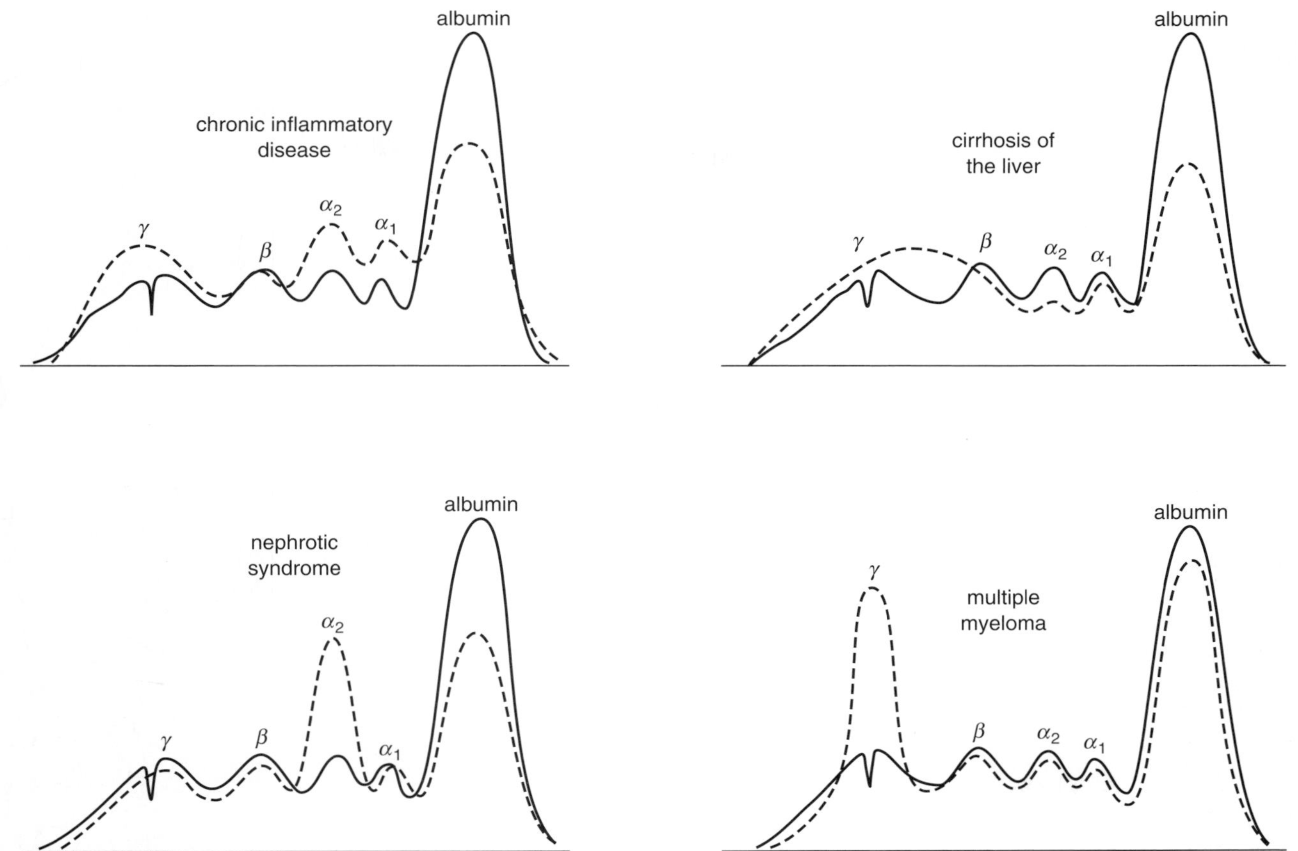


**figure 2.2**

Abnormal serum electrophoresis patterns associated with specific disorders. Dotted lines indicate the normal densitometer pattern of serum.

not spill over into other areas of the circulatory system. The most important fibrinolytic factor is known as **plasminogen,** which is normally classified as an alpha$_2$-globulin.

**4. Immunoglobulins**

As previously mentioned, the plasma contains a large number of **antibodies,** or immunoglobulins. These plasma proteins are produced by certain lymphocytes and play a major role in the body's defense against invading pathogens. Immunoglobulins are found mainly in the **gamma-globulin fraction** of the plasma.

**5. Inflammation**

Plasma also contains a number of proteins that play a major role in the production of inflammatory responses resulting from infection, allergy, or autoimmune disease. Many of these proteins are present in an inactive form but can be activated quickly by appropriate chemical stimuli.

## BLOOD CELLS AND PLATELETS: THE SOLID PHASE OF BLOOD

The solid particles of the blood may be divided into three major categories: (1) red blood cells, or erythrocytes, (2) white blood cells, or leukocytes, and (3) platelets, or thrombocytes (figure 2.3). The term thrombocyte is really a misnomer, since platelets are not cells at all but are small packages of cytoplasm nipped off from large mother cells in the bone marrow, known as megakaryocytes. They are much smaller than regular blood cells, only about 2 μm in diameter, and much more fragile than the average blood cell. Blood cells and platelets are uniquely designed for their functions in the organism.

### Red Blood Cells

Erythrocytes may be described as biconcave discs—that is, they are thicker at the edges than in the middle (figure 2.4). The average red blood cell is about 7.5 μm in diameter and about 2 μm thick at the edges. The essential role of erythrocytes is to transport oxygen to the respiring tissues and to carry carbon dioxide away from the tissues. Red blood cells are ideally designed both in form and content for these tasks.

Since the exchange of gases is the prime function of erythrocytes, it is clear a large cell surface area will facilitate this gas exchange. The biconcave shape allows a much larger ratio of surface area to content than, for instance, a round or globular shape. It is possible to design an even more efficient surface to volume ratio for the red blood cell by producing many

## Table 2.3 Summary of the Major Plasma Proteins

| Plasma Protein | Plasma Concentration (g/L) | Molecular Weight (daltons) | Functions |
|---|---|---|---|
| *Albumin* | 40.0 | 69,000 | Regulation and maintenance of plasma volume and distribution of extracellular fluid. Also functions as a carrier protein. |
| *$\alpha_1$-Globulins* | | | |
| $\alpha_1$-Antitrypsin | 3.0 | 45,000 | Anticoagulant effect |
| HDLs (high-density lipoproteins) | 0.5 | 200,000 | Lipid transport |
| *$\alpha_2$-Globulins* | | | |
| Ceruloplasmin | 0.4 | 160,000 | Copper transport |
| Haptoglobins | 1.2 | 95,000 | Types of glycoproteins that bind free hemoglobin to conserve iron. |
| $\alpha_2$-Macroglobulin | 3.0 | 800,000 | Anticoagulant effect |
| VLDLs (very low-density lipoproteins) | 1.0 | 10,000,000 | Lipid transport |
| *β-Globulins* | | | |
| Transferrin | 3.0 | 90,000 | Iron transport |
| Hemopexin | 1.0 | 80,000 | Binds ferriheme |
| $C_3$ ($\beta_{1C}$-globulin) | 1.2 | 220,000 | Component of complement pathway |
| $C_4$ ($\beta_{1E}$-globulin) | 0.4 | 240,000 | Component of complement pathway |
| Plasminogen | 0.7 | 140,000 | Fibrinolysis |
| Fibrinogen | 3.0 | 350,000 | Fibrin formation |
| LDLs (low-density lipoproteins) | 1.0 | 2,300,000 | Lipid transport |
| *γ-Globulins* | | | |
| Immunoglobulins | | | |
| IgA | 2.5 | 170,000 | Each Ig group has different antibody functions. They move mainly as γ-globulins on zone electrophoresis, but some migrate as β-globulins or as $\alpha_2$-globulins. |
| IgD | 0.03 | 180,000 | |
| IgE | Trace | 200,000 | |
| IgG | 10.0 | 150,000 | |
| IgM | 1.0 | 900,000 | |

folds and invaginations of the membrane. However this shape would interfere with the smooth passage of red blood cells through capillary blood vessels. The smooth round shape of erythrocytes produces little friction as they are squeezed through the narrow capillary tubes.

Regarding content, red blood cells are also ideally suited for their function. Each erythrocyte contains large quantities of hemoglobin. Hemoglobin is the respiratory pigment that allows oxygen and carbon dioxide to be carried in red blood cells, as is explained later.

Finally, not only the shape and content of red blood cells ensures an efficient exchange of gases in the tissues and lungs but also the number of erythrocytes present in the bloodstream. As can be seen in table 2.4, there are about 1,000 times more red blood cells than white blood cells in the bloodstream. The average woman has about 4.0 to 5.5 million erythrocytes per microliter (=1 $mm^3$) of blood. The average man usually has a little higher count (from 4.5 to 6.0 million per microliter of blood), and newborn infants usually have the highest count (5.0 to 6.5 red blood cells per microliter of blood).

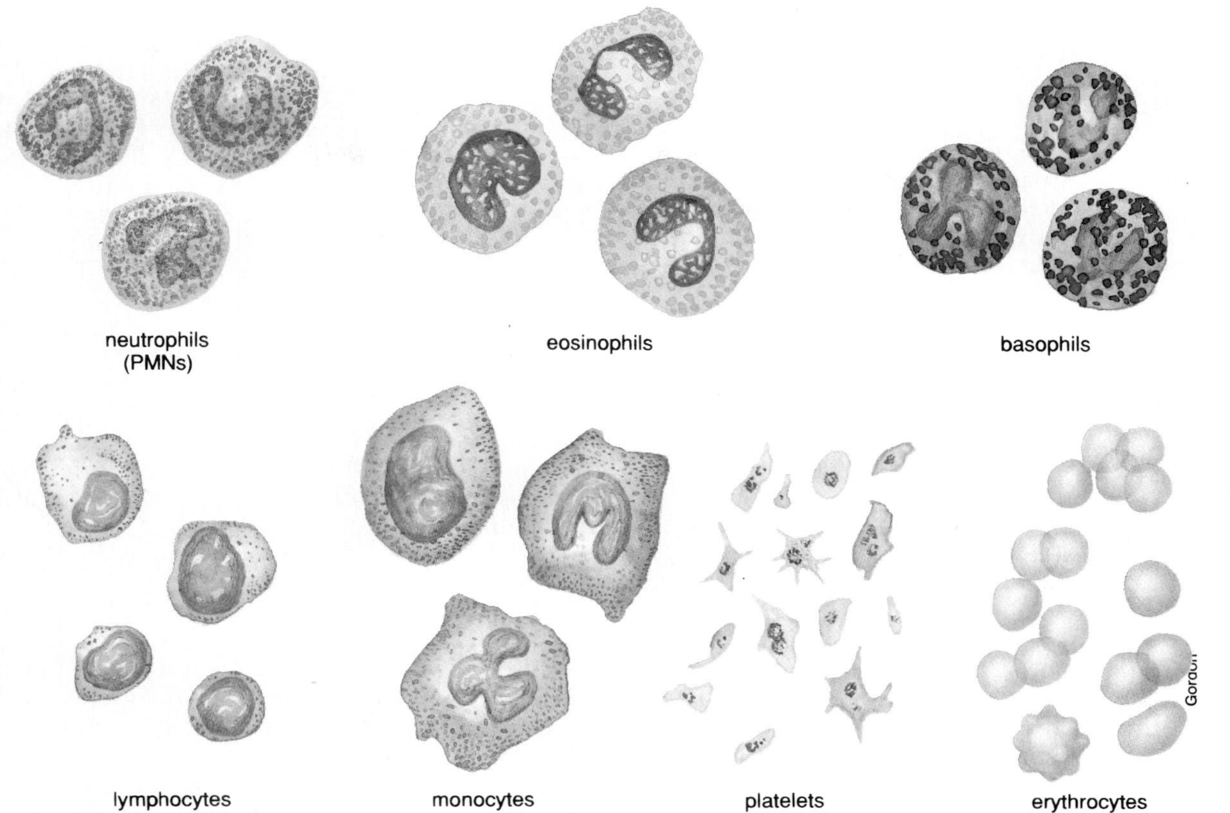


figure 2.3

The normal shapes of the formed elements of the blood: the various white blood cells (leukocytes), the platelets, and the red blood cells (erythrocytes).

## White Blood Cells

White blood cells, or leukocytes, form a heterogeneous group of cells. Although they are much less numerous in the blood than red blood cells, each microliter of blood is estimated to contain between 5,000 and 10,000 white blood cells. From the morphological point of view, leukocytes are traditionally divided into three major categories: granulocytes, lymphocytes, and monocytes (table 2.4).

### *Granulocytes*

Granulocytes are usually further subdivided into three groups according to the staining characteristics of their cytoplasmic granules. The three categories are neutrophils, eosinophils, and basophils—and each of these types of cells has its own unique functions.

**Neutrophils** are the most numerous white blood cells and comprise about 55 to 75% of the total leukocyte population in normal individuals. They comprise 90 to 95% of all granulocytes. Neutrophils are easily recognized in a stained blood smear (using Wright's

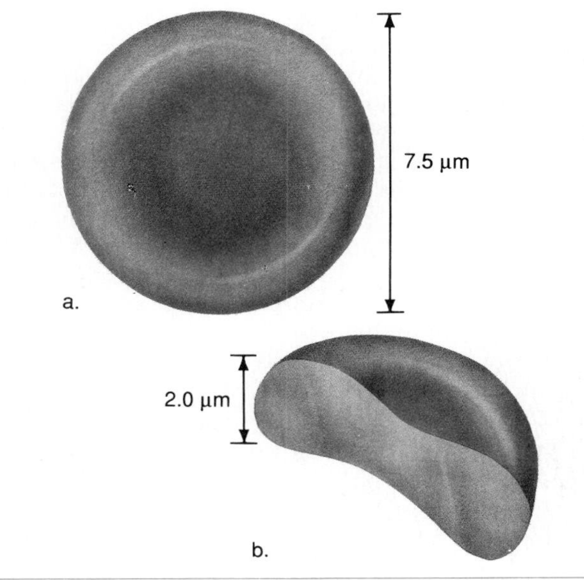


figure 2.4

The shape and size of a normal red blood cell. (*a*) Top view; (*b*) cross section.

**Table 2.4 Normal Values of Blood Cells and Platelets**

| | | |
|---|---|---|
| Erythrocytes | Women | 4.0–5.5 million/µL |
| | Men | 4.5–6.0 million/µL |
| | Newborn infants | 5.0–6.5 million/µL |
| Leukocytes | 5,000–10,000/µL | |
| Granulocytes | | |
| Neutrophils | | 55–75% |
| Eosinophils | | 1–3% |
| Basophils | | 0–1% |
| Lymphocytes | | 20–35% |
| Monocytes | | 2–6% |
| Platelets | 140,000–340,000/µL | |

stain) by numerous small purple granules in the cytoplasm and by a nucleus that is divided into three to five interconnected lobes (figure 2.3). Great variation exists in the shape of the neutrophil nucleus, and for that reason these leukocytes are also commonly referred to as **polymorphonuclear neutrophils,** or **PMNs.** They are phagocytic cells and consequently have amoeboid characteristics, although at rest they usually have a round to ovoid shape. Their function is to ingest and destroy invading pathogenic organisms. They are considered to be the first line of defense against microbes. PMNs are most active early in an infection, during the acute inflammation stage. Because of their relative short life span, they are not very efficient in destroying infectious agents. Their major task is to slow down the infection and to hold it locally, until the specific immune system, made up of lymphocytes and macrophages, is mobilized to effectively eradicate the pathogenic organisms.

The second group of granulocytes has been coined **eosinophils** because of the presence of many large red granules in their cytoplasm. These granules contain many basic proteins with a great affinity for the red acidic dye eosin, present in Wright's stain. Hence these granules stain bright orange-red. The nucleus of eosinophils does not stain as deeply as that of neutrophils, nor does it lobulate as much. Usually the nucleus of an eosinophil has only two lobes (figure 2.3). This type of granulocyte is also capable of phagocytosis, although ingestion of bacteria is not its major function. The main role of eosinophils is to attack parasitic worms, specifically the larval forms of schistosomes or blood flukes, one of the major groups of helminth parasites.

As shown in table 2.4, the number of eosinophils per microliter is much lower than that of neutrophils. In every 100 white blood cells counted in a blood smear, only 1 to 3 are classified as eosinophils.

The last group of granulocytes are the **basophils.** They are so named because their cytoplasmic granules stain a deep blue-black with the basic dye methylene blue, due to the fact that these granules contain many acidic molecules. Again the nucleus of basophils is not as obvious as in the PMNs, mainly because of the presence of many densely staining blue-black granules. In many cases the nucleus appears to be deeply indented rather than lobulated (figure 2.3). Basophils are the least common white blood cells in the bloodstream, making up less than 1% of the total leukocyte population. Basophils are not considered to be phagocytic as are the other types of granulocytes. The basophil has been called a **mediator cell,** together with a closely related and similar looking cell, the **mast cell,** found in connective tissue. Both these cell types have granules containing vasoactive amines, such as histamine and serotonin. Basophils and mast cells play a major role in infections and in certain types of allergies by producing states of acute inflammation.

### *Lymphocytes*

The second category of white blood cells is called the lymphocytes. They are the second most common group of white blood cells in the blood, making up 20 to 35% of the total leukocyte population. Lymphocytes are relatively small cells, only slightly larger than red blood cells. They are easily recognized in a stained blood smear by the presence of a large, densely stained nucleus surrounded only by a tiny amount of cytoplasm. In other words, the large round or oval nucleus almost completely fills the cell (figure 2.3). Usually a small number of larger lymphocytes are also present—that is, there is more cytoplasm surrounding the nucleus, and in those cases the cytoplasm may contain a number of densely stained reddish-purple granules. These lymphocytes are known as **large granular lymphocytes (LGLs)** and belong to a category of lymphocytes known as **natural killer cells (NK cells).** Inexperienced observers may sometimes confuse these LGLs with monocytes, another category of white blood cells, since they may look morphologically rather similar.

The functions of the lymphocytes are all concerned with the immune system. From the functional point of view, lymphocytes may be divided into three major groups: **B lymphocytes** (or **B cells**), **T lymphocytes** (or **T cells**), and **natural killer cells (NK cells).** Distinguishing between B cells and T cells in a blood smear is impossible from the morphological point of view. Most lymphocytes in the peripheral bloodstream may be classified as T cells; they make up 75 to 80% of all lymphocytes in the blood. B cells make up 10 to 15% of the blood lymphocytes, whereas less than 10% are NK cells. Simple techniques may be used to separate the various lymphocyte populations based on the nature of their surface markers. For instance, it was found accidentally that T cells have CD2 surface receptors for sheep red blood cells (SRBCs), B cells do not. When SRBCs are mixed with populations of lymphocytes, the sheep red blood cells will form a **rosette** around T cells but not around B cells

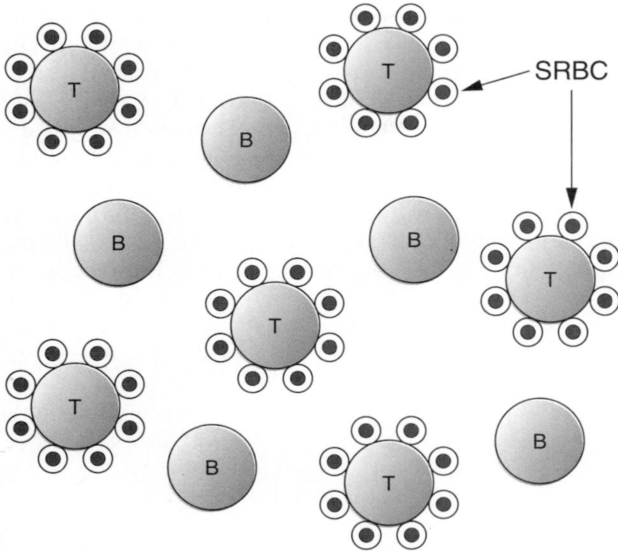


**figure 2.5**
T-cell rosettes that form when sheep red blood cells (SRBCs) are added to a mixture of T and B lymphocytes.

(figure 2.5). By counting the number of rosettes in the first 100 lymphocytes encountered under the microscope, one can find the exact percentage of T cells in a given blood sample.

The major function of B cells is to produce **antibodies.** They can be stimulated to produce antibodies against extracellular pathogenic organisms such as bacteria and certain viruses. The major functions of these antibodies are to kill and help in the elimination of these extracellular parasites. Hence the function of B cells is mainly in the area of **humoral immunity.** The reason why this group is referred to as B cells is that these lymphocytes were first described as originating from the bursa of Fabricius in birds. Later it was discovered that in mammals the bursal equivalent was the bone marrow.

There are two major types of T cells. One group of T lymphocytes, known as **cytotoxic T ($T_c$) cells**, is mainly responsible for **cell-mediated immunity (CMI)**—that is, they attack and destroy cells that have been invaded by viruses and other pathogenic organisms. They also attack neoplastic (=cancerous) cells. The other major group of T cells is known as **helper T ($T_H$) cells.** They play a major role in activating B cells and $T_c$ cells. The reason why this group of lymphocytes is called T cells is that they are all originally derived from the thymus gland.

Natural killer cells lack specific B- and T-cell receptors. They are called natural killer cells because they will attack most virus-infected and neoplastic cells without needing to be primed for attack by specific antigens, as is the case with B and T lymphocytes. As previously mentioned, NK cells can often be distinguished from B and T cells in a blood smear by the fact the NK cells frequently are larger and contain granules—hence the name large granular lymphocytes. They are best thought of as being a more primitive type of lymphocyte. NK cells are also derived from the bone marrow.

*Monocytes*

**Monocytes** are the final group of leukocytes. They are the largest of the circulatory white blood cells, with a diameter of 17 to 24 μm, which is two to three times that of erythrocytes. As previously mentioned, morphologically a monocyte may look quite similar to a large, immature lymphocyte. Both have a large nucleus, pale cytoplasm, and a number of granules. However the shape of the nucleus of the monocyte is usually quite distinctive from that of an immature lymphocyte. The large lymphocyte usually has a round or oval nucleus, whereas the nucleus of the monocyte is normally twisted into a C or S shape (figure 2.3).

The peripheral bloodstream usually contains relatively few monocytes. They make up only 2 to 6% of all white blood cells. Although they are few in number in the blood, they are plentiful in the tissues, where they are known as **macrophages.** Their role in the defense of the organism against the onslaught of invading pathogens is extremely important.

The functions of monocytes are many and varied. They are long-lived cells that are extremely efficient in the phagocytosis and killing of microbial organisms. They play a role in the activation of the immune system by engulfing and presenting the antigens (parts of digested pathogens) to the appropriate B cells and T cells, so that they can become activated and start producing antibodies and lymphokines (=products produced by T cells). These antibodies and lymphokines are usually most effective in killing the invading parasites. Because of this function, monocytes and macrophages are also known as **antigen-presenting cells,** or **APCs.** Another role of the monocytes is to clean up and remove dead pathogens and destroyed cells after an infection is over.

Monocytes are called macrophages in contrast with **neutrophils,** which are known as **microphages,** since they are usually much smaller. The PMNs normally have a diameter of about 12 μm and may be half the size of a macrophage or monocyte. Macrophages are also more efficient and effective as phagocytic agents than PMNs.

### Platelets

As mentioned earlier in this chapter, platelets are not cells but fragile packages of cytoplasm, much smaller than blood cells (figure 2.3). The average person has about 250,000 (or 150,000 to 400,000) platelets per microliter of blood. Platelets are usually round or oval and have a very invaginated surface area, which allows for the adherence of many compounds, especially clotting factors.

Platelets have a dual role in the circulatory system. First, they are instrumental in maintaining the integrity of the endothelial lining of the blood vessels. When the number of platelets falls below normal, the endothelial cells tend to shrivel up, and as a result the basement membrane surrounding the endothelium becomes exposed, thus promoting clotting. It seems that platelets play a nutrient role that allow the endothelial cells to maintain tight intracellular junctions.

Second, platelets play a major role in repairing any breaks in the vascular system, especially at the microcirculation level—that is, where the capillaries, arterioles, and venules are found. Any break in the endothelial lining of the blood vessel will expose the underlying basement membrane. Platelets have a great affinity for this membrane, and when given a chance they will adhere to it in large numbers. This starts the clotting process and will plug up the hole, as is described in a later chapter.

## *Further Readings*

Diggs, L. W., Sturm, D., and Bell, A. *The Morphology of Human Blood Cells,* 5th ed., Abbott Laboratories, Abbott Park, Ill., 1985.

## *Review Questions*

1. Define the following terms: blood, plasma, serum, albumins, globulins.
2. What percentage of blood plasma is water? List three important functions of the water in the plasma.
3. Name the major electrolytes in plasma. Describe their major function.
4. What percentage of plasma is made up of proteins? How would you classify plasma proteins according to their structure and their functions?
5. Define electrophoresis. What is the aim of this technique? Draw a normal serum electrophoretogram and its densitometer pattern. Label each of the peaks.
6. Mention some proteins associated with each of the major protein fractions of an electrophoretogram.
7. List five major functions of plasma proteins.
8. In a normal individual, what percentage of plasma proteins may be classified as albumins? What is the major function of albumins?
9. Describe the size and shape of a normal erythrocyte.
10. What are the normal levels of erythrocytes in (a) an adult male, (b) an adult female, (c) a newborn infant?
11. Explain why the shape of the red blood cell is ideally suited for its function.
12. Define hemoglobin. Describe its function.
13. List the three major types of white blood cells.
14. What is the normal level of white blood cells in the blood of a healthy adult?
15. What is the normal percentage of (a) neutrophils, (b) eosinophils, (c) basophils, (d) lymphocytes, and (e) monocytes in the blood of a normal healthy adult?
16. What are PMNs? What does the abbreviation stand for?
17. How can you distinguish between neutrophils, eosinophils, and basophils in a blood smear?
18. Describe the major function of (a) neutrophils, (b) eosinophils, and (c) basophils.
19. List the three major groups of lymphocytes. What is the normal level of each in the bloodstream?
20. What is the major function of B cells? What do they produce? What is the major function of these products?
21. Name the two major types of T cells. What is the function of each?
22. What is another name for NK cells? What is their major function?
23. How can we separate T cells from non-T lymphocytes in the blood?
24. How does the morphological structure of a monocyte differ from that of a neutrophil and a lymphocyte?
25. List some major functions of monocytes and macrophages.
26. What are platelets? From what cell are they derived? Where in the body are they formed? Why is the term thrombocyte a misnomer?
27. What is the normal level of platelets in the blood?
28. Describe the two major functions of platelets.

# Chapter Three

## *Process of Hematopoiesis*

## VARIOUS ASPECTS OF HEMATOPOIESIS

The process of the development of blood cells is known as **hematopoiesis.** This term is derived from the Greek words "haima," meaning blood, and "poiesis," which denotes making or creating. The morphogenesis of blood cells can be discussed from a variety of perspectives. One way to look at hematopoiesis is from the embryonic point of view **(ontogeny);** another way is to look at the evolution of blood cells in the animal kingdom **(phylogeny).** The development of blood is also commonly considered from the **histological** point of view and described through a series of maturational steps from primitive precursor cells to the final mature product that is released into the circulation. Since each of the different cell types follows its own specific pattern of differentiation, each of these cell lines will be discussed separately. The proliferation and differentiation of red blood cells is known as **erythropoiesis;** the development of white blood cells is commonly referred to as **leukopoiesis;** and the development of platelets is called **thrombopoiesis.**

All blood cells are ultimately derived from a common ancestral cell in the bone marrow known as the **pluripotent stem cell (PPSC).** This original precursor cell can move in either of two directions. It can develop into either a **lymphoid stem cell,** from which all lymphocytes are ultimately derived, or it can develop into a **myeloid stem cell,** which is considered to be the precursor of all other blood cells. The development of the early lymphocytes takes place in either the bone marrow (for B cells and NK cells) or in the thymus gland (for T cells). From these ancestral organs the more mature lymphocytes migrate to the spleen, the lymph nodes, and other lymphoid organs. These mature lymphocytes in the peripheral lymphoid organs can be stimulated to multiply and differentiate when they are exposed to foreign antigens. All other blood cells have in normal situations only one point of origin: the bone marrow. Studying the structure of the bone marrow, the thymus gland, and the peripheral lymphoid tissues is important for a good understanding of the various aspects of hematopoiesis. In this chapter the microanatomy of the bone marrow is discussed, and the structure of the thymus gland and other lymphoid tissues is deferred to later in this book, when the process of lymphopoiesis is discussed in more detail.

### Ontogeny of Blood Cells

All human blood cells are ultimately derived from the primitive **mesenchymal** cells present in the **yolk sac** of the embryo. During the third week of gestation, these mesodermal cells aggregate into clusters of cells known as **"blood islands."** As the number of blood islands increases, the cells on the outside of these islands join up to form a primitive vascular system, and the cells on the inside of these clusters become detached and develop into primitive yolk sac stem cells. As more and more of these islands become connected, increasing numbers of these primitive stem cells are carried off in the mounting stream of plasma. As this primitive vascular system develops, it begins to invade the developing embryo. The growth of the vascular system parallels the growth of the embryo. To state it another way, the growth of the embryo is limited by the growth of the vascular system. As different tissues and organs develop during the second and third months of gestation, they become infiltrated by blood vessels, which carry the migrating stem cells in their plasma. These primitive stem cells will use some of these organs as a base for further development. They move out of the bloodstream and start to colonize these organs. The first organ they migrate to is the developing **liver,** which will become the main source of hematopoiesis after the yolk sac becomes exhausted and begins to disappear. As can be seen in figure 3.1, the liver becomes the main organ of blood cell formation sometime during the third lunar month of pregnancy and will remain that way until the end of the sixth month of gestation, when the **bone marrow** takes over as the major organ of hematopoiesis.

The **spleen** also becomes an organ of hematopoiesis around the third month of gestation. However it is active in that capacity for only a few months, and it stops being an organ in which blood cells are formed by the seventh month of embryonic development (except for lymphocytes). Developing bone marrow becomes colonized by stem cells at the beginning of the fourth month of gestation and gradually increases in importance as a hematopoietic tissue. By the sixth month of embryogenesis, it surpasses the liver as the main source of blood cell production. By the time birth takes place, the bone marrow has become the sole source of blood cell formation (except for the lymphoid organs producing lymphocytes), as the liver ceases to act as an organ of hematopoiesis.

The **lymph nodes** and the **thymus** gland are invaded by stem cells around the fourth month of gestation. They remain the secondary source of lymphocyte production throughout the life of the individual, just as the bone marrow remains the source of the other blood elements.

It is interesting to note that the liver and the spleen retain a residual capacity for hematopoiesis. In times of great demand, when bone marrow is no longer able to supply the necessary blood cells, these two organs may revert to their embryonic role and start producing blood cells again.

With respect to blood cell formation, the terms "medullary" and "extramedullary" hematopoiesis are often used. **Medullary** hematopoiesis refers to blood cell production in the bone marrow, and **extramedullary** hematopoiesis refers to blood cell production outside the bone marrow. In normal situations this means only the lymphoid organs, but the liver and the spleen are also potential extramedullary organs of hematopoiesis.

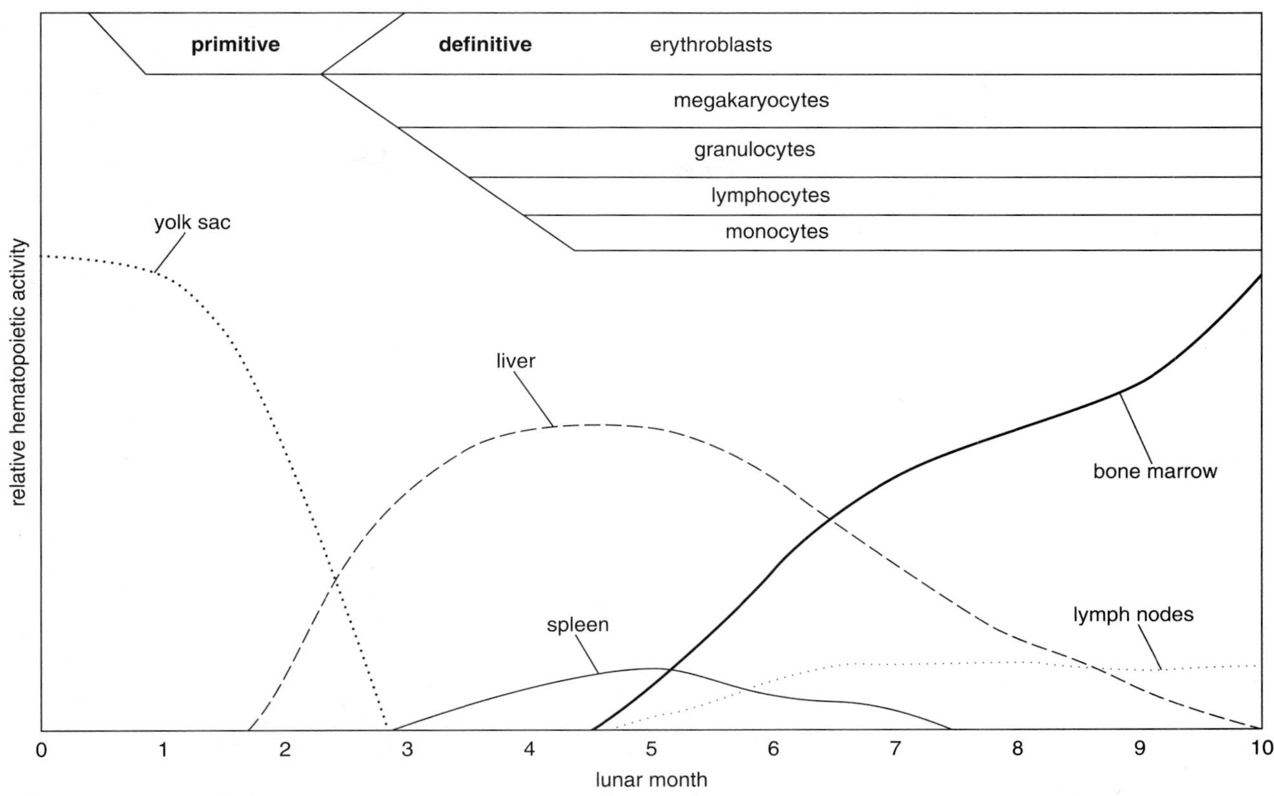


**figure 3.1**
The relative activity of various hematopoietic organs during antenatal life.

At birth medullary hematopoiesis occurs in almost every bone of the body. Gradually the long bones stop producing blood cells. By the time an individual reaches 25 years of age, blood cell formation takes place almost exclusively in the flat bones, especially the sternum, ribs, skull, vertebrae, and innominates. These organs retain hematopoietic capacity throughout life, but as figure 3.2 shows, even these organs diminish their output with advancing age.

## Phylogeny of Blood Cells

The evolution of blood cells is still not very well understood, although many individual facts are presently known. Blood cells exist in both the vertebrate and invertebrate groups of animals, although the number and diversity of cells in the blood circulation is much greater in the vertebrate group. In most invertebrates the only blood cells present are phagocytic cells that play a role in the defense of the organism against invading pathogens.

Red blood cells (i.e., cells containing hemoglobin) are rarely found in invertebrate phyla, although there are some exceptions, such as the nemerteans (e.g., proboscis worms) and some molluscs (e.g., *Arca*), and some echinoderms. There are other invertebrates that have hemoglobin. However it is not concentrated in blood cells but occurs freely in the hemolymph as high-molecular-weight structures made up of large polymers of hemoglobin. In all vertebrates, in contrast with invertebrates, the oxygen-carrying pigment hemoglobin is present within specialized blood cells. The question can be asked, What is the evolutionary advantage of carrying hemoglobin inside a cell, rather than having it diffused inside the plasma? The answer is chiefly one of efficient concentration and circulation. If hemoglobin were carried freely in the blood, these large molecules would render the fluid highly viscous and difficult to circulate, thus placing great stress on the heart. By concentrating these hemoglobin molecules within cellular packages, the overall viscosity is reduced. Red blood cells are sufficiently small and flexible to be squeezed through the fine capillaries where the exchange of gases takes place.

The mature red blood cells of most vertebrates are nucleated and contain metabolically active ribosomes and mitochondria. However in mammals the nucleus is squeezed out of the cell just before it enters the peripheral circulation.

The locus of blood-forming organs also differs in the various classes of vertebrates. For example, in most bony fishes and amphibia, the kidneys are the major organs of hematopoiesis, whereas in other fishes the gonads perform that function. In turtles the liver remains the major organ of hematopoiesis throughout adulthood.

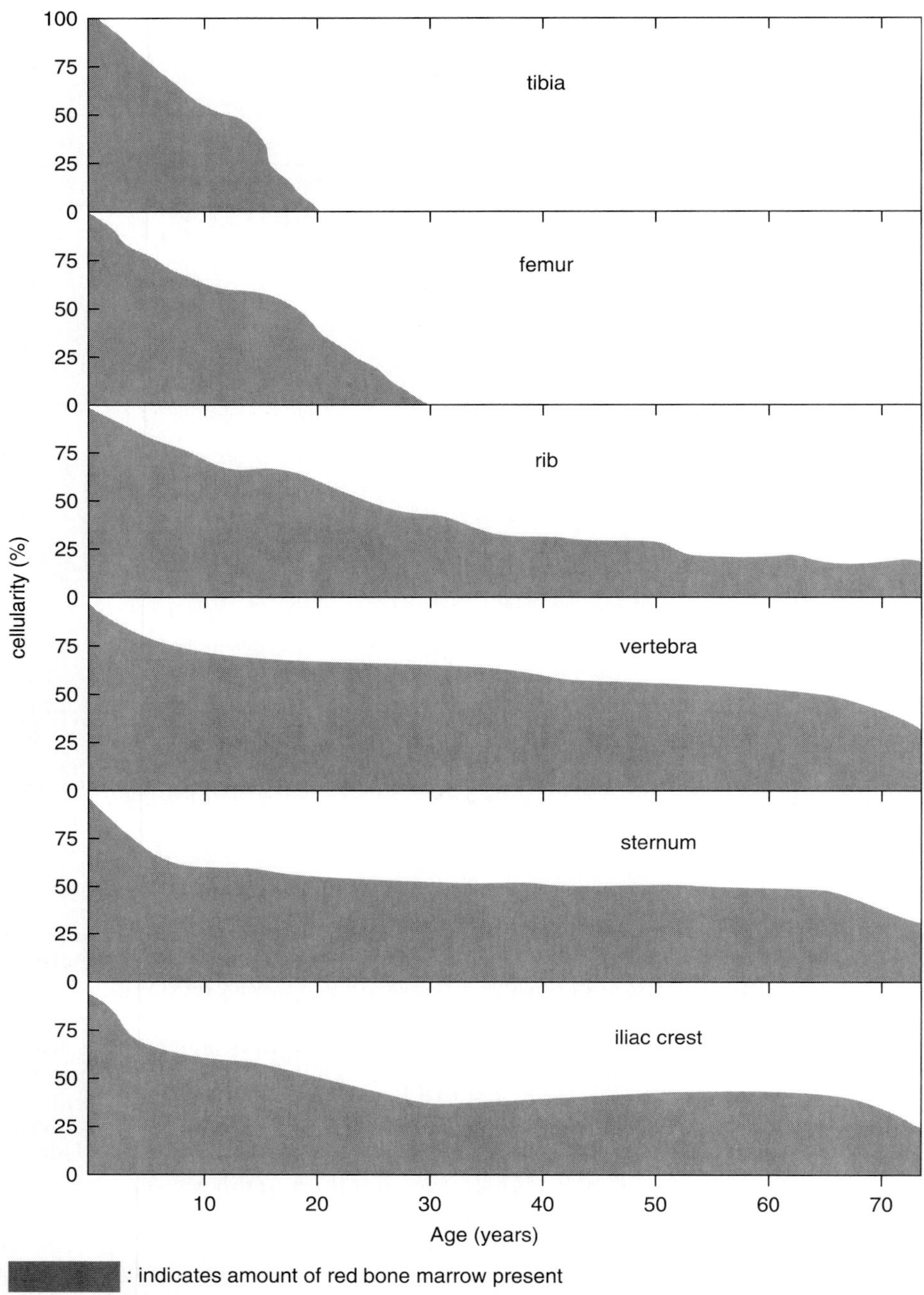


**figure 3.2**

Age and variation in amount of red bone marrow in major bones.

## MICROANATOMY OF BONE MARROW

All blood cells are derived from the mesenchymal tissue of the yolk sac. These primitive mesenchymal cells are not only the precursor cells for all blood cells but are also the ancestral cells for all connective tissue cells including fibroblasts, chondroblasts, and osteoblasts (figure 3.3). Some of these undifferentiated mesenchymal cells evolve into definitive stem cells as soon as the liver, spleen, and bone marrow become colonized and begin to function as organs of hematopoiesis. The essence of these stem cells, or **hemocytoblasts,** is that they are pluripotent—that is, they have the capacity to give rise to all blood cell lines. After birth, the only organ that contains these blood stem cells is the bone marrow.

The bone marrow is the largest, most widely dispersed, and least homogeneous organ of the body. Its volume is between 1,600 and 3,000 cc in an adult. Nutrient arteries enter the bone through hard bone

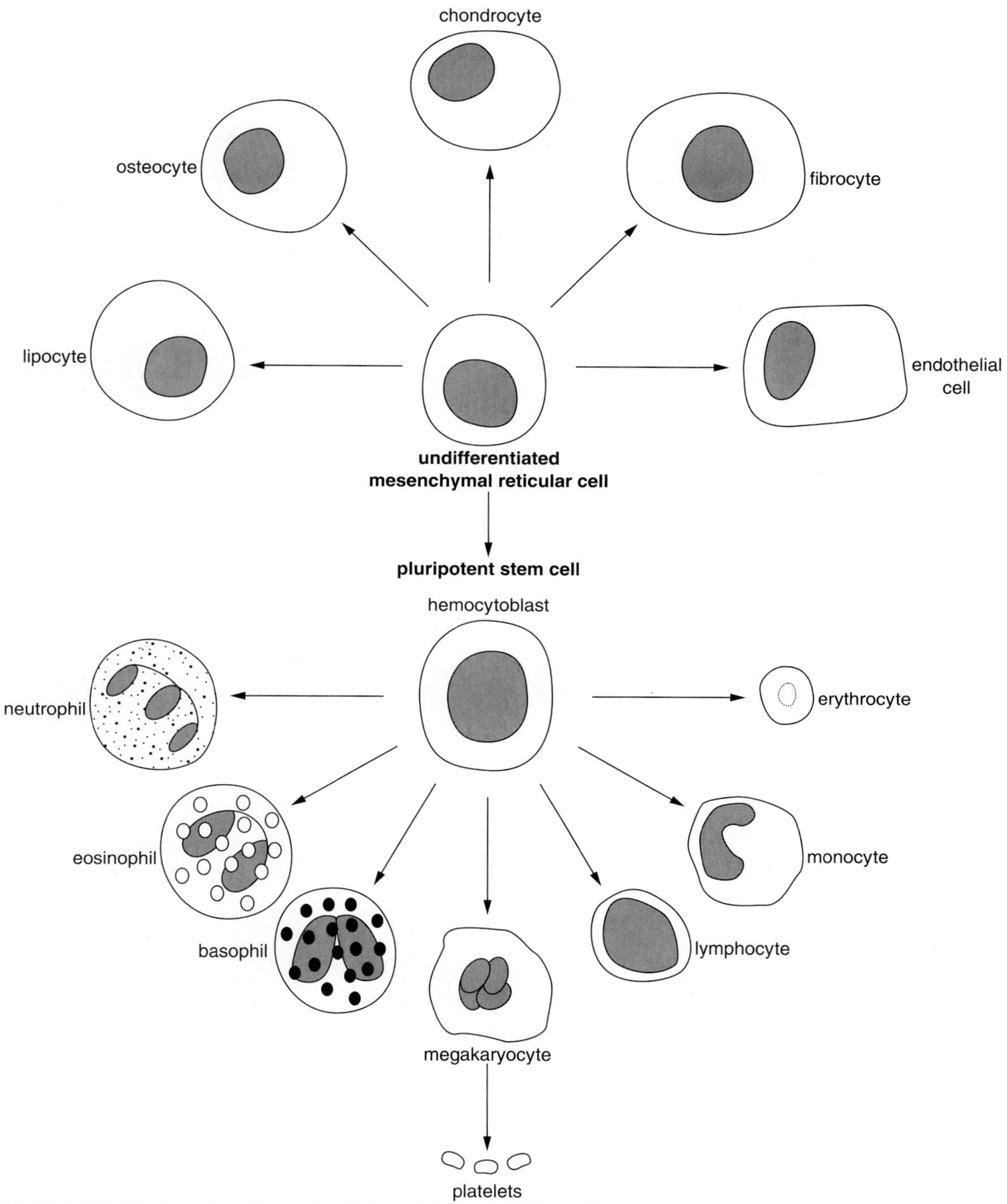


**figure 3.3**

The origin of blood cells, platelets, and other body cells derived from the undifferentiated mesenchymal reticular cells.

ducts known as foramina. These arteries branch into arterioles, which in turn give rise to a bed of sinusoids. These sinusoids substitute for capillary beds found in most other organs. In cross section, the bone marrow structures have been compared to wheels with spokes and a central hub. The marrow sinusoids form the spokes and travel in a radial direction toward a central longitudinal vein (the hub) lying in the long axis of the bone (figure 3.4). No lymphatic vessels are present. Hematopoietic tissues lie outside the sinusoids, and blood cell formation takes place extravascularly in the marrow stroma. The sinusoidal walls are made up of three layers: (1) an internal layer of **endothelial** cells; (2) a middle layer composed of **basement membrane,** which contains many collagen fibers; and (3) an outer layer of **adventitial reticular** cells. Both adventitial and endothelial cells are mobile, capable of phagocytosis, and can

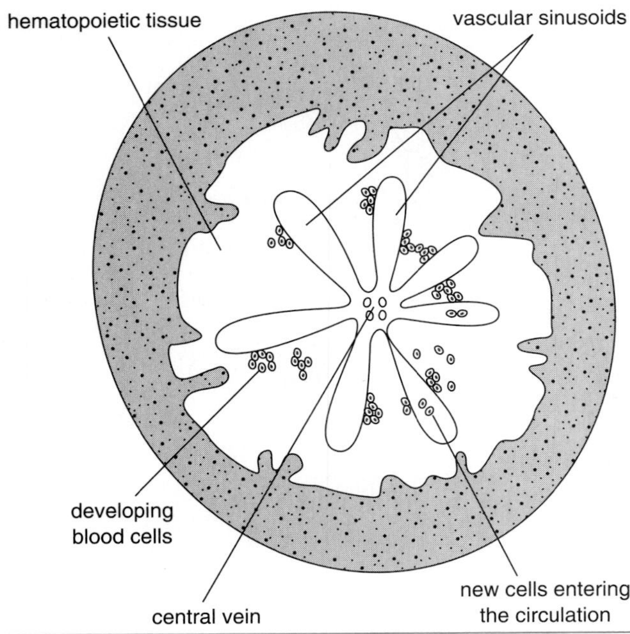


**figure 3.4**

Cross-sectional diagram of bone.

From *Blood, Pure and Eloquent,* edited by Maxwell M. Wintrobe. Copyright © 1980 McGraw-Hill Book Company, New York, NY. Reprinted by permission of McGraw-Hill, Inc.

differentiate into other cell types. They often leave their position in the sinusoid, thus making a gap in the wall through which the formed blood cells can move into the circulation. Bone marrow tissue is a very dynamic structure, changing continually in response to increased or decreased demands of hematopoiesis. New sinusoids are constantly formed, others collapse and become inactive, and others disappear when the reticular and endothelial cells move away.

Developing blood cells—having undergone proliferation and maturation outside the vascular compartment—gain access to the sinusoids at a critical moment in their maturation sequence and move from there into the peripheral circulation. The mechanism for this event is poorly understood.

Adult bone marrow varies from being made up almost exclusively of fatty tissue **(yellow marrow),** as can be seen in the long bones of adults, to almost total hematopoietic tissue **(red marrow),** as seen in the ribs and sternum of young people. In general, the older the person, the more the yellow marrow (=fatty tissue) begins to dominate. Figure 3.5 shows a sequence of bone marrow during the different stages of life. It should be remembered, however, that in case of need, much of the yellow marrow may revert to red marrow, which is the hallmark of active hematopoiesis.

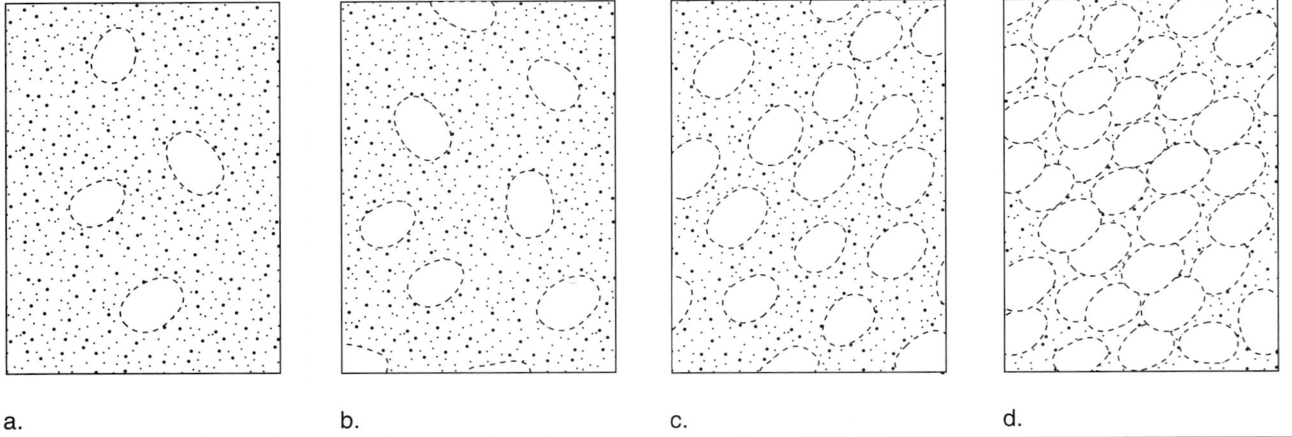


**figure 3.5**

Histology of the bone marrow at various stages of the human life cycle: (*a*) newborn; (*b*) juvenile; (*c*) adult; (*d*) elderly. Notice that the number of fat cells increases with age, while that of the hematopoietic cells decreases.

### Techniques for Bone Marrow Study

Two methods are commonly employed to remove bone marrow cells for further study and analysis. One technique is known as **aspiration** and the other as **biopsy.** Each of these two methods has its advantages and disadvantages. The most common is the aspiration technique. In this method a needle is inserted into the soft center of the bone and a small quantity of bone marrow tissue is aspirated. The contents are then smeared on a microscope slide, then fixed, stained, and examined. In adults, aspiration is usually done from the iliac crest (figure 3.6). In children or infants the tibia of the leg is most commonly used. Since aspiration disturbs the bone marrow architecture, it is used primarily to determine the types of cells present and their relative numbers.

In the biopsy method a piece of bone marrow is removed intact without disturbing the architecture of the bone. Biopsies are done with a fairly thick, hollow biopsy needle, and a core sample of the bone marrow is taken. Once the tissue is removed, the core is carefully extruded from the needle; it is then fixed, embedded in paraffin, sectioned, stained, and examined under the microscope. Normally biopsies are also performed on the iliac crest bone. This technique is much more traumatic, and also more dangerous, since a much wider needle is used. The advantage of this technique is that it gives a better picture of the real structure of the bone marrow.

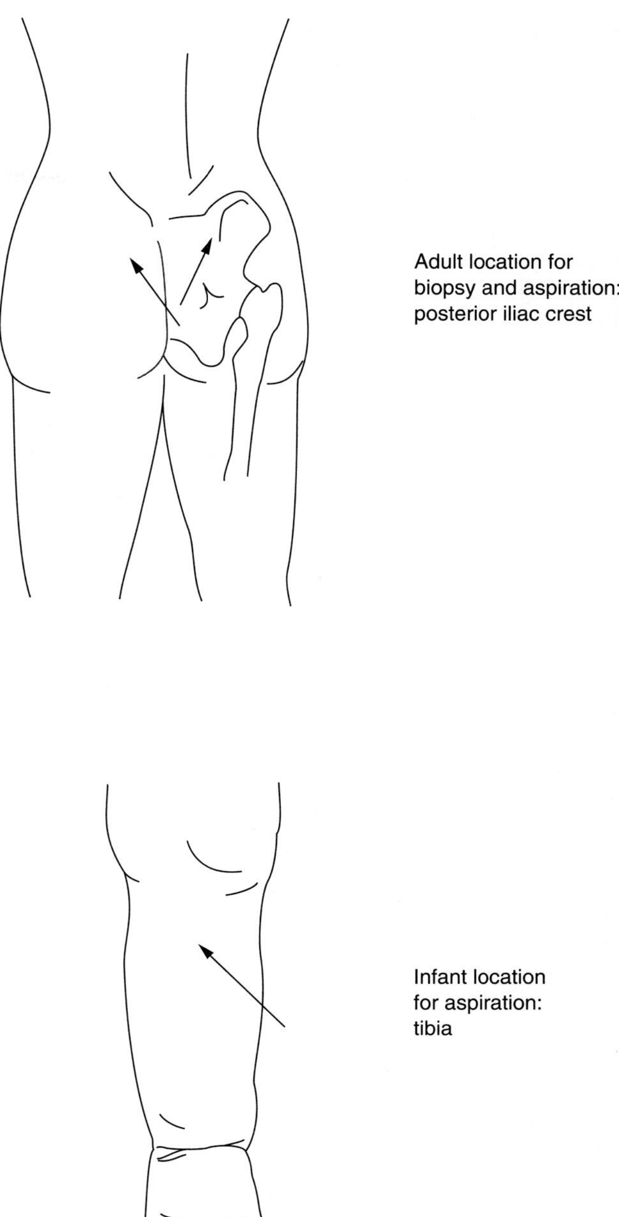


**figure 3.6**
Bone marrow aspiration and biopsy sites.

## PLURIPOTENT STEM CELL

### Nature of Stem Cells

The concept of a single pluripotent stem cell as the ancestral cell for all solid blood elements has long been controversial. The concept has always been very attractive, but until fairly recently, nobody was able to verify the existence of such a cell and hence the truth of this theory. During the last hundred years or so, a great variety of hypotheses regarding the origin of blood cells developed. Around the turn of the century, the German scientist Paul Ehrlich postulated that there were two different cell lines: the myeloid line, from which all cells produced in the bone marrow were derived, and a lymphoid line, from which the lymphocytes originated. Other histologists proposed the existence of three separate cell lines: lymphoid, erythroid, and myeloid lines. Again others believed that there was a different stem cell for each of the major blood cell lines. This polyphyletic point of view was suggested in the histology text by Osgood and Sabin, among others.

The monophyletic school of thought holds that all blood cells develop from a single ancestral cell. This view was espoused in the still famous histology text of Maximov and Bloom, published more than 50 years ago. Only recently was this elusive pluripotent stem cell unequivocally identified. The reason why it took such a long time may be that they are present only in small numbers in bone marrow and in even smaller numbers in blood. Estimates are that blood contains only 1 to 5 stem cells per 100,000 nucleated cells. However at certain times (e.g., after whole-body radiation) their number increases both in the marrow and in the bloodstream. Under normal circumstances they are almost impossible to detect, also because, morphologically speaking they look very much like small lymphocytes.

### Stem Cells as Colony-Forming Units

An easier way to identify and count stem cells is to make use of their colonizing capacity. In this technique suspensions of bone marrow cells are injected intravenously into lethally irradiated mice, in which the spleen and the marrow have been reduced to stroma and are hematologically empty. After 10 days discrete colonies are observed in the animal's spleen. Each stem cell proliferates actively and produces a small clonal colony of undifferentiated stem cells, which some 5 days later differentiate into either myeloid or lymphoid stem cells. Counting each of these colonies provides a valid quantitative assay of the number of stem cells. Stem cells may therefore be defined from the operational point of view as **colony-forming units,** commonly abbreviated **CFUs.** Whether a colony of stem cells will differentiate into an erythrocytic, granulocytic, or other cell line seems to depend on the position of the colony in the hematopoietic tissue. This **hematopoietic inductive microenvironment,** commonly abbreviated **HIM,** seems to determine to a certain extent the further development of these stem cells. The nature of this HIM is still very uncertain and is the object of intensive study.

### Stem Cell Kinetics

Another way to study the pluripotent stem cell is from the kinetic point of view. A true stem cell has the following two basic characteristics: (a) it is self-maintaining—that is, it has the ability to divide and give rise to daughter cells that have the same capabilities as the parent cell; and (b) it has the ability to give rise to further differentiated cells. Under normal conditions the number of stem cells in each person remains more or less constant. How this is achieved is not known, but several mechanisms have been proposed. When stimulated, the stem cell could divide asymmetrically, with one offspring remaining within the stem cell compartment, while the other leaves and becomes a blast cell for a specific blood cell line. In an alternative scheme a stem cell leaves its compartment under a differentiating stimulus and is replaced by the progeny of another stem cell. These views are illustrated in figure 3.7.

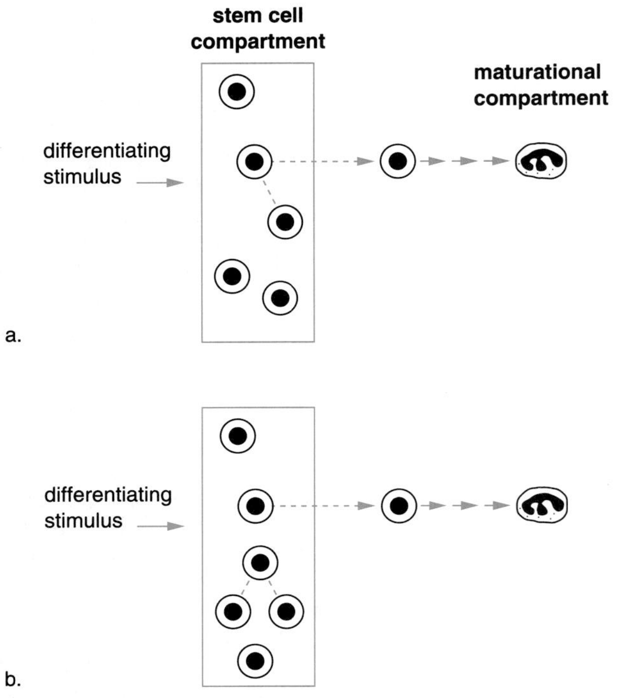


**figure 3.7**

Proposed mechanisms of pluripotent stem cell renewal and maintenance. (*a*) Asymmetric cell division; (*b*) replacement by progeny of other pluripotent stem cells.

## RELATIONSHIPS OF VARIOUS BLOOD CELLS

With the development of new and sophisticated tissue culture techniques, as well as the growth of our understanding and use of cell markers, it has been possible to study the development and relationships of various blood cell lines in great detail. Although a number of questions remain unsolved, a general pattern has emerged over the last few years. A number of different committed stem cells (colony-forming units, or CFUs) have been described or postulated. However, there is only one truly pluripotent stem cell. This cell is currently named **LM-CFU,** which stands for **lymphoid-myeloid colony-forming unit.** This totipotent hematopoietic stem cell can develop in either of two directions (figure 3.8). It can become a lymphoid stem cell known as **lymphoid-CFU (L-CFU)** or a myeloid stem cell, which is now commonly referred to as a **GEMM-CFU (granulocyte-erythrocyte-megakaryocyte-monocyte colony-forming unit).** All nonlymphocytic blood cells are derived from this pluripotent stem cell. Other stem cells develop from this GEMM-CFU with more restricted potentials.

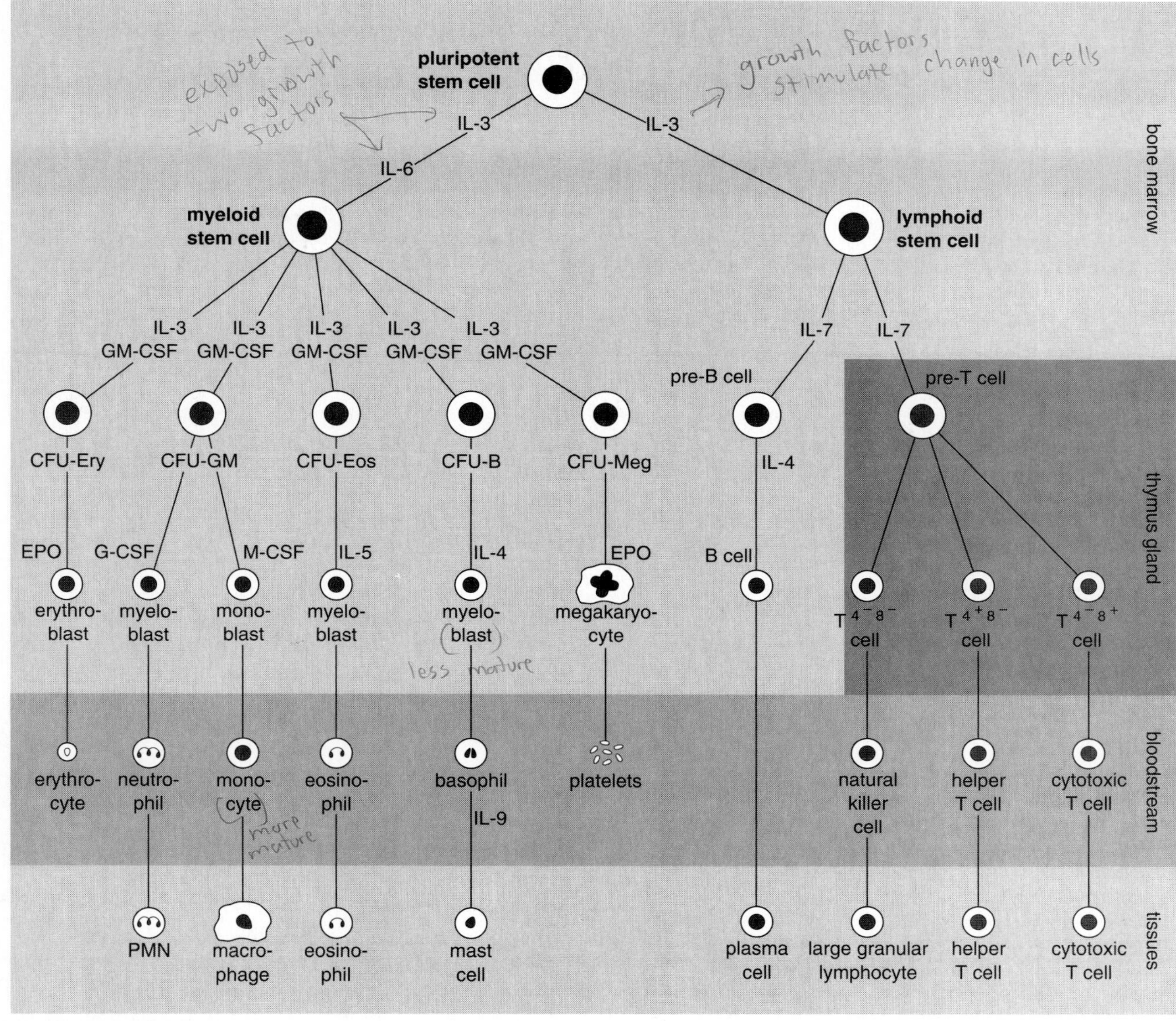


**figure 3.8**

The origin and relationships between the various blood cell lines. The development and maturation of the various blood cells are regulated by a number of growth factors, including colony-stimulating factors (CSF), interleukins (IL), and erythropoietin (EPO).

Currently five different types of committed stem cells are recognized. These are (1) erythroid progenitor cells (erythroid-CFUs), (2) megakaryocyte progenitor cells (megakaryocyte-CFUs), (3) granulocyte (=neutrophil) monocyte progenitor cells (GM-CFUs), (4) eosinophilic progenitors (eosinophil-CFUs), and (5) basophil colony-forming units (basophil-CFUs) (figure 3.8).

The GM-CFUs will eventually give rise to two types of unipotent stem cells—the granulocyte-CFU, which is the progenitor cell for neutrophils, and the monocyte-CFU, which will give rise to monocytes and macrophages.

## REGULATION OF HEMATOPOIESIS

The study of the methods by which cell populations proliferate is known as **cytokinetics.** Cytokinetics is concerned with such phenomena as mass or size (i.e., the total number of cells in a population), maturation time, life span, and turnover rate. Maturation and survival times of major blood cell types are given in table 3.1.

From the cytokinetic point of view, all body cells fit into one of three categories of proliferation. Some cells are constantly dividing (e.g., cells of skin epidermis, sperm cells). They divide to replace mature cells constantly lost by attrition. Another category may be

**Table 3.1** Maturation and Survival Times of Blood Cells and Platelets

| Cell Types | Maturation Time (from stem cell until mature cell, in days) | Survival Time (including tissue phase, in days) |
|---|---|---|
| Erythrocytes | 3–5 | 120 |
| Granulocytes | 5–6 | 9–10 |
| Monocytes | 5–6 | Months to years |
| Lymphocytes | Variable (hours to days) | Months to years |
| Platelets | 4–5 | 10 |

**Table 3.2** Effects of Cytokines on Hematopoietic Stem Cells

| | Cytokines | | | | | | | | | |
|---|---|---|---|---|---|---|---|---|---|---|
| **Target Cells** | **EPO** | **GM-CSF** | **G-CSF** | **M-CSF** | **IL-3** | **IL-4** | **IL-5** | **IL-6** | **IL-7** | **IL-9** |
| Pluripotent stem cell | – | + | – | – | + | – | – | – | – | – |
| Myeloid stem cell | – | + | – | – | + | – | – | + | – | – |
| Lymphoid stem cell | | | | | | | | | | |
| B progenitor | – | – | – | – | – | + | – | – | + | – |
| T progenitor (thymus) | – | – | – | – | – | – | – | – | + | – |
| Granulocyte–monocyte progenitor | – | + | + | + | + | – | – | – | – | – |
| Monocyte progenitor | – | + | – | + | + | – | – | – | – | – |
| Neutrophil progenitor | – | + | + | – | + | – | – | – | – | – |
| Eosinophil progenitor | – | + | – | – | + | – | + | – | – | – |
| Basophil progenitor | – | + | – | – | – | + | – | – | – | – |
| Mast cell | – | + | – | – | + | + | – | – | – | + |
| Megakaryocyte | + | + | – | – | + | – | – | – | – | – |
| Erythroid progenitor | + | + | – | – | + | – | – | – | – | – |

(+) indicates cytokine stimulates the proliferation and differentiation of the target cell. (–) indicates the cytokine has no effect on the cell type.

defined as nonproliferating cells—that is, cells that will not divide under normal circumstances. Most nerve cells and muscle cells fit into this category. Finally, some cells proliferate only on demand and divide only when called upon to do so by an external stimulus.

The hematopoietic cells of the bone marrow and the lymphocytes not only belong to this category, but also belong to the first category. Most have fairly limited life spans and need to be replaced constantly.

Bone marrow cells, like many other continuously proliferating tissues, have two patterns of behavior: the **steady-state** pattern and the **non-steady-state** mode. In the steady-state pattern, proliferation is at a constant rate—that is, new cells are produced at a rate equaling the number of old cells being removed from the blood, thus keeping the total population constant. To maintain a normal quantity of erythrocytes, for example, new mature cells must enter the circulation at the astonishing rate of 2 million per second. This assumes a constant balance between production and destruction of red blood cells (RBCs). In the non-steady-state mode, proliferative behavior is changed and results in either increased or decreased cell production, depending on the physiological or pathological condition of the individual. Both types of proliferation imply the existence of feedback mechanisms. The bone marrow is either stimulated or inhibited by some biological factor(s). Factors that stimulate bone marrow appear to be humoral factors, or hormones, present in the blood. They are secreted by some endocrine cell or organ when the level of a given cell type drops. Currently many such **growth factors** have been identified and undoubtedly more factors will be elucidated in the future.

Steady-state levels of hematopoiesis are maintained by a series of growth factors or cytokines produced by the stromal cells of the bone marrow (table 3.2). A number of these growth factors are known as **colony-stimulating factors (CSFs),** and others are classified as **interleukins (ILs)** as they are also produced by a variety of white blood cells. In addition, the proliferation of red blood cells and platelets is also influenced by a factor produced in the kidneys during hypoxia. This latter growth factor is known as **erythropoietin,** frequently abbreviated as **EPO.**

The earliest-acting growth factor is known as **multi-CSF** (also known as **IL-3**) and has not yet been detected in stromal bone marrow cells but only in helper T cells ($T_H$). Bone marrow cells, however, are thought to produce an as-yet-unidentified factor that maintains the steady-state levels of the pluripotent stem cells. Other growth factors that influence the development of the early stages of hematopoiesis include **granulocyte-monocyte colony-stimulating factor (GM-CSF)** and **IL-6.** The development of B- and T-lymphocyte progenitor cells is influenced by **IL-7** and probably by other factors as well.

Non-steady-state conditions also result in the production of a number of growth factors more specific in their actions that are usually produced by activated white blood cells or fibroblasts. More specific-acting growth factors include **IL-4,** which stimulates the development of basophils and mast cells; **IL-5,** which activates eosinophil progenitor cells; **IL-9,** which stimulates mast cell production; **granulocyte-CSF (G-CSF),** which stimulates neutrophil production; **monocyte-CSF (M-CSF),** which regulates monocyte and macrophage proliferation specifically; and EPO, which stimulates erythrocyte and megakaryocyte production. A summary of these growth factors is given in table 3.2.

The regulation of blood cell production is not only a function of increased or decreased production of various growth factors but also is a function of the presence of receptors on the progenitor cells for these stimulatory or inhibitory factors. Generally the more receptors present on the cell membrane of these progenitor cells, the more likely they are to be stimulated.

## *Further Readings*

Clark, S. C., and Kamen, R. The Human Hematopoietic Colony-Stimulating Factors. *Science,* 236:1229 (1987).

Dexter, T. M., and Spooncer, E. Growth and Differentiation in the Hematopoietic System. *Ann. Rev. Cell Biol.* 3:423 (1987).

Dorshkind, K. Regulation of Hematopoiesis by Bone Marrow Stromal Cells and Their Products. *Ann. Rev. Immunol.* 8:111 (1990).

Golde, D. W., and Gasson J. C. Hormones that Stimulate the Growth of Blood Cells. *Scientific American,* July 1988.

Sachs, L. The Molecular Control of Blood Cell Development. *Science,* 238:1374 (1987).

Spangrude, G. J., Heimfeld, S., and Weissman, I. Purification and Characterization of Mouse Hematopoietic Stem Cells. *Science,* 241:58 (1988).

## *Review Questions*

1. Define the term hematopoiesis. From what words is it derived?
2. From what cell are all blood cells ultimately derived?
3. In which two directions can the original blood cell go?
4. Where do the following cells originate: (a) red blood cells, (b) granulocytes, (c) monocytes, (d) B lymphocytes, (e) T cells, (f) monocytes, and (g) megakaryocytes?
5. Where do the first blood cells originate in the early embryo?
6. When do the following organs begin to function as loci of hematopoiesis: (a) liver, (b) spleen, (c) bone marrow, and (d) lymph nodes?
7. When do the liver and the spleen cease to be organs of hematopoiesis?
8. When does the bone marrow become the major organ of hematopoiesis?
9. Distinguish between medullary and extramedullary organs of hematopoiesis. Give examples of each.
10. When do the long bones stop being hematopoietic organs? What are the major organs of hematopoiesis for a 30-year-old person?
11. Name the major types of blood cells found in invertebrates.
12. Why is hemoglobin concentrated in blood cells in vertebrates?
13. Describe in general terms the structure of bone marrow. Where in the bone marrow are blood cells formed?
14. How do blood cells get into the bloodstream?
15. Distinguish between red and yellow bone marrow. How are these tissues related to age?
16. Name the two major techniques for bone marrow study. Give a major advantage and disadvantage of each.
17. Why did it take such a long time before the existence of the pluripotent stem cells could be demonstrated?
18. Describe the morphological appearance of a PPSC.

19. How does one count the number of PPSCs in a bone marrow sample?
20. What are the two major characteristics of a PPSC from the kinetic point of view?
21. What is another name for the PPSC? Into what two types of stem cells can it develop?
22. What is a GEMM-CFU? Into what types of blood cells can it develop?
23. Draw a diagram that illustrates the relationships between the various blood cells.
24. How is the process of hematopoiesis regulated? Distinguish between steady-state and non-steady-state patterns.
25. What are CSFs and ILs? Where are they produced?
26. Name some general early-acting growth factors.
27. What are the major specific-acting growth factors? On what blood cell type does each of these factors act?
28. How many blood cells does a normal healthy individual produce each day to replace those that are destroyed?
29. What other factor determines the responsiveness of an immature blood cell to a given growth factor?
30. Define EPO. What is its function? When and where is it produced?

# SECTION TWO

## *Red Blood Cells*

# Chapter Four

## *Erythropoiesis and the Morphology of Red Blood Cells*

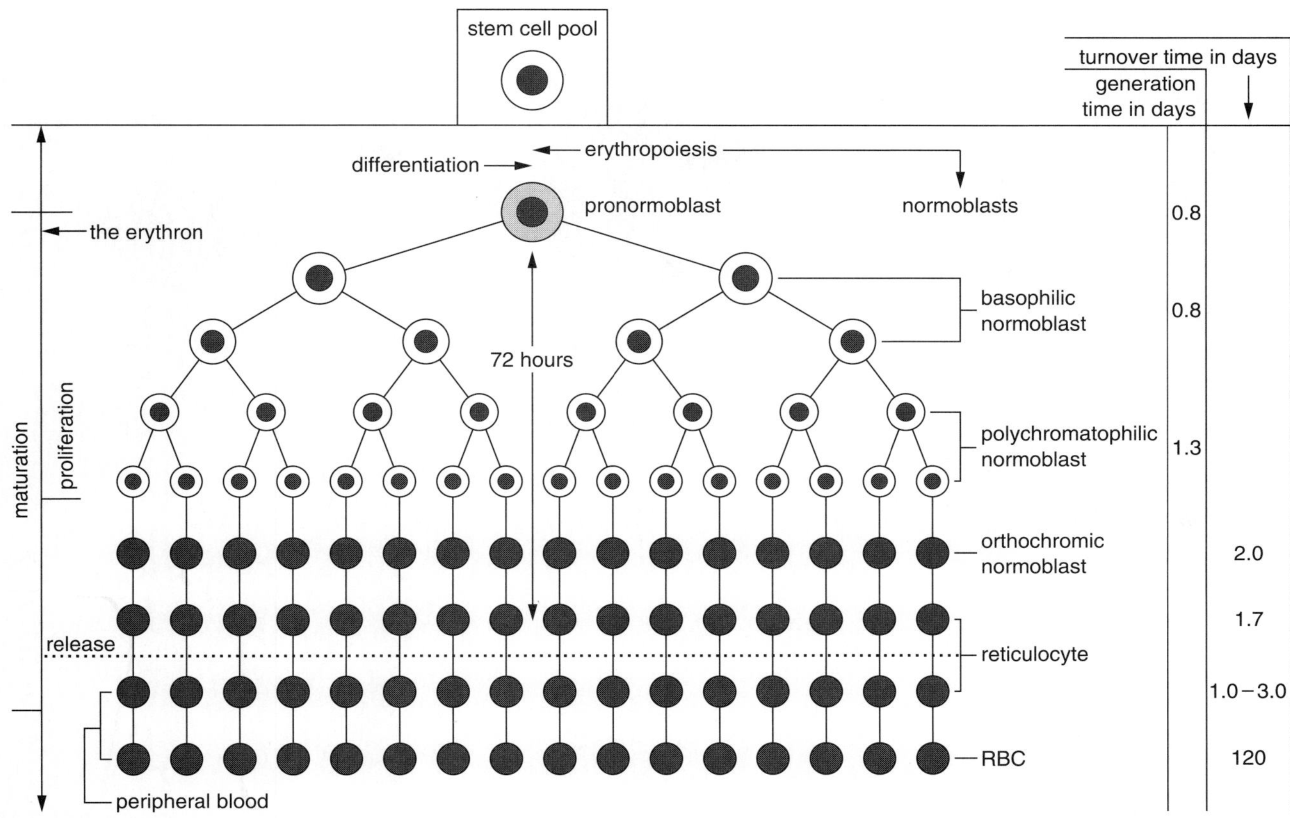


**figure 4.1**
Erythropoiesis, the development and maturation of the red blood cells.

## KINETICS OF ERYTHROPOIESIS

Each committed erythroid stem cell may give rise to up to an estimated 16 mature red blood cells. As can be seen in figure 4.1, each committed stem cell, or pronormoblast, usually goes through four cell divisions to produce a total of 16 daughter cells. However, not all maturing cells will become erythrocytes. Normally a small percentage—less than 10%—die before they reach the final mature stage. To the extent that the process of erythropoiesis fails to deliver cells to the circulating blood, it is called **ineffective.** In certain diseases, such as megaloblastic anemias, the extent of ineffective erythropoiesis is abnormally high (sometimes more than 50%), and relatively few mature red blood cells reach the circulation, despite intense erythropoietic activity in the bone marrow.

The normal process of proliferation and maturation lasts from 3 to 5 days, depending on the urgency of red blood cell needs of the circulating blood. During that time four mitotic divisions usually occur. Each subsequent division is accompanied by a nuclear and cytoplasmic maturation; hence they are commonly referred to as **maturational divisions.** Each successive division yields a smaller cell in which the nucleus becomes progressively smaller and more compact until it is ejected from the cell in the final stages. The cytoplasm also shows a gradual decrease in basophilia—that is, the deep blue cytoplasm found in the early stages gradually disappears and is progressively replaced by an increase in pink color due to the growing levels of hemoglobin. In the final stages the bright pink of the mature red blood cells shows no traces of the original blue cytoplasm.

### Stages of Erythrocyte Development

Usually six stages in the development of red blood cells are recognized. The different stages are pictured in figure 4.2 and are described as follows:

1. The **pronormoblast** (or **proerythroblast**) is the earliest stage in erythrocyte development and is the name given to a committed erythroid stem cell. This pronormoblast is a fairly large cell, varying in diameter from 12 to 14 µm. The nucleus is prominent; it contains coarse chromatin and one or more nucleoli. The nucleus is large, relative to the cytoplasm, and is usually round with a smooth nuclear membrane. When stained with Wright's stain, the cytoplasm is a deep blue and lacks inclusions. This cell stage is also known as the **rubriblast** stage.

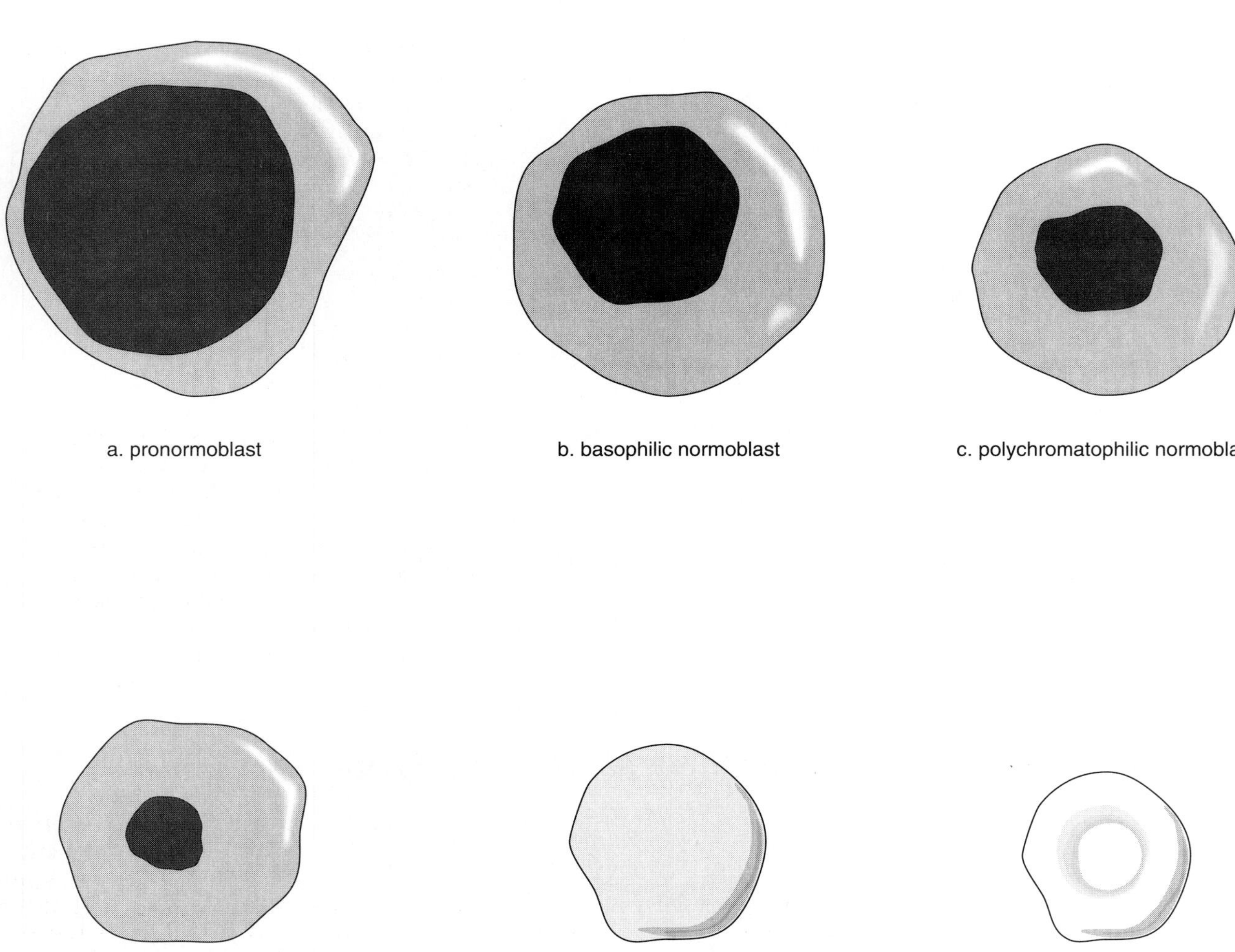


**figure 4.2**

The six stages of erythrocyte development. Note the progressive decrease in cell size and the condensation of the nucleus in the first four stages (*a*) through (*d*).

2. **Basophilic normoblast (basophilic erythroblast).** This cell may be similar in size or smaller than the pronormoblast, and the chromatin of the nucleus is coarser. Generally speaking, the two cells look remarkably similar, especially since the cytoplasm of the basophilic normoblast also stains a deep blue with Wright's stain. This is why the cell is called a basophilic normoblast—it has a great affinity for the basic dye methylene blue. What distinguishes this cell from the pronormoblast is the absence of nucleoli in the nucleus. The presence of one or more well-defined nucleoli establishes the identity of a pronormoblast, whereas the absence of nucleoli indicates the presence of a basophilic normoblast. This cell stage is also known as the **prorubricyte stage.**

3. **Polychromatophilic normoblast (polychromic erythroblast).** This third stage in the development of red blood cells is usually smaller than the previous two cell types, having a diameter of about 10 µm. It can be easily distinguished from the basophilic normoblast by two major changes. First, the nucleus is much more condensed and stains much darker, as the chromatin is now extremely coarse and shows a checkerboard appearance. Second, the cytoplasm is no longer deep blue but appears much paler and shows a variable mixture of pink and blue, producing a kind of lavender color when stained with Wright's stain. This is the stage in which the hemoglobin appears for the first time in the cytoplasm; this hemoglobin causes the pinkish shine

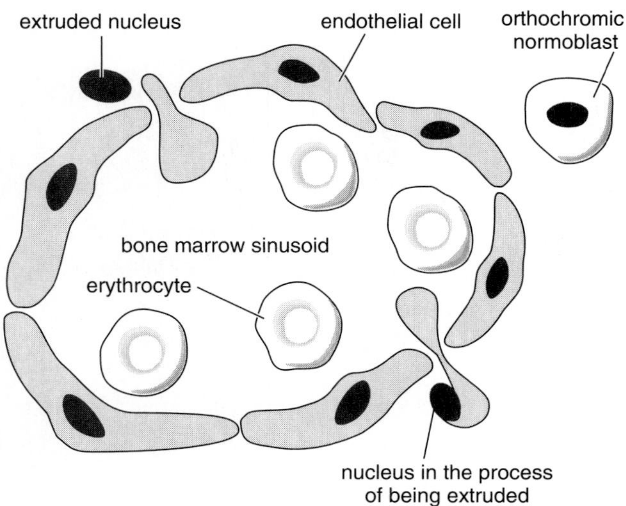


**figure 4.3**
The entrance of the newly formed red blood cells from the bone marrow into the circulatory system.

between the lighter blue of the cytoplasm. This cell is called a polychromatophilic normoblast (or a polychromic erythroblast) because of the mixture of hues in the cytoplasm. This cell stage is also known as the rubricyte stage.

4. **Orthochromic normoblast (or pyknotic erythroblast)** is the fourth cell in the erythrocyte maturation series. It is called "orthochromic" since the cytoplasm is now almost completely salmon-pink, and practically all traces of blue have disappeared. (*Orthos* in Greek means *right,* hence orthochromic is having the right color.)

   The other distinguishing feature of this cell is the shape and size of the nucleus. The nucleus appears as a blue-black sphere, indicating an extremely dense mass of chromatin. Such a dense nucleus is known as a **pyknotic** nucleus and is the most characteristic feature of this stage of development—hence this stage is also known as the "pyknotic erythroblast." Normally no further cell division takes place during this stage (figure 4.1), although this cell retains its capacity to undergo cell division. In certain situations (e.g., in iron deficiency anemia), cell division will occur in this phase, as is explained later. Hence this cell type is still called a blast cell—that is, capable of undergoing blast formation and mitosis. At the end of this phase, the nucleus is extruded, and the cell enters the circulation by squeezing itself through an opening in the endothelial lining of the bone marrow sinusoid (figure 4.3). The nucleus is then phagocytosed and digested by one of the bone marrow macrophages. This cell type is also known as a **metarubricyte.**

5. **Reticulocyte.** The last immature erythrocyte in the maturation series is the reticulocyte. It is given this name because when stained with supravital dyes, such as methylene blue or brilliant cresyl blue, a reticular network of strands can be observed inside the cell (figure 4.4). These threads are remnants of RNA strands. This irregular tangle of threadlike material disappears in a day or two, and the cell now becomes a fully mature red blood cell. The reticulocyte is slightly larger and less regular in shape than the erythrocyte.

6. **Erythrocyte.** This is the final stage in red blood cell development. The maturation of a reticulocyte to an adult erythrocyte takes about 24 to 48 hours. In the course of maturation, the ribosomes and mitochondria disappear, and the cell loses its capacity for hemoglobin synthesis and oxidative metabolism. The ribosomal RNA is also gradually degraded to extracellular ribonucleosides. The maturing red blood cell enters the circulation as a reticulocyte. Therefore the reticulocyte level of the blood is the most common clinical index used to measure erythropoietic activity. Under normal conditions approximately 1% of the red blood cells need to be replaced each day, since the normal erythrocyte life span is about 120 days. Hence the normal reticulocyte count is about 1% of the total erythrocyte count.

### Regulation of Erythrocyte Production

The rate of erythropoiesis is governed by the rate of oxygen transport to the respiring tissues. This latter rate in turn is a function of the amount of hemoglobin present and the cardiac output. Since most of the hemoglobin present in the blood is inside the red blood cells, there is a direct correlation between the number and the size of the erythrocytes and the availability of oxygen for the respiring tissues. When the amount of oxygen reaching the tissues decreases, the rate of erythropoiesis is stepped up. This feedback loop between the oxidizing tissues and the bone marrow is regulated by a humoral agent known as **erythropoietin (EPO).** This hormone was first discovered in the blood and urine of anemic rats by Erslev in 1953. Erythropoietin has recently been isolated and purified. EPO has a molecular weight of 39,000 daltons, a carbohydrate content of 50 to 60%, and a sialic acid content of about 10%. The exact structure of erythropoietin is still unknown. Removal of the sialic acid residues will result in loss of biological activity. Erythropoietin is found in both blood and urine, produced mainly in the kidneys, but extrarenal production of this hormone has also been established. The exact nature of the cells that produce erythropoietin is still a subject of intense discussion.

The function of erythropoietin is to stimulate committed stem cells to differentiate into erythroblasts (or normoblasts). How this is done is not yet known in detail, but evidence indicates that erythropoietin stimulates the synthesis of messenger RNA. In the steady-state pattern of erythropoiesis, the levels of erythropoietin are rather constant, and the regulation

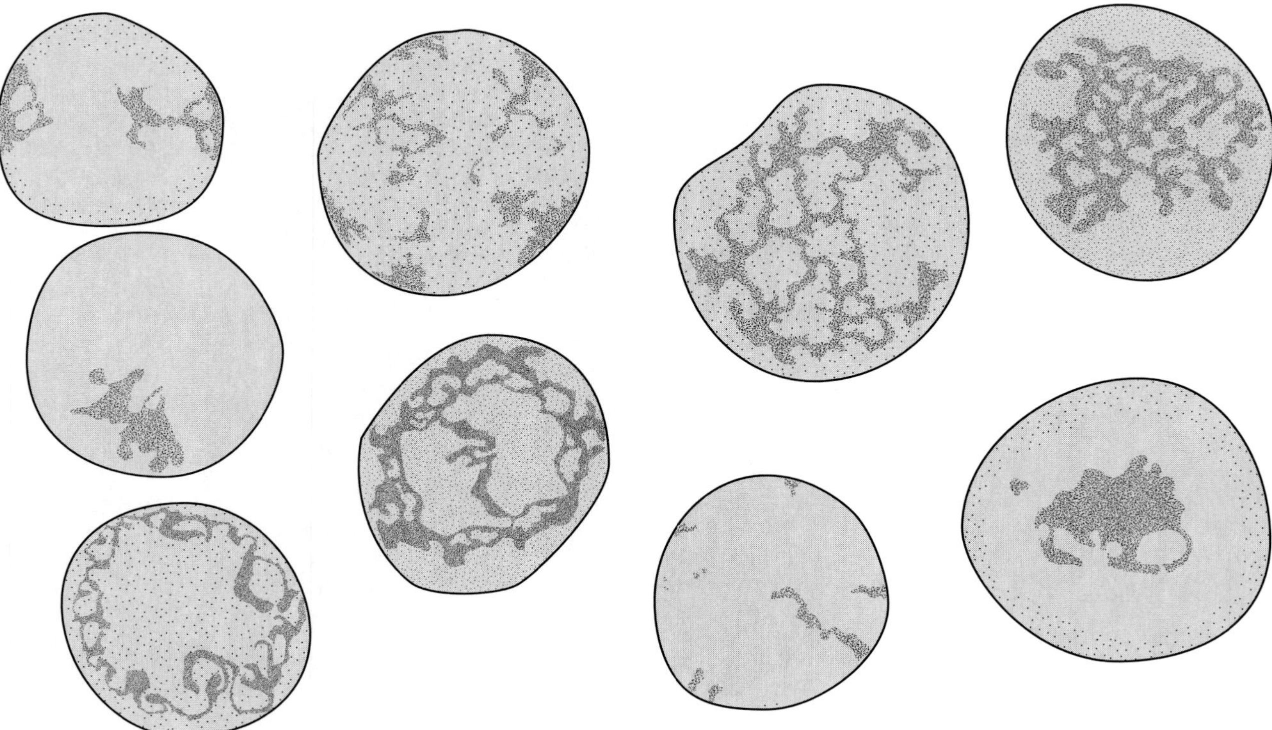

**figure 4.4**
Reticulocytes in a blood smear stained with methylene blue.

of the production of new red blood cells may also be controlled by secretions of bone marrow stromal cells. Recent developments seem to indicate that apart from erythropoietin, IL-1, IL-3, IL-4, insulinlike growth factor (IGF), and platelet-derived growth factor (PDGF) also may play a role in the maintenance of normal red blood cell levels. In non-steady-state conditions, which occur due to changed physiological situations (e.g., high altitude; anemia), erythropoietin production increases sharply and results in increased erythropoietic activity in the bone marrow. Bone marrow has a tremendous capacity to increase its activity, up to eight times its normal level. Obviously in non-steady-state modes, the role of EPO is of greater significance than in steady-state conditions.

## MORPHOLOGY OF RED BLOOD CELLS

Red blood cell structure can be readily determined from a thin blood film stained with Romanowsky dyes. The normal erythrocyte stains salmon-pink with Wright's stain because of the hemoglobin present within the cell. The three basic features of a red blood cell are its size, its shape, and its content. Each of these is discussed in the following material.

### Size of Erythrocytes

Normal erythrocytes are nearly uniform in size, with diameters of 7.2 to 7.9 μm (figure 4.5*b* and figure 4.6). The average red blood cell is about 2 μm thick at its thickest part. However erythrocytes with much larger and much smaller sizes may be found in certain disease states.

An increase in the size of a red blood cell is known as **macrocytosis.** An erythrocyte is referred to as a **macrocyte** when its diameter exceeds 9 μm (figure 4.5*c*). Macrocytosis is commonly encountered in the blood of healthy newborn infants. In older children, however, it is always associated with disease. This phenomenon is especially common in megaloblastic anemias.

The opposite situation is known as **microcytosis.** A red blood cell is called a **microcyte** if its diameter is smaller than 6 μm. Microcytosis is usually caused by a lack of hemoglobin in the red blood cells. This condition is common in iron deficiency anemia and in certain hemolytic anemias (figure 4.5*a*).

In other types of anemia red blood cells of various sizes may be encountered in a blood smear (figure 4.5*e*). This condition is known as **anisocytosis** and is commonly found in persons suffering from both megaloblastic and iron deficiency anemias.

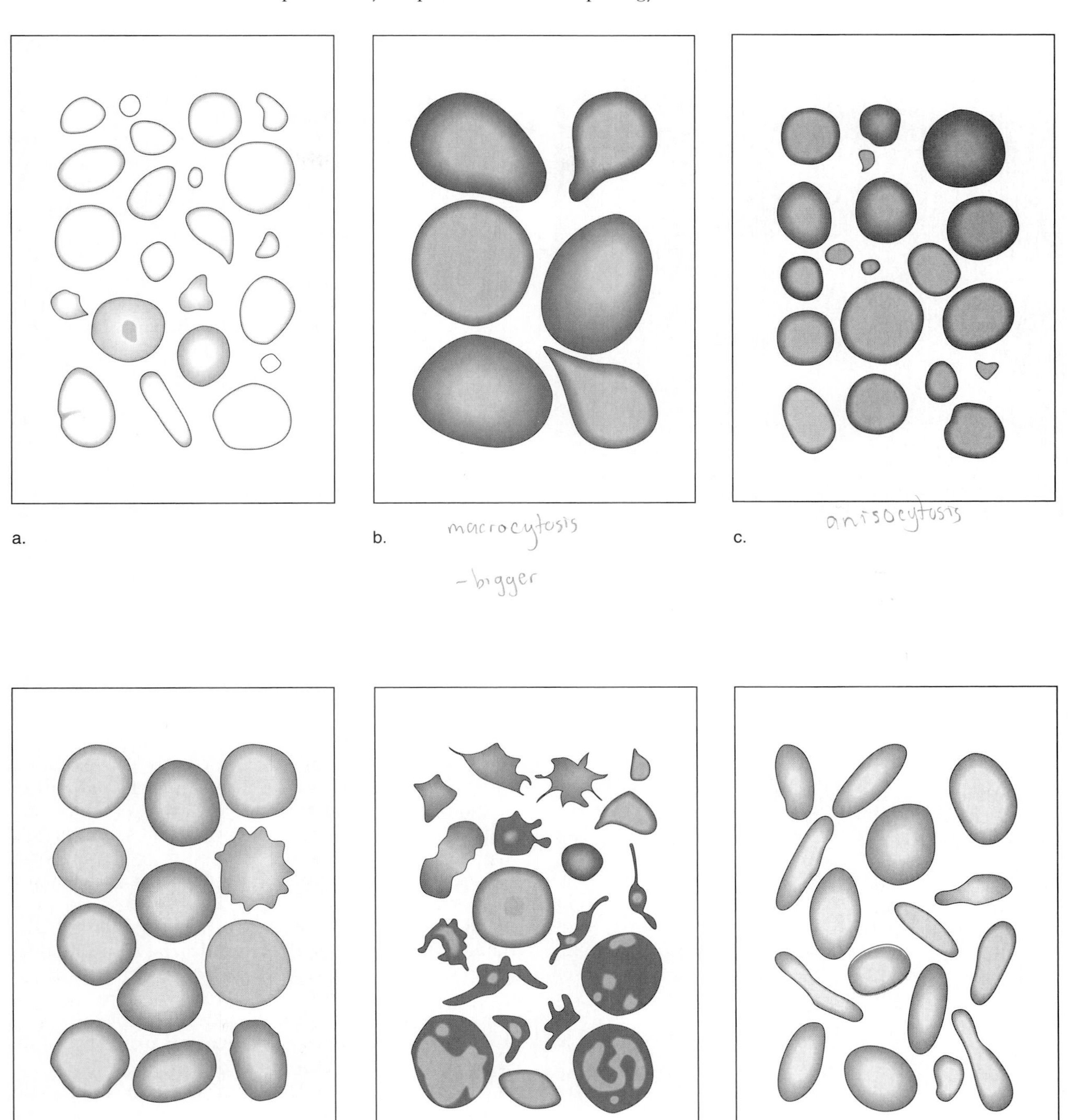


**figure 4.5**
Variations in erythrocyte morphology: (*a*) microcytosis, hypochromia; (*b*) normocytosis, normochromia; (*c*) macrocytosis; (*d*) poikilocytosis; (*e*) anisocytosis, spherocytosis; (*f*) hereditary ovalocytosis.

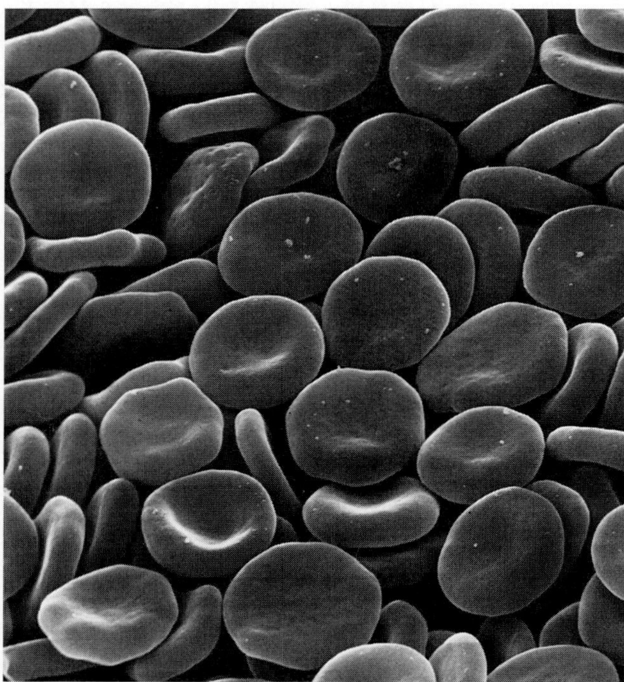

figure 4.6
Scanning electron micrograph of normal red blood cells, showing their biconcave shape.

### Shape of Erythrocytes

The shape of the normal erythrocyte is a biconcave disc, which is thickest at its edges (2 µm) (figure 4.6). This configuration allows for maximum surface contact of hemoglobin within the cell, thus greatly facilitating the exchange of blood gases. Further, this shape gives the red blood cell great flexibility and elasticity and allows it to be folded when it has to move through very narrow blood capillaries. Also, the smooth, round edges reduce the amount of friction the cell may encounter in the microcirculation.

However not all red blood cells are perfectly round, flat, and smooth. On the contrary, a great number of variations in the shape of the erythrocyte may be observed under the microscope. The presence of many abnormally shaped erythrocytes indicates the presence of a certain disorder. The following types of abnormal red blood cells are commonly observed in blood smears.

The presence of many abnormal cell shapes on a blood smear is known as **poikilocytosis,** and all the abnormal cells may be grouped as **poikilocytes** (figure 4.5*d*). However, many different terms have been associated with specific abnormal shapes. The most common of these poikilocyte groups are as follows:

**Ovalocytes** (also known as **elliptocytes**) are oval-shaped red blood cells (figure 4.5*f*). Ovalocytosis is frequently observed in a few cells in a normal blood film. However, when present in large numbers, they are often an indication of an abnormal state. Large numbers of ovalocytes are commonly encountered in hemolytic anemias. They are especially common and numerous in an anemia called hereditary elliptocytic anemia (HEA), also known as hereditary elliptocytosis, where they make up 50 to 90% of the total red blood cell population in an individual.

**Spherocytes** are small spherical erythrocytes. These ball-shaped cells are much thicker in the center than normal red blood cells, and they usually stain uniformly dark (figure 4.5*e*). Spherocytosis is seen in an inherited disorder known as hereditary spherocytic anemia (HSA) but also may be observed in other hemolytic anemias.

**Teardrop cells** are erythrocytes that have assumed the shape of a teardrop (figure 4.7*a*). This type of erythrocyte is commonly associated with such diseases as myelofibrosis and thalassemia and is also seen in many hemolytic anemias. Teardrop cells are also known as dacrocytes.

**Sickle cells** occur predominantly in patients with sickle-cell anemia. This is an inherited abnormality found mainly in people who are homozygous for hemoglobin S (HbS). Sickle cells are usually shaped like a crescent or sickle but may assume a wide variety of shapes, as is illustrated in figure 4.7*b*. Sickle cells are also known as **drepanocytes** or **meniscocytes.**

**Target cells** are an abnormal form of reticulocyte in which there is a spot or disc of hemoglobin in the center of the cell surrounded by a clear area, which in turn is surrounded by a rim of hemoglobin at the outer edge (figure 4.7*c*). The inner spot gives the cell the appearance of a target. These target cells are also known as **codocytes** and are commonly found in such diseases as thalassemia major and sickle-cell anemia.

**Schistocytes** or schizocytes, are fragmented cells that appear in various shapes and sizes, from small triangular forms to normally sized cells with grossly distorted outlines (figure 4.7*d*). They are seen in a number of different anemias but are especially common in hemolytic anemias, in severe burns, and in DIC (=disseminated intravascular coagulation).

**Acanthocytes,** or "thorny" cells, are abnormally shaped erythrocytes with many spiny projections on their outer surface (figure 4.7*e*). They are also known as **spur cells,** as they have spikes of varying length extending from the cell surface. Since they have no basic cell shape, they may be considered a subclass of schistocytes. These abnormal red blood cells are commonly seen in alcohol liver disease, disorders of lipid metabolism, and people who have undergone splenectomy.

**Keratocytes** are horn-shaped cells—that is, they are red blood cells with one or two projections that look like horns on a helmet. Hence these cells are also known as **helmet cells.** This condition is

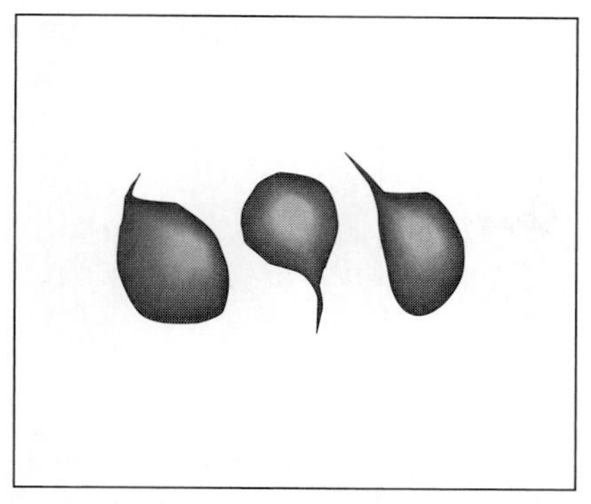

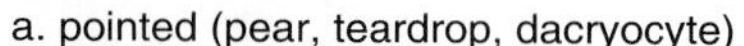

a. pointed (pear, teardrop, dacryocyte)

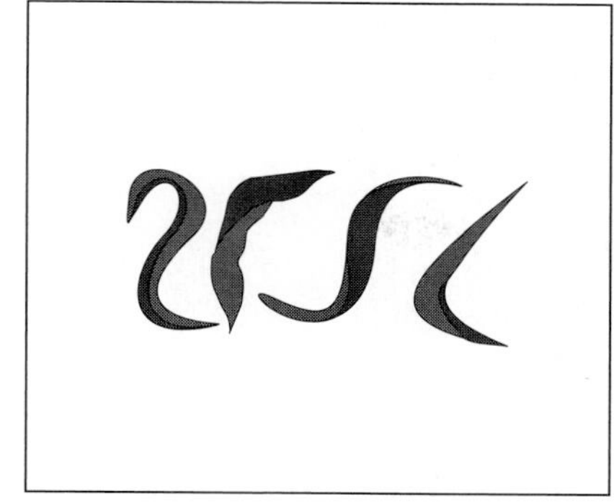

b. sickled (drepanocyte, meniscocyte)

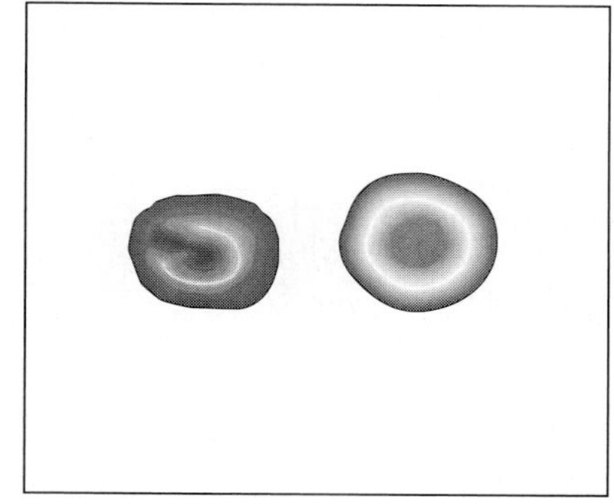

c. target (codocyte)

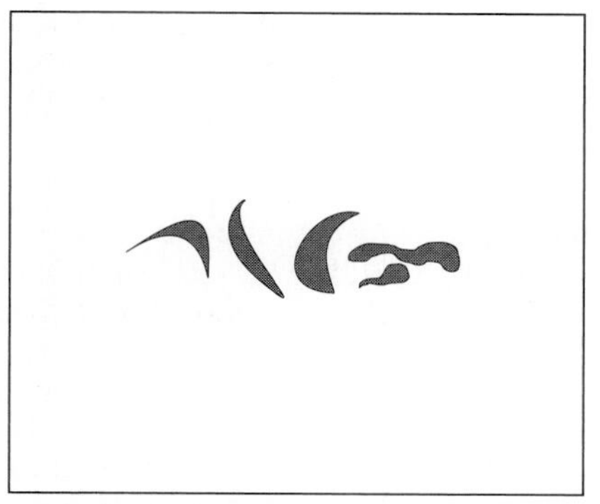

d. schistocyte (schizocyte)

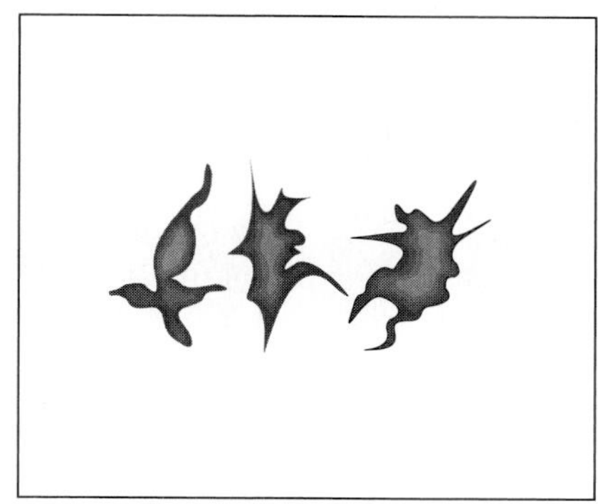

e. acanthocyte (thorn, spur, spicule)

f. helmet (keratocyte)

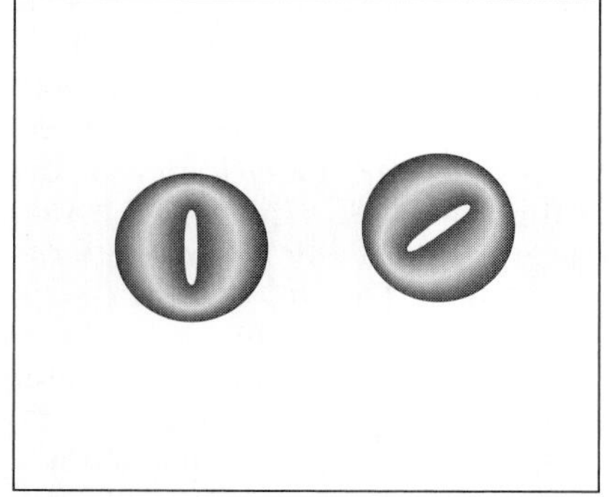

g. stomatocyte

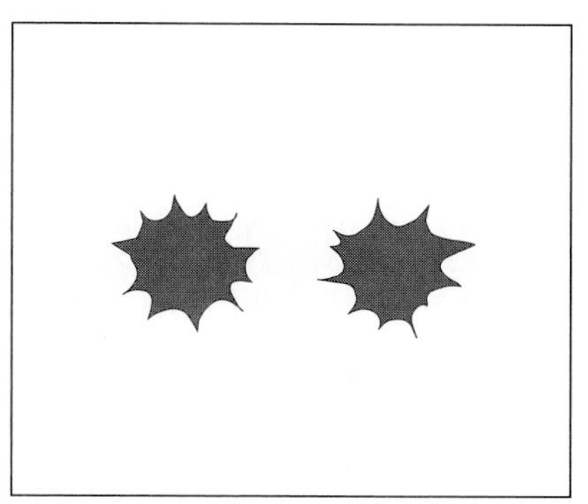

h. burr (echinocyte)

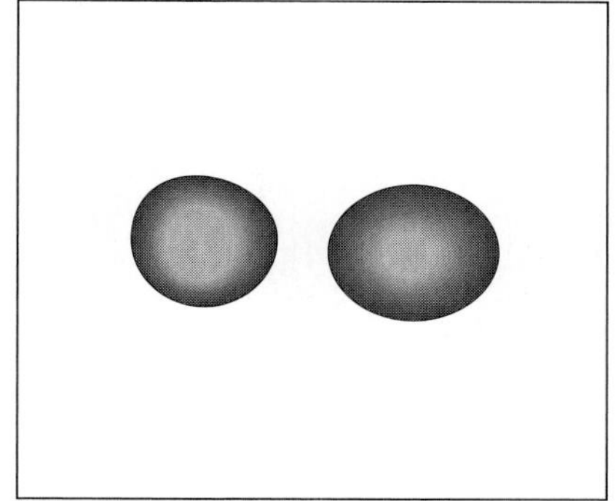

i. normal (discocyte)

**figure 4.7**

Poikilocytosis (*a*) through (*h*), and normal red blood cells (*i*).

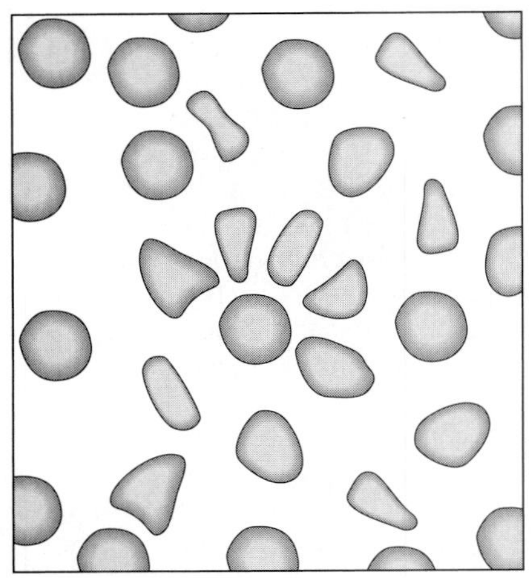
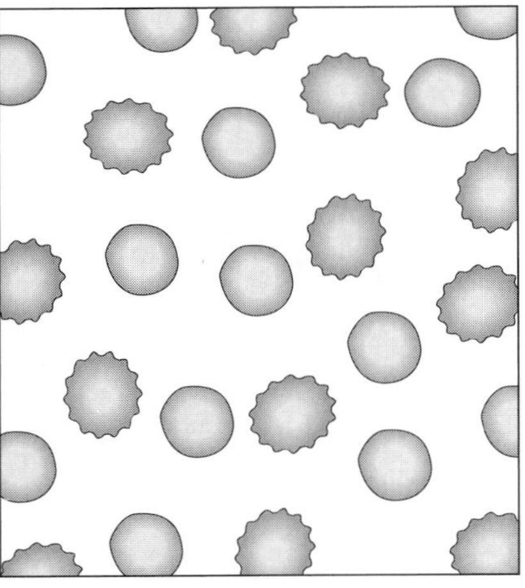
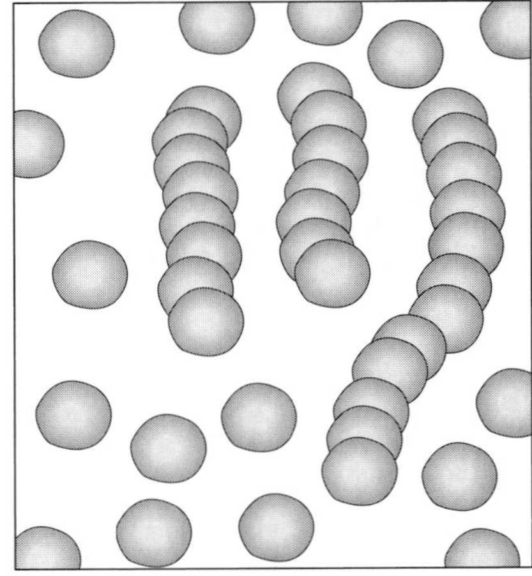

a. design formation b. crenated erythrocytes c. rouleau formation

**figure 4.8**

Normal red blood cell variations frequently encountered in blood smears: (*a*) design formation; (*b*) crenation; (*c*) rouleau formation.

frequently found in microangiopathic hemolytic anemias (figure 4.7*f*).

**Stomatocytes** or **mouth cells** are uniconcave cells with the shape of a cup. On a stained blood smear they have a slitlike area of central pallor. The presence of many of these cells may be the result of an inherited condition or may be caused by cirrhosis of the liver or lead poisoning. They are also frequently seen in a number of neoplastic diseases (figure 4.7*g*).

**Burr cells** are characterized by the presence of many short regular spines extending from the outer edge of the red blood cells, as shown in figure 4.7*h*. They are also known as **echinocytes.** Burr cells are commonly seen in cases of uremia and neonatal liver disease.

**Leptocytes** are also known as **thin cells,** as they appear as very thin, flat cells with only a thin rim of hemoglobin coloring the edges. They are seen in a variety of anemias including iron deficiency anemia, thalassemia, and other abnormal hemoglobin disorders (figure 4.5*a*).

**Design formation** is another common artifact. This condition occurs when a number of erythrocytes get bunched up and are pushed out of shape by the surrounding red blood cells, as illustrated in figure 4.8*a*. This may be caused by fat or oil on the slide applying too much pressure on the red blood cells while preparing a blood smear, or by bad technique in smear preparation.

**Crenated cells** have many projections on their outer surface, which may be either sharp or blunt (figure 4.8*b*). These artifacts are due to the shrinkage of the cells in hypertonic media or in conditions of increased pH. They may be seen in both wet and dry preparations and are frequently the result of improper laboratory smear preparation.

**Rouleau formation** is another artifact often encountered in blood smears. It is not an abnormality of red blood cell shape, but red cells that align face to face in columns, resembling a "stack of coins" (figure 4.8*c*). A few rouleau formations are commonly seen in a blood film and may be considered artifacts of the smearing technique. If, however, we see numerous rouleau formations in a blood smear, it may indicate the presence of increased plasma globulin, which may be indicative of a disease known as multiple myeloma. It may also indicate that the blood sample is getting old.

### Content of Erythrocytes

The normal red blood cell is a biconcave-shaped cell filled mainly with **hemoglobin.** Because the erythrocyte is thicker at the edges than in the middle, more hemoglobin is found on the outside of the cell, and for that reason a normal erythrocyte stains pink to red on the outside with a pale area in the center. However not all red blood cells contain equal amounts of hemoglobin. In some disease states hemoglobin production is below normal—consequently those cells show an increased central pallor with only a thin rim of stained hemoglobin visible. This phenomenon is known as **hypochromia** and is illustrated in figure 4.5*a*. Common in iron deficiency anemia, hypochromia may also be encountered in other anemias.

In certain situations staining with Romanowsky dyes does not result in the production of uniformly pinkish-red erythrocytes. Sometimes a number of red blood cells in a blood smear will stain various shades of blue or gray with tinges of pink. This is known as **polychromatophilia** and indicates the presence of immature red blood cells. This condition is especially common in a number of hemolytic anemias.

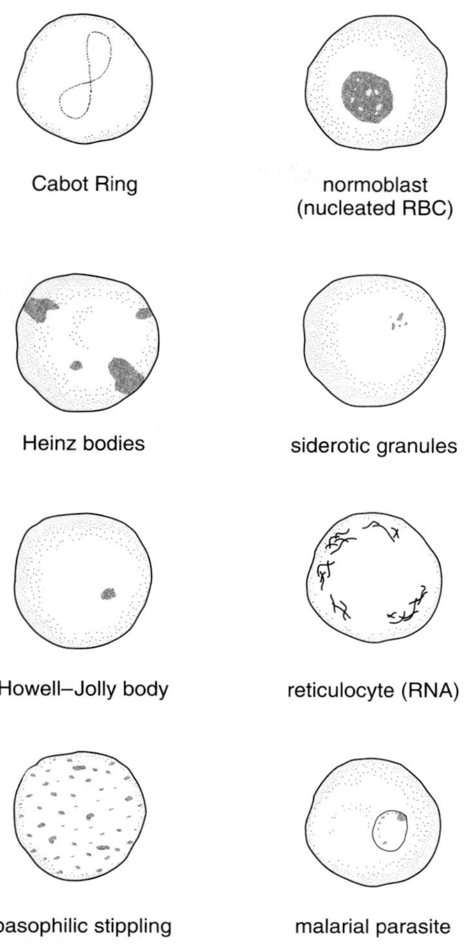


**figure 4.9**

Red blood cell inclusions that may be encountered in blood smears.

Sometimes blood films will show red blood cells with deeply colored spots or rings inside their cytoplasm. Often these inclusions are indications of disease patterns. The most common of these erythrocyte inclusions are as follows:

**Howell-Jolly bodies.** These are small, well-defined particles in the red blood cell, usually about 1 μm in diameter, that stain either deep blue, purple, or red, depending on the staining technique used (figure 4.9). Normally they occur singly, but sometimes occur in multiples. Howell-Jolly bodies are thought to be nuclear fragments and occur in cases of severe anemia and also after splenectomies.

**Heinz bodies.** These particles are not visible with the normal Wright's stain but can readily be seen with special supravital stains such as methyl violet (figure 4.9). They appear as irregular purplish particles usually found near the edge of the erythrocyte and are generally not bigger than 1 μm in diameter. Heinz bodies are granules of precipitated hemoglobin, and their presence is thought to indicate some oxidative injury to the red blood cell. They are commonly found in several types of hemolytic anemia.

**Cabot rings.** These are purple or reddish-staining ringlike inclusions. Cabot rings appear as thin threads in the form of a ring or a figure-eight (figure 4.9). They are thought to be remnants of fused microtubules that earlier formed the spindle apparatus associated with the process of mitosis. They occur only rarely and then only in severe cases of anemia.

**Pappenheimer bodies** are dark-blue staining granules found at the edge of red blood cells. They are iron-containing bodies and are commonly seen in sideroblastic anemia, thalassemia, and a number of other anemias (figure 4.9).

**Basophilic stippling.** This term refers to the presence of coarse or fine punctate basophilic inclusions in red blood cells (figure 4.9). The granules may be blue, gray, or brownish. In many cases these multiple diffuse dots represent deposits of aggregated ribosomes. When basophilic stippling involves a large number of erythrocytes (1 to 2%), it usually indicates lead poisoning.

Last, **malaria parasites** may occasionally be encountered in the red blood cell film. These protozoan parasites may be present in several shapes but are commonly found in the ring form, with a blue ring and a red dot, as shown in figure 4.9.

The untrained eye may mistake a platelet for an inclusion body. In a blood smear a platelet quite frequently will be present on top of a red blood cell. However, distinguishing between a platelet and an inclusion body is not difficult since a platelet is usually much larger and much more irregular in shape than the average inclusion particle.

## COMPONENTS OF RED BLOOD CELLS

One way to look at the structure of a red blood cell is to discuss its size, shape, and content, as was previously done. Another way to look at the structure is to discuss the various components that make up the red blood cell. The average cell is composed of three major parts: a nucleus, a membrane, and the cytoplasm. The red blood cell has only two of these three components, as the nucleus is extruded before the erythrocyte enters the peripheral circulation.

Red blood cells are essentially containers full of hemoglobin. The structure and the functions of hemoglobin are discussed in some detail in the following chapters. The biochemical machinery of the cytoplasm is discussed in the chapter dealing with the fate of red blood cells. Hence the only component that is discussed here is the red blood cell membrane.

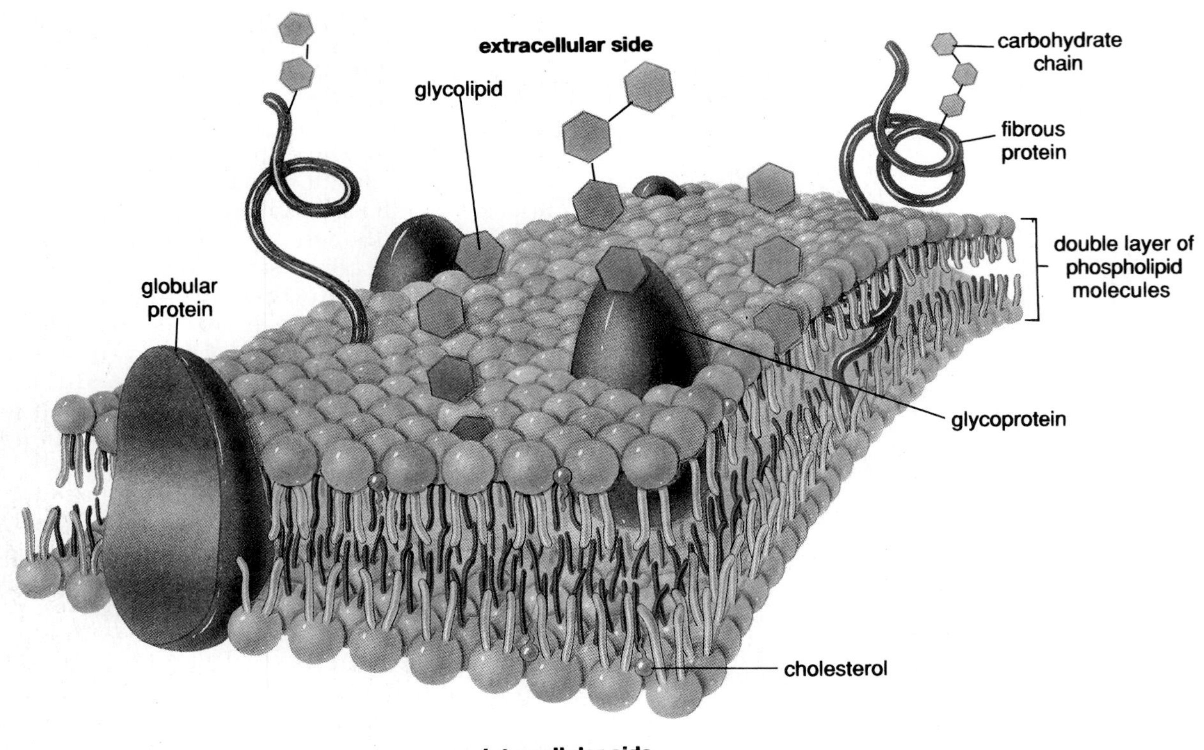


figure 4.10

The fluid-mosaic model of the cell membrane as proposed by Singer and Nicholson.

**Structure of the Erythrocyte Membrane**

The erythrocyte membrane is a dynamic structure, composed of many different chemical compounds, all of which play a role in maintaining the integrity or functions of the membrane. The structural complexity of this cell membrane has become increasingly evident with the development of new techniques designed to study membrane ultrastructure. A current model of the cell membrane incorporating many of the new findings is given in figure 4.10.

In this model the membrane consists of a double layer of phospholipid (the matrix) in which many proteins are present, forming a kind of mosaic when viewed from the outside. Some of these proteins are stationary, held in place by elements on the inside of the membrane; other proteins float in and on this fluid lipid membrane. Hence this model is often called the "fluid-mosaic" model of the cell membrane.

The chemical composition of the substances making up the cell membrane is as follows: about 50% of the membrane is protein, 40% is fat, and up to 10% is carbohydrate. Each of these components is discussed in the following material.

*Lipid Matrix*

The lipid matrix is made up of several different groups of fats. The major group is the phospholipids, which constitute about 60% of the total lipids, and the rest is made up of about 30% neutral lipids, mainly cholesterol, and 10% glycolipids.

The lipid membrane structure is composed of a double layer of phospholipids. Phospholipids are made up of a polar phosphate head and a nonpolar hydrophobic tail, made up of two fatty acids. The polar heads are always found on the outside and on the inside of the membrane, whereas the nonpolar hydrocarbon tails point toward the center of the membrane (figure 4.11). This lipid bipolar layer must have a certain degree of rigidity but also be sufficiently fluid to permit lateral movement of both

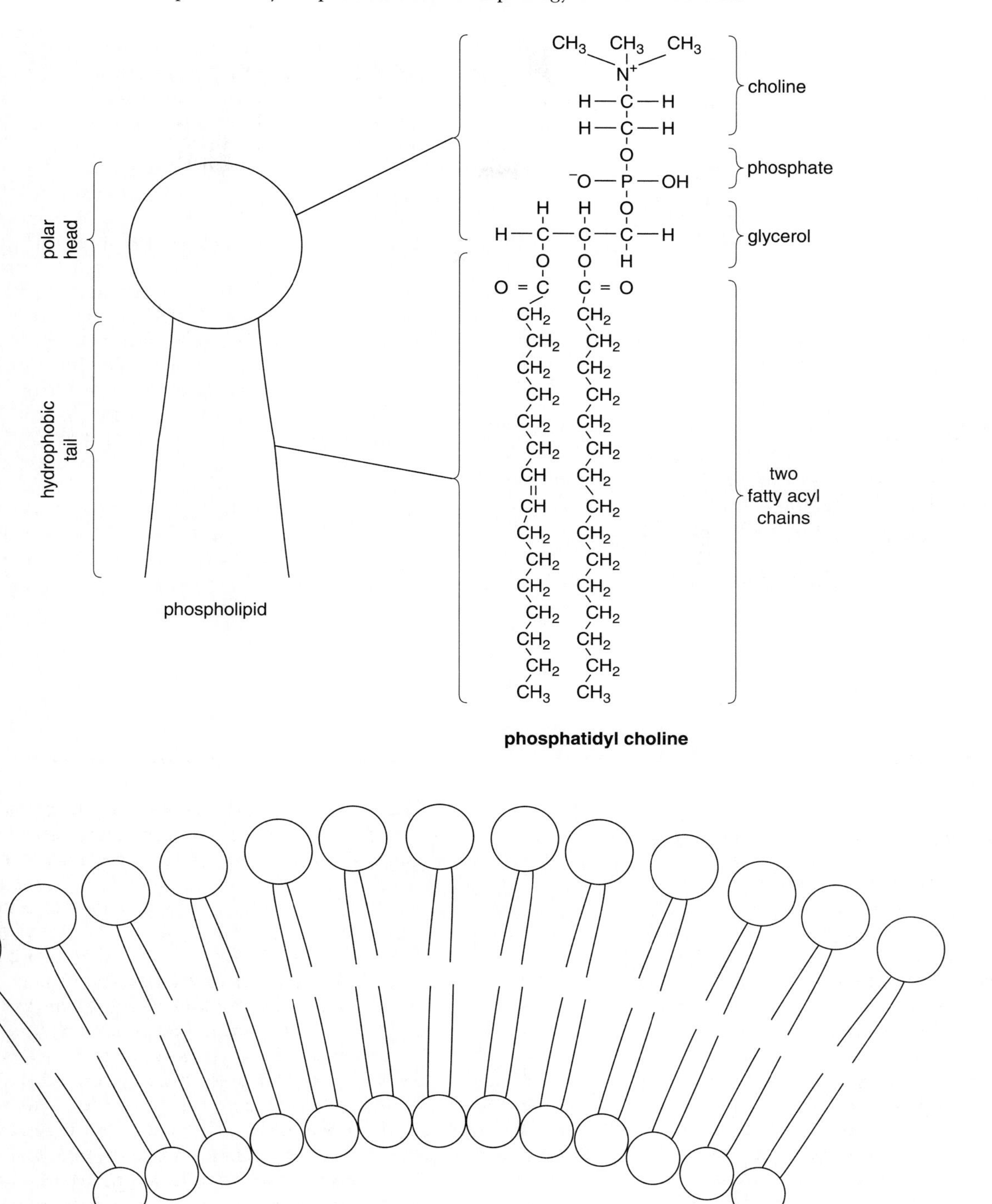


**figure 4.11**
The structure of phospholipid and its arrangement in the cell-membrane bilayer.

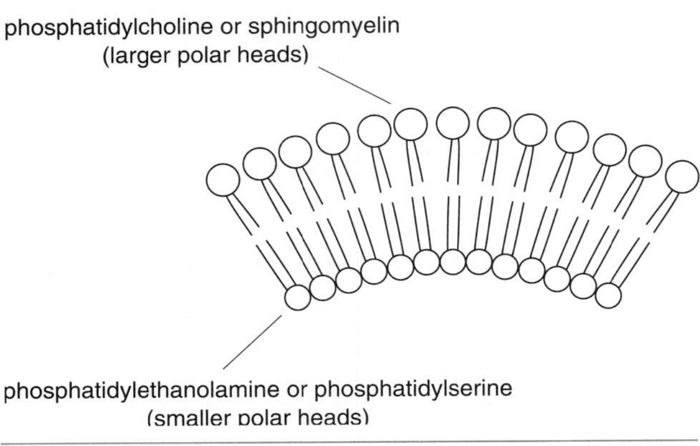


figure 4.12

Phospholipid placement in the cell membrane. To enable proper cell membrane curvature, phospholipids with larger polar heads tend to localize on the external part of the bilayer, while phospholipids with smaller polar heads tend to localize on the internal part.

the constituent lipids and the many proteins. Current thinking is that the lipid matrix is not a homogeneous mass, but rather that it consists of a more stable gel-like phase separated by more fluid patches. This lipid matrix must also conform to the curvature of the red blood cell membrane. The particular requirements may be met by alterations in the relative proportions of the different lipid components or by alterations in the fatty acid composition of particular kinds of lipids. The relative proportions of certain phospholipids appear to depend on the radius of the curvature of the membrane. For instance, phosphatidylcholine and sphingomyelin have larger polar head groups than phosphatidylethanolamine and phosphatidylserine and therefore tend to occur on the outer surfaces of the lipid core, where there is more area per residue, as illustrated in figure 4.12.

The two polar layers consist almost exclusively of phospholipids and glycolipids, whereas cholesterol is found almost exclusively in the center of the membrane. It appears to fit neatly between the hydrocarbon tails of these structural lipids, and this may explain why cholesterol stabilizes the two-layer structure.

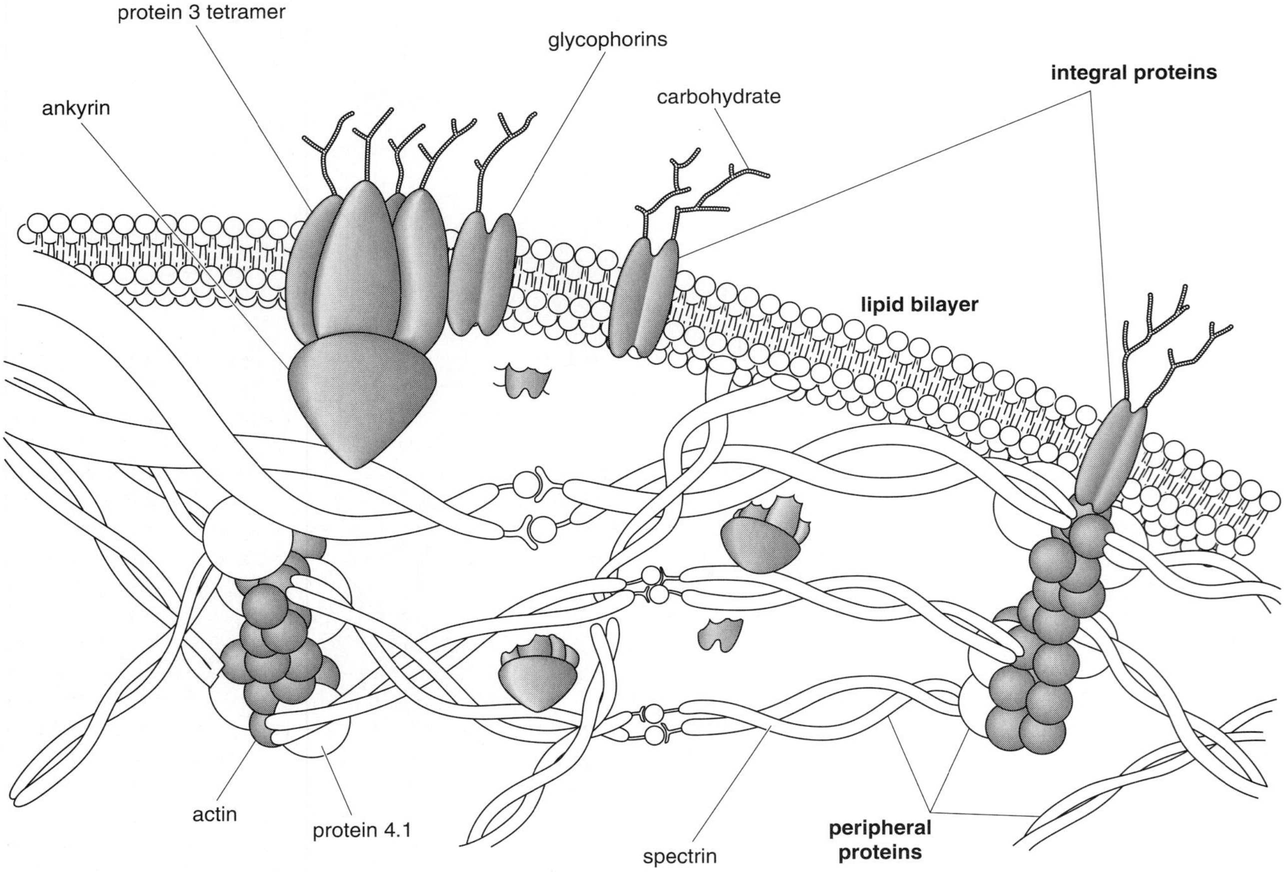

figure 4.13

Diagram of a cell membrane with integral and peripheral proteins.

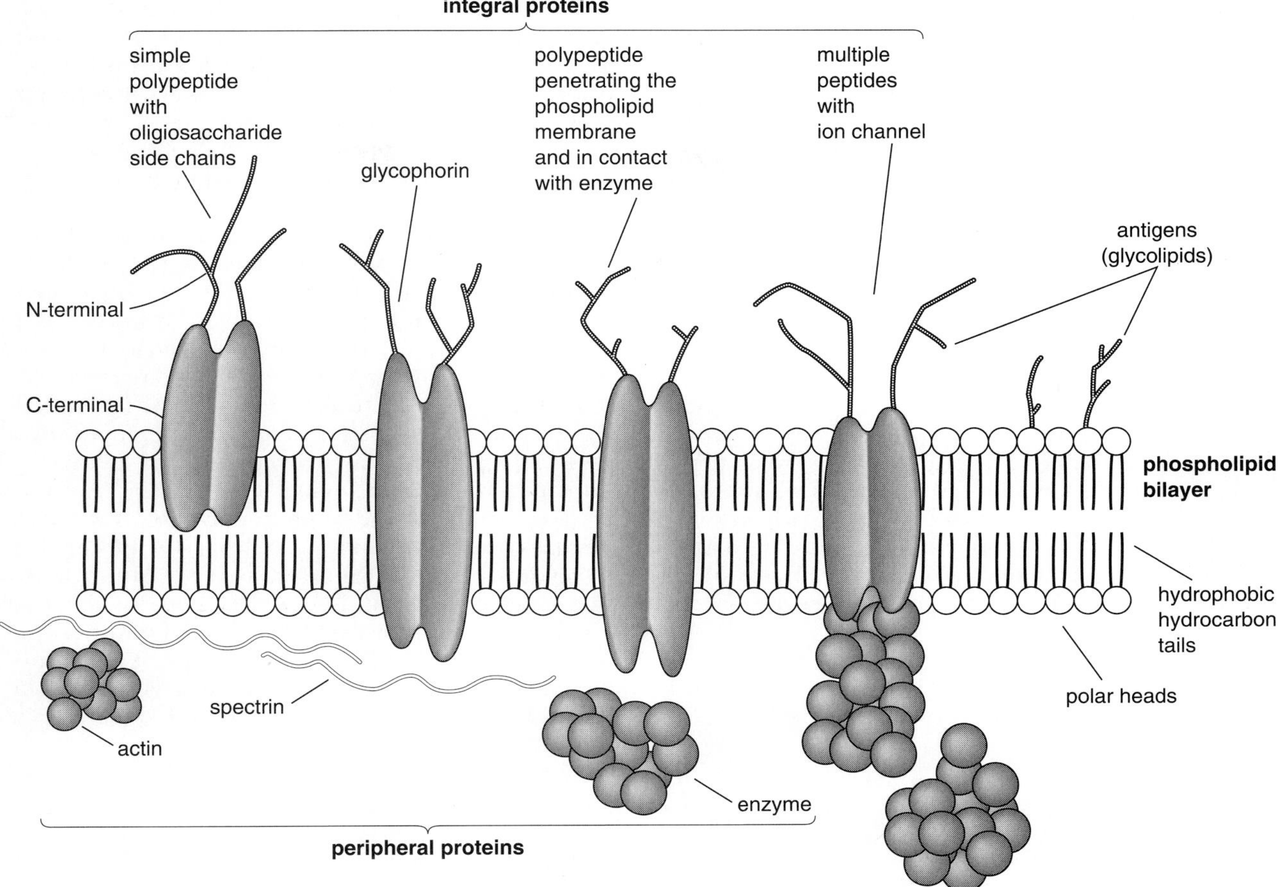


**figure 4.14**

Diagram of a cell membrane with various glycoproteins and glycolipids.

*Membrane Proteins*

Proteins associated with the red blood cell membrane may be classified according to their position, structure, or function. As regards their position, they may be grouped into peripheral and integral proteins. **Peripheral proteins** are proteins that are only loosely connected with the membrane and may easily be separated from it by ultracentrifugation and other methods. To this group belong the structural proteins, such as spectrin and actin, which are found on the inside of the membrane and play a major role in giving the erythrocyte its shape (figure 4.13). The **integral proteins,** on the other hand, penetrate the phospholipid bilayer and cannot be separated from it without destroying the membrane. This group includes receptor proteins and transmembranal transport proteins.

From a structural point of view, membrane proteins may be divided into glycoproteins, lipoproteins, glycolipoproteins, and simple polypeptides. The great majority of membrane proteins are thought to be glycoproteins—that is, proteins that have one or more carbohydrate moieties attached to them. Three major types of glycoproteins are generally recognized, all of which have their carbohydrate groups on the outside of the cell membrane (figure 4.14). The first type is a simple glycoprotein embedded in the lipid matrix with oligosaccharide side-chains attached to the N-terminal portion of the polypeptide, whereas the C-terminal end is hidden in the fluid matrix. The second group is made up of large glycoproteins that protrude through the inner surface of the cell membrane and may be connected to some structural protein on the inside of the membrane (e.g., glycophorin A, which accounts for about 10% of all the integral proteins). Last, are those protruding glycoproteins that consist of two or more polypeptide chains; these appear to be responsible for ion transport through the membrane by making ion channels.

Membrane proteins may also be divided according to their function into structural proteins, transport proteins, and receptor proteins. Structural proteins play a role in maintaining the shape of the red blood cell, such as spectrin, which accounts for about 20% of all membrane proteins. Transport proteins are multiple polypeptide structures that regulate the entrance of materials into the cell, whereas receptor proteins bind to hormones and other messenger molecules, thus forming complexes that initiate changes in the metabolic processes of the cell.

*Carbohydrate Moieties*

The red blood cell membrane does not contain independent carbohydrate molecules. All carbohydrates are present as part of other membrane molecules—such as proteins (forming glycoproteins), lipids (glycolipids), or mixtures of both (glycolipoproteins) (figure 4.14). These carbohydrate groups are almost all found on the outside of the cell membrane. Many of the carbohydrate moieties on the red blood cell act as antigenic determinants and play a role in the recognition of these red blood cells as belonging to the "self." For instance, the ABO blood group antigens are carbohydrate structures attached to lipids (glycolipids).

Individuals receiving blood transfusions should get red blood cells that match the antigens found on their own blood cells; otherwise the transfused red blood cells will be rapidly destroyed by the recipient's immune system.

**Importance of the Erythrocyte Membrane**

The red blood cell membrane must be composed of the proper compounds and each compound must be present in the right quantities. Defects in proteins may result in abnormalities in cell shape. For instance, defects in the structure of spectrin are said to be one of the major causes of an anemia known as hereditary spherocytosis. In this condition most of the red blood cells are spherical rather than biconcave. This leads to early destruction of the erythrocytes and may lead to anemia.

Alterations in the lipid composition due to congenital or acquired abnormalities in the plasma, cholesterol, or phospholipid levels may also result in membrane abnormalities and hence early hemolysis. For instance, an increased cholesterol and phospholipid level has been suggested as one of the causes of target cell anemia, whereas a large selective increase of cholesterol alone may account for acanthocyte formation.

## *Further Readings*


Diggs, L. W., Sturm, D., and Bell, A. *The Morphology of Human Blood Cells,* 5th ed., Abbott Laboratories, Abbott Park, Ill., 1985.

Elgsdaeter, A., Stokke, B. T., Mikkelsen, A., and Branton, D. The Molecular Basis of Erythrocyte Shape. *Science,* 234:1217 (1986).

Marchesi, V. T. The Cytoskeletal System of Red Blood Cells. *Hospital Practice,* Nov. 1985.

Shohet, S. B., and Lux, S. E. The Erythrocyte Membrane Skeleton: Pathophysiology. *Hospital Practice,* Nov. 1984.


## *Review Questions*

1. Distinguish between effective and ineffective erythropoiesis. What is the normal level of effective erythropoiesis?
2. What is the length of the proliferation and maturation time of an erythrocyte? What is the life span of a red blood cell?
3. Name the six stages of development of red blood cells. Describe the distinguishing characteristics of each stage.
4. Describe a pyknotic nucleus.
5. What is the normal reticulocyte level in the bloodstream? What does it mean when that level is increased?
6. Name the major growth factors that regulate erythrocyte production in steady-state and non-steady-state situations.
7. What is the normal diameter and thickness of a red blood cell?
8. When do we refer to an erythrocyte as a macrocyte and a microcyte?
9. What is meant by the term anisocytosis? When could this condition develop?
10. Define the term poikilocytosis.
11. Describe the following poikilocytes including when they normally occur, and provide alternate name(s) for these cells.

    echinocyte, acanthocyte, elliptocyte
    drepanocyte, dacrocyte, codocyte
    schistocyte, keratocyte, spherocyte
    stomatocyte, leptocyte

12. What is meant by (a) design formation, (b) rouleau formation, and (c) crenated red blood cells? What causes these conditions?
13. What is meant by hypochromia and polychromatophilia? What causes these conditions?
14. Define the following erythrocyte inclusion bodies: (a) Howell-Jolly bodies, (b) Heinz bodies, (c) Pappenheimer bodies, and (d) basophilic stippling. What conditions are associated with the presence of these inclusions?
15. Describe the fluid-mosaic model of a cell membrane.
16. What is the chemical composition of the red cell membrane?
17. Name the major groups of lipids found in the red blood cell membrane. In what proportions are they present? What is the function of each of these lipids?
18. Distinguish between integral and peripheral membrane proteins.
19. List the major groups of functional proteins in the cell membrane.
20. How would you classify the ABO blood group antigens on the red blood cell membrane?
21. Explain why all constituents of the red blood cell membrane need to be present in the right proportions. Give an example.

# Chapter Five

## *Structure of Hemoglobin*

## STRUCTURE OF HEMOGLOBIN

The major function of red blood cells is to carry blood gases. Erythrocytes are the principal agents that transport oxygen from the lungs to the respiring tissues and also transport most of the carbon dioxide from the tissues back to the lungs. These functions are accomplished mainly by the protein hemoglobin, which is ideally suited to carry oxygen and also is a great help in the transport of carbon dioxide by acting as a buffer system. To understand how hemoglobin performs these functions in such an efficient manner, we need an awareness of the chemical structure of this remarkable molecule.

The hemoglobin molecule is a complex structure made up of four subunits. Stated another way, hemoglobin is a tetramer made up of four monomers, as shown in figure 5.1. Each monomer consists of a heme and a globin unit. The globin units are made up of polypeptide chains of two distinct types. In human adults one type is referred to as the alpha-chains (α-chains), and the other two are called beta-chains (β-chains).

In discussing the actual structure of the hemoglobin molecule, we customarily describe the heme fraction first, then account for the structure of the globin unit; next we consider the monomer made up of a heme and a globin, and finally consider the structure of the tetramer.

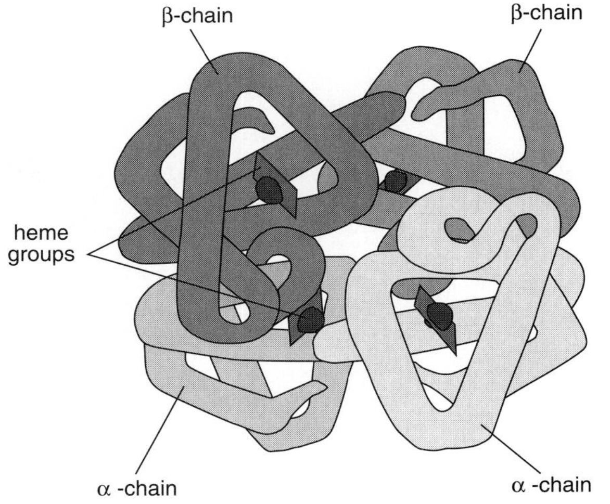


**figure 5.1**

The three-dimensional structure of the hemoglobin molecule.

### Structure of Heme

**Heme** is a chemical structure made up of a porphyrin ring with an iron atom inserted in the center, which is attached to four nitrogen atoms, as illustrated in figure 5.2. The **porphyrin ring** structure is present in some form in almost all living organisms. The porphyrin skeleton is found in hemoglobin, myoglobin, cytochrome, chlorophyll, and in many oxidases and catalases. Almost all living organisms have the ability to synthesize these porphyrin structures. Among the rare exceptions are the *Hemophilus* group of bacteria (e.g., *H. pertussis* and *H. influenzae*), which require hemoglobin derivatives in their culture media to compensate for this metabolic defect and are thus appropriately named *Hemophilus*. Porphyrin structures have the ability to form coordination compounds with metallic elements—in heme the porphyrin binds to iron, in chlorophyll to magnesium, and in vitamin $B_{12}$ to cobalt.

The porphyrin ring of the heme structure can bind with either bivalent or trivalent iron. In the former case ($Fe^{2+}$) it is known as **ferroheme** (or **ferrous heme**); in the latter situation ($Fe^{3+}$) it is called **ferriheme** (or **ferric heme**). Heme compounds bind readily to various nitrogen bases to form a class of compounds known as hemochromogens, or hemochromes. Thus hemoglobin may be classified as a hemochrome in which the protein globin is the nitrogenous base.

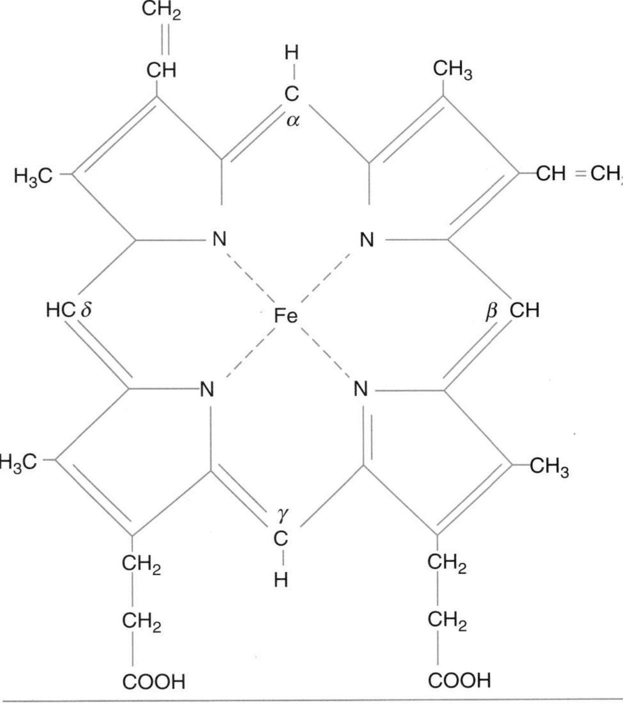


**figure 5.2**

The heme structure.

The biosynthesis of heme involves a number of steps, each catalyzed by a different enzyme, as illustrated in figure 5.3.

The porphyrin ring is made up of four pyrrole units. The **pyrrole** structure as such does not exist in nature, but humans have the ability to produce it from a related compound known as **porphobilinogen**

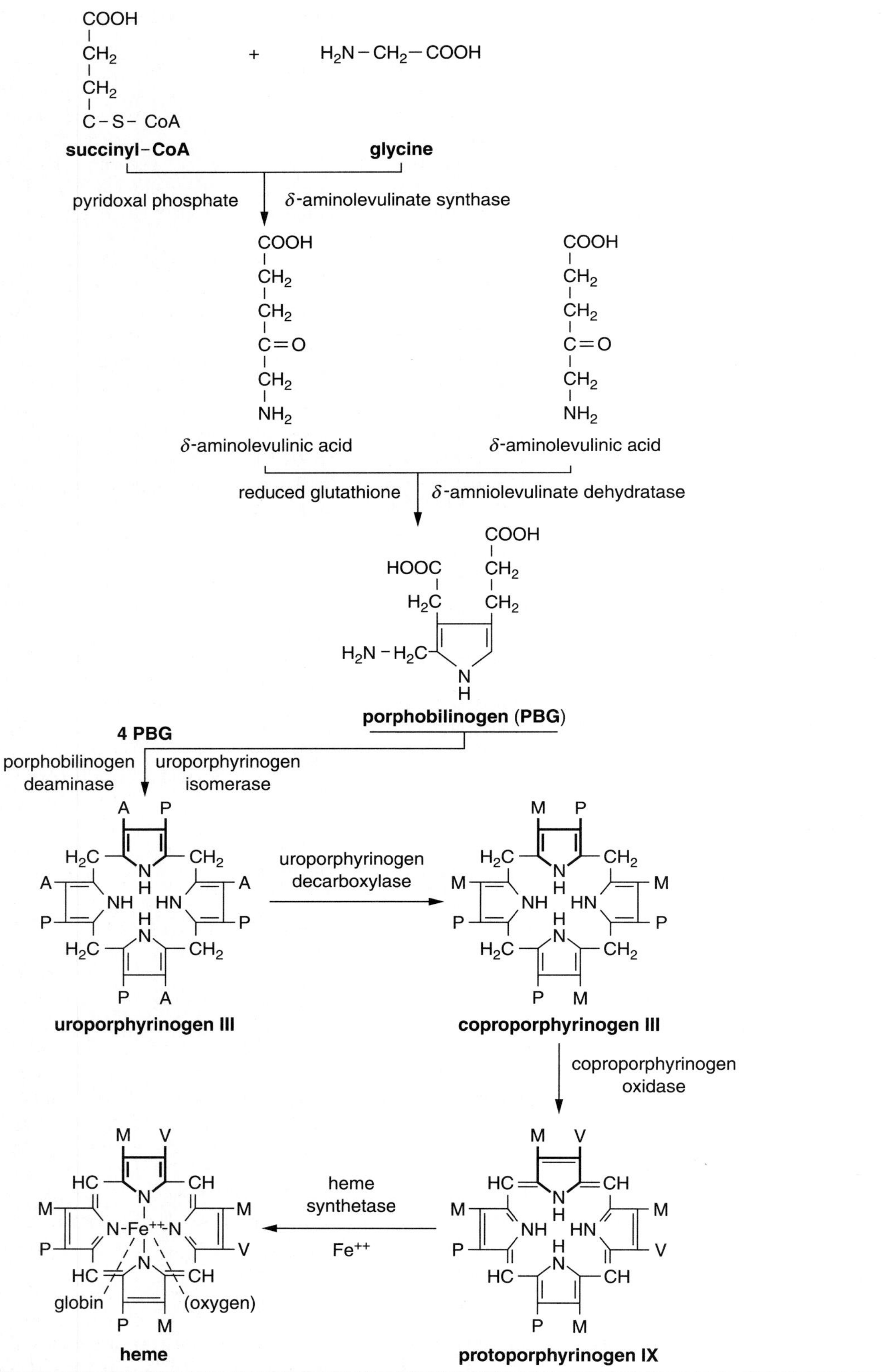


**figure 5.3**

The biosynthesis of heme. A = acetyl (-$CH_2$-COOH); P = propionyl (-$CH_2$-$CH_2$-COOH); M = methyl (-$CH_3$); V = vinyl (-CH=$CH_2$).

**(PBG).** The synthesis of porphobilinogen involves the condensation of succinyl coenzyme A (formed in the tricarboxylic acid cycle) and glycine into Δ-aminolevulinic acid (Δ-ALA). The enzyme catalyzing this reaction is called Δ-ALA synthetase, and evidence indicates that this enzyme is rate-limiting in the pathway of heme synthesis. Two molecules of Δ-ALA condense to form porphobilinogen. This reaction is catalyzed by the enzyme Δ-ALA dehydrase in the presence of reduced glutathione. In turn four molecules of PBG condense to form uroporphyrinogen. Since it is possible to interchange the external side-chains of each pyrrole fragment (i.e., the acetyl and propionyl groups), several different stereoisomers of the same compound can be produced. However for heme synthesis, only **uroporphyrinogen III** is important. This tetrapyrrole is then changed to **coproporphyrinogen III** by the decarboxylation of acetic side-chains to methyl groups on each pyrrole unit. Coproporphyrinogen is then converted into **protoporphyrinogen IX** by oxidation of the propionic acid side-chains on two of its pyrroles to vinyl groups. Because protoporphyrinogen has three types of side-chains—methyl, vinyl, and propionic acid—15 different possible isomers of protoporphyrinogen could be produced, but only **protoporphyrinogen IX** is utilized in the formation of heme and is the only isomer important in biology. The final step in heme synthesis is the formation of **protoporphyrin IX** from protoporphyrinogen IX by the removal of six hydrogen atoms. Simultaneously a ferrous iron is inserted into the porphyrin ring by the enzyme heme synthetase to produce heme, the final product. All these events take place in the mitochondria.

### Structure of Globin

The globin portion of the hemoglobin monomer is a protein. Generally speaking, proteins consist of chains of **amino acids.** Amino acids are small molecules with molecular weights of 100 to 200 daltons. About 20 different amino acids occur in nature, and they may be present in almost any sequence, opening up the possibility of virtually an unlimited number of different protein molecules. The general structure of an amino acid is as follows:

$$
\begin{array}{c}
R \\
| \\
H_2N - C - COOH \\
| \\
H
\end{array}
$$

where R is the variable. R can be just another H atom, as in glycine, or may consist of a much more complex side-chain as in phenylalanine (figure 5.4). The $NH_2$ fraction of the molecule is known as the amino group, and the COOH part is called the carboxyl group. Apart from glycine, all amino acids may exist in either D (dextro, or right) and L (levo, or left) forms. In other words, mirror images of the molecules can exist that will rotate polarized light in either the D or L direction. This is a consequence of the so-called alpha-carbon in the center of the molecule being asymmetric and giving the molecule two alternative stable positional forms, which are mirror images of each other. Although both forms of most amino acids occur in each cell, only L forms are utilized in the structure of proteins. A mixture of L and D forms would result in unstable compounds, since proper peptide bonds could not form.

A peptide bond is formed when two amino acids link up in such a way that the amino group of one amino acid is next to the carboxyl group of the next amino acid. When this happens, a water molecule is eliminated and a peptide bond is formed, as illustrated in figure 5.5. This is the reason why only one type of amino acid configuration (the L form) is used in the formation of polypeptides.

Although proteins may be defined as simple polypeptide chains made up of strings of amino acids, the actual structure of the protein is usually much more complicated. Normally three or four different levels of complexity are used to describe the actual configuration of a protein molecule.

The **primary structure** of a protein molecule refers to the basic number and sequence of amino acids in a polypeptide chain. On one end it starts with the amino group of the first amino acid (this is known as the N-terminal), and the polypeptide ends with the carboxyl group of the last amino acid (the C-terminal) (figure 5.6). Most globin structures are made up of about 150 amino acid residues, but the number varies for each globin chain. For instance, the human α-chain globin has 141 amino acids, and the β- and γ-chains each have 146 amino acid residues. Myoglobin has 153 amino acids in its primary structure.

The **secondary structure** of a protein molecule refers to the basic configuration of a polypeptide chain. The structure is almost invariably a simple coil or helix and is referred to as the **alpha-helix.** Each full turn of the helix involves a length of about four amino acids. The alpha-helix is normally stabilized by some bonding pattern between amino acids lying above and below each other in adjacent turns (figure 5.6*b*).

However not all parts of the polypeptide chain are twisted; some are simply strung out and uncoiled. About 75% of all amino acids in the globin structure are helical.

The **tertiary structure** of a protein involves the overall shape of each polypeptide chain. Some proteins occur in fairly straight coils, which includes the fibrous and the contractile proteins. The majority of proteins, however, are more or less globular, and the globins belong to this category. In order to obtain a

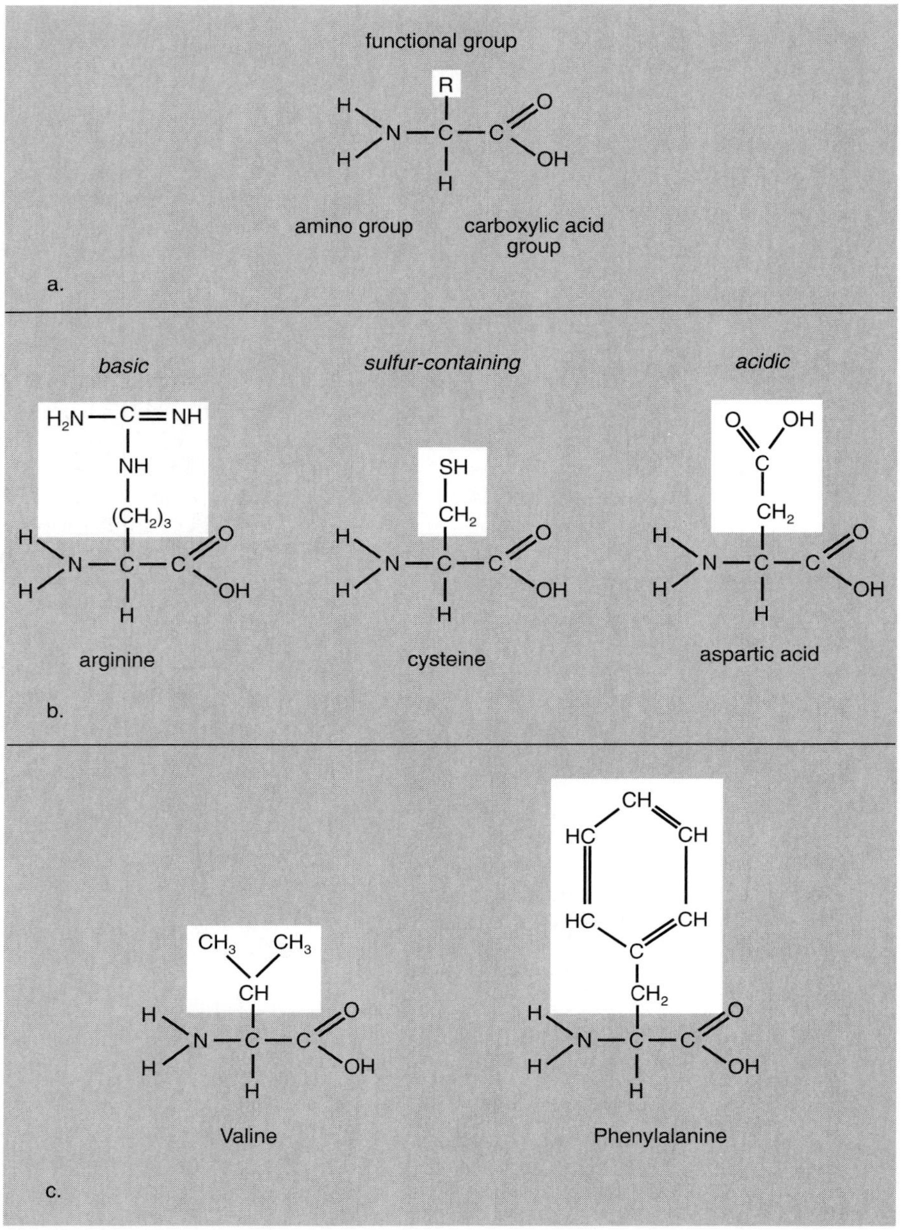

**figure 5.4**

Amino acid structure and examples. (*a*) All amino acids have a central carbon, an amino group (-$NH_2$), a carboxyl group (-COOH), and a functional (-R) group that differs in each amino acid. (*b*) Some R groups are either charged or polar. (*c*) Others are nonpolar.

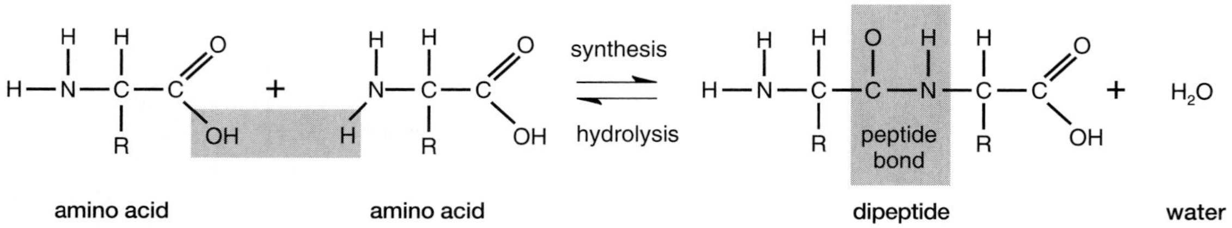

**figure 5.5**

The formation of peptide bonds. The amino group of one amino acid is linked to the carboxyl group of the next amino acid.

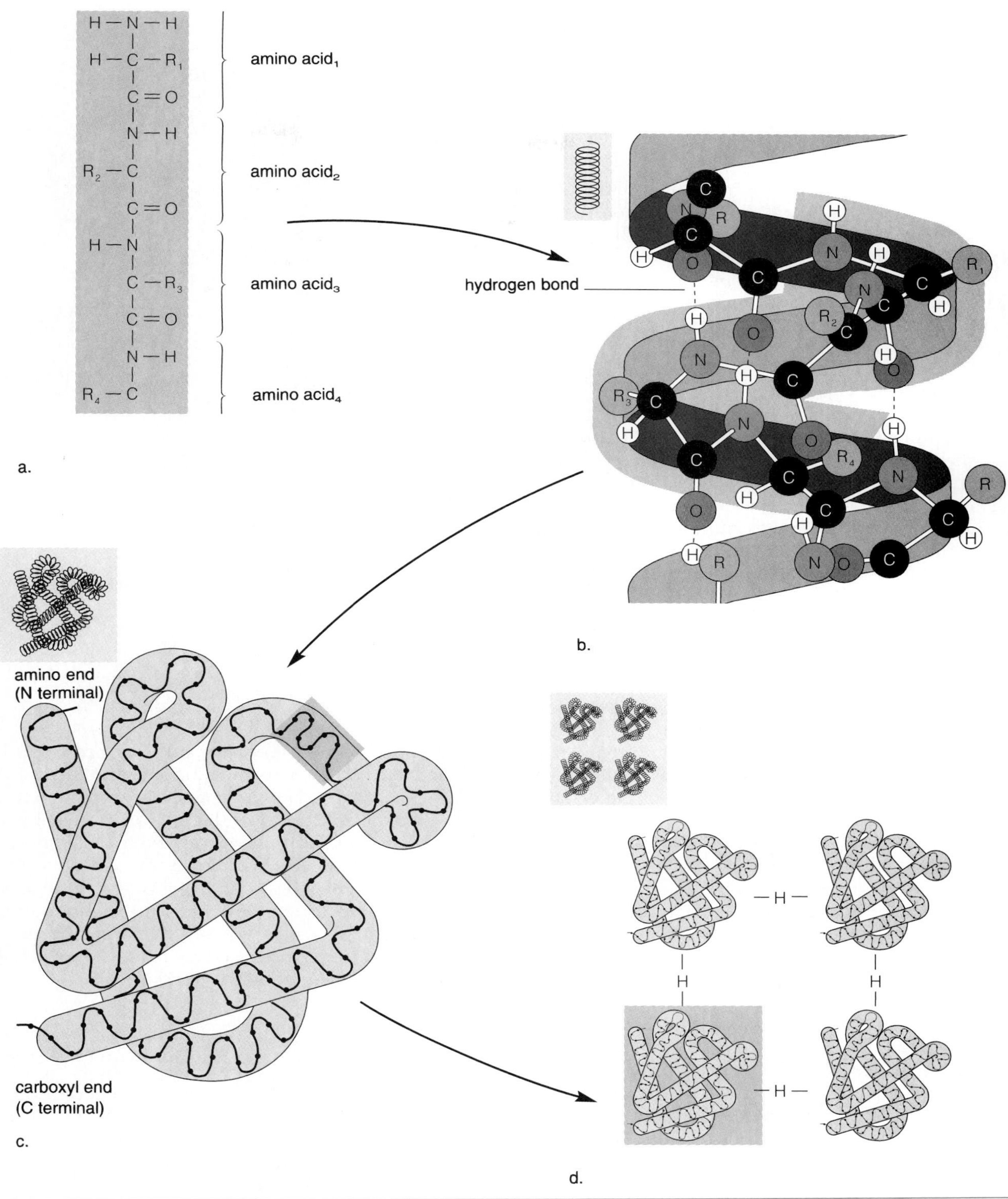


**figure 5.6**

Diagram showing (*a*) the primary structure of a protein; (*b*) the secondary structure (the α-helix); (*c*) the tertiary structure of a globin chain; and (*d*) the quaternary structure of hemoglobin.

globular form, the protein coil has to fold up onto itself. The form taken by the folded coil is absolutely precise and stabilized by bonds developed between adjacent folded sections (figure 5.7). To fold like this, some parts have to bend more than others, and luckily some amino acids are particularly prone to bending (e.g., proline).

Each globin monomer is made up of eight helical segments labeled A to H, which are joined together by nonhelical segments designated AB, BC, CD, and so on, as illustrated in figure 5.7.

When the composition of the amino acid sequences of the various globin monomers are compared, a rather confusing picture is produced if only the absolute sequence (primary structure) is taken into account. If, however, the amino acid sequences of the various helical segments are compared, certain key amino acids will always occupy the same position. For instance, the important amino acid histidine, which links the iron atom in the heme structure to the globin part, is the eighty-ninth amino acid in the α-chain and the ninety-second amino acid in the β-chain—but it is the eighth amino acid of the F helical segment in both the α- and the β-chains. In the modern nomenclature of the hemoglobin molecule, each amino acid is referred to by its numerical position in each helical segment (e.g., F8, E6, and A9).

### Hemoglobin Monomer

The heme fraction is present as a flat plate within a pocket of the folded globin (figure 5.7). The iron atom is attached to the four nitrogen atoms of the pyrrole subunits and is also covalently bonded to the F8 histidine, and the sixth bond should be occupied by oxygen (figure 5.8).

Approximately 21 amino acids of the globin chains come within 4 Å of the heme plate and hold it in position via some 60 atomic interactions. All but one contact in the α-chain and two in the β-chain are **nonpolar.** It is of vital importance that amino acids lining the internal heme pocket of each hemoglobin monomer are nonpolar in nature. Any substitution of a nonpolar amino acid with a polar amino acid will make the heme pocket less hydrophobic, and thus make it more difficult for the heme to bind with oxygen. As a matter of fact, any replacement of the 21 amino acids that bind to the heme group, even by another nonpolar amino acid, will interfere not only with the stability of the heme group but also with the tertiary structure of the globin chain.

The tertiary structure is the true three-dimensional representation of a protein. Most proteins need a given configuration to be able to perform their function. They require a certain shape to be able to interact with other substances. If they lose their proper shape (e.g., by heating), they are no longer able to perform their biological function, despite the fact that they still possess the right sequence of amino acids.

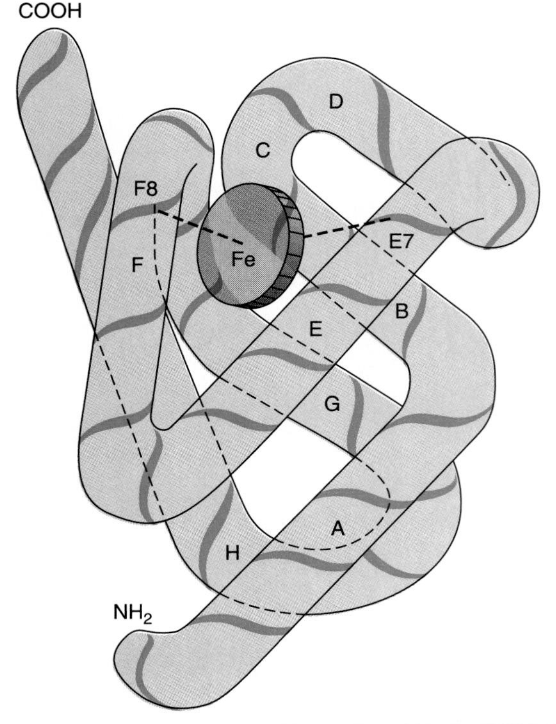


**figure 5.7**
The structure of a globin monomer showing the eight helical segments and the position of the heme plate in the heme pocket of the globin chain.

The correct tertiary structure of a protein is important not only for its function but also for its solubility. To stay in a soluble state, amino acids found on the outside of the globular protein must be **hydrophilic,** thus **polar** in nature. This is also important for hemoglobin, which is dissolved in the cytoplasm of the red blood cell. Any changes in the composition of amino acids on the outside of the hemoglobin molecule may make it less soluble, and increase its chance of precipitation and crystallization, thus rendering it useless as an oxygen carrier.

### Hemoglobin Tetramer

The **quaternary structure** of a protein is limited to complex molecules—that is, those made up of two or more polypeptide chains. Each of these subunits must be conjugated in an exact way to form a complex protein with the precise configuration to fulfill its function. The hemoglobin molecule is made up of four monomers, as illustrated in figure 5.6*d.* These monomers are held together by **hydrophobic bonds** between the adjacent parts. These bonds play a vital role in the physiological allostery displayed by the molecular complex. It is important to remember that not all subunits in the hemoglobin molecule are identical, although such hemoglobin molecules may

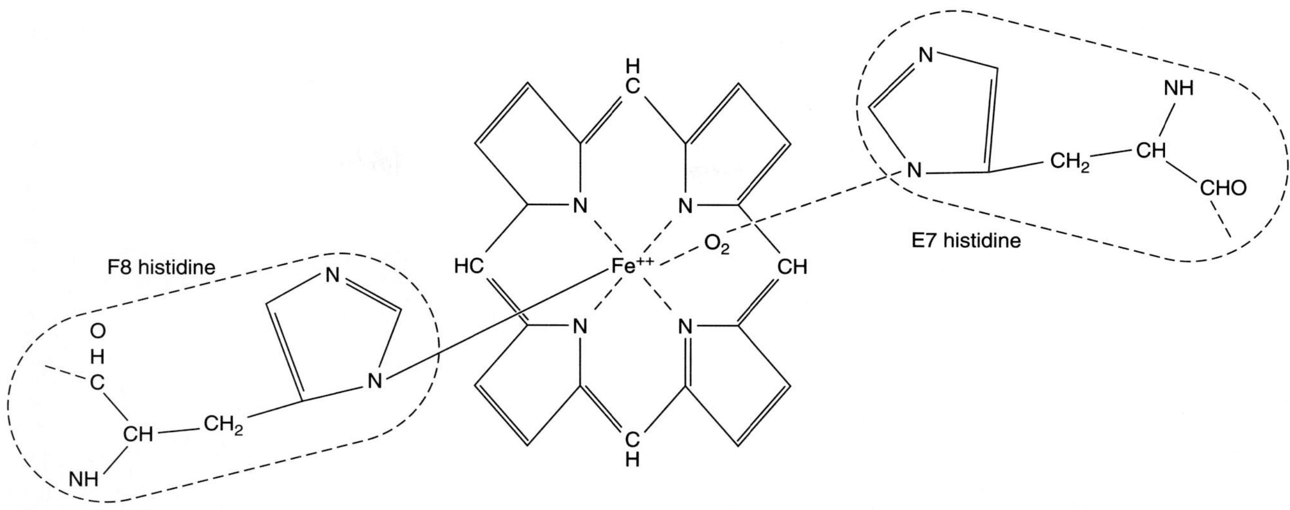


**figure 5.8**

The six valences of iron (Fe) in the globin monomer.

be formed in the laboratory and do occasionally occur in nature. The normal adult tetramer is made up of two α- and two β-chains. These chains are held together by two pairs of contacts called $\alpha_1\beta_1 + \alpha_2\beta_2$ and $\alpha_1\beta_2 + \alpha_2\beta_1$, as illustrated in figure 5.9. The first pair of contacts is large and is formed by 20 interactions between the α- and β-chains, mainly from the B and H helices. These stabilizing contacts prevent the dissociation of the dimers $\alpha_1\beta_1$ and $\alpha_2\beta_2$ into free monomers.

The $\alpha_1\beta_2 + \alpha_2\beta_1$ contacts are functional and comprise only nine amino acids of either chain, mainly from the C, F, and G segments. Large movement takes place across these contacts upon oxygenation and deoxygenation, resulting in changes in the quaternary structure.

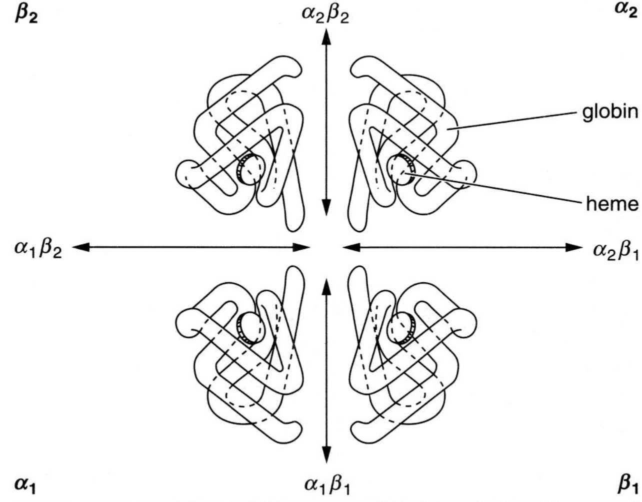


**figure 5.9**

The quaternary structure of hemoglobin showing the contacts between monomers.

## VARIATIONS IN THE STRUCTURE OF HEMOGLOBIN

It is understandable that a large and complex molecule such as hemoglobin, which is controlled by several different genes, has been subjected to evolutionary change more frequently than a small simple compound controlled by only one gene. Currently more than 300 different variations of the human hemoglobin molecule have been substantiated and described. These variations may be classified under two broad headings: normal and abnormal variations. Normal variations may be described as those types of hemoglobin that occur in most if not all human beings during some stage of their development. All the rest are classified as abnormal. The abnormalities may be either in the heme or the globin structure, but most frequently they are due to changes in the amino acid sequences of the globin chains. These variations may be caused by a number of factors, such as chromosomal deletions or elongations, or by hybrid hemoglobin formation—but most commonly they are due to single point mutations. Over 90% of all abnormal hemoglobin variations are due to single amino acid substitutions caused by single base substitutions in the corresponding triplet codon.

### Normal Variations

Normal hemoglobin variations correspond with different stages of human development. They fall neatly into three categories: embryonic, fetal, and adult

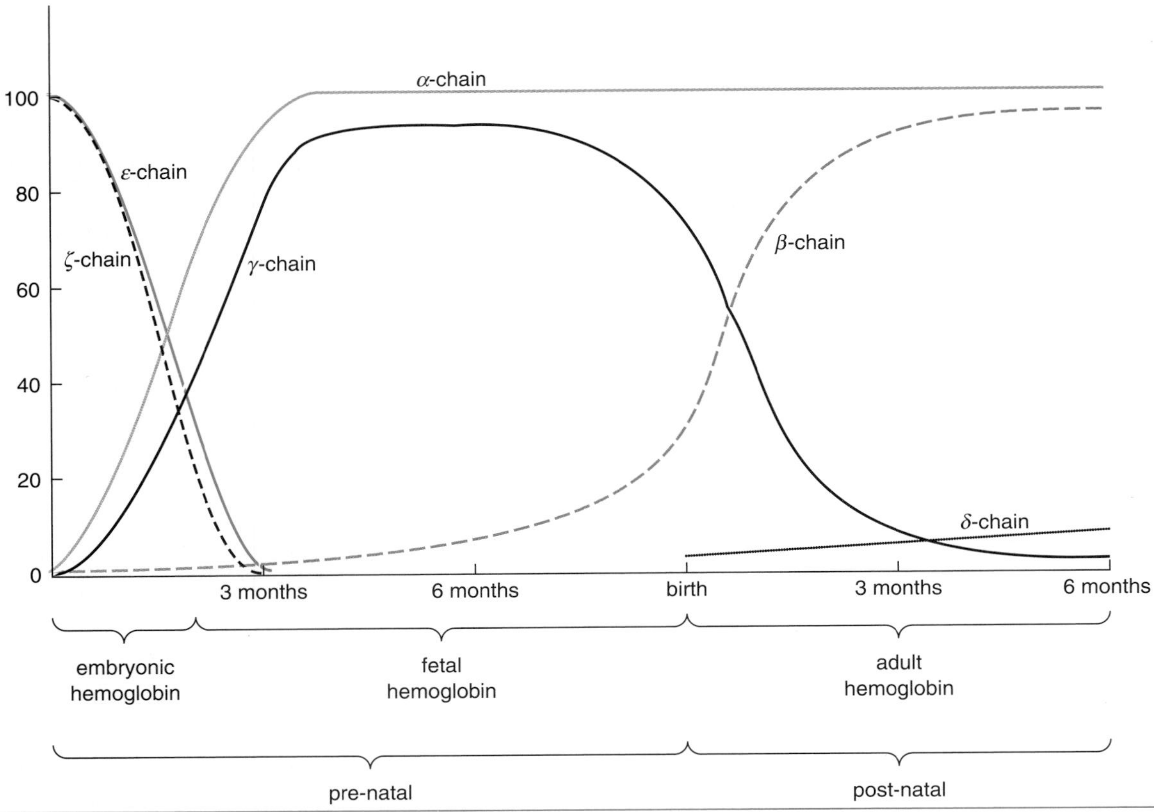


**figure 5.10**

The relative rates of globin chain synthesis during embryonic, fetal, and neonatal life.

From D. L. Rucknagel, *Clinical Obstetrics and Gynecology,* 12:49, 1969. Copyright © 1969 J. B. Lippincott Company, Philadelphia, PA. Reprinted by permission of the publisher and the author.

hemoglobins. Apparently there is a need for change from one type to the next as the human organism moves through successive stages of growth. This phenomenon is not confined to humans but is also found in many other higher vertebrates.

There are three kinds of **embryonic** hemoglobin: **Gower 1,** which is made up of two ζ- and two ϵ-chains; **Gower 2,** composed of two α- and two ϵ-chains; and hemoglobin **Portland,** which has two ζ- and two γ-chains. These hemoglobins occur early in gestation and are soon replaced by **fetal hemoglobin (HbF),** which consists of two α- and two γ-chains. Before birth, this fetal type is gradually replaced by **adult hemoglobin (HbA),** of which there are two types: **HbA$_1$,** made up of two α- and two β-chains; and **HbA$_2$,** which consists of two α- and two δ-chains. The most important of these is HbA$_1$, which accounts for about 97% of the total HbA. The schedule of synthesis of the various types of normal globin chains during prenatal and postnatal life is given in figure 5.10.

Recent research in the genetics of the hemoglobin molecule has resulted in new information regarding the development, gene organization, and evolution of these normal hemoglobins. It was found that globin genes could be divided into two families, each family forming a cluster of genes on different chromosomes. One group, labeled the **alpha-family,** includes the genes for the α- and ζ-globins and is found in the long arm of the sixteenth pair of chromosomes. The other category, called the **beta-family,** includes the genes for the other globins (β,γ,δ, and ϵ), and is found in the long arm of the eleventh pair of chromosomes. Further detailed chromosome mapping revealed that on both sets of chromosomes the genes of each family are lined up in developmental order—that is, the embryonic genes come first, then the fetal genes, and last the adult genes (figure 5.11). For most globin chains there is only one gene per chromosome, except for γ and α, both of which are represented by two active genes. Both the α-genes are identical, but there is a small difference in the two γ-genes—hence they are labeled γ-G and γ-A. Also of note is that in between these functional globin genes are a number of nonfunctional "pseudogenes."

Apart from these six types of hemoglobin, several other types, also considered normal, should be mentioned. One of these, commonly referred to as **glycosylated hemoglobin (HbG)** consists of a minor variation of HbA$_1$. It differs from normal HbA$_1$ in that it has a molecule of glucose attached to the N-terminal valine of the beta-chains, as illustrated in figure 5.12. The normal level of glycosylated hemoglobin in adults is about 5% of the total hemoglobin. In diabetics this level is elevated to two to three times this amount. Recent investigations have demonstrated that this glucose is not added during the biosynthesis of HbA$_1$, but rather becomes nonenzymatically attached to the hemoglobin molecule anytime in the life span of the

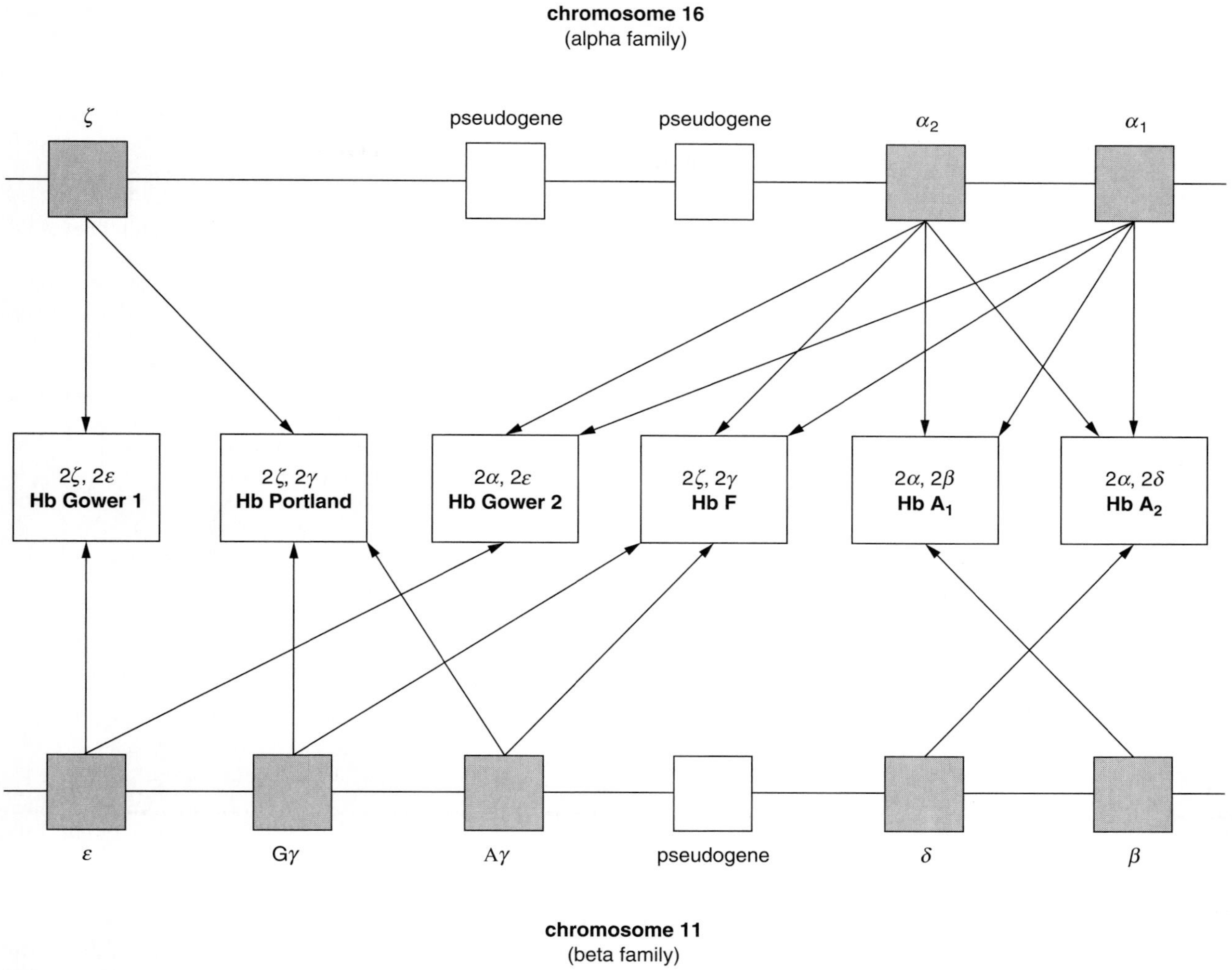

chromosome 16
(alpha family)
ζ
pseudogene
pseudogene
α2
α1
2ζ, 2ε
Hb Gower 1
2ζ, 2γ
Hb Portland
2α, 2ε
Hb Gower 2
2ζ, 2γ
Hb F
2α, 2β
Hb A1
2α, 2δ
Hb A2
ε
Gγ
Aγ
pseudogene
δ
β
chromosome 11
(beta family)

**figure 5.11**
The location of the globin genes on human chromosomes 11 and 16, and the results of their possible combinations.

β2
α2
glucose
globin
heme
glucose
α1
β1

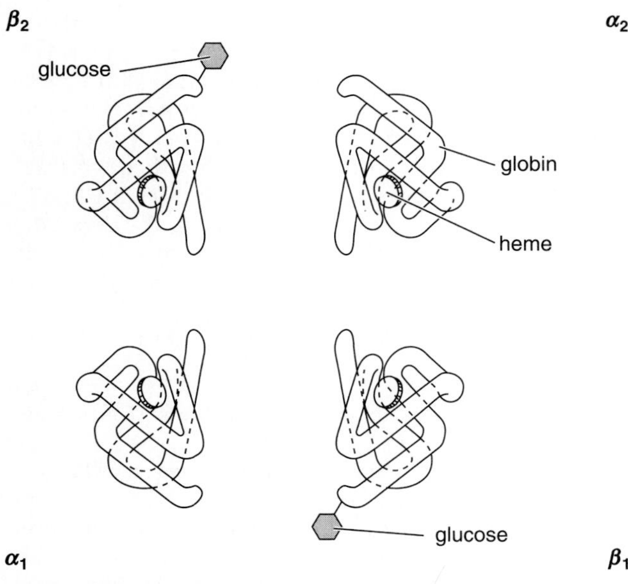

**figure 5.12**
Glycosylated hemoglobin.

erythrocyte. Since the attachment occurs slowly and depends on the circulating levels of blood glucose, the glycosylated hemoglobin level is thought to represent the time-averaged blood glucose level. Consequently it provides a new and valuable assay of the status of glucose metabolism in normal and diabetic patients. One advantage of this test over the older, and more well-known, glucose tolerance test (GTT) is that it is much less time-consuming. A single glycosylated hemoglobin determination can substitute for several glucose determinations at different time intervals as is needed for the GTT. Further, since HbG levels do not vary immediately after meals or exercise, samples can be taken anytime of the day, and fasting is no longer necessary.

Two other variations of hemoglobin made up of normal globin chains occur in certain disorders. They are abnormal variations since they are associated with clinical syndromes but can also be included in the normal category, since they are made up of globin chains which occur in "normal" hemoglobins. One variation is known as **HbH.** It consists of four β-chains and is commonly found in α-thalassemia.

**Table 5.1 Summary of Normal Human Hemoglobins**

| Hemoglobin | Structure | Comments |
|---|---|---|
| A | $\alpha_2\beta_2$ | Comprises 92% of adult hemoglobin |
| $A_{1c}$ | $\alpha_2$ ($\beta$-NH-glucose)$_2$ | Comprises 5% of adult hemoglobin<br>Increased in patients with diabetes |
| $A_2$ | $\alpha_2\delta_2$ | Comprises about 2% of adult hemoglobin<br>Elevated in β-thalassemia |
| F | $\alpha_2\gamma_2$ | Predominant hemoglobin in fetus from the 3rd through 9th month of gestation<br>Facilitates transfer of oxygen across placenta<br>Increased in β-thalassemia and other disorders |
| Gower 1 | $\epsilon_4$ or $\zeta_2\,\epsilon_2$ | Present in early embryo<br>Function not known |
| Gower 2 | $\alpha_2\epsilon_2$ | Present in early embryo<br>Function not known |
| Portland | $\zeta_2\gamma_2$ | Present in early embryo<br>Function not known |
| H | $\beta_4$ | Found in α-thalassemia<br>Low solubility<br>Nonfunctional |
| Barts | $\gamma_4$ | Trace present in newborns<br>May comprise 100% of hemoglobin in homozygous α-thalassemia<br>Nonfunctional |

Finally, there is **Hb Barts,** which consists of four γ-chains and may be considered to be another fetal hemoglobin. A summary of all the normal hemoglobins is given in table 5.1.

### Abnormal Variations

Adult hemoglobins do not normally exhibit any differences in the amino acid sequences of their globin chains. However, a number of mutations of HbA have been found. Currently more than 300 abnormal variants have been described. Most of the mutations are rather rare and have been found accidentally as a result of clinical or anthropological investigations. However a small number of abnormal hemoglobins are fairly widespread (e.g., HbS). All abnormal hemoglobins may be classified according to their cause into four major categories: (a) single point mutations, (b) deletions, (c) elongations, and (d) hybrid hemoglobins.

#### *Single Point Mutations*

A single point mutation refers to the substitution of a single base pair for another base pair in the DNA sequence that codes for a given protein. As a result the normal triplet codon is changed, and consequently a different amino acid will be inserted into the polypeptide chain that forms the globin fragment of the hemoglobin molecule. No single site in the chains seems to be more susceptible to mutations than others. However more β-chain mutations are known than α-chain substitutions. There are two kinds of single base pair mutations: transition mutations and transversion mutations. A **transition mutation** is the replacement of one purine by another purine (e.g., adenine by guanine) or of one pyrimidine by another pyrimidine (e.g., cytosine by thymine). In contrast a **transversion mutation** is the replacement of a purine by a pyrimidine or vice versa (e.g., adenine by cytosine). Of all the variants described, about 30% are guanine-adenine mutations.

The best-known example of a single point mutation is hemoglobin S (HbS) found in sickle-cell anemia. In HbS the sixth amino acid of each β-chain has been changed from glutamic acid to **valine.** This situation could have arisen only if the adenine in the codons for glutamic acid (GAA and GAG) has been replaced by the pyrimidine uracil, thus producing the codons GUA and GUG, which code for valine, thus manufacturing HbS.

Some other well-known single point mutations are as follows:

**HbC Lysine substitutes for glutamic acid at position A6 in the β-chain.**

**HbD Glutamine substitutes for glutamic acid at position 121 in the β-chain.**

**HbE Lysine substitutes for glutamic acid at position 26 in the β-chain.**

**Hb Sydney Alanine substitutes for valine at position E11 in the β-chain.**

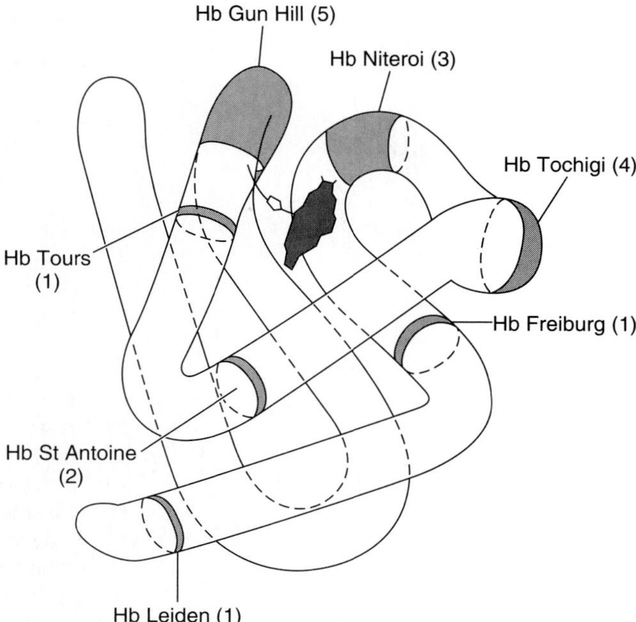


**figure 5.13**

Abnormal, and frequently unstable hemoglobins resulting from amino acid deletions. Numbers refer to the number of amino acids deleted.

**Hb Milwaukee Glutamic acid substitutes for valine at position E11 in the β-chain.**
**Hb Bristol Aspartic acid substitutes for valine at position E11 in the β-chain.**

*Deletions*

At least eight types of hemoglobin variants have arisen from the deletion of one or more amino acids from the globin chains; to a certain degree all are unstable hemoglobins. Interestingly multiple deletions always have taken place in an interhelical segment (figure 5.13). One possible explanation is that multiple deletions in a helical segment would make the hemoglobin nonviable. Deletions cause instability because they interfere with the ordered α-helical structure of hemoglobin and also with heme binding. Most deletions are single amino acid deletions such as Hb Leiden, Hb Freiburg, and Hb Tours. Examples of multiple deletions are Hb Gun Hill (five amino acids), Hb Tochigi (four amino acids), and Hb Niteroi (three amino acids).

*Elongations*

Several hemoglobins have longer subunits than normal. They were discovered rather recently and have caused a great deal of interest. The first to be described was found in a West Indian family and was labeled Hb Constant Spring, for the place where it was discovered. The abnormality was in the α-chain. In 1972 another elongation was discovered, again of the α-chain (5 amino acids longer). This was called Hb Wayne. Recently an elongation of the β-chain was discovered (10 extra amino acids) and labeled Hb Tak.

These elongations can be explained by the fact that they are the result of a deletion of a nucleotide in the stop codon. This results in the formation of a codon for another amino acid. The gene is then read and translated into amino acids until another stop codon is reached.

*Hybrid Hemoglobins*

Hybrid hemoglobins are those in which the non-α-chains are mixtures of either the δ- and β-chains or the γ- and β-chains. They are thought to have arisen from a misplaced synapse during meiosis, followed by a nonhomologous crossing-over. The most common are the Lepore hemoglobins, of which four varieties are known. They all have an N-terminal sequence of the δ-chain and a C-terminal sequence of the β-chain. They vary only in positions where the crossing-over has taken place (figure 5.14).

Although more than 300 structural variants of human adult hemoglobin have been described, only about 30% cause disease. Which of these will be clinically significant can be predicted if the site and the nature of the mutation are known. For instance, mutations of the surface polar amino acids often result in the production of a less soluble form of hemoglobin that will precipitate under certain circumstances (e.g., HbS, HbC, HbD, and HbE). Further, mutations of internal nonpolar amino acids, which are important to the stability of the tertiary structure, and mutations of amino acids lining the heme pocket, will often result in some type of clinically significant abnormality. These variations in hemoglobin structure can produce three different categories of clinical abnormalities.

First, the abnormal hemoglobin may cause increased red blood cell hemolysis because abnormal hemoglobin is less soluble and tends to precipitate out readily. These precipitated hemoglobins form aggregates called Heinz bodies in red blood cells. The presence of these large insoluble inclusions creates rigid areas in red blood cells, causing reduced deformability; when they reach the spleen they will be taken out prematurely.

Second, these mutations may cause a change in the tertiary or quaternary structure of the hemoglobin molecule, which may result in the abnormal function of hemoglobin—that is, it may develop higher or lower oxygen affinity levels and thus interfere with oxygen transport.

Last are variants resulting from elongation, deletion, or hybridization—possibly producing lower than normal levels of hemoglobin, thus also resulting in anemia.

Careful analysis of the structure of a large number of hemoglobins in a wide variety of

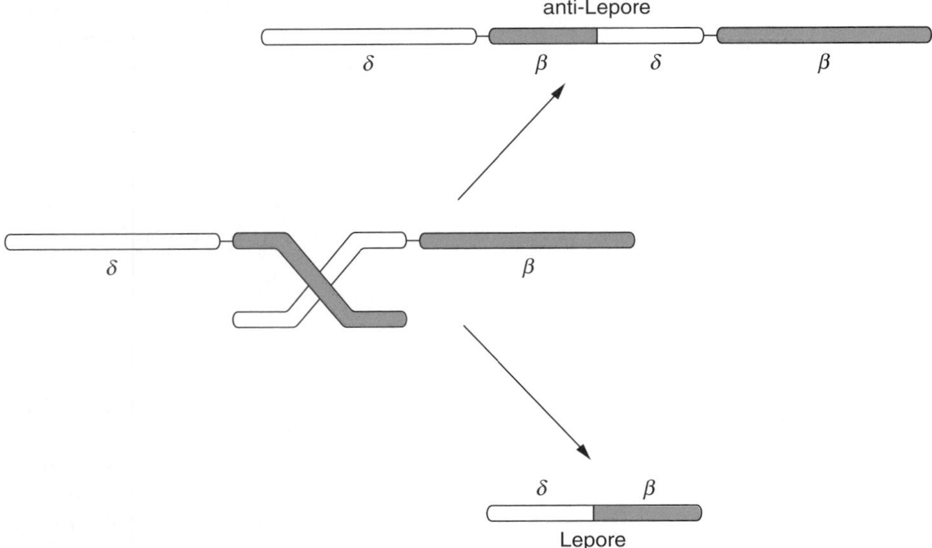


**figure 5.14**
Abnormal hybrid globin genes.

species has shown that certain amino acids are always found in the same position on the globin chain, whereas others vary with great frequency. Species that are higher on the evolutionary level have more invariant amino acids than those lower on the ladder. Obviously the invariant amino acids appear to be more important to the structure and function of the hemoglobin molecule than those amino acids that vary with great frequency. The severity of the abnormality can be postulated to be directly correlated with the variance or invariance of the amino acid that is replaced. In humans the invariant amino acids make the heme contacts and form the major helical contacts. Replacement of one of these is likely to alter the stability and function of the hemoglobin molecule.

## Further Readings

Jeffreys, A. J., et al. Evolution of Gene Families: The Globin Genes. In: *Evolution: From Molecules to Men,* D. S. Bendall, ed. Cambridge University Press, 1983.

Perutz, M. F. Hemoglobin Structure and Respiratory Transport. *Scientific American,* Dec. 1978.

White, J. M. Haemoglobin Structure and Function: Its Relevance to Biochemistry and Medicine. *Molecular Aspects of Medicine,* Vol. 1, No. 2, 1977.

## Review Questions

1. Distinguish between the terms monomer and tetramer when discussing the hemoglobin molecule. What does each monomer consist of?
2. Define the term heme. Distinguish between ferroheme (or ferrous heme) and ferriheme (or ferric heme). Mention other biological molecules that contain the heme structure.
3. Describe the appearance of a porphyrin ring structure. What are the building blocks of a porphyrin ring? What other nonheme biological compounds employ the porphyrin ring structure?
4. What pyrrolelike structure occurs in humans? Where is it produced? From what precursor compounds is it produced?
5. How many protoporphyrinogens are produced in humans? Which one is important for heme production?
6. What are globins composed of? Compare the length of the α- and β-globins.
7. Describe the (a) primary, (b) secondary, and (c) tertiary structure of a protein molecule.
8. How many amino acids are needed to make one full turn of the alpha-helix? Name an amino acid that does not lend itself to coiling. How many helical segments are in each hemoglobin monomer? How are these segments labeled?

9. How are the various amino acids in each hemoglobin monomer numbered? Give examples. What is the significance of F8 histidine?
10. What is the position of the heme plate in each globin monomer? How many amino acids are in close proximity to the heme plate? How many atomic interactions keep the heme plate in place?
11. Describe the nature of the amino acids lining the heme pocket. Describe the nature of the amino acids lining the outside of the hemoglobin molecule. Why is it important that these amino acids are not substituted by other amino acids?
12. Describe the quaternary structure of the hemoglobin molecule. Name the two categories of contacts that keep the four monomers together. Which are the tight contacts? Which are the loose functional contacts?
13. Distinguish between normal and abnormal hemoglobins in humans. How many hemoglobins belong to each category?
14. List the three groups of normal hemoglobins. Which hemoglobins belong to each group? How would you distinguish between each of these normal hemoglobins?
15. How many globin gene families occur in humans? On what chromosome is each family found? Which globin genes belong to what family?
16. What is unique about the position of each of the globin genes on their respective chromosomes? What separates the α- and γ-genes from the β- and δ-genes?
17. What is glycosylated hemoglobin? What is its normal level? What does an increase in glycosylated hemoglobin indicate?
18. What are the normal levels of $HbA_1$ and $HbA_2$ as percentage of total hemoglobin?
19. Describe the nature of HbH and Hb Barts. When do they occur?
20. List the four classes of abnormal hemoglobins. Which of these classes is the most common?
21. How are abnormal hemoglobins usually discovered? What percentage of these abnormal hemoglobins is associated with clinical abnormalities?
22. Which abnormal hemoglobins are most likely to cause clinical disorders? Give three reasons why these abnormalities usually result in conditions of anemia.
23. Provide two examples of each of the four classes of abnormal hemoglobins. Describe the nature of the abnormality.

# Chapter Six

## *Functions of Red Blood Cells*

## TRANSPORT OF BLOOD GASES

The major function of erythrocytes is to carry **blood gases**—that is, to bring oxygen from the lungs to the actively metabolizing tissues and carry carbon dioxide back from the tissues to the lungs. This process is quite complex, but red blood cells are ideally suited for this task. The erythrocyte consists of 65% water and 35% solids; 90 to 95% of the solids are hemoglobin molecules (figure 6.1). Hemoglobin makes red blood cells such efficient carriers of blood gases. The basic requirements for an efficient oxygen carrier are as follows:

1. The combination of the carrier with $O_2$ must be reversible, if the carrier is to take on $O_2$ in the lungs and release $O_2$ to the respiring tissues.
2. The carrier should be able to take on a full load of $O_2$ in the lungs and should be able to unload a large fraction of this $O_2$ in the respiring tissues, which have $Po_2$ of 20 to 40 mm Hg. It will be shown that hemoglobin fills both requirements perfectly.

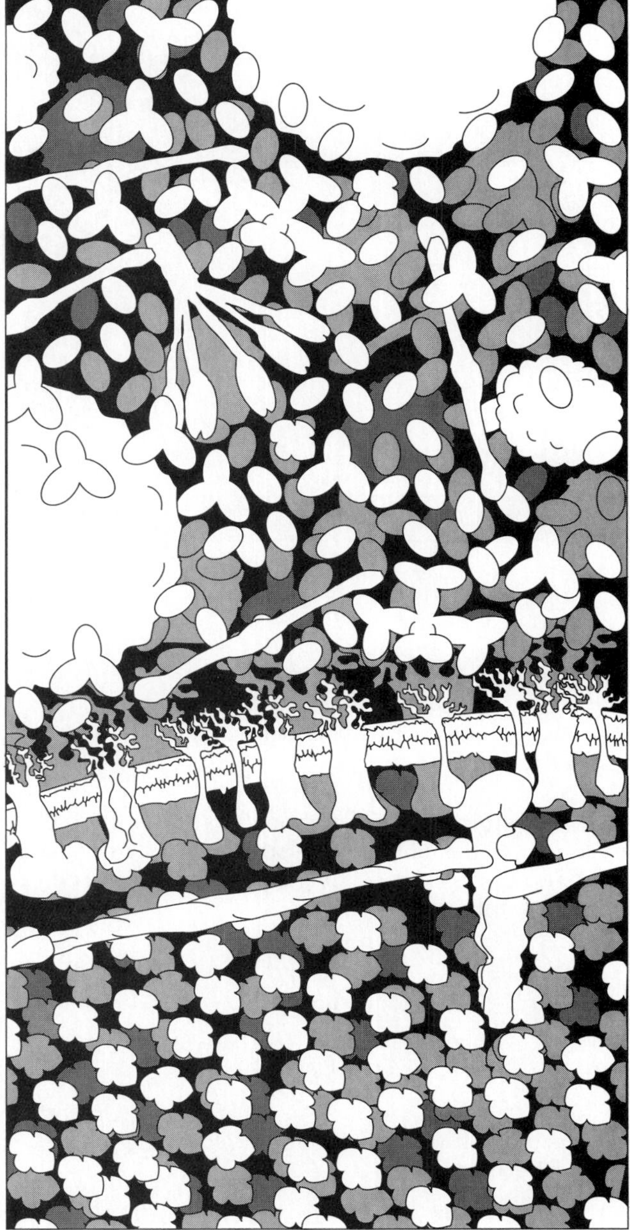

**figure 6.1**

Diagram of part of a red blood cell (lower half) and the surrounding blood plasma (upper half) magnified approximately a million times. The red blood cell is almost entirely filled with hemoglobin molecules. The plasma contains large spherical globulins; smaller oval-shaped albumins; long, fibrous proteins such as fibrin; Y-shaped antibodies; and some less common proteins.

### Law of Partial Pressure of Gases

Electrolytes and simple solutes remain in solution in a relatively simple way. Gases, on the other hand, remain in solution only under pressure. **Henry's law** states that the solubility of a gas is directly proportional to the pressure on this gas, causing it to dissolve. For instance, if a certain amount of oxygen will dissolve in water at a pressure of 250 mm Hg, then twice that amount can be dissolved if the pressure is increased to 500 mm Hg. However this relationship holds only if there is only one gas present. The air we breathe is a mixture of gases, mainly nitrogen, oxygen, and carbon dioxide. When we deal with a mixture of gases, **Dalton's law** of the partial pressure of gases comes into play. This law states that the total pressure exerted by a mixture of gases is equal to the sum of the partial pressures. The partial pressure of a gas is denoted by P, hence the partial pressure of oxygen is the $Po_2$, and $Pco_2$ indicates the partial pressure of carbon dioxide. The pressure of air in the atmosphere known as barometric pressure, or $P_B$, is the sum of the partial pressures of oxygen, carbon dioxide, and nitrogen of the air. The ratio of the partial pressure of oxygen to the barometric pressure is equal to the ratio of the number of moles of oxygen to the total number of moles of the air. Equal volumes of gas contain an equal number of moles. Therefore a fraction of a gas by volume is equal to the fraction of a gas by moles. The fraction of a gas by volume is the percentage divided by 100. Thus the fraction of oxygen in a gas mixture equals the percentage of oxygen divided by 100. Air consists of

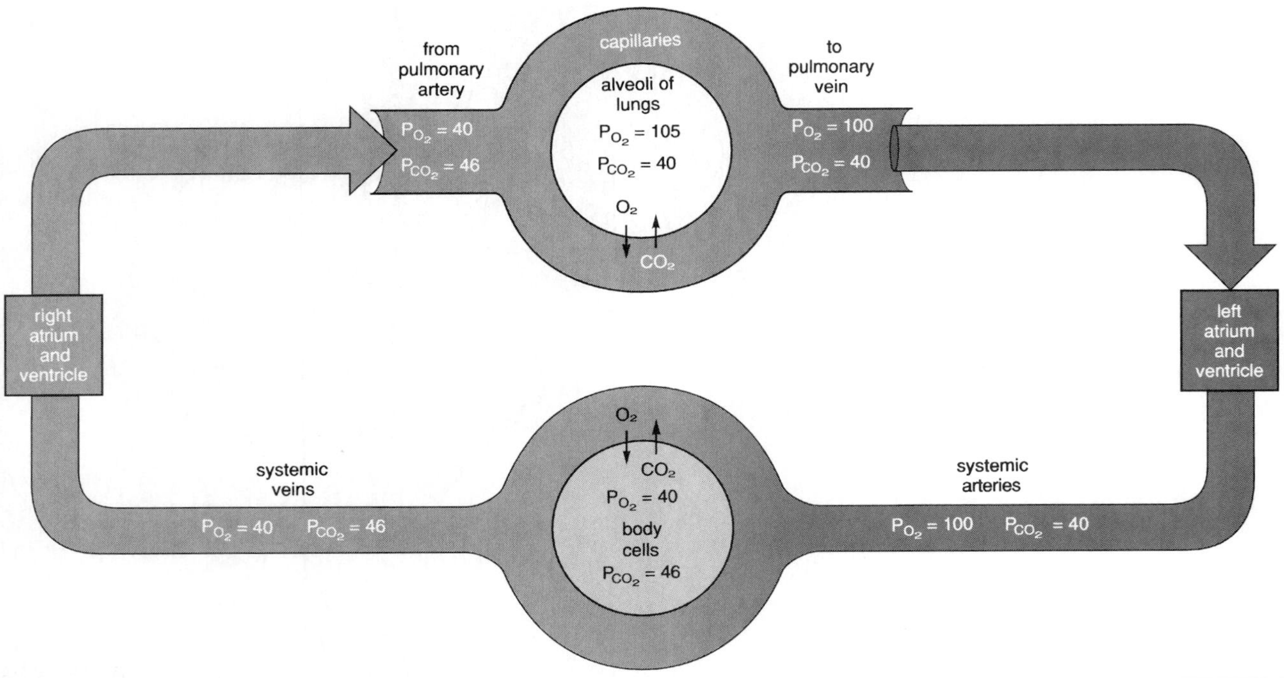


figure 6.2
The changes in the partial pressures of oxygen and carbon dioxide in the alveoli of the lungs, the arteries, the body cells, and the veins.

20.93% oxygen, 0.03% carbon dioxide, and 79.04% nitrogen. If the barometric pressure is known, then the partial pressures of the gases in the air can be calculated as follows:

$$P_{O_2} = \frac{P_B\ (\%O_2)}{100}$$

$$P_{CO_2} = \frac{P_B\ (\%CO_2)}{100}$$

$$P_{N_2} = \frac{P_B\ (\%N_2)}{100}$$

For instance, if the $P_B$ is 760 mm Hg, then the $P_{O_2}$ of atmospheric air is

$$P_{O_2} = \frac{760\text{ mm Hg} \times 20.93 = 159\text{ mm Hg}}{100}$$

### Composition of Alveolar Air

Inhaled air mixes with the air already in the trachea, bronchi, and bronchioli of the lungs. Some of this mixture enters the alveoli, where it comes into contact with the lung capillaries. The gas mixture in the alveolar spaces is known as alveolar air. From this mixture oxygen passes into the blood, and carbon dioxide is released from the blood into the alveolar air. Under resting conditions, the amount of gas flowing in and out of the alveolar spaces during breathing is more or less steady—that is, in general as much oxygen enters the alveoli as is released into the blood, and as much carbon dioxide is released into the bronchioli as enters the alveoli from the blood.

Alveolar air contains all three gases that are present in the atmosphere. However alveolar air is also saturated with water vapor from the tissues. Water vapor exerts the same partial pressure in a mixture as all other gases. At the normal body temperature of 37°C, the partial pressure of water vapor is 47 mm Hg. Given that the total pressure of gases in alveolar air is the same as barometric pressure, it follows that the pressure of the other gases is $P_B$ minus 47 mm Hg.

To calculate the partial pressure of oxygen in alveolar air, multiply the fraction of $O_2$ in dry air with the $P_B$ minus 47 mm Hg and again divide that product by 100. In a normal adult human at rest (at sea level), the $P_{O_2}$ of alveolar air is around 100 mm Hg and the $P_{CO_2}$ would be about 40 mm Hg. Hyperventilation raises the alveolar $P_{O_2}$ and lowers the $P_{CO_2}$, whereas hypoventilation does just the opposite.

### Partial Pressures of $O_2$ and $CO_2$ in the Blood

When the partial pressure of a gas differs in two parts of a system, diffusion occurs from the area with the higher partial pressure to the place of lower partial pressure. The rate of diffusion depends on the steepness of the diffusion gradient. The greater the difference in partial pressure between the two places, the faster the diffusion rate. The rate of diffusion also depends on the nature of the barrier between the two parts of the system. In the lungs the alveolar membrane separating the air from the blood contains an aqueous phase, which is a considerable barrier to the diffusion of oxygen because oxygen is relatively poorly soluble in water. However, normally venous blood in the lung capillaries has a much lower $P_{O_2}$ (40 mm Hg) than alveolar air (100 mm Hg). Consequently oxygen will diffuse readily into the venous blood. Figure 6.2 gives a summary of the partial pressure of

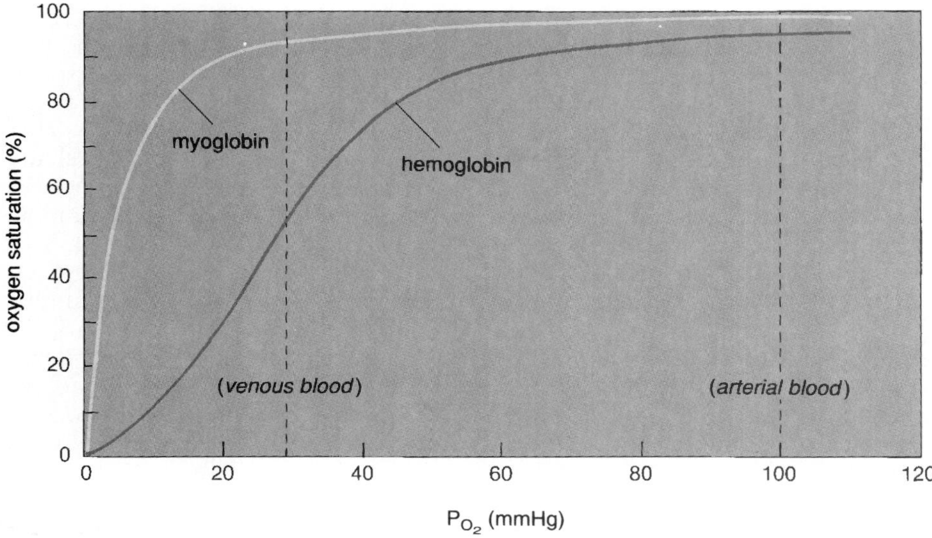


figure 6.3
The oxygen dissociation curves for hemoglobin and myoglobin.

gases in the different compartments of the body and that of inspired and expired air; note that the $Po_2$ of arterial blood is 5 mm Hg lower than the $Po_2$ of alveolar air. This may be explained in part by the fact that a small amount of venous blood is shunted from the right ventricle to the left ventricle without getting into the lungs (about 2%). This blood supplies the lung tissues via the bronchial arteries. There is a further dilution caused by blood flowing from the coronary arteries directly back to the left chamber of the heart. These physiological shunts cause the blood in the systemic arteries to contain less oxygen than would be expected from the exchange in the lungs. In the tissues oxygen is given off at a great rate because the $Po_2$ of the tissues is much lower than the $Po_2$ of the blood. Sometimes the difference between the $Po_2$ of blood and that of surrounding tissues may be enormous—for instance, in exercising muscle the $Po_2$ may be near zero.

Movement of carbon dioxide takes place in the reverse direction. The $Pco_2$ of venous blood is 46 mm Hg, whereas the $Pco_2$ of alveolar air is only 40 mm Hg. Consequently carbon dioxide will move out of the blood into the alveoli since arterial blood contains carbon dioxide at a partial pressure of 40 mm Hg, which is similar to that of alveolar air. In the tissues carbon dioxide is absorbed since the $Pco_2$ in the blood is lower than that of the tissues.

## ROLE OF HEMOGLOBIN IN $O_2$ TRANSPORT

### Oxygen Binding by Hemoglobin

Hemoglobin has a great affinity for oxygen. If we plot the percentage saturation of hemoglobin with oxygen versus the oxygen concentration expressed as the partial pressure of oxygen, we get a sigmoidal, or S-shaped curve, as illustrated in figure 6.3. This same figure also shows the oxygen-saturation curve of myoglobin, which in comparison shows a completely different curve. This hyperbolic curve of myoglobin is the type of curve we would expect for a simple association reaction. The S-shaped curve of hemoglobin is fundamentally different—which is of great significance for the oxygen transport. Both myoglobin and hemoglobin meet the first requirement of a good oxygen carrier, since both proteins bind oxygen when the $O_2$ concentration is very high and release oxygen when the $O_2$ concentration is very low. Regarding the second basic requirement—that the carrier take on a full load of oxygen in the lungs and release a large fraction of it at the delivery site where there are intermediate $Po_2$s—only hemoglobin qualifies as a good carrier. Inspection of the myoglobin and hemoglobin curves reveals that while both are essentially saturated in the lungs, only hemoglobin would release an appreciable amount of its bound oxygen to the respiring tissues, such as a working muscle with a $Po_2$ of 20 mm Hg. At the same $Po_2$ of 20 mm Hg, most of the oxygen carried by myoglobin is still bound, and most of the oxygen carried by hemoglobin is given off to the metabolizing tissues. In other words, since hemoglobin is capable of delivering a much larger fraction of its bound oxygen to respiring tissues (most of which have intermediate $Po_2$s of 20 to 40 mm Hg) than myoglobin, hemoglobin is much more efficient as a transport system for oxygen. Hence the significance of the S-shaped curve for hemoglobin.

### Relationship Between the Structure and Function of Hemoglobin

Since each hemoglobin molecule consists of four monomers, each of which has a heme plate with a ferrous iron in the center, it follows that hemoglobin can combine with four molecules of oxygen. Although this oxygenation is a rapid phenomenon taking less than 0.01 second it should be realized that the binding of oxygen is a sequential process—that is, first one iron atom is oxygenated, then the second and the third, and finally the fourth. This sequential binding is the result of the quaternary structure of the hemoglobin molecule and the relationships between its four component globin chains. By shifting its four component peptide chains, the hemoglobin molecule fosters either the uptake or the release of oxygen. In other words, although the oxygen binds to the ferrous iron of each heme nucleus, the surrounding globin chains also play an essential role in binding and releasing oxygen molecules.

To understand this phenomenon more clearly, we need to compare the structure and action of hemoglobin with that of myoglobin. The myoglobin molecule is essentially that of a hemoglobin monomer. It also consists of one heme and one globin. When this monomeric molecule is exposed to high oxygen tension, it will bind readily to one oxygen molecule. This oxygen molecule will remain attached to the myoglobin structure as long as the oxygen tension remains fairly high. Only when this oxygenated monomeric molecule is transported to an area of extremely low oxygen tension will the bond become unstable and the oxygen molecule released from the heme group. This phenomenon is illustrated in figure 6.3 by the hyperbolic curve.

Hemoglobin is essentially a polymer of four myoglobinlike molecules. It developed late in the evolution of animals, but when it was produced it resulted in a great breakthrough in the efficiency of a hemochrome as an oxygen carrier. With the formation of the hemoglobin tetramer, the oxygen-dissociation curve (also known as the oxygen-saturation curve) for hemoglobin changed from a hyperbolic to a sigmoid shape. This change allowed animals to bind and release oxygen over a much wider range of oxygen tensions. The question to be asked is, How did this cooperation between hemoglobin monomers actually modify the oxygen-binding characteristics of the individual oxygen molecules? The explanation lies in a phenomenon known as **allostery,** a term denoting that a change in shape alters the properties of a molecule. When a single heme in a tetrameric hemoglobin molecule binds to an oxygen molecule, not only does the monomer with the bound oxygen change shape slightly, but it also alters the total configuration of the whole tetrameric complex. When the second oxygen binds, it will result in a further shift of the whole molecule, as does the binding of the third oxygen molecule. As a result of these allosteric changes, the binding characteristics of the last oxygen molecule are completely different from the first.

The physical movement involved in the hemoglobin allostery is minute, involving only small shifts of a few angstroms (1 Å is 0.1 nm, which is one ten-millionth of a millimeter); however the effect on the binding characteristics of the hemoglobin molecule is profound. The binding of the first oxygen requires more energy than that of the second and the third, whereas the binding of the last oxygen molecule requires almost no energy at all. This phenomenon is the result of the differences in shape of the hemoglobin molecule when it is oxygenated or deoxygenated. A fully oxygenated hemoglobin molecule is said to be in a relaxed (R) state. The deoxygenated hemoglobin molecule is in a tense (T) state. In the T state salt bridges between the globin monomers keep the heme plate in a tight pocket (figure 6.4). Binding of the first oxygen molecule involves breaking several salt bridges. Movement in the monomer that occurs as a result of opening the heme pocket and binding the first oxygen will break more salt bridges between adjacent monomers. This allows the second and third oxygen molecules to bind more readily to their heme plates. In the process of binding the second and third oxygens, more changes take place in the monomers—resulting in breaking the last salt bridges, so that the final oxygen molecule can bind very easily without any expenditure of energy. In other words, the affinity of the hemoglobin molecule for the fourth oxygen molecule is much greater than for the first one. The following analogy helps to explain the various energy expenditures involved in the sequential oxygenation of the hemoglobin molecule: More energy is required to remove the first postage stamp from a block of four than to remove the next two. As a result the last stamp is then automatically free (figure 6.5). A similar situation exists in the hemoglobin molecule. The tight relationships that exist between the four monomers of a deoxygenated hemoglobin molecule is similar to that of postage stamps in a block of four. The energy required to bind the first oxygen is greater than that of the next two, and practically no energy is needed to bind the final oxygen. The energy required to bind the second and third oxygen molecules is somewhere between that of the first and the fourth.

The reverse is true for the release of oxygen from a fully oxygenated hemoglobin molecule. The first oxygen is readily given up to the respiring tissues, but each succeeding oxygen molecule is given up with increasing difficulty and requires reduced oxygen tensions. Hence the shape of the sigmoid curve.

### Factors Affecting Oxygen Binding by Hemoglobin

Several factors have a profound effect on the oxygen-saturation curve of hemoglobin. The three most important of these factors are (1) the pH, (2) 2,3-biphosphoglycerate (2,3-BPG), and (3) temperature.

#### *pH*

The pH of the solution has a marked effect on the position of the oxygen-saturation curve of hemoglobin, as illustrated in figure 6.6. The pH of plasma

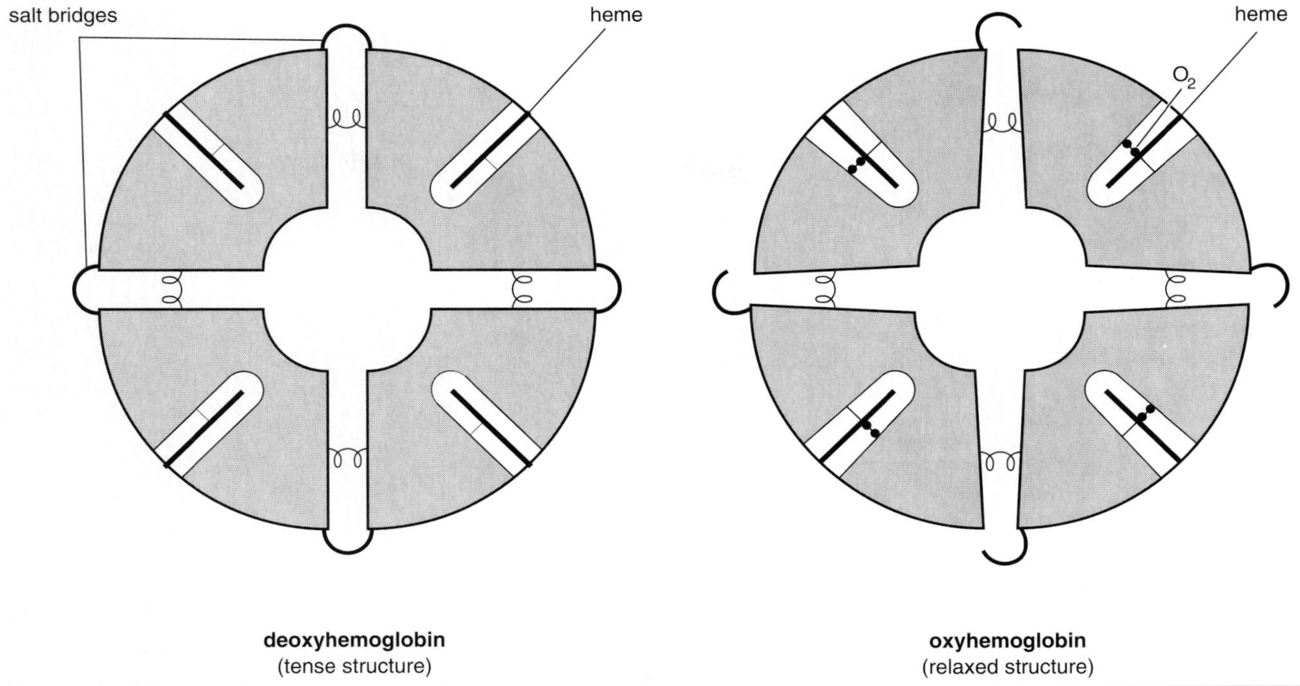


**figure 6.4**

Steric structure model: The conformational differences between deoxyhemoglobin and oxyhemoglobin.

**figure 6.5**

The postage stamp analogy of the energy involved in the oxygenation of hemoglobin.

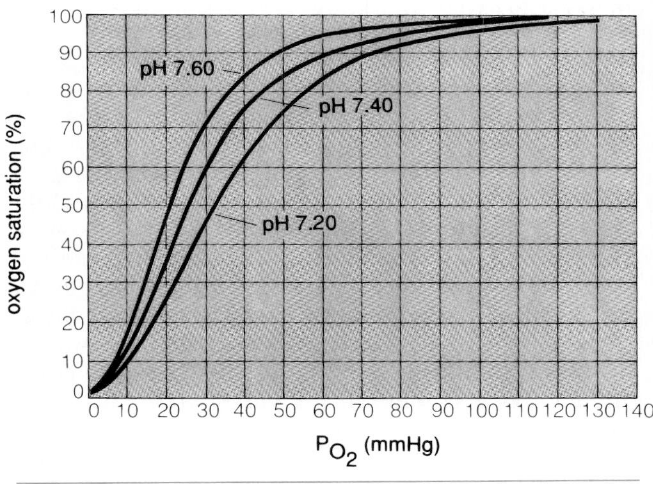


**figure 6.6**

The effect of pH on the oxygen dissociation curve of hemoglobin.

and the cytoplasm of red blood cells is normally about 7.4. A drop in pH shifts the curve to the right—that is, at a given $Po_2$, the amount of oxygen bound by hemoglobin decreases as the pH is lowered. A rise in pH shifts the curve to the left, which means that at a given $Po_2$ the amount of oxygen bound to hemoglobin is increased. To put it more simply, when there is a drop in pH, the hemoglobin releases its oxygen more readily. This is important, especially in a working muscle, which needs large quantities of oxygen. Active muscle cells produce large amounts of lactic acid, resulting in a drop in pH in the muscle tissue—consequently more oxygen is freely given off to these respiring muscle cells. This effect of changes in pH on the oxygen-carrying capacity of hemoglobin is known as the Bohr effect, and it has important consequences not only for the carrying of oxygen by red blood cells but also for the transport of carbon dioxide.

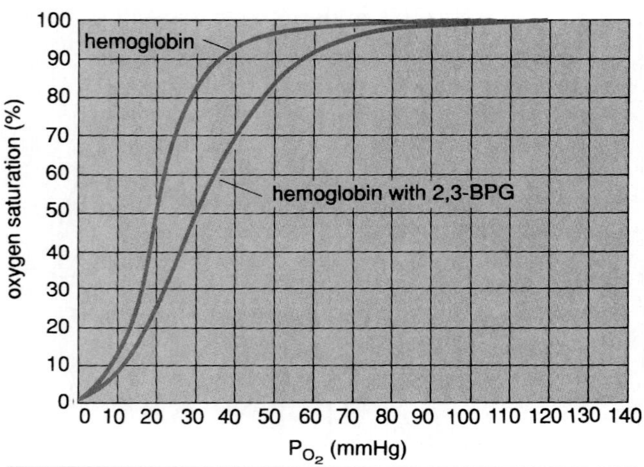


figure 6.7

The effect of 2,3-biphosphoglycerate (2,3-BPG) on the oxygen dissociation curve of hemoglobin.

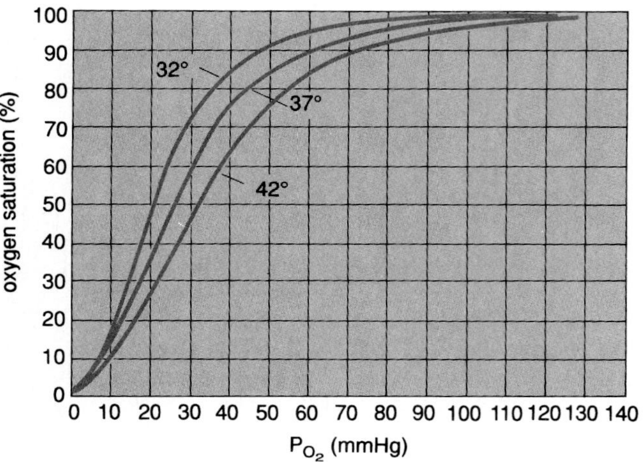


figure 6.8

The effect of temperature (°C) on the oxygen dissociation curve of hemoglobin.

*2,3-BPG*

The compound 2,3-biphosphoglycerate (2,3-BPG), previously known as 2,3-diphosphoglycerate (2,3-DPG), is quite plentiful in red blood cells. It is formed from 3-phosphoglyceraldehyde, which is a product of glycolysis via the Embden-Meyerhof pathway (figure 7.2). It is a highly charged anion that is bound by electrostatic bonds in the cavity between two beta subunits of the deoxyhemoglobin molecule but not in the oxyhemoglobin structure. This bound 2,3-BPG has a marked allosteric effect on the oxygen affinity of deoxyhemoglobin. As can be seen in figure 6.7, removal of 2,3-BPG from deoxyhemoglobin shifts the oxygen-saturation curve to the left—that is, it increases the affinity of hemoglobin for oxygen and makes the curve look remarkably like the one for myoglobin. In other words, hemoglobin would be little better than myoglobin as an oxygen carrier were it not for the allosteric effect of 2,3-BPG—which decreases the oxygen affinity of hemoglobin, thus allowing oxygen to diffuse into these respiring tissues even at relatively high $Po_2$s. Increasing the amount of 2,3-BPG will shift the curve to the right, thus lowering the oxygen affinity of hemoglobin. This is an important compensatory mechanism in various types of hypoxia, such as that experienced in high altitudes, or in cases of congestive heart failure, pulmonary insufficiency, or severe anemias. In all these conditions increased amounts of 2,3-BPG ensures that the respiring tissues will be given the same amounts of oxygen at lower $Po_2$s.

The concentration of 2,3-BPG is strongly affected by the pH, since a lowering of pH (acidosis) inhibits erythrocyte glycolysis and consequently inhibits the production of 2,3-BPG. The level of 2,3-BPG is further influenced by a number of hormones: growth hormone, thyroid hormones, and androgens—all of which cause an increase in 2,3-BPG.

*Temperature*

A rise in temperature shifts the curve to the right, and a drop in temperature shifts the oxygen-saturation curve to the left (figure 6.8). Again, this has important consequences from the physiological point of view. For instance, the temperature of exercising muscle rises several degrees, and the effect of this rise in temperature is to make hemoglobin unload more oxygen at a given $Po_2$. On the other hand, cold skin may be bright pink in light-skinned people, not only because cold reduces the rate of oxygen consumption but also because hemoglobin binds oxygen more tightly in the cold.

## ROLE OF HEMOGLOBIN IN $CO_2$ TRANSPORT

Hemoglobin plays an essential role in the transport of carbon dioxide from the tissues to the lungs. In the tissues $CO_2$ is taken up by the passing blood, since the $Pco_2$ of the metabolizing tissues is higher than that of blood (46 mm Hg vs. 40 mm Hg). If all $CO_2$ were to be carried in the plasma as a simple aqueous solution, we would expect very wide shifts in the pH of the blood due to the formation of carbonic acid, following the reactions:

$$CO_2 + H_2O \rightleftharpoons H_2CO_3 \rightleftharpoons HCO_3^- + H^+$$

However the changes in pH in the plasma are very small, indeed negligible, because most of the carbon dioxide is taken up into the red blood cells. Less than 3% of the total $CO_2$ in blood is present in plasma. In erythrocytes, hemoglobin acts as a **buffer** system to control the level of the pH. This buffering action is mainly a function of oxyhemoglobin, since oxyhemoglobin is a stronger acid than deoxyhemoglobin.

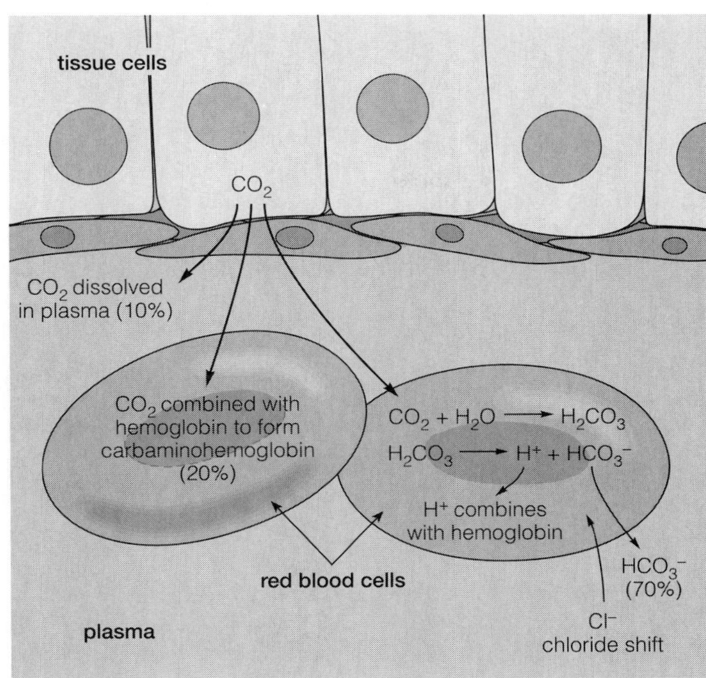


**figure 6.9**
The three ways carbon dioxide is transported in the blood.

Consider the following equilibria:

$$\text{Deoxyhemoglobin} + O_2 \rightleftharpoons \text{Oxyhemoglobin} \rightleftharpoons \text{Oxyhemoglobin}^- + H^+ \text{ (Reaction A)}$$

If these reactions are at equilibrium, and the concentration of $H^+$ is increased, then the reactions will be driven to the left according to the law of mass action. This shift to the left decreases the concentration of oxyhemoglobin$^-$ and increases the concentration of deoxyhemoglobin, which in turn will drive the reaction farther to the left, resulting in the release of oxygen. This phenomenon is of great physiological significance because it enables hemoglobin to play an essential role in the transport of carbon dioxide.

For every mole of oxygen consumed in the oxidation of glucose in the tissues, a mole of $CO_2$ is produced, since the overall equation for oxidation of glucose is

$$C_6H_{12}O_6 + 6O_2 \rightarrow 6CO_2 + 6H_2O + \text{energy}$$

Therefore an actively metabolizing tissue produces large quantities of $CO_2$ that must be transported to the lungs via the venous circulation. Most of this $CO_2$ diffuses rapidly into the red blood cells, where it is hydrated to carbonic acid by the enzyme carbonic anhydrase. The $H_2CO_3$ produced is a weak acid, which dissociates to yield $H^+$ and $HCO_3^-$:

$$CO_2 + H_2O \rightleftharpoons H_2CO_3 \rightleftharpoons H^+ + HCO_3^- \text{ (Reaction B)}$$

The law of electrical neutrality states that there must be equal numbers of positive and negative electric charges in a solution, or at least all systems tend to move toward an equilibrium. This applies also to red blood cells. Hemoglobin has a number of negative charges, and these are normally balanced by positive charges of cations present within the erythrocytes—the cations being chiefly potassium ($K^+$) and sodium ($Na^+$) ions. As a result of changes occurring in the blood at the tissue level with the removal of $O_2$ and the uptake of $CO_2$, there is a sudden increase in hydrogen cations ($H^+$) and bicarbonate anions ($HCO_3^-$) within the red blood cells. Normally $HCO_3^-$ within the erythrocyte is in equilibrium with the $HCO_3^-$ in plasma. But as a result of the sudden high bicarbonate concentration generated inside the red blood cells after the uptake of $CO_2$, some of this $HCO_3^-$ diffuses out of the red blood cells into the plasma. Since $HCO_3^-$ ions are negatively charged, the electrical neutrality of the erythrocytes and the plasma would be disturbed unless one of two things happens: either an equal number of positive cations could diffuse from the red blood cells into the plasma, or an equal number of negatively charged anions could diffuse from the plasma into the erythrocytes. The membrane of erythrocytes is impermeable to cations, at least over the very brief period in which these exchanges occur. For that reason, an inward diffusion of anions takes place. The anions available in the plasma are chloride ions ($Cl^-$), which diffuse into the red blood cells while the $HCO_3^-$ ions diffuse out until an equilibrium is reached. This process is known as the **chloride shift** (figure 6.9).

The fate of the hydrogen ions ($H^+$) produced as a result of the dissociation of carbonic acid is quite different. As was previously discussed, the pH of the

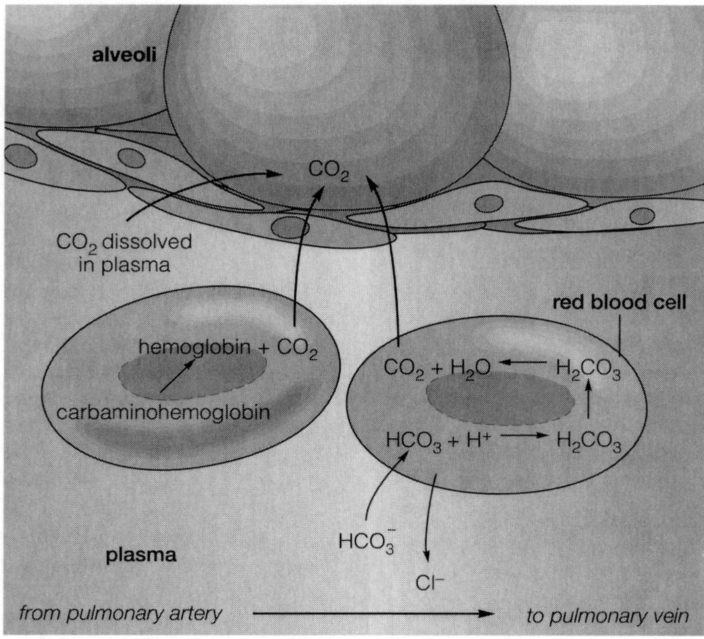


**figure 6.10**

Carbon dioxide release from the blood into the lungs.

blood is about 7.4 and must be maintained within very narrow limits. Relatively small increases or decreases of plasma $H^+$ ions (known respectively as acidosis and alkalosis) could be life-threatening. Considering the large amounts of $CO_2$ that must be carried to the lungs, we might expect an accumulation of toxic concentrations of $H^+$. This does not occur. For every mole of $CO_2$ produced by the respiring tissues, a mole of $O_2$ must be delivered to these tissues. This is done by converting a mole of oxyhemoglobin to deoxyhemoglobin. The latter process requires the uptake of $H^+$, as was shown in reaction A. Hence the production of $H^+$ as a result of ionization of $H_2CO_3$ (reaction B) is largely taken up by the conversion of oxyhemoglobin to deoxyhemoglobin and $O_2$ via reaction A. As a result virtually no change in $H^+$ concentration occurs. This phenomenon, where a combination of reactions involving large quantities of $H^+$ occurs with a minimal change in pH, is known as **isohydric shift.** An overall equation for the process occurring in the capillaries of the metabolizing tissues can be obtained by summing up reaction A and reaction B to yield the following formula:

$$CO_2 + H_2O + \text{Oxyhemoglobin}^- \rightleftharpoons HCO_3^- + \text{Deoxyhemoglobin} + O_2$$

This reaction is driven to the right by high $P_{CO_2}$s and low $P_{O_2}$s in the metabolizing tissues. This is known as the **Bohr effect,** and it allows the red blood cells to take up a large fraction of $CO_2$ produced by the tissues and transport it to the lungs as $HCO_3^-$ and $H^+$ without a major change in the pH of the blood. In the lungs the high $P_{O_2}$ and the low $P_{CO_2}$ of the alveoli drives the reaction in the reverse direction:

$$O_2 + \text{Deoxyhemoglobin} + HCO_3^- \rightleftharpoons CO_2 + \text{Oxyhemoglobin} + H_2O$$

Here the $CO_2$ is removed via the lungs to the outside. The $H^+$ released in the conversion of deoxyhemoglobin to oxyhemoglobin$^-$ (reaction A) is taken up by the conversion of $HCO_3^-$ to $CO_2$ (reaction B), enabling the entire process to occur without a significant change in the pH of the blood (figure 6.10).

Although most of the $CO_2$ is transported from the tissues to the lungs as $HCO_3^-$, approximately 20% is transported as **carbaminohemoglobin.** The free hemoglobin groups of the N-terminal valine residues of the hemoglobin molecule combine reversibly with $CO_2$ to form carbaminohemoglobin according to the following formula:

$$CO_2 + H_2N(\text{terminal valine})\,P \rightleftharpoons H^+ + {}^-OOC\text{-}NH(\text{terminal valine})P$$

where P represents the rest of the polypeptide chain. In a rapidly metabolizing tissue, where the $P_{CO_2}$ is high, the reaction is driven to the right. In the lungs, where the $P_{CO_2}$ is low, the reaction is driven to the left and $CO_2$ is removed. Thus the hemoglobin molecule is a very finely tuned and efficient transport system, which not only carries molecular $O_2$ to the tissues but also participates actively in the removal of $CO_2$ to the outside.

## Further Readings

Maclean, N. Haemoglobin. *Studies in Biology,* No. 93. Edward Arnold, London, 1978.

Perutz, M. Hemoglobin Structure and Respiratory Transport. *Scientific American,* Dec. 1978.

White, J. M. Haemoglobin Structure and Function: Its Relevance to Biochemistry and Medicine. *Molecular Aspects of Medicine,* Vol. 1, No. 2, 1977.

## Review Questions

1. Describe the composition of a red blood cell.
2. Give two major functions of hemoglobin.
3. What are the two basic requirements of an efficient oxygen carrier?
4. What does Henry's law state? What does Dalton's law of the partial pressure of gases state?
5. What is meant by $P_{O_2}$ and $P_{CO_2}$? How are they calculated?
6. Give the average $P_{O_2}$ and $P_{CO_2}$ of (a) atmospheric air, (b) alveolar air, (c) arteries, (d) normal respiring tissues, and (e) veins.
7. Explain why the $P_{O_2}$ and the $P_{CO_2}$ changes in each of the compartments in question 6.
8. Draw an oxygen-saturation curve for hemoglobin and for myoglobin. What is the name of each curve? What is the significance of both curves?
9. Distinguish between the allostery of an oxygenated and a deoxygenated hemoglobin molecule.
10. Explain why more energy is required to bind the first than the last oxygen molecule to a hemoglobin tetramer.
11. Explain why hemoglobin is a more efficient oxygen carrier than myoglobin.
12. List three major factors that affect oxygen binding to hemoglobin.
13. What is the effect of a drop in pH on the oxygen-saturation curve of hemoglobin? Why is this important?
14. Describe the effect of an increase in 2,3-BPG on the saturation curve of hemoglobin. Why is this important?
15. Describe the effect of an increase in temperature on the oxygen-dissociation curve of hemoglobin. Why is this important?
16. What are the three ways in which $CO_2$ is carried in the blood? What is the percentage of each of these?
17. What happens when $CO_2$ is released into the blood? What would happen if large amounts of $CO_2$ were carried in the blood plasma? Why is that dangerous?
18. Explain how hemoglobin acts as a buffer system. Describe the buffering action of hemoglobin. Explain what is meant by isohydric drift.
19. Explain what is meant by the chloride shift. Why is such a chloride shift necessary?
20. What is carbaminohemoglobin? How is it formed? How much of the $CO_2$ is carried by carbaminohemoglobin?

# Chapter Seven

## *Metabolism and Fate of Red Blood Cells*

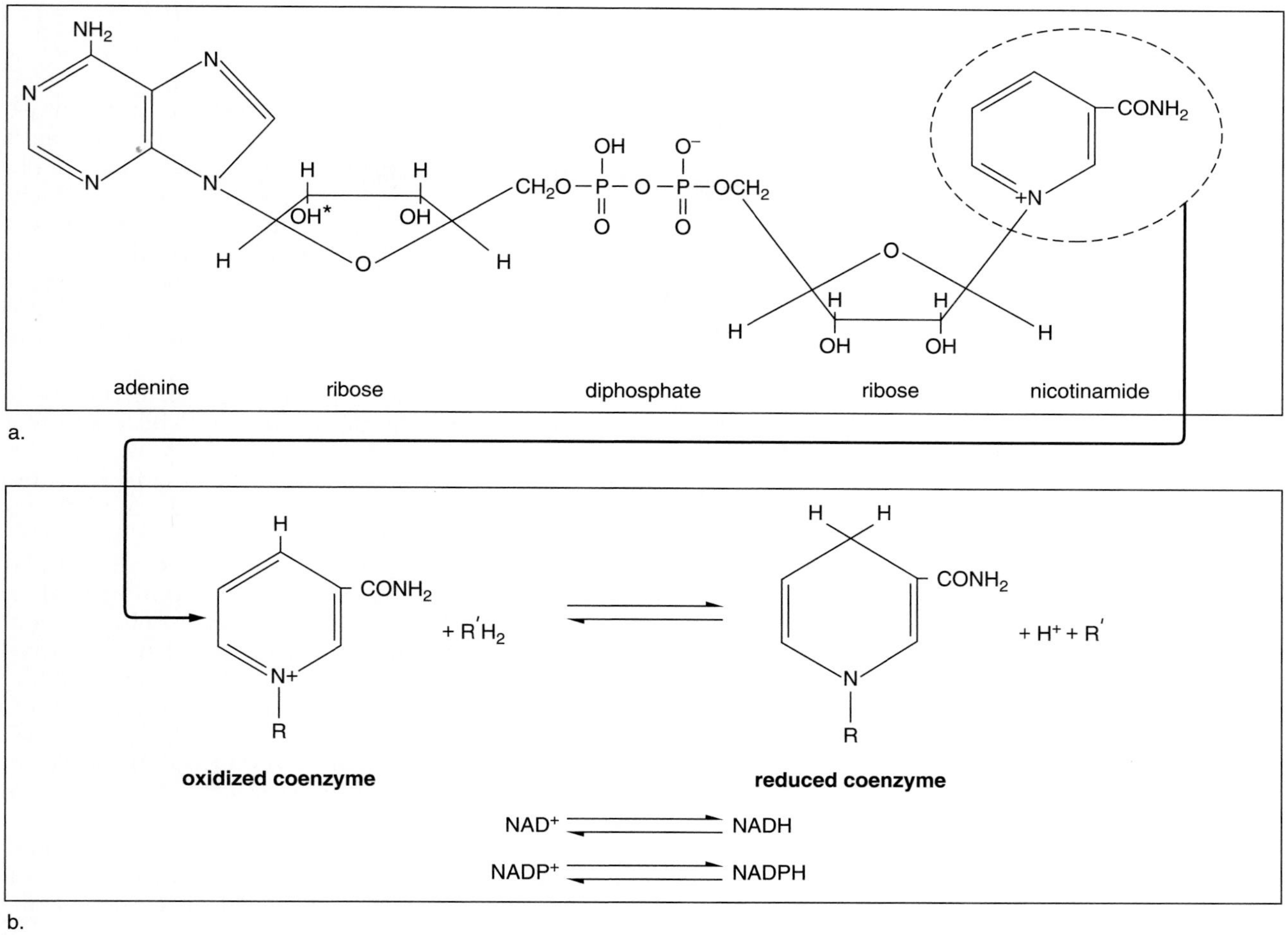

**figure 7.1**

(*a*) The structure of nicotinamide adenine dinucleotide ($NAD^+$). Nicotinamide adenine dinucleotide phosphate ($NADP^+$) has an additional phosphate group at the site of the asterisk. (*b*) The reduction of $NAD^+$ and $NADP^+$ to NADH and NADPH (R = rest of molecule; R´= hydrogen donor).

## ENERGY REQUIREMENTS OF RED BLOOD CELLS

Energy requirements of red blood cells are small but critical. Energy is required for a number of processes. First, it is needed to keep the cation ($Na^+$, $Ca^+$) pumps going; in that way the cell's osmotic stability is maintained. Second, energy is needed to generate $NAD^+$ (=nicotinamide adenine dinucleotide), which is needed to produce NADH (figure 7.1). NADH provides the energy needed for the conversion of pyruvate to lactate and also helps in the regulation of the methemoglobin reduction system. Third, energy is required to produce $NADP^+$ (=nicotinamide adenine dinucleotide phosphate), which is needed to produce NADPH (figure 7.1). NADPH is generated via the pentose monophosphate shunt (also known as the hexose monophosphate shunt) and is also a major energy source in keeping hemoglobin in its reduced state. Fourth, energy is needed to generate 2,3-BPG, which is also an essential ingredient for the proper functioning of hemoglobin as an oxygen carrier. Last, red blood cells need to maintain their shape as biconcave discs, despite the fact that they have to change their shape many times a day as they pass through the capillaries of the microcirculation, which may have a diameter as small as 3.5 µm.

## GLUCOSE METABOLISM IN RED BLOOD CELLS

Energy needed for all these functions is provided by glucose, which is broken down via several routes to yield ATP, NADH, and NADPH. All three energy molecules are needed by red blood cells to maintain their metabolic activities. The major roles of ATP as an energy molecule include (1) running the cationic pumps ($Na^+$ and $Ca^+$) to maintain the osmotic balance of erythrocytes; (2) providing the energy to keep the cell membrane in good shape by ensuring proper lipid turnover and by phosphorylation of membrane proteins; (3) providing the phosphates needed to prime the Embden-Meyerhof pathway, thus starting the process of anaerobic glycolysis (figure 7.2); and (4) contributing the active phosphates that allow the generation of NADPH from NADH. The energy functions of NADH and NADPH are to provide reducing power, thus preventing excess oxidation of many compounds.

Most of the energy in red blood cells is produced by the conversion of glucose via two pathways: the **Embden-Meyerhof (EM) pathway** (anaerobic glycolysis) and the **hexose monophosphate shunt (HMS)** (figure 7.2). The EM pathway utilizes about 90 to 95% of all the glucose and supplies about 75% of the cell's energy. The HMS utilizes about 5 to 10% of the glucose and supplies about 25% of the potential energy of the cell. The EM pathway produces all the ATP for the cell (a net of 2 ATPs for each glucose molecule); it also produces NADH and 2,3-BPG. The HMS produces all the NADPH for the red blood cell. Ironically despite the presence of large amounts of oxygen in red blood cells, most energy production is achieved via the anaerobic pathway. This paradox may be explained by the fact that mature erythrocytes lack mitochondria, thus preventing the aerobic oxidation of glucose and pyruvate. This anaerobic breakdown of sugar is a process that results in the splitting of the six-carbon glucose molecule into two three-carbon molecules of pyruvic acid. This process involves a series of enzymatically controlled steps that do not require oxygen. Before a molecule of glucose is broken down, it is first activated by ATP—that is, there is a transfer of a phosphate group from an ATP molecule to glucose as ATP becomes ADP. This process is known as **phosphorylation.** The end result of the breakdown of one glucose molecule via the EM pathway is the net gain of two ATP molecules and two molecules of reduced, energy-carrying NADH. The process of glycolysis is summarized in figure 7.2. This diagram also shows the hexose monophosphate shunt. This shunt is also known as the **pentose phosphate pathway** because here the hexose glucose 6-phosphate is converted to the pentose ribulose 5-phosphate. The importance of this HMS lies in the production of **NADPH** from $NADP^+$. This step involves the enzyme **glucose-6-phosphate dehydrogenase (G-6-PD).** NADPH is of vital importance to the integrity of red blood cells. NADPH is needed to keep glutathione in its reduced state. **Reduced glutathione** plays a major role in prolonging the life of red blood cells. Any defect in the hexose monophosphate shunt will result in the early hemolysis of red blood cells. The mean lifetime of an erythrocyte is 120 days, and most of the constituent hemoglobin molecules survive unscathed for that time.

## SURVIVAL OF RED BLOOD CELLS

### Formation of Methemoglobin

Although hemoglobin is a remarkably stable molecule, it is slowly but continually oxidized to **methemoglobin,** which is the form of hemoglobin that contains ferric iron ($Fe^{3+}$) rather than ferrous hemoglobin ($Fe^{2+}$). Methemoglobin is useless as an oxygen carrier. The processes involved in this oxidation are very slow compared with enzymatically catalyzed metabolic events, but their cumulative effects would be disabling nonetheless if there was not some means of counteraction.

A small amount of methemoglobin is present in the blood of all normal individuals. This quantity is about 0.3 g/dL blood, or about 1.7% of the total hemoglobin. An increased amount of methemoglobin may result in a clinical condition known as **methemoglobinemia,** which is a type of anemia. This anemia may be caused by a failure of erythrocyte cellular mechanisms to reconvert methemoglobin back to normal hemoglobin or it may be the result of a more rapid production of methemoglobin than the normal cellular mechanisms can handle. The latter condition is caused by a number of oxidant drugs such as the antimalarials chloroquine and primaquine or by such antibacterial drugs as penicillin and sulfonamides. It may also be caused by certain viruses and bacteria, as well as by certain foods such as fava beans.

The red blood cell has an efficient mechanism to minimize the accumulation of methemoglobin. This method involves the help of NADH, according to the following reaction:

Hemoglobin ($Fe^{2+}$) ⟲ $NAD^+$
Oxidant drugs, etc. ( ) ( Methemoglobin reductase
Methemoglobin ($Fe^{3+}$) NADH

### Formation of Hydrogen Peroxide

The oxidation of hemoglobin to methemoglobin results in the formation of the **superoxide** radical ($O_2^-$), by the transfer of a single electron:

$$\text{Hemoglobin } (Fe^{2+}) + O_2 \rightarrow \text{Methemoglobin } (Fe^{3+}) + O_2^-$$

The enzyme **superoxide dismutase** (also known as hemocuprein) present in all red blood cells catalyzes the conversion of superoxide to **hydrogen peroxide** ($H_2O_2$):

$$2O_2^- + 2H^+ \xrightarrow{\text{superoxide dismutase}} H_2O_2 + O_2$$

Large quantities of hydrogen peroxide in the red blood cells cause an irreversible oxidation of the sulfhydryl (—SH) groups in hemoglobins (called **sulfhemoglobins**). The $H_2O_2$ also oxidizes the —SH groups on the membrane proteins.

If sulfhemoglobin accumulates in large amounts in red blood cells, it tends to precipitate out in discrete granules known as **Heinz bodies.** These Heinz bodies are removed from red blood cells as they

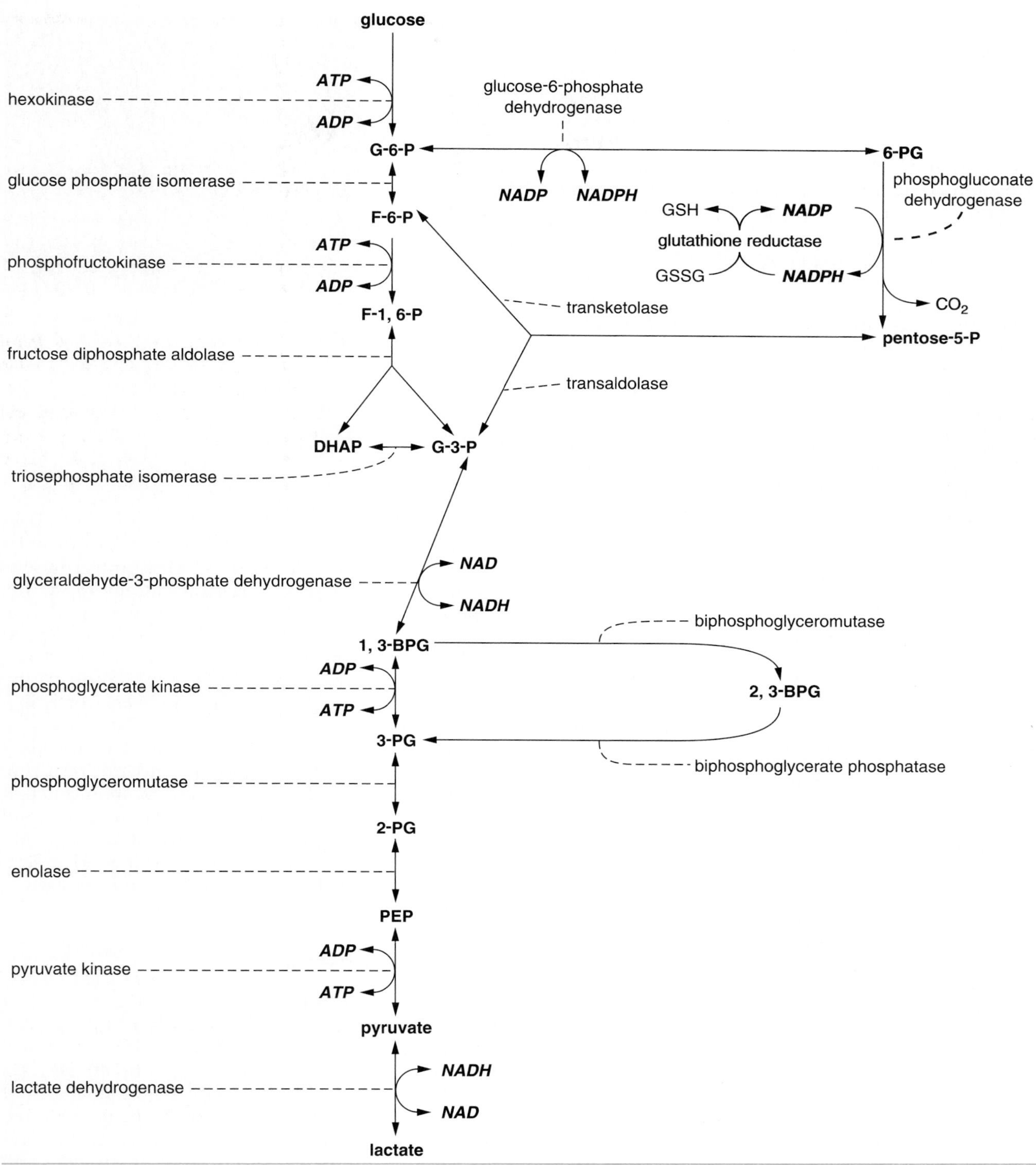


**figure 7.2**

The Embden-Meyerhof (EM) pathway (also known as the anaerobic glycolytic pathway), and the hexose monophosphate shunt [HMS], (also known as the pentose phosphate pathway) in the red blood cell. G-6-P = glucose 6-phosphate; F-6-P = fructose 6-phosphate; F1,6-P = fructose 1,6-diphosphate; DHAP = dihydroxyacetone phosphate; G-3-P = glyceraldehyde-3-phosphate; 1,3-BPG = 1,3-biphosphoglycerate; 3-PG = 3-phosphoglycerate; 2-PG = 2-phosphoglycerate; PEP = 2-phosphoenolpyruvate; 6-PG = 6-phosphogluconate; 2,3-BPG = 2,3-biphosphoglycerate; GSH = reduced glutathione; GSSG = oxidized glutathione.

percolate through the spleen by a process known as **pitting.** Each time an erythrocyte is pitted, it becomes smaller and more fragile and more distorted in shape. As a result it will be removed from the circulation long before it reaches its potential life span of 120 days. If this happens on a large scale it will result in anemia. Therefore maintaining low levels of hydrogen peroxide in red blood cells is extremely important.

### Breakdown of Hydrogen Peroxide

$H_2O_2$ is decomposed in two ways. One method involves the enzyme **catalase** in the following reaction:

$$2H_2O_2 \xrightarrow{\text{catalase}} 2H_2O + O_2$$

This is a slow process. A more efficient route involves the **glutathione peroxidase** system, which changes reduced glutathione to oxidized glutathione. Glutathione is a tripeptide made up of glutamic acid, cysteine, and glycine, as illustrated in figure 7.3. The removal of hydrogen peroxide by **reduced glutathione,** usually abbreviated as GSH (glutathione with its sulfhydryl radical in reduced form), is catalyzed by the enzyme **glutathione peroxidase** in the following reaction:

$$2GSH + H_2O_2 \xrightarrow{\text{glutathione peroxidase}} GSSG + H_2O$$

This **oxidized glutathione** (GSSG) is then reduced again with the aid of NADPH, produced in the pentose phosphate pathway, in the following reaction involving the enzyme **glutathione reductase:**

$$GSSG + NADPH + NADPH + H^+ \xrightarrow{\text{glutathione reductase}} 2GSH + NADP^+$$

Any malfunctioning of the hexose monophosphate shunt—for example, lack of glucose-6-phosphate dehydrogenase (G-6-PD) will result in a lack of NADPH. This in turn prevents the return of oxidized glutathione to its reduced form. Since the cell has only a limited amount of glutathione, this will result in an accumulation of hydrogen peroxide and in turn produce increasing amounts of sulfhemoglobin in the red blood cell.

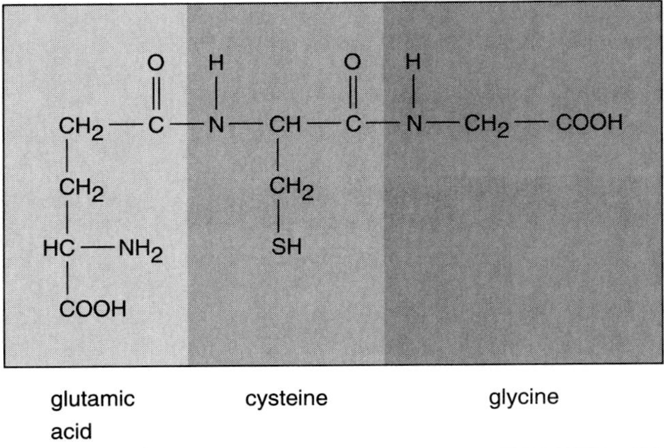


**figure 7.3**
Glutathione (reduced form). This tripeptide is made up of the amino acids glutamic acid, cysteine, and glycine.

## DESTRUCTION OF RED BLOOD CELLS

### Intravascular Versus Extravascular Hemolysis

Normal red blood cells circulate in the bloodstream for about 120 days. After that time the senescent erythrocyte is selectively removed and destroyed by monocytic-phagocytic cells that line the cords and sinusoids in the red pulp area of the spleen. Although the spleen is the normal organ of destruction, other organs containing mononuclear phagocytic cells, notably the liver, are also competent to assume that role. Although most old red blood cells are broken down in the spleen, a small amount of intravascular hemolysis also takes place. Probably less than 10% of the senescent erythrocytes break down while still in the systemic circulation. When red blood cells hemolyze outside the cells of the mononuclear phagocytic system, two things can happen to the hemoglobin released from the cell after the membrane is ruptured. Some hemoglobin is changed at once to methemoglobin by oxidation of the iron to its ferric state. This methemoglobin is later taken up by macrophages of the spleen and the liver, where it is broken down by hydrolytic and proteolytic enzymes. The rest of the free hemoglobin is linked to an $alpha_2$-glycoprotein called **haptoglobin,** and the resulting complex is eventually also removed by mononuclear phagocytic cells and broken down. The normal level of haptoglobin in the blood is about 200 mg/dL, and a sharp drop of this substance is a good indication that a great deal of intravascular hemolysis is taking place.

### Degradation of Hemoglobin

Whether present as hemoglobin in an intact cell, or as methemoglobin or a hemoglobin-haptoglobin complex after hemolysis, all hemoglobin that is not lost to the exterior by bleeding is eventually taken up by phagocytic cells. In these cells the tetramer is split up into monomers again, and the heme is separated from the globin. The globin fractions are further broken down to amino acids by proteolytic enzymes and returned to the bloodstream where they are taken up again by the liver and other somatic cells to be used again in new protein synthesis. The heme fragments are also broken down further. First the iron is removed and reused for the synthesis of new hemoglobin in the bone marrow. The porphyrin moiety of heme, on the other hand, is not reused. The porphyrin ring is opened, and the resulting tetrapyrrole, known as **biliverdin,** is released into the bloodstream.

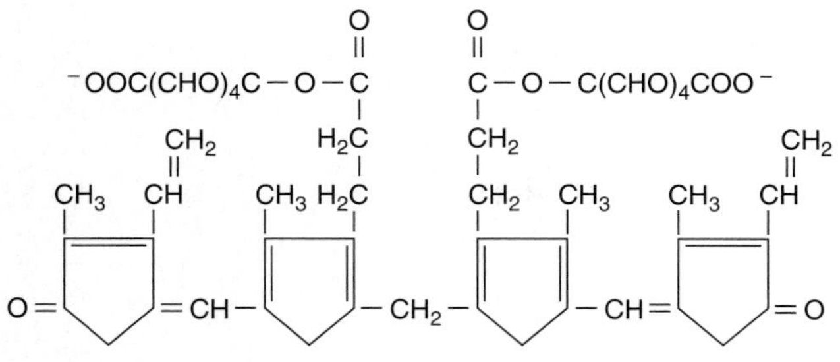

**figure 7.4**
The structure of bilirubin diglucuronide (conjugated bilirubin).

In the blood this biliverdin is oxidized to **bilirubin** and attached to albumin. As the blood percolates through the liver, the bilirubin is again detached from the albumin and taken into the hepatic cells. From here it is secreted into the bile. Since bilirubin is only sparingly soluble in plasma and water, the liver converts bilirubin first to a water-soluble form, which can then be readily secreted into the bile. The process of increasing the water solubility (osmolarity) of bilirubin is achieved by conjugating it with glucuronic acid. The resulting compound, known as **bilirubin diglucuronide** (figure 7.4), is then secreted into the bile and passed on into the gastrointestinal tract. As the conjugated bilirubin reaches the beginning of the large intestine, the glucuronides are removed by specific bacterial enzymes, and the remaining compound is subsequently reduced by the fecal flora to a group of tetrapyrrolic compounds known as **urobilinogens.** A small portion of the urobilinogen is reabsorbed, but since the venous drainage of the intestine passes directly back to the liver, virtually all the reabsorbed urobilinogen is eventually reexcreted by the liver (figure 7.5). However the great majority of urobilinogens escape reabsorption and are then converted to either mesobilirubin or stercobilin, and the remainder is oxidized by the fecal flora to urobilin. These end-products are colored compounds that give the stools their characteristic color. When obstruction of the bile duct prevents the entry of bilirubin into the gastrointestinal tract, the stools become pale or clay-colored. The process of the degradation of hemoglobin is summarized in figure 7.6.

## Hyperbilirubinemia and Jaundice

Since bilirubin is relatively insoluble and is carried in the blood attached to albumin, urine does not normally contain any detectable amounts of bilirubin. So the measurement of bilirubin is done from the plasma. The bilirubin level of the blood is an excellent indicator of the level of breakdown of hemoglobin and consequently of the amount of red blood cell hemolysis. The average adult loses about 6 g of hemoglobin per day, thus producing about 200 mg of bilirubin, since each gram of hemoglobin yields about 35 mg of bilirubin. The normal serum level of unconjugated bilirubin in transit to the liver does not exceed 0.6 mg/dL of blood. However increased hemolysis will result in excess bilirubin production, with values of 1.0 mg/dL or more, producing a condition known as **hyperbilirubinemia.**

The bilirubin in the blood is bound to albumin and cannot escape in the urine. Since small quantities of albumin do leak through the capillary walls, hyperbilirubinemia will result in the movement of increased amounts of bilirubin into the tissues, eventually coloring them yellow. This condition is known as jaundice. The lighter the patient's skin, the sooner the yellow coloring will become apparent. In dark-skinned people, jaundice is most readily detected by the yellow discoloration of the whites of the eyes.

The appearance of jaundice does not always indicate increased red blood cell destruction. If the bile duct is obstructed by an impacted gallstone, or by the encroachment of a tumor, the conjugated bilirubin cannot reach the intestines and diffuses back into the bloodstream. As a result the plasma concentration of bilirubin rises, but this is a different form of bilirubin. Since it is in its conjugated form, it is highly soluble and does not need to be bound to albumin. This has two important consequences. First, jaundice is likely to appear much quicker, since bilirubin will diffuse rapidly into the tissues in its soluble form. Second, this conjugated bilirubin will also appear in the urine, which is an important indicator regarding the cause of the jaundice.

Jaundice may also appear when the function of the liver is impaired by excess alcohol, infection (e.g., viral hepatitis), or by hepatotoxins (e.g., carbon tetrachloride). In these cases the liver is incapable of removing the albumin-bound bilirubin at its usual rate, and since the breakdown of erythrocytes continues at a normal rate, the concentration of unconjugated albumin-linked bilirubin in the plasma rises and will cause jaundice as it diffuses out of the blood with the albumin. Since the injured liver still functions at a low level, variable amounts of bilirubin will still be removed, and conjugated and excreted into the gallbladder together with the bile. Therefore the stools will retain their normal color.

Moreover, the urobilinogen carried to the liver via the portal veins of the intestines is not completely removed and escapes into the general circulation. This urobilinogen is soon carried to the kidneys and will appear in the urine. This may be used as another parameter to discover the cause of jaundice.

## Direct and Indirect Bilirubin Tests

In the clinical diagnosis of the cause of jaundice, assessing the type of bilirubin (albumin-bound or conjugated) is of paramount importance. Two simple tests will help in the diagnosis. To discover the presence of conjugated bilirubin in the serum (or urine), diazotized sulfanilic acid is added and will produce a

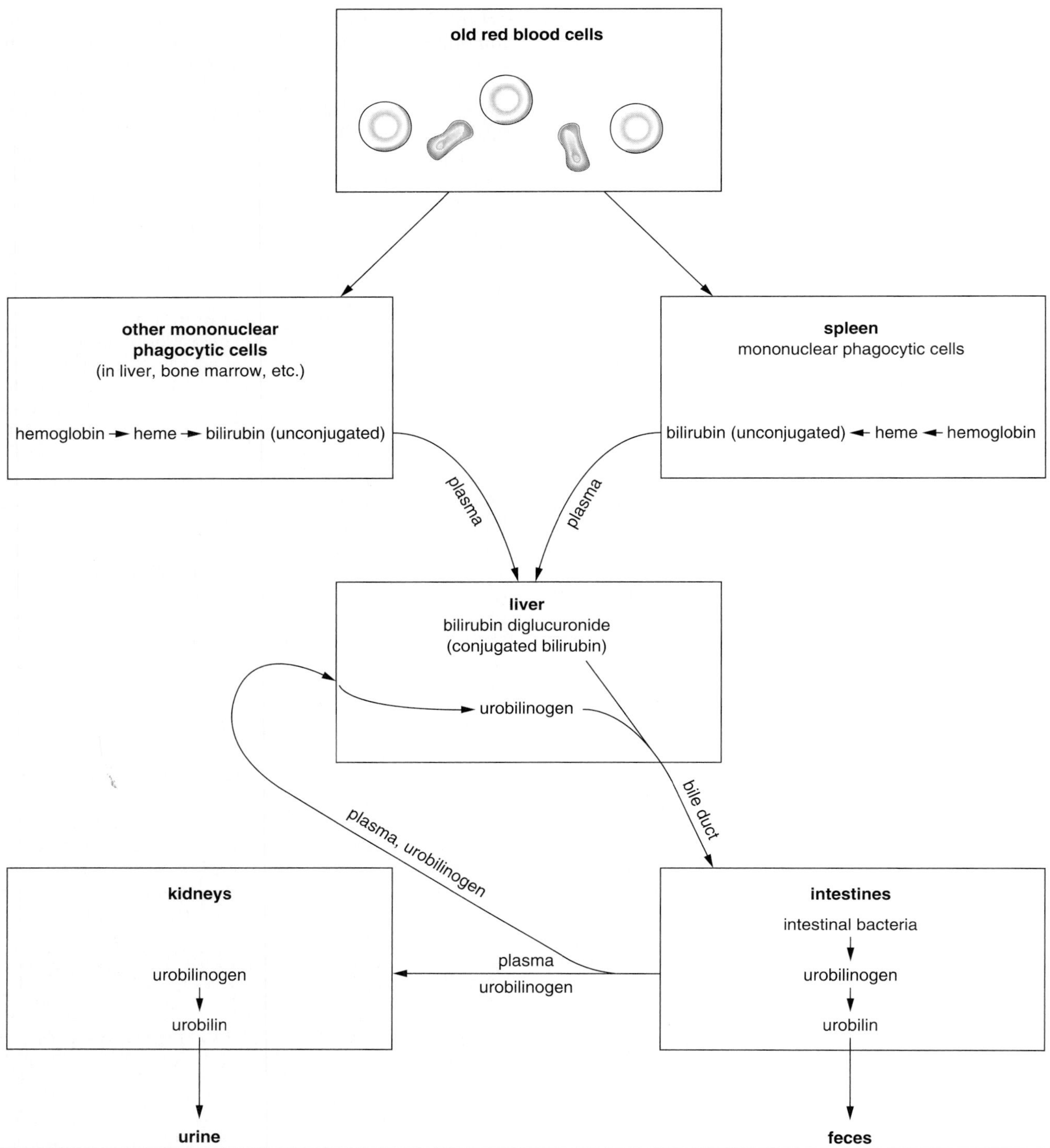


**figure 7.5**

Bilirubin metabolism and excretion.

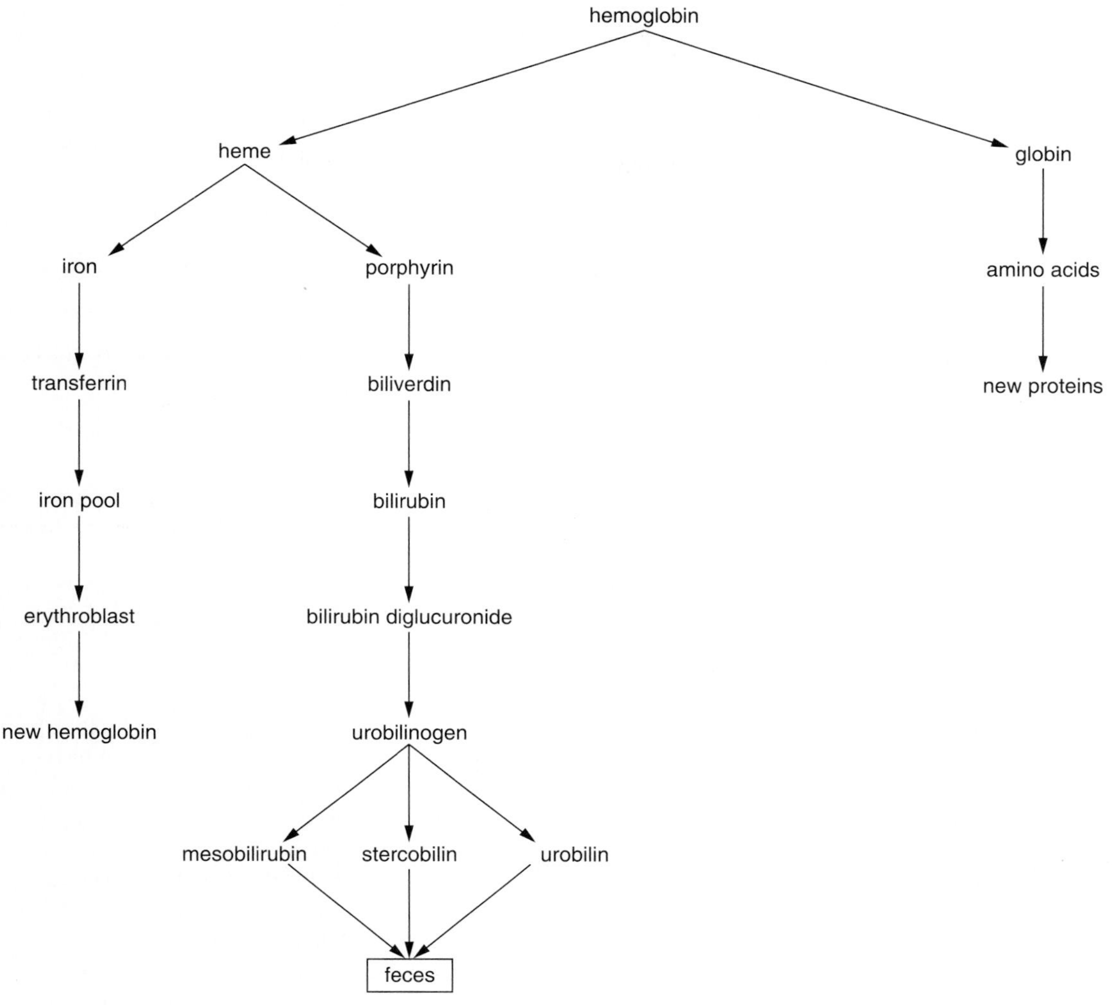


**figure 7.6**
Summary of the breakdown of the hemoglobin molecule.

reddish-purple azo compound when this type of bilirubin is present. The deeper the color, the more bilirubin has been produced. This test is known as the **direct bilirubin test,** since it needs only one incubation.

In the case of albumin-bound bilirubin, a different procedure is followed. Since this type of bilirubin is not water soluble, it requires methanol to initiate the coupling with the diazo reagent. This test requires two steps: (1) the addition of methanol and (2) the addition of the diazo sulfanilic acid, to produce a change in color when bilirubin is present. For that reason it is known as the **indirect bilirubin test.**

The typical findings for all three major causes of jaundice are summarized in table 7.1.

**Table 7.1** Typical Findings Associated with the Three Major Causes of Jaundice

| Test or Sign | Cause of Jaundice | | |
|---|---|---|---|
| | *Intravascular Hemolysis* | *Biliary Obstruction* | *Cirrhosis of the Liver* |
| Diazo reaction | Indirect | Direct | Indirect |
| Bilirubin in urine | No | Yes | No |
| Urobilinogen in urine | No | No | Yes |
| Color of stool | Normal | Pale | Normal |

# Further Readings


Cello, J. P. Diagnostic Approaches to Jaundice. *Hospital Practice,* Feb. 1982.

Gollan, J. L., and Knapp, A. B. Bilirubin Metabolism and Congenital Jaundice. *Hospital Practice,* Feb. 1985.


# Review Questions

1. List the five major energy requirements of a red blood cell.
2. What are the major energy molecules available for metabolic reactions in erythrocytes? What are the major function(s) of each?
3. Name the two major biochemical pathways used for the generation of energy molecules.
4. Why is the major pathway for energy in the red blood cell the anaerobic glycolytic pathway?
5. Compare the EM pathway and the HMS regarding (a) utilization of glucose, (b) production of energy, and (c) production of energy compounds.
6. What happens to glucose in the EM pathway? How many net molecules of ATP and NADH are generated?
7. Give another name for the HMS. What is the major importance of the HMS? Why is this important?
8. Define methemoglobin. What is the normal level of methemoglobin in the blood?
9. List some major causes of an increase in methemoglobin. Why is such an increase dangerous?
10. How is superoxide generated? What is it catalyzed into? By what enzyme?
11. Why are high levels of $H_2O_2$ dangerous? What does it produce in the red blood cell?
12. Describe Heinz bodies. How are they formed? Why do Heinz bodies shorten the red blood cell life span?
13. Give two ways red blood cells reduce the amount of $H_2O_2$. Which of these two methods is the most efficient?
14. What is the function of reduced glutathione? How is oxidized glutathione reduced again? Why is this important? What is the role of the HMS in this process?
15. Distinguish between intravascular and extravascular hemolysis. What is the major method of hemolysis in the normal human? What are the percentages of each?
16. What happens to hemoglobin when red blood cells disintegrate within the bloodstream?
17. What is the normal level of haptoglobin in the blood? What do low levels of haptoglobin indicate?
18. Where does all free and bound hemoglobin end up eventually? What happens to the hemoglobin in these cells? What happens to the globin? The heme? The iron? The porphyrin ring?
19. Distinguish between biliverdin and bilirubin. What are the normal levels of bilirubin in the blood?
20. Why is bilirubin glucuronated in the liver? What happens to this complex?
21. Distinguish between conjugated and unconjugated bilirubin in the blood. What do increased levels of each indicate?
22. Why are increased levels of bilirubin frequently associated with jaundice? Which type of bilirubin produces jaundice slowly, which rapidly? Why?
23. Which bilirubin type occurs in the urine? What does that normally indicate?
24. Explain why gallstones are usually associated with pale stools.
25. Distinguish between direct and indirect bilirubin tests. What does each indicate?

# Chapter Eight

## *Introduction to Anemias*

## MAJOR CAUSES AND EFFECTS OF ANEMIA

There are many causes that may upset the structure and function of red blood cells and as a result may produce clinical manifestations in the person experiencing those changes. These causes may be broadly categorized into four major groups. First are **genetic factors,** which may cause abnormalities in the structure of many of the molecules present in red blood cells such as hemoglobin, the enzymes needed for red blood cell metabolism, and the proteins associated with the red blood cell membrane. Genetic factors may slow the development of red blood cells or shorten their life span or reduce their effectiveness.

Second, a number of **environmental factors** may interfere with the normal development and functioning of erythrocytes. For instance, exposure to radioactive materials, ingestion of toxic substances, or lack of essential minerals and vitamins in the diet may all severely harm the process of hematopoiesis and reduce the integrity of mature red blood cells.

Third are a number of **pathological factors.** For instance, some infectious organisms (e.g., malaria parasites) may invade the red blood cells and destroy them prematurely. Other parasites (e.g., tapeworms) may take away essential minerals and vitamins needed by red blood cells to develop properly.

Last are a number of **iatrogenic factors**—which are the result of side-effects of legitimate medical treatment—that may interfere with the normal development and function of red blood cells. For instance, to combat certain cancers, it may be necessary to prescribe certain drugs (chemotherapy) or X-ray treatment (radiotherapy) that may have adverse effects on the production or function of erythrocytes. Other drugs may bind to the surface of the red blood cells, thus forming neoantigens, which may activate the body's immune system. The latter may start to produce antibodies that will eventually destroy the affected red blood cells.

Whatever the nature of the factors involved, ultimately all effects can be reduced to either or both of the following two phenomena: (1) reduced red blood cell production and/or (2) shortened erythrocyte life span. Either case results in lower amounts of hemoglobin in the peripheral circulation, which will affect the amount of oxygen reaching the respiring tissues. When that level falls below a critical threshold, it may have severe effects on the functioning of the organism. People who lack hemoglobin in their blood may suffer from many different symptoms, all of which are ultimately caused by lack of oxygen to the tissues. Such people are diagnosed as being anemic—that is, as not having enough oxygen-carrying pigment in their bloodstream as a result of too few red blood cells. So for all practical purposes, all diseases of red blood cells may be reduced to the condition of anemia.

## DEFINITION OF ANEMIA

### Traditional Definition

It is not easy to give an exact definition of anemia. Traditionally anemia has been defined as a clinical condition associated with either a decrease in the number of total red blood cells, or a decrease in the concentration of hemoglobin, or a decrease in the hematocrit, as compared with a normal person. The difficulty of this definition lies in the word "normal." What is normal varies from one group to the next. For instance, a red blood cell count of 5 million per microliter is normal for an adult man at sea level but would be quite insufficient for another adult man living at high altitude. A hemoglobin level of 12 g/dL of blood is still within the normal range for an adult woman but is much too low for an infant. A hematocrit value of 35% is too low for a nonpregnant woman but is acceptable for a pregnant woman. Therefore any definition of normal should always include the relevant group.

### Concept of the Erythron

Another way to approach the definition of anemia is to start with the concept of the **erythron,** which may be defined as the size of the total red blood cell mass of the body. This includes the erythrocytes in the systemic circulation, the developing red blood cells in the bone marrow, and the elements involved in their production. Table 8.1 gives the components of the erythron of a normal 150 lb man. Table 8.1 shows that the circulating erythrocyte compartment is the largest, and that the pool of the reticulocytes in the bone marrow is about the same size as the pool of reticulocytes in the peripheral blood. Since the average red blood cell has a life span of about 120 days, about 0.8% of the erythrocytes have to be replaced daily.

In any situation—whether normal or abnormal—the size of the erythron is simply a function of the amount of new red blood cells produced per unit of time, and their life span. This relationship may be formulated as follows:

$$M = I \times T$$

where M is the size of the red blood cell mass, I is the input of new erythrocytes, and T is time, representing the life span of the red blood cell. By definition, anemia exists when M is less than normal. Therefore anemia may result from diminished red blood cell production or from diminished red blood cell life span or both. A decreased red blood cell life span may occur as a result of blood loss or as a consequence of increased hemolysis.

There is no physiological mechanism in the body that can prolong the life span of a red blood cell beyond its 120 days. However, the body does possess a powerful mechanism that can compensate for the loss of erythrocytes that are lysed prematurely. The

**Table 8.1** Components of the Erythron in the Average 150 lb Adult

| | Cell Number/lb | Relative Number | Average Volume per Cell ($\mu m^3$) | Total Cellular Volume of Each Compartment (mL) | Total Erythron Components in Each Compartment | Average Transit Time (days) |
|---|---|---|---|---|---|---|
| *Marrow* | | | | | | |
| Nucleated cells | $11.7 \times 10^9$ | 1.7 | 250 | 88 | 6% | 5.0 |
| Reticulocytes | $18.1 \times 10^9$ | 2.7 | 120 | 44 | | 3.0 |
| *Blood* | | | | | | |
| Reticulocytes | $6.8 \times 10^9$ | 1.0 | 100 | 23 | 94% | 1.0 |
| Erythrocytes | $67.5 \times 10^9$ | 100 | 90 | 2,000 | | 120.0 |

Source: Data from C. A. Finch, *Blood* 50:699, 1977.

hormone **erythropoietin** is able to stimulate the bone marrow to produce up to eight times the normal daily amount of red blood cells to compensate for the premature destruction of erythrocytes in the peripheral circulation.

Normally, when the red blood cell life span is reduced in half, the output is increased to twice the daily amount: $M = 2I \times T/2$. When it is reduced to one-fourth, four times as many red blood cells are put into the circulation: $M = 4I \times T/4$. The maximum that the bone marrow can handle is a cell with a life span of about 15 days—that is, one-eighth of the normal life span: $M = 8I \times T/8$.

Beyond that point the marrow's ability to compensate is usually exhausted. Thus when the red blood cell life span is reduced to about 1 week—that is, one-sixteenth of normal, we can predict a halving of the erythron: $M = 8I \times T/16$, and anemia would obviously result.

The formula $M = I \times T$ makes it easy to understand why, up to a point, accelerated hemolysis may be compensated for so effectively that no overt anemia is detected. It also makes clear why a combination of mild hemolysis and relatively mild marrow failure together can cause severe anemia.

## CLASSIFICATION OF ANEMIAS

Anemias can be classified in many ways. However, traditionally they are grouped by their morphological characteristics or by their etiology. The morphological classification deals with the size of the red blood cells and their hemoglobin concentration. The etiological classification is concerned with the causes of the anemias—that is, the pathophysiological mechanisms responsible for the red blood cell deficit. Neither scheme is wholly satisfactory since certain types of anemia may show characteristics of more than one category.

### Morphological Classification

This is the most useful type of classification from the diagnostic point of view, since the initial laboratory data on red blood cell size and hemoglobin content can rule out certain types of anemia and indicate other types. Second, these morphological features may suggest an underlying cause. For instance, macrocytosis is usually associated with folate or vitamin $B_{12}$ deficiency. Since the classification of the anemias based on red blood cell morphology deals with cell size and hemoglobin content, the following two parameters are used in this classification: (1) the **MCV,** or **mean corpuscular volume,** which is normally 85 to 100 femtoliters (fL) or $\mu m^3$ per red blood cell and (2) the **MCHC,** or the **mean corpuscular hemoglobin concentration,** which is normally about 31 to 35 g/dL of red blood cells. Based on these two parameters, all the anemias may be divided into the following three major categories:

1. **Normocytic-normochromic anemias** have an MCV of 85 to 100 fL and an MCHC of 31 to 35 g/dL. In other words, in these anemias the cells have a normal size and a normal hemoglobin content. This category includes several types of anemia. First, are anemias caused by bleeding as a result of either external or internal injury. Second, are anemias associated with hypoproliferation of the hematopoietic stem cells (bone marrow failure). This group includes aplastic anemias, anemia of endocrine disease, anemia of chronic renal failure, myelophthisic anemia, and anemia caused by toxic depression of the bone marrow. This group also includes many hemolytic anemias.
2. **Microcytic-hypochromic anemias** exhibit an MCV of less than 85 fL and an MCHC of less than 30 g/dL. This means that both cell size and hemoglobin content are below normal. This category also includes several different types of anemia, such as iron deficiency anemia, which is the most common anemia, and sideroblastic anemias.

3. **Macrocytic-normochromic anemias** have an MCV of more than 100 fL and an MCHC of 30 to 35 g/dL. This means that the cell size is much larger, but the hemoglobin concentration is normal. This group includes megaloblastic anemias and certain hemolytic anemias.

In some normal physiological situations the MCV may be outside the normal adult range. For instance, in newborns the MCV may be high for a few weeks. In young children the MCV may be low (e.g., a level below 70 fL at 1 year of age is normal), but it rises slowly throughout childhood to the normal adult range. Pregnancy also tends to be associated with a rise in MCV, even in the absence of other causes, such as lack of folic acid in the diet.

### Etiological Classification

The advantage of this classification is that it fosters an understanding of the disease process in kinetic terms. Its major shortcoming lies in the fact that some anemias are due to more than one pathophysiological mechanism. Often one mechanism predominates early in the development of certain anemias, and others supervene later. As was said in the beginning of this chapter, basically there are only two causes of anemia: (1) decreased erythrocyte production or (2) increased erythrocyte loss. Further analysis reveals that decreased red blood cell production may be caused by either decreased proliferation of new erythrocytes or by impairment in the development of red blood cells. When the cause is decreased production of new red blood cells in the bone marrow, this may be due to either lack of erythropoietin or to bone marrow injury or defects. Lack of erythropoietin may result from the fact that the source of erythropoietin—the kidneys—are damaged in some way. This results in an anemia known as **anemia of renal failure.** Anemia may also be caused by low oxygen requirements by the body, as frequently happens in people suffering from hypothyroidism. Erythropoietin production is regulated by the need for oxygen. If oxygen needs are low, little erythropoietin is produced; hence few new red blood cells are produced, and anemia may result. This type of anemia is commonly referred to as **anemia of endocrine disease.**

It may also be that nothing is wrong with the kidneys or with the oxygen levels, and normal amounts of erythropoietin are produced; however the stem cells are unable to respond to the erythropoietin stimulus. They may lack receptors for this hormone on their membrane, or some unknown factor may interfere with hormone binding to the stem cells. This occurs fairly frequently in persons suffering from chronic disease. Hence this anemia is known as **anemia of chronic disease.**

Decreased proliferation of new red blood cells may be also due to bone marrow damage. A number of causes may produce defective bone marrow. First, normal marrow may be crowded out by a tumor of the bone or other tissues. The resulting anemia is called **myelophthisic anemia.** Normal, healthy bone marrow may also be replaced by a cancer of one of the blood cell types. Many types of leukemia are associated with anemia. Third, bone marrow may be damaged by some physical factor (e.g., radiation) or a chemical factor (e.g., benzene) or some infectious agent (e.g., a virus), all of which may kill the stem cells and thus stop further cell proliferation. This type of anemia is known as **aplastic anemia.** Finally, there is a rare type of inherited anemia associated with hypoproliferation of bone marrow cells. This condition is known as **Fanconi's anemia.**

The second cause of decreased erythrocyte production is due to failure in red blood cell maturation. Not much is wrong with the stimulation of the stem cells, which produce many normoblasts. But few mature red blood cells reach the peripheral circulation. Most of them die within the bone marrow before they reach the final stage of development. In other words, there is a high level of ineffective hematopoiesis resulting from a number of different causes such as a lack of folic acid or vitamin $B_{12}$ in the diet. This will result in the production of very low numbers of macrocytic-normochromic red blood cells. This type of anemia is known as **megaloblastic anemia.** Another cause of impairment in maturation may be the lack of iron in the diet. This will result in the production of many small (microcytic) and hypochromic red blood cells and produce an anemia known as **iron deficiency anemia.** Sometimes sufficient amounts of iron are available in the diet, but for some reason this iron is unable to reach the developing red blood cells. This happens frequently in persons suffering from chronic disease; hence it is known as **anemia of chronic disease.** Finally, there may be some impairment in the enzymes involved in hemoglobin synthesis. If something is wrong with the production of the heme fraction, this will result in an anemia known as **sideroblastic anemia.** If the lack of hemoglobin is due to impairment in the globin synthesis, it produces **thalassemia syndromes.** A summary of these causes is given in table 8.2.

The second major cause of anemia is increased erythrocyte loss. Nothing is wrong with the red blood cell production, but the cells produced have a shorter than normal life span. There are two major causes for this reduced life span: red blood cells may be lost prematurely due to either bleeding or increased intravascular hemolysis. Bleeding may be acute or chronic. Acute bleeding is usually a result of trauma. Chronic bleeding is usually due to gastrointestinal ulcers or to increased menstrual loss. Both types may result in **anemia of hemorrhage.** Increased intravascular hemolysis may be from hereditary causes, or it may be acquired later in life as a result of some external factor. All cases result in **hemolytic anemia.** There are three major causes of **hereditary hemolytic anemia.** It may be the result of inherited defects in the red blood cell membrane. This may produce such diseases as **hereditary spherocytic anemia** or **hereditary elliptocytic anemia.** The problem may also be caused

**Table 8.2** Etiology of Anemia

| Cause | Mechanism | Anemia |
|---|---|---|
| Decreased erythrocyte production, due to: | | |
| Decreased proliferation of new erythrocytes, due to: | | |
| Decreased erythropoietin, due to: | Impaired production by the kidneys | Anemia of renal failure |
| | Low oxygen requirements | Anemia of endocrine disease |
| | Impaired stem cell response to erythropoietin | Anemia of chronic disease |
| Bone marrow damage or defect, due to: | Replacement of marrow by tumor | Myelophthisic anemia |
| | Replacement of normal marrow by cancerous cell line | Anemia associated with myeloproliferative disease |
| | Damage to bone marrow by physical or chemical agents, or infections | Aplastic anemia |
| | Inherited bone marrow defect | Fanconi's anemia |
| Impairment in the maturation of new erythrocytes, resulting in: | | |
| Macrocytic-normochromic erythrocytes, due to: | Folic acid deficiency | Megaloblastic anemia |
| | Vitamin $B_{12}$ deficiency | Megaloblastic anemia |
| Microcytic-hypochromic erythrocytes, due to: | Iron deficiency | Iron deficiency anemia |
| | Unavailability of iron to blast cells | Anemia of chronic disease |
| | Impairment of heme synthesis | Sideroblastic anemia |
| | Impairment of globin synthesis | Thalassemia syndromes |
| Increased erythrocyte loss, due to: | | |
| Hemorrhage | | |
| Acute | | |
| Chronic | | |
| Intravascular hemolysis, as a result of: | | |
| Hereditary factors, resulting in: | Defects in the erythrocyte membrane | Hereditary spherocytosis<br>Hereditary elliptocytosis |
| | Defects in erythrocyte metabolism | G-6-PD deficiency anemia |
| | Abnormal hemoglobin production | Sickle-cell anemia<br>HbC disease, HbD disease<br>HbE disease |
| Acquired accelerated hemolysis, due to: | Activation of the immune system | Immunohemolytic anemia |
| | Physical factors | Red cell fragmentation syndromes |
| | Chemical agents | Various forms of hemolytic anemia |
| | Microorganisms | Various forms of anemia (e.g., anemia of malaria) |
| | Secondary to other diseases | Various forms of anemia (e.g., anemia of hepatic failure) |
| | Sensitivity to complement | Paroxysmal nocturnal hemoglobinuria |

by defects in the enzymes associated with red blood cell metabolism. For instance, lack of G-6-PD will result in a special type of hemolytic anemia due to the buildup of sulfhemoglobin **(sulfhemoglobinemia).** Finally, increased hemolysis may be caused by the presence of abnormal hemoglobin in the red blood cell, as seen in **sickle-cell anemia.**

In **acquired hemolytic anemias,** hemolysis is accelerated because of a number of external factors. A summary of the major factors is given in table 8.2.

## SIGNS AND SYMPTOMS OF ANEMIA

### Overt Symptoms

Overt symptoms commonly associated with anemia are shortness of breath (especially during exercise), weakness, lethargy, and headaches. The headaches are often accompanied by dizziness and nausea; a lack of interest in food is also common. The latter may result in flatulence, constipation, or diarrhea.

Other obvious symptoms include pallor of the skin and mucous membranes. The paleness of the mucous membranes is a more reliable indicator of anemia, since blanching of the skin may also be caused by a sensitivity to cold. In younger people the anemia symptoms usually also include tachycardia and palpitation, and in older persons signs of cardiac failure, angina pectoris, and claudication may be experienced.

Moderate and severe anemias may also be associated with visual disturbances, which are caused by retinal hemorrhages. This sign is particularly evident in rapid-onset anemias. Further, severe anemias often show delayed wound healing, especially when caused by acute or chronic bleeding.

In women of childbearing age, anemia may be associated with menstrual irregularity or even complete amenorrhea.

Table 8.3 gives a summary of the signs and symptoms that may be associated with anemia, organized according to the body systems affected. However, it should be noted that people with severe cases of anemia may show few overt symptoms and others with only mild forms of anemia may be severely incapacitated.

### Basic Truisms Regarding Anemia

The presence or absence of clinical features may be predicted by the following factors:

1. Anemias that develop rapidly usually cause more symptoms than anemias that develop slowly, because the body has less time for adaptation in the cardiovascular system and for compensation by the bone marrow.
2. The elderly tolerate anemias less well than the young, because their bone marrow is less able to compensate for the lack of hemoglobin and red blood cells. Further, in the young the body also can compensate by increasing the cardiac output. In the elderly this cardiovascular compensatory mechanism is often impaired, and as a result they are unable to pump the blood faster throughout the circulatory system.
3. People who eat well-balanced diets will usually show fewer overt symptoms than persons who are on diets, eat junk foods, or otherwise have impaired nutrition.

Finally, it should be remembered that mild anemias usually show very few if any symptoms. Only when hemoglobin levels fall below 10 g/dL of blood will the first symptoms arrive. Even severe anemias—that is, those with hemoglobin levels of 5 to 6 g/dL produce remarkably few symptoms, when the onset is gradual, and the individual is otherwise healthy.

Anemia is usually also associated with a rise in 2,3-BPG in the red blood cells. This results in a shift of the oxygen-dissociation curve to the right, so that oxygen is more readily given up to the tissues. This adaptation occurs more readily in some anemias than in others.

**Table 8.3 Signs and Symptoms of Anemia**

*Cardiovascular*

Tachycardia
Palpitations
Strong arterial pulses
Cardiac enlargement
Murmurs
Bleeding in the retina

*Respiratory*

Difficult or painful breathing (dyspnea)
Problem with breathing when lying down (orthopnea)
Increased depth and rate of respiration

*Neuromuscular*

Headaches
Dizziness (vertigo)
Faintness
Inability to concentrate
Feeling of fatigue
Sensitivity to cold
Ringing in the ear (tinnitis)

*Integumentary*

Paleness (pallor) of the skin, especially the mucous membranes and nailbeds
Delayed wound healing

*Gastrointestinal*

Anorexia
Nausea
Flatulence
Constipation
Diarrhea

*Genitourinary*

Menstrual irregularity
Amenorrhea
Increased urination

Apart from these general symptoms, which are common to all anemias, often specific signs are associated with particular types of anemia. For instance, koilonychia (flat nails) is unique to iron deficiency anemia, jaundice with hemolytic anemia, leg ulcers with sickle cell anemia, bone deformities with beta-thalassemia major. These and other specific signs are discussed in detail in the following chapters.

## DIAGNOSIS OF ANEMIA

### Routine Laboratory Investigations

A number of laboratory tests are routinely performed in the diagnosis of anemia. Not only do they give us an indication whether anemia is present, but they will also help in pinpointing the type of anemia involved.

The following is a series of common tests used in the diagnosis of anemia:

**1. Red Blood Cell Count**
The aim of this test is to count the number of red blood cells per microliter (μL) (= per cubic millimeter, or $mm^3$). The normal values are 4.5 to 6.5 million per microliter in men and 4.0 to 5.5 million per microliter in women.

**2. Determination of the Hemoglobin Concentration**
The purpose of this test is to provide a quantitative determination of concentration of hemoglobin in the blood, expressed in grams per deciliter (100 mL per liter) of blood. The normal values are 13.5 to 18.0 g/dL in men and 11.5 to 16.5 g/dL in women. To perform this test, the red blood cells first have to be hemolyzed.

**3. Determination of the Hematocrit Value**
The purpose of this test is to determine the ratio of blood cells to plasma, expressed as a percent of the whole blood volume. This ratio can be found by centrifuging a small amount of blood in a small capillary tube. When whole blood is centrifuged at high speed, the blood cells will be packed at the lower end of the tube and can be reported as a percentage of the whole blood volume. The normal values in men are from 40 to 54% and in women from 37 to 47%.

**4. MCV and MCHC**
To calculate the MCV and the MCHC, we need to know the red blood cell count, the hemoglobin determination, and the hematocrit value. These two parameters are crucial in determining the type of anemia from the morphological point of view. The normal levels of MCV and MCHC were mentioned earlier in the chapter.

**5. White Blood Cell Count**
This test is another useful instrument to determine whether the anemia is a "pure" anemia or whether it is the result of a general bone marrow failure. In the latter case, **pancytopenia** will be noted. Pancytopenia involves not only a drop in red blood cells but also a drop in white blood cells and even platelets. On the other hand, a rise in the number of neutrophils is often associated with hemolytic anemia, infection, or hemorrhage.

**6. Reticulocyte Count**
The reticulocyte count is another important test in the investigation of anemia. The normal reticulocyte count is about 1% of the total number of red blood cells (normal range is 0.5 to 2.0%). However in cases of anemia this percentage could be much higher—especially in chronic hemolytic anemias where the body has had time to adjust for the loss of red blood cells. In case of an acute hemorrhage, an erythropoietin response occurs within 6 hours. Within 2 to 3 days the reticulocyte count rises dramatically, reaching a maximum about 6 to 10 days later, and this high level remains until the red blood cell level returns to normal (provided the cell can do so—e.g., iron deficiency is not present). If the reticulocyte count is not raised in an anemic person, this suggests impaired bone marrow function or lack of erythropoietin.

**7. Blood Smear**
The blood smear is another important tool in the diagnosis of the cause(s) of anemia. Abnormally shaped red blood cells, erythrocyte inclusions, and abnormally shaped white blood cells and platelets give ready clues regarding the etiology of the anemia. For instance, the presence of target cells may indicate thalassemia, whereas the presence of Heinz bodies may point toward methemoglobinemia. The occurrence of normoblasts in the blood film suggests hyperproliferation of the bone marrow frequently associated with hemolytic anemia. Hypersegmented neutrophils are often encountered in megaloblastic anemia, whereas the presence of large "shift" platelets suggests hemorrhage.

**8. Bilirubin Measurement**
The purpose of the bilirubin test is to determine the level of bilirubin in the blood. The normal values of unconjugated, albumin-bound bilirubin are between 0.5 to 0.6 mg/dL of serum. In cases of accelerated hemolysis, these unconjugated bilirubin levels may increase eight-to tenfold (4.0 to 5.0 mg/dL) depending on the conjugating capacity of the liver. Elevated levels of unconjugated bilirubin may result from either increased hemolysis or increased ineffective hematopoiesis. Unconjugated bilirubin is usually measured by the indirect bilirubin test. As explained in the previous chapter, bilirubin is usually taken up by the liver where it is conjugated to glucuronic acid to make it more soluble in bile for transport to the gastrointestinal tract. If the bile duct is blocked, much of this conjugated bilirubin may diffuse back into the bloodstream from where it may be secreted into the urine. The level of conjugated bilirubin in blood or urine is measured by the direct bilirubin test.

**9. Estimation of Serum Haptoglobin**
The measurement of the serum haptoglobin level is another test to estimate the presence of increased hemolysis in the circulating blood. The normal level of serum haptoglobin is about 200 mg/dL of blood. If that level drops sharply, it may indicate the presence of large amounts of free hemoglobin in the blood as a result of intravascular hemolysis.

One way to measure the amount of haptoglobin is by electrophoresis of the serum sample. Haptoglobin is found in the alpha$_2$-fraction of the globulin. By comparing the alpha$_2$-fraction of the sample with that of a standard serum, we can estimate the amount of haptoglobin present.
If the haptoglobin fraction is completely saturated with hemoglobin, free hemoglobin will appear in the serum electrophoretogram in the beta-globulin region of the separation.

**10. Plasma Iron Levels and TIBC**

The effectiveness of red blood cell production by the bone marrow can be assessed by measuring the plasma iron concentration and total iron-binding capacity (TIBC) of the blood. The extent of erythropoiesis can be assessed by calculating the plasma iron content and by measuring the rate of clearance of transferrin-bound iron from the plasma. As most of the iron leaving the plasma is taken up by erythroblasts, the iron turnover is related to the total erythropoietic activity. For example, a plasma iron turnover of three times normal suggests a threefold expansion of erythropoiesis. Reduced plasma turnover, on the other hand, indicates erythropoietic hypoplasia. The normal values of plasma iron levels are 65 to 173 μg/dL, whereas the normal TIBC levels are between 224 to 366 μg/dL.

## Special Tests

The preceding 10 tests are routinely performed in clinical and office laboratories to determine or confirm the nature and cause of anemia in a given patient. However these routine tests may not give the physician sufficient information to determine the underlying cause of the anemia, which makes decisions regarding treatment rather difficult. In those situations frequently one or more special tests are ordered that will give the clinician a better insight as to the underlying cause of the anemia.

A large number of specialized tests are available, some of them more useful than others. Frequently they are quite expensive as they require special instrumentation or invasive surgery such as bone marrow biopsy. Some of the more common special tests include the following:

**1. Bone Marrow Examination**

Sometimes it is necessary to supplement the blood film analysis with a bone marrow examination. The techniques used in bone marrow extraction (the bone marrow smear and marrow biopsy) were discussed in chapter 3. Examining the bone marrow is useful in providing information regarding the proportion of the different cell lines, especially the ratio of erythroid versus myeloid cells, which normally should be 1:3. It will also show the presence of foreign cells in the bone marrow, as would be the case in secondary carcinoma. A bone marrow smear would also give information about the size and shape of the developing cells—for example, whether they are normoblastic or megaloblastic. Finally, a bone marrow biopsy slide would give a good indication regarding the architecture of the bone marrow, providing a clear picture of the ratio of hematopoietic cells, fat cells, and fibroblasts.

**2. Use of Radioactive Iron**

Sometimes the incorporation of radioactive iron ($^{59}Fe$) into the circulating cells is used to measure effective erythropoiesis. In this test labeled iron is injected into the bloodstream. In healthy persons the $^{59}Fe$ disappears from the circulation within 2 to 3 hours. In cases of erythroblastic hypoplasia (i.e., below normal production of red blood cells), the clearance of $^{59}Fe$ is delayed, whereas in erythrocytic hyperplasia (i.e., increased production of red blood cells), the clearance is more rapid (figure 8.1).

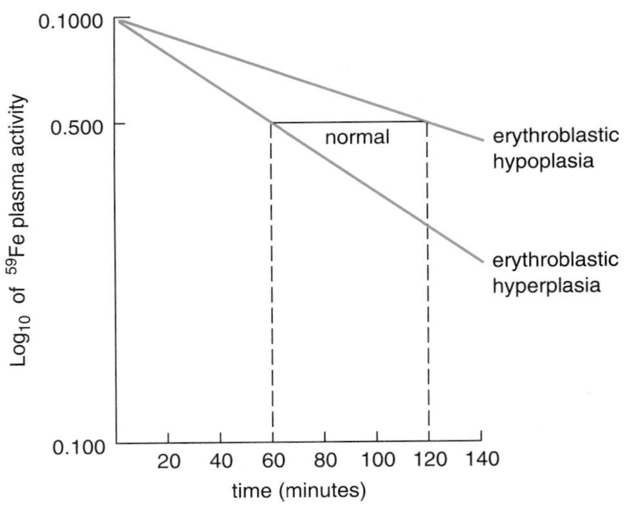


**figure 8.1**

$^{59}Fe$ clearance from the blood. Note that in normal individuals, half of the $^{59}Fe$ is cleared within 60 to 120 minutes. In erythroblastic hyperplasia, clearance is more rapid; while in erythroblastic hypoplasia, clearance is delayed.

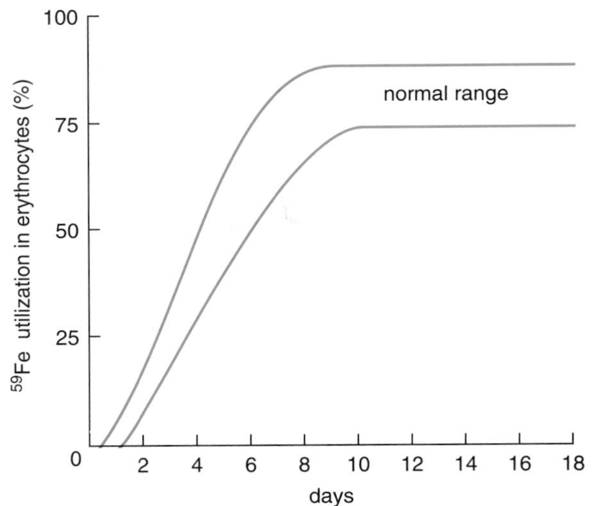


**figure 8.2**

$^{59}Fe$ incorporation into the red blood cells over two weeks. Note that on day 10, 75 to 90 percent of the $^{59}Fe$ has been incorporated into the circulating erythrocytes.

In normal individuals 70 to 80% of the injected $^{59}Fe$ reappears again in the circulation, incorporated in the hemoglobin of the red blood cells, within 10 days (figure 8.2). Red blood cell $^{59}Fe$ incorporation values of less than 70% between the tenth and fourteenth day indicate diminished or ineffective erythropoiesis.

**3. Red Blood Cell Life Span**

Measurement of the life span of the red blood cell is another good diagnostic tool in the investigation of hemolytic anemias. Erythrocyte life span is measured by $^{51}Cr$-labeled red blood cell survival. A sample of the individual's blood is taken and incubated with a chemically stable form of $^{51}Cr$: $^{51}CrO_4^-$. This chromate anion rapidly penetrates the red blood cells, where it is reduced to $^{51}Cr^{3+}$, which can no longer escape through the cell membrane. Most of the trapped $^{51}Cr^{3+}$ forms ionic bonds with negatively charged groups of hemoglobin radicals and other intracellular proteins. These labeled cells are then reinjected into the circulation, where they diffuse rapidly throughout the systemic circulation. The disappearance of $^{51}Cr$ from the blood is measured sequentially over 3 weeks. In a normal person about 1% of the intracellular $^{51}Cr$ is eluted every day from the blood and is excreted primarily in the urine. In anemic persons, especially those suffering from hemolytic anemia, this process takes place much more rapidly.

**4. Osmotic Fragility Test**

The osmotic fragility test is commonly performed in patients suspected of suffering from a hereditary form of hemolytic anemia. The test consists of exposing drops of blood to increasing dilutions of NaCl and observing when hemolysis begins to occur. In a normal individual large-scale hemolysis begins to occur at dilutions of 0.3 to 0.5% NaCl. In hereditary spherocytosis, on the other hand, hemolysis starts at 0.7% NaCl (figure 8.3).

**5. Glycosylated Hemoglobin Determination**

Although this is not a routine test, it is commonly performed in the laboratory on patients suspected of having diabetes mellitus or who are considered to be prediabetic. The test determines the amount of glycosylated hemoglobin as a percentage of total hemoglobin. In this test glycosylated hemoglobin is separated from normal hemoglobin by chromatography, and the resulting fractions are measured in a spectrophotometer. The normal level of glycosylated hemoglobin is around 5%.

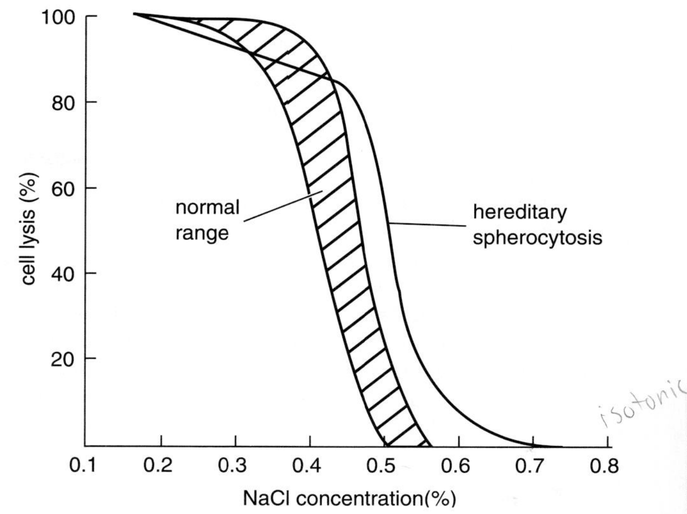


**figure 8.3**

The osmotic fragility test. The shaded area represents the normal osmotic fragility curve of red blood cells. The single line represents the fragility curve seen with hereditary spherocytosis. Note that the curve is shifted significantly to the right.

Many other special tests may be performed on occasion to investigate the structure and functions of red blood cells. These include the sickle-cell preparation, sedimentation rate test, total protein determination, G-6-PD deficiency test, hemoglobin electrophoresis, and fetal hemoglobin determination.

## *Further Readings*

Hillman, R. S., and Finch, C. A. General Characteristics of the Erythron. Chapter 1 in *Red Cell Manual,* 5th ed. F. A. Davis, Philadelphia, 1985.

———. The Detection of Anemia. Chapter 2 in *Red Cell Manual,* 5th ed. F. A. Davis, Philadelphia, 1985.

Seiverd, C. E. *Hematology for Medical Technologists,* 5th ed. Lea & Febiger, Philadelphia, 1983.

## *Review Questions*

1. Give some traditional definitions of anemia. What is the problem with these definitions?
2. What is meant by the erythron? What is the formula? List the components of the erythron in a normal 150 lb man.
3. Explain why a person with severe hemolysis may show only a mild form of anemia. Explain why in other situations mild hemolysis may result in severe anemia. What is the ultimate compensation factor of bone marrow?
4. List the three categories of anemia from the morphological point of view. Give the two major parameters for each category. What are the numbers of each parameter for each category?
5. Provide two examples of each morphological category of anemia.
6. What are the two major categories of anemia from the etiological point of view?
7. What are the two major reasons for the decrease in red blood cell production?
8. Explain the cause(s) of (a) anemia of renal failure, (b) anemia of chronic disease, (c) anemia of leukemia, (d) myelophthisic anemia, (e) aplastic anemia, and (f) Fanconi's anemia.

9. What are two major causes of megaloblastic anemia? How would you classify megaloblastic anemia from the morphological point of view?
10. List four important groups of microcytic-hypochromic anemias. What causes each of these types of anemia?
11. What are the two major causes of erythrocyte loss?
12. What are the two major forms of hemorrhagic anemia? Give an example of each.
13. List the three major categories of hereditary hemolytic anemias. Give an example of each.
14. List five different causes of acquired hemolytic anemia.
15. Name five overt symptoms associated with mild forms of anemia.
16. Name five overt symptoms frequently associated with moderate and severe anemias.
17. From the clinical point of view, when is an anemia classified as mild, moderate, or severe? (Express your answers in hemoglobin levels.)
18. What are three basic truisms associated with all anemias?
19. Mention 10 different laboratory tests routinely used to diagnose anemia.
20. Explain why the white blood cell count is useful in the determination of anemia.
21. What is a normal reticulocyte count? What does a high reticulocyte count indicate?
22. What is a normal bilirubin level? What does a high bilirubin level indicate?
23. What is a normal haptoglobin level? What does a low haptoglobin level indicate?
24. What is a normal plasma iron level? What does a low plasma iron level indicate? What may a high plasma iron level mean?
25. What is the normal level of glycosylated hemoglobin? What does a high level of glycosylated hemoglobin indicate?
26. At what NaCl dilutions do normal red blood cells usually hemolyze? At what dilutions do they hemolyze in hereditary spherocytic anemia?
27. Describe how you could determine the red blood cell life span.

# Chapter Nine

## *Aplastic Anemias*

## CLASSIFICATION OF APLASTIC ANEMIA

The term "aplastic anemia" is a bit misleading, since it seems probable that prolonged complete *aplasia*—that is, no blood cell production at all, would be incompatible with life. A better name would be "hypoplastic anemia"; however the original term **aplastic anemia** has been used for a long time. The main characteristics of this group of poorly understood disorders is hypoplasia of the bone marrow due to decreased erythroblast formation.

Although many different causes of aplastic anemia have been identified in more than half of patients with depressed bone marrow function no specific causes have been found. These cases are termed **idiopathic,** though the great majority are likely due to exposure to some as-yet-unknown environmental pollutant. In the meantime, however, **idiopathic aplastic anemia** is the most common type of decreased bone marrow activity. The known causes of hypoplasia of the bone marrow may be divided into two major categories. First are those that are *inherited.* These are often referred to as **primary aplastic anemias.** The most important of these primary, or inherited, marrow hypoplasias is **Fanconi's anemia.** The symptoms of this disorder appear in the first decade of life and are often accompanied by multiple congenital abnormalities—that is, not just the bone marrow is affected but also the kidneys, the skin, and the spleen. Many victims are also mentally retarded and later develop leukemia. Since this form of congenital aplastic anemia is caused by a rare recessive gene, it can be expressed only by inbreeding (e.g., by marriage between first cousins). Luckily the incidence of primary aplastic anemia is very low.

The great majority of cases of aplastic anemia whose cause is known are classified as **secondary,** or **acquired aplastic anemias.** This category includes aplastic anemias acquired later in life due to exposure to some environmental agent. These anemia-causing agents include ionizing radiation, toxic chemicals, or parasitic infections, as well as anemias resulting as a side-effect of other diseases, as explained in the following material.

## ETIOLOGY OF APLASTIC ANEMIA

The known causes of secondary or acquired aplastic anemia may be divided into at least six major categories: (1) decreased production of erythropoietin, (2) decreased effectiveness of erythropoietin, (3) damage to bone marrow by ionizing radiation,(4) damage to bone marrow by toxic chemicals, (5) replacement of normal hematopoietic tissue by tumors, and (6) competition for nutrients between normal and abnormal blood cells.

### Decreased Production of Erythropoietin

Since erythropoietin is produced mainly by the *kidneys,* it follows that any major damage to the kidneys may have severe consequences for the production of erythropoietin. Lack of erythropoietin will result in understimulation of the bone marrow, and hence may result in a decreased production of red blood cells, and thus anemia. This type of aplastic anemia is commonly called **anemia of renal failure.**

Low erythropoietin production also may be due to lack of stimulation to the kidneys to produce this red blood cell growth factor. The kidneys are normally stimulated to produce more erythropoietin by low oxygen tensions **(hypoxia)** in the respiring tissues. In certain endocrine diseases, however, the body's requirements for oxygen are lowered. For instance, in hypothyroidism, the basal metabolic rate (BMR) drops considerably, resulting in lower oxygen needs in the tissues. Consequently the need for hemoglobin drops and fewer red blood cells are produced. Analysis of a blood sample of such a hypothyroidic person will show a much lower hemoglobin level. This type of aplastic anemia is commonly called **anemia of endocrine disease.**

### Decreased Effectiveness of Erythropoietin

Situations may occur in which nothing is wrong with erythropoietin production, but the bone marrow cells are no longer able to respond to the activating stimulus of erythropoietin. The erythroid stem cells may have lost their receptor sites for this growth factor and as a result can no longer be stimulated by this hormone. Also the erythroid stem cells may not produce a sufficient number of receptor sites needed to elicit a full response. In addition, the receptor sites may be slightly changed, so that they can no longer activate the mitotic mechanism of the stem cells, even when they are complexed with erythropoietin. These phenomena of low levels or altered receptor sites are commonly associated with chronic disease conditions such as infectious hepatitis, and hence this type of anemia is called **anemia of chronic disease.**

### Damage to Bone Marrow by Ionizing Radiation

High radiant energies such as X rays, gamma rays, and neutrons released during laboratory or reactor accidents, or released by deliberate therapeutic exposure as a result of radiotherapy, affect all tissues in which cell turnover is rapid. These include the gonadal epithelium, intestinal epithelium, hair cells of the skin, and hematopoietic tissues. Intensive radiation kills cells in each of these tissues, and death of the patient usually follows due to acute marrow aplasia and intestinal ulceration. If the patient survives a critical 3- to 6-week period, surviving stem cells may slowly produce marrow regeneration. If the stem cells are damaged, regeneration may be incomplete and permanent marrow hypoplasia and pancytopenia may result.

**Table 9.1** Chemical Agents Implicated in Aplastic Anemia

**Agents Regularly Producing Marrow Hypoplasia if Dose is Sufficient**

Benzene and derivatives (toluene, etc.)
Cytostatic agents (6-mercaptopurine, busulfan, melphalan, vincristine, etc.)
Other poisons (inorganic arsenic)

**Agents Occasionally Associated with Marrow Hypoplasia**

| *Class* | *Relatively Frequent* | *Infrequent* |
|---|---|---|
| Antimicrobial | Chloramphenicol | Streptomycin |
| | Organic arsenicals | Amphotericin B |
| | Penicillin, tetracyclines | Sulfonamides |
| | | Sulfisoxazole (Gantrisin) |
| Anticonvulsant | Mephenytoin (Mesantoin) | Mephenytoin |
| | Trimethadione (Tridione) | Diphenylhydantoin (Dilantin) |
| | | Primidone |
| Analgesic | Phenylbutazone | Aspirin |
| Antithyroid | | Carbimazole |
| | | Tapazole |
| | | Potassium chlorate ($KClO_4$) |
| Hypoglycemic | | Tolbutamide (Orinase) |
| | | Chlorpropamide (Diabinese) |
| Antianxiety | | Chlorpromazine (Thorazine) |
| | | Chlordiazepoxide (Librium) |
| Insecticide | | DDT |
| | | Parathion |
| Miscellaneous | | Colchicine |
| | | Acetazolamide (Diamox) |
| | | Hair dyes |
| | | Carbon tetrachloride ($CCl_4$) |
| | | Thiocynate (SCN) |

### Damage to Bone Marrow by Toxic Chemicals

Historically, chemical agents were the first recognized causes of aplastic anemia. Chemical substances capable of depressing bone marrow may be roughly divided into two categories: (1) chemicals that regularly produce hypoplasia if the dose is large enough and (2) chemicals that only occasionally produce hypoplasia.

An example of the first group is **benzene.** This chemical was the first clear-cut specific agent identified as a cause of aplastic anemia. For many years, aplastic anemia was a frequent cause of disease in persons working in the rubber tire industry—many workers were exposed to fatal doses of benzene. Since that time, many other industrial chemicals, agricultural compounds, and pharmacological drugs have been identified as causative agents of bone marrow suppression. A list of chemical agents that are regularly or occasionally associated with aplastic anemia is given in table 9.1. Clearly most of the listed agents are drugs rather than industrial or agricultural compounds.

Presently the antibiotic chloramphenicol leads the list as being the most common chemical agent causing aplastic anemia. This drug is a nitrobenzene derivative and usually causes a brief, reversible suppression of the bone marrow in most, if not all, patients exposed to it. Chloramphenicol rarely causes prolonged self-sustaining marrow aplasia (1 in every 30,000 treated persons). Usually the marrow reverts to normal after the patient stops taking the drug. (In this instance at least, the disorder seems to be dose-dependent—many individuals with aplastic anemia were previously exposed to this antibiotic drug.) Table 9.1 shows many common agents that have been implicated, from analgesics (e.g., phenylbutazone and even aspirin) to anticonvulsants and antianxiety drugs. Further, insecticides and hair dyes have been named as infrequent causes of aplastic anemia.

### Replacement of Normal Hematopoietic Tissue by Tumors

Bone marrow may be further damaged by infiltration of the marrow space by cancerous cells. These metastasizing tumor cells may gradually crowd out hematopoietic tissues. The condition of hematopoietic elements being replaced by nonhematopoietic elements is known as **myelophthisis,** and the corresponding anemia is known as **myelophthisic anemia.**

### Competition for Nutrients Between Normal and Abnormal Blood Cells

Rapid production of one or more abnormal blood cell lines may interfere with the normal production of hematopoietic cell lines. This happens in leukemias and in infectious granulomas. However in most cases the crowding out is not complete, and bone marrow examinations reveal the presence of both normal and abnormal hematopoietic tissue side by side. Even when sufficient normal hematopoietic tissue is present, low new blood cell production is common in the presence of these cancerous cell lines. This suggests that other mechanisms are at work. It has been proposed that the decreased proliferation may be due to local competition between normal and abnormal cells for the available essential nutrients. Some evidence indicates that these abnormal cell lines secrete substances that may inhibit the growth and development of new normal blood cells. Anemia that develops as a result of leukemia and other myeloproliferative disorders is often called **anemia of myeloproliferative disease.**

## INCIDENCE AND CLINICAL FEATURES OF APLASTIC ANEMIA

Aplastic anemia may occur at any age and its onset can be either insidious or acute. However onset is usually insidious and is often discovered as a result of an investigation into the cause of a prolonged infection (low leukocyte count), prolonged bleeding (low platelet count), or because of complaints of weakness and fatigue (low red blood cell count).

A peak incidence occurs around 30 years of age with a slight male predominance. This may reflect the fact that more relatively young males are exposed to dangerous chemicals and radiation in the workplace than are females.

Infections of the mouth, throat, and lungs are common, and systemic infections may be life-threatening. Bleeding is another common symptom, especially bleeding from the gums, and prolonged menstrual bleeding, as well as the presence of clear-cut ecchymosis after small vessel rupture.

Finally, all the common symptoms of anemia may be observed in hypoplasia of the bone marrow. Symptoms of weakness, fatigue, headaches, and dizziness are usually more common in this type of anemia, as in many other types, because of the inability of the marrow to compensate for low hemoglobin levels.

## LABORATORY FINDINGS OF APLASTIC ANEMIA

Laboratory findings invariably show **pancytopenia**—that is, low levels of red blood cells (RBCs), white blood cells (WBCs), and platelets. The reticulocyte count is also always low. The anemia, however, is normocytic and normochromic; hence the red blood cell indices—mean corpuscular volume (MCV) and mean corpuscular hemoglobin concentration (MCHC)—are normal. Hemolysis is not common, as is the case with many other normocytic-normochromic anemias, and hence few abnormal cells will be found in the peripheral blood.

Bone marrow analysis shows a distinct loss of hematopoietic tissue and replacement of that tissue by fat cells, as illustrated in figure 9.1. The bone marrow biopsy also will show strands of reticular cells (remnants of cords and sinuses) with small clusters of lymphocytes. But the greater part of the bone marrow stroma is empty and contains fat cells.

The serum iron levels are always elevated, often to the point of 100% saturation of the transferrin—hence the total iron-binding capacity (TIBC) is always low. Although the iron content of the mononuclear (=monocyte-macrophage) cells in the bone marrow is also increased, most of the extra iron is taken up by the parenchymal cells of the liver, where it is stored.

## PROGNOSIS OF APLASTIC ANEMIA

The prognosis for aplastic anemia is usually not very good. Generally speaking, about 50% of patients die as a direct consequence of anemia, infection, or bleeding, or a combination of these. Another 20% require continued blood transfusions, and perhaps 30% recover sufficiently to require no further support, even though the blood counts may remain subnormal indefinitely.

## TREATMENT OF APLASTIC ANEMIA

The therapy for aplastic anemia includes five different measures: (1) discontinuance or avoidance of the suspected toxic agent, (2) careful personal hygiene to prevent infection or bleeding, (3) blood transfusions, (4) bone marrow stimulants, and (5) bone marrow transplantations.

The first two measures are self-evident. The only solution for this type of anemia is to avoid the physical or chemical agent that caused the condition in the first place. Second, since all blood cell levels are down, the patient is much more susceptible to infectious disease and bleeding. Hence the emphasis is on careful personal hygiene until the anemia no longer exists.

### Blood Transfusions

Blood transfusions are used mainly to "buy time" in the hopes that a remission will occur. However transfusions should be kept to a minimum since they increase the opportunity for the development of immune reactions. These immune reactions would also increase the chance of rejection of the transplanted bone marrow cells if this course of treatment is subsequently elected.

Most individuals tolerate chronic anemia rather well, and it is rarely necessary to raise the hemoglobin levels above 8 to 9 g/dL. In severe aplastic anemia, this

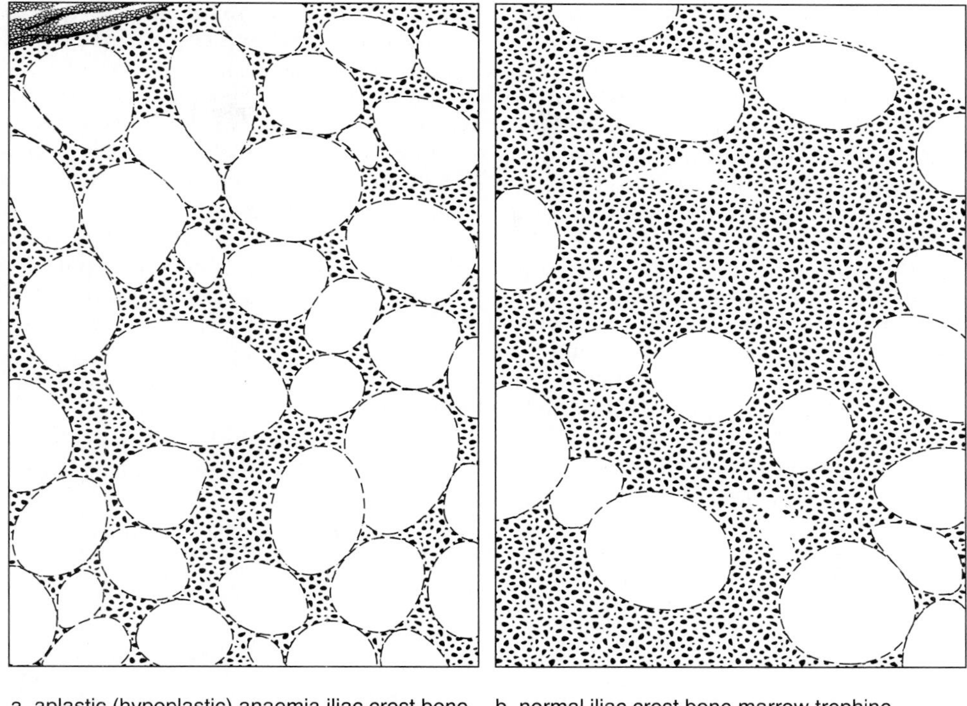


**figure 9.1**
Bone marrow in (*a*) normal and (*b*) in aplastic anemia. (matched ages).

goal can be met with about 2 U of blood every 2 weeks. Usually only washed or frozen red blood cells are used in the transfusion to avoid immunization to transfused white blood cells and platelets. Sometimes, however, it may be necessary to transfuse platelet concentrates—for instance, in the case of life-threatening bleeding. For minor bleeding, corticosteroids should be used first. Corticosteroids cause the tightening of the vessels of the microcirculation and thus minimize capillary bleeding. They do not stimulate hematopoiesis, although sometimes these steroids may increase white blood cell production. What happens most likely is that the tightening of small blood vessels prevents white blood cells from leaving the circulation, and thus it appears that their number is increased.

Sometimes it may be necessary to transfuse white blood cells also—for instance, in case of a serious infection. Again this should be done only as a last resort in order to minimize immunization. Usually infectious complications can be handled effectively (once the offending pathogen is known) by an appropriate antibiotic regimen.

### Bone Marrow Stimulants

Bone marrow stimulants have been used with considerable success, especially in instances in which the marrow still shows a residual capacity for hematopoiesis. High doses of androgenic hormones somehow stimulate erythropoiesis and sometimes also leukopoiesis and thrombopoiesis. It is not understood how these androgens act on the bone marrow. They are not thought to act by stimulating the production of erythropoietin, since the level of erythropoietin is already very high in people suffering from most types of aplastic anemia.

### Bone Marrow Transplantations

Finally, bone marrow transplantation must also be considered. Marrow transplantation would be the ideal therapy if it could be done successfully. The chances of a successful transplantation are still rather restricted, since host-versus-graft reactions of the immune system usually prevent the injected bone marrow cells from establishing themselves. However the number of successful bone marrow transplants is increasing yearly. **Syngeneic grafts**—that is, where the donor and recipients are identical twins—have long been used successfully. **Allogeneic grafts**—that is, donor and recipient are genetically dissimilar—are more difficult to perform because of the presence of varying histocompatibility antigens or human leukocyte antigens (HLA) on the blood cell membranes. These cell surface antigens are recognized as "nonself" by the immune system of the recipient, and therefore the injected marrow cells are attacked by white blood cells, especially lymphocytes. New techniques have made HLA compatibility matching much easier, and consequently allogeneic marrow transplantations are becoming more common. But even where the HLA types are matched, half the patients develop immune reactions such as skin rash, diarrhea, and ultimately rejection of the transplanted cells. However new and strong immunosuppressive drugs such as cyclosporine and antilymphocyte serum (ALS) have helped to increase the success rate of bone marrow transplantation.

Finally, we should note that pure red blood cell hypoplasia is rather rare. However it may occur as a transient phase in many hemolytic anemias, as is explained in another chapter. A chronic form of red blood cell hypoplasia is also known, which is thought to be caused by an aberration in the immune system, resulting in a constant low level of attack on committed erythroid precursor cells. In the great majority of cases, however, erythroid hypoplasia is just one aspect of overall hypoplasia of the bone marrow, resulting in decreased numbers of white blood cells and platelets also. This condition is known as **pancytopenia** and is a common feature in most forms of aplastic anemia.

## *Further Readings*


Beck, W. S. Normocytic Anemias. Lecture 3 in *Hematology,* 4th ed., W. S. Beck, ed. The MIT Press, Cambridge, Mass., 1985.

Crosby, W. H. Red Cell Mass: Its Precursors and Its Perturbations. *Hospital Practice,* Feb. 1980.

Goldstein, M. The Aplastic Anemias. *Hospital Practice,* May 1980.

Katz, A. J. Transfusion Therapy: Its Role in the Anemias. *Hospital Practice,* June 1980.


## *Review Questions*

1. Distinguish between idiopathic, primary, and secondary aplastic anemia.
2. What is the incidence of aplastic anemia in each of these categories?
3. Give the six major causes of acquired aplastic anemia. What are the names of the anemias associated with each of the six causes?
4. List five different chemical agents that are regularly implicated as the cause of bone marrow hypoplasia.
5. Mention five chemical agents that are occasionally associated with causing aplastic anemia.
6. List five different chemical agents that are infrequently implicated as the cause of aplastic anemia.
7. Why is aplastic anemia slightly more common in males than in females? Why is it most common in people 30 to 40 years old?
8. Explain why aplastic anemia is usually an insidious disease.
9. Why are infections and bleeding common symptoms of aplastic anemia?
10. Mention five important laboratory findings commonly associated with aplastic anemia.
11. Explain the term pancytopenia. Why is it always a feature of aplastic anemia?
12. What is the prognosis for most cases of aplastic anemia?
13. List five general therapeutic measures commonly employed to combat aplastic anemia.
14. Why is good personal hygiene important in the treatment of aplastic anemia?
15. Why should blood transfusions be kept to a minimum?
16. Name a good bone marrow stimulant. In what situation can bone marrow stimulants cause unwarranted complications?
17. Describe some difficulties associated with bone marrow transplants. List some measures currently used to overcome these difficulties.

# Chapter Ten

## *Megaloblastic Anemias*

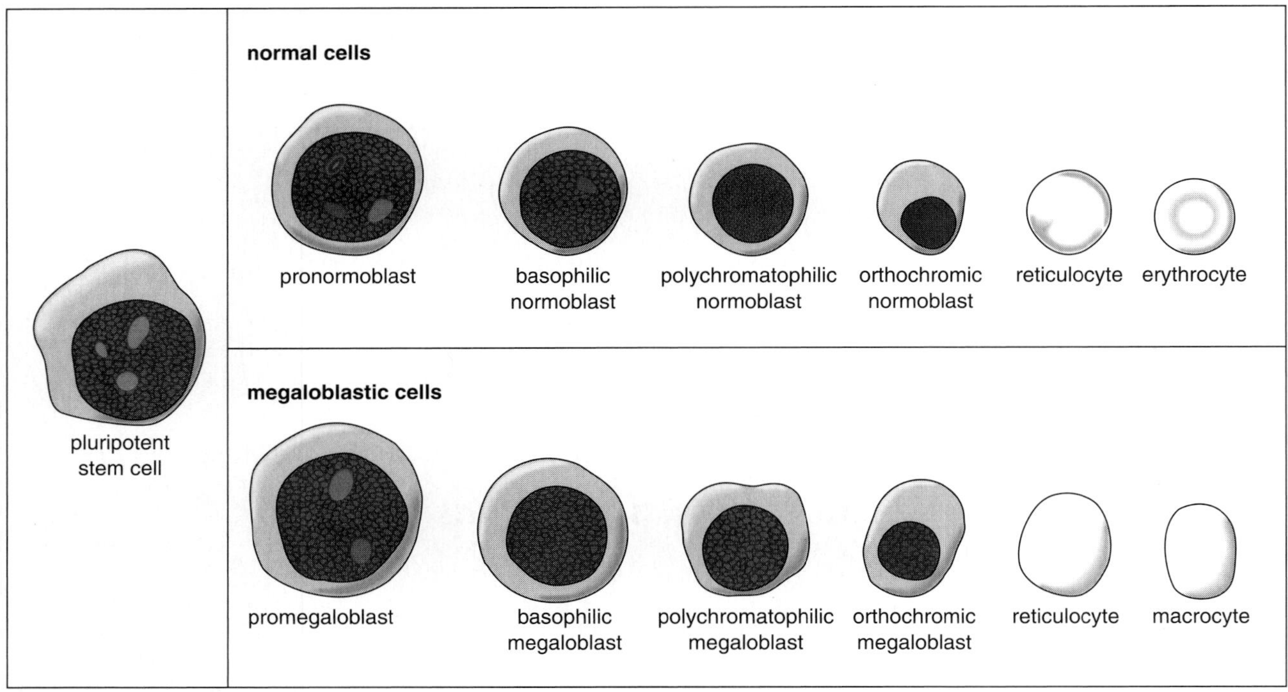


figure 10.1
The six stages of erythrocyte development of normal and megaloblastic red blood cells.

## INTRODUCTION TO MEGALOBLASTIC ANEMIA

A number of different anemias are the result of maturational defects of the red blood cells. In all cases these anemias have been associated with various forms of **anisocytosis** and **poikilocytosis.** As was shown in table 8.2, they are generally classified as belonging to either of two major categories: those usually producing larger than normal erythrocytes **(macrocytic anemias),** and those normally producing smaller than normal red blood cells **(microcytic anemias).**

### Definition of Megaloblastic Anemia

*Megaloblastosis*

The term "macrocytic anemia" is often used synonymously with the term **megaloblastic anemia,** since the large red blood cells in the circulation are usually the products of larger than normal blast cells (or **megaloblasts**) in the bone marrow. However there are other causes of macrocytosis that are not associated with megaloblastic anemia, as is explained at the end of this chapter. Megaloblastic marrow cells are the result of abnormalities in the DNA synthesis of the blast cells, due to the lack of one or more ingredients needed for its synthesis. The mechanism for hemoglobin production in these developing erythrocytes is not affected, and consequently these red blood cells will manufacture normal amounts of hemoglobin. This is why megaloblastic anemias have an increased mean corpuscular volume (MCV) but a normal mean corpuscular hemoglobin concentration (MCHC).

*Nuclear-Cytoplasmic Asynchronism*

The exact biochemical mechanism(s) causing megaloblastosis of the bone marrow cells is not known. It appears, however, that the period preceding each cell division is prolonged in order to complete DNA synthesis, resulting in an increased growth of each blast cell. As can be seen from figure 10.1, the megaloblastic red blood cell precursors are at all stages larger than the corresponding cells in a normal erythrocytic series. They also have a higher than normal ratio of cytoplasmic to nuclear area. In addition, the nucleus of the megaloblastic erythroid cell fails to become pyknotic (completely condensed) but retains its granular structure, although the cytoplasm starts to mature and begins to produce hemoglobin. This morphogenic abnormality is referred to as the **nuclear-cytoplasmic asynchronism** (or "nuclear-cytoplasmic dissociation")—that is, the nucleus retains its primitive appearance, and the cytoplasm matures normally.

*Ineffective Hematopoiesis*

Many megaloblasts die in the bone marrow, thus failing to produce mature red blood cells that enter the general circulation. This is known as **ineffective erythropoiesis,** and the rate may be as high as 50%

or more—that is, 50% or more of all developing erythrocytes do not reach the peripheral circulation. In an effort to counteract the premature death and removal by macrophages of these immature red blood cells, the bone marrow is stimulated to produce more erythroid percursor cells, many of which will undergo the same fate. As a result few reticulocytes are produced, and consequently relatively few red blood cells are found in the peripheral circulation. Erythrocytes that do reach the bloodstream are usually large and oval (macro-ovalocytes) in a blood smear. The reason for this oval shape is not known.

In general megaloblastosis is not limited to the erythroid cell series only. As a matter of fact, all tissues with a high rate of mitosis are affected. This includes the production of white blood cells and platelets. Hence megaloblastic anemia is commonly associated with low numbers of white blood cells in the peripheral blood also **(leukopenia)**, as well as reduced numbers of platelets **(thrombocytopenia)**.

These megaloblastic leukocytic cells of the bone marrow show the same nuclear-cytoplasmic asynchronism and enlargement as the erythroid series. In many cases of megaloblastic anemia, extra-large band cells may be found in the peripheral circulation, as well as **hypersegmented neutrophils.** The megakaryocytes of the bone marrow may become truly gigantic in megaloblastic anemia, and as a result the platelets produced are also much larger than normal. The granulation of the cytoplasm of these giant cells is usually deficient, and the nucleus is sometimes bizarre, showing numerous unattached lobes, giving the cell an exploded appearance.

### Etiology of Megaloblastic Anemia

The maturation defects resulting in megaloblastic anemia are usually associated with five different causes. These are as follows:

1. **Lack of folic acid.** This compound, also known as **folate**, is an essential vitamin necessary for the synthesis of DNA. Lack of folate is by far the most common cause of megaloblastic anemia.
2. **Lack of vitamin $B_{12}$.** This compound, also known as **cobalamin**, is another essential vitamin needed for DNA synthesis. It is the second most frequent cause of megaloblastic anemia.
3. **Abnormalities of folate or vitamin $B_{12}$ metabolism.** Usually these compounds are stored in the body in an inactive form, and special enzymes are needed to reactivate them. Any interference with this activation process may also result in the lack of the active form of these compounds needed for DNA synthesis. Luckily this is a rare cause.
4. **Congenital deficiencies of some enzyme involved in purine or pyrimidine synthesis** may also cause megaloblastic anemia, since it also inhibits DNA synthesis. Again this is a rather rare occurrence.
5. Finally, **therapy with drugs that inhibit purine or pyrimidine synthesis** such as hydroxyurea, cytosine arabinoside, and 6-mercaptopurine also cause megaloblastic anemia. These drugs are commonly used as part of a chemotherapeutic regimen in the fight against certain cancers. Hence this cause of megaloblastic anemia is much more common. However the anemia is usually limited, since the drug cycle used to combat these tumors is usually limited to a certain timespan, after which the bone marrow can again start its normal production of blood cells.

### Classification of Megaloblastic Anemia

Classification of megaloblastic anemias is closely related to the causes of these disorders. Customarily megaloblastic anemias are divided into the following three groups:

1. Megaloblastic anemias caused by *lack of folic acid* in the diet. This is by far the largest group.
2. Megaloblastic anemias caused by *lack of vitamin $B_{12}$*. It is rare that the diet does not provide enough vitamin $B_{12}$. The quantities needed are so small, and we have such great storage capacity for this essential vitamin, that in the majority of cases, the problem may be found in the inability to absorb vitamin $B_{12}$ from the food in the intestines into the body. An important form of megaloblastic anemia, **pernicious anemia,** is due to the inability of the body to absorb vitamin $B_{12}$ and is discussed later in this chapter.
3. Megaloblastic anemias *not caused by lack of folate or vitamin $B_{12}$*. Compared to the previous two causes, these causes of megaloblastic anemia are relatively rare and include deficiencies or abnormalities in enzyme or pyrimidine and purine synthesis.

### Clinical Features of Megaloblastic Anemia

The onset of megaloblastic anemia is usually *insidious,* with gradually progressing signs and symptoms of anemia. Usually the diagnosis of megaloblastic anemia is made accidentally as a result of routine laboratory investigations of the blood of people who have symptoms associated with anemia such as weakness, fatigue, palpitations, shortness of breath, headache, or dizziness. Specific overt features are sometimes also present. For instance, the condition of **atrophic glossitis,** which means that the tongue appears red, beefy, and swollen (figure 10.2) occasionally occurs in megaloblastic anemia. This condition is caused by the lack of vitamin $B_{12}$ or folate, which affects all rapidly dividing tissues, which includes the gastro-intestinal tract, and thus the tongue. Macrocytic changes in the epithelial lining of the tongue causes this atrophic glossitis.

**Purpura,** or the presence of purple patches in the skin, is sometimes also associated with moderate or severe megaloblastosis. This condition is caused by small amounts of bleeding under the skin, due to lack of platelets in the circulation. Consequently small leaks in the microcirculation are not repaired fast enough, and bleeding occurs. The patient may also be slightly **jaundiced.**

This phenomenon may be explained be the excess breakdown of hemoglobin, mainly from immature erythroid bone marrow cells that have been removed by macrophages. Also contributing to the jaundice may be premature hemolysis of macro-ovalocytes in the peripheral circulation. These large red blood cells are more easily damaged than normal erythrocytes and hence have a shortened life span.

### Laboratory Findings in Megaloblastic Anemia

Laboratory investigations will usually show red blood cells that are normochromic but macrocytic, with MCVs ranging from 100 to 140 fL per RBC or more. Macro-ovalocytes up to 14 µm in diameter are often present.

The red blood cell count and hematocrit value are usually very low. When the hematocrit reaches a value of less than 20%, nucleated red blood cells begin to appear in the blood.

The white blood cell count may also be below normal. White blood cells on a blood smear will appear larger than normal. Many neutrophils will show a **hypersegmented nucleus**—that is, they have more than five lobes. This phenomenon may already be observed in early stages of megaloblastic anemia and is often one of the most clear-cut diagnostic features, since 98% of all cases of megaloblastic anemia will exhibit this hypersegmentation of the polymorphonuclear neutrophil (PMN) nucleus. In severe cases 5 to 10% of all neutrophils will have hyperlobulated nuclei with 6 to 16 segments. Hypersegmentation of the PMNs may also be seen in other anemias such as iron deficiency anemia, anemia of renal failure, and in anemias associated with myeloproliferative disorders. However in these instances there is an absence of large-scale macro-ovalocytosis. Platelets are also fewer in number, but platelets observed in the peripheral blood are usually larger than normal. Bone marrow analysis will show the presence of many megaloblastic cells with distinct nuclear-cytoplasmic asynchronism.

Serum analysis of patients with milder forms of megaloblastic anemia often exhibit a moderate increase in unconjugated bilirubin levels, as well as decreased serum haptoglobin levels. These changes are due to the increased release of hemoglobin into the blood from erythroid cells destroyed in the bone marrow and in the systemic circulation. Studies with $^{51}$Cr-labeled red blood cells from patients with megaloblastic anemia have shown that the life span of such erythrocytes has been reduced to one-third to one-half of the normal life span.

The increased destruction of red blood cells in the bone marrow and in the systemic blood compartment also produces increased serum iron levels. The plasma iron turnover rate, however, is three to five times normal, due to increased production of new precursor cells.

For unknown reasons, megaloblastic erythroid precursor cells contain large amounts of the enzyme lactate dehydrogenase (LDH). The increased destruction of these cells produces elevated levels of LDH in the serum.

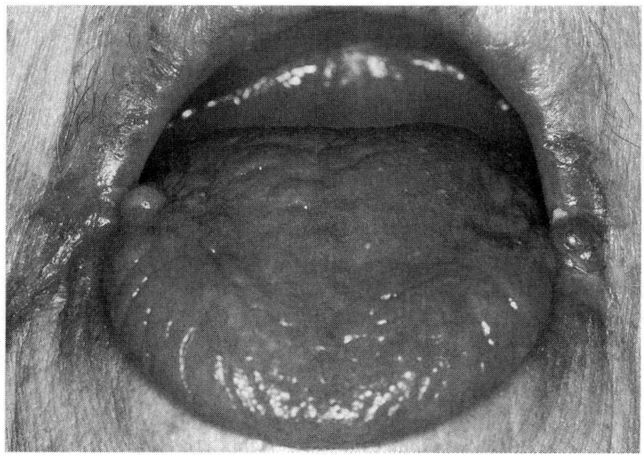

figure 10.2

Atrophic glossitis, the characteristic beefy tongue of a patient with megaloblastic anemia.

Many of the laboratory findings previously mentioned may also be seen in other anemias, especially hemolytic anemias. However a distinction can readily be made since a combination of these signs, together with macro-ovalocytosis and neutrophil hypersegmentation, are unique for megaloblastic anemia.

## MEGALOBLASTIC ANEMIA DUE TO FOLIC ACID DEFICIENCY

### Chemical Aspects of Folic Acid

Folic acid (or folate) is the common name for pteroylmonoglutamate (or pteroylmonoglutamic acid). As shown in figure 10.3, the molecule consists of three parts: (1) a pteridine double ring (left), (2) *para*-aminobenzoic acid (middle), and (3) glutamic acid (right).

Folic acid is the parent molecule of a large group of compounds known as **folates.** The folates are derived from the original molecule in three ways: (1) by the addition of extra glutamic acid residues, attached to the preceding glutamic acids via peptide linkages, producing so-called **folate polyglutamates;** (2) by the addition of hydrogen atoms at positions 7 and 8 **(dihydrofolate)** or at positions 5, 6, 7, and 8 **(tetrahydrofolate,** or **THF)** of the pteridine double ring; or (3) by the addition of single carbon units—for example, a methyl (—$CH_3$) group at N-5, or a formyl (—CHO) group at N-5 or N-10.

### Dietary Aspects of Folic Acid

Humans are unable to synthesize the folate structure and thus require preformed folate as a vitamin in their diet. Luckily folate compounds are widely distributed in nature. Green leafy vegetables are rich sources of this vitamin, especially spinach, lettuce, broccoli, and asparagus. Each of these contains more than 1 mg of folate per 100 g of dry weight. Folates

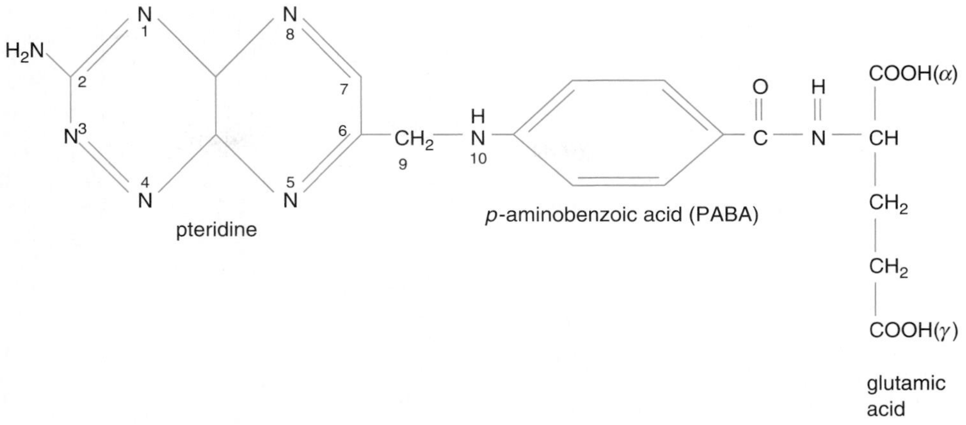

**figure 10.3**

The structure of folic acid, also known as pteroylmonoglutamic acid. The structure consists of three parts: pteridine, *para*-aminobenzoic acid (PABA), and glutamic acid.

**Table 10.1** Comparison of the Nutritional Aspects of Vitamin $B_{12}$ and Folic Acid

| | Vitamin $B_{12}$ | Folate |
|---|---|---|
| Normal dietary intake | 7–30 μg | 600–1,000 μg |
| Main sources | Animal produce | Liver, greens, and yeast |
| Cooking effect | Minimal | Easily destroyed |
| Minimal daily requirement | 1–2 μg | 100–200 μg |
| Body stores | 2–3 mg (sufficient for 2–4 years) | 10–12 mg (sufficient for 4 months) |
| Absorption site | Ileum | Duodenum and jejunum |
| Mechanism | Intrinsic factor | Conversion to methyltetrahydrofolate |
| Limit | 2–3 μg daily | 50–80% of dietary content |
| Intracellular physiological forms | Methyl- and adenosylcobalamin | Reduced polyglutamate derivatives |
| Therapeutic form | Hydroxycobalamin | Folic (pteroylglutamic) acid |

are also found in liver, kidneys, yeast, and mushrooms. Further, this vitamin may be synthesized de novo from the three basic compounds by many bacteria, some of which live as native microflora in the human intestinal tract. Sulfonamides block the incorporation of *p*-aminobenzoic acid and thus inhibit bacterial folate synthesis.

The average Western diet contains about 200 μg of folate per day. Excessive cooking, especially with large amounts of water, can remove or destroy the greater part of the folates in these foods. The minimum daily requirement (MDR) for folic acid is about 100 to 200 μg, and the average diet contains about two to four times that amount. Table 10.1 shows the nutritional aspects of both folate and vitamin $B_{12}$; additionally it can be seen that the body stores of folate are much smaller than those of vitamin $B_{12}$. The average person has a folate reserve that would last only 4 months. However a number of situations occur in which the need for folate is greatly increased—for example, during pregnancy, growth, and in certain disease states. In these situations the stores of folate will be much smaller.

### Metabolic Aspects of Folic Acid

Details of the mechanisms of absorption of folic acid are not well understood, but it is believed that the jejunum is the principal site of folate absorption. Folic acid in food is present mainly in **polyglutamate** forms. But in the upper small intestine, all these dietary forms are converted to the **monoglutamate** state. Folate is absorbed and transported in the blood in monoglutamate form. Whether a portion of this folic acid in the blood is protein-bound is unclear, since serum folate is largely dialyzable. However evidence

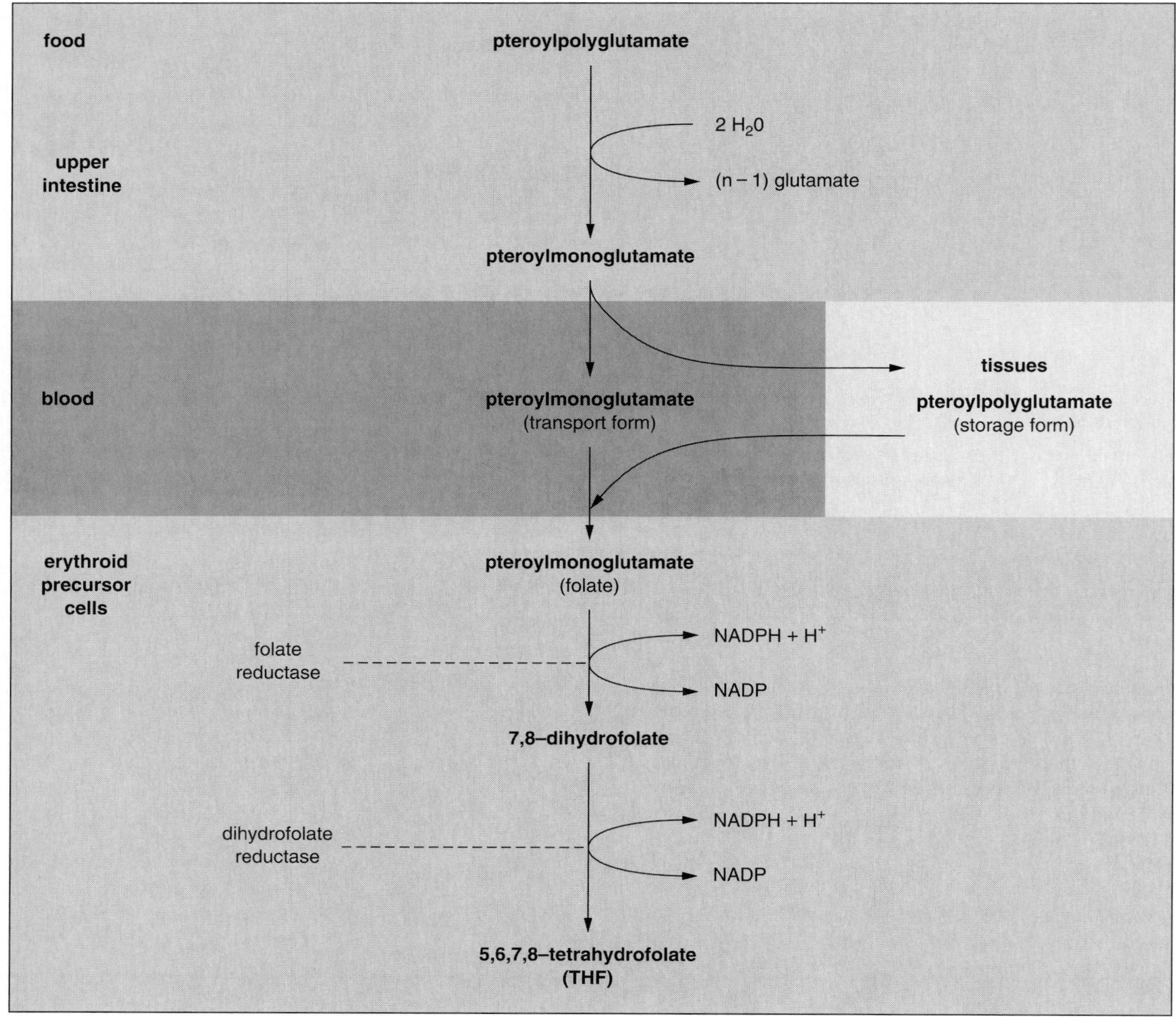


figure 10.4
The biosynthetic pathway of tetrahydrofolate (THF).

indicates that some transport protein may exist. In the tissues this monoglutamate is converted back to and stored in polyglutamate forms.

For metabolic activity, folate must be reduced again to the monoglutamate form. Further, it must be made biologically active by the stepwise additions of four hydrogen atoms at positions 5, 6, 7, and 8 to form THF. This reaction is catalyzed by the enzymes folate reductase and dihydrofolate reductase (figure 10.4).

Folic acid antagonists used in cancer chemotherapy (e.g., methotrexate), or those used in fighting bacterial infections (e.g., trimethoprim), or parasitic infections (e.g., pentamidine), inhibit these enzymes and deprive the cell of reduced folate and thus interfere with new DNA synthesis. This may cause megaloblastic anemia. To counteract these effects by folate antagonists, massive doses of reduced folate are given to these patients.

### Functional Aspects of Folic Acid

Folates act as coenzymes in a number of biochemical reactions in the body involving single-carbon transfers. Three of these reactions are involved in DNA synthesis: two in the synthesis of purines and one in thymidylate synthesis. The others are largely concerned with amino acid interconversions. Folate deficiency is thought to cause megaloblastic anemia by inhibiting the enzyme thymidylate synthetase, which is needed to produce thymidine, one of the two pyrimidine bases of DNA. In bone marrow cells THF plays a role in the synthesis of thymidine monophosphate (=thymidylate), from its precursor compound deoxyuridine monophosphate (=deoxyuridylate), as illustrated in figure 10.5. This reaction is catalyzed by the enzyme thymidylate synthetase, with 5,10-methylene THF as a **coenzyme.**

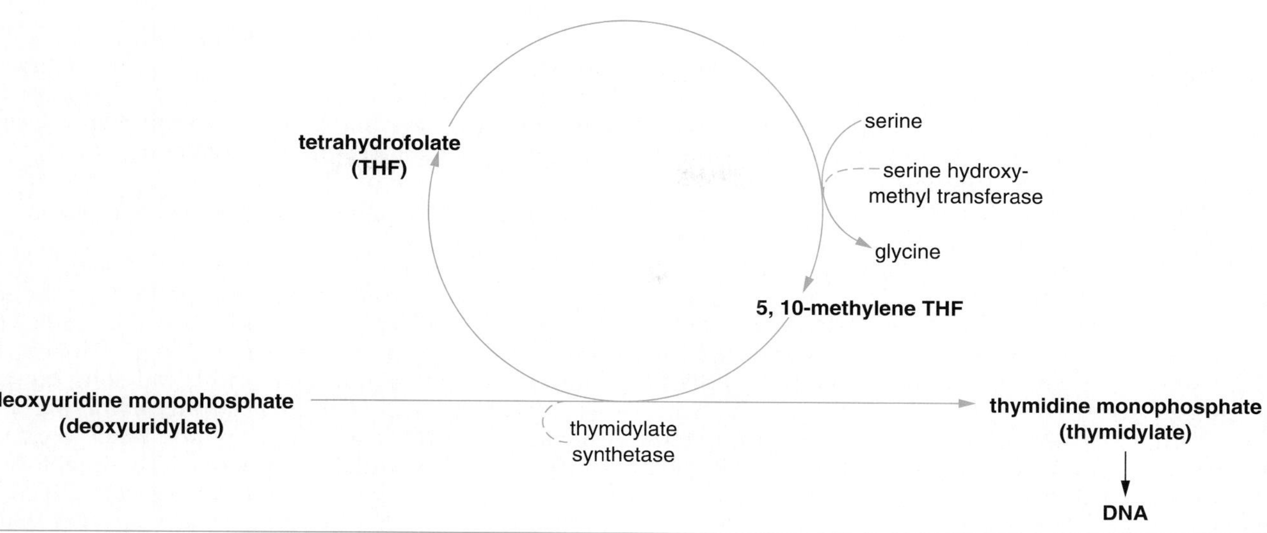


figure 10.5

The role of tetrahydrofolate (THF) in DNA synthesis.

The latter serves as a one-carbon donor in the methylation of deoxyuridylate to thymidylate. This reaction is thought to be rate-limiting for DNA synthesis. Thus if the cell is deprived of folate, there will be reduced synthesis of thymidylate and therefore of DNA. The inability of the cell to produce enough 5,10-methylene THF could be either due to a depletion of folate or the inability to make THF, as happens during therapy with dihydrofolate reductase inhibitors.

### Causes of Folic Acid Deficiency

There are many different causes for folate deficiency but all can be reduced to belonging to one or more of the following three categories: (1) decreased intake, (2) increased requirements, (3) blocked activation (table 10.2).

#### *Decreased Intake*

Because folate in the diet is not greatly in excess of the nutritional requirements, and because the body folate reserves are relatively small, it follows that folic acid deficiency shows up rapidly in individuals with poor diet. Loss of folate due to excessive cooking may also cause deficiency. Further, prolonged and excessive use of alcohol is also often associated with folate deficiency. Not only do alcoholics tend to eat poorly, the ethyl alcohol itself acts as a folate antagonist for some unknown reason. Since excess alcohol also causes liver damage, and the liver is one of the major storage organs for folate, clearly hepatic impairment due to alcohol consumption is another major cause of megaloblastic anemia. However alcohol is not the only cause of liver disease. Any factor that causes a serious impairment of the liver may contribute to a shortage of folate and hence be instrumental in the development of megaloblastic anemia (e.g., viral hepatitis).

**Table 10.2** Causes of Folate (Folic Acid) Deficiency

| |
|---|
| *Decreased Intake* |
| Poor diet |
| Excessive cooking |
| Alcohol |
| Liver disease |
| Malabsorption (e.g., sprue, intestinal lymphoma) |
| *Increased Requirements* |
| Pregnancy |
| Childhood and adolescence |
| Hyperactive hematopoiesis |
| Neoplastic disease |
| *Blocked Activation* |
| Drugs (e.g., methotrexate, trimethoprim, pentamidine, phenytoin, sulfasalazine) |
| Oral contraceptives |
| Vitamin C deficiency |

Finally, a decreased intake of folate may also be due to malabsorption of folic acid from the gastrointestinal tract. Such intestinal syndromes as tropical sprue, celiac disease, intestinal lymphoma, or massive resection of the jejunum may all result in malabsorption of folate.

#### *Increased Requirements*

Increased requirements of folate are obvious in all situations where new DNA is needed for growth. This includes, first of all, the condition of pregnancy. The daily requirements of pregnant women are 5 to 10 times the normal daily needs. This is especially true during the last trimester of pregnancy and also when multiple fetuses are present. More than two-thirds of pregnant women who show signs of anemia may do so because of folate deficiency. Many pregnant women tend to eat poorly due to feelings of nausea and morning sickness. Although we do not yet know why, the fetus can absorb folic acid and other nutrients at the expense of the mother, even when the available supply is greatly reduced. For these reasons, pregnant women are advised to supplement their daily intake of food with extra folate and other essential vitamins and minerals.

Second, growing children need increased amounts of folate for normal physical development. Children who are poor eaters will often show signs of megaloblastosis.

Third, folate requirements rise sharply in cases of hyperactive hematopoiesis due to excessive bleeding or increased hemolysis. Indeed, megaloblastic changes in the bone marrow often occur shortly after the onset of severe hemolytic disorders.

Last, moderate to severe folate deficiency is often observed in patients with neoplastic disease and is especially common in leukemias and metastatic cancers. The deficiency presumably reflects a competitive utilization of this vitamin by the tumor cells. This phenomenon resembles the preemption of maternal nutrients by the fetus.

*Blocked Activation*

Folate deficiency is often caused by drug therapy. One of the side-effects of certain therapeutic drugs is that they block the action of folic acid. As previously explained, many drugs act as dihydrofolate reductase inhibitors, thus preventing the production of methylene THF, which is the active agent needed for production of new DNA for the erythroid and other body cells.

Other drugs may cause a deficiency of folic acid by inhibiting absorption of folate from the intestines. This group includes several anticonvulsive drugs such as phenytoin and anti-inflammatory drugs such as sulfasalazine. Recently certain oral contraceptives have also been shown to inhibit folate absorption from the intestines. In addition, lack of ascorbic acid (vitamin C) may influence the uptake of folate from the intestinal tract, since many cases of scurvy are associated with megaloblastic anemia due to folate deficiency.

### Diagnosis of Folic Acid Deficiency

The diagnosis of folic acid deficiency involves the measurement of serum folate and red blood cell folate, both of which are low in megaloblastic anemia. Once a deficiency of folate has been established, further investigations are needed to discover the cause of a lack of this vitamin in the body. In this respect an accurate diet history is important in identifying the cause. If a poor diet is considered the causative agent, then a full therapeutic response could occur within a few weeks after the administration of daily physiological doses of folate. If a proper response has not occurred within that time, other causes should be suspected. An in-depth medical history may show underlying disease or the use of drugs antagonistic to the absorption or action of folate. Further, tests for intestinal malabsorption, or a jejunal biopsy may be required.

### Treatment of Megaloblastic Anemia Due to Folic Acid Deficiency

In most cases the only therapy needed is the administration of physiological doses of folate. The usual regimen is to give the patient orally 5 mg of folate per day for 4 months. In instances where lifelong therapy is needed (e.g., in inherited chronic hemolytic anemias or in myelosclerosis) a maintenance regimen ensuring sufficient daily folate should be established. In cases of blocked activation due to side-effects of certain drugs, the offending drug should be withdrawn; if that is impossible, high doses of methyl-THF should be given to overcome side-effects.

In most cases requiring only oral folate therapy, the patient should feel better within a few days after the regimen has started. A strong reticulocyte response may be observed after 7 or 8 days along with an increase in the red blood cell count. The hemoglobin content will increase approximately 2 g/100 mL every 2 weeks until normal. White blood cell and platelet counts will also reach normal limits within 10 days, and the bone marrow will return to normoblastosis within the same period. Sometimes the response will remain inadequate after proper folate treatment. In that case a deficiency of other vitamins (e.g., vitamin $B_{12}$) or iron should be suspected. Blood transfusions should be avoided as much as possible, since they may cause iron overload and immunization problems.

## MEGALOBLASTIC ANEMIA DUE TO VITAMIN $B_{12}$ DEFICIENCY

### Chemical Aspects of Vitamin $B_{12}$

Vitamin $B_{12}$ is synthesized in nature only by certain microorganisms. Animals, including humans, ultimately depend on microbial synthesis for their vitamin $B_{12}$ supply. Foods that contain vitamin $B_{12}$ are those of animal origin (e.g., liver, fish, meat, eggs, and milk).

Vitamin $B_{12}$ consists of a small group of compounds known as **cobalamins,** all of which have the same basic structure. Figure 10.6 shows a cobalt atom in the center of a porphyrin ring structure, which is attached to a nucleotide portion. Four major forms of the compound exist as follows:

1. **Methylcobalamin (methyl-$B_{12}$)** has a methyl group (—$CH_3$) attached to the central cobalt atom. This is the main form of vitamin $B_{12}$ in plasma (figure 10.7).
2. **Deoxyadenosylcobalamin (ado-$B_{12}$)** has a deoxyadenosyl attached to the central cobalt. This is the major form in which vitamin $B_{12}$ is stored in the tissues.

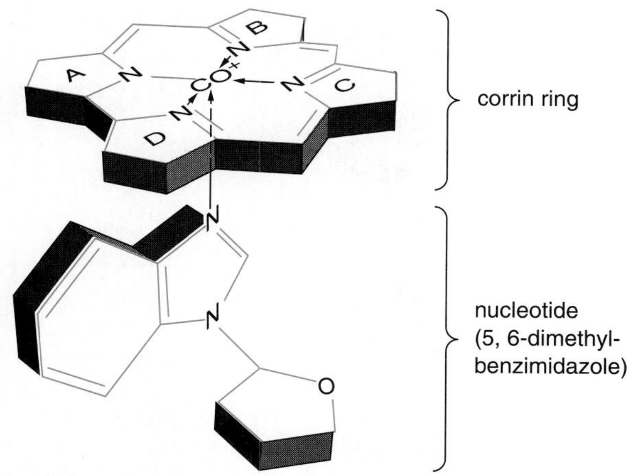


**figure 10.6**

The structure of vitamin $B_{12}$ (cobalamin).

3. **Hydroxycobalamin (OH-$B_{12}$)** has a hydroxyl (—OH) group attached to the central cobalt atom. This is the main form of cobalamin used in medical treatment.
4. **Cyanocobalamin (cyano-$B_{12}$)** has a cyanide (—CN) radical attached to the central cobalt. This type of cobalamin is used in laboratory tests to study the absorption and metabolism of vitamin $B_{12}$ using radioactive labeled cobalt ($^{57}Co$ or $^{58}Co$).

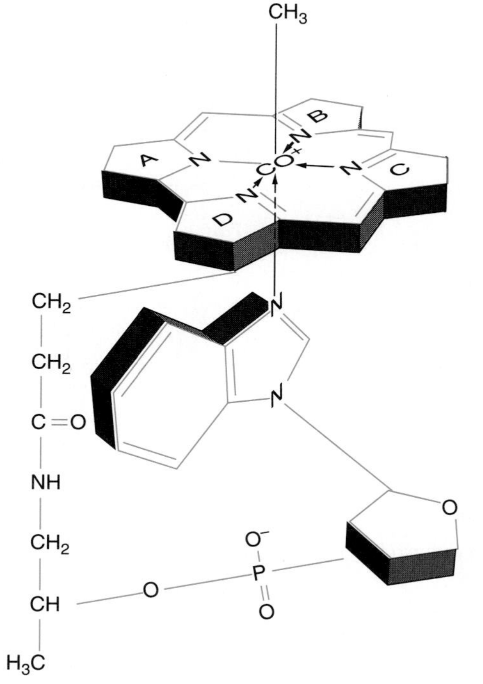


**figure 10.7**

The structure of methylcobalamin.

### Dietary Aspects of Vitamin $B_{12}$

Vitamin $B_{12}$ is found in many foods of animal origin such as liver and other organ meats, seafood, including both fish and shellfish, and in dairy products, especially in milk and in fermented cheeses.

The average daily diet in Western countries contains about 5 to 30 µg of vitamin $B_{12}$, and of this amount 1 to 5 µg are absorbed. The total body content in an adult male is between 2 to 3 mg, most of which is found in the liver and kidneys. Daily losses through excretion are about 2 to 5 µg. This is about 0.1% of the total body pool. Consequently this amount must be replaced daily (table 10.1).

Since the minimum daily requirements (MDRs) are so small, and the body stores so large, it takes several years to develop a deficiency of cobalamin and consequently megaloblastic anemia.

### Metabolic Aspects of Vitamin $B_{12}$

Normally vitamin $B_{12}$ is attached to protein complexes in food and must be released from these complexes to be absorbed. This occurs in the stomach and upper small intestine. In order to absorb this vitamin into the body, it must be attached to a glycoprotein called **intrinsic factor (IF).** This IF is produced by the parietal cells in the wall of the stomach. The resulting complex is then attached to surface receptors on the wall of the ileum. From there it is taken into the mucosal cells lining the ileal wall by the process of pinocytosis. In the mucosal cells the complex dissociates again, and vitamin $B_{12}$ is transferred to the blood vessels. The IF is not absorbed. In the blood vitamin $B_{12}$ binds to either one of two special plasma transport proteins called **transcobalamin I** and **II (TC I** and **TC II).** Vitamin $B_{12}$ attached to TC II is delivered to the bone marrow and other tissues (figure 10.8). Cobalamin attached to TC I does not transfer readily to the hematopoietic tissues and appears to be functionally "dead." TC I is thought to be largely synthesized by the granulocytes, since in myeloproliferative diseases (where granulocyte production is greatly increased), the TC levels are also increased considerably. This fact is used as a diagnostic tool in identifying myeloproliferative disease.

### Functional Aspects of Vitamin $B_{12}$

Very little is known about the exact role of vitamin $B_{12}$ in the human organism. Thus far, only two biochemical reactions have been clearly shown to require vitamin $B_{12}$ as coenzyme, and neither is directly involved with DNA synthesis. One involves the peripheral nervous system, and the other appears to play a role in the regulation of folate synthesis and thus may be indirectly involved in DNA synthesis. This latter function may result in megaloblastic anemia when vitamin $B_{12}$ is deficient. The cobalamin involved appears to be methyl-$B_{12}$. Vitamin $B_{12}$ acts as a coenzyme to transfer a methyl (—$CH_3$) group from 5-methyl-THF to homocysteine to form methionine, in the following reaction:

$$\text{5-methyl-THF} + \text{homocysteine} \xrightarrow{\text{vitamin } B_{12}} \text{THF} + \text{methionine}$$

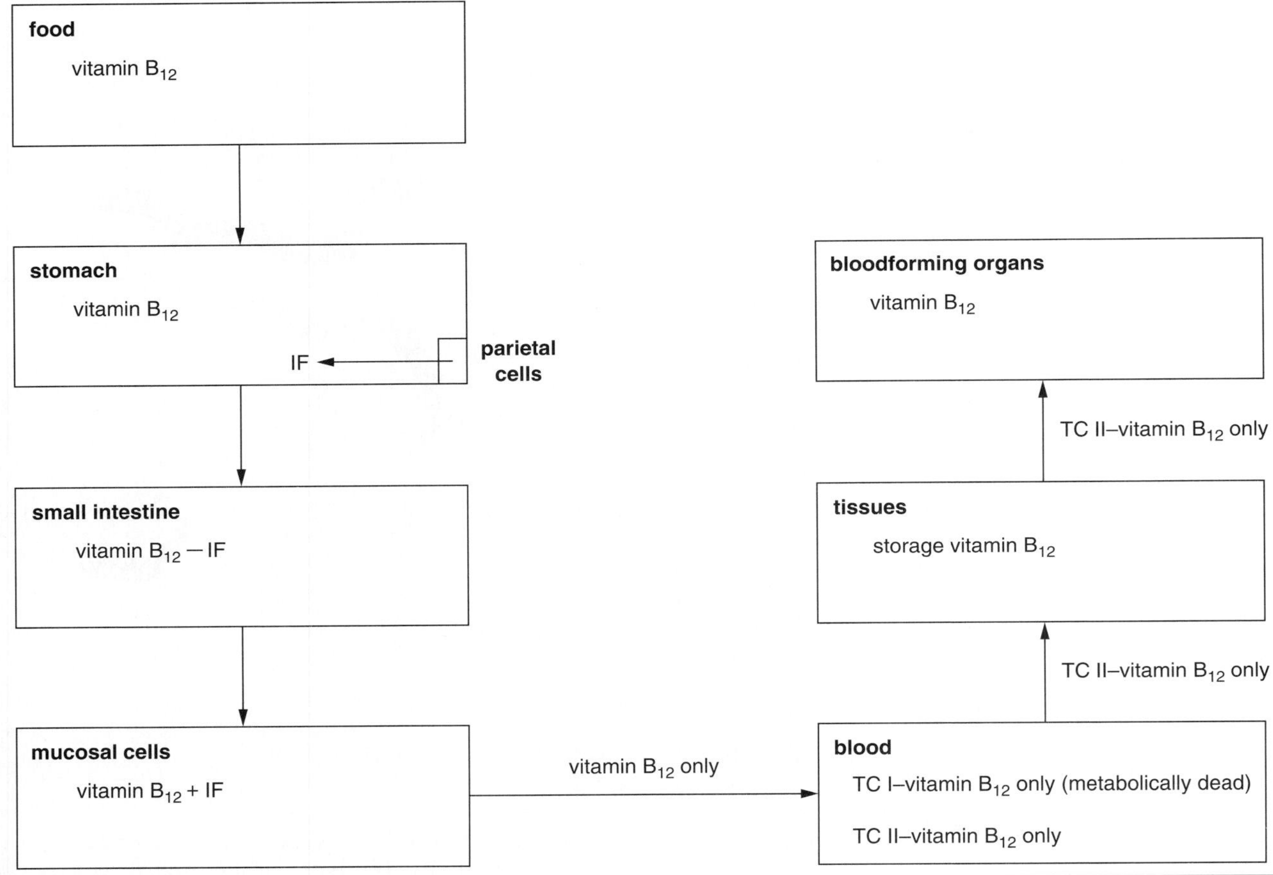


**figure 10.8**

The metabolic aspects of vitamin $B_{12}$. IF = intrinsic factor; TC = transcobalamins.

The main storage form of folate in the cell is 5-methyl-THF. Evidence indicates that this storage form is metabolically inactive and must be demethylated to THF in order to become active as a coenzyme needed for DNA synthesis (figure 10.9). If the sole pathway to convert 5-methyl-THF to THF is the preceding reaction, then the availability of methyl-$B_{12}$ will control the cellular supply of THF. In vitamin $B_{12}$ deficiency states, cells will be deprived of THF, and most folate will be trapped in an unusable storage form: 5-methyl-THF. This is the so-called **methyl-folate trap hypothesis.** According to this hypothesis, DNA synthesis is impaired because the lack of vitamin $B_{12}$ causes a secondary intracellular deficiency of metabolically active folate. This hypothesis would also explain why many patients with a cobalamin deficiency have an elevated serum folate level.

## Causes of Vitamin $B_{12}$ Deficiency

Deficiency of vitamin $B_{12}$ is rarely caused by inadequate dietary intake, although it has been reported as a cause in severe cases of malnutrition and in total vegetarianism. In the great majority of cases, cobalamin deficiency is due to diminished intestinal absorption. This malabsorption may be caused by any one or more of the following four factors:

**1. Lack of intrinsic factor**
This is the most common cause of vitamin $B_{12}$ deficiency in the body. Although the lack of IF may be caused by partial or total gastrectomy, much more frequently it is due to disease of the stomach, resulting in failure of IF secretion. Megaloblastic anemia caused by a defect in the secretion of IF by the parietal cells of the stomach is known as **pernicious anemia (PA).** Strong evidence suggests that most cases of pernicious anemia are due to a genetic defect or an inherited condition. Because of the importance of this type of anemia, it is discussed separately, later in this chapter.

**2. Jejunal bacterial growth**
Cobalamin deficiency may also occur in patients who have a bacterial overgrowth in the jejunum as a result of impaired intestinal motility. This is known as the **stagnant loop syndrome.** Several enteric bacteria have the ability to take up vitamin $B_{12}$, thus preventing intestinal absorption.

**3. Fish tapeworm infestation**
Infestation of the jejunum by the fish tapeworm *Diphyllobothrium latum* also may lead to cobalamin deficiency, due to competitive consumption of this vitamin by the parasite.

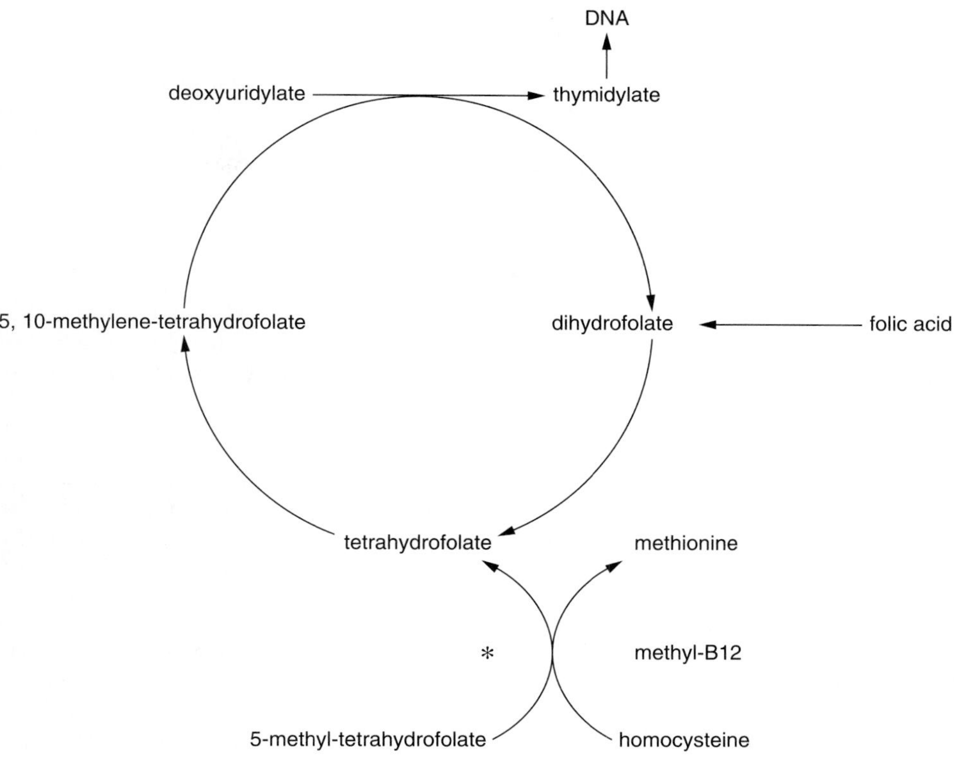


**figure 10.9**
The role of vitamin $B_{12}$ in the production of THF.

**4. Ileal dysfunction**
Any inflammation of the ileum **(ileal enteritis)** or infiltration disease of that part of the small intestine (such as occurs in ileal tuberculosis and lymphoma of the ileum) may lead to vitamin $B_{12}$ deficiency. This same effect may also be caused by tropical sprue or by surgical resection of the ileum.

### Diagnosis of Vitamin $B_{12}$ Deficiency

To discover whether lack of vitamin $B_{12}$ is the causative agent for megaloblastic anemia in a given patient, the serum level of cobalamin is measured. Normal plasma contains about 150 to 450 pg/mL of vitamin $B_{12}$, depending on the size of the individual. If the level is found to be subnormal, the cause of the vitamin $B_{12}$ deficiency should be established. This is primarily done by the **Schilling test.** In this test a small radioactive tracer dose of cyanocobalamin is given by mouth and a 24-hour urine sample is collected and assayed for radioactivity. Since absorbed vitamin $B_{12}$ may be stored in the liver, or bound by serum transcobalamins, a parenteral (intramuscular) dose of nonradioactive vitamin $B_{12}$ is given previously to saturate both the body stores and the transcobalamins. As a result almost all of the absorbed labeled vitamin $B_{12}$ will be excreted in the urine. In patients with cobalamin deficiency due to malabsorption, this will result in abnormally low readings of the urine sample. This is known as part I of the Schilling test.

To determine whether this decrease is due to lack of IF or some other factor, a second test is performed, known as part II of the Schilling test. In this test the patient is given another dose of radioactive vitamin $B_{12}$, together with a dose of IF concentrate. If the readings of the urine sample show high levels of vitamin $B_{12}$, it may be concluded that the megaloblastic anemia is caused by lack of IF. If the readings remain low, it must be concluded that the condition is caused by other factors such as jejunal overgrowth or sprue or that tapeworms are involved. To find out whether it is caused by jejunal bacterial overgrowth, the test is repeated after a 5-day course of antibiotics. If after the antibiotic treatment the readings are high again, it may be concluded that bacteria were the cause.

Other tests, including endoscopy or an X-ray analysis of the intestines following a barium meal, are sometimes performed. Finally a good diet history should also not be neglected.

### Treatment of Megaloblastic Anemia Due to Vitamin $B_{12}$ Deficiency

In most cases the only therapy needed is treatment with extra vitamin $B_{12}$. This is usually given parenterally, since the major cause for deficiency is malabsorption. The injected compound is **hydroxycobalamin** and is given initially in six doses of 100 µg of vitamin $B_{12}$ over a period of 2 to 3 weeks. Usually the patient begins to feel better after a few days, and the blood cell values should be back to normal within 2 to 3 weeks.

In cases where the body is unable to assimilate vitamin $B_{12}$ on a permanent basis, a maintenance program of 1,000 mg every 3 months is initiated.

If the deficiency is caused by bacterial overgrowth or parasitic infection, appropriate antibiotic or other pharmacological regimens should be used to eliminate these causes.

### Pernicious Anemia

As was previously explained, the term pernicious anemia is reserved for those cases of megaloblastic anemia due to vitamin $B_{12}$ deficiency that are caused by lack of intrinsic factor (IF).

There are two types of pernicious anemia: an adult type and a childhood type of PA. **Adult-type PA** is by far the most common. It is caused by inflammation and atrophy of the stomach and is known as **chronic atrophic gastritis.** Patients suffering from this condition have very thin stomach walls, which become inflamed and infiltrated by many small lymphocytes and plasma cells. The gastritis not only affects the production of IF but also the secretion of hydrochloric acid. Hence **achlorhydria** is also common in patients with pernicious anemia.

The disease affects more women than men: the ratio is three women for every two men. It usually begins to show up in later adulthood, with a peak occurrence around 60 years of age. It has been suggested that there is a genetic base for this form of PA, since it is most common in people of Northern European descent, people with fair complexions, blue eyes, and blood type A. This group also tends to have a greater incidence of cancer of the stomach.

Pernicious anemia is probably **autoimmune** in origin, since 90% of all patients with adult-type PA have antibodies in their serum against the parietal cells of their stomach. More than 50% of all affected people also have antibodies against the IF glycoprotein. Two types of anti-IF antibodies may be present as follows:

1. Type I, or **blocking antibodies,** prevent the IF proteins from binding to vitamin $B_{12}$.
2. Type II, or **binding antibodies,** can bind to both the IF-$B_{12}$ complex as well as to free IF. This type of antibody does not interfere with the binding site for vitamin $B_{12}$ (as does type I), but it prevents the complexes from binding to the ileal wall. Both types of antibodies are found not only in serum, but also may be present in the gastric juice, where they inhibit the small amounts of IF that may remain in the stomach juices and thus further contribute to the malabsorption of vitamin $B_{12}$.

**Childhood pernicious anemia** consists of either a lack of IF or abnormalities of IF with an otherwise normal stomach with normal HCl secretion. This is a rare inherited disease caused by an autosomal recessive gene.

If this congenital PA is present, it will not be obvious until the child is about 2 years of age when the vitamin $B_{12}$ stores that were derived from its mother *in utero* have been depleted.

Treatment of both types of PA require lifelong administration of vitamin $B_{12}$, given parenterally at 3-month intervals.

## OTHER MEGALOBLASTIC ANEMIAS

In rare instances megaloblastic anemia is not caused by either folate or vitamin $B_{12}$ deficiency. First of all, a number of unusual congenital deficiencies of enzymes are involved in cobalamin or folate synthesis, resulting in the production of abnormal, nonfunctional, vitamins. Other congenital enzyme abnormalities interfere with the production of transcobalamin II (TC II). As was previously explained, the latter is needed to transport vitamin $B_{12}$ to the body cells undergoing mitosis, including the bone marrow cells. Second, a number of defects in DNA synthesis do not involve folate and vitamin $B_{12}$. Congenital deficiencies of one or more enzymes concerned with purine or pyrimidine synthesis can also cause megaloblastic anemia (e.g., hereditary orotic aciduria).

Last, cancer chemotherapy with drugs that inhibit purine or pyrimidine synthesis such as hydroxyurea, 6-mercaptopurine, and cytosine arabinoside also cause megaloblastic anemia. These antimetabolites are used in chemotherapy for leukemia, lymphoma, and solid tumors since they all block DNA synthesis nonspecifically.

## CAUSES OF MACROCYTOSIS OTHER THAN MEGALOBLASTIC ANEMIA

Finally, not all macrocytosis is associated with megaloblastic anemia. The presence of a significant number of larger than normal red blood cells in a blood smear is frequently seen in the following conditions: (1) alcoholism, (2) liver disease, (3) myxedema, (4) pregnancy, (5) myeloma, and (6) increased reticulocyte formation.

1. **Alcoholism** will eventually lead to megaloblastic anemia due to (1) the toxic effect of alcohol on bone marrow cells, (2) the deficiency of folate, due to bad eating habits of alcoholics, and (3) destruction of the liver (cirrhosis) and the inability to store folate and vitamin $B_{12}$. However macrocytosis has been observed in alcoholics that still have functional livers and that eat balanced diets—so it may result from the toxic effect of alcohol on the hematopoietic process.
2. **Liver disease** has many causes such as alcoholism, infection (hepatitis), and recreational drugs that destroy liver cells. All these conditions appear occasionally in macrocytosis.
3. **Myxedema** is a syndrome that results from hypothyroidism. The latter results in a low basal metabolic rate (BMR). This may slow the development of red blood cells and hence may cause an increase in the size of the cells.

4. **Pregnancy** may also produce macrocytosis of red blood cells, without causing full-blown megaloblastic anemia. This may be due to the fact that the fetus competes with the mother for the available folate and vitamin $B_{12}$ and hence less of these essential vitamins may remain for the mother. This may slow the process of blood cell development and maturation and will result in larger cells.
5. **Myeloma** is a cancer of the lymphocytes. Cancer cells also undergo rapid cell division. These cancer cells may also compete with regular body cells for the available folate and cobalamin. Hence less than normal amounts may be available for the bone marrow, which may slow the blood cell development process. Slower cell division and maturation will lead to larger red blood cells.
6. **Increased reticulocyte formation** may be a cause for transient macrocytosis. Reticulocytes are usually larger than normal red blood cells. If many reticulocytes enter the general circulation in a short period of time, temporary macrocytosis may be observed. Increased reticulocytosis may be observed in cases of heavy blood loss due to trauma or intravascular hemolysis.

## *Further Readings*

Beck, W. S. Megaloblastic Anemias I. Vitamin $B_{12}$ Deficiency. In *Hematology,* 4th ed., W. S. Beck, Ed., The MIT Press, Cambridge, Mass., 1985.

Beck, W. S. Megaloblastic Anemias II. Folic Acid Deficiency. In *Hematology,* 4th ed., W. S. Beck, Ed., The MIT Press, Cambridge, Mass., 1985.

Herbert, V. Megaloblastic Anemias. In *Cecil Textbook of Medicine,* P. B. Beeson, W. McDermott, and J. B. Wijngaarden, Eds., W. B. Saunders, Philadelphia, 1979.

Herbert, V. The Nutritional Anemias. *Hospital Practice,* March 1980.

## *Review Questions*

1. Define the terms megaloblastosis and nuclear-cytoplasmic asynchronism.
2. What is the ultimate cause of macrocytosis associated with megaloblastic anemia?
3. Give the five proximate causes of megaloblastic anemia.
4. Explain why megaloblastic anemia is a classic example of ineffective hematopoiesis.
5. List the three major classes of megaloblastic anemia.
6. Describe some of the unique overt clinical features associated with megaloblastic anemia.
7. What major laboratory features are associated with megaloblastic anemia?
8. Give the official name for folic acid. What are the three components of this vitamin?
9. Define (a) polyglutamates, (b) hydrofolates, and (c) methylhydrofolate.
10. What is the minimum daily requirement (MDR) for folic acid? What is the normal dietary level of folic acid? What are the total body stores of folic acid? How long can they last without being replenished?
11. Describe the normal food sources of folic acid.
12. Where in the intestines is folic acid absorbed? In what form? In what form is it transported in the blood? In what form is it stored in the tissues? Which are the major storage tissues for folic acid?
13. What is the active form of folic acid? Where is the active form produced?
14. Explain the metabolic function of the active form of folic acid.
15. List the three major categories of causes of folic acid deficiency.
16. Mention five different reasons for a decreased intake of folic acid.
17. Describe several different situations that require an increased intake of folate.
18. List three ways in which activation of folic acid can be blocked.
19. How would you diagnose folic acid deficiency in the body?
20. What is the normal treatment for folic acid deficiency?
21. Give another name for vitamin $B_{12}$. Describe the chemical structure of vitamin $B_{12}$.
22. List the four common forms of vitamin $B_{12}$. What is the significance of each form?
23. What is the MDR of vitamin $B_{12}$? What is the normal dietary intake of vitamin $B_{12}$? What are the total body stores of vitamin $B_{12}$? How long will they last if they are not replenished?
24. Give the major food sources for vitamin $B_{12}$.
25. Where and how is vitamin $B_{12}$ absorbed into the body? Where is it stored?
26. Describe the major metabolic function of vitamin $B_{12}$.
27. List the four major causes of vitamin $B_{12}$ deficiency.
28. How is vitamin $B_{12}$ deficiency diagnosed? How would you know if this lack of vitamin $B_{12}$ was caused by lack of IF or by some other factor?
29. Describe the normal treatment for vitamin $B_{12}$ deficiency.

30. Define pernicious anemia. What are its two forms? Which of the two is the most common?
31. Provide evidence that pernicious anemia is an inherited disorder. Which group of people is most prone to develop pernicious anemia?
32. What evidence is there that pernicious anemia may be an autoimmune disease? Which two types of antibodies have been found in this anemia?
33. Describe the treatment for pernicious anemia.
34. Mention at least five different causes of macrocytosis that are not associated with megaloblastic anemia. Explain each of these causes.

# Chapter Eleven

## *Iron Deficiency Anemia*

## ETIOLOGY OF MICROCYTOSIS

Maturation defects of the red blood cells result in the production of abnormally shaped erythrocytes. Although various forms of anisocytosis and poikilocytosis are normal in these maturational anemias, they can ultimately be grouped according to their size into two categories: those usually producing smaller than normal red blood cells **(microcytic anemias)** and those generally producing larger than normal erythrocytes **(macrocytic anemias).** Microcytic anemias show both a reduced mean corpuscular volume (MCV) and a lower mean corpuscular hemoglobin concentration (MCHC)—that is, in addition to microcytosis, they also exhibit hypochromia. Apparently the condition of hypochromia causes microcytosis. It has been suggested that in the course of erythroid maturation, cell divisions continue to occur until the MCHC of the developing cell reaches a critical value. In these hypochromic anemias, one or more extra cell divisions occur because the MCHC is depressed. Since each cell division reduces the cell size, the extra cell divisions yield red blood cells of a smaller size—that is, having a smaller MCV. The hypochromia is caused by an impairment in the hemoglobin synthesis due to abnormalities in either the heme or globin synthesis, or both. The most common cause of impairment in hemoglobin synthesis is the lack of iron. If the body lacks iron, it cannot produce heme, and hence no hemoglobin—resulting in a type of anemia known as **iron deficiency anemia (IDA).** Ironically, although iron is one of the most common elements in the earth's crust, iron deficiency is by far the most common cause of all anemias. An estimated one-quarter of the world's population is deficient in iron. This high incidence occurs because the human body has a limited ability to absorb and store iron and also because excess iron is commonly lost as a result of bleeding, especially in women of childbearing age.

## DIETARY ASPECTS OF IRON

Iron, as part of heme compounds, is found not only in the hemoglobin of red blood cells but also is present in virtually all body cells as part of heme-containing enzymes, such as cytochromes and catalases. Further, iron is an integral part of the myoglobin of muscle cells. The normal distribution of body iron is given in table 11.1. From this table we can see that the total body stores of iron for adults is between 2 to 4 g (depending on the size of the adult), and that approximately two-thirds is present in hemoglobin. Consequently when the iron reserves of the body are depleted, the most sensitive indicator will be the hemoglobin concentration in the blood. Every day about 0.8% of the total red blood cell mass must be replaced. This process requires about 120 g of hemoglobin; about 20 mg of iron is actually needed. The majority of iron is provided by the old red blood cells destroyed by the reticuloendothelial (RE) cells of the spleen and liver. The body has a remarkable capacity to use the same iron over and over again. For that reason, iron stores do not need to be large. Approximately 1 mg of iron is lost each day through excretion from the epithelium of the skin and through the intestinal wall. Even in cases of massive iron overload, the amount of iron lost via the physiological routes is limited to about 2 mg per day maximum. The only other way that iron can escape from the body is by bleeding. This is the reason why menstruating women have a greater minimum daily requirement (MDR) of iron than men. The average adult male needs 1.0 to 1.5 mg of dietary iron each day to replace the amount lost through excretion, whereas the average adult female needs 1.5 to 2.5 mg per day.

**Table 11.1 Distribution of Iron in the Adult Body**

| | Male (g) | Female (g) | % of Total |
|---|---|---|---|
| Hemoglobin | 2.4 | 1.7 | 65 |
| Ferritin and hemosiderin | 1.0 (0.3–1.5) | 0.3 (0–1.0) | 30 |
| Myoglobin | 0.15 | 0.12 | 3.5 |
| Heme enzymes (e.g., cytochromes, catalase, peroxidases, flavoproteins) | 0.02 | 0.015 | 0.5 |
| Transferrin-bound iron | 0.004 | 0.003 | 0.1 |

The average daily Western diet contains about 10 to 30 mg of iron, of which 5 to 10% is absorbed. However, in conditions of pregnancy or in iron deficiency, that percentage may increase to 20 to 30%. But even in those situations the majority of dietary iron remains unabsorbed.

## METABOLIC ASPECTS OF IRON

### Absorption of Iron

Dietary iron is absorbed through the mucosal cells of the duodenum and to a lesser extent through the jejunum. Iron is present in food in both inorganic form—as ferric ($Fe^{3+}$) and as ferrous ($Fe^{2+}$) iron—and in organic form as heme. Although ferric iron is the most common form of dietary iron, heme iron predominates in some foods, such as red meats. In general, iron in foods of animal origin is better absorbed than that of plant origin. There are two reasons for this: (1) apparently many plant foods contain substances that will chelate iron and form insoluble

**Table 11.2 Factors that Promote Iron Absorption and Factors that Reduce Absorption**

| Promote Absorption | Reduce Absorption |
|---|---|
| Ferrous form | Ferric form |
| Inorganic iron | Organic iron |
| Acids: HCl, vitamin C | Alkalis: antacids, pancreatic secretions |
| Solubilizing agents: sugars, amino acids | Precipitating agents: phosphates |
| Iron deficiency | Iron excess |
| Increased erythropoiesis | Decreased erythropoiesis |
| | Infection |

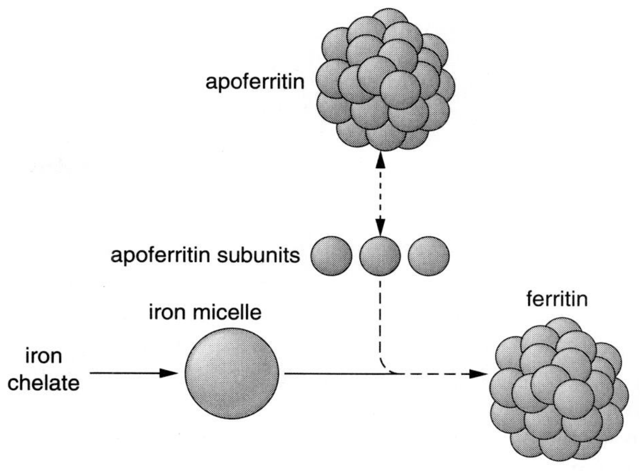


figure 11.1
Structures of apoferritin and ferritin.

precipitates in the lumen of the gastrointestinal tract and (2) ferric iron forms insoluble ferric hydroxide polymers in water at pH levels greater than 2.5. Since the pH level of the upper intestines is between 4.0 to 6.0, ferric iron will not be in absorbable form unless it has been previously solubilized at the lower pH of the stomach. Patients with achlorhydria as a result of either atrophic gastritis or surgical removal of the stomach will have great difficulty absorbing ferric iron from food. They are also more prone to develop iron deficiency anemia. Ferric iron in food can be bound to sugar, ascorbic acid, or certain amino acids and may be absorbed in the more alkaline environment of the intestines. Ferrous iron and heme iron remain soluble in food and are well absorbed at the pH ranges encountered in the duodenum and the jejunum. Table 11.2 tabulates the major factors that either favor or reduce the absorption of dietary iron from the intestines. From that list we can see that in situations of iron deficiency and increased hematopoiesis, more iron tends to be absorbed by the intestines. In cases of iron excess and decreased hematopoiesis, iron uptake tends to be reduced. Alkalis such as antacid medications also tend to inhibit iron absorption.

## Transport of Iron

Generally speaking, the amount of iron absorbed equals the amount lost by excretion. If it were greater, iron overload would develop since the body's storage capacity is limited. Iron absorption is an active process, and control of the amount of absorption lies with the mucosal cells of the upper intestines. Any iron absorbed as part of a heme structure is removed from the porphyrin ring in these mucosal cells. Most inorganic iron is absorbed in the ferrous form but inside the mucosal cells is oxidized to ferric iron ($Fe^{3+}$). The mucosal cells pass some of the iron directly into the bloodstream, but most of it is bound to a specific storage protein known as **apoferritin,** which is found in many tissues and combines with iron to form **ferritin.** Apoferritin is a globular protein made up of 24 subunits. Iron forms a micelle of ferric hydroxyphosphate that surrounds each subunit, as shown in figure 11.1. Each ferritin molecule can contain as many as 4,000 atoms of iron. Apoferritin without iron tends to dissociate into apoferritin monomers. However the addition of iron to each monomer tends to polymerize these monomers to form a large complex of ferritin made up of 24 subunits. Ferritin is the principal storage form of iron in the tissues.

Table 11.1 shows that 25 to 30% of the total body iron is stored in ferritin. Ferritin iron is in equilibrium with plasma iron. Although some ferritin is also found in the plasma, the greater part of iron in the blood is transported bound to a β-globulin called **transferrin,** or siderophilin. Transferrin is a globular protein with a molecular weight (MW) of 80,000 daltons. It is synthesized in the liver, has a half-life of 8 to 10 days, and is capable of binding two iron atoms per molecule. Transferrin is reutilized after it has given its iron to the mononuclear phagocytic cells of the bone marrow and other tissues. Normally transferrin is about 35% saturated with iron; the normal plasma iron level is about 130 μg/dL in men and 110 μg/dL in women. However there appears to be a diurnal variation, with higher values occurring in the morning and lower levels in the evening.

Transferrin receives only a small amount of its iron from the mucosal cells of the intestines. The greater part of its iron is gained from the phagocytic cells of the spleen and other tissues. As can be seen from table 11.1, the total plasma iron bound to transferrin is only 4 mg. Since the body requires approximately 20 mg of iron each day to produce 120 mg of new hemoglobin (0.8% of the total red blood cell mass), it follows that total plasma iron turns over about five times a day.

### Regulation of Iron Uptake

Iron absorption from the intestines into the bloodstream is increased when iron stores are depleted or when erythropoiesis is increased. When large quantities of iron are ingested, more iron is bound in the mucosal cells, but the absorption into the bloodstream is increased very little. Iron in the mucosal cells stays bound to ferritin and is lost when the cells are shed into the lumen of the intestine, from where it is passed in the stool. The mucosal cells also become loaded after parenteral administration of iron; this has an effect on the uptake of dietary iron, which will be reduced in such situations. Thus one factor regulating iron uptake in mucosal cells is the amount of iron present in these cells. This is known as the **mucosal block** and refers to the ability of the mucosa to prevent excess iron from being absorbed.

### Iron Overload

The normal functioning of the homeostatic mechanism that maintains iron balance in the body is essential for health. If more iron is absorbed than excreted, iron overload results. Development of iron overload can result from a number of causes but the most common one is an abnormal functioning of the mucosal cells of the small intestine, which behave as if iron deficiency is present continuously absorbing large quantities of iron. First, the transferrin of the blood is saturated; next, visible aggregates of ferritin with various amounts of porphyrin appear in RE cells such as macrophages and other mononuclear phagocytic cells. These aggregates are known as **hemosiderin.** Small amounts of hemosiderin in these RE cells are normal, but large amounts are harmful since they interfere with the normal functioning of the cells. This produces a condition known as **hemosiderosis.** If iron overloading continues, iron will be deposited in other cells of the body, especially in the cells of the skin, endocrine organs, and liver. These large deposits of ferritin and hemosiderin will cause bronzing of the skin and result in a condition known as **hemochromatosis.** Excess deposits of iron in the liver may result in cirrhosis of the liver. The endocrine tissues affected by iron overload are the islets of Langerhans and the gonads; destruction of the islets of Langerhans may result in diabetes mellitus, and the gonads, which may result in sterility.

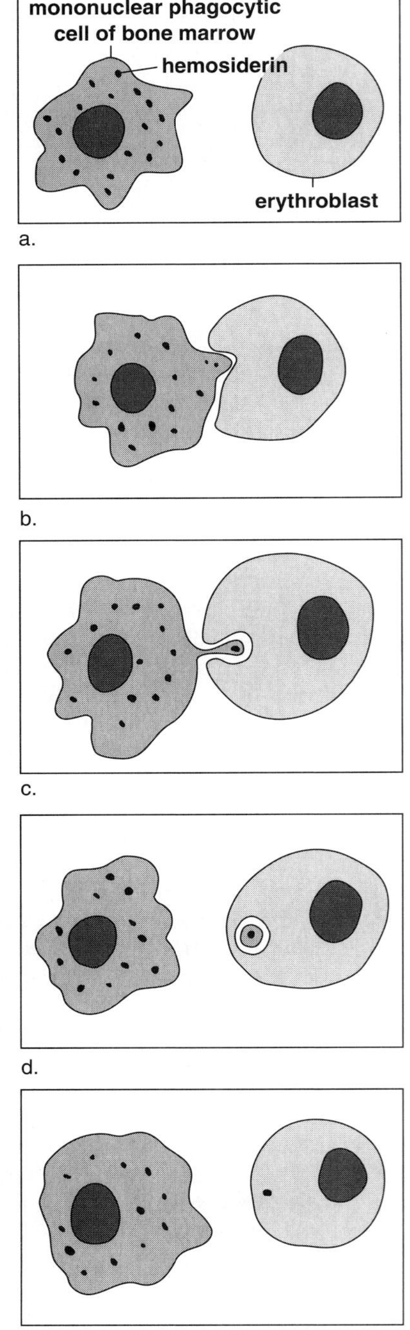


**figure 11.2**
Ropheocytosis.

### Incorporation of Iron into Red Blood Cells

Under normal circumstances, transferrin-bound iron in the blood is taken up by mononuclear phagocytic (RE) cells that line the sinuses in the bone marrow. Once in these cells, iron binds once again to apoferritin and is stored as ferritin or as granules of hemosiderin. Electron microscopy has revealed that erythrocytic blast cells cluster around these phagocytic cells and engulf small granules of hemosiderin and ferritin by a process known as **ropheocytosis.** In this process a thin strand of RE cytoplasm, containing several granules of iron-containing compounds, protrudes into the wall of the erythroblast, and the invagination thus produced is pinched off to form a vacuole within the developing red blood cell (figure 11.2). In this way the erythroblastic cells receive large quantities of iron in discrete packets. The cytoplasmic fluid in the vacuole disappears rapidly, leaving the iron-containing

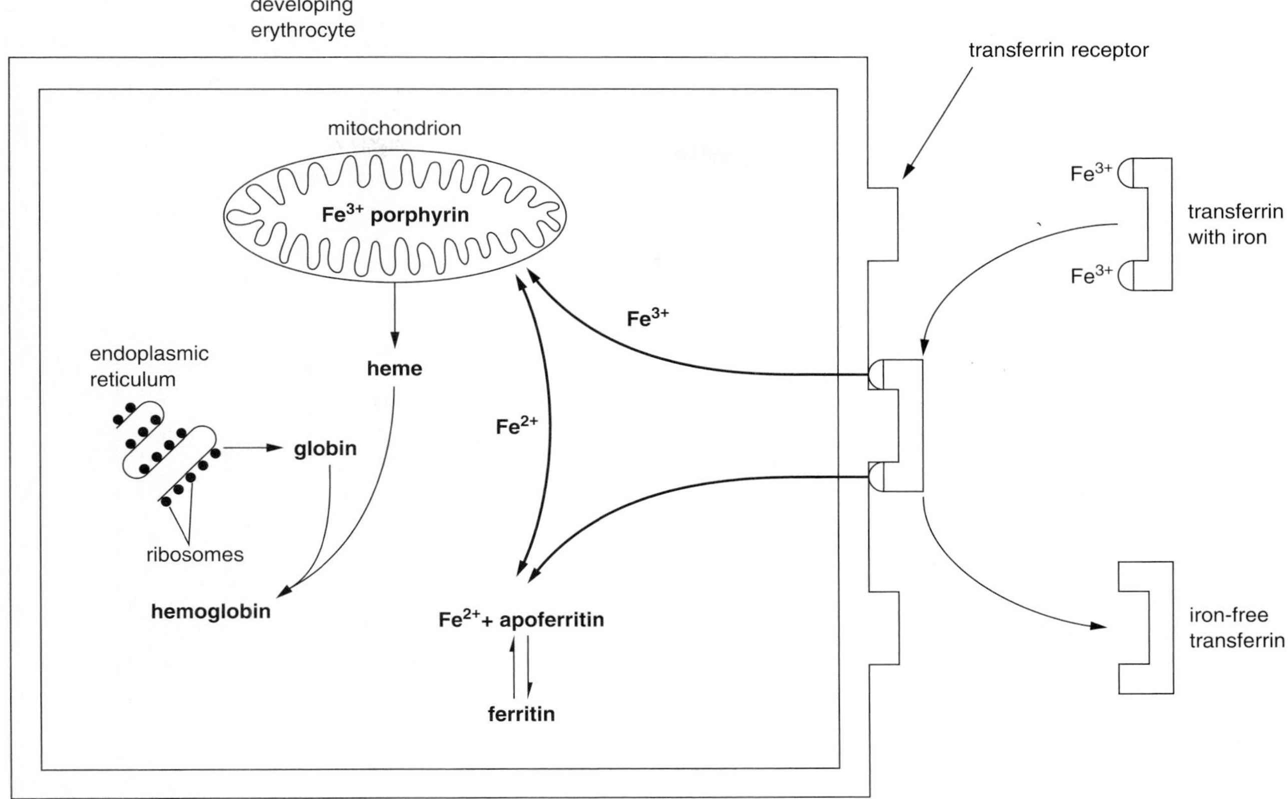


**figure 11.3**

The incorporation of iron into the developing red blood cell.

compounds free in the cytoplasm of the developing red blood cell. These young cells containing stainable granules of iron (stained with Prussian blue) are called **sideroblasts.** Often stainable particles of iron persist into the reticulocyte stage and sometimes even in the mature red blood cells. Stained granules in reticulocytes and mature erythrocytes are called **Pappenheimer bodies,** and the red blood cells are often referred to as **siderocytes.** When iron stores are depleted, neither sideroblasts nor siderocytes will be found in the bone marrow. This is a good diagnostic tool to confirm an iron deficiency.

Within developing red blood cells, iron remains present in the form of ferritin until it is delivered to the mitochondria. In the mitochondria iron is dissociated from the apoferritin and incorporated into the protoporphyrin ring to form heme. The heme is then attached to globin at the ribosomes to form hemoglobin monomers, which then combine to form the tetramer structure known as hemoglobin (figure 11.3).

Each of these steps requires the presence of certain enzymes. Interference with this enzymatic activity by toxic substances (e.g., lead poisoning) or by hereditary deficiencies (e.g., thalassemia) will result in metabolic bottlenecks, producing red blood cells with lots of iron and little hemoglobin. A summary of iron metabolism is given in figure 11.4.

## CAUSES OF IRON DEFICIENCY ANEMIA

As is the case with many anemias, the causes of IDA are many and varied. However they may be grouped into four major categories: (1) blood loss, (2) increased demand, (3) malabsorption, and (4) poor diet.

### Blood Loss

Bleeding is by far the most common cause of iron deficiency anemia. In adult men and in postmenopausal women, bleeding is mainly from the gastrointestinal tract. Blood loss is usually in the form of a chronic hemorrhage caused by gastric or duodenal ulcers, excessive aspirin ingestion, hiatus hernia, carcinoma of the stomach or intestines, colitis, diverticulitis, or hemorrhoids.

In women of childbearing age, IDA is mostly due to blood loss caused by menstruation, pregnancy, and childbirth.

In very rare instances, bleeding may be from the kidneys and bladder, resulting in hematuria and hemoglobinuria.

### Increased Demand

Three groups of people need increased amounts of iron. First, pregnant women need more iron, since their body iron is preferentially taken by the developing

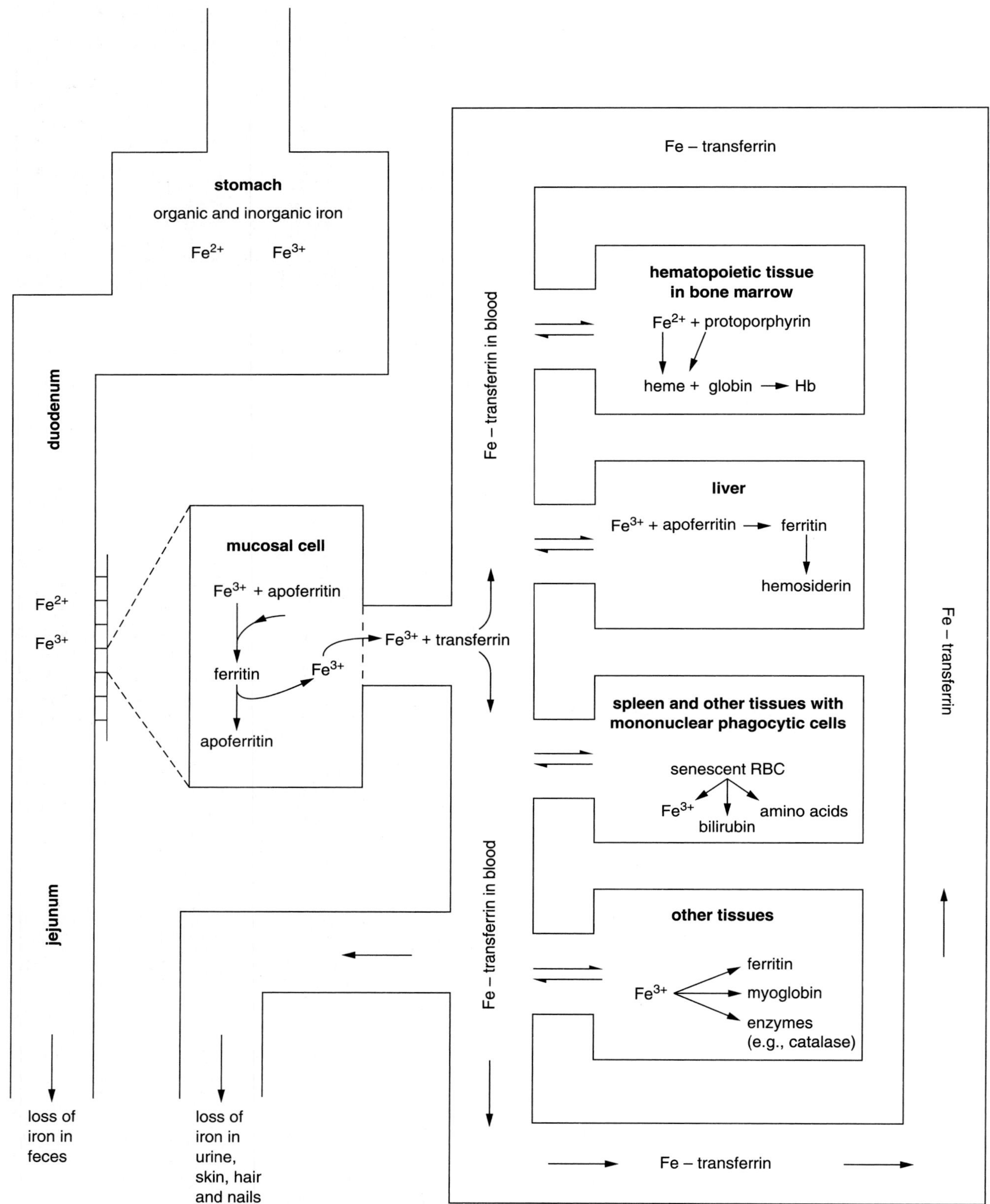


figure 11.4
Iron metabolism in the body.

fetus at the expense of the mother. Further, during pregnancy the blood volume of the mother increases by two-thirds.

Second, infants and young children need extra iron, especially those that are almost exclusively breast- or bottle-fed over a long period of time. Milk does not contain iron, and iron deficiency may develop unless supplemental iron is given.

Third, older children and adolescents need extra iron since they are experiencing rapid growth. In addition, many children lack a well-balanced diet that provides them with sufficient ferrous iron and heme.

### Malabsorption

Celiac disease, partial or total gastrectomy, and atrophic gastritis all may result in achlorhydria and thus predispose the person to iron deficiency.

Second, intestinal parasites, such as hookworms, may take up dietary iron at the expense of the infested person and thus may contribute to IDA. Also tropical sprue will result in the inability to absorb iron. **Sprue** is a chronic celiac disease associated with malabsorption of food elements such as fat, xylose, and vitamin $B_{12}$ from the small intestine.

### Poor Diet

Although poor diet is rarely the sole cause of iron deficiency, it is an important contributing factor for persons living in many underdeveloped countries. Their poor-quality, largely vegetable diet may produce a background of latent iron deficiency, especially when combined with repeated pregnancies, prolonged lactation, or parasitic infection.

## CLINICAL FEATURES OF IRON DEFICIENCY ANEMIA

The first symptoms of IDA are usually those associated with general anemia such as weakness, fatigue, inability to concentrate, headaches, and dizziness. No other overt signs are present at the early stages of IDA. Later phases, however, show a number of highly specific signs that may indicate the presence of this type of anemia. These symptoms are mainly epithelial changes caused by a deficiency of intracellular iron-containing enzymes. They include **angular stomatitis** (sores at the corners of the mouth), **glossitis** (inflammation of the tongue, often associated with open sores), **koilonychia** (flat or spoon-shaped nails, as illustrated in figure 11.5), **esophageal webbing** (lesions of the esophagus), **dysphagia** (not eating, perhaps due to pain in the esophagus), and **pica** (craving for unusual foods).

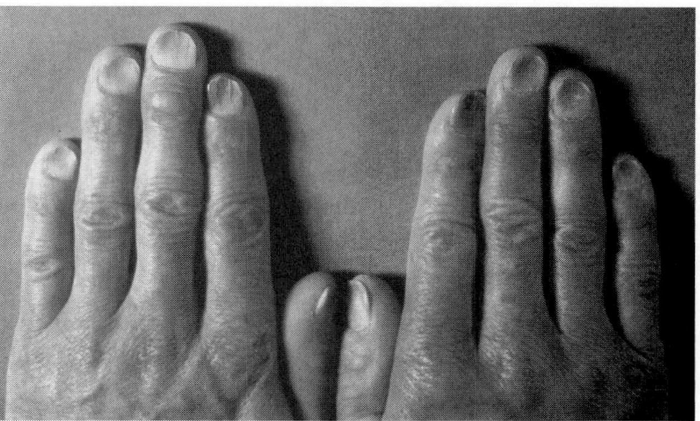

figure 11.5
Koilonychia, the development of spoon-shaped nails that may occur in severe cases of iron deficiency.

## LABORATORY FINDINGS OF IRON DEFICIENCY ANEMIA

The laboratory findings associated with IDA show a low MCV and low MCHC. A blood smear will confirm microcytosis and hypochromia. In severe cases of IDA, poikilocytosis may be observed. If iron deficiency is associated with lack of folate or vitamin $B_{12}$, anisocytosis may be noted, with both microcytic and macrocytic cells present in the blood film. In the latter case the MCV and MCHC may be fairly normal.

The reticulocyte count is always low in IDA, whereas the platelet count is often moderately increased, especially when low-level chronic bleeding is present.

The bone marrow may show erythroid hyperplasia, but the developing red blood cells will be hypochromic. Staining of a bone marrow smear with Prussian blue will show a complete absence of hemosiderin granules, thus indicating the absence of any iron stores in these erythrocyte precursor cells.

The serum iron levels are decreased, and the unbound iron-binding capacity (UIBC) rises sharply to give a less than 10% transferrin saturation, compared with a normal level of 35%, as illustrated in figure 11.6. The serum ferritin level may be reduced to zero, as indicated in table 11.3.

### Treatment of Iron Deficiency Anemia

Once the prognosis of iron deficiency anemia has been established, the underlying cause should be investigated. In case of a simple lack of iron in the diet, a therapeutic response should follow a course of treatment with oral iron. Many different oral iron preparations are available. The method of choice, which is

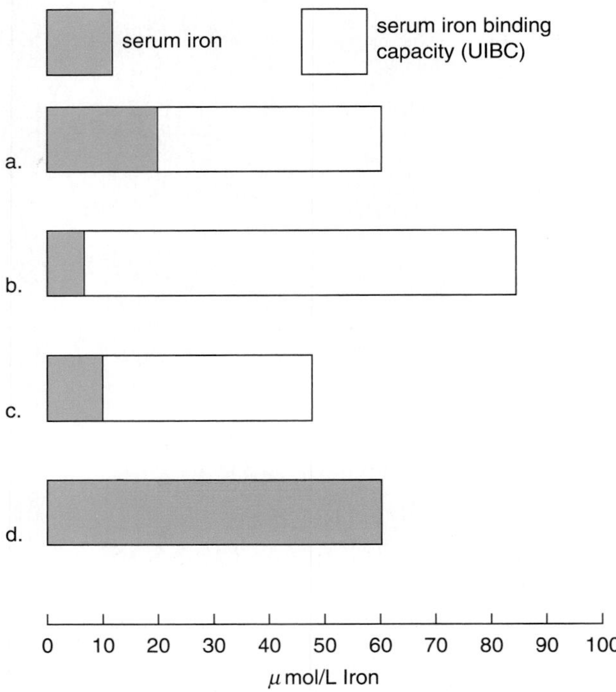


figure 11.6

Serum iron and unsaturated iron-binding capacity (UIBC). (*a*) Normal individuals; (*b*) iron deficiency anemia; (*c*) anemia of chronic disease; and (*d*) iron overload.

also the cheapest, is the administration of **ferrous sulphate.** A standard regimen of 300 mg ferrous sulphate tablets containing 60 mg of elemental iron, given three times daily (with or after meals), is usually more than adequate. Mild gastrointestinal intolerance manifested by nausea, abdominal pain, constipation, or diarrhea usually is a dose-related effect of the therapy and can often be alleviated by dose reduction or by using ferrous gluconate, which contains only 37 mg of elemental iron per 300 mg tablet, or by using the more expensive preparations such as ferrous succinate, ferrous fumarate, or ferrous lactate. An elixir (a liquid containing iron plus a flavoring substance) is available for children.

Oral iron therapy should be given long enough to ensure both the correction of the anemia and the building up of iron stores. This means that the average replacement therapy should continue for about 4 to 6 months. The patient should begin to feel better, however, within days after the oral regimen has started. The hemoglobin level should rise by 2 mg/dL of blood every 3 weeks. Marked reticulocytosis should occur within a week, but it should take 1 or 2 months for the reticulocyte level to return to normal.

**Table 11.3 Serum Ferritin Levels in Normal Individuals and in Cases of Iron Deficiency Anemia and Iron Overload**

| Normal levels | μg/L |
|---|---|
| Male | 40–340 |
| Female | 14–148 |
| Children | 7–142 |
| Iron deficiency | 0–12 |
| Iron overload | 340– > 20,000 |

When the patient is intolerant of all kinds of oral iron or when a rapid response is required (e.g., in late pregnancy) or when the patient has a gastrointestinal disorder that may be aggravated by iron, **parenteral iron** is given. This may be done by injections of iron dextran or iron sorbitol citrate. The hematological response to parenteral iron is not faster than with oral iron therapy, but the iron stores are replenished at a much quicker rate.

Failure to respond to iron therapy (i.e., no improvement within 2 to 3 weeks) may be due to a variety of causes. The most prominent among these are (1) continued active blood loss; (2) failure to take iron tablets as indicated (patient failure); (3) the wrong diagnosis—that is, there are other microcytic-hypochromic anemias that may be mistaken for IDA, such as sideroblastic anemia or anemia of chronic diseases; and (4) failure to absorb oral iron due to inflammation of the intestines, intestinal malignancy, sprue, or parasitic infections. In these instances the underlying causes must be investigated and treated where possible. Meanwhile, iron replacement therapy should be continued.

Since the body has only a limited capacity to store iron, iron therapy brings with it the danger of iron overload. As previously explained, iron overload can seriously damage the liver, the endocrine organs, and even the heart. The danger of iron overload is especially high in patients with repeated blood transfusions, since the body has little ability to remove the excess iron after red blood cells are broken down. When this phenomenon happens, the patient is said to be suffering from **transfusional hemosiderosis.** Iron overloads are also commonly encountered in people suffering from sideroblastic anemia, aplastic anemia, and β-thalassemia major. Finally, iron poisoning is one of the most common causes of fatal poisoning in children who may ingest large quantities of therapeutic iron left carelessly in the home.

## Further Readings

Baum, H., Gergely, J., and Fanburg, B. L. Iron Overload. *Molecular Aspects of Medicine* 6 (1) 1983.

Beck, W. S. Hypochromic Anemias I. Iron Deficiency and Excess. Lecture 6 in *Hematology*, 4th ed., W. S. Beck, Ed., The MIT Press, Cambridge, Mass., 1985.

Bothwell, T. H., et al. *Iron Metabolism in Man*. Blackwell, Oxford, 1979.

Muss, H. B., and White, Douglas R. Iron Deficiency anemia in Adults. *American Family Physician* 117 (2) (Feb. 1978).

## Review Questions

1. Explain why iron deficiency anemia causes microcytosis.
2. What are the total body stores of iron in adult males and females? How much of that iron is present in hemoglobin? How much in ferritin? How much in transferrin?
3. How much iron does the average Western diet contain? How much of it is absorbed in the body?
4. Name the major factors that favor absorption of iron from the gut. Name the major factors that reduce absorption of iron from the gut.
5. How much iron is needed to replenish the normal loss of red blood cells each day? Where does most of the iron come from? Where does the rest of the iron come from?
6. What is the MDR of iron in adult males? In adult females? Why the difference?
7. Exactly where in the body does most iron absorption take place? What is the first storage place for iron once it enters the interior milieu of the body? In what form is it stored?
8. Describe the nature of apoferritin. How many iron atoms can it hold?
9. In what form is iron transported in the blood? How many atoms of iron can be carried by a single transferrin molecule? What is the normal saturation level of transferrin?
10. Where is iron stored in the hematopoietic tissues? In what form is it stored? What is hemosiderin?
11. How is iron taken into the developing red blood cells? What is that process called?
12. Define sideroblasts, siderocytes, and Pappenheimer bodies.
13. What happens when excess iron enters the interior milieu of the body? Where is the excess iron stored? What disorders may this produce?
14. List the four major causes of iron deficiency.
15. Mention some unique clinical features of IDA.
16. Describe the major laboratory findings associated with IDA.
17. What is the normal treatment for IDA? In what form is iron given?
18. In what situations is iron given parenterally? In what form?
19. Describe some of the major causes of failure to respond to iron replacement therapy.
20. List the most common causes of iron overload.

# Chapter Twelve

## *Other Microcytic-Hypochromic Anemias*

## ETIOLOGY OF MICROCYTOSIS AND HYPOCHROMIA

bnormalities in either heme or globin synthesis will result in a decreased production of hemoglobin, and this in turn may lead to more than the four usual maturational divisions of the erythroblast. This will result in the production of smaller than normal red blood cells since each subsequent cell division reduces the cell size.

Abnormalities in heme synthesis may be the result of the three following causes:

1. **Lack of iron**
This results in iron deficiency anemia, which was discussed in chapter 11.
2. **Inability of the erythrocytic precursor cells to take up iron**
In this situation sufficient iron is present in the body, but the erythroblasts are unable to incorporate this iron into their cytoplasm. This condition is known as **anemia of chronic disease** and is discussed in the beginning of this chapter.
3. **Inability to incorporate iron into the porphyrin ring structure**
In this situation the individual has sufficient iron and is also able to take iron into the developing red blood cell; however this iron cannot be incorporated into an abnormal porphyrin ring structure. This results in iron overloads in the developing erythrocytes, producing a condition known as **sideroblastic anemia,** which is also discussed in this chapter.

Abnormalities in globin synthesis frequently result in insufficient synthesis of globin chains. In these situations nothing is wrong with either the availability of iron, with the ability to take up iron, or with the formation of the heme structure. The individual is unable to produce sufficient quantities of either alpha-globin chains or beta-globin chains. This results in a group of anemias known as the **thalassemias,** which are discussed later in this chapter.

## ANEMIA OF CHRONIC DISEASE

Anemia of chronic disease is also known as **anemia of inflammation** since it always is associated with chronic inflammatory or malignant diseases. It is a common disorder and is second only to iron deficiency anemia as the leading cause of anemia. This anemia is a complex disease and may be complicated by the presence of other factors that also cause anemia such as folate deficiency, liver disease, renal failure, and endocrine abnormalities.

Anemia of chronic disease is usually a mild type of anemia—that is, the hemoglobin levels rarely go below 9 to 10 g/dL of blood, and it is nonprogressive. In general the severity of the anemia is directly related to the severity of the chronic inflammatory or malignant disease.

**Table 12.1 Causes of Anemia of Chronic Disease**

*Malignant Diseases*
- Carcinoma
- Lymphoma
- Sarcoma

*Chronic Inflammatory Diseases*
- Infectious tuberculosis, pneumonia, osteomyelitis, etc.
- Noninfectious rheumatoid arthritis, systemic lupus erythematosus (SLE), Crohn's disease, etc.

### Causes of Anemia of Chronic Disease

Syndromes associated with anemia of chronic disease may be grouped into two categories: (1) **malignant diseases** such as carcinomas, sarcomas, and lymphomas and (2) **chronic inflammatory diseases.** The latter may be further subdivided into infectious and noninfectious diseases, as illustrated in table 12.1.

Defining the exact cause(s) of this type of anemia is difficult since it is associated with many different diseases, each with its own etiology. However the following three mechanisms are generally recognized as playing a major role in the production of anemia of chronic disease:

1. The most important cause is *decreased iron release* from the mononuclear phagocytic cells of the spleen and other tissues after they break down old red blood cells and hemoglobin. This iron is normally released into the plasma and recycled over and over again in hemoglobin synthesis. As a result of this decreased iron release, less iron is available for new hemoglobin formation. This will result in hypochromia and consequently microcytosis of the newly formed erythrocytes. The reason for this decreased iron release by the phagocytic cells is not understood.
2. A second mechanism responsible for the production of anemia of chronic disease is a *moderate increase in the destruction of circulating red blood cells.* The life span of erythrocytes of people with this anemia is usually shortened to about 60 to 90 days. The defect does not lie in the red blood cells themselves, since transfused cells from a normal donor also have a shortened survival rate. It has been suggested that the diseases mentioned in table 12.1 are often associated with **vasculitis**—that is, inflammation of the blood vessels—and this condition may promote intravascular clotting and hemolysis. However normal bone marrow would easily compensate for such mild hemolytic states by increasing the output of new red blood cells. In anemia of chronic disease this does not happen—that is, the reticulocyte count is not elevated, and the anemia develops due to a failure to compensate for this increased hemolysis. One explanation for this phenomenon has been previously given: a decreased release of iron from the mononuclear phagocytic cells. A second explanation is given in the following material.

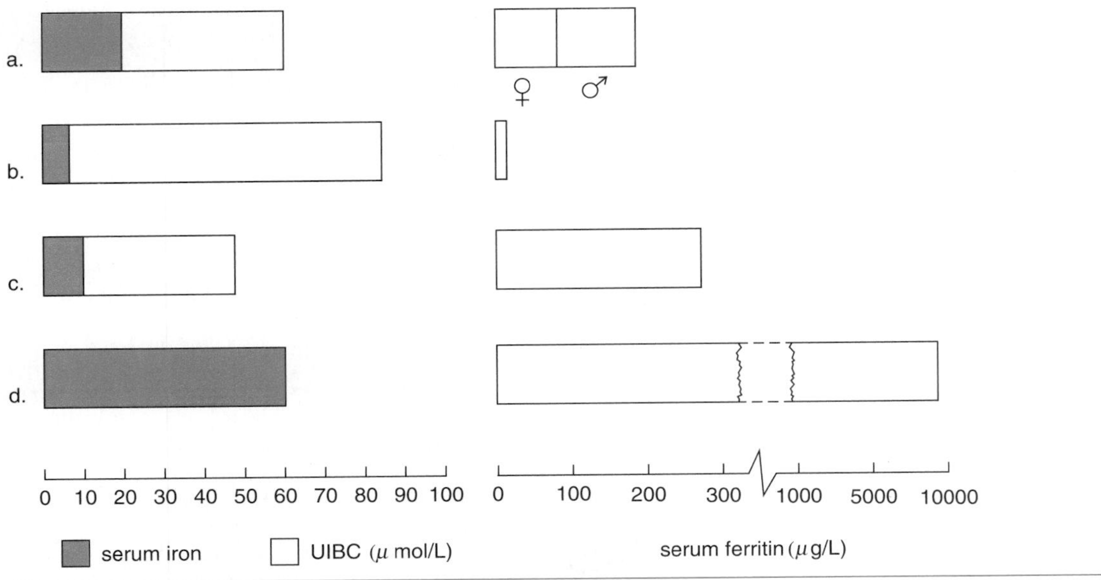


**figure 12.1**

(*a*) The serum iron, UIBC, and serum ferritin levels in normal subjects, (*b*) iron deficiency, (*c*) the anemia of chronic diseases, and (*d*) iron overload. The total iron-binding capacity (TIBC) comprises the serum iron and the unsaturated iron-binding capacity.

3. It has been suggested that anemia of chronic disease may also be caused by a *decreased production of erythropoietin,* since many patients with this anemia have lower plasma and urinary erythropoietin levels.

### Diagnosis of Anemia of Chronic Disease

Anemia of chronic disease is usually a mild form of anemia that develops slowly, with hematocrit values falling to 25 to 35%. Only rarely does the hematocrit decrease below 20%, and in those situations it is usually due to complications produced by factors that are the basis for other types of anemia.

In the majority of cases, the mean corpuscular volume (MCV) and mean corpuscular hemoglobin concentration (MCHC) are fairly normal, but in at least 25% of the patients microcytosis and hypochromia are clearly present. The total iron-binding capacity (TIBC) comprises the serum iron and the unsaturated iron-binding capacity (UIBC).

In anemia of chronic disease, both serum iron levels and the UIBC are reduced, whereas serum ferritin levels are normal or slightly elevated. The TIBC level is a good diagnostic tool to distinguish anemia of chronic disease from iron deficiency anemia (IDA). As illustrated in figure 12.1, the TIBC is usually increased in IDA, whereas it is decreased in anemia of chronic disease. Also, serum ferritin levels are low or absent in IDA and are normal or slightly increased in anemia of chronic disease.

Bone marrow analysis will show that the mononuclear phagocytic cells that line the sinuses contain abundant hemosiderin, and iron stores in the erythroblasts are reduced—that is, very few **sideroblasts** are present.

### Treatment of Anemia of Chronic Disease

This anemia is always associated with malignant or inflammatory disease, and it follows that anemia of chronic disease can be corrected only by successful treatment of the underlying disease.

Since one of the basic features of this anemia is accumulation of iron in macrophages and other mononuclear phagocytic cells, which will not give it up readily to the developing erythroblasts, treatment with oral iron clearly will not alleviate the anemia despite the low serum iron levels.

## SIDEROBLASTIC ANEMIA

Sideroblastic anemias are a group of heterogeneous disorders associated with hypochromia and in some cases also with microcytosis, as a result of defective heme synthesis. In contrast with anemia of chronic disease—where the block is in the movement of iron from the storage cells to the developing red blood cells—in this group of diseases the developing erythroblasts can incorporate iron properly but are unable to bind this iron to the protoporphyrin structures in the mitochondria. Also, in cases of anemia of chronic disease sideroblasts contain a few (between 1 to 5) granules of hemosiderin when stained with Prussian blue.

In sideroblastic anemia, on the other hand, the erythrocytic blast cells have many more granules. These abnormal sideroblasts, containing many granules of hemosiderin, are plentiful in bone marrow smears of patients with this type of anemia.

Often these granules are abnormally large and are frequently arranged in a ring around the nucleus. Such sideroblasts are referred to as **ring sideroblasts,** and

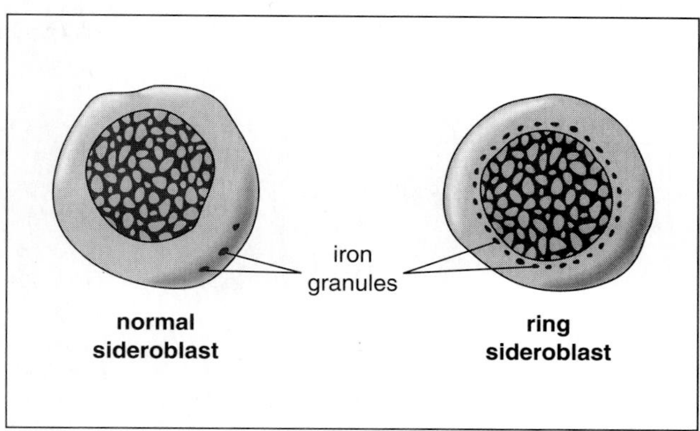


figure 12.2
Comparison of a normal sideroblast and a ring sideroblast when stained for iron granules (hemosiderin granules).

this phenomenon is illustrated in figure 12.2. The reason for the development of this ring structure may lie in the fact that the cell—in an effort to produce hemoglobin—deposits large amounts of iron between the cristae of the mitochondria that are normally present around the nucleus of erythroblasts.

### Causes of Sideroblastic Anemia

The accumulation of iron in the mitochondria may be due to problems with heme synthesis, since most patients with sideroblastic anemia also have an accumulation of abnormal amounts of protoporphyrin in their erythroblastic cells. Apparently one of the crucial enzymes or coenzymes needed for proper heme synthesis is missing. Two crucial steps in heme synthesis occur in the mitochondria, and problems with these two steps have been implicated as the major causes of this type of anemia.

As illustrated in figure 12.3, these steps are (1) the first rate-limiting step that combines glycine and succinyl-CoA to form Δ-aminolevulinic acid (Δ-ALA) and (2) the final step that results in the incorporation of iron into the protoporphyrin ring. The first step requires the enzyme Δ-ALA synthetase, with pyridoxal phosphate (vitamin $B_6$) as coenzyme, and the last step requires heme synthetase (also known as ferrochelatase) as a catalyzing agent. Other enzymes are involved in heme synthesis also, and the lack of any of these may also cause sideroblastic anemia. However, the two causes just mentioned appear to be the most common. Many patients with sideroblastic anemia respond favorably to pyridoxine or pyridoxine phosphate therapy. This implies that in most instances the block may be found in the first step of the process of heme synthesis. Other patients accumulate large quantities of protoporphyrin and nonheme iron in their erythroblasts. This indicates that in other situations the block may be in the last step of the heme formation process. Since the defect in heme production frequently results in accumulations of various forms of porphyrin structures in red blood cells or the liver, or both, certain forms of this disorder have also been coined **porphyrias,** as they are the result of inborn errors of porphyrin metabolism.

### Classification of Sideroblastic Anemias

Formation of abnormal sideroblasts may be due to either an inherited condition or to some acquired environmental factors. Hence all sideroblastic anemias can be divided into the hereditary sideroblastic anemias and the acquired sideroblastic anemias.

#### *Hereditary Sideroblastic Anemias*

These anemias are due to reduced production or lack of either Δ-ALA synthetase or heme synthetase. Most cases are caused by lack of Δ-ALA synthetase. Hereditary sideroblastic anemia appears to be sex-linked since it occurs mainly in males but is transmitted by females. This disease may be manifested either during childhood or in adults.

#### *Acquired Sideroblastic Anemias*

The two basic types of acquired sideroblastic anemia are primary and secondary (table 12.2).

**1. Primary acquired sideroblastic anemia.**
This type of sideroblastic anemia is most frequently found in elderly patients, and it is rarely encountered in individuals younger than 60 years of age. The actual etiology is unknown, but it appears to be a metabolic disorder resulting in defective hemoglobin synthesis. Hemoglobin formation is reduced in many erythroid blast cells, and a variable portion of red blood cells are hypochromic, despite the presence of large quantities of iron. Other studies seem to indicate that enzymatic reactions of heme synthesis in the mitochondria may be at fault.

**2. Secondary acquired sideroblastic anemia.**
This type of sideroblastic anemia can occur as a result of the following four different circumstances:

a. **In association with other disorders.** Sideroblastic anemia has been observed in association with a wide variety of hematological diseases, especially with myeloproliferative disorders such as myeloid leukemia. However it has also been associated with rheumatoid arthritis, periarteritis nodosa, and different forms of carcinoma. Why it occurs in some patients with these chronic diseases and not in others remains a mystery.

b. **Excess alcohol.** Chronic alcoholism is perhaps the most common cause of sideroblastic anemia. Alcoholic patients with this type of anemia usually have low serum pyridoxal phosphate levels, and they seem to respond to pyridoxine (vitamin $B_6$) treatment. Apparently the first product of alcohol metabolism, acetaldehyde, enhances the destruction of pyridoxine in red blood cells.

c. **Drug-induced.** Occasionally patients with tuberculosis will develop sideroblastic anemia after 4 to 6 months of therapy with such antituberculosis drugs as isoniazid, cycloserine, and pyrazinamide. All these are pyridoxine antagonists. Other drugs occasionally implicated include chloramphenicol and azathioprine.

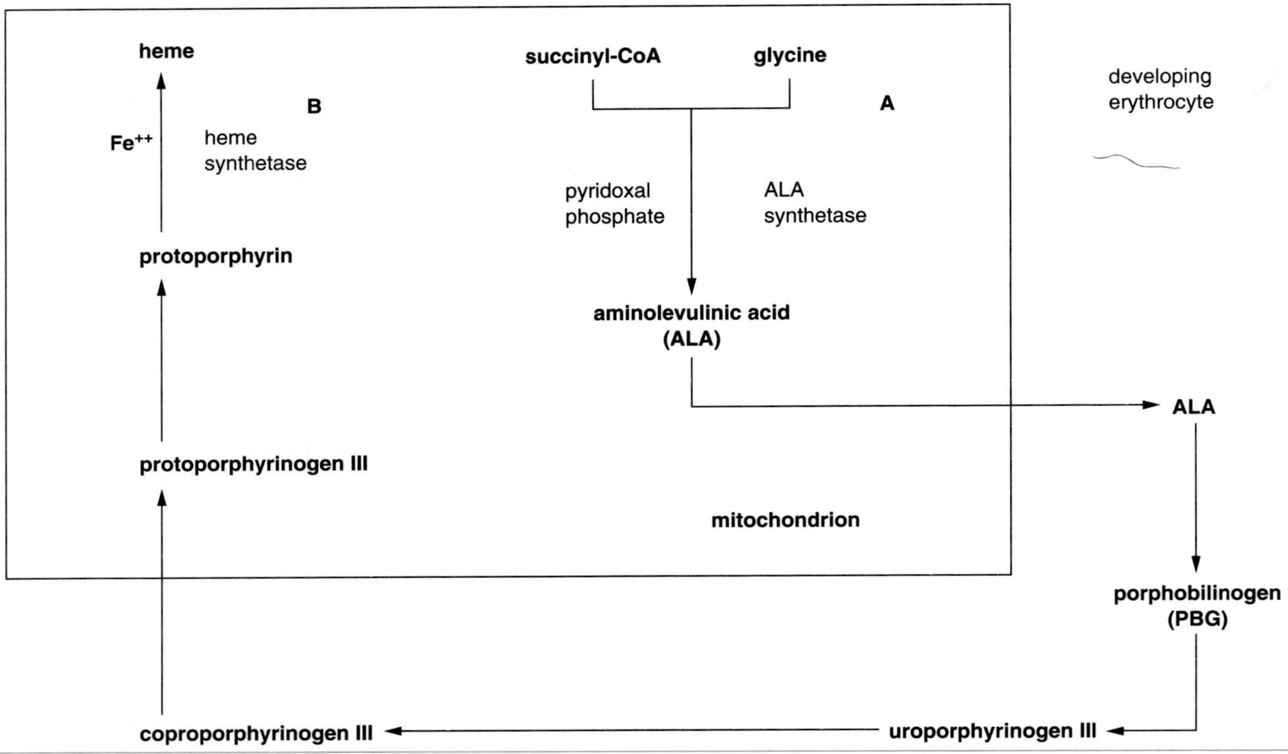


figure 12.3

Abbreviated version of the biosynthetic pathway of heme. In sideroblastic anemia there is frequently decreased activity of ALA synthesis (A) heme synthesis (B), or both.

**Table 12.2 Classification of Sideroblastic Anemias**

*Hereditary Sideroblastic Anemia*

*Acquired Sideroblastic Anemia*

- Primary acquired sideroblastic anemia
- Secondary acquired sideroblastic anemia
  - Those associated with other disorders (e.g., myelosclerosis, myeloid leukemia, periarteritis nodosa, and carcinoma)
  - Excess alcohol
  - Drug-induced (e.g., isoniazid, cycloserine, pyrazinamide)
  - Lead poisoning

d. **Lead poisoning.** Sideroblastic anemia due to lead intoxication is seldom severe in adults, and the hemoglobin levels will rarely go below 10 g/dL. In children, on the other hand, the anemia may be quite severe, with red blood cells more clearly hypochromic. Ringed sideroblasts are common in this type of sideroblastic anemia. Apparently both heme and globin synthesis may be impaired, due to interference by lead in the functioning of enzymes needed for both heme and globin synthesis.

### Diagnosis of Sideroblastic Anemia

Since the onset of sideroblastic anemia is insidious, it is usually discovered only after a routine blood examination, after complaints of weakness, fatigue, dyspnea (shortness of breath), headaches, and the observation of paleness of the skin and mucous membranes.

Laboratory investigations will show many signs of ineffective erythropoiesis. The bone marrow smear will contain increased numbers of erythroid precursor cells (erythroid hyperplasia), since many cells fail to mature properly and die in the bone marrow (intramedullary hemolysis), where they are engulfed by macrophages. Consequently the reticulocyte count is always low. Many of the erythroid blast cells will show excess granules of iron (hemosiderin) when stained with Prussian blue. Many of these abnormal sideroblasts may be ring sideroblasts. Iron-containing granules, also called Pappenheimer bodies, may also be found in the circulating red blood cells.

In hereditary sideroblastic anemia the MCV and the MCHC are always low, but in acquired sideroblastic anemias the MCV is often normal, and the cells may even be slightly macrocytic. In certain forms of this anemia, a dimorphic blood smear may be encountered, showing both microcytic and macrocytic red blood cells. The macrocytosis may result from lack of folic acid, which is often the case in alcoholic patients.

In lead poisoning the red blood cells will show distinct **basophilic stippling** with ordinary Wright's stain. This punctate basophilia is caused by accumulation of denatured ribonucleic acid (RNA) in the erythrocyte due to the interference of lead with the enzymes that normally break down RNA.

### Treatment of Sideroblastic Anemia

Since our knowledge of the underlying causes of sideroblastic anemia is still rather limited, proper treatment is not always easy and in some cases unsuccessful. For instance, the primary acquired type of sideroblastic anemia does not respond to any treatment, and repeated blood transfusions may be necessary to maintain a minimum hemoglobin level. The danger of this type of therapy is, of course, that it may eventually result in fatal hemochromatosis, since more and more iron enters the body, and hardly any is excreted. In some cases this primary type of sideroblastic anemia evolves into leukemia, suggesting a relationship with the myeloproliferative disorders.

Secondarily acquired types of sideroblastic anemia are more easily reversed, once the offending agent is removed (e.g., lead, alcohol, drugs). Some cases, especially those caused by excess alcohol and some inherited types of sideroblastic anemia, show a positive response following vitamin $B_6$ therapy and/or folate treatment.

## THALASSEMIA

Another major group of microcytic-hypochromic anemias due to maturation defects are the thalassemias. The term **thalassemia** comes from the Greek words "thalas," meaning sea, and "emia," which stands for blood. This compound word came into being because this anemia was originally found only in people living in areas bordering the Mediterranean Sea. Thalassemia syndromes are a heterogeneous and complex group of inherited diseases that share certain clinical manifestations such as a reduced MCV and MCHC, ineffective hematopoiesis, and accelerated hemolysis. These symptoms are the result of abnormal hemoglobin synthesis. However thalassemias differ from other microcytic-hypochromic anemias in that the anemia is not caused by abnormalities in the heme synthesis, but the anemia is due to abnormalities in globin synthesis.

### Classification of Thalassemias

In each of the thalassemias the principal biochemical manifestation is a partial or complete (but always selective) deficiency in the production of one of the globin chains.

Normal adult hemoglobin contains two α- and two β-chains. In healthy red blood cell precursors, the α- and β-globin chains are produced in roughly equal amounts. In the thalassemias production of one of these globin chains is deficient, but the formation of the other chain is not affected. Consequently less than the normal amount of adult hemoglobin is produced, and the erythroblasts will have an excess of the unaffected globin chains. Since the two main types of adult globin chains are α- and β-globins, thalassemias may be divided into two major categories—according to the globin chain that is deficient—into **α-thalassemias** and **β-thalassemias.** Each of these categories includes a number of variants that differ in the severity of the anemia they produce.

### Population Patterns of the Thalassemias

The thalassemia trait occurs with high frequency in certain populations. β-Thalassemia is most prevalent in populations that border the Mediterranean Sea (e.g., people from southern Italy, Sicily, Sardinia, Greece, Lebanon, Turkey, and Armenia). α-Thalassemia is more common in the Far East (e.g., China, India, Thailand, Vietnam, and the Philippines). Both α- and β-thalassemia are common in certain African populations as well as in American blacks.

The incidence of these thalassemia syndromes also varies from area to area. In affected Greek and Italian populations, 5 to 10% may be heterozygous for β-thalassemia, whereas in Indochina the gene frequency for heterozygous α-thalassemia may be as high as 20%. The high frequency of thalassemia in certain geographic areas may be attributable to the fact that the heterozygous state for both α- and β-thalassemia confers a biological advantage to the carrier, namely resistance to malarial infection. Thalassemia is inherited as an autosomal dominant trait. If mating occurs between two individuals who are heterozygous for thalassemia, there is a 50% chance that the offspring will also be heterozygous (i.e., have the thalassemia trait), a 25% chance that it will be normal, and 25% chance that it will be fully homozygous for the particular thalassemia, with all its consequences (figure 12.4).

### α-Thalassemias

In this group of diseases, the deficiency is in the synthesis of α-globin. The β-globins are not affected and are produced at their normal rate. The human genome contains two sets of two α-genes—that is, two α-genes on the chromosome inherited from the father, and two α-genes on the chromosomes inherited from the mother. All four of these genes are normally expressed in developing red blood cells, and suppression of all four genes is needed to completely suppress α-chain synthesis. The normal mechanism of suppression of α-genes is by deletion.

#### *Causes of α-Thalassemias*

All α-thalassemias can best be explained by deletion of genes due to the process of **unequal crossing-over.** Since both sets of α-genes are tandemly

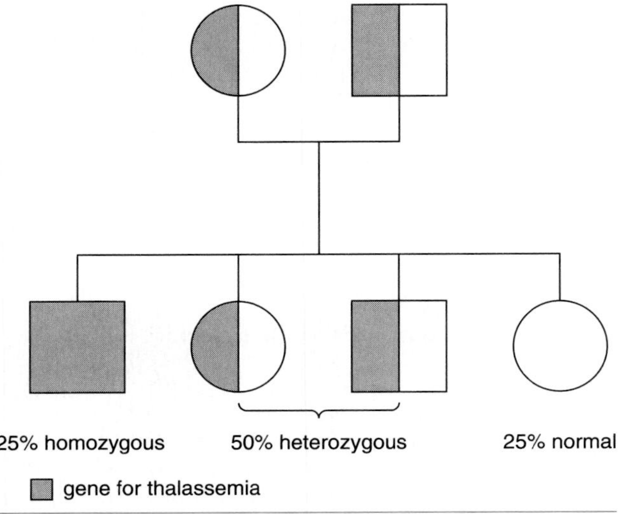


**figure 12.4**
Inheritance of thalassemia as an autosomal dominant trait.

repeated on chromosome 16 (figure 12.5), unequal crossing-over can easily account for the deletion of one or more of these genes. In individuals with one α-globin gene missing, the left-hand portion of the first α-gene is present, the right-hand portion of the second α-gene is present, and the middle of the two-gene sequence is deleted. The simplest explanation for such a deletion pattern is unequal crossing-over (figure 12.6). If unequal crossing-over causes the deletion of one α-globin gene in some individuals, then other individuals should have three α-globin genes on one chromosome. Indeed, a number of such people have been found.

*Classification of α-Thalassemias*

Since people normally have four genes for α-globins, the following four different patterns of α-thalassemia can occur:

1. **One-gene deletion: (–α/αα)**
This pattern is known as the **silent carrier.**
2. **Two-gene deletion: (–α/–α) or (–/αα)**
This pattern is known as possessing the **α-thalassemia trait.**
3. **Three-gene deletion: (–α/–)**
This pattern is known as **HbH disease.**
4. **Four-gene deletion: (–/–)**
This fatal condition is known as **hydrops fetalis.**

A summary of these four types of α-thalassemia is given in table 12.3.

**Silent Carriers** The silent carrier is missing only one functional gene. The three remaining α-genes can direct the production of sufficient α-globin to produce normal quantities of HbF and HbA. In affected newborn infants about 1 to 2% of Hb Barts may be detected, but this disappears within 3 months after birth. Reliable methods for the diagnosis of this silent carrier genotype are not available. As many as 30% of African-Americans may be missing one α-gene.

**α-Thalassemia Trait** This condition occurs when two α-genes are missing. The missing genes may be either on the same or on homologous chromosomes. This condition is common in Southeast Asia, West Africa, and the Mediterranean area. It occurs in about 2% of African-Americans.

The α-thalassemia trait is also asymptomatic as hemoglobin levels usually reach 10 to 12 g/dL. The peripheral blood smear will show considerable microcytosis with many cells having an MCV of 60 to 70 fL. The red blood cells may also appear slightly hypochromic. The imbalance between the α- and β-chain synthesis may result in a slightly excess production of β-globin chains, which may then form tetrameres of HbH. Neonates often show 5 to 6% levels of Hb Barts, which may be a useful diagnostic feature in pinpointing the presence of the trait. Usually Hb Barts disappears after about 3 months. The only persistent hematological feature of the α-thalassemia trait is the microcytic-hypochromic condition, which may lead in some instances to a mild form of anemia.

**Hemoglobin H Disease** In this disorder, also known as **α-thalassemia proper,** three out of four globin genes are absent, resulting in an underproduction of α-globin chains and an overproduction of β-globin polypeptides. These excess β-globin monomers tend to combine to form β-chain tetrameres. Such hemoglobins are classified as HbH; hence the name of this anemia. People with this disorder will have from 5 to 40% of HbH in their blood. In fetuses with this condition, there will be an excess production of γ-chains, since HbF is normally made up of two α- and two γ-chains. Many of these excess γ-chains will also form tetramers known as Hb Barts.

Both Hb Barts and HbH are thermolabile proteins with an oxygen affinity 10 times that of HbA. This high oxygen affinity is due to the lack of heme-heme interaction and the absence of the Bohr effect. These two tetramers mimic the action of myoglobin, which readily takes up oxygen but does not give it up easily to the respiring tissues. Many patients with HbH disease produce a significant amount of normal HbA, and thus the shift is not as extreme as that of myoglobin.

In addition to its high oxygen affinity, HbH is also less stable than HbA and tends to precipitate more readily. This results in the production of Heinz bodies. Such red blood cells are then "pitted" by the mononuclear phagocytic cells of the spleen. This results in considerable poikilocytosis as well as in a shortened life span of these red blood cells.

As less than normal hemoglobin is produced, this anemia is always associated with hypochromia and microcytosis. This is expressed in a lowered MCV and lower MCHC. Hemoglobin levels are usually somewhere between 7 and 11 g/dL, indicating conditions that range from mild to fairly severe anemia. Anemia frequently worsens during pregnancy and during bouts of infectious disease. Splenomegaly is common due to the increased activity of the spleen associated with pitting and destruction of poikilocytes. In a few cases skeletal changes may be seen in

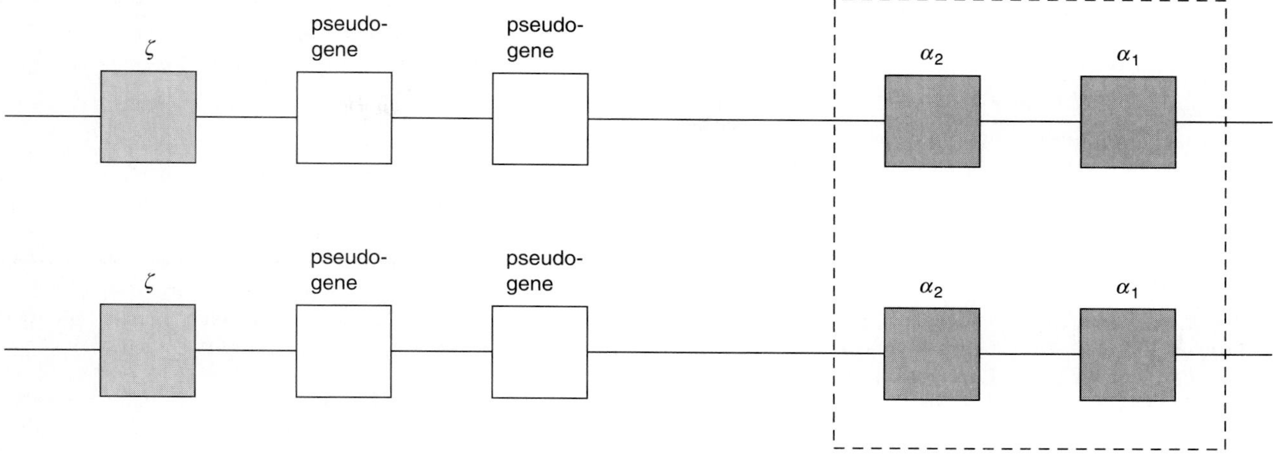


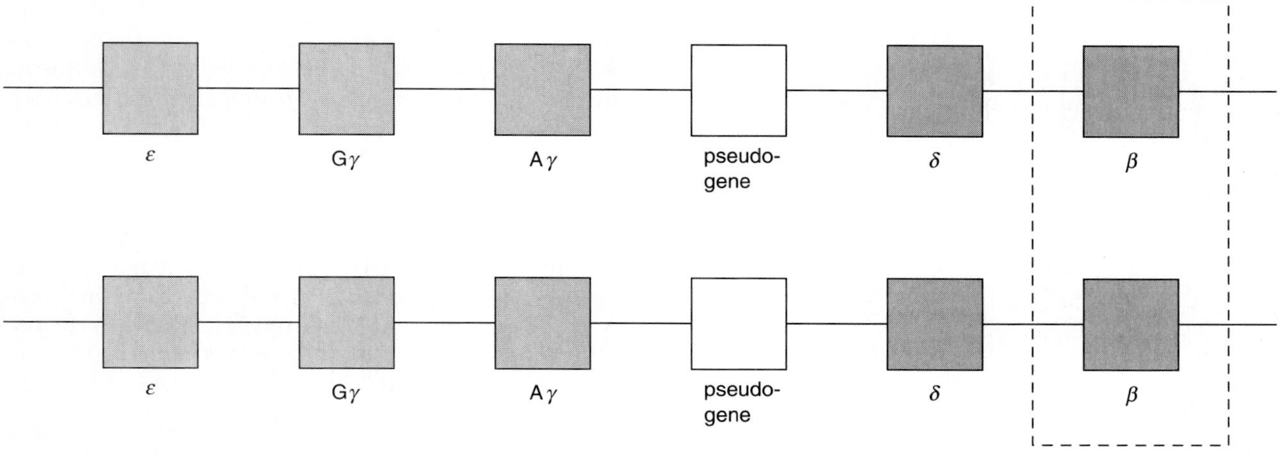


**figure 12.5**

The two families of globin genes: The α-globin family (*a*) and the β-globin family (*b*).

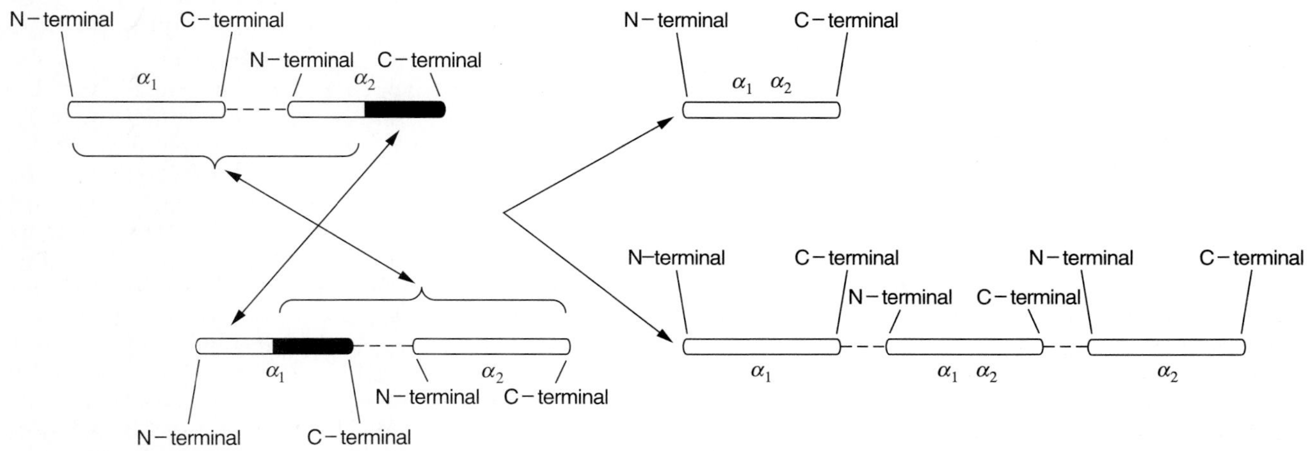


**figure 12.6**

The unequal crossing-over that occurs between $\alpha_1$- and $\alpha_2$-globin genes in α-thalassemia. N and C represent the positions on the DNA that will ultimately code for the N-terminal and the C-terminal amino acids of the protein.

**Table 12.3 Four Forms of Alpha-Thalassemia**

| Genetic Abnormality | | Clinical Syndrome |
|---|---|---|
| *α-thalassemias* | | |
| 4-Gene deletion: | hydrops fetalis | Lethal *in utero* |
| 3-Gene deletion: | hemoglobin H disease | Hemolytic anemia |
| 2-Gene deletion<br>1-Gene deletion | α-Thalassemia trait | Microcytic-hypochromic blood cells but usually no anemia |

the bone marrow due to its hyperactivity. However the latter feature is not common in HbH disease.

The blood smear will show many target cells, poikilocytes, and microcytes. The total red blood cell (RBC) count may be fairly normal due to a moderate increase in reticulocytes (5 to 10% of total RBCs).

A good way to detect the presence of HbH in the blood is by means of electrophoresis of the blood hemoglobin. HbH will show up as a distinct band different from HbA. Another method to detect the presence of HbH is to promote its precipitation in the red blood cell. This can be done *in vitro* by exposing the erythrocytes to the oxidant dye brilliant cresyl blue. Under the microscope the red blood cells will show many punctate (stippled) inclusion bodies.

**Hydrops Fetalis** Since α-globin is needed for both adult (HbA) and fetal (HbF) hemoglobin, it follows that the total deletion of all four α-globin genes leads to a complete failure of HbA and HbF synthesis. This condition is incompatible with life and leads to death either *in utero* or very soon after birth. This phenomenon is known as hydrops fetalis and is seen almost exclusively in Southeast Asia. The hemoglobin of such fetuses consists almost exclusively of Hb Barts with some Hb Portland and HbH. HbF and HbA are completely absent. Hb Barts has a very high oxygen affinity; hence it is almost useless as an oxygen carrier. Thus the fetus usually dies of hypoxia.

*Treatment of α-Thalassemia*

People who are silent carriers for α-thalassemia, as well as those with the α-thalassemia trait, usually do not require any special treatment because no overt anemia is associated with these conditions. Patients who suffer from HbH disease, on the other hand, frequently exhibit mild to moderate anemia as their hemoglobin levels are usually between 7 and 11 g/dL. The body can adjust readily to these levels in the absence of strenuous exercise. Blood transfusions are usually not necessary, but may be required in a sudden hemolytic crisis due to the ingestion of an oxidant drug. Transfusions may also be necessary to prevent mental and physical retardation when young children experience moderate to severe anemia.

Because of the presence of many abnormal red blood cells, splenomegaly is common, and splenectomy may be called for. However this procedure is risky when it is done early in life because the spleen is an important immune organ in young children.

## β-Thalassemias

β-Thalassemias are characterized by a deficiency in β-globin chain synthesis. Since humans have only one set of β-globin genes, total or partial suppression of the expression of either or both β-genes will also result in low hemoglobin production. The underlying mechanisms causing β-thalassemia are completely different from those causing α-thalassemia. α-Thalassemias are the result of gene deletions, whereas β-thalassemias are caused by deficiencies in the regulation of the β-genes.

*Causes of β-Thalassemias*

More and more evidence is becoming available showing that β-thalassemia is usually caused by mutations affecting the normal transcription of β-globin genes. A number of patients with severe β-thalassemia produce normal β-globins but in abnormally small quantities. This type of β-thalassemia is referred to as **$\beta^+$**-thalassemia. The anemia associated with this type of thalassemia may range from mild to severe, depending on how much β-globin is made. Stated another way, the degree of anemia depends on how severely the regulation of β-globin gene transcription is disturbed. In other patients with β-thalassemia, the transcription process is so disturbed that no viable messenger RNA (mRNA) is produced. This type of β-thalassemia is known as **$\beta^0$-thalassemia.**

β-Thalassemia appears to result from defects in the intervening sequences that are normally present between two or more regions of deoxyribonucleic acid (DNA) that make up a gene. When a gene is expressed, the entire DNA segment, including the intervening sequences, is transcribed into large mRNA precursors. These large mRNA precursors are then processed into functional mRNA by a number of enzymes that remove the intervening sequences and splice the proper pieces together (figure 12.7). Apparently this process is not very efficient or does not work at all in β-thalassemia. In $B^+$-thalassemia precursor mRNA copies of the β-globin gene with their intervening sequences pile up in the developing red blood cells, but few functional mRNA copies are produced. Consequently few β-globin proteins are made.

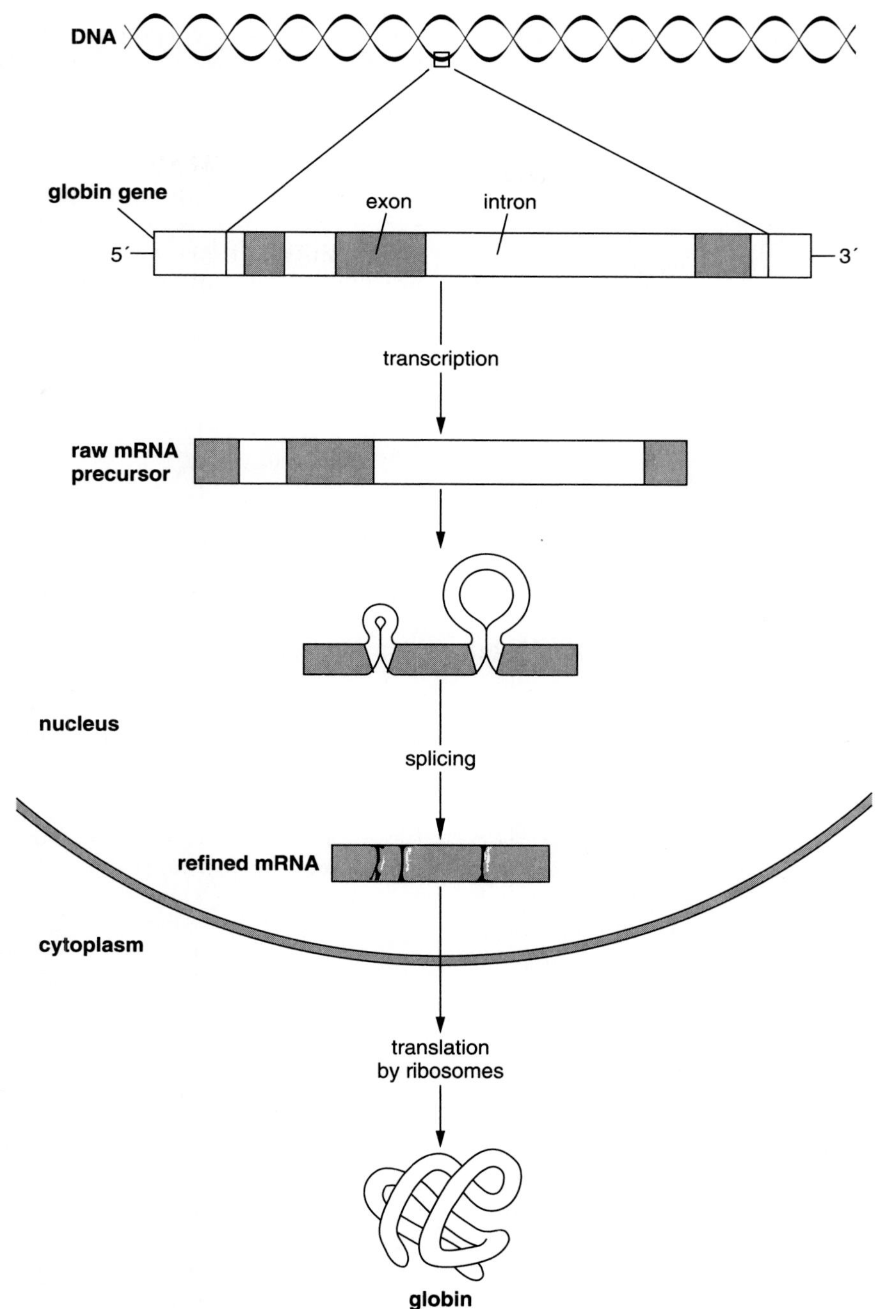


**figure 12.7**

The schematic representation of the biosynthesis of a globin chain. Note that the process begins with the transcription of the entire gene, including the exons and introns. During further processing, the introns are removed to form refined mRNA, which is then translated by the ribosomes into proteins.

The difficulty then is in processing the precursor mRNA into the proper mRNA needed for β-globin protein production. Presumably there are mutations in the intervening sequences of the β-globin genes at sites normally recognized by the enzymes that cut and splice these precursor mRNAs. Because of these mutations the cutting and splicing process has become inefficient. It has been postulated that a number of different mutations affect mRNA processing to various degrees, which would explain why some patients are more anemic than others.

Patients with $\beta^0$-thalassemia may have intervening sequence mutations that completely prevent cutting and splicing the mRNA precursor. However $\beta^0$-thalassemia may also be associated with gene deletions, errors in transcription, or premature termination of protein synthesis. The latter situation will result in the production of abnormal nonfunctional β-globin chains.

*Classification of β-Thalassemias*

Customarily the following three different forms of β-thalassemia are recognized:

1. **β-Thalassemia major** (or $\beta^0$-thalassemia). In this form of the disorder, β-globin chains are not produced, resulting in a complete absence of $HbA_1$.
2. **β-Thalassemia intermedia** (or $\beta^+$-thalassemia). In this condition some β-chains are produced, resulting in varying amounts of $HbA_1$.
3. **β-Thalassemia minor** (or heterozygous β-thalassemia). In this condition at least one normal β-globin gene is present. Thus considerable amounts of $HbA_1$ are being produced.

   In those instances in which very few or no β-globins are produced, very little $HbA_1$ will be present: these conditions are always associated with severe anemia. The resulting anemia is also known as **Cooley's anemia** or **Mediterranean anemia.**

**β-Thalassemia Major** β-Thalassemia major is associated with a complete lack of β-globin production. It is normally detected very early in childhood, when production of γ-chains (which are part of HbF) subsides, and the β-globin chain production normally starts to produce $HbA_1$.

As a result of low hemoglobin production, severe anemia will result rapidly. This hemoglobin deficiency produces characteristic changes in the morphology of red blood cells: microcytosis, hypochromia, anisocytosis, and poikilocytosis. The presence of many target cells and teardrop cells is also common. Further, many red blood cells in the circulation show numerous Heinz bodies as a result of a continued synthesis of α-chains, in the absence of β-chain synthesis. This results in an accumulation of free α-chains, which do not form tetramers as the excess β- and γ-chains do. On the contrary, free α-chains are unstable, denature easily, tend to precipitate, and readily aggregate into Heinz bodies. These are then "pitted" from the red blood cells by mononuclear phagocytes of the spleen, and in doing so they produce abnormally shaped red blood cells with much shorter life spans. Since these inclusion bodies are found not only in mature red blood cells but also in immature precursor cells, phagocytic cells of the bone marrow may damage these developing cells just as phagocytic cells of the spleen do. As a result many developing cells are destroyed before they reach the circulation. Thus this anemia is due not only to accelerated hemolysis of circulating erythrocytes but also to destruction associated with ineffective erythropoiesis.

**β-Thalassemia Intermedia** This condition is due to the partial suppression of the β-globin genes; hence it is also called $\beta^+$-thalassemia. Symptoms may be major and mimic the symptoms of β-thalassemia major, or they may be relatively minor, depending on the severity of the defect in β-chain production. The more β-globin produced, the more $HbA_1$ will be present in the red blood cells, and fewer overt symptoms of the anemia will be present. The degree of anemia will also depend on the extent of compensation by HbF and $HbA_2$, as well as the severity of the unbalanced iron metabolism resulting from blood transfusions.

**β-Thalassemia Minor** The heterozygous state of β-thalassemia is a condition in which the individual has at least one normal functioning β-globin gene. Hence sufficient quantities of $HbA_1$ are produced to prevent overt symptoms of anemia. In this condition hemoglobin levels will be only slightly below normal. The blood smear, however, will show a similar morphology as in β-thalassemia major or intermedia: hypochromia, microcytosis, poikilocytosis, target cells, teardrop cells, and the presence of stippled erythrocytes.

The principal diagnostic feature of this β-thalassemia trait is the elevated level of $HbA_2$, which is about 5% compared with the normal 2%. This feature also distinguishes β-thalassemia from α-thalassemia, where the $HbA_2$ levels are normal.

*Compensatory Mechanisms in β-Thalassemias*

Since anemia associated with $\beta^0$-thalassemia and most cases of $\beta^+$-thalassemia is so severe, the body tries to counteract this potential lethal condition in several ways as follows:

1. Many patients with β-thalassemia major and intermedia *continue to produce HbF after birth.* Normally the production of γ-globin is switched off soon after birth, and hence the production of HbF stops. For some reason, patients with β-thalassemia major and intermedia are able to continue to produce some HbF after birth, although the total amount of HbF produced is insufficient to replace all the missing $HbA_1$. Much current research is attempting to find out what causes the production of γ-chains to be turned on and off. If scientists understood those mechanisms, they perhaps could instruct the body to keep on producing high levels of γ-chains in people with these β-thalassemia disorders and thus could eliminate the severe anemia associated with this disease.
2. The body tries to compensate further by increasing the amount of δ-globin chains, which would result in *increased amounts of $HbA_2$.* Normally the amount of $HbA_2$ is only 2% of the total amount of hemoglobin, but in β-thalassemia this amount may be increased to 5 to 6% of the total hemoglobin content. However not enough δ-globin chains are produced to prevent the precipitation of excess α-chains in the form of Heinz bodies.
3. In an effort to produce more red blood cells, the *bone marrow undergoes erythroid hyperplasia.* This process is so intense that it may cause severe medullary expansion and remodeling of the bone, especially the bones of the skull. The facial appearance of a child with homozygous β-thalassemia commonly shows enlargement of the frontal, parietal, and maxillary bones, giving the face an almost mongoloid appearance, as shown in figure 12.8.

4. Since the bone marrow is unable to produce sufficient numbers of red blood cells, the body reverts back to *extramedullary hematopoiesis,* resulting in an enlargement of the liver and the spleen. The spleen is also enlarged due to accelerated red blood cell destruction.

*Treatment of β-Thalassemia*

Since the anemia associated with $\beta^0$-thalassemia and most cases of $\beta^+$-thalassemia usually starts soon after birth, poor growth and development of these infants will result due to severe chronic anemia. The only way this stunted growth and mental retardation may be prevented is by regular blood transfusions. However this treatment accelerates the development of life-threatening hemosiderosis, as a result of the increased iron levels in the body. Each 500 mL of transfused blood contains 250 mg of iron. Our bodies have little ability to eliminate this excess iron. This excess iron is deposited in many vital organs such as (1) the liver, causing hepatic cirrhosis; (2) the islets of Langerhans, causing diabetes mellitus; (3) the gonads, causing delayed or absent puberty; and (4) the heart, causing arrhythmias and congestive heart failure. The latter effect—heart failure—is the normal cause of death in individuals who have survived early childhood with blood transfusions.

Despite these dangers regular blood transfusions are presently the only treatment to save these children from severe mental and physical defects and death. Hopefully in the near future, bone marrow replacement in fetuses and newborns with normal hematopoietic tissue will be a more satisfactory solution for this condition. But presently regular blood transfusions are the only way to keep hemoglobin levels to about 12 to 14 g/dL in these young children. This requires 2 to 3 U of blood every 4 to 6 weeks. Normally only washed blood—that is, blood containing only red blood cells and no platelets or white blood cells—is given to reduce the possibility of immune reactions. If blood transfusions are not given, death in early childhood will be almost inevitable. If the syndrome is discovered early in life, many of the symptoms just mentioned can be avoided. If the hemoglobin level can be maintained at about 12 to 14 g/dL by repeated transfusions, marrow hyperplasia is repressed and facial changes do not develop. Also there will be better growth and development, less splenomegaly and hepatomegaly, and a better quality of life. However frequent blood transfusions inevitably lead to iron overload, and this may also cause death eventually due to hemosiderosis. Recently new techniques and treatments have been developed to counteract this iron overload. These treatments involve the use of **iron chelation therapy.** Iron is chelated by such compounds as **desferrioxamine,** which can be given with the blood transfusion or by subcutaneous infusions. In persons with heavy iron overload, excretion rates of up to 200 mg of iron per day can be achieved. (Remember the normal loss is 1 to 2 mg of iron per day.) It appears that vitamin C increases the excretion rate of iron by desferrioxamine. With an intense iron chelation regimen, the outlook for these children has markedly improved.

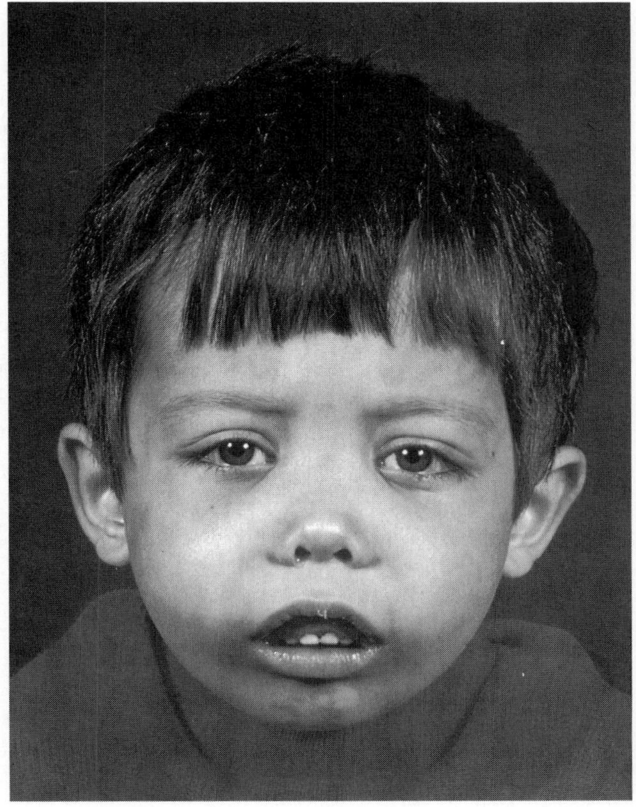

figure 12.8

The facial manifestations of β-thalassemia.

One of the major side-effects of β-thalassemia major is splenomegaly. This enlargement of the spleen increases the total blood volume, and splenectomy may be needed at some time to reduce the blood requirements. This procedure is usually not performed before the child is at least 6 years of age, since it increases the risk of infection, as the spleen is an important organ of the immune system.

The heterozygous state of β-thalassemia (β-thalassemia minor) requires no special medical attention since anemia is usually absent and if present is always very mild. However those with the β-thalassemia trait are carriers and can produce homozygous states in their offspring. If two carriers marry, their offspring have a 25% chance of being homozygous for either β-thalassemia major or intermedia. The transmission of this disease can be halted only by genetic counseling.

**Variations of α- and β-Thalassemias**

Both α- and β-thalassemia conditions may be exacerbated by the presence of abnormal hemoglobins such as HbS, HbC, HbD, and HbE, as well as by nutritional

**Table 12.4 Summary of Laboratory Features of Various Forms of Microcytic Anemia**

| | Iron Deficiency | Chronic Inflammation or Malignancy | Thalassemia Trait (α or β) | Sideroblastic Anemia |
|---|---|---|---|---|
| MCV, MCH, MCHC | All reduced in relation to severity of anemia | Low normal or mild reduction | All reduced very low for degree of anemia | Low in congenital type but MCV often raised in acquired type |
| Serum iron | Reduced | Reduced | Normal | Raised |
| TIBC | Raised | Reduced | Normal | Normal |
| Serum ferritin | Reduced | Normal | Normal | Raised |
| Bone marrow iron stores | Absent | Present | Present | Present |
| Erythroblast iron | Absent | Absent | Present | Ring forms |
| Hemoglobin electrophoresis | Normal | Normal | $HbA_2$ raised in β form | Normal |

deficiencies such as lack of folic acid or vitamin $B_{12}$, or by the presence of abnormal glucose-6-phosphate dehydrogenase (G-6-PD) genes. All these conditions may increase the condition of anemia.

## DIFFERENTIAL DIAGNOSIS OF MICROCYTIC-HYPOCHROMIC ANEMIAS

Since microcytosis and hypochromia, and therefore low MCVs and low MCHCs, are common features in all these anemias, further investigations are often needed to pinpoint the specific type of anemia. Because MCV and MCHC values are not very helpful in this situation, other parameters are used to determine the type of microcytic anemia. Three iron indices have been very helpful: (1) the serum iron level; (2) the UIBC and TIBC; and (3) the serum ferritin levels. All three are excellent parameters to distinguish among the four major types of microcytic-hypochromic anemias discussed in this and the previous chapter.

Table 12.4 provides a summary of the laboratory features associated with the various forms of microcytic anemia. We can see that the serum iron level is reduced in IDA and anemia of chronic disease and raised in sideroblastic anemia. The TIBC, on the other hand, is increased only in IDA and reduced in anemia of chronic disease. Serum ferritin levels are reduced in IDA, raised in sideroblastic anemia, and are normal in anemia of chronic disease and thalassemias.

## *Further Readings*

Bank, A. Genetic Disorders of Hemoglobin Synthesis. *Hospital Practice,* Sep. 1985.

Beck, W. S. Hypochromic Anemias II. Heme Metabolism and the Porphyrias. Lecture 7 in *Hematology,* 4th ed., W. S. Beck, Ed., The MIT Press, Cambridge, Mass., 1985.

Crosby, W. S. Hemochromatosis: Current Concepts and Management. *Hospital Practice,* Feb. 1987.

Kolata, G. B. Thalassemias: Models of Genetic Diseases. *Science,* 210 (Oct. 1980).

McCullough, J., and Jacob, H. S. The Chronically Transfused Patient: Advances in Treatment. *Hospital Practice,* July 1985.

Nathan, D. G. The Thalassemias. Lecture 10 in *Hematology,* 4th ed., W. S. Beck, Ed., The MIT Press, Cambridge, Mass., 1985.

Pearson, H. A. Splenectomy: Its Risks and Its Roles. *Hospital Practice,* Aug. 1980.

# Review Questions

1. List the three major causes of microcytosis and hypochromia associated with heme synthesis. What anemia is associated with each cause?
2. Name the anemias associated with abnormalities in globin synthesis.
3. How common is anemia of chronic disease? How severe is this anemia usually?
4. Name the major disorders that cause anemia of chronic disease.
5. List the three major underlying mechanisms that produce anemia of chronic disease.
6. Describe some common laboratory features unique to anemia of chronic disease.
7. What is the best treatment for anemia of chronic disease?
8. How would you define sideroblastic anemia? What is another name for sideroblastic anemia?
9. Name the two major causes of sideroblastic anemia.
10. What are the two major categories of sideroblastic anemias?
11. How would you distinguish between primary and secondary acquired sideroblastic anemia?
12. Name the major causes of secondary acquired sideroblastic anemia.
13. Describe some of the unique laboratory features of sideroblastic anemia.
14. What are some of the more common treatments for sideroblastic anemia?
15. What is the origin of the term thalassemia?
16. Name the two major classes of thalassemia. In what populations does each class predominate?
17. List the four variations of α-thalassemia. Why are there four possible variations? What is the major cause of these variations? Which variations do not normally result in anemia? Which of these variations is incompatible with life? Why? What is it called?
18. Define HbH disease. Why is it called HbH disease?
19. Name some of the common laboratory features associated with HbH disease.
20. Why is special treatment rarely required for HbH disease?
21. Under what circumstances may blood transfusions be needed in HbH disease?
22. List the three major variations of β-thalassemia. Which of these are clinically significant?
23. Give the major cause for β-thalassemia syndromes.
24. Explain why β-thalassemia major is a much more severe disorder than HbH disease.
25. What are some of the major compensatory mechanisms produced by individuals with $\beta^0$- and $\beta^+$-thalassemia to increase their hemoglobin levels?
26. At what age does β-thalassemia major become apparent (i.e., begin to show overt symptoms)? Why?
27. Name the current treatment for patients with β-thalassemia major. Give a possible future treatment.
28. What problem is associated with repeated blood transfusions? How can that problem be overcome?

# Chapter Thirteen

## *Anemias Due to Increased Erythrocyte Loss: Hemorrhagic and Hemolytic Anemias*

**Table 13.1** Compensatory Reactions that Occur in the Body After an Episode of Severe Acute Bleeding

| Reaction | |
|---|---|
| Tachycardia | Increased |
| Vasoconstriction (especially of veins) | Increased |
| Thoracic pumping | Increased |
| Movement of interstitial fluid into the capillaries | Increased |
| Secretion of catecholamines (especially epinephrine) | Increased |
| Secretion of vasopressin | Increased |
| Secretion of glucocorticoids | Increased |
| Secretion of renin and aldosterone | Increased |
| Plasma protein synthesis | Increased |
| Formation of erythropoietin | Increased |

There are two major causes for a shortened red blood cell life span as a result of increased erythrocyte loss: one is **bleeding** and the other is **accelerated hemolysis.** The former results in a disorder known as **hemorrhagic anemia,** the latter is defined as **hemolytic anemia.** This chapter first discusses anemia associated with excess bleeding and is followed by an introduction of the general features of hemolytic anemia. The actual disorders that produce hemolytic anemia are discussed in chapter 14 (hereditary hemolytic anemias) and chapter 15 (acquired hemolytic anemias).

## HEMORRHAGIC ANEMIAS

The first question that should be asked when anemia is suspected is, Is it caused by bleeding? Bleeding, or hemorrhage, may be divided into two major categories: acute and chronic bleeding. Each of these conditions presents different problems and needs to be approached differently from the clinical point of view.

### Acute Bleeding

Acute hemorrhage is associated with physical trauma and is easily observed when the injury is external. Internal bleeding, however, may be more difficult to spot immediately. Acute hemorrhage results in a sudden sharp drop of the total blood volume, which may produce immediate severe anemia. However this type of anemia cannot be detected in the laboratory until several hours after the bleeding has stopped, since both plasma and blood cells have been lost proportionately. On the contrary, many overt physical signs may be observed readily because cardiovascular functions will change almost immediately in an effort to compensate for the loss of blood volume.

A large loss of blood is always followed by a sharp drop in blood pressure, increased heart rate, rapid breathing, and intense thirst—all of which are symptoms of what is officially known as the **hypovolemic** or **hemorrhagic shock syndrome.** A complete list of compensatory reactions as a result of acute bleeding is given in table 13.1. Within hours after the hemorrhage, there will be a great influx of interstitial fluid into the bloodstream, but it may take several days after a single acute hemorrhage for the blood volume to return to normal. During this adjustment period the hemoglobin levels, the hematocrit, and the red blood cell values will fall progressively. Replenishment of circulating red blood cells will start within a few days, but it may take many weeks before the blood cells are back to their original level. When acute traumatic blood loss is great, it is advisable to compensate for this drastic loss by the transfusion of one or more units of donor red blood cells. One unit of concentrated packed red blood cells will elevate the hemoglobin level about 1 g/dL and increase the hematocrit value about 3%. A patient may be capable of tolerating low levels of hemoglobin for short periods of time (8 to 10 g/dL) provided the individual does not exercise strenuously. Moreover the low levels of hemoglobin will act as a stimulus for the production of new red blood cells. A sudden drop in red blood cells will result in a sharp decrease in oxygen delivery to the body tissues (hypoxia). Decreased levels of oxygen in the blood will stimulate the kidneys to produce more erythropoietin, which then will stimulate the bone marrow to produce more erythrocytes. Increased erythropoietin (EPO) levels are usually evident within hours after the blood loss (figure 13.1), but evidence of hyperproliferation of the bone marrow does not appear until 7 to 10 days later. Increased numbers of reticulocytes begin to appear 4 to 7 days later, achieving a maximum concentration about 10 days after the initial stimulation by erythropoietin (figure 13.2). The reticulocyte response is also dependent on the presence of iron stores, which vary greatly with the age, health, and sex of the patients. The total iron stores in humans are quite limited, and it may be necessary to give iron replacement therapy during recovery. Iron-poor patients will produce hypochromic-microcytic reticulocytes and red blood cells.

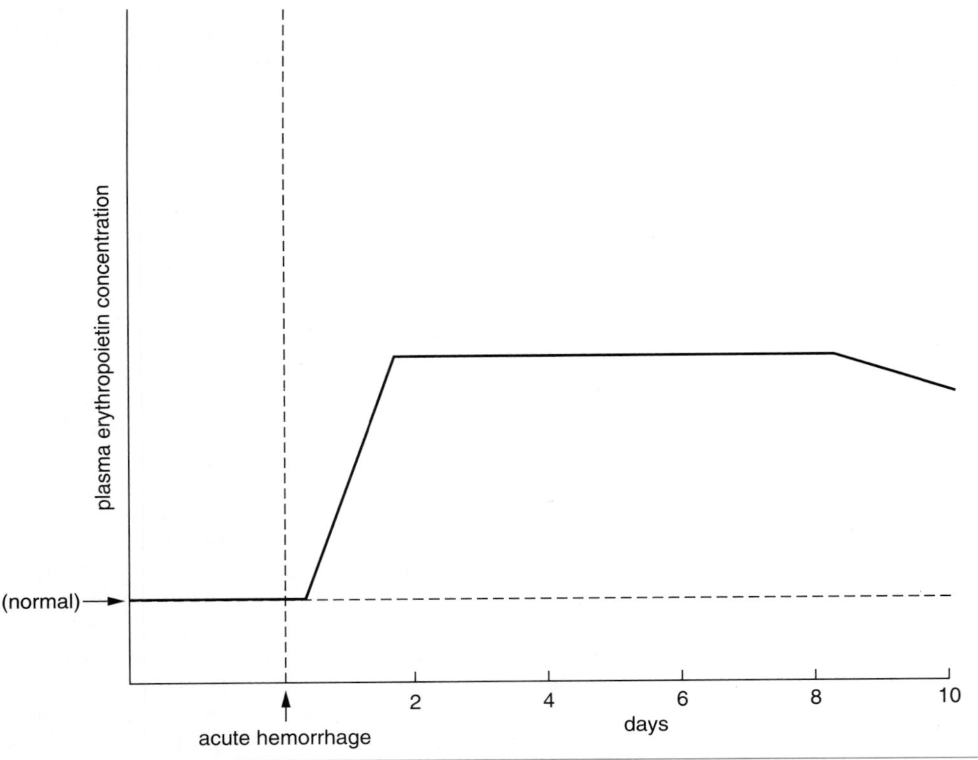


**figure 13.1**

Erythropoietin (EPO) response after severe, acute, uncorrected blood loss.

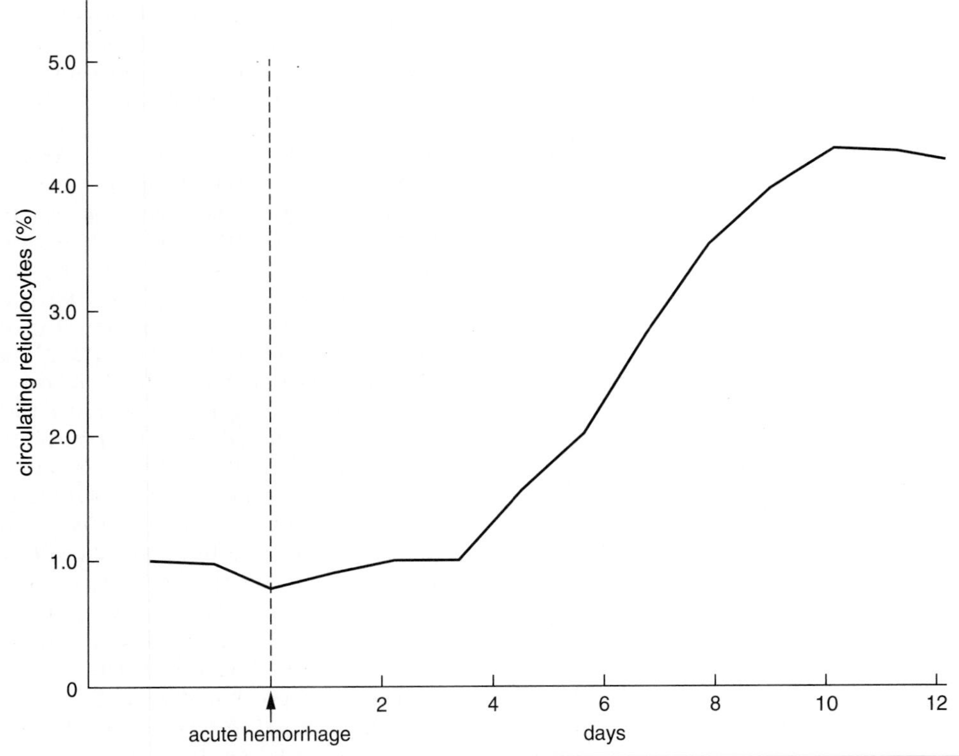


**figure 13.2**

Reticulocyte levels after uncorrected acute blood loss.

When the hemorrhage is internal—that is, into the tissues—it is likely that macrophages will digest the red blood cells, break down the hemoglobin, and recycle the iron. Thus in internal bleeding there is usually a better response by the bone marrow because of the availability of iron.

However acute bleeding does not result only in loss of red blood cells but also white blood cells and platelets. Usually the spleen contains a reserve of platelets, and if the spleen is not damaged, the clotting functions of the body should quickly respond back to normal. Loss of white blood cells, however, may take longer to replace and hence may make the patient slightly more prone to infections during that period.

### Chronic Bleeding

Chronic low-grade bleeding is one of the major causes of anemia. The three most common types are (1) hemorrhaging into the gastrointestinal tract from gastric or duodenal ulcers, (2) excessive bleeding associated with severe and prolonged menstruation, and (3) bleeding associated with intestinal malignancies. These phenomena may go unnoticed for a long time by the person suffering from this chronic blood loss. Although excessive menstrual bleeding is obvious, many women tend to disregard its seriousness. Similarly, although gastric and duodenal ulcers are frequently associated with considerable pain, many sufferers will try to ignore them for long periods of time. To determine whether someone has intestinal bleeding, examination of stool specimens for traces of blood is necessary.

Chronic blood loss is usually associated with a depletion of the iron stores, especially in women. Consequently the symptoms of iron deficiency anemia will be present, including microcytosis and hypochromia, resulting in low red blood cell indices. Chronic bleeding is usually diagnosed by physicians as a result of individuals describing such symptoms as fatigue, dizziness, and headaches.

## INTRODUCTION TO HEMOLYTIC ANEMIAS

Anemias due to an accelerated destruction of the peripheral red blood cells are commonly referred to as **hemolytic anemias.** They may be defined as anemias resulting from a shortened life span of erythrocytes in the circulation. We should note, however, that this definition is not an exact one. Many of the anemias discussed in previous chapters also show accelerated hemolysis—for example, megaloblastic anemia, sideroblastic anemia, and thalassemia. However increased hemolysis of the anemias discussed previously is primarily due to ineffective erythropoiesis. In these cases accelerated hemolysis is only a secondary result of primary developmental defects. In true hemolytic anemias, the bone marrow produces sufficient numbers of normal-looking red blood cells with adequate levels of hemoglobin to satisfy the normal needs of the body. However for some intrinsic or extrinsic reason, large numbers of red blood cells are destroyed, sometimes suddenly, long before the end of their normal life span. Because the body has a tremendous capacity for compensatory action, many cases of hemolytic anemia are frequently without major overt symptoms. Although red blood cell counts may be fairly normal, the reticulocyte count drastically increases. This is known as **compensated hemolytic disease,** a sign of increased bone marrow activity. At some point, however, the bone marrow will no longer be able to compensate, either because it lacks the essential building components or because the breakdown rate becomes greater than the production rate of new erythrocytes. In these situations overt hemolytic anemia will eventually become apparent.

### Classification of Hemolytic Anemias

All hemolytic anemias may be classified according to their etiology into two major groups: (1) **Hereditary hemolytic anemias,** which are the result of some intrinsic defect in the red blood cells and (2) **Acquired hemolytic anemias,** which are the result of one or more extrinsic factors acting on erythrocytes in the peripheral circulation thus shortening their life span.

### Etiology of Hemolytic Anemia

The causes of hemolytic anemia may be hereditary or acquired. The causes of hereditary hemolytic anemia are grouped into three categories: (1) those caused by defects in the red blood cell membrane; (2) those due to defects in erythrocyte metabolism; and (3) those caused by the presence of abnormal hemoglobins. As shown in table 13.2, a number of different anemias are associated with each of these categories of inherited red blood cell abnormality.

Classifying the causes of the acquired hemolytic anemias is not as easy, as they are many and varied. In this text the causes have been summarized into the following six major categories:

1. One of the major causes of acquired hemolytic anemia is the *inappropriate activation of the body's immune system.* These anemias are referred to as **immunohemolytic anemias** and can be caused by either alloantibodies or autoantibodies, as is explained later.
2. Another common cause is the *ingestion of drugs and chemicals* which may interfere with the structure and functions of red blood cells and thus prepare them for early destruction.
3. A number of *physical factors* may result in premature erythrocyte hemolysis. For instance, increased resistance and friction in the circulatory system due to inflammation of blood vessels (vasculitis) or the presence of blood clots (phlebitis) may result in accelerated hemolysis of red blood cells. The same thing could happen in the presence of defective vascular protheses, such as faulty heart valves.

## Table 13.2 Classification of Hemolytic Anemias

*Hereditary Hemolytic Anemias*
- Defects of the cell membrane
  - Hereditary spherocytic anemia (HSA)
  - Hereditary elliptocytic anemia (HEA)
- Defects in erythrocyte metabolism
  - G-6-DP deficiency anemia
  - Pyruvate kinase (PK) deficiency anemia
- Abnormal hemoglobins
  - Sickle cell anemia
  - Hemoglobin C disease
  - Hemoglobin D disease
  - Hemoglobin E disease

*Acquired Hemolytic Anemias*
- Immunological destruction of red blood cells
  - Transfusion with incompatible blood
  - Hemolytic disease of the newborn
  - Autoimmune hemolytic anemia (AIHA)
    (warm-active AIHA and cold-active AIHA)
- Physical destruction of red blood cells
  - March hemoglobinuria
  - Traumatic cardiac hemolytic anemia
  - Microangiopathic hemolytic anemia
- Hemolytic anemia induced by chemical agents
- Hemolytic anemia caused by microorganisms
  - Anemia of malaria
  - Anemia of clostridia
- Hemolytic anemia secondary to other diseases
- Paroxysmal nocturnal hemoglobinuria (PNH)

4. *Infectious diseases* may also be the cause of accelerated erythrocyte hemolysis. Such infectious agents as the malaria parasite *Plasmodium falciparum* routinely use red blood cells for their propagation and in the process destroy them.
5. Hemolytic anemia could also develop as a *secondary effect of certain clinical conditions* such as splenomegaly, liver disease, and renal failure.
6. Last, the enigmatic disorder known as *paroxysmal nocturnal hemoglobinuria* (PNH) is considered by many to be an acquired hemolytic anemia. The etiological basis for PNH is still unknown, but evidence points to a disturbance of the hematopoietic stem cell because all blood cells show evidence of a similar membrane defect. PNH may be due to some unknown environmental influence or drugs that produce a certain somatic mutation.

### Clinical and Laboratory Features of Hemolytic Anemia

Although there are many different kinds of hemolytic anemia, the overt symptoms are the same as most other anemias and include such features as weakness, fatigue, dyspnea, palpitations, dizziness, headaches, and inability to concentrate. Frequently these anemias are also associated with pallor of the skin and mucous membranes. Often some degree of jaundice and splenomegaly is present.

The following laboratory features are indicative of hemolytic anemia:

**1. Accelerated hemolysis**
Indications of accelerated hemolysis are usually shown by increased levels of bilirubin in the blood, and high levels of urobilinogen in the urine.

**2. Increased erythropoiesis**
Higher levels of reticulocytes in the peripheral blood reflect bone marrow hyperplasia.

**3. Damaged red blood cells**
Many hemolytic anemias will exhibit abnormally shaped red blood cells such as sickled cells, spherocytes, and poikilocytes. Many of the erythrocytes have an increased osmotic fragility.

**4. Intravascular hemolysis**
In the normal individual less than 10% of red blood cells are destroyed in the blood vessels. Most destruction takes place in the spleen. In many hemolytic anemias, however, the level of intravascular hemolysis is drastically increased. When this happens we see the following unique symptoms:

a. **Absence of haptoglobin in the blood.** In moderate to severe hemolytic anemia, all haptoglobin will be bound to the free hemoglobin in the blood. The resulting haptoglobin-hemoglobin complexes are rapidly removed from the circulation by the cells of the reticuloendothelial system.

b. **Presence of free hemoglobin in the blood.** This condition is known as **hemoglobinuria.**

c. **Presence of free hemoglobin in the urine.** This condition is known as **hemoglobinuria.**

d. **Presence of methemalbuminemia.** Some of the free plasma hemoglobin is taken up by the liver cells, where it is oxidized to the trivalent form and released back into the bloodstream. This methemoglobin binds readily to plasma albumin, thus forming methemalbumin.

e. **Decreased red blood cell life span** as determined by the $^{51}Cr$ test. This test may be a useful indicator in certain patients who do not show obvious changes in the preceding parameters.

## Further Readings

Jandl, J. Hemolytic Anemias I. Introduction. Lecture 11 in *Hematology,* 4th ed., W. S. Beck, Ed., The MIT Press, Cambridge, Mass., 1985.

Swedberg, J., Driggers, D., and Johnson, R. Hemorrhagic Shock. *American Family Physician,* July 1983.

## Review Questions

1. Name the two major causes of increased erythrocyte loss.
2. Explain why hemorrhagic anemia cannot be detected in the laboratory until several hours after blood loss has occurred.
3. Explain what is meant by the hemorrhagic shock syndrome. What major compensatory mechanisms are employed by the body to counteract the effects of rapid blood loss?
4. Describe the levels of erythropoietin and reticulocytes in the days following a drastic blood loss.
5. How many units of packed red blood cells are needed to raise the blood hemoglobin level by 3 g/dL?
6. Explain why the body can compensate more readily for internal blood loss than for external blood loss.
7. Explain why people who have had a severe blood loss are much more prone to infections.
8. Name the two most common types of chronic bleeding.
9. Explain why chronic blood loss often results in hypochromic-microcytic anemia.
10. Explain why severe chronic blood loss may show few overt symptoms of anemia, and other cases of mild chronic blood loss may result in clear overt symptoms of anemia.
11. Name the two major categories of hemolytic anemia.
12. List the three major categories of hereditary hemolytic anemia.
13. Name the major causes of acquired hemolytic anemia.
14. List four important laboratory findings commonly associated with hemolytic anemia.
15. What are the normal serum levels of haptoglobin? Why are low haptoglobin levels a clear indication of hemolytic anemia?

# Chapter Fourteen

## *Hereditary Hemolytic Anemias*

The three major categories of hereditary hemolytic anemia are caused by (1) defects in the red blood cell membrane, (2) deficiencies in erythrocyte metabolism, and (3) the presence of abnormal hemoglobins. The most important anemias belonging to each of these categories are discussed in the following material.

## HEMOLYTIC ANEMIAS CAUSED BY DEFECTS IN THE RED BLOOD CELL MEMBRANE

Two major types of hemolytic anemia belong to this category: (1) **hereditary spherocytosis,** also known as **hereditary spherocytic anemia (HSA)** and (2) **hereditary elliptocytosis,** or **hereditary elliptocytic anemia (HEA).** Both disorders have similar clinical and laboratory features. The only major difference is that in HSA the red blood cells appear to be round in a blood smear, whereas in HEA they appear to be oval or elliptical. Further, HEA usually produces a milder type of anemia than HSA.

### Hereditary Spherocytic Anemia (HSA)

Hereditary spherocytic anemia is an inherited disorder that is transmitted as an autosomal dominant trait. It is the most common form of inherited hemolytic anemia in Northern Europeans and people of Northern European descent. HSA is always associated with splenomegaly and with the presence of microspherocytes.

#### *Causes of HSA*

The normal biconcave shape of the red blood cell is characterized by a surface area of membrane that is greater than necessary for the volume of cellular contents. This high ratio of surface area to volume permits the cell to twist and bend and allows for considerable movement of the intracellular contents. The biconcave shape allows the red blood cell to withstand the considerable mechanical stress to which it is constantly subjected in the microcirculation, where the vessel diameter is at times smaller than the diameter of the normal erythrocyte. Red blood cell flexibility is maintained by the unique structure of its membrane. On the inside of the membrane are a number of important structural proteins such as actin, spectrin, and ankyrin—all of which play a major role in the maintenance of red blood cell shape. People suffering from HSA have a defect in the production or shape of one or more of these structural proteins. It appears that the majority of people with HSA have an *abnormal (mutant) gene for the protein spectrin.* Lack of spectrin, or a defective form of spectrin, weakens the structural scaffolding of the red blood cell membrane. The bone marrow releases more or less biconcave red blood cells into the peripheral circulation; however each time these red blood cells percolate through the sinuses of the spleen they tend to lose a small part of their membrane. This results in progressively smaller red blood cells or microcytes as more and more of the plasma membrane is lost. Eventually the microcytes become too rigid due to the constant loss of membrane while enveloping the same volume and are destroyed by the mononuclear phagocytic cells of the spleen and other organs. Because the spleen has to work overtime to get rid of these abnormal red blood cells, it tends to enlarge, and thus the phenomenon of splenomegaly occurs.

There is another reason for the occurrence of HSA. The preceding reason mainly explains the development of microcytosis but does not completely explain the condition of spherocytosis. A normally structured red blood cell membrane also aids in the proper functioning of the membrane. These functions include regulating the transport of vitamins, minerals, and nutrients across the cell membrane. One especially important mechanism is regulating the exchange of sodium and potassium across the membrane. This requires the presence of active ATP-dependent ion channels. These channels actively move sodium out of the cell in exchange for transporting potassium into the cell. The unique shape of the red blood cell and its surface to volume ratio are very closely linked to the maintenance of these electrolyte exchanges. In people suffering from hereditary spherocytic anemia, the *sodium pumps seem not to be able to function properly,* and hence there will be a net increase of sodium into the cell. This will upset the osmotic equilibrium and lead to an influx of water into the red blood cell. As a result the ratio of surface to volume is further reduced and the cell becomes more spherical. The round shape makes the red blood cell less flexible and makes it more susceptible to entrapment in the spleen. It also becomes more prone to osmotic lysis within the circulation itself. The above two reasons explain the presence of numerous microspherocytes in the circulation of people with HSA.

However a third mechanism has been implicated in HSA that explains the rapid deterioration of the microspherocytic red blood cells in the spleen. Spherocytes *use more energy than normal blood cells* in an effort to maintain their osmotic equilibrium. When spherocytes are incubated for 24 hours at 37°C in the absence of glucose, they show a much greater degree of hemolysis than do normal red blood cells. If this *in vitro* phenomenon also occurs *in vivo,* then it seems logical that the low glucose environment in the spleen will accelerate hemolysis of entrapped microspherocytes. In the absence of a spleen, abnormal cation flux, glucose dependency, and microspherocytosis persist, but there is much less hemolysis (figure 14.1).

#### *Diagnosis of HSA*

The clinical effects of HSA vary greatly from person to person. Severe anemia is unusual and normally occurs only as a transient phenomenon associated with

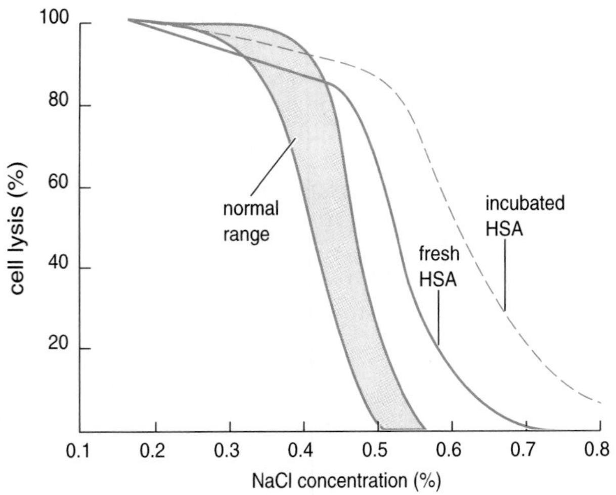


figure 14.1

Osmotic fragility curves of normal blood (shaded area), fresh blood of a patient with hereditary spherocytic anemia (HSA), and for HSA blood after 24 hours of incubation in the absence of glucose.

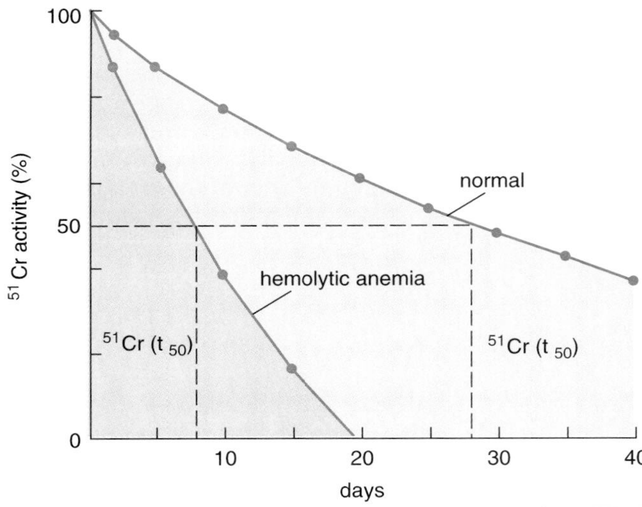


figure 14.2

Survival rates of $^{51}$Cr-labeled red blood cells in normal individuals and in patients suffering from hemolytic anemia. Note that the mean half-life ($t_{50}$) of $^{51}$Cr-labeled red cells in normal individuals is about 30 days, while that of patients with hemolytic anemia is about 10 days.

the onset of another cause of anemia, such as iron or folate deficiency, or as a result of an "aplastic crisis" due to the action of a medicinal drug or toxic chemical. Normally the anemia is fairly mild, because the reduced life span of the red blood cells is compensated for by an increased erythrocyte production in the bone marrow.

The increased red blood cell destruction does, however, result in higher levels of unconjugated bilirubin (rarely more than 3 to 4 mg/dL) and decreased levels of haptoglobin. The liver will have to work harder to get rid of the increased amounts of bilirubin, possibly leading to the formation of gallstones in the gallbladder, which may result in biliary obstruction and the pains associated with it. Gallstones occur most frequently in older overweight individuals. When gallstones are present in young adults, accelerated red blood cell destruction should be suspected.

Other hematological findings include a sharp increase in the number of reticulocytes (from 5 to 20%) and the presence of spherocytes and microspherocytes in the blood smear. These spherocytes are more sensitive to osmotic hemolysis than normal red blood cells. This is the basis for the **osmotic fragility test,** a unique test which is a primary diagnostic tool in identifying HSA. In this test red blood cells are suspended in a series of saline solutions of decreasing concentrations. Normal red blood cells swell and lyse relatively abruptly between 0.3 and 0.5% NaCl (physiological saline is 0.85% NaCl). Red blood cells of people with HSA show a moderately increased osmotic fragility plus a small number of highly fragile cells that may start to hemolyze at 0.7% NaCl (figure 14.1). These sensitive cells are spherocytes that have been severely injured by mononuclear phagocytic cells of the spleen and have lost large fragments of their plasma membranes. When cells of suspected HSA patients are first incubated for 24 to 48 hours in a sterile, non-glucose-containing medium and then tested for osmotic fragility, the increased sensitivity is much more marked. This test is a sensitive diagnostic tool for those cases that do not show large-scale spherocytosis in the blood smear.

Another test that is sometimes performed to assess the severity of red blood cell destruction is the use of **$^{51}$Cr-labeled red blood cells.** As shown in figure 14.2, the half-life of normal $^{51}$Cr-labeled erythrocytes is about 30 days, whereas in most hemolytic anemias, including HSA, the half-life has been reduced to less than 15 days. The same $^{51}$Cr test can also be used to document the fact that in HSA most red blood cell destruction takes place in the spleen. Figure 14.3 illustrates the surface counting pattern of $^{51}$Cr-labeled red cells in a patient with HSA. This graph shows that the majority of labeled erythrocytes gather in the spleen, rather than in the liver or lungs or other organs. This is an indication that most red blood cells in HSA are destroyed in the spleen. Finally, in HSA splenomegaly is almost always present due to an overactive red pulp, full of sequestered microspherocytic red blood cells.

### *Treatment of HSA*

Clearly the therapy of choice is **splenectomy** since the spleen causes most of the damage and destruction to the red blood cells. Such an operation should only be performed on children that have reached the age of 5 or 6 years, because the spleen is also an important organ of the immune system that filters out bacteria

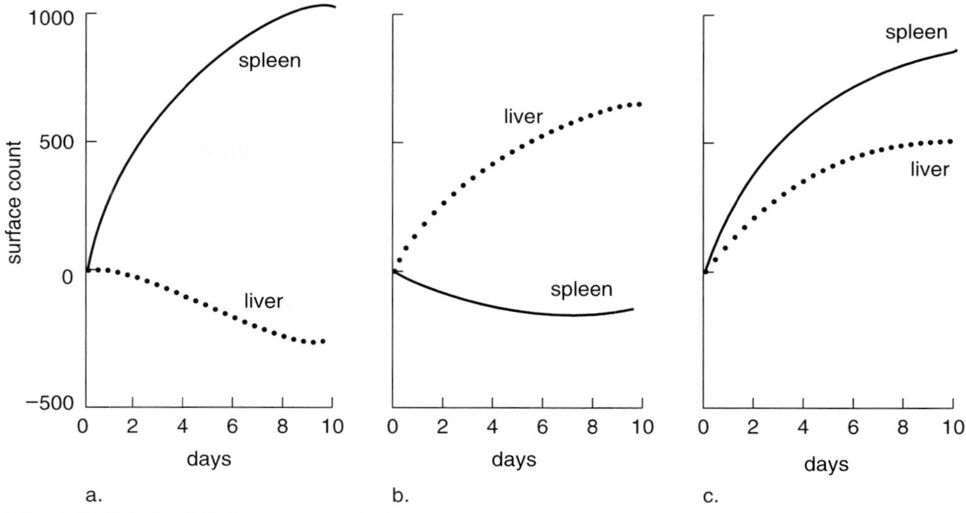


**figure 14.3**

Surface counting patterns of $^{51}Cr$-labeled red blood cells in hemolytic anemias. (*a*) In hereditary spherocytosis, most of the radiolabeled red cells are destroyed in the spleen; (*b*) in sickle cell anemia, most of these radiolabeled red cells are destroyed in the liver; (*c*) in autoimmune hemolytic anemia, both organs are involved in red cell destruction.

and other microorganisms that enter the body. The spleen also contains many lymphocytes in the white pulp, which can activate the specific acquired immune system against pathogens. Its function as an organ of the immune system is especially important during early childhood; for that reason, splenectomy is usually postponed until later childhood. Splenectomy removes only the major site of red blood cell destruction. Damage to the red blood cells continues, however, by other mononuclear phagocytic cells in the liver and other organs. Hence the presence of microspherocytes will persist after splenectomy. However the life span of these small red blood cells will be considerably increased.

### Hereditary Elliptocytic Anemia (HEA)

Hereditary elliptocytic anemia is also an inherited disorder that is transmitted by an autosomal dominant gene. It occurs in a wide variety of populations with a fairly high frequency of about 40 per 100,000. However it is usually not fully expressed. In most instances hemolysis is almost absent or very mild. Hence laboratory features such as reticulocytosis and increased bilirubin levels are only slightly above normal. In people who express the dominant elliptocytic gene, 15 to 60% of all red blood cells are oval. A small number of elliptical cells are fairly common—however in normal people that number never reaches above 10% of all red blood cells. Because of the low levels of premature hemolysis, overt anemia is rarely present since the bone marrow can readily compensate. Only the presence of complicating factors, such as lack of folic acid or lack of iron, will produce overt moderate anemia. Few will ever experience severe anemia because of the elliptocytic condition. In those who exhibit moderate to severe overt anemia, the blood smear will show not only elliptocytes but also microcytes and many bizarre poikilocytes. In those situations splenomegaly will be common as well. Hereditary elliptocytic anemia is caused by a membrane defect associated with the actin-spectrin-ankyrin cytoskeleton, since ghost cells and isolated cytoskeletons retain their elliptical shape.

The diagnostic features of overt HEA are similar to those of HSA. Since the effect of the membrane abnormality is minimal in most cases, no treatment is necessary. In cases of severe anemia, splenectomy may be necessary, just as in overt HSA.

## HEMOLYTIC ANEMIAS CAUSED BY DEFICIENCIES IN RED BLOOD CELL METABOLISM

The second group of hereditary hemolytic anemias are those involving abnormalities in red blood cell metabolism. These abnormalities are caused by deficiencies of certain enzymes needed in the anaerobic glycolytic pathway (the Embden-Meyerhof pathway) and in the hexose monophosphate shunt (also known as the pentose phosphate pathway). Although many functionally abnormal mutant forms of enzymes in both pathways are known, 95% or more of the clinically important inherited metabolic defects are due to deficient **glucose-6-phosphate dehydrogenase (G-6-PD)** activity in the hexose monophosphate shunt. The next most common red blood cell enzyme deficiency is the **pyruvate kinase (PK)** deficiency in the Embden-Meyerhof pathway. Both disorders are discussed in the following material. We should note, however, that compared with G-6-PD deficiency PK deficiency is a rare occurrence.

## G-6-PD Deficiency Anemia

As explained in chapter 7, many substances known as **oxidants** may cause the production of hydrogen peroxide ($H_2O_2$). The production of $H_2O_2$ is associated with the oxidation of the free sulfhydryl (—SH) groups of the hemoglobin molecule. Hemoglobin that contains many oxidized sulfhydryl groups is known as **sulfhemoglobin.** Sulfhemoglobin molecules tend to bind to one another, thus forming aggregates that readily precipitate into distinct granules known as Heinz bodies. These Heinz bodies in turn tend to attach to the red blood cell membrane. When the red blood cell percolates through the red pulp of the spleen, mononuclear phagocyte cells of the spleen tend to remove these Heinz bodies by nipping off part of the cell membrane containing these bodies. This process is known as "pitting," and the resulting cells will appear "eaten-out." This process may be repeated many times over, and as a result many red blood cells become misshapen—that is, display poikilocytosis. Such pitted cells are more fragile than normal red blood cells and are more prone to cation leakage, thus increasing the osmotic pressure within the cell. As a result the cells tend to become less flexible and more spherical. They tend to become sequestered in the spleen, where they are destroyed prematurely by phagocytic cells that line the sinuses and cords of the red pulp of the spleen.

In normal individuals the level of oxidant activity, and hence the level of sulfhemoglobin, is carefully controlled by the compound known as **reduced glutathione (GSH),** which neutralizes the activity of the oxidant drugs and oxidant products of infections according to the following formula:

$$\underset{\text{(reduced glutathione)}}{2GSH} + \text{oxidant compound} \rightarrow \underset{\text{(oxidized glutathione)}}{GSSG} + 2H_2O$$

Once glutathione is oxidized, it can no longer bind to other oxidants; only reduced glutathione can interact with these toxic compounds. Since there is a continuous onslaught of oxidants on the red blood cells, the body needs a steady source of reduced glutathione to eliminate them. Each red blood cell has only a limited amount of glutathione, so the oxidized glutathione (GSSG) must become reduced again. This is achieved with the aid of NADPH in the following reaction:

$$GSSG + NADPH \xrightarrow{\text{glutathione reductase}} 2GSH + NADP^+$$

The NADPH needed for this reaction is provided by the hexose monophosphate shunt (HMS) via the enzyme glucose-6-phosphate dehydrogenase (G-6-PD), which changes glucose 6-phosphate (G-6-P) to 6-phosphogluconate (6-PG) in the following reaction:

$$\text{G-6-P} \xrightarrow{\text{G-6-PD}} \text{6-PG}$$
$$\text{NADP} \rightarrow \text{NADPH}$$

Persons who lack G-6-PD or have an abnormal form of G-6-PD will be unable to generate sufficient amounts of NADPH. Hence they will be unable to reconvert oxidized glutathione back to reduced glutathione. As a consequence there will be a steady increase in the amount of oxidants in the red blood cells, resulting in the formation of sulfhemoglobin, which will precipitate out as Heinz bodies. A summary of this process is given in figure 14.4.

People with abnormal or deficient G-6-PD do not automatically suffer from hemolytic anemia. The body has other ways of eliminating excess oxidants, but these other processes are much slower. However they are generally sufficient to deal with low levels of oxidant stress. Only in situations in which the body is suddenly exposed to large amounts of oxidants will such individuals be unable to cope with the activity of these toxic compounds, resulting in a rapid outbreak of hemolytic anemia. Individuals with a normal functioning G-6-PD system are able to respond to an increased demand of reduced glutathione because of sudden severe oxidant stress by speeding up the production of NADPH in the hexose monophosphate shunt.

### *Causes of G-6-PD Deficiency Anemia*

Deficiency of G-6-PD is an inherited disorder affecting millions of people throughout the world. The gene for G-6-PD is located on the X chromosome; thus the inheritance of G-6-PD is sex-linked. As a result males have only one gene for this enzyme and females have two. However in females only one of the two genes is expressed in any given cell. Although more than 100 different allelic variants of this enzyme are known, the individual red blood cells of females produce only one type of G-6-PD. Any heterozygous deficient female—that is, a female with one normal and one abnormal gene—has red blood cells that are either normal or lack the active G-6-PD enzyme. Therefore mean enzymatic activity in females who carry a gene for G-6-PD deficiency may be normal, moderately reduced (which is the usual case), or grossly deficient. Generally speaking, overt signs of anemia are rare in heterozygous deficient females. On the other hand, in males (who have only one gene) the presence of a G-6-PD deficient gene will result in the absence of any normal G-6-PD, and this will invariably result in an accelerated hemolysis of the red blood cells. Consequently such males are much more prone to overt hemolytic anemia.

The great majority of the more than 100 known G-6-PD protein variants are enzymatically normal and thus do not cause any clinical problems. Most of them are discovered accidentally as a result of routine electrophoresis or other biochemical analysis. The overwhelming majority of people worldwide have a normal gene for G-6-PD known as **type B G-6-PD** or in the newer terminology **GdB.** In blacks there is another normal gene known as **type A G-6-PD** or **GdA.**

Some of the common abnormal enzyme types include the following:

1. **GdA$^-$(type A$^-$)** is found in about 10% of American blacks. Since this variant is commonly associated with accelerated hemolysis, these people have a greater chance of suffering from hemolytic anemia.

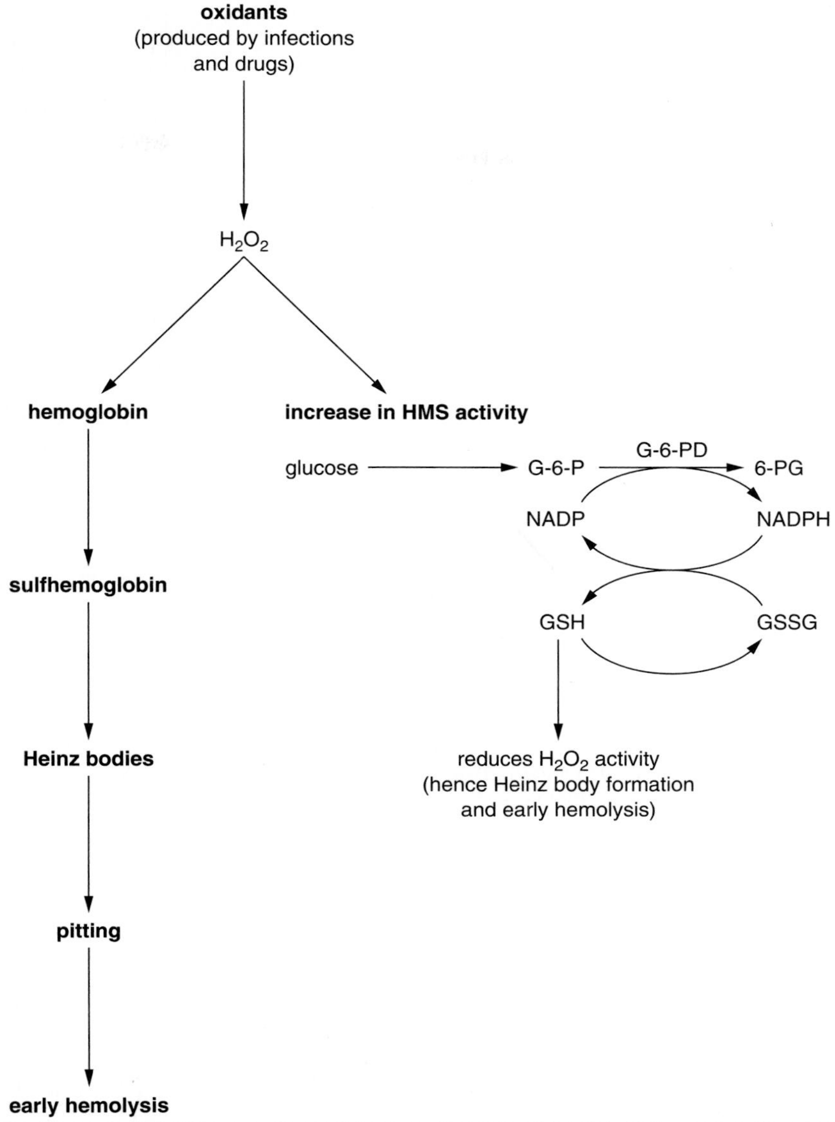


**figure 14.4**

In G-6-PD deficiency anemia, the reduced glutathione levels (GSH) cannot be maintained because insufficient NADPH is generated in the hexose monophosphate shunt (HMS). Consequently, hemoglobin is oxidized to sulfhemoglobin, which readily precipitates. The resulting Heinz bodies are pitted by the cells in the spleen, which "eat" these inclusions. This leads to the formation of poikilocytes and early hemolysis.

2. **Gd Mediterranean (GdMed)** is the second most common abnormal variant. It is found mainly in populations that border the Mediterranean Sea such as Sicilians, Sardinians, Greeks, Turks, Lebanese, Sephardic Jews, and Arabs.
3. **Gd Canton** is an abnormal variant commonly found in people of Asian origin.

Most of these enzyme variants are thought to differ only in one amino acid substitution, in a matter analogous to abnormal hemoglobins like HbS. Despite the disadvantages of having an abnormal gene for G-6-PD, they remain common in many areas. The reason for the continued existence of numerous harmful genes is that they may confer protection against the malaria parasite, *Plasmodium falciparum,* since the incidence of these abnormal genes is most common in areas where malaria is endemic.

The severity of hemolytic anemia associated with G-6-PD deficiency depends not only on the type of G-6-PD alleles carried, but also on the age of the red blood cells and on the quantity and quality of the oxidizing agent affecting the system. As shown in figure 14.5, G-6-PD activity—even in normal red blood cells—deteriorates progressively with age. Older red blood cells are more sensitive to oxidative denaturation, especially if the enzyme activity is already deficient in these cells, as in the case with $GdA^-$ and

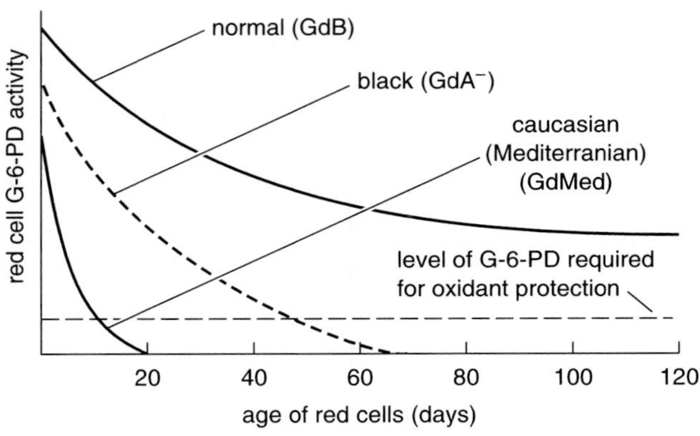


**figure 14.5**

Decay of G-6-PD in the red blood cells of normal individuals (GdB) and those of the $GdA^-$ and GdMed variants. Note that the latter two show a much faster decay of G-6-PD than that of normal individuals.

GdMed. Figure 14.5 also shows that normal red blood cells retain enough G-6-PD activity to produce sufficient NADPH to maintain enough glutathione in a reduced state in the face of normal oxidant stress throughout their life span. The diagram also shows that in the $GdA^-$ syndrome, the young red blood cells have normal enzyme activity. Only older red blood cells become grossly deficient in G-6-PD and are therefore abnormally sensitive to oxidant stress. This is even more pronounced in GdMed. Here even very young red blood cells become deficient in G-6-PD. Consequently the entire red blood cell population of individuals with GdMed is susceptible to oxidant stress. In most individuals with either $GdA^-$ or GdMed enzyme types, red blood cell survival in the absence of oxidant chemicals or infection is long enough to be replaced by fresh erythrocytes from the bone marrow. Consequently in the absence of oxidants, no overt signs of anemia are present.

Clearly the type and dose of oxidant agent plays a major role in determining whether hemolytic anemia will occur in susceptible individuals. The original observation that oxidant drugs play a major role in producing this type of anemia in the presence of abnormal G-6-PD genes was made after black soldiers developed severe hemolytic crises after receiving primaquine as a prophylactic agent against malaria during World War Two. Since that time many other oxidant drugs have been implicated as causative agents. Table 14.1 lists a number of common drugs that frequently will lead to sudden occurrences of hemolytic anemia in susceptible people. The list also mentions other agents that may cause damage to G-6-PD deficient red blood cells.

The most common cause of accelerated hemolysis due to G-6-PD deficiency is not oxidant drugs, but infection. Virtually every type of infection has been implicated. Although the exact mechanism of action is unknown, it has been suggested that production of superoxide anions and hydrogen peroxide by white blood cells in response to phagocytosis of pathogens are the primary agents damaging G-6-PD deficient red blood cells. Finally, there is the unique phenomenon of **favism.** Ingestion of the fava bean, or even inhalation of fava pollen, may result in a severe hemolytic crisis in a small portion of people carrying the GdMed gene. Since only a small fraction of people with GdMed are involved, other factors may be involved as well. Fava beans are very rich in L-dopa, and a metabolite of that compound, dopaquinone, is a potent oxidant. Thus the limited sensitivity to fava beans may be due to differences in L-dopa metabolism.

One last point: a rare group of alleles found almost exclusively in whites produces an abnormal type of G-6-PD that does not depend on the presence of oxidants to produce its effects. Consequently these people suffer from chronic hemolytic anemia, since the enzymatic activity of these variants appears to be inadequate to protect the red blood cells from normal wear and tear.

**Table 14.1** Agents Shown to be Involved in the Onset of Hemolytic Anemia in Individuals with G-6-PD Deficiency Syndromes

*Drugs*
- Antimalarials
  - Quinine
  - Chloroquine
  - Primaquine
  - Pyrimethamine
- Analgesics
  - Phenacetin (acetophenetidin)
  - Acetylsalicylic acid
  - Paracetamol
- Antibacterials
  - Sulfonamides
  - Penicillin
  - Nitrofurazones
  - Isoniazid
  - Streptomycin
- Miscellaneous
  - Vitamin K
  - Probenecid
  - Naphthalene
  - Methylene blue

*Infections*
- Respiratory viruses
- Infectious mononucleosis
- Bacterial pneumonia
- Septicemia
- Hepatitis

*Other Factors*
- Fava beans (and its pollen)
- Diabetic ketoacidosis
- Uremia

*Diagnosis of G-6-PD Deficiency Anemia*

If the anemia is acute and associated with large-scale intravascular hemolysis as evidenced by high levels of bilirubin, absence of haptoglobin, hemoglobinemia, and hemoglobinuria, then G-6-PD deficiency may be suspected. The suspicion that G-6-PD deficiency is involved will be increased if the sudden onset of anemia is associated with acute illness or infection, the taking of a new drug, or the ingestion of fava beans.

This diagnosis should be confirmed, however, by assays of G-6-PD activity in red blood cells. Several tests for this diagnosis are available. Some are fairly general screening tests based on NADPH-mediated dye coloration; others are based on the production of methemoglobin in the presence of methylene blue; still others are direct assays of enzyme activity. A definitive test of enzyme activity depends on spectrophotometric measurement of NADPH production. Cytochemical estimation of G-6-PD activity in individual red blood cells can be performed with the aid of tetrazolium dyes. This test can detect the presence of less than 5% G-6-PD deficient cells. Hence it is useful in identifying female heterozygote carriers.

*Treatment of G-6-PD Deficiency Anemia*

Once it has been established that the anemia is caused by an abnormal variant of the G-6-PD gene, the agent causing the oxidant stress should be discovered. This can usually be done by a detailed analysis of the medical history and eating habits of the patient. If the presence of a certain oxidant is suspected, it should be removed immediately. In most cases the anemia will disappear shortly afterward.

In cases involving a severe hemolytic crisis, blood transfusions may be necessary. Blood transfusions may also be needed in those rare cases of G-6-PD deficiency that are not subject to oxidant stress.

**Pyruvate Kinase (PK) Deficiency Anemia**

PK deficiency is a disorder of the Embden-Meyerhof glycolytic pathway. In contrast with G-6-PD deficiency, which affects many millions of people, hemolysis due to PK deficiency is rare, affecting only several hundred patients. Although abnormalities of other enzymes in the glycolytic pathway have been described, most cases are due to PK deficiency.

*Causes of PK Deficiency Anemia*

PK deficiency is inherited as an autosomal recessive gene, which means that hemolytic anemia will occur only in the homozygous state. People with the heterozygous condition have a normal metabolism, although their red blood cells may have slightly lower levels of enzyme activity.

Pyruvate kinase is needed to catalyze the formation of pyruvate from phosphoenolpyruvate, with the generation of ATP (see figure 7.2). Deficiency of this enzyme will alter the concentration of glycolytic intermediate metabolites and decrease the production of ATP to inadequate levels to meet the requirements of the red blood cell. When that happens, erythrocytes will lose their integrity and will hemolyze readily, causing anemia in the affected individual.

Reticulocytes have residual ATP-generating capacity in the Krebs cycle, hence reticulocytes are better able to withstand PK deficiencies than more mature red blood cells.

*Diagnosis of PK Deficiency Anemia*

Overt symptoms of anemia may be quite severe in homozygous newborns and young children. Many will have jaundice and damage to the central nervous system due to the deposition of bilirubin in the basal ganglia (kernicterus). This may lead to spasticity, mental deficiency, deafness, and epilepsy. Later in life the anemia usually becomes much less severe, even though hemoglobin levels may stay low (5 to 10 g/dL) because of a gradual rise in the production of intracellular 2,3-biphosphoglycerate (2,3-BPG), which causes the oxygen-dissociation curve to shift to the right. The exact diagnosis for this disorder requires specific enzyme assays, since no other unique features are associated with this form of hemolytic anemia. Drugs are not implicated in the pathogenesis of this disease. Splenomegaly is usually present due to the sequestration of many fragile red blood cells in the spleen.

*Treatment of PK Deficiency Anemia*

Splenectomy is usually performed since it alleviates many of the severe symptoms of this anemia, although it does not cure it. In cases of severe anemia, blood transfusions may be needed, although, generally speaking, splenectomy will reduce the need for frequent blood transfusions. As explained earlier in this chapter, physicians are reluctant to remove the spleen early in life because of its importance in the immune system.

Children that survive the hemolytic crisis of early childhood will be able to tolerate the relatively mild anemia of later youth and adulthood quite well, because the increased presence of 2,3-BPG enhances oxygen release to the tissues and because the reticulocytes are able to function normally.

## HEMOLYTIC ANEMIAS CAUSED BY ABNORMAL HEMOGLOBINS

This last group of hereditary hemolytic anemias is caused by abnormal hemoglobins. As mentioned in chapter 5, more than 300 genetic variants of the human hemoglobin molecule are known. Only about 10 of these are considered to be normal hemoglobins.

The majority of mutant hemoglobin forms are single point mutations—that is, they differ only in one amino acid from the normal globin chains. Most of these abnormal variants are harmless, since they do not interfere with the normal structure and functioning of the hemoglobin molecule. Most of them were discovered accidentally as a result of routine hemoglobin electrophoresis or by other tests used to separate different forms of normal hemoglobins.

### Effects of Abnormal Hemoglobins

A small number of hemoglobin variants may be considered pathogenic since they interfere with the normal structure and function of erythrocytes. These pathological variations may cause anemia in three different ways, which are not mutually exclusive:

1. Some variants, especially those with substitutions of amino acids at the surface of the molecule, have *reduced solubility* and a tendency to combine with other similar hemoglobin molecules; these complexes ultimately form precipitates in the red blood cell.
2. Other variants, especially those with amino acid substitutions around the heme pocket, will *interfere with the normal oxygen transport* function of the hemoglobin molecule. Some of these mutant forms have an increased oxygen affinity and will not release oxygen at the normal tissue $Po_2$. Others have decreased oxygen affinity and do not transport oxygen as readily as normal hemoglobins.
3. Last, some variants cause instability to the quaternary structure of the hemoglobin molecule, which may result in spontaneous or drug-induced *denaturation of the hemoglobin.*

Table 14.2 lists some examples of each of these types of pathological variants. Most hemoglobin variants are designated by either a letter (e.g., HbS) or more commonly by the name of the town in which they were first discovered (e.g., Hb Sydney) or by a combination of both (HbM Boston). From a global point of view, only HbS—and to a lesser extent HbC, HbD, and HbE—are able to produce hemolytic anemia on a significant scale. The other mutant hemoglobins have limited clinical importance, but they have played a major role in the elucidation of the structure-function relationship of the hemoglobin molecule. In this section only the hemolytic anemia caused by HbS (sickle cell anemia) is discussed, and a few remarks about the anemias associated with HbC, HbD, and HbE are made at the end of this chapter.

### Sickle Cell Anemia

The name **sickle cell** was first coined in 1910 by the American physician James B. Herrick, who noticed the presence of peculiar elongated or sickled cells in the bloodstream of a black student suffering from hemolytic anemia. Since that time, the study of the structure of HbS and its relationship to the formation of sickled cells has been the subject of continuous investigations. Currently a great deal is known about the unique behavior of this abnormal hemoglobin.

The structure of HbS is similar to that of HbA except for one amino acid substitution on the surface of the β-globin chains. In HbS the nonpolar amino acid valine replaces the polar glutamic acid at position A6 of the β-chains. This results in a molecule with a different isoelectric point (7.2 rather than the 7.0 for HbA). Consequently HbS can easily be separated from HbA by the process of electrophoresis (figure 14.6). Further, HbS in the deoxygenated state is 50 times less soluble than deoxygenated HbA.

#### *Causes of Sickle Cell Anemia*

Sickle cell anemia is inherited as an autosomal codominant trait. As illustrated in figure 14.7, the gene frequency for HbS is widely distributed around the world. In some western and central African countries, the gene frequency may reach as high as 30%. Approximately 10% of the American black population has been estimated to be carriers of this trait. The reason why this deleterious gene has persisted so long in so many different populations is that the presence of the heterozygous genotype protects the carriers against the fatal complications caused by the malaria parasite *P. falciparum.*

People who are heterozygous for HbS normally produce more than 50% of their hemoglobin in the form of HbA. As a matter of fact, usually the amount of HbS is only 25 to 35% of the total hemoglobin present. Since the formation of sickled cells is dependent on the amount of deoxy-HbS present, the availability of large quantities of HbA inhibits the sickling phenomenon so much that under normal circumstances the red blood cells of people heterozygous for HbS (*A/S* type) will not sickle at all. Only under conditions of extreme stress, such as severe hypoxia or severe infections, does some degree of sickling occur. The sickling trait is almost invariably asymptomatic, and for that reason individuals with the sickle cell trait should not be placed in high-risk categories by insurance companies and other organizations. The only common side-effect that may occur in people heterozygous for HbS is an occasional episode of painless hematuria. This is due to the fact that the low oxygen tension in the kidneys may at times make the red blood cells susceptible to sickling, which may result in some kind of renal infarction, resulting in painless hematuria.

#### *Diagnosis of Sickle Cell Trait*

The presence of the sickle cell trait may be diagnosed in a number of ways. Three of the more common tests are (1) **hemoglobin electrophoresis**—individuals with the sickle cell trait will have about 35% HbS and about 60% HbA; (2) **sickling preparation**—sickled cells may be observed under the microscope after

## Table 14.2 Some Examples of Human Hemoglobin Variants Associated with Hemolytic Anemia

| Features | Hemoglobin Variant | Alteration |
|---|---|---|
| Reduced solubility | | |
| | S (sickle) | $\beta^{6\ GLU \to VAL}$ |
| | C | $\beta^{6\ GLU \to LYS}$ |
| | C Harlem | $\beta^{6\ GLU \to VAL}$ and $\beta^{73\ ASP \to ASN}$ |
| Abnormal $O_2$ transport | | |
| Increased $O_2$ affinity | Chesapeake | $\alpha^{92\ ARG \to LEU}$ |
| | Rainer | $\beta^{145\ TYR \to HIS}$ |
| | Hiroshima | $\beta^{143\ HIS \to ASP}$ |
| | Tacoma | $\beta^{30\ ARG \to SER}$ |
| | Zurich | $\beta^{63\ HIS \to ARG}$ |
| | Gun Hill | Deletion: $\beta^{93\text{–}97 \to 0}$ |
| | Freiburg | Deletion: $\beta^{23\ VAL \to 0}$ |
| Decreased $O_2$ affinity | Kansas | $\beta^{102\ ASN \to THR}$ |
| | Hammersmith | $\beta^{42\ PHE \to SER}$ |
| | Seattle | $\beta^{76\ ALA \to GLU}$ |
| | Torino | $\alpha^{43\ PHE \to VAL}$ |
| Methemoglobins | M Boston | $\alpha^{58\ HIS \to TYR}$ |
| | M Iwate | $\alpha^{47\ HIS \to TYR}$ |
| | M Saskatoon | $\beta^{63\ HIS \to TYR}$ |
| | M Hyde Park | $\beta^{92\ HIS \to TYR}$ |
| Unstable molecules | | |
| α-Chain mutants | Torino | $\alpha^{43\ PHE \to VAL}$ |
| β-Chain mutants | Zurich | $\beta^{63\ HIS \to ARG}$ |
| | Seattle | $\beta^{76\ ALA \to GLU}$ |
| | Geneva | $\beta^{28\ LEU \to PRO}$ |
| | Hammersmith | $\beta^{42\ PHE \to SER}$ |
| | Tacoma | $\beta^{30\ ARG \to SER}$ |
| | Gun Hill | Deletion: $\beta^{93\text{–}97 \to 0}$ |
| | Freiburg | Deletion: $\beta^{23\ VAL \to 0}$ |

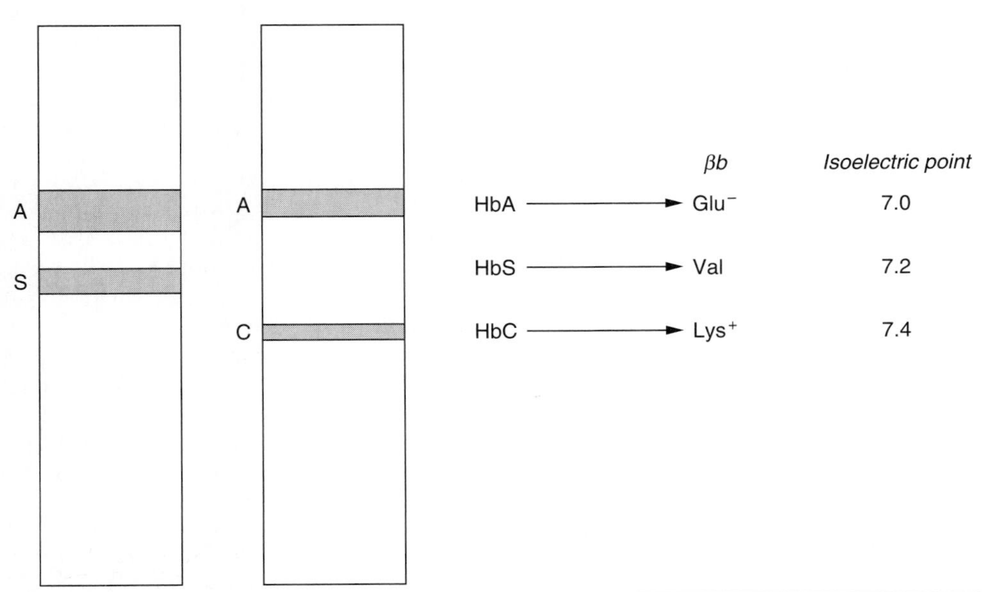


**figure 14.6**

Electrophoresis patterns of hemoglobins from individuals with the sickle cell trait (*A/S*) and with the HbC trait (*A/C*).

figure 14.7

The geographical distribution of thalassemia and of HbS, HbC, HbD, and HbE diseases.

mixing the blood sample with an oxygen-consuming agent such as metabisulfite; and (3) **solubility tests**—these tests depend on the fact that deoxy-HbS has a lower solubility at high ionic strength solutions than does HbA (e.g., the dithionite tube test).

*Molecular Basis for Sickling*

The sickling phenomenon of red blood cells is caused by the substitution of valine for glutamic acid at the sixth residue of the β-chains of hemoglobin molecules. Several hypothetical models have been proposed in an attempt to explain the decreased solubility of deoxy-HbS. Although all theories differ on the exact molecular events involved, most agree that the sickling of red blood cells is ultimately caused by the stiff, polymerized aggregates of hemoglobin.

One of the more popular hypotheses regarding the molecular events involved is the theory put forward by Perutz and Lehmann. They proposed that when the HbS molecule is reduced to the deoxy state, a gap is produced between the two β-chains. This allows the hydrophobic valine at position 6 to form a bond with the adjacent α-chain of another deoxy-HbS molecule (figure 14.8). Although the exact binding site of the next molecule is not known, it could be position 23 in the α-chain since Hb Memphis, which is similar to HbS except for an amino acid substitution at position 23 of the α-chain, sickles much less readily than pure HbS. Although this hypothesis has not been proven conclusively, it does give a logical explanation for the process of polymerization of deoxy-HbS.

Recent electron micrographs have shown these polymerized HbS molecules are arranged in bundles of fibers, each with a diameter of 150 Å. Each of these fibers appears to consist of a helical polymer, having at least six molecules of hemoglobin per cross-sectional area (figure 14.9). Both hydrophobic and electrostatic bonds contribute to the stability of the fibers. It appears that the initial step is the aggregation of individual HbS molecules into a single strand. This process is known as **nucleation.** The degree of nucleation depends greatly on the concentration of deoxy-HbS in the erythrocyte. Apparently a certain

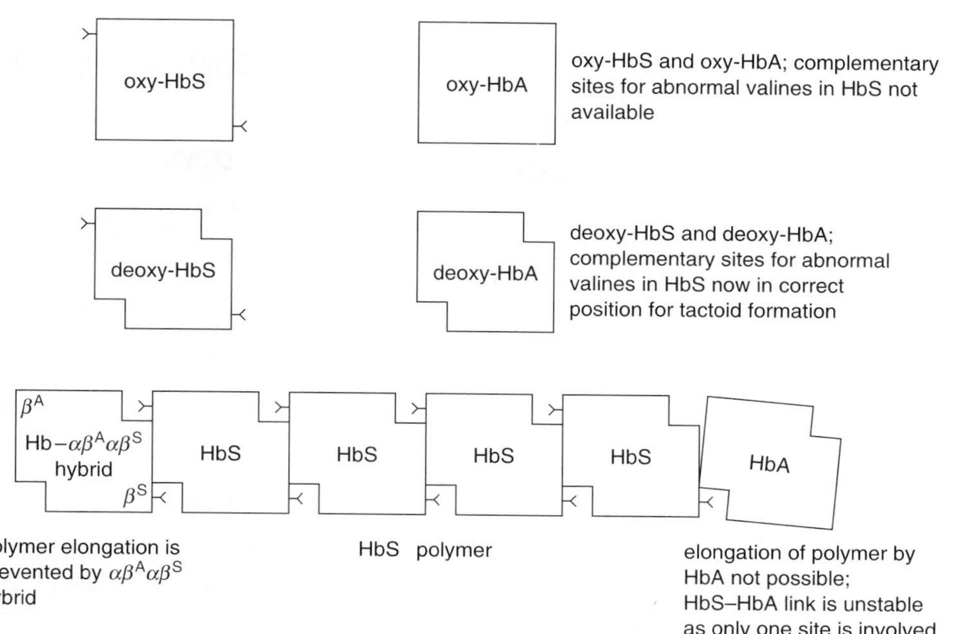


**figure 14.8**

The formation of HbS polymers and their interruptions by the addition of HbA.

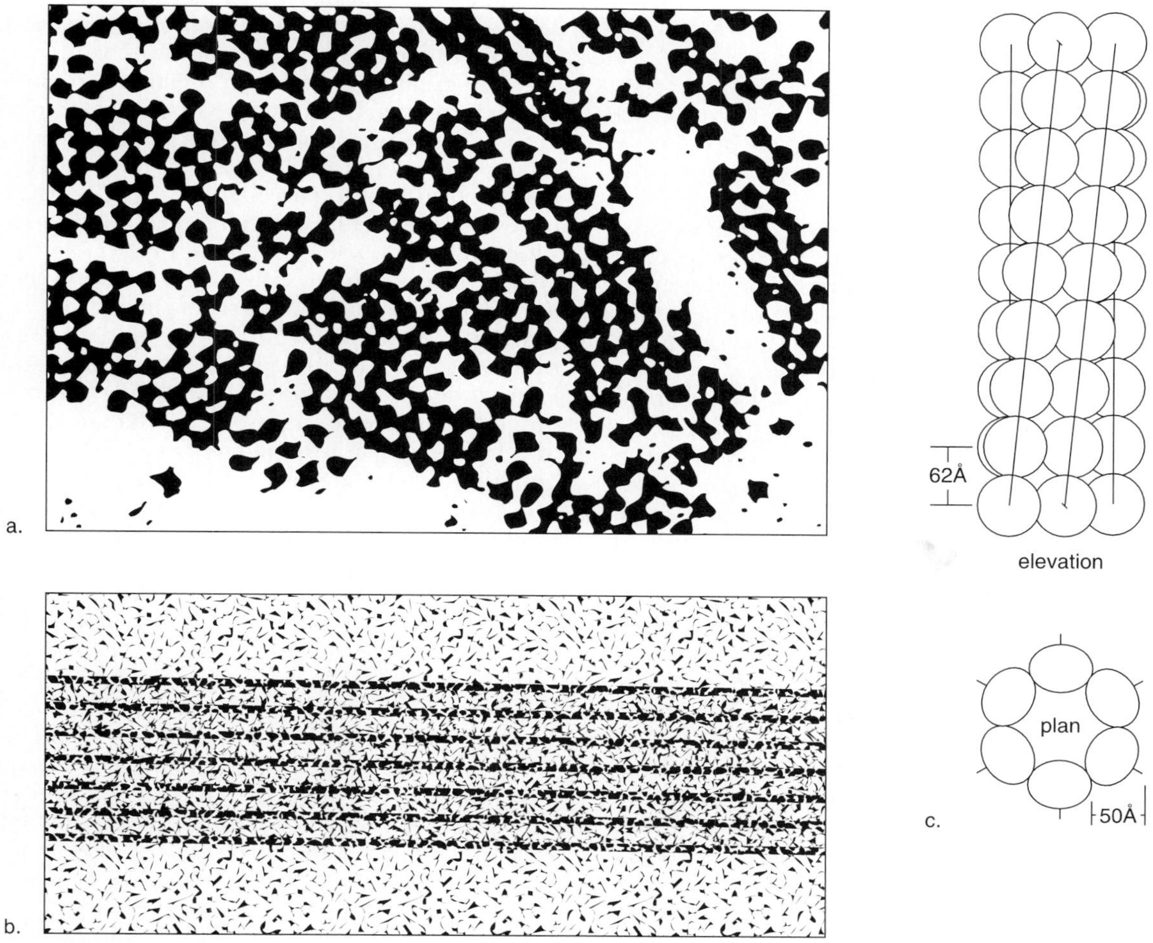

a.

62Å

elevation

plan

50Å

c.

b.

**figure 14.9**

Electron micrographs of concentrated solutions of deoxygenated HbS. (*a*) Transverse section. (*b*) Longitudinal section, showing parallel fibers. (*c*) Model of fiber made up of deoxygenated HbS polymers showing a tubular arrangement.

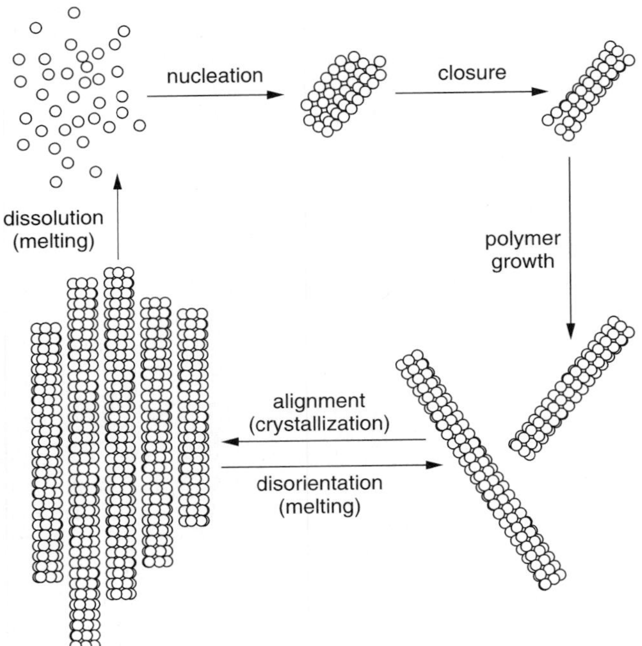


figure 14.10

Polymerization and fiber formation of deoxygenated HbS.

Source: William S. Beck, *Hematology,* 4th ed., 1981 MIT Press, Cambridge, MA.

figure 14.11

Sickled red blood cells.

minimal threshold level is needed to start the process. The next step involves the fusion of six or more strands into a helical fiber which keeps growing, as illustrated in figure 14.10.

In the beginning these fibers are randomly oriented but soon change to a uniform direction and become aligned in a parallel manner. The alignment and the continuing growth of the long fibers are the probable causes for the distortion of the red blood cells into an elongated sickled form.

As the red blood cells sickle, they lose their flexibility and become rigid (figure 14.11), and in this form they may obstruct the capillary flow. This obstruction then leads to local tissue hypoxia, which causes further deoxygenation of the HbS and therefore more sickling. This vicious cycle may amplify a small infarction into a larger one.

Ordinarily, upon reoxygenation in the lungs, the fibers readily dissociate (or "melt") again into individual HbS molecules, and the sickled cells resume their normal biconcave shape. However if the cell is sickled over and over, the membrane loses its flexibility, resulting in the formation of **irreversibly sickled cells (ISCs).** These ISCs are subject to sequestration in the red pulp of the spleen and are usually rapidly destroyed there. Continuous formation and destruction of these ISCs is probably the major cause of the moderate to severe hemolysis associated with sickle cell anemia. Further, the cell membrane at the pointed ends of sickled cells is especially vulnerable to lysis in the microcirculation, which also contributes to hemolysis.

*Symptoms of Sickle Cell Anemia*

Although the degree of hemolysis is quite severe in many cases of homozygous (*S/S*) sickle cell disease, with hemoglobin levels between 6 and 9 g/dL, the symptoms of anemia are often fairly mild because HbS gives up oxygen to the tissues more readily than HbA. Further, in homozygous sickle cell anemia the absence of HbA is often partially compensated for by an increased level of HbF, which may vary from 5 to 15% of the total hemoglobin present. The more HbF present, the milder the symptoms will be.

The greatest problems associated with sickle cell anemia are usually not caused by the lack of hemoglobin but occur as a result of the so-called **painful crisis** that occurs periodically. The unpredictable and intermittent occurrence of these acute painful crises are the result of accelerated localized sickling, which will produce local tissue injury. Tissue injury results from the infarction of the local microcirculation. These vascular obstructions may produce acute pain and tissue damage. These painful crises are often triggered by local infections, dehydration following vigorous exercise, childbirth, surgery, prolonged exposure to intense cold, change of altitude, and similar situations that result in sudden deoxygenation. In normal everyday situations not associated with violent exercise, painful crises will not readily occur. Further, a peculiar time delay occurs between the deoxygenation of HbS and onset of the polymerization of the HbS molecules. Usually that delay is sufficient for the deoxygenated red blood cells to escape to the larger blood vessels before fiber formation starts. When this time delay is shortened due to external circumstances such as local infection or surgery, these painful ischemic crises occur and cause damage to tissues and organs. These tissue infarcts may occur in a wide variety of organs including the spleen, lungs, heart, brain, kidneys, and bones. Cerebrovascular complications are especially common in younger children, and this may result in strokes or even death. In older children and adults, the major damage is usually to the cardiopulmonary system, and this may result in congestive heart failure and lung embolisms. Sometimes impairments to the liver may result in the formation of gallstones in the gallbladder, whereas impairments to the kidneys may result in glomerulonephritis and high levels of hematuria. Damage may also occur to the eyes and the skin. The latter condition may result in chronic ulcers. Finally, estimates are that at least one-quarter of all patients with homozygous sickling disease will eventually develop some kind of neurological disorder.

Next to these painful ischemic crises, patients with homozygous sickle cell anemia may also experience occasional hypoplastic crises. These are transient episodes of bone marrow failure apparently triggered by infections. These hypoplastic crises are most dangerous since a combination of accelerated hemolysis and hypoplasia of the bone marrow may result in severe life-threatening anemia. Luckily these dangerous aplastic crises occur much less frequently than ischemic tissue crises. Under normal circumstances the erythropoietic activity of people with homozygous sickle cell disease is greatly increased. This may lead to a rapid depletion of folic acid stores of the body and may result in the development of megaloblastic anemia as well.

Problems associated with the sickling of red blood cells are especially severe in young children since such sickling may produce infarctions in many vital organs. This would impair normal growth and development. Many young children with homozygous sickle cell disease fail to thrive and remain small and sickly. Infarction of the spleen appears to be especially traumatic since the spleen normally functions as a clearinghouse for foreign substances in the blood. A damaged spleen is less efficient in filtering out pathogenic microorganisms, particularly pneumococci, and this will result in a greater risk of infection in these children. The increased number of infections in turn may cause damaging painful crises in other parts of the body.

### *Laboratory Findings of Sickle Cell Anemia*

Laboratory investigations of persons with sickle cell anemia will show low levels of hemoglobins: 6 to 9 g/dL. Hemoglobin electrophoresis will reveal the presence of large amounts of HbS (80 to 95%), considerable amounts of HbF (5 to 15%), and a total absence of HbA. Bilirubin levels will be increased, and haptoglobin levels will be low.

The blood smear will show many sickled cells as well as other poikilocytes. Many of the red blood cells will show Howell-Jolly bodies, and microcytosis is usually marked as well. These are all indications of a spleen that is not functioning optimally.

### *Treatment of Sickle Cell Anemia*

Many antisickling treatments have been proposed, but none of them has been satisfactory so far. Currently the usual management is mainly prophylactic and supportive. All treatment is aimed at avoiding factors known to bring on painful crises. Since infections are the primary cause of onset of these crises, infections must be detected early and promptly treated with antibiotics and other medications. Good personal hygiene is also of paramount importance to reduce the risk of infection. Further, people who are homozygous for HbS need to eat well-balanced diets with extra vitamins, especially folic acid, to be able to support their hyperactive bone marrow.

The painful crises themselves should be treated with complete rest, analgesics to reduce pain, and lots of liquids to increase blood volume and thus lessen the chance of infarction.

Blood transfusions have only a limited role in the management of sickle cell anemia. They should be given only in those cases associated with hypoplastic or aplastic crises. As mentioned earlier, in the absence of hypoplastic episodes, patients tolerate the anemia quite well. Blood transfusions always carry with them the risk of immunization, iron overload, and the transmission of infectious diseases such as hepatitis or acquired immunodeficiency syndrome (AIDS). Hence this type of treatment should be used sparingly.

Particular care is needed during pregnancy, birth, and surgery, since these are all prime occasions for the induction of painful crises. Before delivery or surgical procedures, patients should be transfused with normal blood, thus reducing the proportion of red blood cells high in HbS.

### *Sickle Cell/ β-Thalassemia Syndrome*

One type of sickle cell anemia requires special mention since it may result in severe hemolytic anemia: **sickle cell/β-thalassemia syndrome.** In this condition the patient has inherited two deficient genes (*S*/β-*thal*): one for producing abnormal β-chains and another for affecting the rate of β-chain synthesis. This disorder is found primarily in people that originate from countries surrounding the Mediterranean Sea. This syndrome varies in its clinical manifestations from moderate to severe (table 14.3). Most cases have high levels of hemolysis and considerable sickling. Microinfarctions are common, but painful crises are less frequent than in homozygous sickle cell disease. In severe cases where red blood cells are sequested in the enlarged spleen, splenectomy may be necessary.

The blood smear will reveal many target cells in addition to sickled cells, as well as many hypochromic and microcytic erythrocytes. Polychromatophilia is also common.

Hemoglobin electrophoresis will show 60 to 90% HbS and 10 to 30% HbF. Some HbA may also be present in cases of sickle cell/$\beta^+$-thalassemia. Hence the effects are usually less marked than in homozygous sickle cell anemia (*S/S*). Most patients require little specific therapy except in cases of infections. Good hygiene and good nutrition are recommended.

## Table 14.3 Variations in Clinical Severity of the Various Forms of Sickle Cell Anemia

| Genotype | % of Hemoglobin S | % of Non-S Hemoglobin | Clinical Severity |
|---|---|---|---|
| *S/A* | 30 | 70 (A) | Mild |
| *S/F** | 70 | 30 (F) | Mild |
| *S/S* | 85 | 15 (F) | Moderate-severe |
| *S*/thalassemia | 80 | 20 (A + F) | Moderate |
| *S/C* | 50 | 50 (C) | Moderate |
| *S/O, S/D* | 30 | 70 (O,D) | Moderate-severe |

*Double heterozygous state for hemoglobin S and hereditary persistence of fetal hemoglobin.

*Prognosis for Sickle Cell Anemia*

The prognosis for people with sickle cell anemia is improving gradually. An increasing number of patients are now surviving into adulthood and are having children of their own. But the disease remains potentially debilitating and frequently the quality of life of individuals with homozygous sickle cell anemia is severely impaired. Eradication of the homozygous HbS condition should be of prime importance. Proper genetic counseling is currently the only way to eliminate this disease. If both marital partners are carriers of the sickling gene, they have a 25% chance that one or more of their offspring will be homozygous for HbS. If one of the partners is homozygous and the other is heterozygous, there will be a 50% chance of producing a child with homozygous sickle cell anemia (figure 14.12). Such people should be persuaded not to reproduce but to adopt children.

### Hemoglobin C (HbC) Disease

HbC is another mutant hemoglobin involving one amino acid substitution. In HbC this substitution occurs at position A6 in the β-chains: lysine instead of glutamic acid (table 14.4). The gene frequency of HbC is about one-quarter that of HbS and is found primarily in West African populations (figure 14.7). It is also frequent in African-Americans. The HbC condition may be present in a homozygous form (*C/C*) or in a heterozygous form together with HbA (*A/C*) or with HbS (*C/S*). The homozygous condition is known as **HbC disease** and is associated with a fairly mild form of anemia. The presence of the HbC trait (*A/C*) is completely asymptomatic, whereas the *C/S* condition may result in moderate anemia.

**HbC disease** is a fairly rare disorder in the United States. HbC is less soluble than HbA and tends to form intracellular crystals, especially in slightly hypertonic solutions. The blood smear will reveal striking target cells, but no sickled cells are present in the *C/C* condition. In contrast with HbS, HbC does not polymerize; hence there are no sickled cells. The osmotic fragility of the red blood cells of patients with HbC disease is decreased. The only significant complication that occurs in patients with HbC disease are eye problems.

The heterozygous *C/S* condition is much more common and may be seen in the United States almost as frequently as the *S/S* genotype. Patients with heterozygous *C/S* will usually show mild to moderate forms of anemia with distinct splenomegaly. The blood smear will show many target cells and some sickled cells. Organ infarcts and painful crises may occur occasionally but are not as frequent as in the homozygous *S/S* condition. As a result the life expectancy of persons with *C/S* is greater than those with *S/S*. Complications may occur during pregnancy, but with careful management most problems can be avoided.

### Hemoglobin D (HbD) Disease

**HbD disease** is also known as Hb Punjab as it is found almost exclusively in the Punjab region of northwest India (figure 14.7). This abnormal hemoglobin also involves one amino acid substitution in the β-chains. In HbD glutamine is substituted for glutamic acid at position 121 of the β-chains.

The heterozygous condition of Hb Punjab (*A/D*) is completely asymptomatic. The homozygous genotype (*D/D*) is extremely rare and is not normally associated with overt hemolytic anemia. Although the blood smear may show some target cells and spherocytes, splenomegaly is not normally present. However it is possible to get hemolytic anemia with HbD when there is a heterozygous condition together with HbS (*D/S*) or with the β-thalassemia gene (*D*/β-*thal.*).

Since HbD migrates with the same speed as HbS upon electrophoresis, it is difficult to distinguish between these two abnormal forms of hemoglobin.

### Hemoglobin E (HbE) Disease

HbE is another mutant of HbA. **HbE disease** is also produced by a single amino acid substitution in the β-globin chains. Here glutamic acid is replaced by lysine at position 26 of the β-globin chain. The HbE gene is fairly widespread in Southeast Asia, especially in Manyar (Burma), Thailand, Cambodia, Malaysia, and Indonesia (figure 14.7). Apparently it does not occur with great frequency in China.

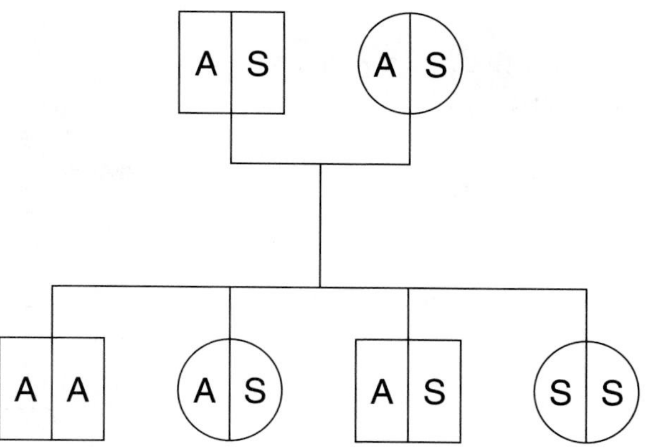


If both partners are heterozygous for HbS, there is a 25% chance that offspring will be homozygous for HbS.
a.

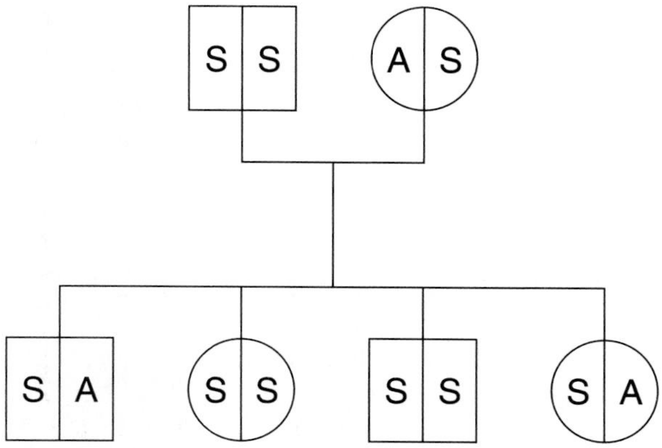


If one partner is homozygous and the other partner heterozygous for HbS, there is a 50% chance that the offspring will be homozygous for HbS.
b.

**figure 14.12**
Chances of inheriting sickle cell anemia (*a*) if both parents are carriers (*A/S*); (*b*) if one is homozygous (*S/S*) and the other is a heterozygous carrier (*A/S*).

**Table 14.4** Effect of an Amino Acid Substitution at Position 6 in the Beta-Chain on the Isoelectric Point

| Hemoglobin | β-chain Sequence |
|---|---|
| A | $H_2N$-Val-His-Leu-Thr-Pro-*Glu*-Glu . . . |
| S | $H_2N$-Val-His-Leu-Thr-Pro-*Val*-Glu . . . |
| C | $H_2N$-Val-His-Leu-Thr-Pro-*Lys*-Glu . . . |

Individuals homozygous for HbE (*E/E*) often exhibit a mild form of hemolytic anemia characterized by the presence of target cells and microcytes. Although there is a lowered osmotic fragility of the red blood cells, overt signs of accelerated hemolysis are minimal.

In the heterozygous condition (*E/A*), HbE makes up about 30 to 45% of all hemoglobin. Such individuals are completely asymptomatic.

# Further Readings


Bunn, F. H. Hemoglobin II: Sickle Cell Anemia and Other Hemoglobinopathies. Lecture 9 in *Hematology,* 4th ed., W. S. Beck, Ed., The MIT Press, Cambridge, Mass., 1985.

Charache, S. Advances in the Understanding of Sickle Cell Anemia. *Hospital Practice,* Feb. 1985.

Lux, S. E. Hemolytic Anemias III. Membrane Disorders. In *Hematology,* 4th ed., W. S. Beck, Ed., The MIT Press, Cambridge, Mass., 1985.

Lux, S. E. Hemolytic Anemias IV. Metabolic Disorders. In *Hematology,* 4th ed., William S. Beck, Ed., The MIT Press, Cambridge, Mass., 1985.

Maugh, T. H. A New Understanding of Sickle Cell Emerges. *Science,* 211 (Jan. 1981).

McCurdy, P. R. The Sickle Cell Trait. *American Family Physician,* Nov. 1974.


# Review Questions

1. List the three major categories of hereditary hemolytic anemia.
2. Name the two major forms of hereditary hemolytic anemia associated with membrane abnormalities. Which one is more severe?
3. Give the three basic mechanisms that cause hereditary spherocytic anemia (HSA).
4. Describe the major laboratory features of HSA.
5. What unique lab test is frequently used to diagnose HSA?
6. Give the treatment of choice for HSA.
7. Where is hereditary elliptocytic anemia found most frequently? What is its incidence?
8. What is an acceptable level of elliptocytosis in a normal blood smear?
9. Define sulfhemoglobin. How is it formed? What happens to sulfhemoglobin in red blood cells?
10. Describe Heinz bodies. Why are they detrimental to red blood cells?
11. How are the levels of sulfhemoglobin kept low in normal blood? What compound keeps the oxidant level low?
12. Why does oxidized glutathione need to be reduced again? How is this accomplished?
13. What does the abbreviation G-6-PD stand for? What is its function? In what pathway is it active?
14. On what chromosome is the G-6-PD gene found? How many alleles of this gene exist? What are two common normal variants?
15. List the three major abnormal variants of the G-6-PD gene. In what populations is each of these variants predominantly found? Why do these abnormal variants still exist in humans?
16. The severity of the G-6-PD deficiency anemia depends on three factors. List them.
17. Mention a number of oxidants that frequently cause acute hemolytic anemia in susceptible individuals.
18. What is the most common cause of hemolytic anemia associated with abnormal G-6-PD enzymes?
19. How would you diagnose the existence of G-6-PD deficiency anemia?
20. Describe the normal treatment for G-6-PD deficiency anemia.
21. When was the term sickle cell first coined? By whom?
22. Compare the structure of HbS with HbA.
23. List three common methods used to elucidate the HbS trait in a person.
24. Compare the severity of the anemia between people with the *A/S* genotype and those with *S/S* genotype.
25. Why does HbS still exist in humans? In which human groups is HbS predominantly found?
26. Describe the molecular basis for sickling.
27. Describe what is meant by (a) nucleation, (b) crystallization, and (c) dissolution.
28. What are irreversibly sickled cells? What is their fate?
29. What is meant by a painful crisis? How is it caused? What initiates such a painful crisis?
30. In which organs can such painful crises occur? What can result from such crises?
31. Define hypoplastic crises. Why are they dangerous? What is the usual cause of such hypoplastic crises?
32. Describe the major preventive measures that can be taken to avoid painful and hypoplastic crises.
33. What can be done if a painful crisis occurs?
34. What can be done if a hypoplastic crisis occurs?
35. In the future how can sickle cell anemia be eliminated?
36. Compare the structure of HbC with HbA.
37. Where does HbC disease mainly occur? Describe HbC disease. Why does HbC disease still exist?
38. Compare the structure of HbD with HbA.
39. In what part of the world is HbD mainly found? What is another name for HbD?
40. Compare the structure of HbE with HbA.
41. In what part of the world is HbE mainly found? Why does it exist in these populations? In what form is it associated with anemia?

# Chapter Fifteen

## *Acquired Hemolytic Anemias*

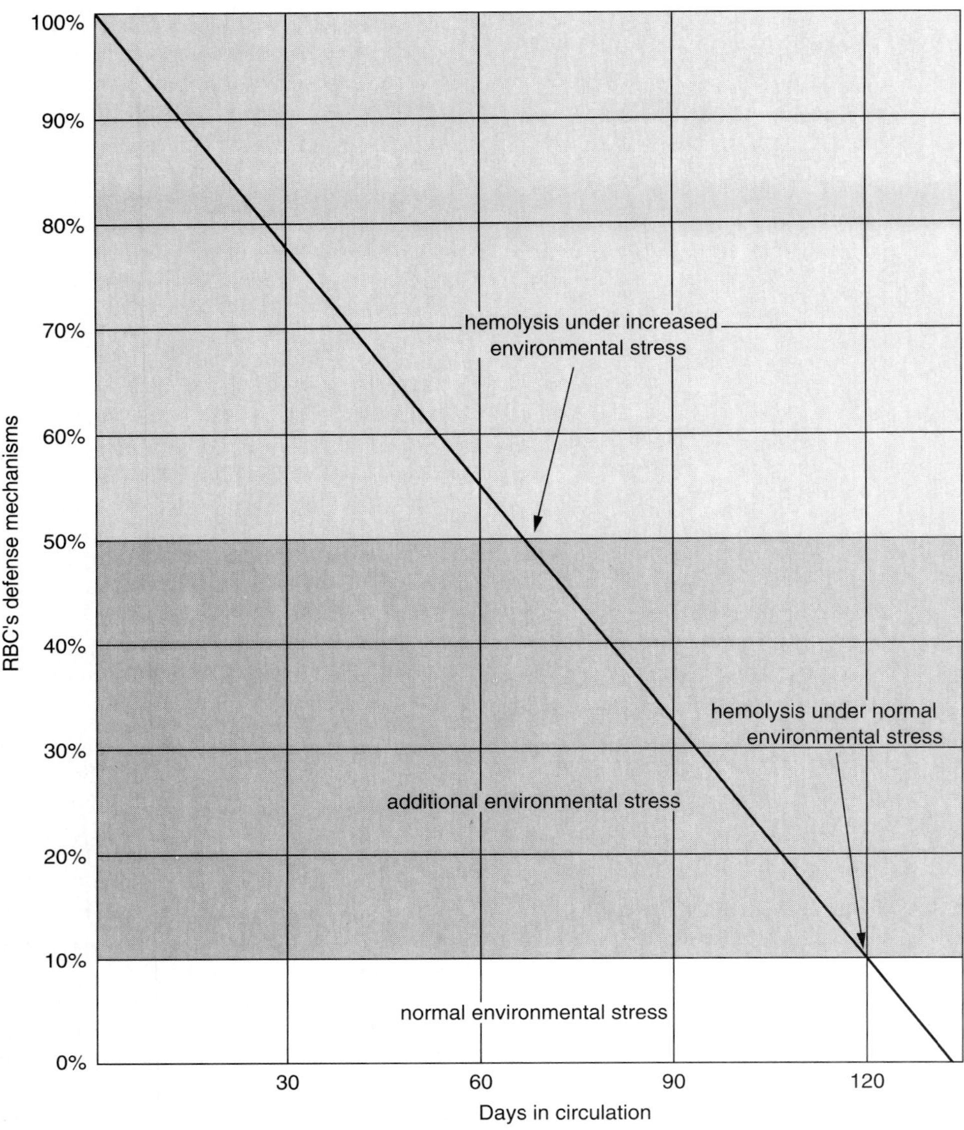


**figure 15.1**

Survival rate of red blood cells exposed to normal environmental stresses and survival rate of red blood cells exposed to increased environmental stresses.

Normal red blood cells have adequate defense mechanisms against all physiological and environmental stresses. However as erythrocytes get older, these defense mechanisms gradually decrease until they reach a point where the cells no longer have the capacity and energy to defend themselves against these stresses, and become subject to intravascular hemolysis or phagocytosis by cells of the mononuclear phagocytic system (figure 15.1). The stresses themselves come from **mechanical factors** (e.g., turbulence or stagnation in the blood flow); **chemical factors** (e.g., the byproducts of metabolic functions, such as oxidants); and **immunological factors** (e.g., destruction by antibodies and complement factors). An increase in these environmental stresses creates a more hostile environment in which the red blood cells must circulate; consequently they spend more energy combating these hostile forces. Since cells have only a finite amount of energy, their life span will be reduced (figure 15.1). If such premature destruction occurs on a large scale, it will result in anemia. Since premature hemolysis is caused by **environmental factors,** it is commonly referred to as **acquired hemolytic anemia.**

## ETIOLOGY OF ACQUIRED HEMOLYTIC ANEMIA

The causes of acquired hemolytic anemia are many and varied. They are grouped into the following six categories according to causation:

1. Activation of the immune system (anemias belonging to this category are known as immunohemolytic anemias)
2. External physical factors
3. Chemical agents
4. Infectious disorders
5. A secondary effect of other diseases
6. Paroxysmal nocturnal hemoglobinuria

## IMMUNOHEMOLYTIC ANEMIAS

Acquired hemolytic anemias caused by activation of the immune system are commonly referred to as **immunohemolytic anemias.** These anemias may be defined as disorders resulting from premature destruction of red blood cells due to inappropriate activation of the immune system. The consequence of this activation is the production secretions by white blood cells, such as antibodies and lymphokines. The production of these cellular secretions results in the activation of other serum factors, such as complement and clotting factors, which make the red blood cells more susceptible to sequestration, phagocytosis, and destruction by cells of the mononuclear phagocytic system.

### Classification of Immunohemolytic Anemia

Immunohemolytic anemias are divided into two major categories: (1) those caused by the action of alloantibodies and (2) those caused by the formation of autoantibodies.

**Alloantibodies** are immunoglobulins produced by the body against antigens that are present on the cells of some individuals but not on the cells of others. Many of the antigens present on red blood cell membranes qualify as alloantigens against which certain individuals can develop alloantibodies. They include the ABO and rhesus (Rh) antigens. Individuals may develop hemolytic anemia in two major ways as a consequence of developing alloantibodies: (1) when a mother with type O blood becomes pregnant and the fetus has type A or B red blood cells and (2) when a Rh-negative mother gets pregnant more than once with a Rh-positive fetus. Both conditions may result in an anemia known as **hemolytic disease of the newborn (HDN).** In the past it was more commonly called erythroblastosis fetalis. HDN occurs in the offspring as a result of antibodies produced by the mother, not by the fetus. Of the two forms of red blood cell incompatibility, ABO-HDN is the more common of the two but rarely causes more than very mild symptoms. The Rh-HDN, however, is the more severe of the two disorders, and is discussed later in greater detail. We should note, however, that many signs and symptoms that occur in Rh incompatibility situations may occur also in certain cases of ABO incompatibility—or for that matter against other incompatible antigens on the red blood cell membrane.

**Autoantibodies** are immunoglobulins synthesized by a person against his or her own self-antigens, including those found on the red blood cells. This type of anemia is commonly referred to as **autoimmune hemolytic anemia** or **AIHA.** People with AIHA produce antibodies that bind to their own red blood cells, which become very susceptible to phagocytosis and premature hemolysis. The two types of autoimmune hemolytic anemia are (1) those caused by **warm-active antibodies**—that is, antibodies active only at body temperature and (2) those caused by **cold-active antibodies**—that is, antibodies active at room temperature or below (between 0 and 10°C). Both types are discussed in the following material.

### Hemolytic Disease of the Newborn (HDN)

#### *ABO Incompatibility*

Blood type O mothers have anti-A and anti-B antibodies in their plasma. Although the maternal and fetal circulations are separate, small blood leaks may occur, and maternal anti-A and anti-B antibodies may enter the fetal circulation where they bind the A or B antigens on the fetal cells. These antibody-antigen complexes activate the complement cascade causing premature lysis of these fetal red blood cells. If this happens, the fetus normally responds with increased erythropoiesis in the bone marrow to compensate for the loss of these cells. If it occurs on a large scale, or when the fetus is unable to compensate properly for the loss of the red blood cells, anemia may occur. Since approximately 23% of all pregnancies involve incompatible HBO systems, it would be logical to conclude that this would frequently result in hemolytic disease in the fetus and newborn. The facts are, however, that such occurrences are very uncommon. In the overwhelming majority of cases, the red blood cell count and the hemoglobin levels of the fetuses and newborns are within normal range, although sometimes a small increase in the reticulocyte count as well as a slightly elevated bilirubin level may be encountered. Normally no overt symptoms of anemia are present, and if erythroblastosis is seen in a blood smear it is usually very mild.

There is no wholly satisfactory explanation for the relatively innocuous nature of the ABO incompatibility syndrome, although several partial answers have been suggested. First, fetal red blood cells have fewer ABO antigens than adult erythrocytes. Therefore they are not as susceptible to phagocytosis and complement activation. Second, it may be that the A and B antigens are not yet well developed in fetuses. This would make it more difficult for the maternal antibodies to attach to fetal erythrocytes. Third, most of the anti-A and anti-B antibodies of the mother belong to the IgM class of immunoglobulins. They are too large to cross the placenta on their own (as the IgG antibodies do), and thus depend on leaks in the maternal and fetal circulations, which appear to be rather rare.

*Rh Incompatibility*

This anemia occurs only in fetuses and newborn infants that are **Rh-positive** and that are born from mothers who are **Rh-negative** and who produce **anti-Rh antibodies.** The rhesus (Rh) factor is a protein present on red blood cell membranes of most people, but not all. A great majority of people possess this rhesus factor, which is inherited: 85% of white populations, 95% of black populations, and 99% of Oriental populations. Consequently less than 15% of the human population does not inherit this Rh factor—that is, it is not expressed on their red blood cell membranes. If an Rh-negative female marries an Rh-positive male, she has a 50% chance of producing an Rh-positive fetus if the father is heterozygous ($Rh^{+}Rh^{-}$), and a 100% chance of producing an Rh-positive fetus if the father is homozygous ($Rh^{+}Rh^{+}$) for the Rh factor. During pregnancy the maternal and fetal circulations are separate, so if the red blood cells of the fetus express the Rh factor, this will not matter to the mother. However at the time of birth, violent contractions occur in the uterus that may cause placental blood vessels to burst, causing some mixing of fetal and maternal blood for a number of hours until the fetus is born. Thus a number of fetal red blood cells will enter the maternal circulation, where they will be recognized as nonself by the mother. She will develop alloantibodies to these fetal red blood cells, which will be destroyed as a consequence within 1 or 2 weeks. But no harm should occur to the fetus since it is born long before that time. Usually the immunity against the fetal Rh antigens is at a low level and consists mainly of IgM antibodies. However when the mother becomes pregnant again with another Rh-positive fetus, small numbers of the fetal red blood cells may get into the maternal circulation as a result of breaks in the fetal circulatory system. This may happen all through her pregnancy. Each time such fetal red blood cells enter the maternal circulation, they act as boosters that stimulate new antibody production. Such secondary, or anamnestic, responses result in the production of higher antibody titers and also result in switching the antibody class from IgM to IgG. These IgG antibodies are much smaller than the IgM antibodies and can readily pass through the placental walls into the fetal circulation, where they attack and destroy the fetal red blood cells (figure 15.2). If the mother has developed a large amount of anti-Rh IgG antibodies, many fetal erythrocytes will be destroyed. In those situations the fetus frequently will die and the mother will experience a miscarriage. However if the mother has only a low level of these anti-Rh antibodies, the fetus may survive in the womb but will be severely anemic when born. Hence this disorder is called hemolytic disease of the newborn. The level and type of maternal antibodies against the Rh antigens on the fetal red blood cells then depends on two factors. First, it depends on how many red blood cells entered the maternal circulation during the initial birth process. Second, once the mother is sensitized against these Rh antigens, and she becomes pregnant with a second Rh-positive child, it depends on how many fetal red blood cells enter the maternal bloodstream and with what frequency they enter.

The term **erythroblastosis fetalis** is also used to describe HDN since blood smears from these babies show the presence of many immature red blood cells or erythroblasts. This is an indication that many red blood cells are destroyed prematurely.

*Symptoms of HDN*

The clinical symptoms of a newborn child with hemolytic disease of the newborn (HDN) vary greatly from mild self-limiting anemia to severe life-threatening situations as a result of cardiac failure, pulmonary congestion, and edema. In many cases the baby's skin turns yellow soon after birth as a result of jaundice. Jaundice does not occur before birth because much of the hemoglobin released into the fetal bloodstream as a result of red blood cell destruction is taken up by the mother's blood and metabolized in her liver. The fetal liver is unable to conjugate bilirubin until well after birth. After the child is born, red blood cell destruction is limited to the residual concentration of the mother's antibody, which has a half-life of about 23 days. So gradually it is cleared from the circulation, and the bone marrow of the newborn will slowly be able to replace the lost red blood cells. However in the first 2 or 3 months, the rapid growth of the fetus requires substantial numbers of erythrocytes, and growth will be slower than normal.

The other major problem is the accumulation of bilirubin in the blood and tissues of the neonate, which no longer has the maternal liver available to help eliminate it. Since its own liver is unable to conjugate all the bilirubin produced, much of the unconjugated bilirubin is deposited in the tissues, where it can do damage. One favorite place of deposition is the central nervous system, especially the basal ganglia. This results in a condition called **kernicterus,** which may result in irreversible brain damage and mental and physical retardation. Another favorite tissue is the skin and hence the condition of **jaundice.** If the mother is able to produce several subsequent Rh-positive babies, the jaundice will be increasingly pronounced in the subsequent babies. Splenomegaly is often present as a result of the spleen's increased action to eliminate excess remnants of old erythrocytes (ghost cells) and conjugated hemoglobin.

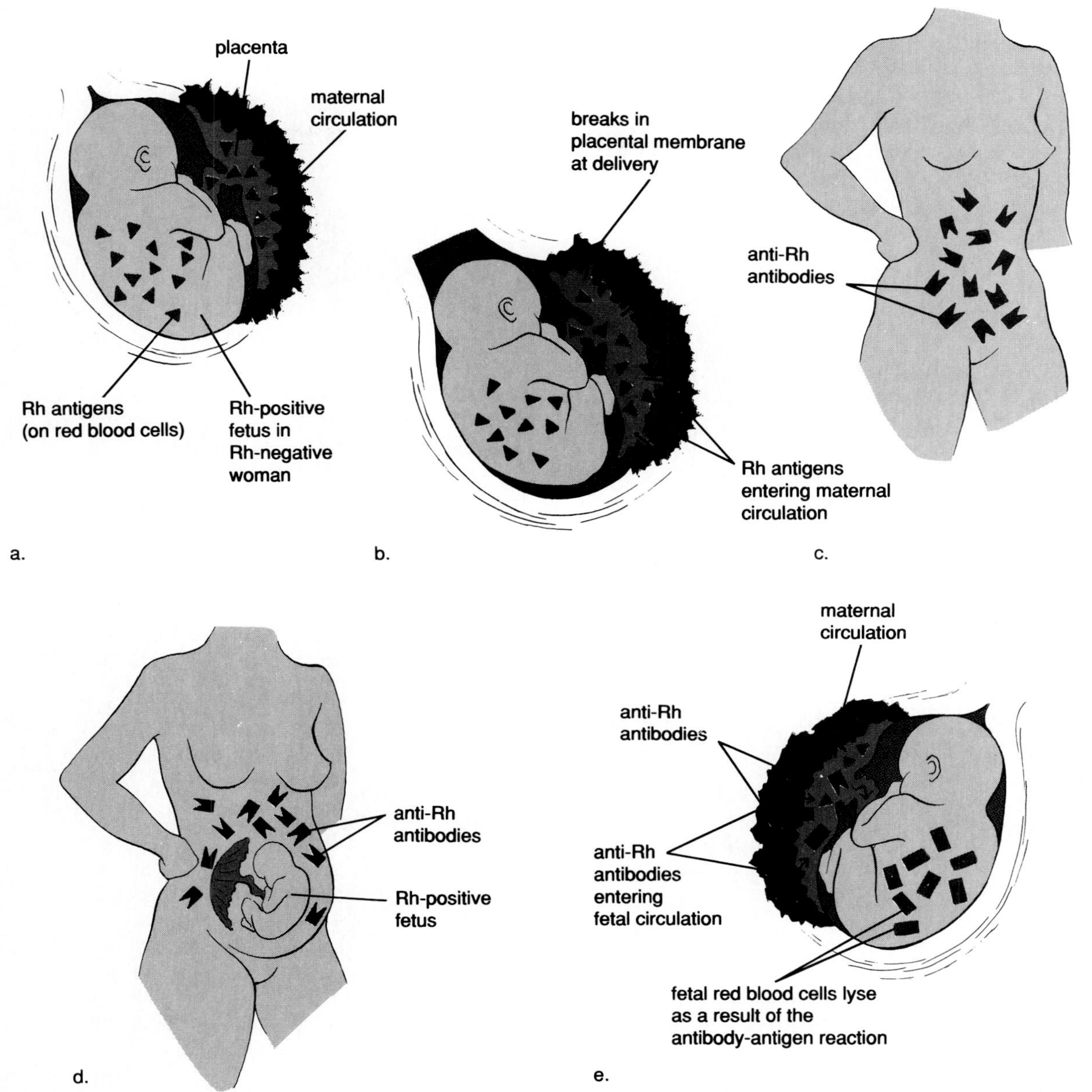


figure 15.2

Development of hemolytic disease of the newborn (HDN). (*a*) Rh-negative woman becomes pregnant with her first Rh-positive child. (*b*) Pregnancy is normal, but during delivery, some Rh-positive cells enter the maternal circulation. (*c*) Woman now produces anti-Rh antibodies. (*d*) After conception of a second Rh-positive child, her anti-Rh antibodies travel through the placenta into the fetal circulation. (*e*) The anti-Rh antibodies attack the fetal red blood cells, causing them to lyse. As a result, the child may be born with severe hemolytic anemia.

### *Laboratory Findings of HDN*

The most obvious laboratory feature is the presence of many erythroblasts (normoblasts) and reticulocytes in the blood smear. The blood film may also show the presence of many ghost cells. The hematocrit value will be low, and levels of unconjugated bilirubin will be high. Haptoglobin levels will be low or absent, and hemoglobinemia and hemoglobinuria will be common. The bone marrow will show erythroid hyperplasia and iron turnover rates will be high.

### *Treatment of HDN*

Blood transfusions may be needed if the level of bilirubin exceeds 10 mg/dL within 24 hours after birth. Transfusion is usually not just infusion of blood but rather an **exchange transfusion,** where the fetal blood containing the maternal anti-Rh antibody is exchanged for compatible blood without the antibody. This procedure drastically reduces the level of maternal antibody, which thereby decreases the level of fetal RBC hemolysis.

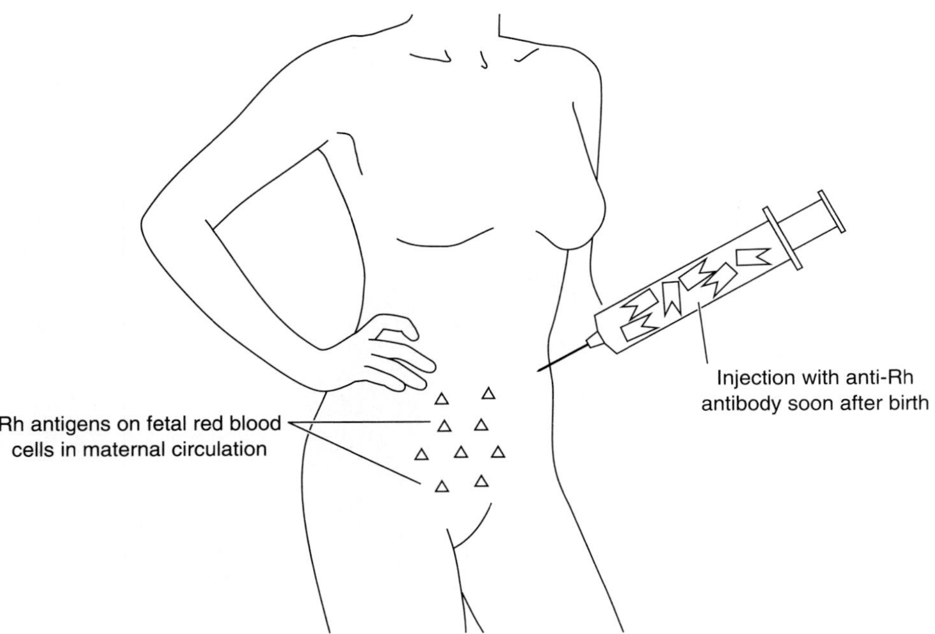


**figure 15.3**

Maternal injection with $Rh_o$ (D) immunoglobulin (RhoGAM, Gamulin Rh), an anti-Rh gamma-globulin, soon after birth will destroy the fetal red blood cells before the mother develops her own antibodies against the fetal Rh antigen.

Amniocentesis may also be used to monitor a fetus known to be compromised with maternal anti-Rh isoantibody. Increased levels of bilirubin can be measured with a spectrophotometer. If the bilirubin level becomes dangerously high late in pregnancy, labor may be induced prematurely. If delivery of the fetus is not possible, intrauterine exchange transfusions may be required.

The best treatment, however, is preventive treatment. If the mother is Rh-negative—and if it was shown immediately after birth of her first child that the baby was Rh-positive (by analyzing the fetal cord blood)—the mother should be given **RhoGAM** within 72 hours after giving birth. RhoGAM is Rh-positive gamma-globulin (=Rh-positive antibody). This antibody will attack and destroy the fetal red blood cells that entered the maternal circulation before the immune system of the mother is properly activated. In other words, the fetal red blood cells will be destroyed before the mother is able to develop her own antibodies against these erythrocytes. Hence she will be able to conceive another Rh-positive child without any complications (figure 15.3).

### Warm-Type Autoimmune Hexmolytic Anemia (AIHA)

**Warm-type AIHA** is caused by the production of autoantibodies of the **IgG class** against antigens present on red blood cell membranes. IgG molecules have maximum activity at 37°C or body temperature; hence they are warm-type. Once these antibodies are produced by the B cells of the affected person, they are released into the bloodstream where they bind to appropriate antigens on red blood cell membranes. The number of antibodies that can bind to a given red blood cell depends on the number of antigens present on the erythrocyte surface. The number usually varies from about 1,000 to 100,000 epitopes or antibody-binding sites per red blood cell (RBC).

Red blood cells coated with IgG are preferentially engulfed by macrophages—especially by the phagocytic cells of the spleen, which have special Fc receptors for IgG on their surfaces. This coating of antibodies to an RBC surface is known as **opsonization,** and the antibodies are often referred to as **opsonins** (figure 15.4). This process of opsonization facilitates the process of phagocytosis. In most cases, however, the RBC is not completely destroyed by the phagocytes of the spleen. Rather a part of its membrane is nipped off by the phagocytic cell in a process called "pitting." As a result these red blood cells become progressively smaller each time they percolate through the spleen, because each time the phagocytic cells remove another part of the RBC membrane. In order to maintain their volume, these red blood cells become more spherical and thus less flexible. Eventually they become so rigid that they can no longer escape the splenic microcirculation because they can no longer squeeze between the phagocytic cells of the cords and sinuses. As a result they become trapped (sequestered) in the spleen and are prematurely destroyed by the mononuclear macrophages of the red pulp of the spleen. If this happens on a large scale, overt anemia will result.

Once the antibodies bind to the antigens on the surface of the red blood cells, they are also able to activate **complement** proteins present in the serum. These activated complement proteins bind on the surface of red blood cells (e.g., C3b, C4b, C5b). This also facilitates the process of phagocytosis of RBCs

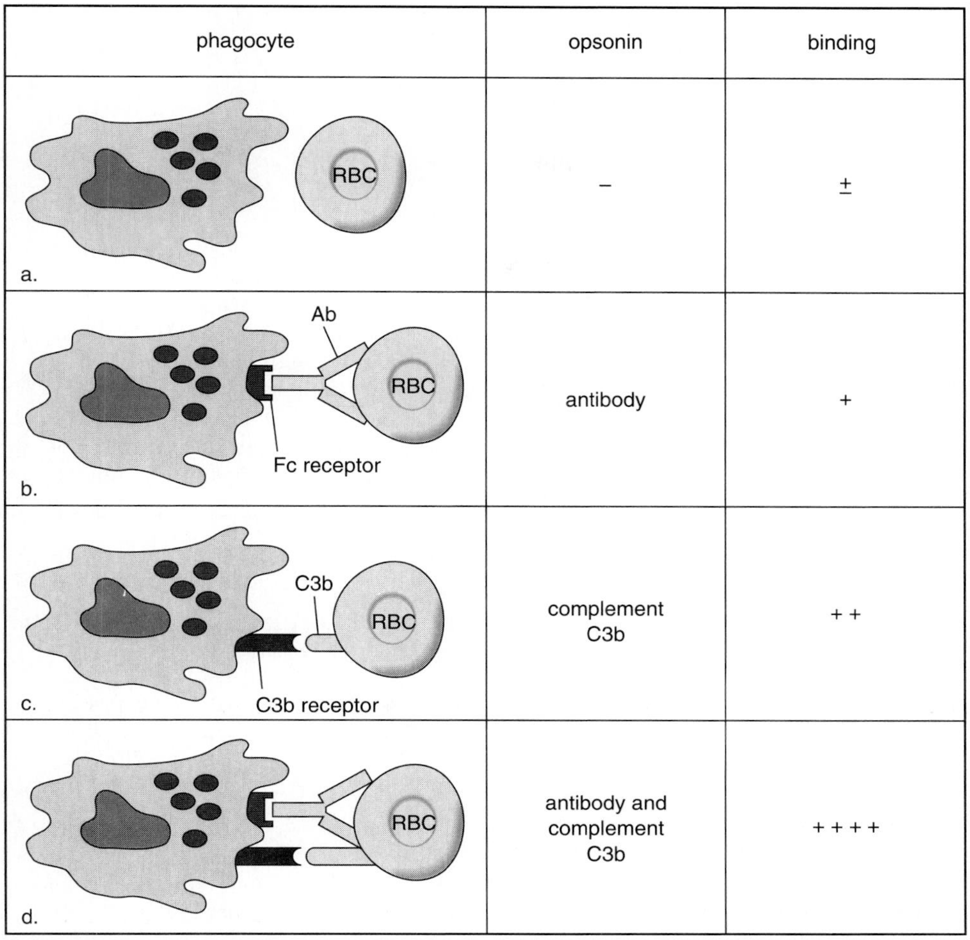

| phagocyte | opsonin | binding |
|---|---|---|
| a. RBC | – | ± |
| b. Ab, RBC, Fc receptor | antibody | + |
| c. C3b, RBC, C3b receptor | complement C3b | + + |
| d. RBC | antibody and complement C3b | + + + + |

**figure 15.4**

Opsonization. When red blood cells are opsonized by (*b*) antibodies and (*c*) C3b proteins, the phagocytic cells can engulf and destroy them much more easily. The phagocyte has receptors for opsonins on its cell membrane, which greatly enhances its ability to bind to and phagocytose the affected red blood cells.

because phagocytic cells also have special receptors for these activated complement factors (figure 15.4). The warm antibodies of the IgG class rarely activate the complete complement cascade. The latter would result in premature intravascular lysis of the red blood cells. This does not normally happen in AIHA, although there are exceptions.

The degree of red blood cell destruction (or hemolysis) by warm-type IgG antibodies is influenced by the following three factors:

1. **Number and distribution of antigens** on the red blood cell surface. Relatively few antigens of a given type are present on the red blood cell membrane, and as a result few antibodies can bind. Generally speaking, the more antibodies bound to the red blood cell membranes, the faster the erythrocyte will hemolyze.
2. **Type of IgG antibody** involved. There are four subclasses of IgG immunoglobulins (or antibodies): IgG1, IgG2, IgG3, and IgG4. Only antibodies of the subclasses IgG1 and IgG3 can bind firmly to the receptors of the phagocytic cells of the spleen. Consequently these two IgG subclasses cause greater damage to red blood cells.
3. **Functional capacity of the mononuclear phagocytes.** The capacity of the mononuclear phagocytes (i.e., monocytes, macrophages, reticular cells, endothelial cells) for destroying antibody-coated red blood cells depends on the number of Fc receptors for IgG on the surface of these phagocytic cells, and on the rate of blood flow through the spleen and other organs, such as the liver, which contain large numbers of phagocytic cells.

### *Causes of Warm-Type AIHA*

Autoimmune hemolytic anemias associated with warm-active antibodies may be classified as idiopathic warm-type AIHA, drug-induced warm-type AIHA, or warm-type AIHA associated with other diseases, according to their causes.

**Idiopathic Warm-type AIHA** About 60% of all warm-type AIHAs may be idiopathic—that is, they occur spontaneously without any known causative agent. Although chemicals or drugs are frequently suspected, no specific chemical compound has been identified.

As long as no specific cause can be determined, these hemolytic anemias are classified as **idiopathic.** Idiopathic warm-type AIHA can occur in both sexes and at all ages, although there seems to be a higher incidence of the middle-aged group found in the workplace.

**Drug-induced Warm-type AIHA** About 20 to 25% of all warm-active AIHAs are drug-induced. Patients treated with certain drugs run the risk of developing AIHA. The first drug implicated was stibophen, a compound used in the treatment of schistosomiasis. This was in 1954; since then many other drugs have been found to have similar effects. Many of these drugs act as **haptens.** A hapten is a small chemical compound that by itself is nonimmunogenic—that is, it cannot elicit an immune response in the body—but when it is bound to a carrier protein, antibodies can be formed against the hapten-protein complex. Once these antibodies are produced, they are able not only to react against the complex but also to bind to the hapten independently. A list of drugs that have been implicated is given in table 15.1.

Drugs implicated in the production of AIHA may be divided into the following three major categories according to the mechanism by which they initiate the process of hemolysis:

**1. Drugs that bind directly to the red blood cell membrane** Certain drugs have the ability to bind to proteins that are normally present on red blood cell membranes. These proteins are self-antigens—that is, they are recognized by the immune system as belonging to the self. Once a foreign drug binds to these proteins, a **neoantigen** is produced, and the body will develop antibodies of the IgG class against this complex. Most of the drugs are small compounds, or haptens, unable to activate the immune system by themselves. They can do so only when bound to a carrier protein, in this case a membrane protein of the RBC, thus producing a neoantigen. Once the antibodies are produced, they will bind to these drug-protein complexes and thus become opsonins. As a result the red blood cells will bind readily to the phagocytic cells in the spleen and elsewhere, where they are pitted constantly, until they become so small and rigid that they are destroyed in the spleen. A good example of this type of drug is penicillin, which may produce acute autoimmune hemolytic anemia, especially when it is used in large doses for prolonged periods of time (figure 15.5*a*).

**2. Drugs that bind to serum proteins** A number of drugs form complexes with normal serum proteins in a manner similar to those that bind to the proteins of the red blood cell membrane. These drugs also act as haptens, and antibodies (IgG, IgM, or both) are also produced against the hapten-protein complex (they are also neoantigens). For some reason, these antibody-antigen complexes are then deposited on the red blood cell membrane, where they tend to activate the complement system (figure 15.5*b*). This results in the production of activated complement factors on the red blood cell surface (e.g., C3b, C4b, C5b). The phagocytic cells have special receptors for these activated complement factors, and hence these red blood cells become much more susceptible to premature hemolysis. If this happens to many red blood cells simultaneously, acute hemolytic anemia will occur. Drugs in this category include stibophen as well as the antimalarials quinine and quinidine, the antibiotic sulfonamides, and the analgesic phenacetin.

**3. Drugs that induce antibody against RBC antigens** α-methyldopa has been implicated in this type of AIHA. It is a commonly used antihypertensive drug that may produce warm-type AIHA in some users (less than 1%). Exactly how this occurs is not known, but α-methyldopa may react with the gene coding for the Rh factor on the red

## Table 15.1 Drugs Implicated in the Production of Autoimmune Hemolytic Anemia (AIHA)

| |
|---|
| Drugs that bind directly to the red blood cell membrane |
| Penicillin (antibiotic) |
| Drugs that bind to serum proteins |
| Stibophen (schistosomicidal) |
| Quinine, quinidine (antimalarials) |
| Sulfonamides (antibiotics) |
| Phenacetin (analgesic) |
| Drugs that induce antibody against RBC antigens |
| Synthetic α-methyldopa (Aldomet—antihypertensive) |

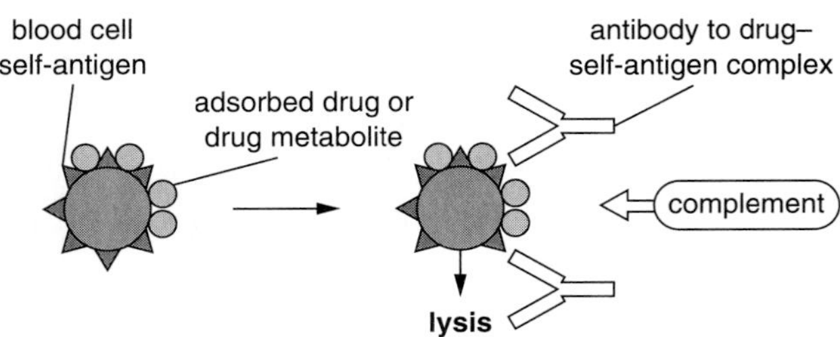


**figure 15.5**
The action of drugs in autoimmune hemolytic anemia. (*a*) Drugs that bind directly to the red blood cell membrane. (*b*) Drugs that bind first to serum proteins, resulting in a complex that binds to the red blood cell membrane. In both cases, the result is the activation of the complement system and, eventually, cell lysis.

blood cells, since analysis shows that introducing this drug results in the production of anti-Rh antibody but not anti-α-methyldopa antibody. Normally the numbers of Rh antigens on red blood cells are very low. The introduction of α-methyldopa, however, may greatly increase the number of Rh antigens on the red blood cell, and this in turn may trigger the production of anti-Rh antibodies by the immune system. The affinity of this antibody is usually very low, and may explain why so few people develop overt AIHA when they take this drug. In some cases, however, greater affinity occurs, resulting in the development of stable antibody-antigen complexes on the red blood cell surface. These complexes then activate the complement system, and the activated complement factors on the erythrocyte surface thus make these cells much more receptive to hemolysis by phagocytic cells of the spleen. As soon as the drug is withdrawn, the level of anti-Rh antibodies begins to drop, disappearing altogether within 3 to 4 months. The body itself produces a natural L-dopa, which does not cause hemolytic anemia. Only the synthetic compound α-methyldopa does so. Presumably other medications may act in a manner similar to α-methyldopa, such as L-dopa, metanemic, and procainamide.

**Warm-type AIHA associated with other diseases**
About 15 to 20% of all warm-active AIHAs occur as a result of other disorders. The major disorders involved are the lymphoproliferative diseases (e.g., lymphocytic leukemia), systemic lupus erythematosus (SLE), and infectious mononucleosis. Certain other infectious disorders have also been implicated.

*Symptoms of Warm-Type AIHA*

Warm-type AIHA may range from a mild chronic hemolytic anemia with few overt symptoms to an acute, explosive disorder with rapid hemolysis and many overt symptoms commonly associated with severe hemolytic anemia. However, in most cases this warm-type AIHA is self-limiting and thus usually shows mild symptoms.

In idiopathic warm-type AIHA, the anemia may vary from a very short transient phenomenon lasting only a few weeks to a chronic condition that may last many years. In drug-related cases the symptoms usually stop soon after the drug is withdrawn. In those instances in which the AIHA is the result of other diseases, the hemolytic anemia will usually stop when the disease goes into remission.

In cases with moderate to severe hemolysis, we can expect the development of splenomegaly, hepatomegaly, and in severe cases also congestive heart failure. When there is only a mild anemia, overt symptoms will usually be absent because the body is able to compensate for the increased loss of red blood cells by increased hematopoietic activity of the bone marrow.

*Laboratory Findings of Warm-Type AIHA*

Laboratory investigations of this disorder will usually exhibit a high percentage of reticulocytes (10 to 30%), and in cases of acute explosive hemolysis, the presence of nucleated red blood cells can be observed in the peripheral blood. Since hemolysis is the result of constant pitting of red blood cells by phagocytic cells of the spleen and liver, we can also expect many spherocytes, microcytes, and microspherocytes. As many as 60% of all red blood cells may reveal spherocytosis. The level of bilirubin will also be higher than normal, and frequently haptoglobin levels will be lower. Hemoglobinemia and hemoglobinuria will also be present in moderate to severe cases of hemolytic anemia. Finally, the bone marrow will show erythroid hyperplasia.

*Diagnosis of Warm-Type AIHA*

A unique test is used to diagnose the presence of AIHA caused by the warm-type IgG class of antibodies. This test is the **direct Coombs' test,** classified as an antiglobulin agglutination test—that is, a test in which the red blood cells clump, or agglutinate, after Coombs' serum is added if the RBCs are covered with IgG. Coombs' serum is anti-IgG (antiserum or antiglobulin) produced in rabbits. IgG is a serum protein found in the γ-globulin fraction of the serum.

If the human γ-globulin serum fraction is injected into a rabbit, the rabbit's immune system will respond to it as foreign proteins and develop its own antibodies against it. Thus Coombs' serum is rabbit serum that contains anti-IgG antibody (antiglobulin). One of the properties of the IgG molecule is that it has two antigen-binding fractions (Fab) and one complement-binding fragment (Fc) (figure 15.6). In theory IgG should be able to cross-link or bind to two different particles at the same time at the Fab sites, causing them to clump or agglutinate. This is true when IgG binds to most microorganisms such as viruses and bacteria. IgG molecules are, however, unable to clump red blood cells directly, because each red blood cell is negatively charged on its surface. Hence similarly charged cells will repel each other. In normal physiological conditions, two red blood cells will always repel each other at a minimum distance of 300 Å. This repulsive force is known as the **zeta potential.** The greatest span of the two antigen-binding sites on an IgG molecule is only 240 Å (figure 15.7). Hence an individual IgG molecule is unable to bridge the zeta potential gap. Coombs' serum, however, contains an antibody against the Fc or tail fraction of the IgG molecule and is able to cross-link the Fc fractions of two IgG molecules of two different erythrocytes, thus causing them to agglutinate (figure 15.8).

In warm-type AIHA the red blood cells are covered with IgG. Therefore if Coombs' serum is added, the erythrocytes will clump readily. This is a positive test for the presence of IgG antibody on red blood cells and thus for the existence of warm-type AIHA.

*Treatment of Warm-Type AIHA*

In most cases of warm-type AIHA, the anemia is so mild that no special treatment is needed. However in certain rapid-onset cases that produce moderate to severe anemia, the treatment of choice is the administration

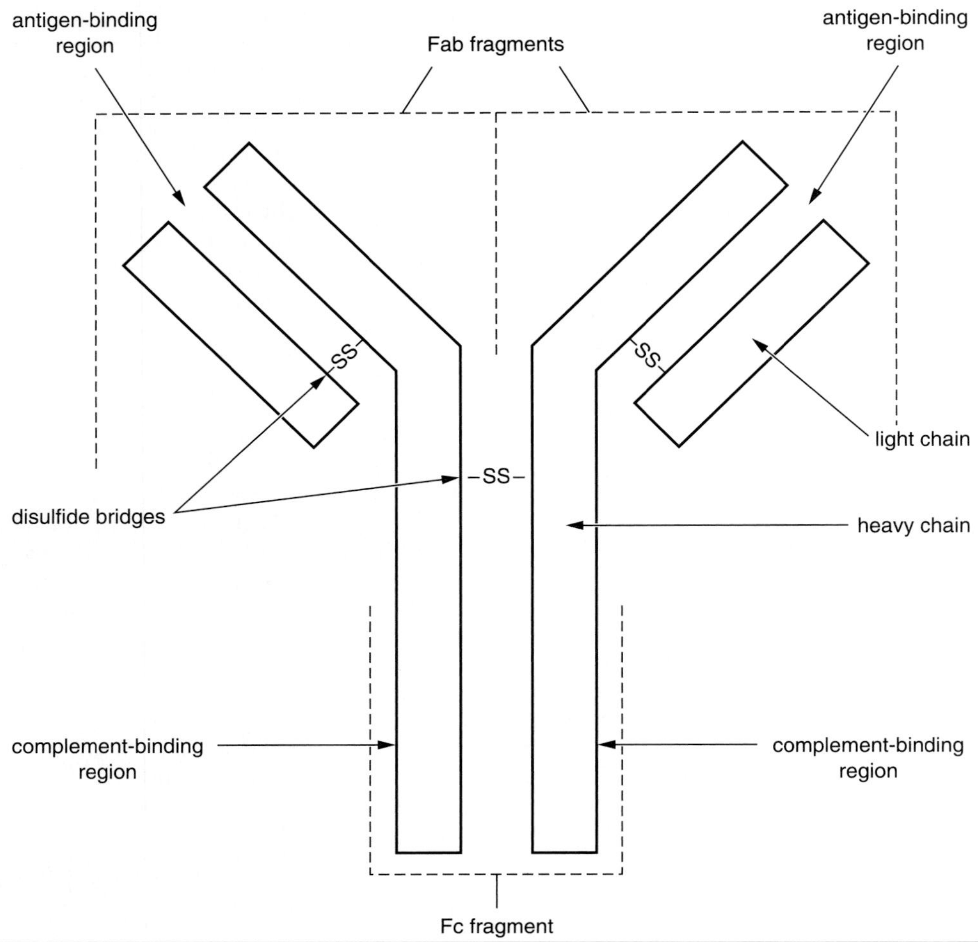


**figure 15.6**

Each antibody molecule consists of two antigen-binding fractions (Fab) and one complement-binding fraction (Fc).

of corticosteroids, since these steroid hormones are effective in at least 80% of all cases of AIHA. Apparently these corticosteroids prevent the premature destruction of red blood cells by inhibiting the activity of mononuclear phagocytic cells. These steroids probably also have some immunosuppressive effect since antibody levels fall rapidly during administration of these hormones.

If steroid therapy fails to control hemolysis, splenectomy may be considered as a next step, since most of the premature destruction of the red blood cells takes place in the spleen. If removal of the spleen is inappropriate or fails to alleviate the hemolysis, a treatment course with immunosuppressive drugs may be needed (e.g., azothioprine).

Blood transfusions are not normally a part of the treatment program because of the dangers of isoimmunization. Such an action is justified only in cases exhibiting a life-threatening explosive level of hemolysis.

### Cold-Type Autoimmune Hemolytic Anemia (AIHA)

In contrast with warm-type antibodies belonging to the IgG class of immunoglobulins, cold-type antibodies belong mainly to the **IgM class.** Maximum activity occurs at temperatures between 0 and 10°C, with an optimum temperature between 2 and 4°C. Above 10°C the effective affinity of the IgM molecule for the red blood cell antigen diminishes and dissociates from the cell-bound antigens. No activity is present above 30°C. This contrasts with the IgG molecules, which have maximum activity at 37°C or body temperature. Also contrary to IgG antibodies, immunoglobulins of the IgM class are able to agglutinate red blood cells directly, because they are much larger and are able to bridge the zeta potential gap between individual erythrocytes. IgM antibodies also activate complement much more rapidly compared to IgG molecules. However, similar to the activation of IgG autoantibodies, IgM autoantibodies rarely induce activation of the full complement cascade but function through the activated complement factors C3b, C4b, and C5b.

#### *Causes of Cold-Type AIHA*

Practically all cases of **cold-type AIHA** are associated with the production of an IgM antibody against the **"I" antigen** normally present on red blood cell membranes of most adults. Also individuals normally possess low levels of this cold-active anti-I IgM class of

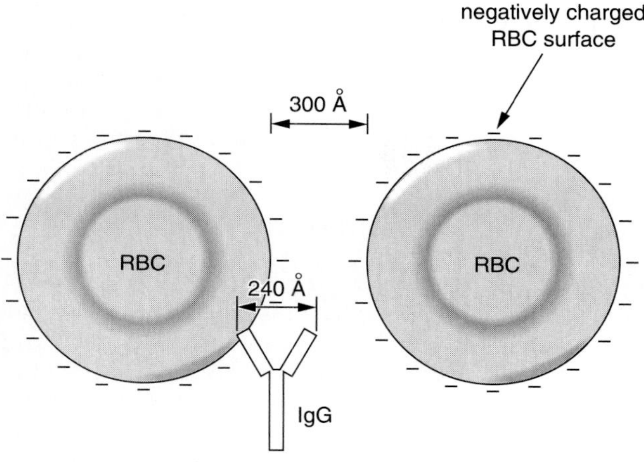


**figure 15.7**
The zeta potential is the repellent force between two negatively charged red blood cells.

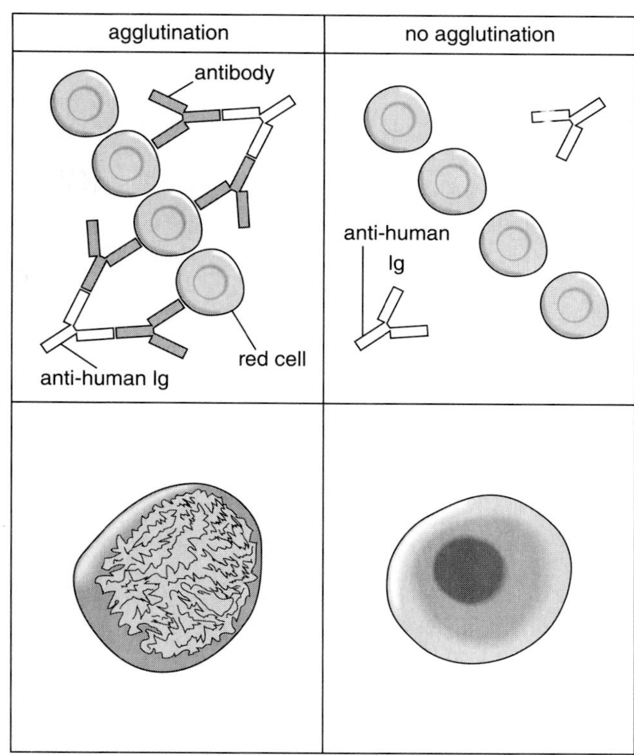


**figure 15.8**
The direct Coombs' test. The red blood cells are directly tested for the presence of antibodies. Individuals suspected of suffering from autoimmune hemolytic anemia will have IgG antibodies attached to the erythrocyte surfaces. If these cells are incubated with serum containing anti-IgG antibodies, the cells will agglutinate. If they have no IgG antibodies on their surface, the anti-IgG antibodies cannot cause agglutination.

antibody in their blood. The I antigen is not present on the RBC membranes of fetuses and very young infants; they have an equivalent **"i" antigen** instead. After birth these i antigens are gradually replaced by I antigens. The reason for the cold reactivity of the IgM class of antibodies does not appear to be a function of the antibody but rather of the Ii antigens on red blood cell membranes. These antigens presumably move to more accessible positions of the membrane surface as the temperature is reduced. In certain situations there may be a sudden increase in IgM against this red blood cell antigen. When such individuals are exposed to cold temperatures, they may develop hemolytic anemia.

The high levels of I antigens on the red blood cell membrane and high titers of anti-I IgM antibodies in the blood will result in the binding of large numbers of IgM molecules to the membrane of erythrocytes. These antibody-antigen complexes can now activate the complement system via the classical pathway and thus cause production of activated complement factors, especially C3b, which also binds to the red blood cell surface. C3b may cause intravascular hemolysis if the complement sequence goes on to completion. Mononuclear phagocytic cells of the liver and spleen have special receptors for this C3b marker and will attach to them. This will result in the pitting of the red blood cells. The latter will result in the production of microspherocytes, which will become sequestered in the splenic microcirculation and become a target for complete phagocytosis and premature hemolysis.

Furthermore, IgM-coated red blood cells are subject to agglutination in the peripheral circulation, because the IgM antibodies can bridge the zeta potential gap (figure 15.9). This clumping process occurs most readily in the blood vessels of the extremities such as the hands, feet, nose, and ears, which are more easily subjected to cold temperatures. This agglutination may cause problems of circulation in these organs; however when red blood cells pass back from colder areas to the warmer internal circulation, antibodies detach again from the red blood cells, thus preventing clumping in the internal circulation. However if activated complement factors have been bound by the time the blood cells return to the warmer internal environment, they will stay bound, and thus subject these cells to premature hemolysis by the phagocytic cells of the liver and spleen.

In about 30 to 40% of all cases of cold-type AIHA, no cause is known for the sudden increase in anti-I antibodies in the blood. These cases are referred to as **idiopathic cold-type AIHA.** In the other 60 to 70% of cases, the anemia is associated with the presence of other disorders. Elevated titers of cold-type IgM antibodies are especially common during the recovery

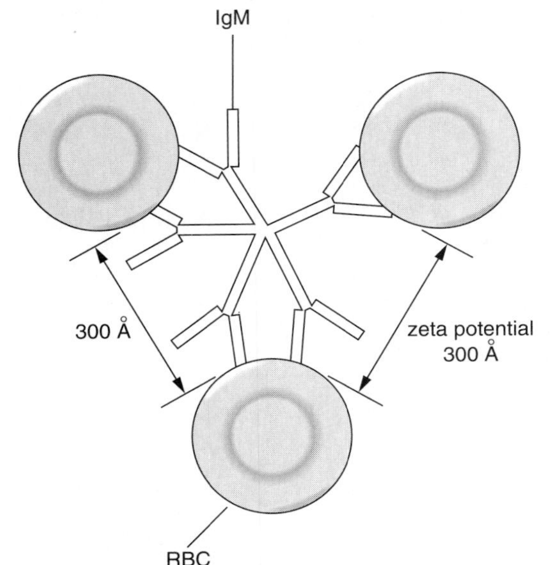


**figure 15.9**

The IgM antibody is a pentameric antibody with a diameter greater than 300 Å; thus it can easily bridge the zeta potential gap.

phase of certain infectious diseases such as mycoplasmal pneumonia and infectious mononucleosis (the latter is frequently associated with a sudden increase in anti-i antibodies). Why infection should stimulate the production of antibodies that can react with antigens on the RBC membrane is unknown. In addition, instances of this cold-type AIHA have occurred in persons suffering from lymphoproliferative disorders.

*Symptoms of Cold-Type AIHA*

Generally speaking, this cold-type AIHA occurs in older people whose circulation has become less efficient. Most overt symptoms occur as a result of the blockage of small blood vessels in the extremities due to agglutination of large numbers of red blood cells. These occlusions may result in sharp pains due to the development of hypoxia and cyanosis in these areas. Another effect is the painful blanching of the skin of the affected extremities.

Most affected individuals have very few overt symptoms of anemia since the level of premature hemolysis is usually rather low. However exceptions are possible, and severe rapid-onset hemolysis may occur, resulting in full-blown cases of hemolytic anemia. Finally, splenomegaly is not uncommon in people with a chronic form of cold-type AIHA.

*Laboratory Findings of Cold-Type AIHA*

Laboratory features of cold-type AIHA are similar to those of warm-type AIHA, except that spherocytosis is usually less marked. There is also a tendency for fresh blood specimens to agglutinate when exposed to room temperature or below. Often it is necessary to draw blood into a preheated syringe and to warm the microscope slides before making the blood smears. The occurrence of large-scale clumping at temperatures below 30°C is usually a diagnostic feature of this disorder.

*Treatment of Cold-Type AIHA*

In cases of idiopathic cold-type AIHA, the best treatment available is preventive: avoidance of cold exposure and use of homely measures such as mittens and warm underwear. In cases where the underlying cause is known, treatment should be directed toward curing the disorder. For instance, when the syndrome is associated with lymphoproliferative disorders, improvement in cold type AIHA symptoms may follow treatment, such as antilymphoma or antileukemic chemotherapy.

In most instances, especially in chronic forms of cold-type AIHA, antibody titers should be lowered. This may be achieved by plasmapheresis. The administration of alkylating drugs, such as chlorambucil, has also been used with some success. Corticosteroid drug treatment has not been very effective in this case. Since splenomegaly is usually mild, splenectomy is normally unnecessary.

## OTHER ACQUIRED HEMOLYTIC ANEMIAS

Besides the anemias associated with the activation of the immune system, other causes of acquired hemolytic anemia include (1) physical factors; (2) chemical agents; (3) microorganisms; (4) as a consequence of other diseases; and (5) paroxysmal nocturnal hemoglobinuria (PNH). Each of these causes will be discussed in the following material.

### Hemolytic Anemia Caused by Physical Factors

This type of acquired hemolytic anemia is the result of mechanical damage to the red blood cells and is therefore commonly referred to as the **red blood cell fragmentation syndrome.** The most common causes for this red blood cell fragmentation syndrome are (1) prolonged violent physical exercise, especially long-distance running or marching, which may result in a condition known as **march hemoglobinuria**; (2) artificial heart valves, which may produce a condition known as **traumatic cardiac hemolytic anemia**; and (3) damage caused by vasculitis or formation of small clots in the microcirculation, which may result in a disease known as **microangiopathic hemolytic anemia.**

*March Hemoglobinuria*

March hemoglobinuria is rather common in dedicated long-distance runners and in people such as soldiers who march for long periods of time. Running and marching may cause destruction of red blood cells in the microcirculation of the feet due to the pressure on the blood cells by repeated forceful contact of the soles of the feet with hard surfaces. These people commonly pass dark urine immediately after such prolonged exercises. This phenomenon is caused by the intravascular destruction of red blood cells, resulting in the release of free hemoglobin in the blood. This hemoglobin is then filtered out by the kidneys, from which it is moved to the urine. Normally no overt anemia is associated with this phenomenon, but it can result in a mild form of anemia in people who lack reserves of iron or folic acid or in those who are otherwise susceptible to some form of anemia. No special laboratory findings are associated with this condition, except perhaps lower than normal levels of plasma haptoglobin and increased levels of free hemoglobin in the blood and urine for a short time after such long-distance exercises.

Softer lining of the running or walking shoes as well as the development of a better stride may reduce the level of hemoglobinuria.

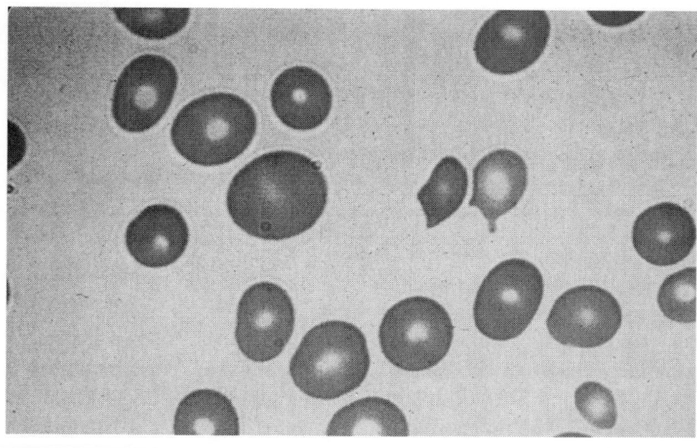

figure 15.10

Blood smear from a patient diagnosed with microangiopathic hemolytic anemia, also known as disseminated intravascular coagulation (DIC).
From The American Society of Hematology Slide Bank, 3rd edition 1990, used with permission.

*Traumatic Cardiac Hemolytic Anemia*

Patients who have had intricate corrective surgery, such as the emplacement of artificial heart valves, may also develop mild cases of hemolytic anemia with hemoglobinuria. Hemolysis is caused by the forceful interaction of the fragile red blood cells with the hard surfaces of these inert structures. The majority of instances of traumatic cardiac hemolysis involve the artificial aortic valve.

The severity of the anemia in patients with hemolytic anemia due to heart valve prostheses is highly variable. Almost all patients with aortic valve replacement experience a mild form of hemolysis that is completely compensated for by a slight increase in bone marrow activity. A few patients, however, experience moderate to severe anemia that may require regular blood transfusions.

Laboratory features of this type of hemolytic anemia include the presence of many poikilocytes and schistocytes such as triangular cells and other fragmented forms with sharp points. If iron deficiency is present, then microcytosis and hypochromia are also common. The reticulocyte count is always elevated, and the hemoglobin levels are usually below normal. Apart from extra iron and vitamin intake, no treatment is usually necessary, except in very severe cases, which will require the replacement of the artificial valve.

*Microangiopathic Hemolytic Anemia*

Abnormalities of the vascular endothelium due to inflammation (vasculitis) or **disseminated intravascular coagulation (DIC)** may also cause premature red blood cell destruction. Such conditions may be caused by infectious diseases such as Rocky Mountain spotted fever, malignant neoplasms, thrombocytopenic purpura, hypertension due to kidney impairment, and various forms of septicemia. All these disorders cause changes in the normal blood vessel diameters, especially in the microcirculation. As a result these blood vessels become narrower. Narrowing of the smaller arteries and arterioles especially damages red blood cells, as the blood is pushed with great force through these vessels.

Laboratory findings are similar to those of other fragmentation syndromes and include the presence of many fragmented cells, poikilocytes, teardrop cells, and anisocytes (figure 15.10), high levels of reticulocytes, increased serum and urine hemoglobin levels, and lower serum haptoglobin levels.

Treatment for this condition involves a three-pronged approach: (a) treatment of the primary disease that originally caused the condition, (b) treatment aimed at removing the intravascular clots, and (c) treatment for the anemia itself. Treatment of the primary condition is often difficult as the causative agent may not be known. However symptoms such as hypertension may be relieved by the administration of antihypertensive drugs, and septicemia may be dealt with by antibiotics. The problem of coagulation may be handled with anticoagulant drugs. The anemia is usually mild and thus warrants no special treatment except for extra iron and vitamins. However in cases of severe anemia, blood transfusions may be needed.

### Hemolytic Anemia Caused by Chemical Agents

A number of industrial chemicals, pharmaceutical drugs, and natural poisons may induce hemolytic anemia in susceptible individuals (table 15.2). The most important of these are as follows:

1. **Industrial chemicals** such as lead, copper salts, sodium and potassium chlorate, nitrobenzene, aniline, arsine, trinitrotoluene (TNT), naphthalene (=mothballs), and methylene blue are well-recognized causes of hemolytic anemia. Each of these compounds acts in its own specific way to produce accelerated hemolysis.

**Table 15.2 Well-known Chemical Agents Implicated in Production of Hemolytic Anemia**

*Industrial Chemicals*
- Lead
- Copper
- Sodium chlorate
- Potassium chlorate
- Nitrobenzene
- Aniline
- Arsine gas
- Methylene blue
- Naphthalene (mothballs)
- Trinitrotoluene (TNT)

*Pharmaceutical Drugs*
- Aspirin
- Phenacetin
- Acetanilid
- Quinine, chloroquine, primaquine
- Nitrofurans
- Sulfonamides
- Chloramphenicol
- Dimercaprol

*Natural Poisons*
- Spider bites
- Insect bites
- Snakebites

2. **Pharmaceutical drugs** such as the analgesics aspirin, phenacetin, and antipyrine; antimalarials such as quinine, chloroquine, and primaquine; sulfonamide drugs; chloramphenicol; probenecid; and dimercaprol. All these drugs act primarily on erythrocytes deficient in G-6-PD, but all red blood cells are potentially sensitive to these oxidant drugs if given at high doses.
3. **Natural poisons** such as spider bites, insect stings, and snakebites have on occasion been associated with severe hemolysis.

### Hemolytic Anemia Caused by Infectious Agents

Certain infectious diseases may also cause hemolytic anemia via a variety of mechanisms. The most important of these and the simplest to explain are those microorganisms that invade the red blood cells, use their contents to multiply, and in the process destroy them. This describes what happens to red blood cells when invaded by the malaria parasites of the genus *Plasmodium.* In addition, certain infectious microorganisms produce substances that lyse the membrane of red blood cells and thus destroy them. This is what happens when the bacterium *Clostridium welchii* produces the potent hemolytic substance known as lysolecithin.

Infectious bacteria may also cause premature destruction of red blood cells indirectly by producing polysaccharides that become adsorbed to the red cell membrane. The body's immune system reacts to these foreign antigens by developing antibodies against them. Once the antibodies are produced, they bind to the foreign polysaccharides on the erythrocyte membrane. These antibody-antigen complexes will then activate the complement system and cause premature lysis of the red blood cells.

**Table 15.3 Microorganisms Implicated in the Production of Hemolytic Anemia**

- *Bartonella bacilliformis*
- *Clostridium welchii*
- *Streptococcus pneumoniae*
- *Escherichia coli*
- *Hemophilus influenzae*
- *Leishmania donovani*
- *Mycobacterium tuberculosis*
- *Neisseria meningitidis*
- *Plasmodium falciparum, P. malariae, P. vivax*
- *Salmonella* species
- *Toxoplasma gondii*
- *Vibrio cholerae*

Table 15.3 provides a short list of some of the major microorganisms implicated in the production of hemolytic anemia. Laboratory findings for these types of acquired hemolytic anemia are similar to those in other hemolytic disorders and include such phenomena as hemoglobinemia, hemoglobinuria, low haptoglobin levels, increased bilirubin levels, and increased reticulocytosis. In some cases such as malaria, the parasite can be identified inside the red blood cell in a stained blood smear.

All treatment should focus on the removal of the underlying infection. In cases of sudden explosive hemolysis, blood transfusions may be necessary.

### Hemolytic Anemia Caused by Other Diseases

The average life span of the red blood cell is considerably shortened in many systemic disorders. Many inflammatory and malignant disorders are associated with an accelerated hemolysis of erythrocytes. Patients suffering from renal failure frequently produce abnormal **burr cells,** which are also subject to early hemolysis. Patients with liver disease often produce acanthrocytes and target cells, which are also preferentially destroyed by phagocytic cells of the spleen. Splenomegaly often associated with this form of anemia because increased numbers of phagocytic cells in the red pulp accelerate hemolysis of abnormal erythrocytes. An enlarged spleen also causes many red blood cells—normally present in the peripheral circulation—to be sequestered in the spleen, thus making fewer cells available for blood gas transport. Further, an enlarged spleen will result in an increase in the plasma volume, producing a dilution of blood cells in the circulation. Treatment for this type of anemia is to reduce the size of the spleen, or if that is impossible, removal of the spleen (splenectomy) may be necessary.

**Paroxysmal Nocturnal Hemoglobinuria (PNH)**

This unique syndrome occurs as a result of the action of complement on red blood cell membranes. The complement system is composed of at least 16 inactive proteins normally found in blood serum. These circulating plasma proteins can be activated sequentially by two major pathways. One pathway is by the presence of antibody-antigen complexes. When complement is activated by these immune complexes, it follows the **classical pathway** of complement activation. The complement system can also be activated by the presence of bacterial endotoxins, certain plant polysaccharides, cobra venom, or yeast extracts. When complement is activated in this manner, it follows the **alternate pathway,** which skips several of the pathway steps (C1, C4, C2) common to the classical pathway. The alternate pathway does not require the presence of specific antibody for initiation. The overall result of this pathway, however, is the same as that of the classical pathway. In PNH complement is activated via the alternate pathway, although the initiating agent is not known.

*Cause of PNH*

Although the cause of the defect that attracts activated complement to the red blood cell membrane is not known, it is not considered to be an inherited factor, since no familial connections have been identified. The defect may be the result of a somatic mutation in some of the hematopoietic stem cells in the bone marrow, since patients with PNH have two distinct populations of red blood cells: those sensitive to the action of complement and those that are not. Other researchers have suggested that environmental factors such as drugs or industrial chemicals may be the cause of this complement activation.

*Symptoms of PNH*

PNH is predominantly a disease of young adults between the ages of 25 and 45 years, affecting both sexes equally. This disorder may affect all blood cell lines produced in the bone marrow, not just red blood cells, since pancytopenia is a common phenomenon in these patients. All the affected blood cells are destroyed prematurely by the action of activated complement.

The presence of PNH may be suspected from a history of dark urine on awakening in the morning. Hence the name **nocturnal hemoglobinuria,** because of an increased intravascular hemolysis during the night. PNH is often associated with iron deficiency since large quantities of iron are lost in the urine. Patients with PNH also have a higher than normal incidence of venous blood clots. Perhaps this results from the release of clotting factors from lysed red blood cells into the venous circulation.

Anemia associated with PNH ranges from none at all to fairly severe. Distinct morphological forms of red blood cells are not associated with this disorder. However if the anemia is moderate to severe, the number of reticulocytes in the blood will show a distinct increase. When the iron stores are depleted, microcytosis and hypochromia will be exhibited.

The severity of this disorder depends on a number of factors including (1) the proportion of red blood cells sensitive to complement, (2) the ability of the bone marrow to compensate for increased erythrocyte loss, and (3) the presence of other complications such as thrombosis. Obviously complications and overt symptoms are more likely to happen in those patients with large numbers of complement-sensitive cells.

*Diagnosis of PNH*

The diagnosis of PNH depends on the demonstration of the presence of complement-sensitive red blood cells. Several tests have been devised for this purpose. These tests are based on the ability of a small amount of activated complement to lyse sensitive red blood cells. Two common tests are as follows:

1. **The Ham test**
In this test fresh serum is acidified, usually with hydrochloric acid to a pH of 6.4. The cells from the patient to be tested are then placed into the acid serum for 30 minutes at 37°C. Lysis of RBCs is then determined and compared with normal controls. Normal red blood cells will not lyse under these conditions, but complement-sensitive cells will rupture readily.
2. **The sucrose lysis test**
In this test red blood cells of patients suspected of having PNH are incubated in a sucrose medium of reduced ionic strength. To this mixture a small amount of complement-containing serum is added. In this medium PNH cells are lysed readily, and normal cells are not.

*Treatment of PNH*

In severe cases of hemolytic anemia, blood transfusions may be needed. The advantage is that complement-sensitive cells are replaced by normal cells. However the always-present danger of isoimmunization or the infection of the patient with hepatitis or acquired immunodeficiency syndrome (AIDS) precludes transfusion as a routine treatment.

Many PNH patients not receiving blood transfusions often become iron deficient because of the constant loss of iron in the urine. Therefore iron replacement therapy is required. Since PNH is also frequently associated with thromboembolic phenomena, anticoagulant therapy may be needed. Finally, androgen therapy is commonly used to treat this disease, since high doses of androgenic hormones stimulate the hematopoietic cells of the bone marrow. However the dangers of excess androgenic drugs in females are obvious.

PNH was once thought to be a serious disease with poor chances of survival. The diagnosis of this disorder was made only in the most severely affected patients who produced large amounts of hemoglobin in the urine due to the presence of large numbers of complement-sensitive cells. Now it is recognized that many people have PNH, most of them experiencing only modest amounts of hemolysis. Consequently most of them can lead fairly normal lives without drastic medical intervention.

## *Further Readings*


Freda, V. J., Pollack, W., and Gorman, J. G. Rh Disease: How Near the End? *Hospital Practice,* June 1978.

Jandl, J. H. Hemolytic Anemias II. Immunohemolytic Anemias. Lecture 12 in *Hematology,* 4th ed., William S. Beck, Ed., The MIT Press, Cambridge, Mass., 1985.

Rose, N. R. Autoimmune Diseases. *Scientific American,* Feb. 1981.

Rosse, W. F. Autoimmune Hemolytic Anemia. *Hospital Practice,* Aug. 1985.


## *Review Questions*

1. List five different causes of acquired hemolytic anemia.
2. Distinguish between alloantibodies and autoantibodies. What anemias are associated with each?
3. What is another name for hemolytic disease of the newborn (HDN)? Why do we use this term?
4. Explain under what two circumstances HDN can develop. Explain why it develops in these situations.
5. Explain why ABO-HDN is usually much milder than Rh-HDN.
6. What is meant by the rhesus factor? What percentage of the general population is Rh-positive?
7. Describe some of the major clinical features of HDN.
8. List the major laboratory features associated with HDN.
9. What treatment is commonly used for HDN?
10. Explain how HDN can be prevented. What is RhoGAM?
11. Distinguish between warm-type and cold-type AIHA regarding (a) types of antibodies involved, (b) temperature of action, (c) basic causes, (d) age group involved, and (e) severity of the anemia.
12. Explain what is meant by opsonization, pitting, and neoantigens.
13. Explain how opsonization of red blood cells can cause hemolytic anemia.
14. The degree of destruction of red blood cells by IgG opsonins depends on which three factors?
15. Describe idiopathic AIHA. What percentages of all warm-type AIHAs can be classified as idiopathic?
16. Give two categories of known causes of warm-type AIHA. What percentage of AIHA belongs to each group?
17. List the three categories of drugs that have been implicated as causes of warm-type AIHA. Give an example of each.
18. Describe the common symptoms associated with warm-type AIHA.
19. List the major laboratory findings associated with warm-type AIHA.
20. Which test is used to diagnose warm-type AIHA? How is this test performed?
21. What is Coombs' serum? How is it produced?
22. Define the zeta potential of red blood cells. What is the distance of the zeta potential between red blood cells?
23. Why are IgG opsonins unable to bridge the zeta potential gap?
24. What is the function of Coombs' serum?
25. List some common treatments for warm-type AIHA. Why are these treatments deemed useful?
26. What is the major cause of cold-type AIHA?
27. What type of antibodies are produced in cold-type AIHA? Why are these antibodies more efficient in agglutinating red blood cells?
28. What percentage of cold-type AIHA is idiopathic?
29. Name some infectious diseases that can produce cold-type AIHA.
30. Describe the common clinical symptoms associated with cold-type AIHA.
31. List the laboratory features associated with cold-type AIHA.
32. Describe the preferred treatment for cold-type AIHA.
33. Name the three types of hemolytic anemia caused by physical factors.
34. Why is the urine of long-distance runners often dark brown after a long race?
35. Define microangiopathic anemia. What are some of the common causes for this condition? Why do these causes frequently produce anemia?
36. Name the three major classes of chemical agents that may induce hemolytic anemia. Give one or two examples for each category.

37. Mention five different microorganisms that frequently cause hemolytic anemia.
38. Define paroxysmal nocturnal hemoglobinuria (PNH). What is a unique feature of this condition?
39. What is the major cause of PNH?
40. Describe the major symptoms of PNH. Which age group is most likely to get PNH?
41. List some common laboratory features associated with PNH.
42. The severity of PNH depends on which three factors?
43. How is PNH usually diagnosed?
44. Describe the common treatment for PNH.

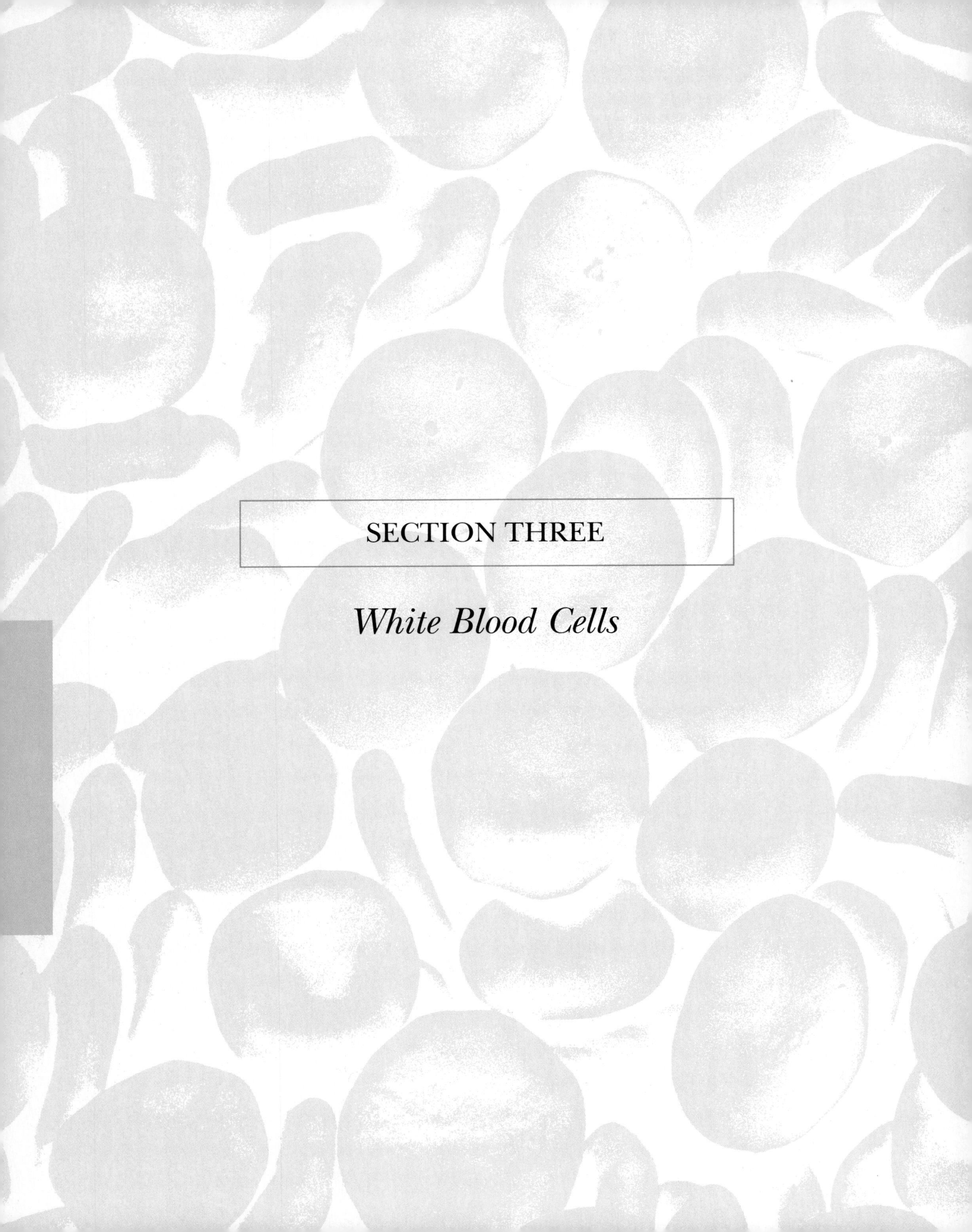

SECTION THREE

# *White Blood Cells*

# Chapter Sixteen

## *Development and Structure of Neutrophils*

## INTRODUCTION TO WHITE BLOOD CELLS

The last twelve chapters have dealt with the structure, functions, and disorders of red blood cells. The next several chapters deal mainly with white blood cells. The term white blood cell, or **leukocyte,** is a catch-all name given to a variety of nucleated cells present in the peripheral bloodstream. The major functions of the white blood cells are to protect the body against invasion of extrinsic pathogens and to save the organism from being overrun by neoplastic cells and other abnormal cells.

### Various Classifications of White Blood Cells

Unlike red blood cells, there are many different types of white blood cells that may be classified in a number of different ways as follows:

1. Based on the *shape of the nucleus,* these leukocytes may be divided into **mononuclear** and **polymorphonuclear** white blood cells. Mononuclear leukocytes have a single, nonsegmented nucleus. This group includes the monocytes and all the various lymphocytes. Polymorphonuclear white blood cells have nuclei of varying shapes. Usually their nuclei are divided into a number of interconnected segments. This group includes neutrophils, eosinophils, and basophils.
2. Based on the *presence or absence of specific staining granules,* white blood cells may be divided into **granulocytes** and **agranulocytes.** Granulocytes may be further divided by the nature of their specific-staining granules into neutrophils, eosinophils, and basophils. Compared with granulocytes, agranulocytes have very few granules that can be readily stained with specific dyes. The latter group includes monocytes and lymphocytes.
3. Based on their *site of origin,* leukocytes may be divided into **myeloid** and **lymphoid** white blood cells. Myeloid leukocytes are those that are produced in the bone marrow. This group includes all white blood cells except the lymphocytes. Although we should realize that ultimately all lymphocytes are also derived from the bone marrow, most lymphocytes found in the peripheral blood are derived from lymphoid tissues and are therefore referred to as lymphoid white blood cells.
4. Based on their *function,* white blood cells may be divided into a number of categories. Some white blood cells may be classified as **phagocytes,** as their major function is to engulf and destroy invading pathogens and neoplastic cells. There are two classes of phagocytes in the blood: **macrophages** and **microphages.** Monocytes are classified as macrophages, and neutrophils are normally referred to as microphages. There are other cells with phagocytic ability in the blood (e.g., eosinophils), but do not play a major role in the phagocytic process.
   Other white blood cells are referred to as **immunocytes,** as they play major roles in the specific immune system. This group includes lymphocytes and macrophages.

Historically white blood cells have been divided into five different categories: (1) neutrophils; (2) eosinophils; (3) basophils; (4) monocytes; and (5) lymphocytes. Each of these groups of leukocytes has a different origin, a different structure, and possesses unique functions. These categories are discussed separately in the next set of chapters.

## STAGES OF DEVELOPMENT OF NEUTROPHILS

As explained in chapter 3, all blood cells are derived from the same pluripotent stem cells in the bone marrow. This precursor cell may become either a myeloid or a lymphoid stem cell. The neutrophil is derived from the myeloid pluripotent stem cell, which commits itself via a common granulocytic-monocytic precursor cell to become a committed neutrophilic stem cell (see figure 3.8). When this neutrophilic stem cell becomes activated and starts to divide, it undergoes six different stages of development before the mature neutrophil is produced. This process of proliferation and maturation lasts about 5 to 6 days. During this time usually about four to five mitotic divisions take place. As illustrated in figure 16.1, cell division normally stops after the cell reaches the fourth stage of development. The process of neutropoiesis goes through the following six phases:

1. When the neutrophilic stem becomes activated, it enlarges and becomes a **myeloblast.** This enlarged stem cell is normally about 20 to 25 μm in diameter. As can be seen in figure 16.2, the nucleus of this myeloblast is more or less round or oval and relatively large when compared with the volume of the cytoplasm. The nuclear membrane is usually quite smooth. The nuclear material consists of fine open chromatin, and several distinct nucleoli are always present. The cytoplasm is deep blue when stained with Wright's stain but tends to be darker near the periphery of the cell. Generally no, or very few, cytoplasmic inclusions are observed.
2. The second stage in this proliferation and maturation series is called the **neutrophilic promyelocyte.** This cell is quite similar in shape to the neutrophilic myeloblast, except for the fact that the cytoplasm now contains numerous large azurophilic granules. The nucleus is still round or oval, but the chromatin has become a bit coarser, and the nucleoli have all but disappeared. This cell may also be classified as a blast cell, since this type of cell actively divides.

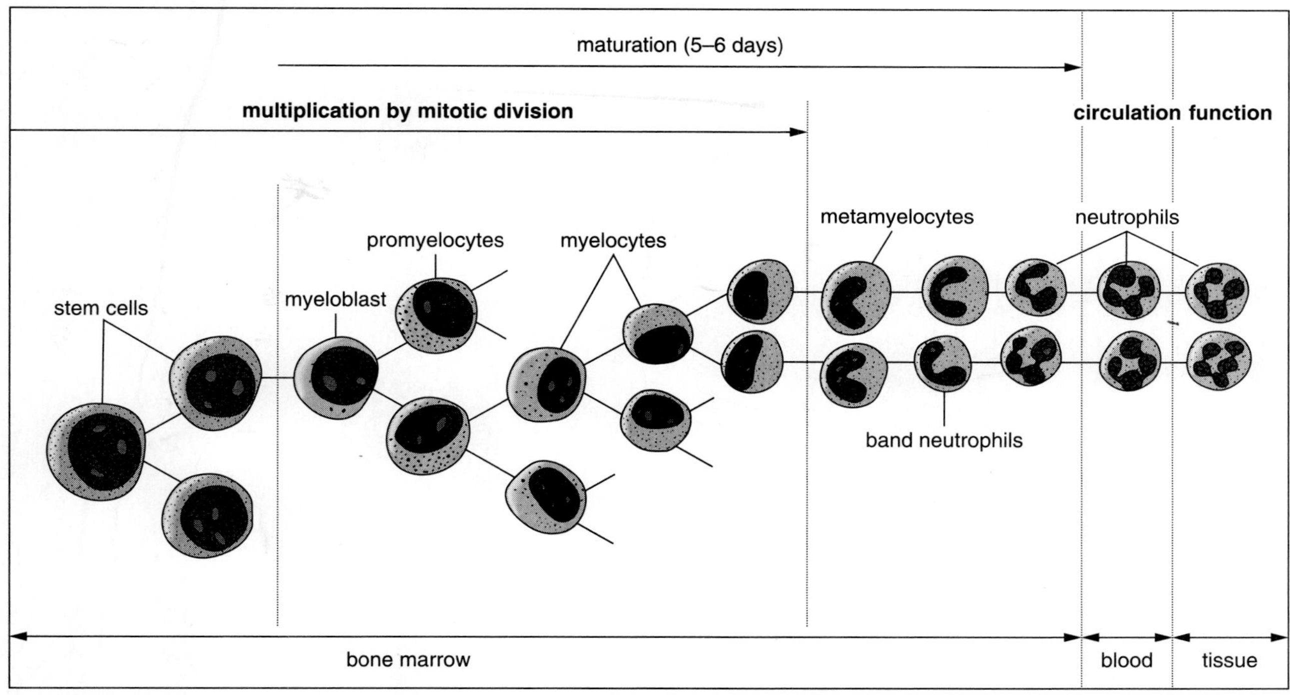


**figure 16.1**
The kinetics of granulopoiesis.

3. The third stage in this neutrophilic series is called the **neutrophilic myelocyte,** and it differs in a number of ways from the promyelocyte. First, this cell is normally somewhat smaller than the previous stages, averaging only 15 μm in diameter. Second, the nucleus is now distinctly oval and has become eccentric—that is, it has clearly moved to one side of the cell. The nuclear chromatin is much coarser, and all nucleoli are gone. Third, the cytoplasm has lost its unique basophilic (=blue) coloration and has become more acidophilic (=pink) with Wright's stain. In this stage the specific or secondary neutrophilic granules appear for the first time. Gradually they become more numerous than the original azurophilic granules. However these primary granules remain present throughout the life of the cell, although in relatively smaller quantities. The primary azurophilic granules become scarcer than the specific-staining secondary granules because they are produced only during the promyelocyte stage of neutrophil development, and with each cell division they become less numerous in each of the subsequent daughter cells.
4. The fourth stage in this series is the **neutrophilic metamyelocyte.** This cell is no longer considered to be a blast cell since mitosis has stopped. From here on only maturation takes place. This cell can easily be distinguished from the previous myelocyte stage by the shape of its nucleus. The nucleus of the neutrophilic metamyelocyte has taken on a distinct kidney- or bean-shaped configuration. The nuclear material has by now become extremely coarse and stains deeply basophilic. The small secondary cytoplasmic granules characteristic of neutrophils have become so numerous that the primary azurophilic granules are almost hidden. Finally, this cell stage is again considerably smaller than the previous myelocyte phase.
5. The fifth stage is commonly known as the **neutrophilic band cell.** It is also known as the **stab cell** and is the last juvenile form of the neutrophilic series. This cell is easily recognized by its sausage-shaped nucleus. Its size is again somewhat smaller than the metamyelocyte, and its numerous secondary granules do not stain as prominently as in previous stages. This is the stage in which neutrophils enter the bloodstream. Normally 2 to 6% of all neutrophils in the peripheral circulation are neutrophilic band cells. If that number is higher, it is an indication that the bone marrow is producing higher than normal levels of neutrophils (e.g., during severe infections).
6. The final stage of development is the **neutrophilic segmented cell.** This cell is also frequently referred to as the **polymorphonuclear neutrophil,** or more commonly called **PMN,** since the nucleus has now become segmented into several nuclear lobes connected by thin strands of chromatin. The normal PMN has three to five lobes, although in certain diseases such as megaloblastic anemia, this number may increase to as many as 10 to 15 lobes. This final neutrophilic cell is the smallest cell of the whole series, being between 12 and 15 μm in diameter in the normal bloodsmear.

A summary of the developmental changes of the neutrophilic series is given in figure 16.3.

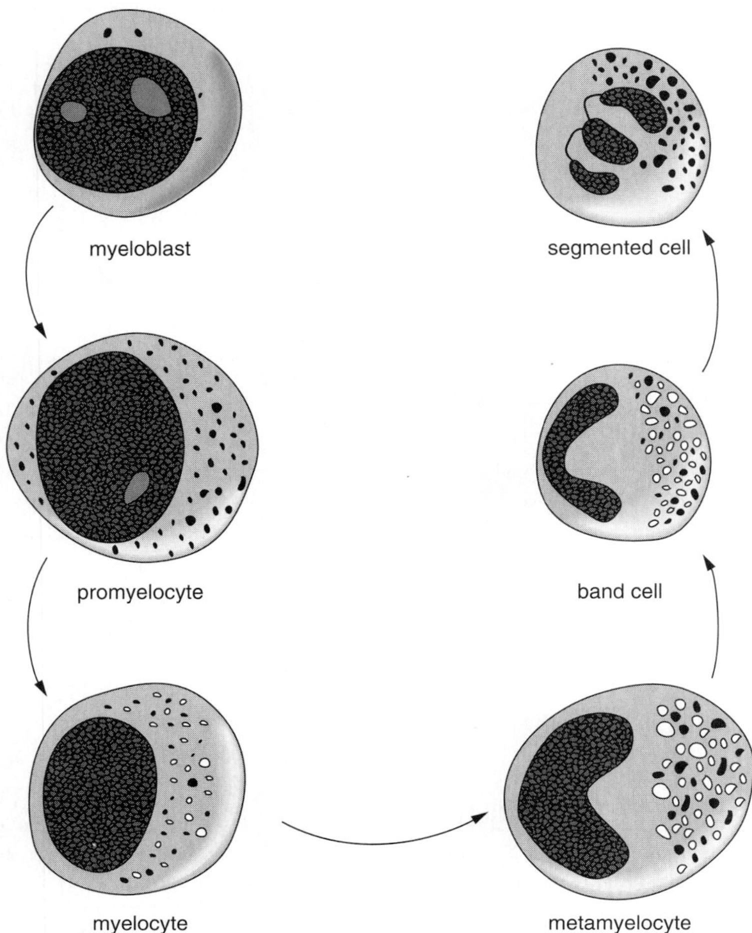


**figure 16.2**

The maturation sequence of the neutrophilic granulocytes. The black dots represent the primary granules, the white circles represent the secondary granules.

## REGULATION OF NEUTROPHIL PRODUCTION

The life span of the neutrophils can easily be detected by employing radioactive labels either *in vivo* or *in vitro.* The most common label used is radioactive phosphorus ($^{32}P$), which attaches readily to the serine residues of some cellular proteins. Studies with $^{32}P$ have produced interesting insights into the kinetics of the PMN. First, the total granulocyte pool includes two subpools: the **circulating granulocyte pool** and the **marginating granulocyte pool.** The term "granulocyte" is used synonymously with neutrophil since the overwhelming majority of granulocytes (more than 97%) in the blood are neutrophils. Moreover what is described here for PMNs also is true for eosinophils and basophils. The marginating granulocyte pool (MGP) consists of granulocytes that have "marginated"—that is, attached themselves to the walls of the blood vessels of the microcirculation. Most of these white blood cells are in the process of leaving the circulatory system. This means that the neutrophil counts performed in the laboratory measure only the number of cells in the circulating granulocyte pool (CGP).

Second, when radioactive decline is measured, it becomes clear that the normal half-life of PMNs in the blood is only about 6.7 hours. Thus the average time that a neutrophil spends in the circulation is only a little more than 10 hours, certainly not more than 24 hours. Other studies have revealed that it takes about 5 to 6 days before a labeled neutrophil enters the bloodstream. Hence this may be presumed to be the time it takes for a myeloblast to become a mature neutrophil. Finally, radiolabeled studies have shown that the total life span of a neutrophil from early myeloblast to death in the tissues is about 9 to 10 days, thus after the PMN leaves the circulation it lives for another 2 to 3 days in the tissues before it dies. In discussing the kinetics of the neutrophil, we need to take into account three different areas of action: the bone marrow, the bloodstream, and the tissue. Each of these phases may have several divisions. For instance, bone marrow may be divided into a mitotic compartment and a maturation phase. The blood compartment may be divided into a circulating and a marginating pool. The tissue compartment may be divided into areas in which neutrophils can move freely and into areas in which movement is restricted—for example, in an area of infection (figure 16.4).

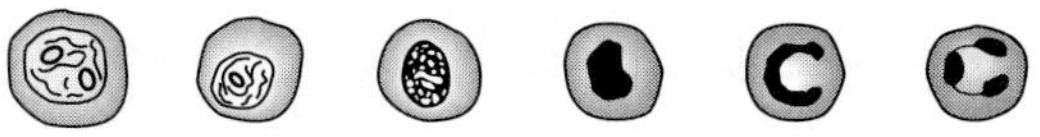

| | myeloblast | promyelocyte | myelocyte | metamyelocyte | band form | mature segmented form |
|---|---|---|---|---|---|---|
| cytoplasmic basophilia | +++ | + | – | – | – | – |
| azurophilic granules | –/+ | +++ | ++ | + | + | + |
| specific-staining granules | – | – | +++ | ++ | ++ | ++ |
| nucleoli | +++ | + | – | – | – | – |
| nuclear chromatin | dispersed ——► progressively coarser and more basophilic | | | | | |
| nuclear size | large ——► progressively smaller and then indented | | | | | |
| nuclear shape | mononuclear ——► ——► ——► indented ——► segmented | | | | | |
| cell size | progressively smaller | | | | | |
| capacity to divide | + | + | + | – | – | – |

**figure 16.3**

Characteristics of the various stages of neutrophilic granulocyte development and maturation.
Source: William S. Beck, *Hematology*, 4th ed., 1981 MIT Press, Cambridge, MA.

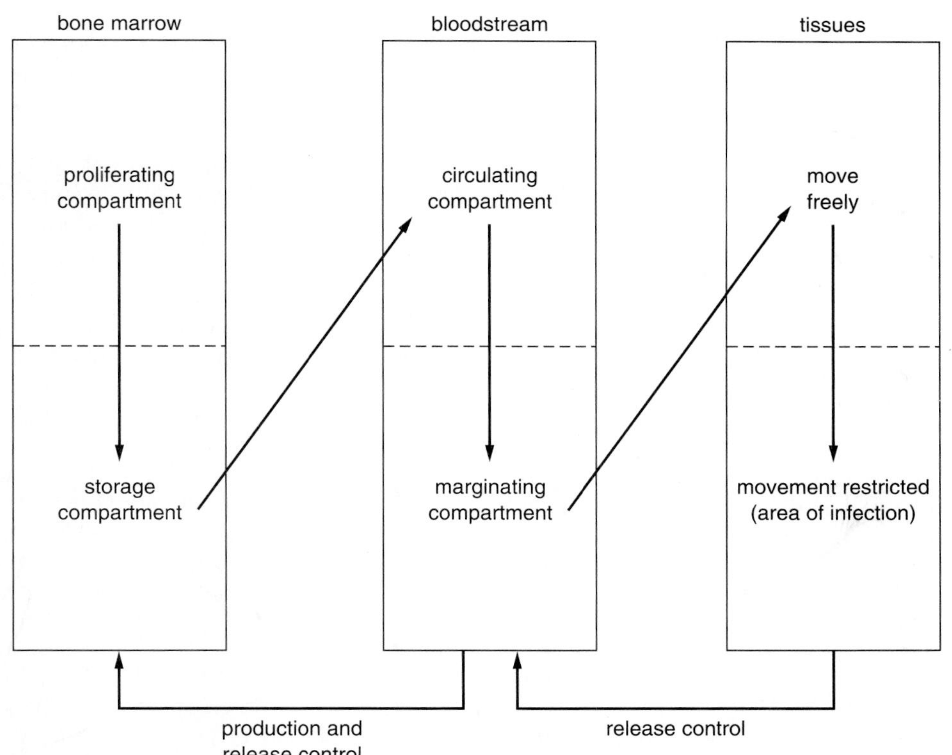


**figure 16.4**

The movement and control of the neutrophils in the various body compartments.

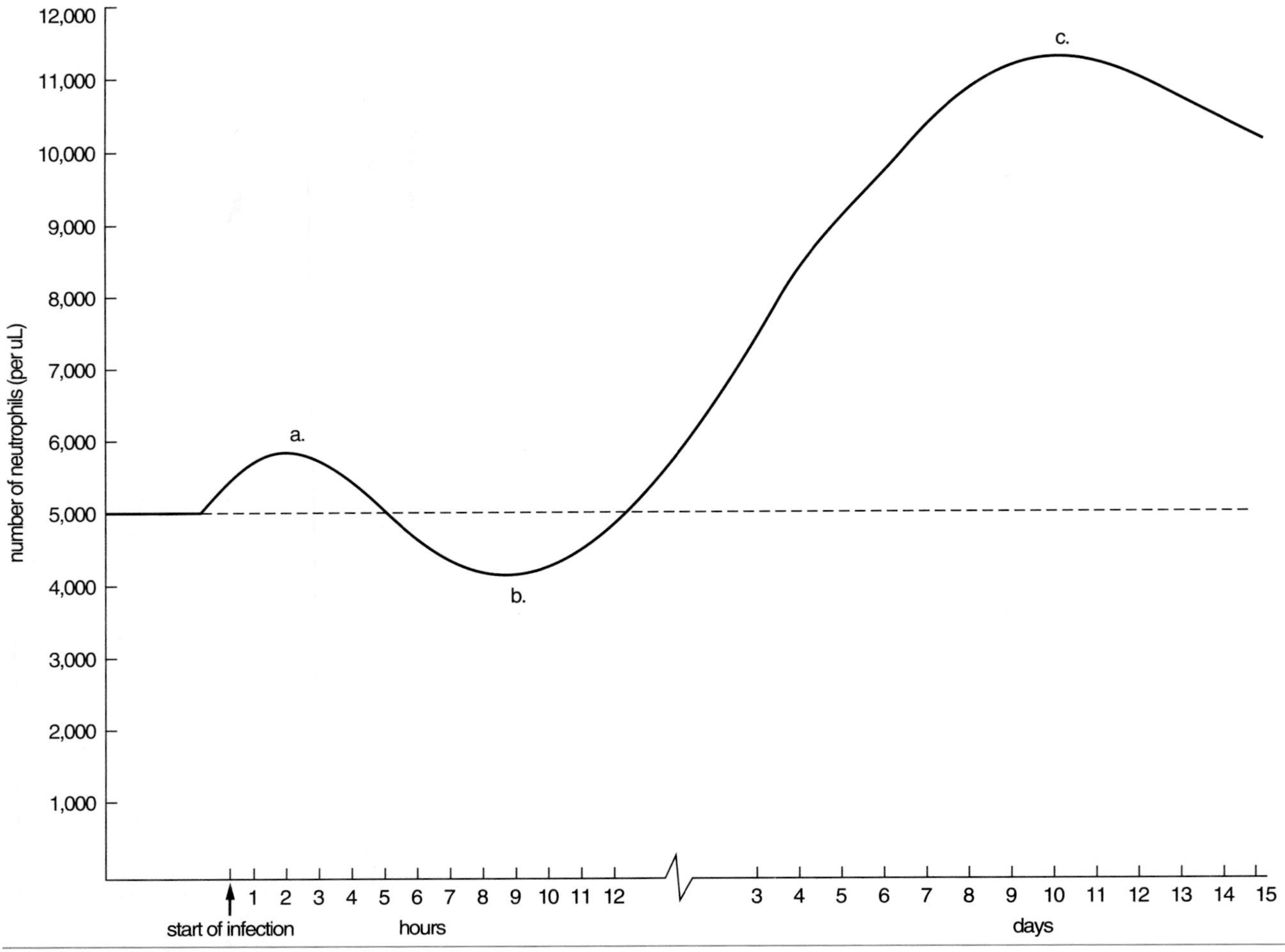


**figure 16.5**

Circulating neutrophils during infection. (*a*) Temporary neutrophilia due to the movement of PMNs from the bone marrow storage compartment to the bloodstream. (*b*) Temporary neutropenia due to the massive movement of PMNs from the bloodstream to the area of infection. (*c*) Prolonged neutrophilia due to the bone marrow response to the infection.

In the steady-state pattern, the turnover rate from one compartment to the next is rather stable. In non-steady-state patterns frequently associated with infectious diseases and leukemias, these compartments may be in a state of flux. For example, in acute inflammatory conditions, the PMNs are mobilized from the blood compartment to the infected site within hours. However the total blood granulocyte pool may actually increase temporarily due to release of neutrophils from the bone marrow reserve. Thereafter granulocyte turnover rates increase in all compartments until the infection is brought under control. However, if the infection is massive, lack of blood neutrophils—a condition known as **neutropenia**—may occur for a short while as a result of the excessive demand for neutrophils by the tissues. Later in the infection the bone marrow will respond properly to the increased need for PMNs, and **neutrophilia** will occur for a while. Neutrophilia indicates that there is a higher than normal level of PMNs in the bloodstream (figure 16.5).

The concentration of circulating neutrophils stays more or less constant in healthy people and increases in disease conditions, suggesting a feedback mechanism similar to the one regulating red blood cell production. The existence of several neutrophil production factors, sometimes referred to as **granulopoietins** or **granulocytic growth factors** has been demonstrated in the last few decades. The most important of these are as follows:

**1. Multi-colony stimulating factor (Multi-CSF)**
This nonspecific growth factor is also known as **interleukin 3 (IL-3).** Multi-CSF only acts early in the process of differentiation, possibly at the level of pluripotent stem cells, but certainly at the level of myeloid progenitor cells. It is produced by T lymphocytes, specifically by helper T cells.

**2. Granulocyte-monocyte colony-stimulating factor (GM-CSF)**
This growth factor is produced by a number of mediator cells including T cells, endothelial cells, and fibroblasts. GM-CSF mimics many of the functions of multi-CSF. It also stimulates the development and differentiation of myeloid progenitor cells. Multi-CSF can compete with GM-CSF for its receptors on the developing myeloid blood cells. Most progenitor cells have receptors for both multi-CSF and GM-CSF. Multi-CSF (IL-3) can bind to both receptors, whereas GM-CSF can bind only to its own receptors.

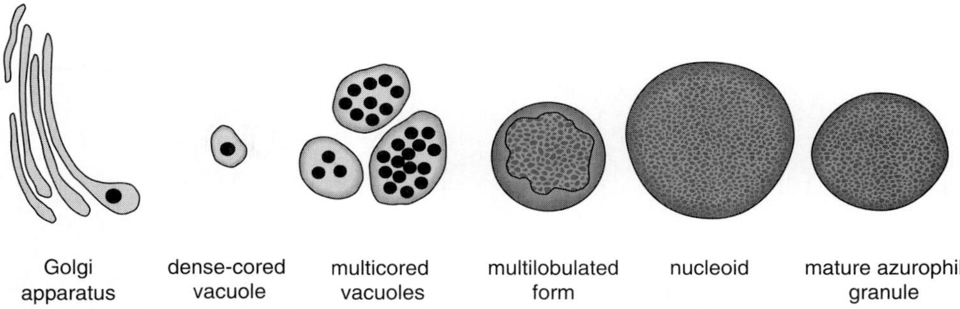


**figure 16.6**

Stages of primary (azurophilic) granule formation in the neutrophilic granulocytes.

**3. Granulocyte colony-stimulating factor (G-CSF)**
This growth factor is uniquely involved with the development and differentiation of neutrophils. G-CSF is produced by monocytes, macrophages, and fibroblasts.

**4. Interleukin 6 (IL-6)**
This growth factor appears to be produced by the bone marrow stromal cells. It is also involved in the development and differentiation of early myeloid progenitor cells.

The action of these growth factors is summarized in table 3.2.

## STRUCTURE OF NEUTROPHILS

As previously mentioned, two types of neutrophils are normally present in the peripheral circulation: the neutrophilic band cell (which is commonly referred to as the stab cell) and the segmented neutrophil (which is commonly referred to as the PMN). The stab cell is considered to be a juvenile neutrophil and is similar in structure to the PMN except for the shape of the nucleus (figure 16.2). Both types of neutrophils contain many granules. Most of these granules are secondary or specific-staining granules (80 to 90%), although a considerable number of primary, or azurophilic, granules are also present (10 to 20%). Electron microscope analysis reveals that the endoplasmic reticulum is not well developed in these latter stages of neutrophil maturation. Free and bound ribosomes are also much less common. These are all indications that little RNA and protein synthesis takes place in the mature neutrophil. Similarly, the Golgi apparatus seems to be atrophic, which means that new granule formation has ceased in the PMN. Further, few mitochondria are present in these two final stages of the neutrophilic series, while the glycogen stores of these circulating neutrophils are quite large. This indicates that the cells have switched from aerobic to mainly anaerobic respiration when they enter the bloodstream.

### Production and Content of Primary Granules

Production of primary, or azurophilic, granules occurs mainly in the promyelocyte stage of neutrophil development. Further cell divisions reduce the concentration of these primary granules as they are parceled out to the subsequent daughter cells.

Formation of azurophilic granules is illustrated in figure 16.6. The production of primary granules starts with the appearance of electron-dense spherules in the cisternae of the Golgi complex. Each individual spherule has a diameter of about 0.1 μm. Once formed, these particles are pinched off from the Golgi apparatus to form a tiny, dense-cored vacuole, several of which merge to form a multicored vacuole. The core material then aggregates and becomes a multilobed mass that subsequently forms a more compact sphere. The vacuolar space surrounding this sphere then disappears, and a mature azurophilic granule results.

The content of these primary granules has been analyzed and contains the following enzymes:

**1. Myeloperoxidase**
This enzyme is present only in primary granules and hence is used as an enzymatic marker to distinguish these granules from secondary granules.

**2. Arginine-rich basic proteins**
Many of these cationic enzymes are in the form of lysozyme.

**3. Sulfated mucopolysaccharides**
These compounds contribute greatly to the azurophilic coloring of the primary granules.

**4. Acid hydrolases**
A variety of acid hydrolases have been found in the primary granules of the human neutrophil including β-galactosidase, β-glucuronidase, mannosidase, arylsulphatase, esterase, and acid phosphatase. The latter enzymatic marker has been used to trace the origin of the primary granules to the Golgi apparatus.

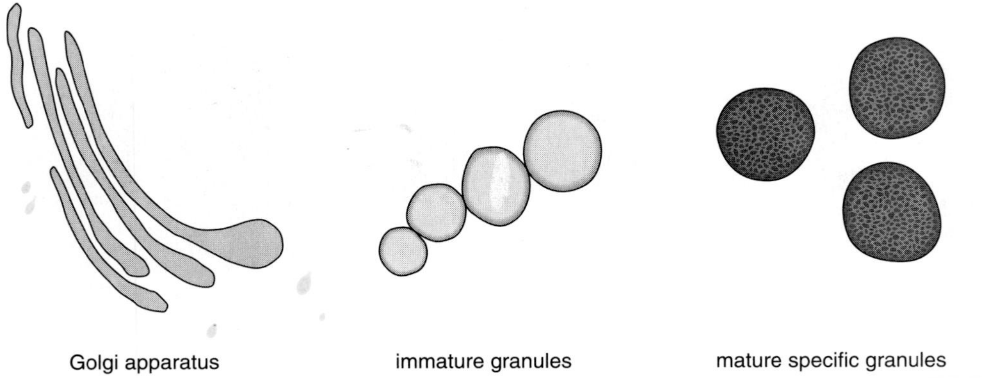


**figure 16.7**

Stages of secondary (specific-staining) granule formation in the neutrophilic granulocytes. Note that several small granules condense to form larger aggregates that gradually form into mature granules. Most mature granules are spherical and measure between 300 and 500 nm (0.3–0.5 μm) in diameter.

## Production and Content of Secondary Granules

Production of secondary granules occurs mainly during the intermediate stages of neutrophil development—that is, during the myelocyte and metamyelocyte phases. With the start of the myelocyte stage, production of primary granules ceases and the neutrophilic blast cell switches over to the development of secondary, or specific-staining, granules. Since formation of secondary granules continues after the cells stop dividing, these secondary granules continue to increase and eventually become more numerous than the primary granules. The Golgi apparatus is also the source of these secondary granules. These specific-staining granules first appear as small structures in the lateral margins of the Golgi and associated vesicles, as illustrated in figure 16.7. Once they have reached the size of about 0.09 μm, they begin to bud off from the outer Golgi cisternae. Several then merge to form larger aggregates that further condense, forming mature secondary granules that are more or less spherical, with diameters of 0.3 to 0.5 μm. Analysis of the contents of these secondary granules has shown that they mainly consist of the following three enzymes:

1. **Alkaline phosphatase**
This distinctive marker occurs only in secondary granules and is therefore used to distinguish the specific granules from primary azurophilic granules.

2. **Basic proteins, mainly in the form of lysozyme**
About two-thirds of the neutrophil lysozyme is in secondary granules, and the other third is in primary granules.

3. **Aminopeptidase**
This enzyme also occurs exclusively in secondary granules.

## Size and Shape of Neutrophils

Fully mature neutrophils tend to be fairly uniform in size with diameters of 12 to 15 μm. In a blood smear they tend to be roundish or ovoid, but have amoeboid characteristics and thus may assume an infinite variety of shapes. They are in constant motion, especially PMNs marginating along the blood vessel walls, which tend to produce numerous pseudopodia of clear cytoplasm in advance of the cell. They pull the granule-containing cytoplasm behind them, and the nucleus is always in the rear during movement.

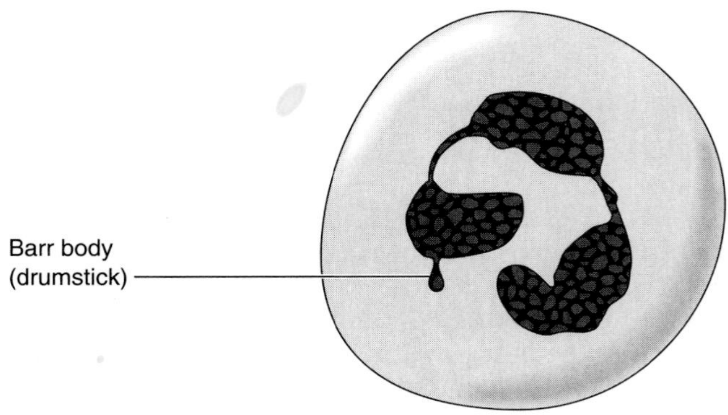


**figure 16.8**

Neutrophil with a "drumstick" or Barr body, commonly seen in human (XX) females.

## Variations in the Structure of Neutrophils

Apart from variation in the shape of the nucleus of the band cell and the segmented cell, one other normal variation is commonly encountered in PMNs. About 5 to 10% of all neutrophils of the human (XX) female contain a special sex chromatin body called a **drumstick** or **Barr body,** which is attached to one of the lobes of the segmented nucleus (figure 16.8).

Apart from this normal variation, a number of abnormal variations may be considered defects in the granulocyte morphology and are often associated with specific diseases. Some of the most important of these abnormalities are as follows:

1. **Macrogranulocytes**
These are larger than normal granulocytes that are commonly found in people suffering from megaloblastic anemia.

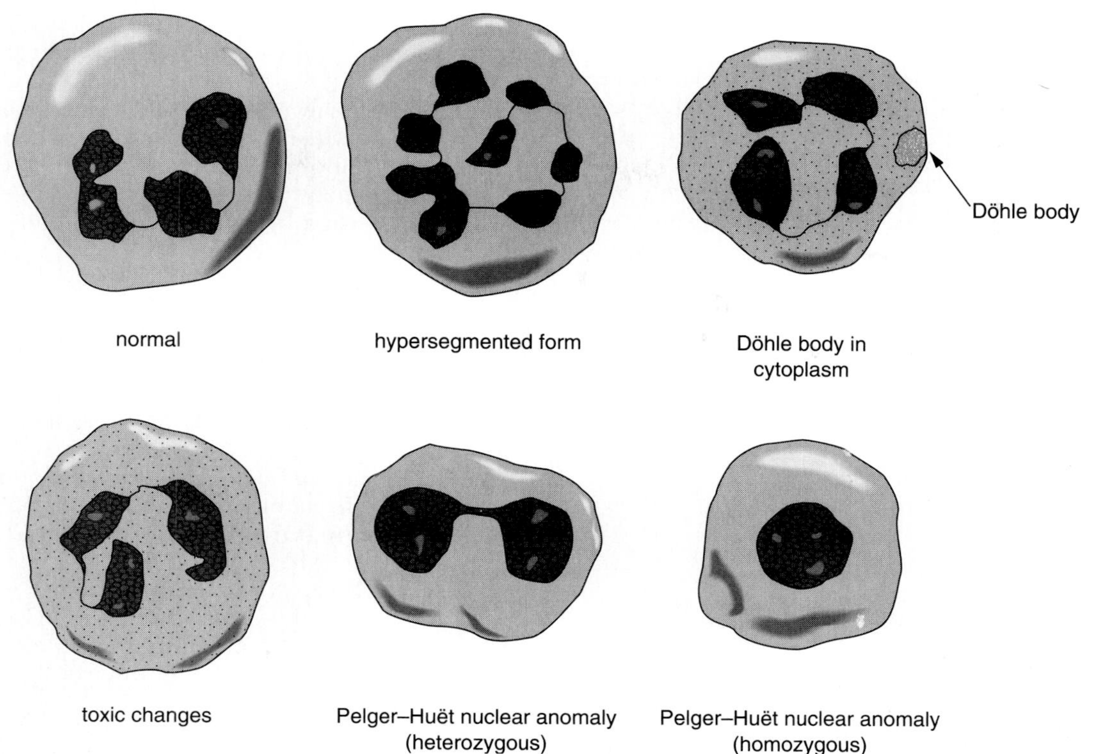


**figure 16.9**
Variations in neutrophil morphology.

**2. Hypersegmented neutrophils**
These PMNs each have more than five lobes in the nucleus and are commonly encountered in pernicious anemia and other megaloblastic anemias. Not all neutrophils in persons with megaloblastic anemia will show hypersegmentation, but a majority will do so (figure 16.9).

**3. Döhle's bodies**
These are single or multiple gray-blue cytoplasmic inclusions usually seen along the outer edges of mature neutrophils (figure 16.9) and are commonly associated with severe infections, burns, and sometimes with pregnancy or cancer. Döhle's bodies are thought to be ribosome-containing remnants of the promyelocyte cytoplasm.

**4. Pelger-Huët anomaly**
This is an inherited abnormality characterized by a failure of normal nuclear segmentation. Neutrophils of individuals who are heterozygous for this anomaly will show typically two-lobed nuclei, or "pince-nez" cells, so-called because these nuclei look like old-fashioned glasses that were pinched onto the nose (figure 16.9). Persons who are homozygous for this condition have round nuclei in their granulocytes. This condition is inherited as a simple autosomal trait and occurs in about 1 of every 6,000 people. It is not associated with any specific disease.

**5. Alder-Reilly anomaly**
This is another, but much more rare, inherited anomaly formed as a result of abnormal polysaccharide metabolism, which prevents the formation of secondary specific-staining granules. Thus in these individuals only primary, azurophilic granules are present. Nuclear maturation is not interfered with, and segmentation of the nucleus is normal. The presence of the Alder-Reilly anomaly is often associated with certain syndromes (e.g., Hurler and Hunter's syndromes) that produce shortened life spans.

**6. Chédiak-Higashi anomaly**
This anomaly is associated with a potentially lethal disorder known as Chédiak-Higashi syndrome. These cells have very few, but gigantic, specific-staining secondary granules that are unable to fuse with the phagosomes (=vacuoles containing engulfed pathogens). Hence these granules are nonfunctional.

## METABOLISM OF NEUTROPHILS

As previously mentioned, the size and number of mitochondria decrease with maturation of the granulocytes. Few functional mitochondria are left in the mature PMN—thus the mature neutrophil obtains most of its energy via anaerobic energy cycles and is not critically dependent on the presence of oxygen for most of its metabolic functions. Key neutrophil functions, such as movement and phagocytosis, continue perfectly well in the absence of oxygen. However for other important functions, such as microbial killing, large amounts of oxygen are needed, as is explained in the next chapter. We should realize, however, that even under aerobic conditions neutrophils continue to rely on anaerobic glycolysis for most of their functions. This is important since many neutrophils must function at low partial pressures of oxygen when infections occur in extravascular tissues.

### Carbohydrate Metabolism

The high levels of energy needed for movement and phagocytosis are provided by the anaerobic breakdown of glucose via the Embden-Meyerhof pathway and the hexose monophosphate shunt (HMS). Glucose needed for these processes is provided by exogenous sugars or endogenous glycogen stores. Some energy also comes from the aerobic pathway via the few mitochondria that are left. Phagocytosis is always associated with increased glycolysis, both under aerobic and anaerobic conditions. Any substance that interferes with the rate of glycolysis will inhibit the process of phagocytosis. For instance, lack of insulin or other hormones involved with sugar metabolism will result in a decrease in the amount of sugar that enters the neutrophil and hence will interfere with the processes of movement and phagocytosis.

### Lipid Metabolism

Actively phagocytosing neutrophils have a much higher rate of lipid turnover than do resting PMNs. This is understandable because new lipid membrane is needed constantly to replace the membrane that was lost with formation of the phagocytic particles (phagosomes); otherwise the cell would get smaller and less flexible.

Although the mature PMN is incapable of new fatty acid synthesis, it can take up lipids such as triglycerides, phospholipids, and cholesterol from the surrounding plasma and use these lipids to replace the outer plasma membrane lost to the phagocytic vacuoles.

### Protein Metabolism

The concentration of amino acids in the neutrophils is more or less similar to that of other somatic cells. Although we could logically presume that the granulocyte would replace the enzymes it lost to the phagocytic vacuoles during the process of particle digestion, this does not happen. This is another indication that the mature PMN is a "one-shot" cell that does not replace its enzymatic arsenal. Nor is there a need for such replacement, as the active life span of a mature neutrophil is only a few days. Once the cell reaches the end of its life span, it disintegrates in the tissues and the dead cells are removed by wandering macrophages.

## *Further Readings*

Abbas, A. K., Lichtman, A. H., and Pober, J. S. The Cells and Tissues of the Immune System. Chapter 2 in *Cellular and Molecular Immunology.* W. B. Saunders Company, Philadelphia, 1991.

Boggs, D. R., and Winkelstein, A. The Phagocyte System. Chapter 4 in *White Cell Manual,* 4th ed. F. A. Davis Company, Philadelphia, 1983.

Cline, M. J. The Normal Granulocyte. Part I A in *The White Cell.* Harvard University Press, Cambridge, Mass., 1975.

## *Review Questions*

1. Distinguish between mononuclear and polymorphonuclear leukocytes. Which white blood cells belong to each group?
2. Distinguish between granulocytes and agranulocytes. Which white blood cells belong to each group?
3. Distinguish between myeloid and lymphoid white blood cells.
4. Distinguish between macrophages and microphages.
5. Provide another name for neutrophils. Where are neutrophils produced?
6. Name the precursor cells for the neutrophilic stem cells.
7. List the six stages in the development of neutrophils.
8. Distinguish between myeloblasts and promyelocytes.
9. Distinguish between myelocytes and metamyelocytes.
10. What is the length of the proliferation and maturation process of PMNs in the bone marrow?
11. How many cell divisions occur in an activated neutrophilic stem cell? How many potential daughter cells are produced?
12. Distinguish between circulating and marginating neutrophils. What is the level of circulating PMNs in the peripheral blood?
13. How long do PMNs stay in the bloodstream? How long do they live in the tissues? What is the total life span of neutrophils?
14. How are the normal levels of PMNs in the blood maintained? List the major neutrophilic growth factors.
15. Where is IL-3 produced? What is another name for IL-3? Describe its function.
16. What does the abbreviation GM-CSF stand for? Where is it produced? Describe its function.

17. Explain why new infections usually start with a temporary neutrophilia. Why is it followed by a temporary neutropenia? Why does it result in a prolonged neutrophilia?
18. Where and how are primary granules produced in neutrophils?
19. List the major components present in primary granules.
20. When are primary granules produced in neutrophils?
21. When, where, and how are secondary granules produced in PMNs?
22. Describe the contents of secondary granules.
23. What is the average diameter of a neutrophil? What shape does a PMN take when it moves?
24. What is a Barr body in a PMN? Describe its significance.
25. What is a hypersegmented neutrophil? What does its presence indicate?
26. What are Döhle's bodies? What is their significance?
27. What is the Pelger-Huët anomaly? Discuss its significance.
28. What is the Alder-Reilly anomaly? Discuss its significance.
29. What is the Chédiak-Higashi anomaly? Discuss its significance.
30. How does the neutrophil get most of its energy? Why is this?
31. Why do active PMNs have a high lipid metabolism?
32. Explain why PMNs do not have a high amino acid turnover. What does this indicate?

# Chapter Seventeen

## *Functions of Neutrophils*

The primary goal of the neutrophils is to act as a first line of defense when the body is invaded by pathogenic bacteria. Their function is to localize and limit the spread of microorganisms until more efficient white blood cells, such as lymphocytes and macrophages, destroy and remove these foreign agents. This may take several days; meanwhile the invasion must be contained. Once these pathogenic bacteria enter the system, some are able to replicate themselves approximately every 20 minutes. If allowed to multiply uncontrolled, they would overrun and kill the infected person within a few days. Neutrophils keep this multiplication process under control. Because neutrophils must act immediately after an infection occurs, it follows that they are the most numerous white blood cells in the bloodstream.

## MAJOR ASPECTS OF NEUTROPHIL ACTIVATION

To achieve this goal of rapid action and frontline defense, several integrated actions take place. First, the area surrounding the infection must be structurally changed to allow large numbers of polymorphonuclear neutrophils (PMNs) to leave the bloodstream and congregate in the affected tissue. This is achieved by the process of **inflammation.** These granulocytes are key participants in all forms of acute inflammation. Second, large numbers of neutrophils are drawn to the area of infection by the process of **chemotaxis.** Producing this large-scale unidirectional migration of PMNs involves many different chemotactic factors. Third, once neutrophils reach the area of infection, they must slow down the reproductive and metabolic activities of invading pathogens, which they do primarily by the process of **phagocytosis.** The ultimate goal of phagocytosis is to kill pathogen microorganisms and thus prevent them from replicating and causing harm to the body.

As is seen later, neutrophils (microphages) are not as efficient in this process of **microbial killing** as are macrophages, but they are efficient in slowing the rate of bacterial multiplication by phagocytosis.

The major mechanisms and the role of the neutrophils in each of these processes are described in the following material.

## PROCESS OF INFLAMMATION

### Signs of Inflammation

The classical symptoms of inflammation were first described by the Roman physician Aulus Cornelius Celsus, who named them **calor** (heat), **rubor** (redness), **tumor** (swelling), and **dolor** (pain). These symptoms result from structural and functional changes occurring in the vasculature of affected tissues in response to invasion by an infectious agent. Redness is caused by the dilation of the blood vessels. Heat is due to the increase of blood in the inflamed area. Swelling is due to enlargement of the blood vessels and movement of fluid into the tissues (edema), and pain is caused by pressure on the nerve endings by the swollen tissues and also by the action of kinins on the nerve endings.

Mounting a normal inflammatory response is of paramount importance to the maintenance of life in each individual. Not only does it play a major role in the prevention of serious infections by blocking the invasion of microorganisms into the internal milieu of the body via wounds or mucosal surfaces, but the process of inflammation also sets into motion a complex series of reactions that result in the healing and repair of damaged tissues.

Although the inflammatory response is an important mechanism in the maintenance of our health, it is a two-edged sword. In many allergic (or hypersensitivity) reactions, this inflammatory process is activated inappropriately and may produce widespread tissue destruction, which may result in severe debility and on occasion may result in death. For instance, rheumatoid arthritis is caused by inappropriate inflammation of the joints, and glomerulonephritis results from inappropriate inflammation and destruction of many kidney glomeruli—which causes thousands of deaths each year.

### Types of Inflammation

The inflammatory response may be divided into the following three major categories according to the cell types predominating in the inflamed tissues: (1) acute inflammation, (2) subacute inflammation, and (3) chronic inflammation. All inflammations start with an acute phase from which they normally progress into a subacute stage. However relatively few acute inflammations will evolve into a chronic stage.

#### *Acute Inflammation*

Most acute inflammations result from infections caused by mechanical injury such as a cut, wood splinter, burn, or damaged stomach wall. For example, when a person steps on a rusty nail, the nail will have millions of bacteria on its surface, which remain behind when the nail is removed. Within hours the area surrounding the spot of infection may become inflamed. Microscopic tissue analysis of the infected area should reveal the infected area is swarming with neutrophils. During the acute inflammatory phase, few lymphocytes or macrophages are encountered at the site of inflammation (the reason for this disparity is not known). Possibly the first chemotactic factors that activate the PMNs may not be able to activate other white blood cells to the same degree, or perhaps they respond at a slower rate. Another reason may be that neutrophils are much more numerous than other white blood cells. This acute phase is usually associated with tissue destruction, pain, and swelling.

*Subacute Inflammation*

A few days after the initial acute inflammatory response, the infected area proceeds to a subacute phase. Swelling becomes less pronounced and pain becomes less throbbing. Microscopic analysis of the infected tissue will reveal that the neutrophils have been largely replaced by macrophages and lymphocytes. This subacute phase is associated with the formation of new tiny blood vessels and connective tissue. Many fibroblasts appear in the area and begin to produce new collagen fibers in an effort to repair the damaged tissues. The acute phase is associated with tissue destruction, and the subacute phase is associated with reconstruction. As mentioned before, the newly arrived white blood cells, macrophages and lymphocytes, are much more efficient in killing and removing pathogenic microorganisms than are neutrophils. Because of their short life span, many neutrophils will die in the infected area and their remnants are also removed by macrophages. Within about a week, the subacute phase disappears and the infection appears to be over.

*Chronic Inflammation*

In a few situations macrophages and lymphocytes may be unable to eradicate the infection because the pathogens multiply faster than the white blood cells can destroy them. In that case additional mononuclear phagocytes migrate to the infected area. Macrophages undergo a change that transforms them into **giant cells.** Such extra-large cells are produced by the process of **endomitosis**—that is, the nucleus keeps dividing within a common cytoplasm. Each time the nucleus divides, the cell becomes larger until it reaches giant proportions. These giant cells wrap themselves around the pockets of infection. They are aided by newly arrived fibroblasts, which will continue to produce vast amounts of dense connective tissue fibers, which also wrap around these loci of infection. These pockets of multiplying microorganisms become completely surrounded by giant cells and fibrous connective tissue. As a result they are separated from nutrients supplied by the blood and hence die inside these clumps of connective tissue called **granulomas.** If many granulomas are produced in vital tissues such as heart muscle, liver, lungs, or kidneys, they may interfere with the normal functioning of the host, which may lead to other complications. However such chronic inflammations are relatively rare and may occur in such infectious diseases as tuberculosis and leprosy.

## Acute Inflammatory Response

Since neutrophils are more closely related to the acute inflammatory response compared to the chronic inflammation response, we should discuss this process in more detail. The acute inflammatory response consists of the following two basic independent events which can occur simultaneously or in sequence: (1) changes in the permeability of the blood vessel walls in the affected area due to blood vessel dilation—a process caused by a number of permeability factors and (2) accumulation of granulocytes near the inflamed area due to the process of chemotaxis.

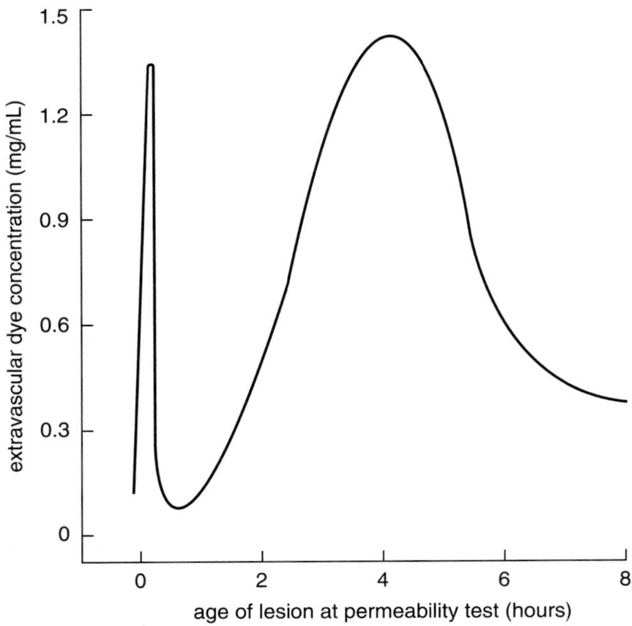


figure 17.1

The two phases of postinjury capillary permeability of a guinea pig. The first phase is short-lived and is completely histamine dependent. The second phase is prolonged and depends upon a number of anaphylactic factors.

*Blood Vessel Dilation*

The two phases in blood vessel dilation in the acute inflammatory response both result in increased permeability. The first phase is short and transient, lasting not more than a few minutes. The second phase begins about 2 to 4 hours after the initial swelling and is more prolonged, often lasting several hours or even days (figure 17.1).

The first phase is exclusively mediated by the vasoactive amine **histamine.** This vasoactive amine does not normally exist freely in the circulation but is stored in granules of the mast cells of the connective tissue, in granulocytes (especially basophils), and platelets of the circulatory system. Histamine is normally released from these cells when they are stimulated by basic (cationic) proteins. These basic peptides are present in large quantities in neutrophils and are released when the PMNs are activated or when they disintegrate at the site of infection.

The second or delayed phase is not exclusively dependent on histamine but is mediated by a variety of chemical compounds. During this prolonged secondary phase, neutrophils begin to migrate in large numbers to the area of infection.

Both the histamine-dependent and histamine-independent phases of blood vessel dilation are associated with fluid movement from the bloodstream into

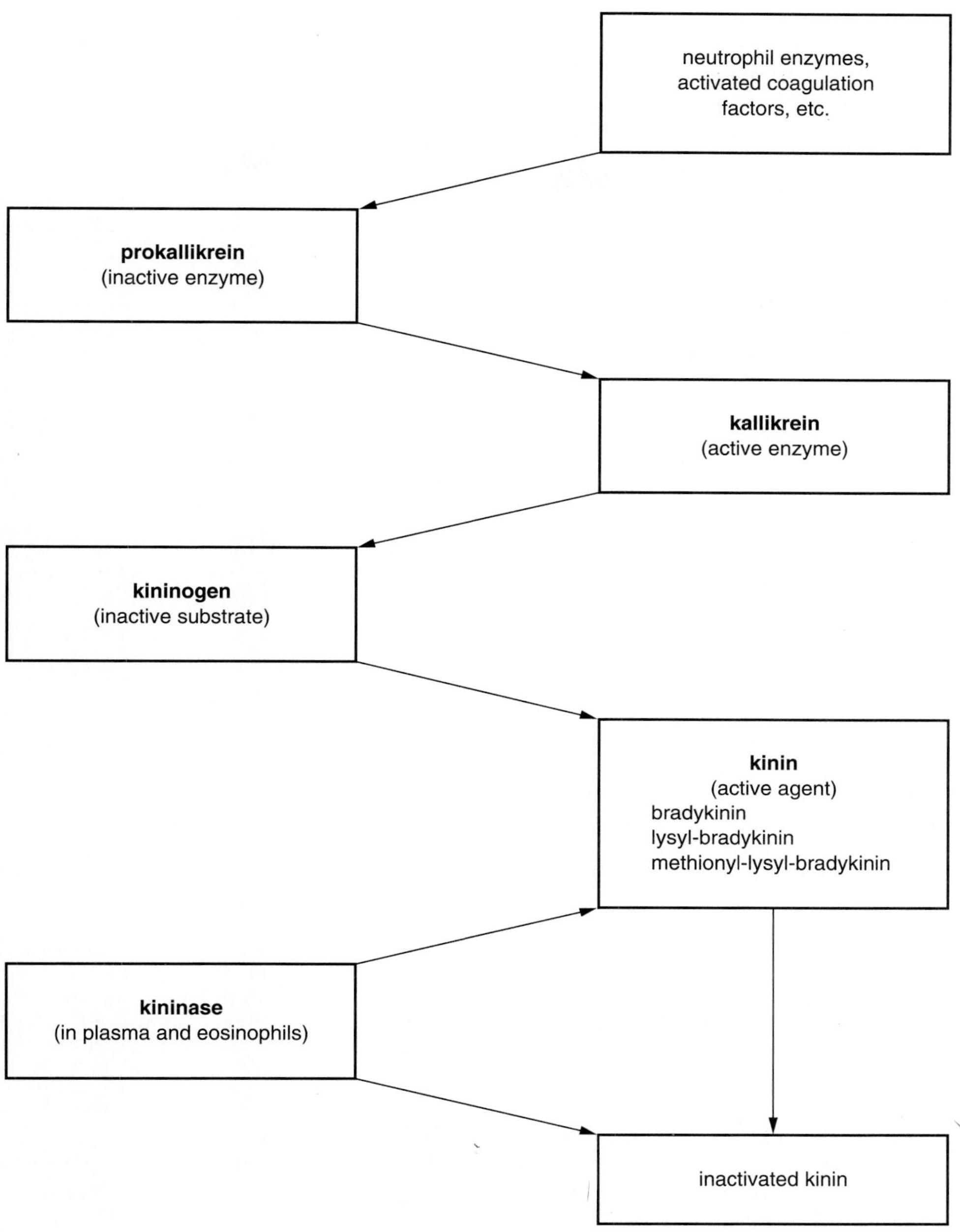


**figure 17.2**

Stages of kinin metabolism.

the tissues through junctional zones between endothelial cells lining the inner walls of the blood vessels, thus causing local edema. Another important point is that these vasodilatory compounds are able to dilate capillaries and postcapillary venules only. The mechanism by which they accomplish this effect is unknown.

*Permeability Factors*

Apart from histamine there are many other permeability factors. The most important of these factors during the second prolonged phase of blood vessel dilation are in the following sections.

**Kinins** Kinins are low-molecular-weight polypeptides present in the α-globulin fraction of the serum. The active enzymatic agents are generated sequentially in a series of steps. This biochemical cascade results in the formation of enzymes, each of which acts sequentially on the next protein. The final result is the formation of three active kinin peptides, the most important of which is **bradykinin.** The latter is made up of nine amino acids and is also the precursor for the other two active kinins: lysyl-bradykinin and methionyl-lysyl-bradykinin (figure 17.2). These kinins are potent constrictors of nonvascular smooth muscle and also have strong

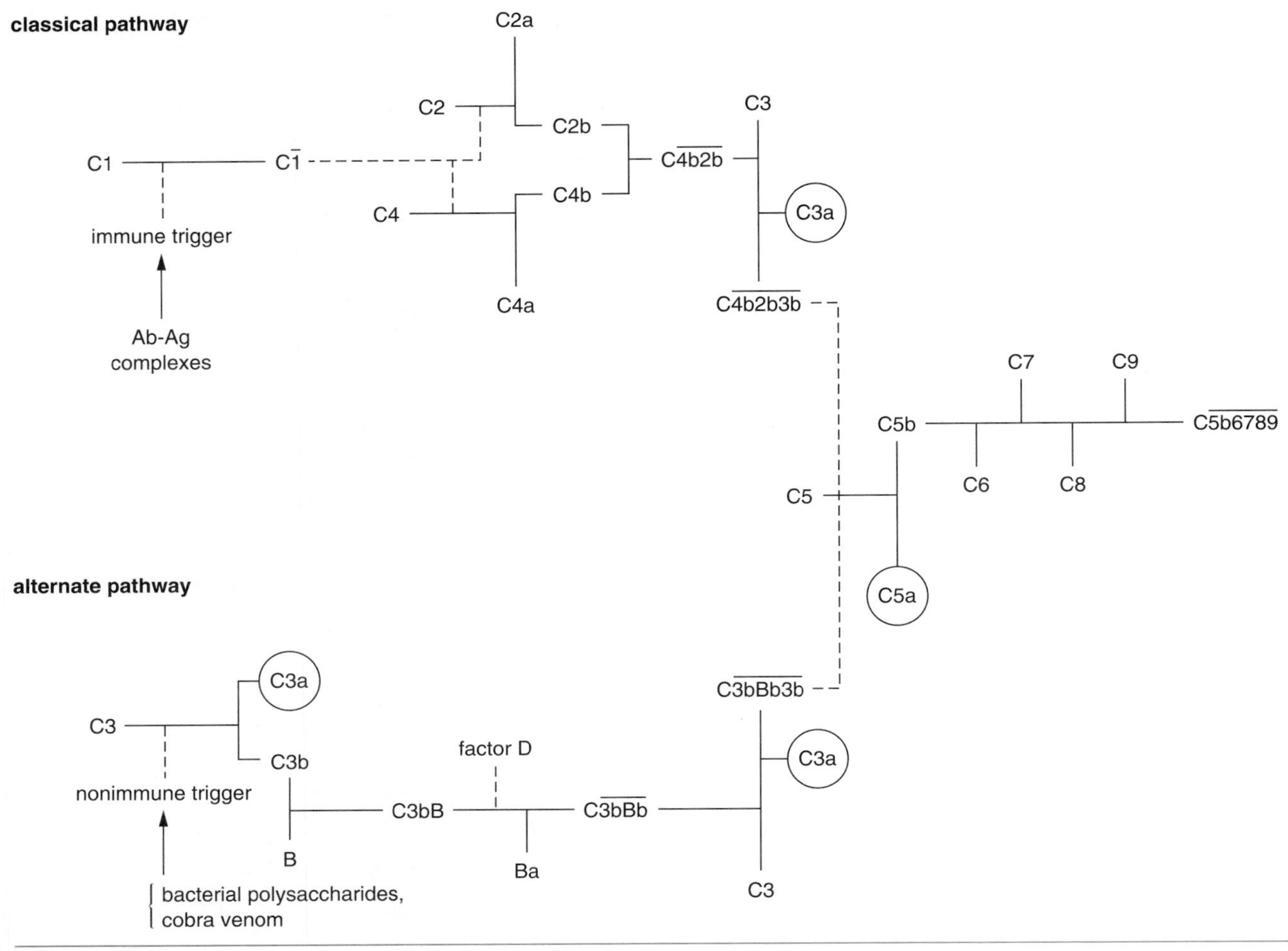


**figure 17.3**
The two major pathways of complement activation, and their convergence into a common pathway with the activation of C5. The major anaphylatoxins and chemotaxins are circled. Bars indicate formed complement complexes.

vasodilating effects. Further, they are also able to cause pain by acting on nerve endings. Granulocytes, especially neutrophils, possess kinin-generating enzymes. Thus PMNs play a major role in initiating the inflammatory response in an area of infection. Activated kinins, however, have a brief life span; they are inactivated within minutes by a series of kininases found in the plasma and certain blood cells, especially eosinophils.

**Activated complement factors** The complement system consists of a series of at least 16 proteins present in inactive form in normal plasma but may be activated by various factors such as antibody-antigen complexes, bacterial polysaccharides, certain yeast products, cobra venom, and aggregates of class A immunoglobulins (IgA). Similar to the kinins these complement factors are activated sequentially. During the cascading process a number of activated products are produced including **C3a** and **C5a,** which are potent vasodilating compounds (figure 17.3). Both C3a and C5a play a major role in the onset of inflammation in many infections, and hence they are commonly referred to as **anaphylatoxins.** Both are small-molecular-weight compounds of approximately 10,000 daltons and act by stimulating mast cells and basophils to release vasoactive amines.

**Prostaglandins** Prostaglandins are fatty acid derivatives produced by many body cells in response to external stimuli such as tissue damage, infectious agents, and hormones. They belong to a family of chemically related compounds known as **eicosanoids,** which are produced exclusively within plasma membranes of the cells of the body. Prostaglandins are derived from a common ancestral compound called **arachidonic acid,** which is released from membrane phospholipids by the action of phospholipases (figure 17.4).

Arachidonic acid is formed from the essential fatty acid, linolenic acid, by elongation and desaturation. The previously mentioned external stimuli may activate phospholipase activity and liberate arachidonic acid to serve as a precursor for further eicosanoid biosynthesis. Depending on the tissue involved and the nature of the external stimulus, arachidonic acid may be utilized in several different ways—that is, it may follow

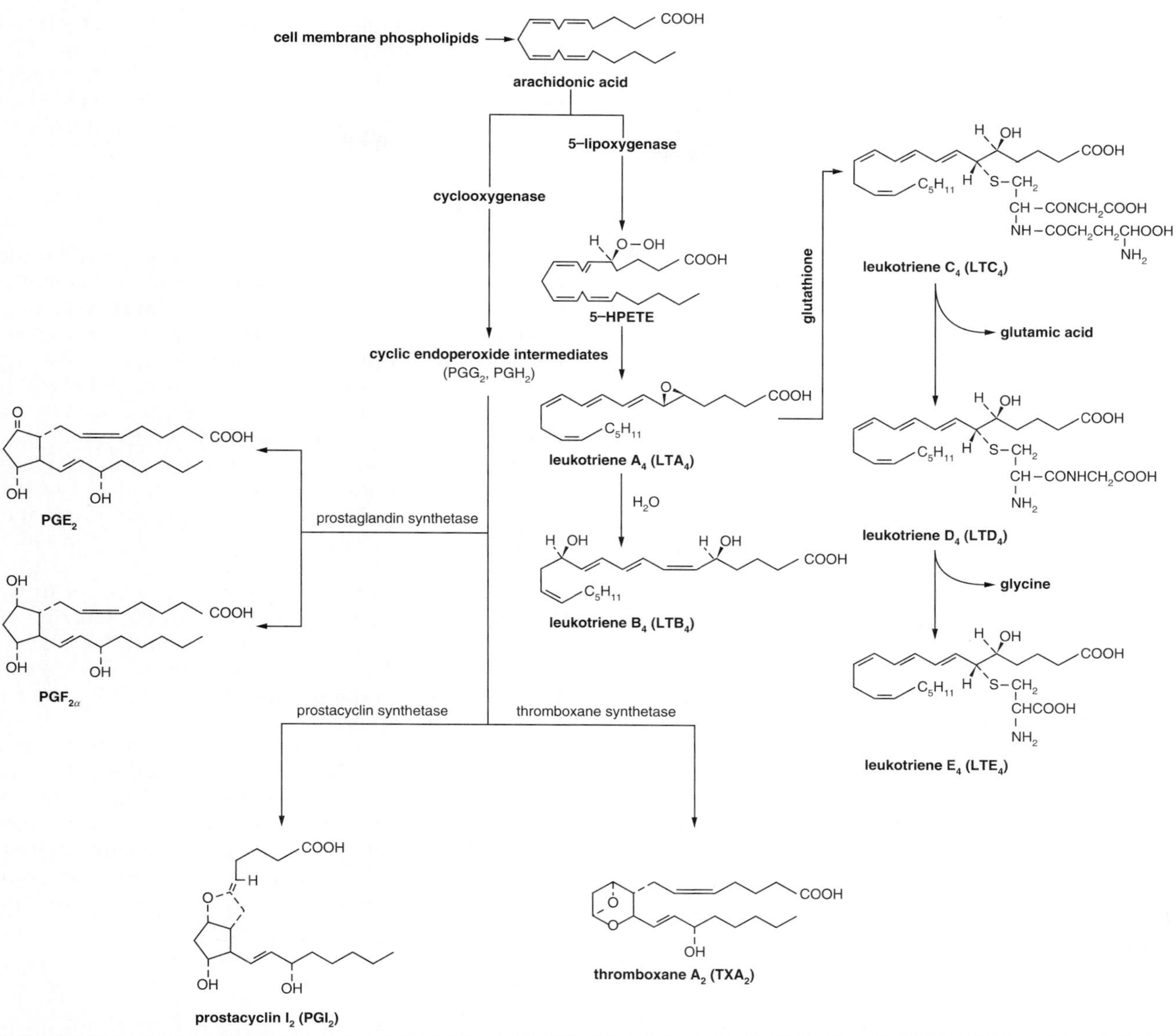


**figure 17.4**

Prostaglandin, prostacyclin, thromboxane, and leukotriene synthesis.

one of two different pathways (figure 17.4). In one pathway arachidonic acid is converted to unstable cyclic endoperoxide intermediates by a group of enzymes collectively called the **cyclooxygenase system.** Then depending on the action of other tissue-specific enzymes, these intermediates may be converted to one of a number of related products including prostaglandin $E_2$ ($PGE_2$), prostaglandin $F_2\alpha$ ($PGF_2\alpha$), prostacyclin $I_2$ ($PGI_2$), and thromboxane $A_2$ ($TXA_2$). Prostaglandins and prostacyclins have vasodilating activity and hence play important roles in the production of local inflammation.

**Leukotrienes** Leukotrienes are so named because they are produced by leukocytes and contain at least three conjugated double bonds. They are also classified as eicosanoids since they are also derived from arachidonic acid. Production of leukotrienes, however, follows another enzymatic pathway. Leukotrienes are formed when the enzyme 5-lipoxygenase converts arachidonic acid to 5-hydroperoxy-6,8,11,14-eicosatetraenoic acid (5-HPETE). The latter compound is then converted to the epoxy acid leukotriene $A_4$ ($LTA_4$), which is subsequently changed to leukotriene $B_4$ ($LTB_4$) by the addition of water, or to leukotriene $C_4$ ($LTC_4$) by the addition of glutathione. This $LTC_4$ is then converted to leukotriene $D_4$ ($LTD_4$) by the elimination of glutamic acid and finally converted to leukotriene $E_4$ ($LTE_4$) by elimination of the amino acid glycine (figure 17.4). The subscript 4 indicates that they all have four double bonds. These leukotrienes are extremely potent vasodilators and greatly increase vascular permeability and local edema. When white blood cells arrive at the site of infection and/or injury, they release these

**Table 17.1 Various Chemical Agents That Influence Membrane Stability**

*Chemical Agents that Cause Lability of Membranes*

| | |
|---|---|
| Antibodies to membranes | Streptolysin S |
| Endotoxin | Vitamin A |
| Etiocholanolone | |
| Progesterone | |
| Streptolysin O | |

*Chemical Agents that Stabilize Membranes or Prevent Enzyme Release*

| | |
|---|---|
| Acetylsalicylic acid | Phenylbutazone |
| Chloroquine | Prostaglandin E |
| Colchicine | Stilbamidines |
| Cyclic AMP | |
| Glucocorticoids | |

leukotrienes into the circulation, thus causing rapid swelling. Since the majority of white blood cells are neutrophils during the acute phase of inflammation, they are mainly responsible for release of these vasodilating compounds.

Previously leukotrienes were considered to be the same compound—commonly referred to as **slow-reacting substance of anaphylaxis (SRS-A)**—because their effects were slower and longer lasting than other substances of anaphylaxis like histamine, serotonin, and bradykinin. Release of these leukotrienes is stimulated by such factors as bacterial products, immune complexes, and certain crystals.

**Basic Protein** The basic (cationic) proteins are another group of permeability factors produced by neutrophils. They are found in a preformed state in lysosomal granules of the PMNs. Some of these basic proteins increase vascular permeability directly, and others activate mast cells and basophils to release vasoactive amines. Each time a neutrophil breaks down and releases its lysosomal contents into the surrounding tissues, it will increase the inflammatory response. A group of substances has been recognized that causes the PMN cell membrane to become more labile and thus promote the release of lysosomal enzymes. This group includes certain bacterial products (endotoxins), immune complexes, poorly soluble crystalline particles, and lipid-soluble compounds that can interact directly with the cell and lysosomal membranes. Table 17.1 provides a list of these agents, along with a list of agents that stabilize PMN membranes and thus prevent the release of enzymes and vasodilatory compounds.

## PROCESS OF CHEMOTAXIS

Permeability changes alone are not enough to cause the PMNs to move out of the circulation *en masse.* Indeed massive inflammation and permeability changes often occur without any noticeable transmigration of leukocytes. This happens in allergic reactions. The wholesale migration of neutrophils and other white blood cells from the circulation to the tissues can take place only if both blood vessel dilation and chemotactic stimuli are present.

### Unidirectional Movement

**Chemotaxis** may be defined as the unidirectional movement of a cell in response to a chemical attractant. The latter is known as a **chemotaxin,** or a chemotactic factor. Neutrophils respond to a gradient concentration of chemotaxins—they move toward a greater concentration of these chemotactic molecules. Nonstimulated neutrophils move randomly. However as soon as they are exposed to a chemotactic stimulus, they begin to move in nearly straight lines to the source of the stimulus. Chemotaxis does not involve a change in speed of these neutrophils but only a change in directional movement.

Locomotion by neutrophils is temperature-dependent; they move best at body temperature. Their motility is also regulated by pH. They move well in fluids with a pH between 6.5 and 7.5. The cytoplasmic flow is continuous and occurs throughout the cell. Pseudopodia produced in the movement are known as lamellipodia.

Transmigration of PMNs to an area of infection is much faster than that of other leukocytes. In an *in vitro* experiment using a chemotactic chamber (i.e., a cell with two compartments separated by a membrane), PMNs averaged 90 minutes or less to respond to a chemotactic stimulus and migrate from one chamber to the next, whereas monocytes and lymphocytes required five or more hours.

### Chemotactic Stimuli

Many different chemotaxins occur in the human organism. The most important of these chemotactic stimuli are as follows:

1. **Bacterial products**
Many bacteria produce low-molecular-weight products that cause white blood cells to move toward that source of chemotaxin (i.e., the bacteria). Recent studies have indicated that many of these chemotactic factors are small peptides such as *N*-formylmethionyl.
2. **Chemical agents produced by neutrophils**
Leukocytes are capable of producing certain factors that attract other white blood cells to the area of infection. This phenomenon may result from active secretion of chemotactic substances produced by these cells or from agents released from the cells after they die and disintegrate at the site of infection.
3. **Damaged tissue**
Many damaged cells produce substances into the body fluids that are capable of attracting leukocytes. They are thought to be mainly prostaglandins and thromboxanes, both of which are produced in the cell membrane.

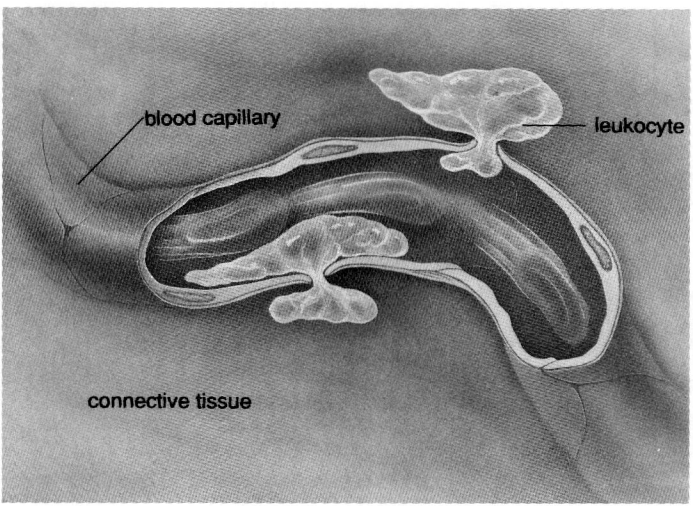


figure 17.5
Diapedesis, the common way in which most white blood cells leave the bloodstream.

**4. Activated complement factors**
Certain activated complement factors also play a major role as chemotaxins. The same complement factors that function as anaphylatoxins also function as chemotaxins: C3a and C5a. Both are considered to be powerful leukoattractants.

**5. Other serum factors**
Several other activated plasma-derived substances have been implicated as chemoattractants for neutrophils. These factors include certain coagulation factors—for example, activated Hageman factor (factor XII)—plasminogen activator substance, and kallikrein.

### Transmigration of Neutrophils

Under normal circumstances white blood cells leave the peripheral circulation only in the blood sinuses. Sinuses may be described as small blood vessels with a large diameter in which the blood flow has slowed considerably. This allows leukocytes to attach themselves against the vessel wall before they move out of the circulation. In most areas of infection, sinusoid vessels are not found; hence the process of inflammation is necessary to create short-term artificial sinuses allowing white blood cells to marginate against the cell wall. This is the first step in the process of leaving the circulation in the vicinity of an infected area. Without inflammation, including local vessel dilation, PMNs and other white blood cells would not be able to attach themselves against the vessel wall as they would constantly be pushed by more numerous red blood cells. Since there are at least 1,000 red blood cells for every white blood cell, the push exerted by the erythrocytes would make it impossible for neutrophils to marginate in a nondilated blood vessel of the microcirculation. The creation of artificial sinuses—due to the process of inflammation—allows red blood cells to move forward at normal speed and allows white blood cells to attach themselves to the vessel wall without interference by other blood particles.

Shortly after the local blood vessels begin to dilate, neutrophils start to cluster along these stretched vessel walls. This process is known as **margination,** or pavementing. Soon after, these marginated neutrophils migrate between the intercellular spaces of adjacent endothelial cells of dilated capillaries and small veins. This process of squeezing through stretched endothelial cells is known as **diapedesis** (figure 17.5). These events accelerate over a period of several hours, and soon the infected area is swarming with vast numbers of extravascular neutrophils, which immediately start to attack and engulf pathogens that entered the interior milieu of the body. Observations of leukocyte transmission with an electron microscope have shown that not all white blood cells move from the bloodstream into the interstitial tissues via the junctions between adjacent cells. Apparently many lymphocytes traverse the blood vessel walls by moving straight through the cytoplasm of endothelial cells. This process is known as **emperipolesis** (figure 17.6). Once the lymphocytes come into contact with the inner endothelial walls of the blood vessel, they become surrounded by cytoplasmic projections from these cells, which engulf them and transport them intact across the cell in a vacuole. However, not all lymphocytes move through the vessel walls by emperipolesis—a significant number move to the extravascular spaces by diapedesis, just as other white blood cells.

## PROCESS OF PHAGOCYTOSIS

Phagocytosis may be defined as the process of **ingestion** and **digestion** of a particle by a cell. Several different stages occur in the process of phagocytosis. First, the phagocytic cell must be attracted to the particles that are to be digested. Second, the cell must make contact with these particles before they can be engulfed. All phagocytic cells have receptors

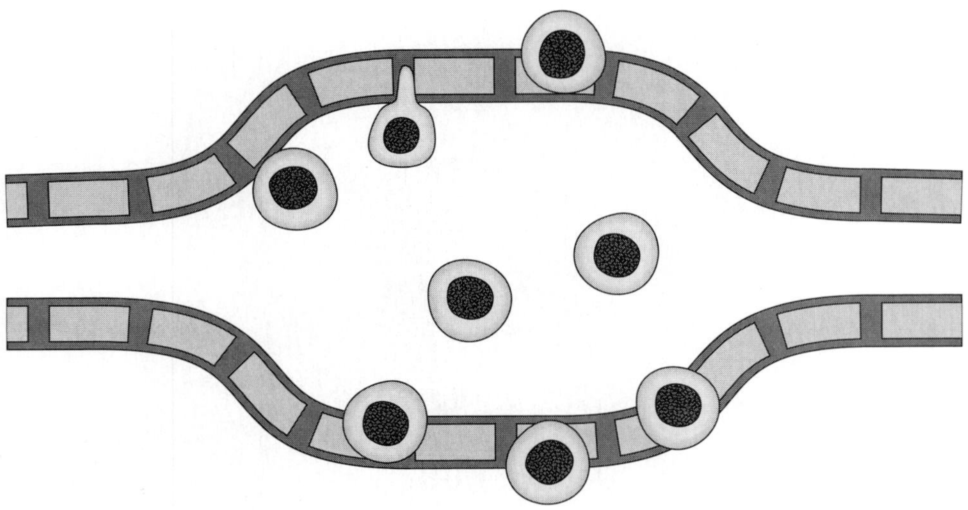

**figure 17.6**

Emperipolesis and diapedesis employed by activated lymphocytes.

on their membranes that can bind to compounds frequently found on the surface of pathogens. Other common receptors are (1) **Fc receptors,** which can bind the Fc fraction of an immunoglobin (Ig) molecule and (2) **C3b receptors,** which can bind to the activated complement factor C3b. Both Ig and C3b molecules—acting as opsonins—are normally present in large numbers on foreign particles that have invaded individuals with an intact immune system. This contact between **opsonins** and receptors allows the phagocyte to get a good grip on the foreign particle, greatly facilitating the process of engulfment. At the point of contact the phagocyte wall starts to invaginate, and as a result the particle is gradually drawn into the cell. Eventually the cell membrane completely surrounds the particle, and the resulting vacuole is pinched off from the external membrane, as illustrated in figure 17.7. The newly formed vacuole containing the engulfed particle(s) is now called a **phagosome.** Within minutes after the phagosome is formed, a number of **lysosomes**—that is, primary and secondary granules containing many digestive enzymes—will fuse with this organelle to form a new structure referred to as a **phagolysosome.** The number of lysosomes that fuse with each phagosome depends on the number of phagosomes produced by the phagocytic cell. Once the phagolysosome is formed, the process of microbial killing and digestion begins.

### Metabolic Events During Phagocytosis

The process of phagocytosis also initiates a series of biochemical events inside the phagocytic cell which optimizes the ability of the cell to destroy the foreign particles and to aid in their digestion. Many rapid changes take place in actively phagocytosing cells; the most important of these are as follows:

**1. Rapid increase in glycolysis**
The energy needed to engulf and digest these particles is mainly derived from an increase in the glycolytic process. This may be demonstrated by the fact that if glycolysis is suppressed, little phagocytosis occurs. Further, the process of phagocytosis results in decreased glycogen levels and increased lactic acid production. Absence of oxygen does not seem to affect the ability of the phagocyte to engulf foreign matter, although it may interfere with the killing process.

**2. Burst in oxygen consumption**
There is a dramatic increase in oxygen consumption within minutes after a particle is ingested. Oxygen consumption may increase from threefold to twentyfold, depending on the amount of foreign material engulfed. This sudden uptake of oxygen is known as the **respiratory burst** (figure 17.8).

**3. Increased hexose monophosphate shunt (HMS) activity**
Phagocytosis also produces a great increase in HMS activity. This sudden increased activity of the hexose monophosphate pathway may be triggered by a sudden increase of lactic acid in the cell as a result of increased glycolysis. This sudden increase in lactic acid causes a drop in the intracellular pH. This lower pH, in turn, acts as a stimulus to release or activate the enzyme NADPH oxidase from the granules. This enzyme converts NADPH to $NADP^+$. An increased supply of $NADP^+$ stimulates the hexose monophosphate pathway to produce more NADPH.

**4. Increased hydrogen peroxide ($H_2O_2$) and superoxide ($O_2^-$) production**
Increased oxygen consumption is associated with a rapid increase in the concentration of $H_2O_2$ and $O_2^-$. Phagocytosing neutrophils produce two to four times as

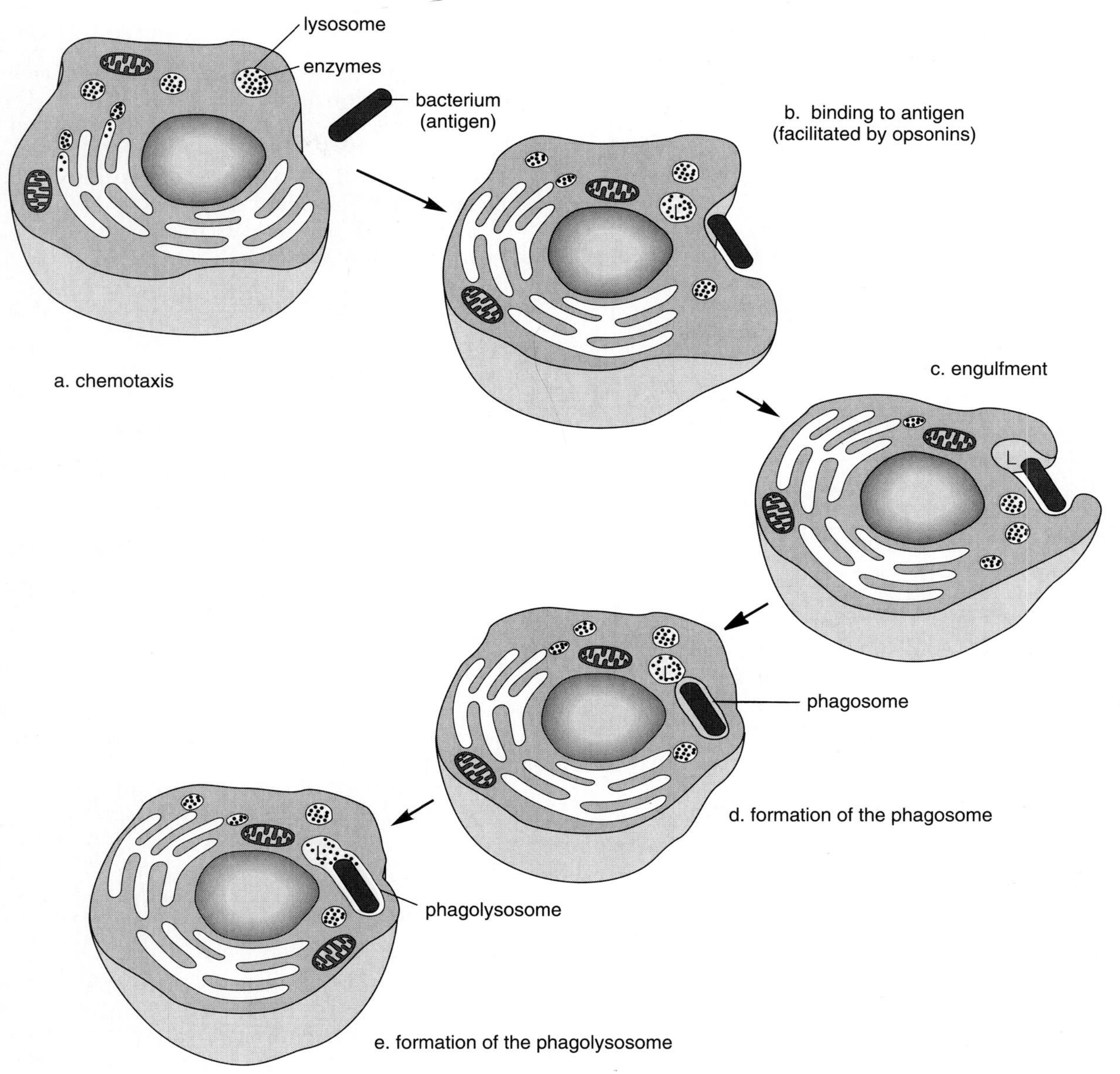


**figure 17.7**

The five steps in the process of phagocytosis.

much $H_2O_2$ as do resting PMNs. These newly generated substances are highly toxic to the ingested particles and play a major role in the process of microbial killing as is explained later.

**5. Increased lipid synthesis**
The actively phagocytosing cell has a much higher rate of lipid synthesis than does the resting cell. This is understandable since new membrane phospholipids are needed to replace the plasma membrane lost to newly created phagosomes; otherwise the cell would become smaller and smaller.

**6. Decreased pH of the phagolysosome**
The pH in phagocytic vacuoles decreases dramatically within minutes after particle digestion. The pH may decrease as low as 4.0 to 4.5. This decrease in pH is needed to activate the acid hydrolases discharged into the phagosomes by the lysosomes and also to activate the enzyme myeloperoxidase. Several mechanisms have been proposed to explain this sudden fall in pH in phagolysosomes. One theory states that the enzyme carbonic anhydrase plays a major role in the acidification of the phagocytic vesicle. In the presence of carbonic

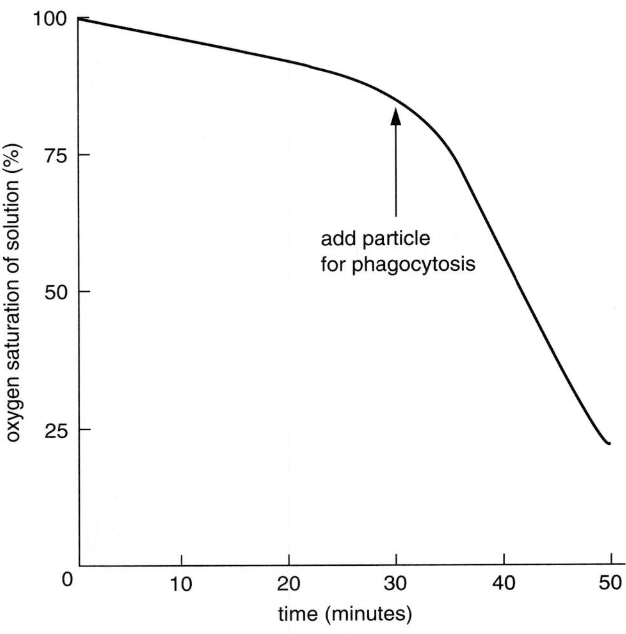


**figure 17.8**
The effect of phagocytosis on neutrophil oxygen consumption. Note that the oxygen saturation of a solution containing suspended neutrophils falls abruptly with the rapid oxygen uptake by the cells after the process of phagocytosis has started.

anhydrase, carbon dioxide is hydrated to form carbonic acid, and a proton pump then restricts these newly created hydrogen ions to the phagolysosome.

$$CO_2 + H_2O \rightleftharpoons H_2CO_3 \rightleftharpoons HCO_3^- + H^+$$

## PROCESS OF MICROBIAL KILLING

A number of different mechanisms are involved in the process of microbial killing. These may be divided into two major categories: (1) oxidative mechanisms and (2) nonoxidative mechanisms.

### Oxidative Mechanisms

As was just mentioned, $H_2O_2$ plays a major role in the process of microbial killing. To produce $H_2O_2$ two oxidative enzymes (oxidases) must be activated that are capable of capturing free $O_2$ molecules and converting them to $H_2O_2$. These enzymes are NADH oxidase and NADPH oxidase.

The activation of **NADH oxidase** results in a direct conversion of $O_2$ to $H_2O_2$:

$$NADH + H^+ + O_2 \xrightarrow{\text{NADH oxidase}} NAD^+ + H_2O_2$$

This $H_2O_2$ can then be converted into water and oxygen with the aid of the enzyme catalase:

$$2H_2O_2 \xrightarrow{\text{catalase}} 2H_2O + O_2$$

or into oxidized glutathione with the aid of the enzyme GSH peroxidase:

$$H_2O_2 + 2GSH \xrightarrow{\text{GSH peroxidase}} 2H_2O + GSSG$$

The latter can be reduced again to GSH by GSSG reductase with the aid of NADPH:

$$GSSG + 2NADPH \xrightarrow{\text{GSSG reductase}} 2GSH + 2NADP^+$$

The $NADP^+$ generated now enters the hexose monophosphate shunt where it is changed again to NADPH (figure 17.9).

In this process glucose is first converted to G-6-P, which is then changed to 6-PG, and finally to pentose 5-phosphate, also known as ribose 5-phosphate (or R-5-P):

$$\text{G-6-P} + 2NADP^+ + H_2O \xrightarrow{\text{G-6-PD}} \text{R-5-P} + 2NADPH + 2H^+ + CO_2$$

The NADPH generated in this reaction now reacts with the second oxidative enzyme, **NADPH oxidase,** which converts $O_2$ into superoxide ($O_2^-$) in the following reaction:

$$NADPH + O_2 \xrightarrow{\text{NADPH oxidase}} NADP^+ + O_2^-$$

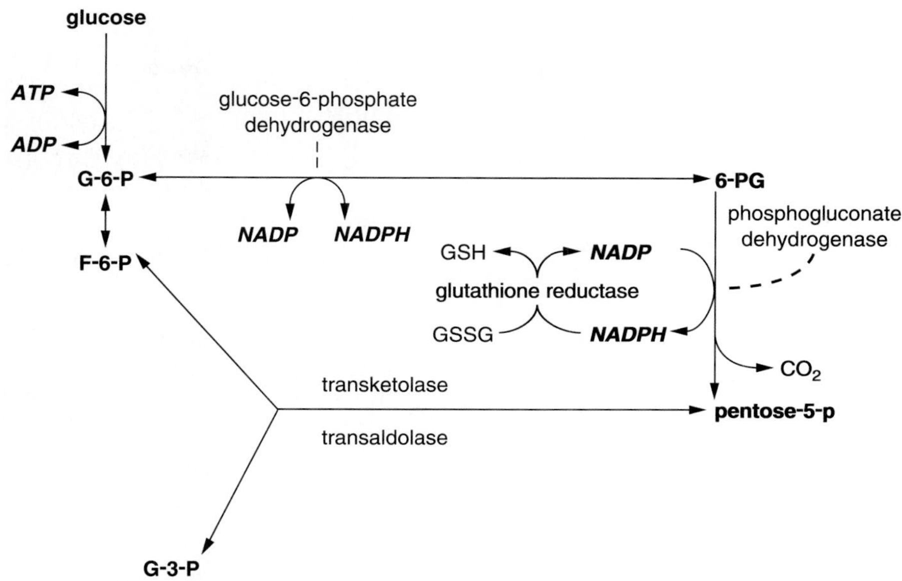


**figure 17.9**

The hexose monophosphate shunt. The F-6-P (fructose-6-phosphate) and G-3-P (glyceraldehyde-3-phosphate) are intermediates in the Embden-Meyerhof pathway for glucose metabolism shown in figure 7.2.

This superoxide ion can then be converted into $H_2O_2$ in several ways, the most important of which are as follows:

$$O_2^- + O_2^- + 2H^+ \xrightarrow{\text{superoxide dismutase}} O_2 + H_2O_2$$

$$O_2^- + O_2^- + 2H^+ \rightarrow {'O_2} + H_2O_2$$

The latter is a spontaneous reaction and produces singlet oxygen ($'O_2$), which can react with any oxidizable molecule—especially any hydrocarbons possessing one or more double bonds—and in doing so will break the carbon chain:

$$R_2C{=}CR_2 + {'O_2} \qquad R_2C{=}O \quad O{=}CR_2$$

During this reaction the singlet oxygen returns from its excited state to its ground state, releasing some energy in the form of light **(chemoluminescence).** This is one characteristic of the respiratory burst. These light flashes are too weak to be seen but can be recorded in a liquid scintillation counter. The foregoing reaction is a major way in which carbohydrates and lipids are broken down inside the phagocytic cell.

The breakdown of proteins is accomplished by the destruction of individual amino acids. For this process $H_2O_2$ is needed also, since it is able to react with chloride, iodide, bromide, and fluoride ions to produce water and hypohalites. This reaction is catalyzed by the enzyme **myeloperoxidase.** The reaction for chloride for example is:

$$H_2O_2 + Cl^- + H^+ \xrightarrow{\text{myeloperoxidase}} HOCl^- + H_2O$$

The hypochlorous acid produced can now react with the amino groups to form **chloramines:**

$$HOCl^- + H\text{-}\underset{COOH}{\overset{R}{C}}\text{-}NH_2 \rightarrow H^+ + H_2O + H\text{-}\underset{COOH}{\overset{R}{C}}\text{-}NHCl \text{ (=chloramine)}$$

These chloramines decompose spontaneously into $NH_3$, $CO_2$, $Cl^-$, and R-CHO (aldehyde), thus producing the decomposition of the particle:

$$H\text{-}\underset{COOH}{\overset{R}{C}}\text{-}NHCl^- + H_2O \rightarrow R\text{-}CHO + NH_4^+ + CO_2 + Cl^-$$

Organic molecules can be destroyed by a third mechanism. When superoxide combines with hydrogen peroxide, it produces a highly unstable **hydroxyl radical** (•OH), which will combine with almost any organic molecule and destroy it:

$$O_2^- + H_2O_2 \rightarrow \bullet OH + OH^- + O_2$$

### Nonoxidative Mechanisms

A number of compounds in neutrophils appear to have antibacterial activity. However few of these factors have been well defined. Some of the more prominent ones include the following:

1. **Lactoferrin,** which is an iron-binding protein, may be bacteriostatic as long as it is not fully saturated with iron. Presumably it acts by limiting the amount of iron available to microorganisms, and iron is required for the growth of many pathogens.
2. **Phagocytin.** This is an inclusive term for a conglomerate of antibacterial compounds produced by phagocytic cells.
3. **Leukin.** This compound is derived from nucleohistone. Apparently it interferes with microbial replication.

The sudden drop in pH in the phagocytic vacuole also functions as a nonoxidative mechanism of microbial destruction. A dramatic decrease in pH may be sufficient to inhibit replication in some of the more sensitive microorganisms such as the pneumococci. In most cases, however, this drop in pH is not sufficient to kill these pathogens, and a variety of oxidative and nonoxidative mechanisms are still required to fulfill that function.

## FATE OF ACTIVATED NEUTROPHILS

Once the bacteria are killed, hydrolytic enzymes present in the phagolysosomes will digest these microorganisms. A list of the major enzymes involved is shown in table 17.2.

Many neutrophils are unable to finish the process of microbial killing and digestion after they have engulfed microorganisms simply because they have reached the end of their life span. However many other PMNs are present to complete the process when the original phagocytic cells begin to disintegrate. Further, the death of many neutrophils in an infected area will result in the release—into that area—of many digestive enzymes previously contained in lysosomal granules. This will effectively lower the pH of the infected area and thus allow the released enzymes to continue their job of destroying and digesting microorganisms found in the infected location. Outside the area of infection, the pH will be too high for these hydrolytic enzymes to function properly. This is also the reason why nonstimulated neutrophils that migrate randomly into the tissues (where they will disintegrate after a few days) will not cause any damage to the surrounding cells upon their death.

The remnants of the disintegrated neutrophils, the end-products of microbial digestion, and the leftovers of necrotic tissue together form a kind of liquefied mass commonly known as **pus.** A localized collection of pus is known as an **abscess.** If the abscess is close to the outer surface of the body, it is frequently drained into the external environment. If there is no exterior access, the material will be hydrolyzed further and ultimately will be removed by the macrophages.

**Table 17.2 Enzymes and Nonenzymatic Substances Found in Neutrophil Granules**

| | |
|---|---|
| *Enzymes* | |
| α-L-Fucosidase | Hyaluronidase |
| α-1,4-Glucosidase | Kininogenase |
| α-Mannosidase | Lipase |
| α-*N*-Acetylgalactosaminidase | Lysozyme |
| α-*N*-Acetylglucosaminidase | Myeloperoxidase |
| Arylsulfatases | Phosphatases |
| β-Galactosidase | Phospholipase |
| β-Glucuronidase | Protease |
| β-*N*-Acetylglucosaminidase | Ribonuclease |
| Cathepsins | |
| Collagenase | |
| Deoxyribonuclease | |
| Elastase | |
| Esterases | |
| *Nonenzymatic Substances* | |
| Cationic proteins: bactericidal proteins, histamine-liberating agent, histamine-independent permeability factors | Histamine |
| | Mucopolysaccharides |
| | Plasminogen activator |
| | Phagocytin and related bactericidal proteins |
| Glycoproteins | |

## Further Readings


Beck, W. S. Leukocytes I. Physiology. Lecture 17 in *Hematology,* 4th ed., W. S. Beck, Ed., The MIT Press, Cambridge, Mass., 1985.

Boggs, D. R., and Winkelstein, A. The Phagocyte System. Chapter 4 in *White Cell Manual,* 4th ed., F. A. Davis Company, Philadelphia, 1983.

Cline, M. J. The Normal Granulocyte. Part I A in *The White Cell.* Harvard University Press, Cambridge, Mass., 1975.

Stossel, T. P. Leukocytes II. Phagocytosis and Its Disorders. Lecture 18 in *Hematology,* 4th ed., W. S. Beck, Ed., The MIT Press, Cambridge, Mass., 1985.


## Review Questions

1. Name the major function of neutrophils.
2. List the four major aspects of neutrophil action.
3. Give the four classical symptoms of inflammation.
4. Why is inflammation an essential ingredient in the process of eliminating pathogens that have invaded the body?
5. Under what circumstances does the inflammatory process become a dangerous mechanism to the body?
6. Mention three important differences between acute and subacute inflammation.
7. When do chronic inflammations occur? What are two unique features of chronic inflammations? Mention some infectious disorders that will produce chronic inflammations.
8. What are the two phases of blood vessel dilation? What chemical factors are associated with each phase?
9. Define kinins. What kinin compounds are the active agents? Describe how they are generated.
10. How are kinins inactivated?
11. What are anaphylotoxins? Which activated complement factors act as anaphylotoxins?
12. What are prostaglandins? To what class of compounds do they belong? From what precursor compound are they derived? Where is that compound normally found?
13. List four classes of compounds derived from arachidonic acid. Which enzyme system produces prostaglandins? Which enzyme system produces leukotrienes?
14. How many different leukotrienes have been described? What were these compounds previously known as? Describe their function.
15. What are basic proteins? Where are they produced? Describe their function.
16. What is chemotaxis? Describe its function.
17. List five important chemotactic stimuli for neutrophils.
18. Define the following terms: margination, diapedesis, emperipolesis.
19. Describe the various stages of phagocytosis. Explain why neutrophils are able to bind readily to foreign particles. Why is this necessary?
20. Distinguish between phagosomes and phagolysosomes.
21. Mention five important metabolic events that occur during the process of phagocytosis.
22. Explain why there is a sudden need for an increase in glycolytic activity.
23. Describe the respiratory burst. Why is it necessary?
24. Explain why the sudden increase in HMS activity is needed in an activated neutrophil.
25. Explain the need for a sudden increase in phospholipid synthesis in the activated neutrophil.
26. Why is a sudden drop in pH in the phagolysosomes needed in the activated neutrophil?
27. Explain the need for the sudden increase in hydrogen peroxide and superoxide in the activated neutrophil.
28. How are $H_2O_2$ and $O_2^-$ generated in the neutrophil?
29. Describe how most carbohydrates and lipids are broken down in the phagolysosomes.
30. How are proteins broken down in the phagolysosomes of activated neutrophils?
31. Mention some nonoxidative mechanisms involved in microbial killing.
32. Define the terms pus and abscess. How is the pus cleared from the system?

# Chapter Eighteen

## *Eosinophils and Basophils*

## DEVELOPMENT, STRUCTURE, AND FUNCTIONS OF EOSINOPHILS

The second most common type of granulocyte in the bloodstream is the eosinophil. However compared with neutrophils, few eosinophils are in the blood. A normal individual has about 3,000 to 7,500 neutrophils per microliter of blood, which means that 55 to 75% of all white blood cells are neutrophils, whereas this same individual has about 50 to 300 eosinophils per microliter, which translates into 1 to 3% of all white blood cells.

Although eosinophils are rather mysterious white blood cells, one of their major functions is to attack and destroy helminth parasites, especially larval blood flukes. Eosinophils possess phagocytic ability, just as neutrophils do. However, when phagocytosis occurs, it is usually associated with degranulation.

### Development of Eosinophils

The proliferation and maturation sequence of eosinophilic granulocytes is similar to that of neutrophils. Both developmental series go through six stages. Both originate from an enlarged committed stem cell known as a **myeloblast.** Distinguishing between a neutrophilic and an eosinophilic myeloblast is impossible—both cells types are large, contain large nuclei with several nucleoli, and have a deep-blue clear cytoplasm.

The second stage produces the **eosinophilic promyelocyte,** which also contains a large round nucleus, but the nucleoli have disappeared. This second stage is associated with the development of primary granules composed of electron-dense material.

In the third stage the **eosinophilic myelocyte** is characterized by a large, oval, eccentric nucleus and secondary granules in the cytoplasm. These secondary, specific-staining granules consist of an electron-dense core made up of crystalloid material and a less dense matrix containing large quantities of special eosinophilic peroxidase enzymes. During this stage another type of granule is also generated, but it is smaller than the specific-staining granules, which stain a bright orange-red in the presence of eosin. These small granules contain acid phosphatase and arylsulfatase.

The fourth stage produces the **eosinophilic metamyelocyte,** which is similar to the myelocyte, except that it is slightly smaller and has a kidney-shaped nucleus, just as the neutrophilic metamyelocyte.

The fifth maturational stage produces the **eosinophilic band cell,** which, just as the PMN, possesses a long sausage-shaped nucleus of dense chromatin and many granules that stain a bright orange-red with Wright's stain.

In the final stage the mature **segmented eosinophil** is released into the bloodstream from the bone marrow. Its nucleus normally contains only two segments or lobes connected by a thin strand of nuclear material, in contrast with the neutrophil, which usually exhibits three to five lobes.

The maturation series of eosinophils is illustrated in figure 18.1.

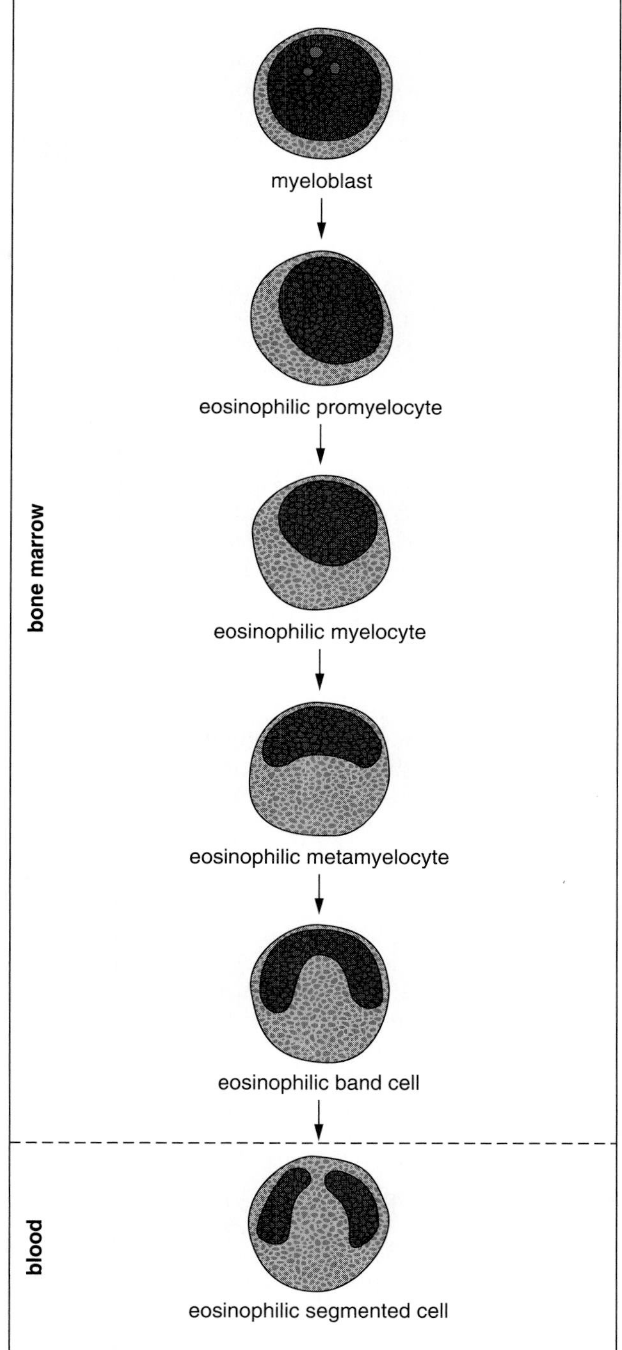


**figure 18.1**

Eosinophil development and maturation.

## Regulation of Eosinophil Production

The peripheral blood contains only a small fraction of the total eosinophil pool. For every circulating eosinophil there are approximately 300 eosinophils in reserve in the bone marrow. Further, for every eosinophil in the blood there are approximately 200 in the tissues, especially in the gut and lungs. Although we do not know exactly how long eosinophils remain in the bloodstream, most experts agree that their residence is short—about 8 to 12 hours.

The maturation time for eosinophils is similar to that of neutrophils, which takes about 5 to 6 days. Eosinophils apparently survive in the tissues somewhat longer than neutrophils. This observation is based on the fact that in tissue culture eosinophils normally survive longer than PMNs—that is, 8 to 12 days compared to 2 to 4 days for neutrophils. Production of eosinophils is also regulated by a humoral feedback system, since a decrease in eosinophils in the bone marrow reserve is replenished in a few days. Several eosinophilic growth factors include **interleukin 3 (IL-3), GM-CSF,** and **interleukin 5 (IL-5).**

In the peripheral bloodstream the level of eosinophils is subject to a diurnal rhythm. The eosinophil count is normally highest at midnight and lowest just before noon. This fluctuation may be correlated with the production of endogenous cortisol by the adrenal cortex. Increased cortisol secretion inhibits the release of eosinophils from the bone marrow compartment. In a normal healthy individual, cortisol release is highest in the morning. Hence it is understandable that this will be followed by a decreased number of eosinophils a few hours later. Corticosteroid drugs and catecholamines will also produce a decrease in the number of circulating eosinophils.

The rate at which eosinophils leave the circulation is regulated by a number of chemotactic factors. The best known of these factors is the **eosinophilic chemotactic factor of anaphylaxis (ECF-A),** which is released from mast cells and basophils.

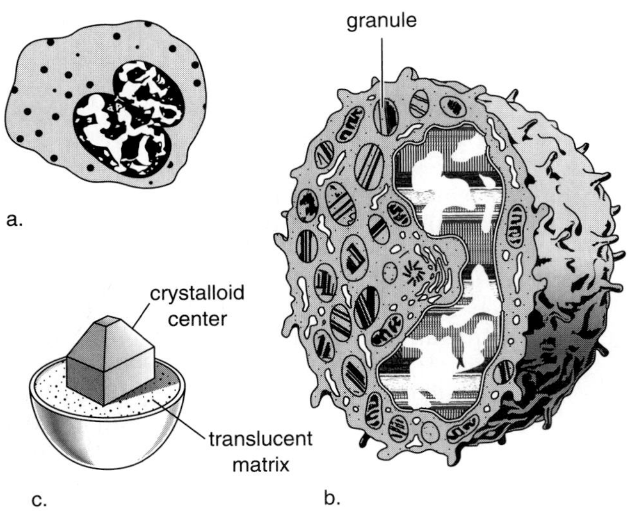


**figure 18.2**

The structure of an eosinophil as seen with (*a*) a light microscope and (*b*) an electron microscope.
(*c*) A representation of a granule with a crystalloid center.
(*a*) Redrawn from M. Bessis: *Blood Smears Reinterpreted.* Springer International, 1977; as appeared in Jan Klein, *Immunology: The Science of Self-Nonself Discrimination.* Copyright © 1984 John Wiley & Sons, Inc., New York, NY. Reprinted by permission of John Wiley & Sons, Inc. (*b* and *c*) Based on R. V. *Krstic: Die Gewebe des Menschen und der Saugetiere.* Springer-Verlag, 1978; as appeared in Jan Klein, *Immunology: The Science of Self-Nonself Discrimination.* Copyright © 1984 John Wiley & Sons, Inc., New York, NY. Reprinted by permission of John Wiley & Sons, Inc.

## Structure of Eosinophils

The structure of eosinophils is in many respects similar to that of neutrophils. The major features distinguishing eosinophils from polymorphonuclear neutrophils (PMNs) are (1) the shape of the nucleus, which usually shows only two lobes; (2) the presence of unique large granules that bind readily to the acidic dye eosin (hence the term eosinophil) and stain them bright orange-red; (3) a more highly developed Golgi apparatus; and (4) the presence of more mitochondria than in mature neutrophils.

Mature eosinophils have two major types of granules: large crystalloid-containing specific-staining granules and smaller less dense granules that contain acid phosphatase and arylsulfatase.

The large granules have two forms: (1) primary immature homogeneous electron-dense granules and (2) mature secondary granules containing an angular crystalloid core surrounded by a less dense translucent matrix (figure 18.2). Primary granules originate in the outer Golgi cisternae, from which small granules bud off. Several of these smaller granules then merge to form larger aggregates that condense to form large homogeneous granules, similar to the primary azurophilic granules of the neutrophils (see figure 16.6). Mature eosinophils contain predominantly secondary granules, but since the Golgi apparatus remains active in mature eosinophils, new immature primary granules may still be produced in the mature cells (figure 18.3).

Analysis of the contents of secondary granules has revealed that they have crystalloid centers, which contain arginine-rich proteins, known as major basic protein (MBP). The affinity of secondary granules for the acidic dye eosin can be attributed to the presence of these arginine-rich basic proteins. The crystalloid core also contains zinc, which may be associated with one or more of the basic proteins. The less dense matrix surrounding the crystalloid core of these specific-staining granules contains eosinophilic peroxidase. This peroxidase is chemically distinct from neutrophilic myeloperoxidase.

## Metabolism of Eosinophils

The metabolism of eosinophils is still poorly understood. This lack of information may stem from the difficulty in obtaining sufficient numbers of eosinophils to do biochemical studies. The presence of relatively large and abundant mitochondria suggests, however, that a significant amount of aerobic metabolic activity takes place within the mature cell.

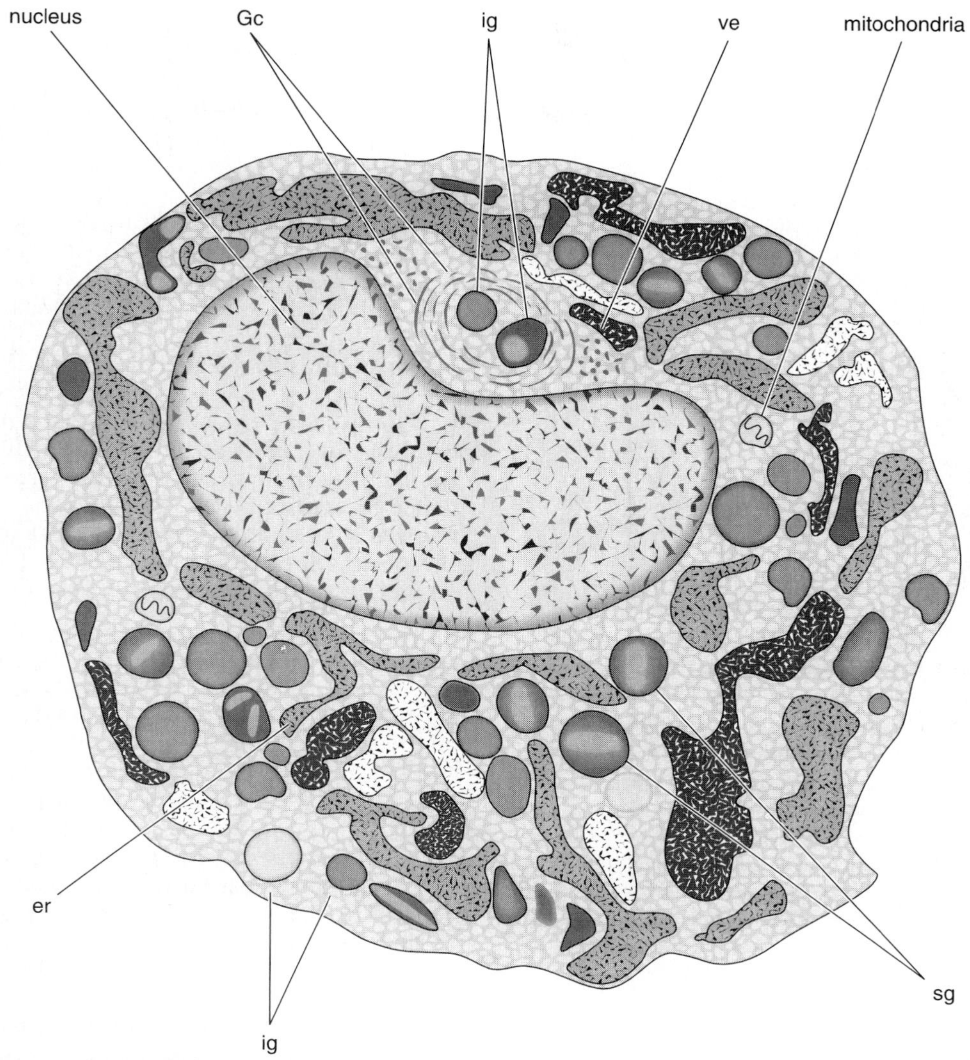


figure 18.3

An eosinophil with mature and immature specific-staining granules. *ig* = immature granules; *sg* = mature granule; *Gc* = Golgi apparatus; *ve* = small vesicles; *er* = endoplasmic reticulum; *m* = mitochondrion.

## Functions of Eosinophils

Although controversy exists regarding the functions of eosinophils, it is now generally agreed that eosinophils play a major role in combating helminth infections and are involved in certain hypersensitivity (allergic) reactions. Both functions are discussed in the following material.

### *Role of Eosinophils in Destruction of Helminth Parasites*

Since the beginning of the twentieth century, infections with helminth parasites have been associated with **eosinophilia**—that is, presence of large numbers of eosinophils in the peripheral bloodstream. Eosinophils seem especially effective against larvae of the genus *Schistosoma* (or blood flukes) of the platyhelminth group of invertebrates. The ability of eosinophils to combat schistosomiasis and other helminth infections is antibody-dependent—that is, they require the presence of IgG molecules before they can interact with these infectious organisms. Eosinophils have large numbers of Fc receptors for antibodies on their cell membranes. Opsonization with IgG molecules allows eosinophils to bind tightly to these parasites and release the contents of their granules and other proteins—which will damage and ultimately kill these organisms. This is an example of ADCC or antibody-dependent cell-mediated cytotoxicity. The major eosinophil secretions involved in helminth destruction are as follows:

**1. Major basic protein (MBP)**
This protein is found in the electron-dense core of secondary specific-staining granules and plays a major role in helminth destruction. MBP is a small protein with a

MW of 9,300 daltons, containing high levels of arginine. MBP is highly destructive to cell membranes of blood flukes (and on occasion also to cell membranes of the host). MBP accounts for about 50% of the total granular protein content.

**2. Charcot-Leyden crystal (CLC) protein**
Also known as **lysophospholipase** this protein has a MW of 13,000, and may also occur in a dimeric form (MW 26,000 daltons). CLC may be a crystalloid fraction of MBP. This protein is not found in the granules but is apparently localized in the plasma membrane of the eosinophil. CLC is also found in basophils and mast cells. The function of this enzyme is to degrade the phospholipid membranes of cells lining the surface of helminth parasites.

**3. Eosinophil peroxidase**
This enzyme is found mainly in the matrix of large specific-staining granules. It is a fairly large protein consisting of a single peptide chain and has a MW of 75,000 daltons. Its function is similar to that of neutrophilic myeloperoxidase and is instrumental in the production of hypohalites such as $HOCl^-$, $HOI^-$, $HOBr^-$, and $HOF^-$ which play a role in the destruction of proteins and thus assist in killing the parasites.

**4. Eosinophil cationic protein (ECP)**
ECP consists of a series of seven cationic proteins, some of which are produced by the eosinophil (proteins 5, 6, and 7). The most important and abundant of these seven proteins is component number 5, which is most often equated with ECP. Component number 5 is a single polypeptide chain with a MW of 21,000 daltons. Its function is to bind to heparin and counteract its anticoagulant activity, thus promoting clotting—which interferes with the activity of the blood flukes, making them easier targets for the white blood cells. ECP also inhibits plasminogen activation by retarding the breakdown of fibrin clots.

**5. Eosinophil-derived neurotoxin**
This protein has a MW of 15,000 daltons and is also found in the matrix of secondary specific-staining granules. Its function is not known, but in experimental animals, it acts as a powerful neurotoxin and causes severe damage to the myelinated sheaths of neurons. This neurotoxic reaction results in stiffness, ataxia, weakness, and muscle wasting. Eosinophil-derived neurotoxin may have a similar function in helminth parasites.

#### *Role of Eosinophils in Hypersensitivity Reactions*

In the past eosinophils were believed to play a major role in containing and reducing inflammation, especially when it occurred as a result of immediate hypersensitivity reactions. This line of reasoning was based on the many eosinophilic enzymes capable of neutralizing such mediators of anaphylaxis as histamine, slow-reacting substance of anaphylaxis (SRS-A), platelet-activating factor (PAF), and heparin—all substances produced by mast cells and basophils when they are activated and release their granules. Subsequently it was found that allergens—that is, substances causing immediate hypersensitivity reactions—stimulate eosinophils to produce secretions inhibiting the release of vasoactive amines from mast cells and basophils. Later studies found that these inhibiting substances were prostaglandins ($PGE_1$ and $PGE_2$). Further, ECP binds to heparin and by doing so neutralizes its anticoagulant activity. The hypohalites (which are products of the interaction of eosinophilic peroxidase and $H_2O_2$ and hypohalites such as $HOCl^-$ and $HOBr^-$) are able to break down leukotrienes (SRS-A), which are also vasoactive compounds.

The presence of eosinophils in low numbers seems to be an important factor in neutralizing anaphylactic (i.e., allergic) reactions. Low numbers of eosinophils inhibit the release of vasoactive amines by mast cells. The presence, however, of large numbers of eosinophils may actually contribute to histamine release by mast cells. High levels of eosinophils during asthma attacks have been implicated as a causative agent in the destruction of bronchial tissues. Eosinophilia has also been associated with the onset of cardiac and lung diseases. Apparently the massive release of major basic protein may result in the destruction of host tissue cells. Thus the role of eosinophils in hypersensitivity reactions remains unclear at this time.

## DEVELOPMENT, STRUCTURE, AND FUNCTIONS OF BASOPHILS

Basophils are the least common of all the blood cells, accounting for less than 1% of the total white blood cell count. Basophils may be closely related to mast cells in the tissues as they are similar in structure and functions. Historically basophils and mast cells were thought to be derived from different ancestral cell lineages, but now many histologists and hematologists believe they are derived from the same precursor cell: the pluripotent stem cell in the bone marrow. It appears that the basophil is merely an early circulatory form of the mast cell. Therefore the mast cell may be a basophil that has moved into the tissues.

### Development of Basophils

Basophils follow the same pattern of development as other granulocytes, which consists of six stages associated with a gradual change in shape of the nucleus, a gradual decrease in cell size, and development of specific granules in the cytoplasm.

The first cell in the basophilic series is again the **myeloblast,** which is a large cell with a large round nucleus with many nucleoli and a clear, deep-blue cytoplasm, similar to neutrophils and eosinophils. The second stage is the **basophilic promyelocyte,** which contains many primary azurophilic granules in the cytoplasm and a nucleus without any prominent nucleoli. The presence of uniquely staining, large blue-black granules becomes obvious in the third stage, the **basophilic myelocyte.** This cell type exhibits an oval, eccentric nucleus. The fourth in the series is the **basophilic metamyelocyte,** which has an eccentric, kidney-shaped nucleus. In this stage the specific-staining granules become more

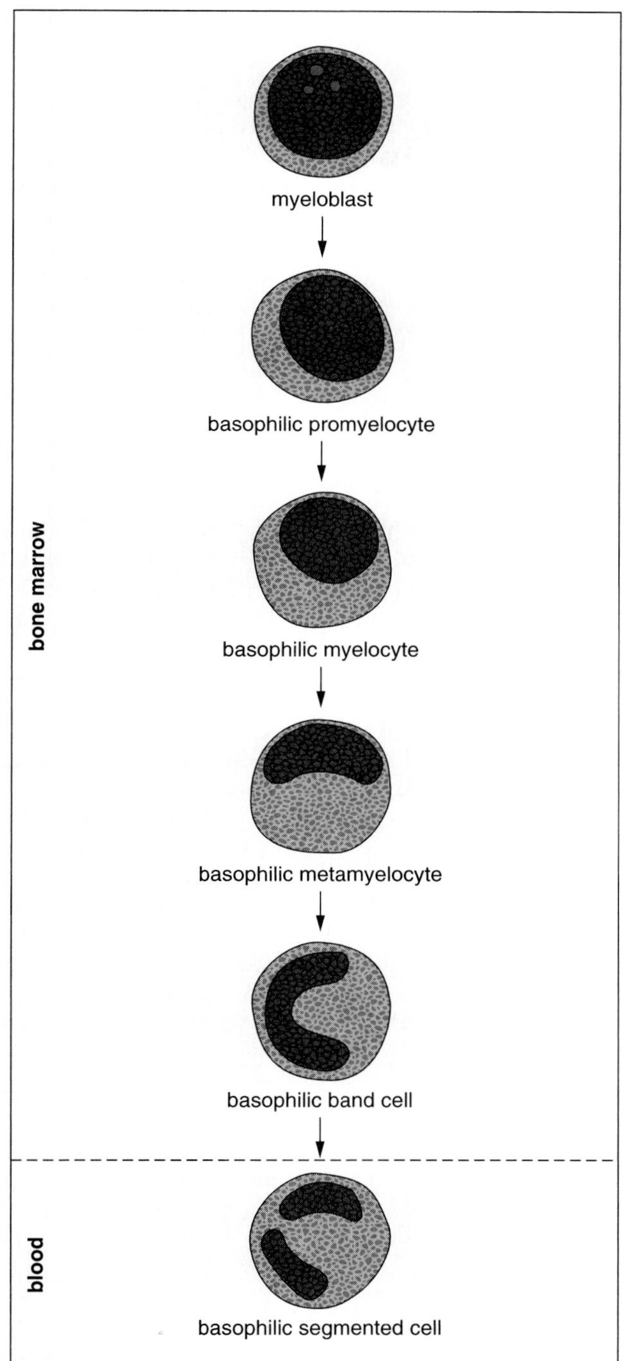


**figure 18.4**

Basophil development and maturation.

numerous and dense, almost concealing the nucleus. The fifth stage is the **basophilic band cell,** and the final cell is the mature basophil itself. The **mature basophil** is the smallest of all granulocytes with a diameter of 10 to 12 µm. The nucleus is usually more indented than segmented, and if segmentation is observed, we rarely see more than two lobes. A summary of the developmental stages of basophils is given in figure 18.4.

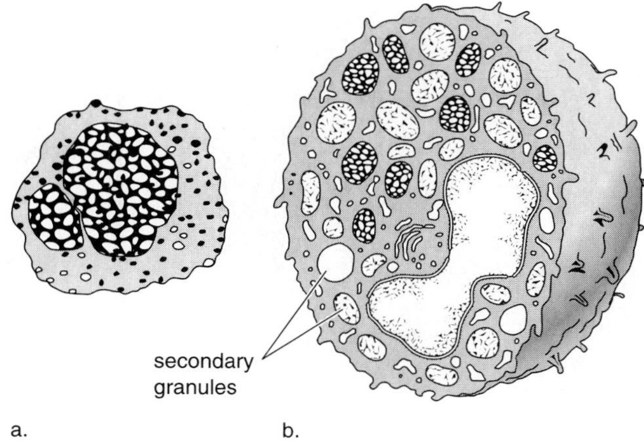


**figure 18.5**

The structure of a basophil as seen with (*a*) a light microscope and (*b*) an electron microscope.

(*a*) Modified from M. Bessis: *Blood Smears Reinterpreted.* Springer International, 1977; as appeared in Jan Klein, *Immunology: The Science of Self-Nonself Discrimination.* Copyright © 1984 John Wiley & Sons, Inc., New York, NY. Reprinted by permission of John Wiley & Sons, Inc.

### Regulation of Basophil Production

Precise information about basophil kinetics is not fully known. Since basophils occur in such low numbers in the peripheral bloodstream and bone marrow, labeling and counting them is difficult. The number of circulating basophils varies with the individual's age. The highest numbers are usually found in newborns and seem to decrease in quantity with age. The normal level of basophils in the blood is regulated by growth factors, just as the other white blood cells. Major growth factors regulating the development and maturation of the basophils are GM-CSF, which is produced by the stromal cells of the bone marrow, and interleukin 4 (IL-4), which is produced by white blood cells including the stromal cells and lymphocytes. Interleukin 9 (IL-9) has also been implicated in the growth and activation of mast cells in the tissues. The level of basophils in the blood is also influenced by certain hormones. For example, administration of glucocorticoids results in a decrease in basophil production, whereas low levels of thyroid hormone (hypothyroidism) appear to be associated with basophilia (i.e., increased numbers of basophils). Gonadal hormones have also been implicated in basophil production. High levels of progesterone may cause basopenia, and high levels of estrogen may produce basophilia. Variations in the levels of basophils can also occur in women of childbearing age.

### Structure of Basophils

Basophils are readily recognized in a blood smear by the presence of many large blue-black granules that fill most of the cytoplasm and virtually cover the nucleus. If the nucleus can be seen, it appears more indented than segmented (figure 18.5). During the early stages of development, basophilic cells produce granules—which in the more mature stages develop into secondary specific-staining granules. These secondary granules consist of electron-dense homogeneous material that

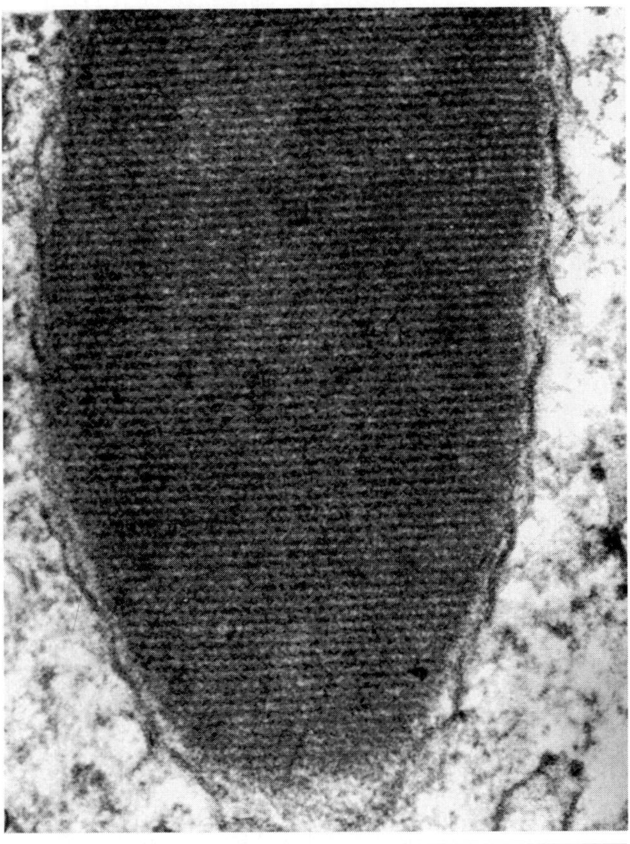

figure 18.6

An electron micrograph of a secondary (specific-staining) granule of a basophil showing the typical banded pattern.

exhibits a typical banded pattern when using electron microscopy (figure 18.6). Just as in other granulocytes, the Golgi apparatus is the primary site of granule formation. Analysis of secondary, specific-staining basophilic granules has revealed that they consist of the following two substances: (1) **acid mucopolysaccharides**—the most important and also the most common of these mucopolysaccharides in the basophilic granules is the anticoagulant **heparin** (blue granule color is due to the affinity of the basic dye methylene blue for acid mucopolysaccharides) and (2) **histamine**—these secondary basophilic granules contain large quantities of the vasoactive compound histamine (as mentioned previously, histamine plays a major role in the production of acute inflammation).

Apart from these two compounds, basophils also possess oxidative enzymes, such as lactate dehydrogenase (LDH), glucose-6-phosphate dehydrogenase (G-6-PD), isocitric dehydrogenase, and succinyl dehydrogenase. However, we do not know if these dehydrogenases are present in the granules or are found freely in the cytoplasm.

### Comparison Between Basophils and Mast Cells

Although mast cells are similar in structure and function to basophils, some subtle differences between the two types of cells may be observed. These include the following:

1. Basophil granules show a marked heterogeneity in size, whereas mast cell granules are more homogeneous in size.
2. Basophil cell membranes appear smooth when observed under a microscope, whereas plasma membranes of mast cells show numerous protrusions (figure 18.7).
3. Basophils have a polymorphonuclear nucleus, whereas the nucleus of the mast cell is usually nonindented and regular in shape.
4. Basophils seem to lack certain enzymes that are conspicuously present in mast cells, especially hydrolases and peroxidases.

### Metabolism of Basophils

As was the case with eosinophils, little is known about the metabolism of basophilic granulocytes. The reason is the same: gathering a sufficient number of basophils for appropriate metabolic studies is not possible at the present time.

However we know that mature basophils possess many large mitochondria. Thus we may presume that they make extensive use of oxidative pathways of energy production. Mature basophils have only a small Golgi region, limited endoplasmic reticulum, and few free polyribosomes—all of which are indications that apparently little DNA and new protein synthesis takes place in these mature cells.

Basophils are not classified as phagocytic cells but possess pinocytic properties and thus can transport materials to the cytoplasmic granules.

### Functions of Basophils

Since basophils and mast cells possess granules containing large amounts of vasoactive amines (mainly histamine) and heparin, their major role is in the production of **acute inflammation** as a result of trauma, infection, and physical and chemical irritants. Mast cells and basophils are also the mediators of inflammation in immediate hypersensitivity reactions that may occur as a result of exposure to allergens.

Histamine exerts a physiological effect by interacting with one of two separate target cell histamine receptors termed H1 and H2. The primary H1 functions of histamine are (1) contraction of the smooth muscle of the bronchi, intestine, and uterus and (2) augmentation of vascular permeability in the blood vessels of the microcirculation. Other H1 actions include pulmonary vasoconstriction, increased production of nasal mucus, increased white blood cell activation, and prostaglandin production in the lung tissues.

Stimulation of H2 receptors increases gastric acid secretion, stimulates production of airway mucus, and inhibits white blood cell activity.

Costimulation of H1 and H2 receptors causes maximal vasodilation, cardiac irritability, and pruritus. A summary of the actions of histamine on various H1 and H2 receptors is given in table 18.1.

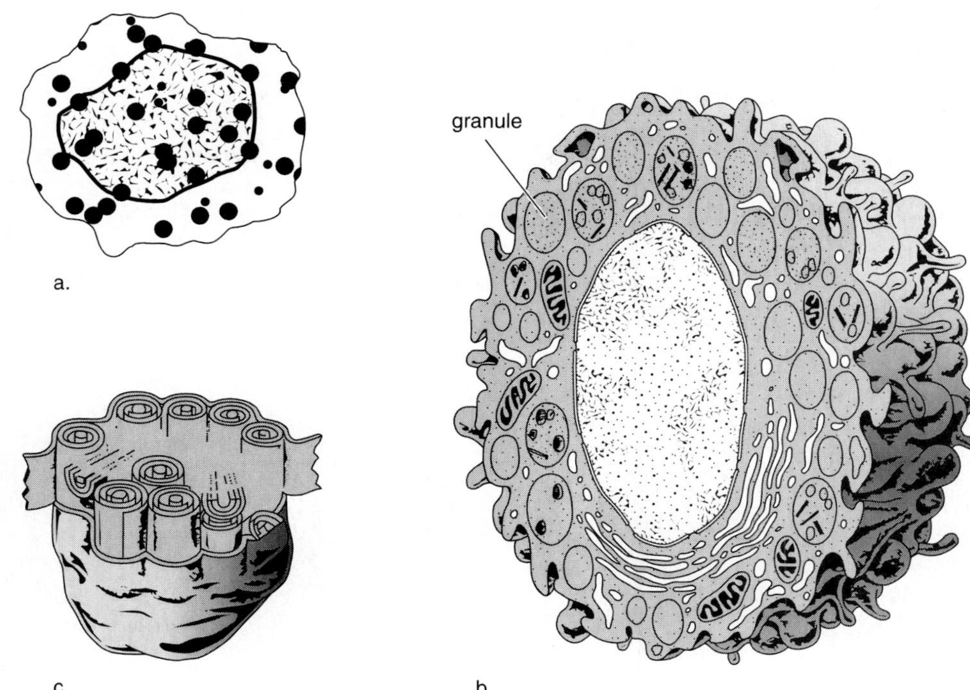


figure 18.7
The structure of a mast cell as seen with (*a*) a light microscope and (*b*) an electron microscope. (*c*) A representation of the structure of a specific-staining granule.

**Table 18.1 Histone Actions with Target Cell Receptors H1 and H2**

| Receptor | Histamine Actions |
|---|---|
| H1 | Increased microcirculation permeability |
| | Smooth muscle contraction |
| | Pulmonary vasoconstriction |
| | Increased intracellular cGMP levels |
| | Increased secretion of mucus |
| | Increased leukocyte activation |
| | Prostaglandin production in lungs |
| H2 | Enhanced gastric acid secretion |
| | Increased secretion of mucus |
| | Increased intracellular cAMP levels |
| | Increased leukocyte activation |
| | Suppressor T cell activation |

### Activation of Basophils

Basophils become activated—that is, release their granules—in a number of different situations. The most important ones are discussed in the following material.

*Trauma and Infection*

A number of factors may stimulate basophil degranulation and thus cause local swelling and blood vessel dilation. One of these factors is produced by neutrophils. PMNs release basic (cationic) proteins when they are activated, and these basic proteins in turn stimulate basophils to release vasoactive amines. Activated plasma kinins, activated complement factors (C3a and C5a), prostaglandin F, and leukotrienes are substances commonly produced in cases of trauma and infection, and all of these may stimulate basophils and mast cells to degranulate and thus produce inflammation.

*Exposure to Physical or Chemical Irritants*

When the airways of our bodies are suddenly exposed to extremely cold temperatures, heavy smoke, or dust or other chemical irritants, the bronchial tubes may become inflamed and make breathing difficult, producing a type of asthmatic attack. Apparently these irritants cause a sudden increase in acetylcholine production by the irritated parasympathetic nerve endings present in these airways. Acetylcholine then stimulates mast cells and basophils to degranulate, producing local swelling and bronchoconstriction.

*Exposure to Allergens*

Exposure to allergens will lead to immediate or acute hypersensitivity reactions. These reactions occur only in people who are allergic to a certain substance (known as an **allergen**) and who have had previous exposures to that allergen. When the immune system of persons allergic to pollen, certain foods, drugs, or other medicines (e.g., vaccines) becomes activated, IgE antibody is produced by B lymphocytes against that allergen.

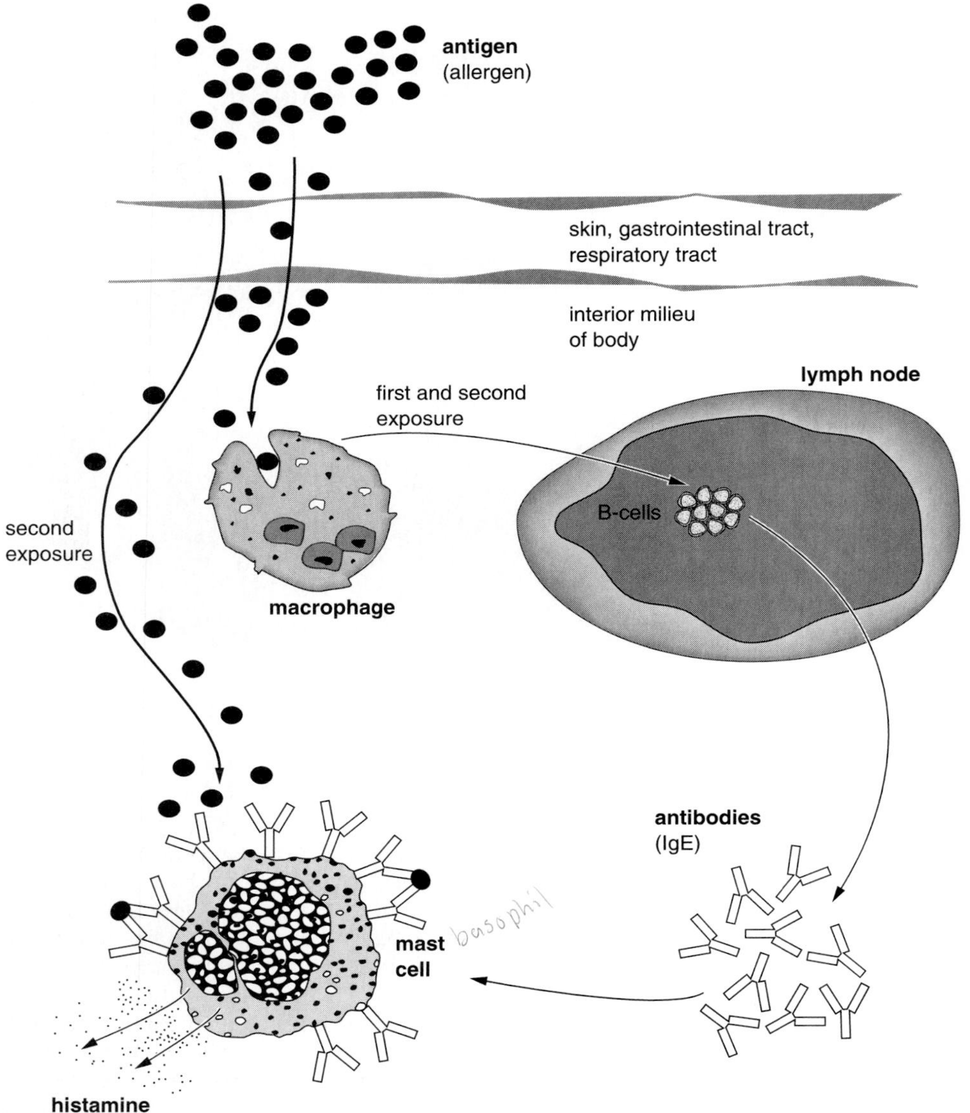


**figure 18.8**

The activation of mast cells (or basophils) by allergens.

The sequence of immediate hypersensitivity reactions is a complex process that can be divided into the following three phases:

1. During the **sensitization phase,** IgE antibody is produced and binds to specific Fc receptors on the mast cells and basophils (figure 18.8).
2. Re-exposure to the allergen triggers the **activation phase,** which causes the mediator cells to respond by degranulation of their contents. Allergic compounds bind to the Fab (fragment, antigen-binding) fraction of the two IgE molecules bound to the mast cells and basophils and cause them to degranulate.
3. The **effector phase,** a complex response, occurs as a result of the effects of many pharmacologically active agents released by mast cells and basophils.

Thus the triggering of basophils and mast cells by bridging between two IgE antibodies initiates a series of events. The intracellular levels of cAMP (3´ 5´-cyclic adenosine monophosphate) and cGMP (cyclic guanosine monophosphate) are known to affect subsequent events and are important in the regulation of these events.

Certain compounds such as basic protein, kinins, prostaglandin F, anaphylatoxins (C3a, C5a), leukotrienes, acetylcholine, and the two IgE-allergen complexes bind to the mediator cell surface where they activate the guanylate cyclase system in the membrane. This causes GTP (guanosine triphosphate) to be converted to cGMP. This cGMP then activates the granule fusion with the external plasma membrane and release of their contents into the surrounding area. The effects

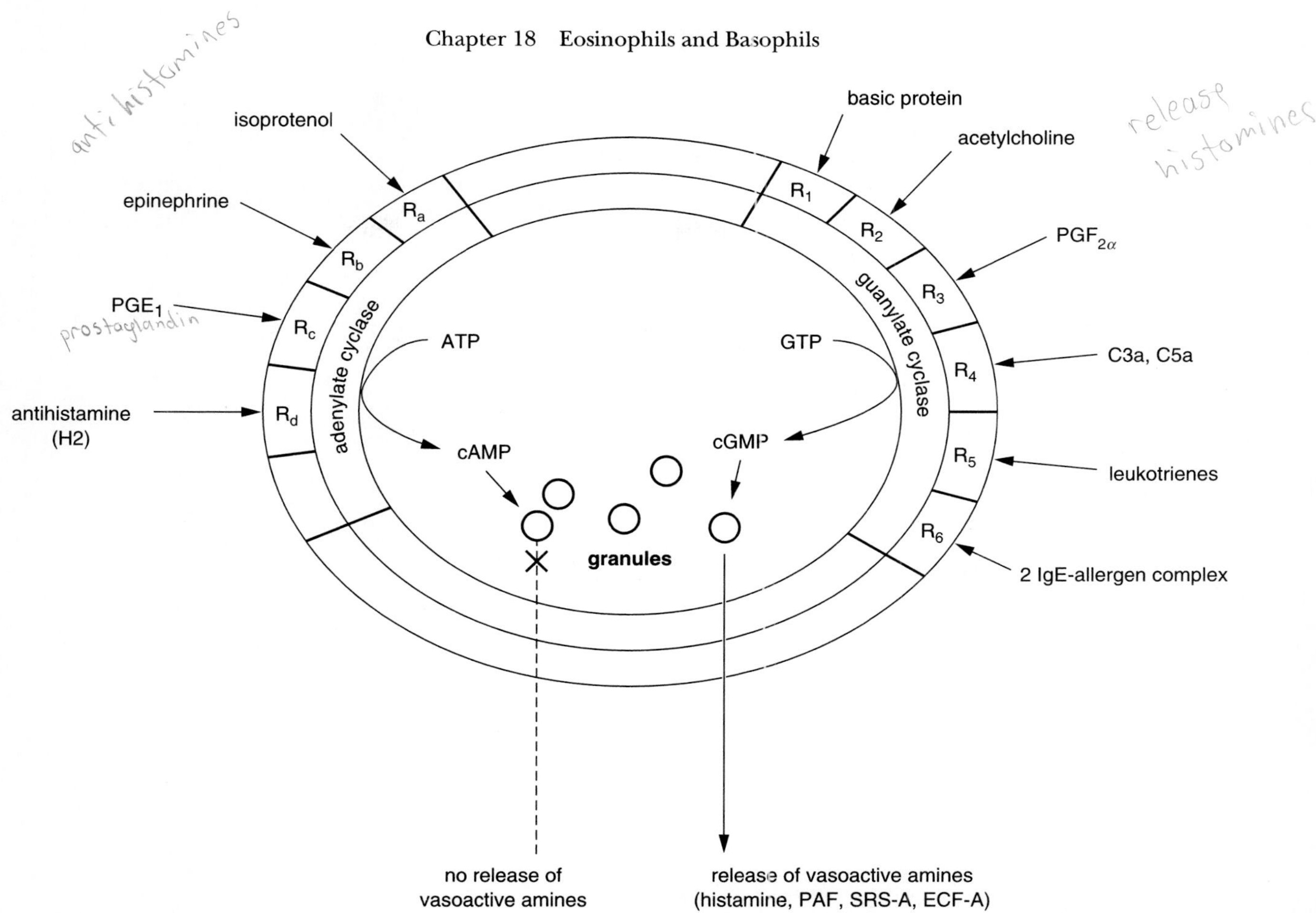


**figure 18.9**

The regulation of vasoactive amine release from mast cells and basophils. The compounds on the right side represent promoters of granule release, those on the left side are inhibitors.

of cGMP are always antagonistic to those of cAMP. cAMP is derived from ATP (adenosine triphosphate) by the action of the membrane enzyme system adenylate cyclase (figure 18.9). If the concentration of cAMP is high, then concentration of cGMP is reduced and vice versa. Therefore if high levels of cAMP are present in the mast cells and basophils, degranulation, and thus inflammation are prevented. This phenomenon explains why emotional upsets may lead to asthmatic attacks in certain individuals. Emotional upsets are associated with high stress levels. The latter causes a sudden increase in the production of norepinephrine in the sympathetic nerve endings in the airways. Norepinephrine then binds to α-adrenergic receptors on the mast cells and basophils and counteracts the action of epinephrine, which normally stimulates the production of cAMP. High levels of norepinephrine inhibit cAMP production, allowing the levels of cGMP to rise. This in turn activates mast cell and basophil degranulation and produces constriction of the airways, thus initiating an asthmatic attack (figure 18.10).

Upon activation of the mast cells and basophils, not only are **preformed mediators** released (e.g., vasoactive amines such as histamine, enzymes such as hydrolases and serine proteases, and proteoglycans such as heparin and chondroitin sulfate), but this activation also results in the production of new mediators. These **newly synthesized mediators** include prostaglandin $D_2$ (which is a powerful vasodilator and bronchoconstrictor), leukotrienes such as $LTC_4$, $LTD_4$, and $LTE_4$—all of which are strong bronchoconstrictors—and platelet-activating factor (PAF), which causes contraction of pulmonary muscles. The biochemical events involved in mast cell and basophil activation and degranulation are summarized in figure 18.9.

## DRUGS THAT COUNTERACT THE ACTIONS OF BASOPHILS AND MAST CELLS

Since inappropriate release of the contents of granules of basophils and mast cells may have a detrimental effect—that is, cause a severe allergic reaction—various drugs have been developed to counteract the anaphylactic reactions caused by these mediator cells.

### 1. Antihistamines

Many negative effects of mast cells and basophils are caused by the actions of histamine, so a number of drugs have been developed to counter its effects. These drugs are collectively known as **antihistamines.** There are two classes of antihistamines: one group binds to H1 receptors, and the other group binds to H2 receptors found on the various body cells. The only antihistamines discussed here are H1 blockers, as H2 blockers do not possess any noticeable anti-inflammatory activity. The H1 blockers are effective in (1) reducing nasal and eye secretions in seasonal rhinitis (or hay fever), (2) treating urticaria (rashes) and angioedema (swelling) associated with hives, and (3) reducing pruritus (itching) and other dermatoses associated with acute hypersensitivity reactions. The principal side-effects of most antihistamines are a sedative effect, dryness of the mucous membranes, and constipation. A list of some commonly used antihistamines is given in table 18.2.

**2. Adrenergic drugs**

This group of drugs—also known as the **sympathomimetic drugs** as they mimic the effects of the sympathetic nervous system—counteracts many of the effects produced by histamine. Of this group **epinephrine** is the drug of choice as it is a natural drug that powerfully counteracts the effects of anaphylaxis. As previously explained, epinephrine (also known as adrenaline) activates the adenylate cyclase system in the membrane of mast cells and basophils. This results in the production of high levels of cAMP, which in turn inhibits production of cGMP and thus prevents degranulation.

**3. Atropine**

This compound inhibits the action of acetylcholine, thus preventing activation of the guanylate cyclase system of mediator cells. Hence it prevents production of cGMP, and thus there will be no degranulation.

**4. Methylxanthines**

This group of compounds, which includes theophylline and caffeine, inhibits the action of the enzyme phosphodiesterase. This enzyme is responsible for the catabolism of cAMP to the inactive form 5´-AMP. If the action of this enzyme is blocked, the levels of cAMP will remain high and thus prevent production of cGMP.

**5. Prostaglandin $E_1$ and $E_2$**

$PGE_1$ and $PGE_2$ are able to stimulate production of cAMP, thus preventing release of histamine from mediator cells.

**Table 18.2 Commonly Used Antihistamines (H1 Blockers)**

*Alkylamines*
- Chlorpheniramine
- Dexchlorpheniramine
- Brompheniramine
- Triprolidine

*Ethanolamines*
- Diphenhydramine
- Dimenhydrate
- Clemastine

*Ethylenediamines*
- Tripelennamine

*Piperazines*
- Hydroxyzine
- Meclizine

*Phenothiazines*
- Promethazine

*Piperidines*
- Cyproheptadine
- Azatidine

*Tricyclics*
- Doxepin
- Amitriptyline
- Imipramine
- Desipramine

**6. Cromolyn sodium**

This compound also inhibits the release of mediator compounds, such as histamine and heparin, from mast cells and basophils. The precise mechanism of action of cromolyn sodium is not known, but evidence shows that it interferes with calcium influx through the cell membrane after these cells become stimulated. Calcium is an essential cofactor for activation of many enzymes that must be activated before degranulation can take place.

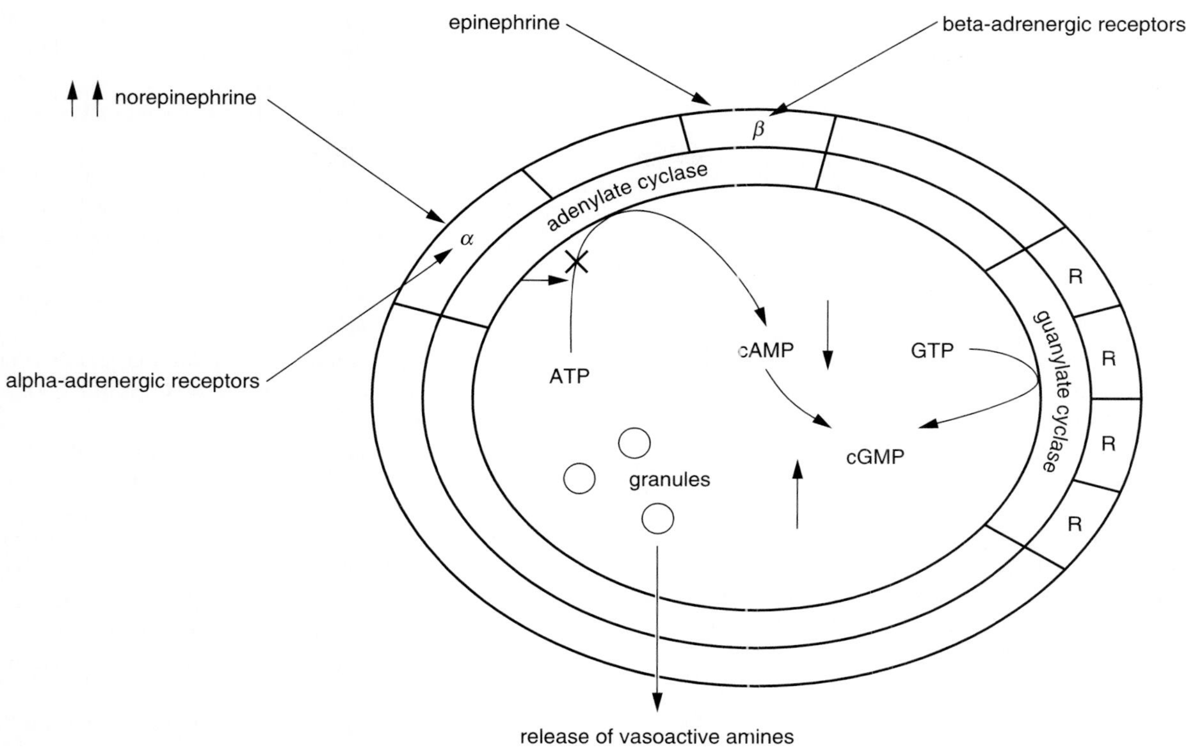


**figure 18.10**

A sudden release of norepinephrine blocks the production of cAMP, thus allowing for an increase in cGMP. This leads to degranulation.

## Further Readings

Boggs, D. R., and Winkelstein, A. The Phagocyte System (Neutrophils, Monocytes, Eosinophils, Basophils). Chapter 4 in *White Cell Manual,* 4th ed., F. A. Davis Company, Philadelphia, 1983.

Goodwin, J. S. Anti-Inflammatory Drugs. Chapter 63 in *Basic and Clinical Immunology,* 7th ed., D. P. Stites and A. I. Terr, Eds., Appleton and Lange, Norwalk, Conn., 1991.

Klein, J. Allergies and Other Antibody-Mediated Hypersensitivities. Chapter 21 in *Immunology.* Blackwell Scientific Publications, Oxford, 1990.

## Review Questions

1. List the six stages of eosinophil development and maturation. Describe the major cell characteristics of each stage.
2. What is the normal number of eosinophils in the blood?
3. What are the normal ratios of eosinophils in blood, bone marrow, and tissues?
4. What is the maturation time for eosinophils? How much time do they spend in the blood? How long do they normally survive in the tissues?
5. List the major growth factors that regulate eosinophil production. Name the hormones that may cause a decrease in the number of blood eosinophils.
6. Explain why the number of eosinophils in the blood is usually higher during the night than during the day.
7. Define ECF-A. What is its function? Where is it produced?

8. Explain the major differences between the structures of a mature eosinophil and a mature neutrophil.
9. Describe the major types of granules found in eosinophils. What is the content of each?
10. List the functions of eosinophils.
11. Explain the statement that eosinophils are the principal effector cells of ADCC against helminth infections. What does ADCC stand for?
12. List the major secretions of eosinophils that cause helminth destruction. How do each of these secretions affect helminth parasites?
13. Why was it originally thought that eosinophils play a role in the containment of immediate hypersensitivity reactions? Why is that role called into question?
14. Provide evidence that eosinophils promote tissue destruction in certain allergic reactions.
15. List the six stages in the development of basophils. Compare and contrast the various stages.
16. What is the normal number or percentage of basophils in the blood? Why is it difficult to find exact data regarding their regulation?
17. Mention a number of drugs and hormones that seem to increase the numbers of basophils in the blood.
18. Compare the structure of a mature basophil with that of a mature neutrophil.
19. Describe the major contents of the secondary specific-staining granules of basophils.
20. Define mast cells. Where are they normally found? What is their origin? What is their relationship to basophils?
21. List some of the major differences between mast cells and basophils.
22. What is the major function of basophils and mast cells?
23. Mention five different situations that may lead to activation and degranulation of basophils and mast cells.
24. Mention five different compounds that may cause basophil and mast cell degranulation.
25. Explain how trauma and infection may result in mast cell and basophil degranulation and thus inflammation.
26. Explain how exposure to physical and chemical stimuli may cause basophil and mast cell degranulation.
27. Explain how exposure to allergens causes basophil and mast cell degranulation.
28. Explain how the release of the granules of basophils and mast cells is regulated at the cellular level.
29. List a number of drugs that counteract the actions of basophils and mast cells.
30. Name the two classes of antihistamines. How do they act?
31. Explain how (a) epinephrine, (b) methylxanthines, (c) prostaglandins, (d) cromolyn sodium, and (e) atropine are able to inhibit the actions of basophils and mast cells.

# Chapter Nineteen

## *Development, Structure, and Functions of Monocytes*

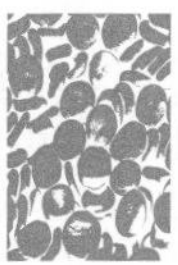

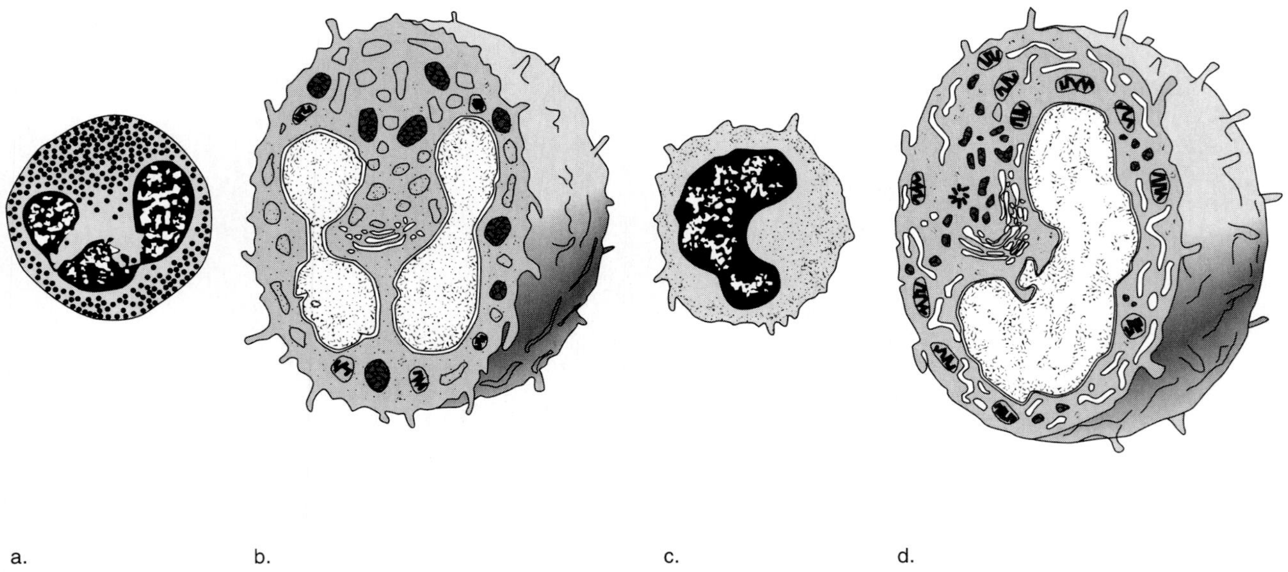


**figure 19.1**

A comparison of the structures of neutrophils and monocytes. The structure of a neutrophil, as seen with (*a*) a light microscope and (*b*) an electron microscope. The structure of a monocyte, as seen with (*c*) a light microscope and (*d*) an electron microscope. Differentiation between neutrophils and monocytes can be determined by the nuclei morphology and cytoplasmic granulation.



Monocytes are the largest blood cells in the bloodstream. Together with neutrophils, they are the major types of blood cells capable of phagocytosis. Because of their size differences, neutrophils are generally classified as **microphages,** whereas monocytes are commonly referred to as **macrophages.** Although monocytes are less numerous in the peripheral bloodstream (2 to 6%) than neutrophils (55 to 75%), they are an important group of cells as they play a major role in the activation and function of the immune system. Monocytes are more closely related to neutrophils than to any other type of blood cells, but despite this close ancestry they have different morphogenic histories. Mature neutrophils are considered end-stage cells—that is, the final forms of differentiation—and blood monocytes are considered an **intermediate stage** in the development of tissue macrophages. Other differences include the shape of the nucleus, which is segmented in neutrophils, whereas the nucleus of monocytes is singular (mononuclear) though frequently twisted in shape. Further, the cytoplasm is distinctly granulated in neutrophils but is much clearer in monocytes, as their granules are smaller and cannot readily be seen under a normal light microscope (figure 19.1).

## DEVELOPMENT OF MONOCYTES AND TISSUE MACROPHAGES

The process of development of mononuclear phagocytes is normally divided into five different stages. In extraordinary circumstances a sixth stage may be added as is explained later. The first stage in the development of the monocyte is the **monoblast.** This cell is indistinguishable from the structures of myeloblasts in that it also has a large nucleus with many nucleoli and a deep-blue, clear cytoplasm when stained with Wright's stain. The second stage is the **promonocyte**—a large cell with a diameter of 15 to 25 μm with a large, often slightly indented nucleus with few if any nucleoli. The moderately abundant cytoplasm stains a pale bluish-gray with Wright's stain, which may or may not contain granules. These first two stages are normally found only in the bone marrow. The third stage is released into the bloodstream and is known as the **monocyte.** This is a large cell with a diameter of 15 to 20 μm. The nucleus appears to be indented or twisted into a C or S shape. The chromatin is fine and usually stains light blue with Wright's stain, compared with the deeper blue staining observed in the nuclei of other white blood cells.

The cytoplasm is pale gray or gray-blue and may contain a few fine pinkish granules. Monocytes are easily distinguished from other white blood cells by

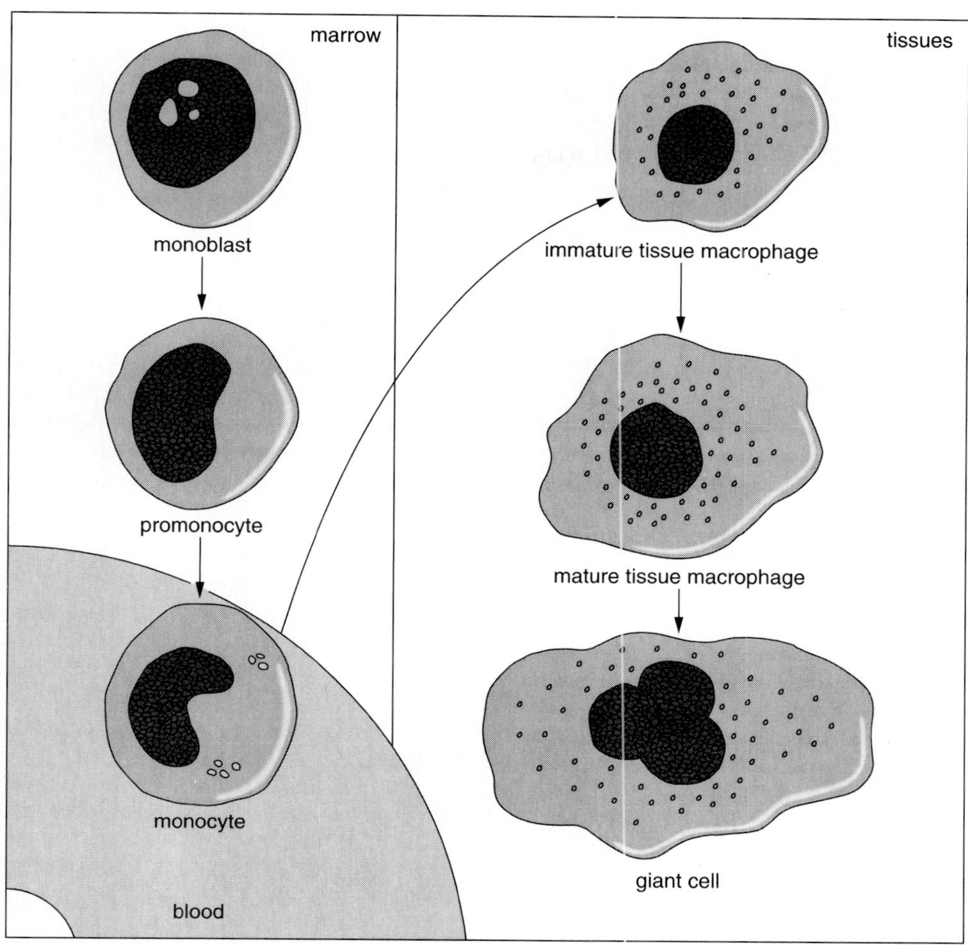


**figure 19.2**

Development of monocytes and tissue macrophages (histiocytes).

their size and shape; however, to the untrained eye, monocytes may resemble large immature lymphocytes. In this case the shape and color of the nucleus should help us distinguish between the two. In the immature lymphocyte the nucleus is always round or oval and is deeply stained, whereas in the monocyte the nucleus is usually twisted and pale.

In contrast with neutrophils and perhaps other granulocytes, the monocyte is not a finite cell but is considered an intermediate phase. The next stage of the monocytic series—the **immature tissue macrophage**—is found in the tissues. They frequently occur in the alveolar, pleural, and peritoneal spaces of the body and are particularly numerous in lymphoid tissues and the spleen. This tissue macrophage, or **histiocyte,** is a large cell, much larger than the average monocyte. These cells vary greatly in size but usually measure 20 to 30 μm in diameter. Young tissue macrophages are considered to be immature cells as they are still capable of DNA synthesis and cell division. Under normal circumstances they produce several generations of daughter cells, until finally a nonreplicating **mature macrophage** is produced. This is the fifth and final stage (figure 19.2). The mature macrophage may be distinguished from the immature histiocyte by its condensed nuclear chromatin and pinkish-red cytoplasm when stained with Romanowsky dyes. Immature macrophages have loose nuclear chromatin and gray-blue cytoplasm.

In case of prolonged chronic inflammations, some of these mature tissue macrophages may change into a sixth stage commonly referred to as **giant cells** because they may become extremely large, with sizes up to 80 μm in diameter. They are able to grow to this size because of **endocytosis**—that is, a process in which the nucleus divides several times within a common cytoplasm. Each time the nucleus divides, the cytoplasm enlarges and the cell grows. The function of these giant cells is to surround bacterial infections and wall them off from the physiological milieu of the body, so they are no longer able to receive nutrients and die within these nodules, which are commonly referred to as **granulomas.**

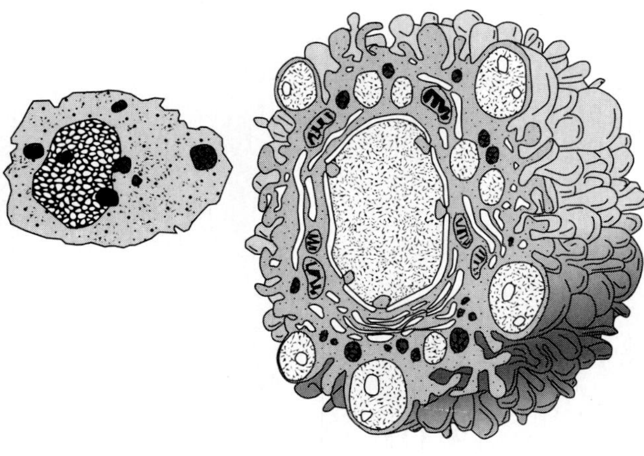


**figure 19.3**

The structure of a tissue macrophage (histiocyte) as seen with (*a*) a light microscope and (*b*) an electron microscope. Note the various cytoplasmic structures and pseudopods evident on macrophage membrane.

(*a*) Modified from M. Bessis: *Blood Smears Reinterpreted.* Springer International; as appeared in Jan Klein, *Immunology: The Science of Self-Nonself Discrimination.* Copyright © 1984 John Wiley & Sons, Inc., New York, NY. Reprinted by permission of John Wiley & Sons, Inc. (b) Modified from R. V. Krstic: *Die Gewebe des Menschen und der Saugetiere.* Springer-Verlag, 1978; as appeared in Jan Klein, *Immunology: The Science of Self-Nonself Discrimination.* Copyright © 1984 John Wiley & Sons, Inc., New York, NY. Reprinted by permission of John Wiley & Sons, Inc.

## REGULATION OF MONOCYTE PRODUCTION

Studies with autoradiographic techniques using tritium-labeled thymidine have revealed some definite information regarding monocyte kinetics. The normal number of cell divisions from activated blast cell to monocyte is about three to four, as occurs with almost all other myeloid cells. The proliferation and maturation process normally takes about 5 to 6 days, similar to neutrophils and other granulocytes. Monocytes remain an estimated 2 to 3 days in the bloodstream before they leave the circulation. Once they leave the blood vessels, they do not return under normal circumstances. In the tissues these monocytes undergo blast formation and develop into immature tissue macrophages. The latter will undergo a number of cell divisions until finally a mature histiocyte is produced (figure 19.3). This mature macrophage may live for many months (and possibly for several years) in the tissues. Mature macrophages may exhibit many different forms depending on where they are found in the body. For example in the bone they become osteoclasts; in the liver, Kupffer cells; in the brain, microglial cells; in the lungs, alveolar macrophages; and in the pleural cavity, serosal macrophages (figure 19.4). Together they form the **mononuclear phagocytic system.**

In normal adults the relative monocyte count is 2 to 6% of the total white blood cell count. The absolute count is between 200 and 500 monocytes per microliter of blood. This number is usually much higher in young children and in certain infectious diseases such as tuberculosis and in some malignancies such as Hodgkin's disease. The increase in the number of monocytes in the blood is known as **monocytosis.** High levels of certain glucocorticoid hormones and steroidogenic drugs may cause a decrease in blood monocytes. This phenomenon can also be caused by certain bacterial endotoxins and other bacterial products. Apart from these external agents, the body also has naturally occurring growth factors that regulate blood monocyte levels. Some of the most important of these regulatory agents are as follows:

**1. Interleukin 3 (IL-3)**
This growth factor—also known as multi-colony-stimulating factor (multi-CSF)—is produced by T lymphocytes and promotes the growth and development of all myeloid progenitor cells including monocytic stem cells. IL-3 also facilitates the self-renewal of the pluripotent granulocyte-erythrocyte-megakaryocyte-monocyte colony-forming unit (GEMM-CFU) stem cells (see chapter 3).

**2. Granulocyte-monocyte colony-stimulating factor (GM-CSF)**
This stimulating agent is not only produced by T lymphocytes but has also been elucidated from endothelial cells and fibroblasts. GM-CSF mimics many of the functions of IL-3. In fact, IL-3 competes with GM-CSF for receptor binding on developing blood cells. Most of the myeloid progenitor cells have receptors for both IL-3 and GM-CSF. Apparently IL-3 can bind to both receptors, whereas GM-CSF can bind only to its own specific markers.

**3. Monocyte (or macrophage) colony-stimulating factor (M-CSF)**
This chemical stimulant has been extracted from other monocytes, fibroblasts, and endothelial cells. Its function is limited to the growth and development of monocytes and macrophages.

## STRUCTURE OF MONOCYTES

The human monocyte is a large cell (15 to 20 μm in diameter) with a large nucleus and a rather clear grayish-blue cytoplasm when stained with Wright's stain. The nucleus is usually twisted or indented, giving it a C or S shape. The nuclear material is only moderately condensed, resulting in a transparent or "lacy" appearance (figure 19.5).

When viewed under an electron microscope, monocytes show a well-developed Golgi apparatus with a few nearby clusters of azurophilic granules. The endoplasmic reticulum is not prominent, but mitochondria are frequently observed. Along the surface are a few **pseudopodia,** indicating that these cells are usually mobile.

The immature tissue macrophage can be easily distinguished from the blood monocyte by its much larger size. Further, its nucleus is large and contains many nucleoli. Its cytoplasm contains many distinct

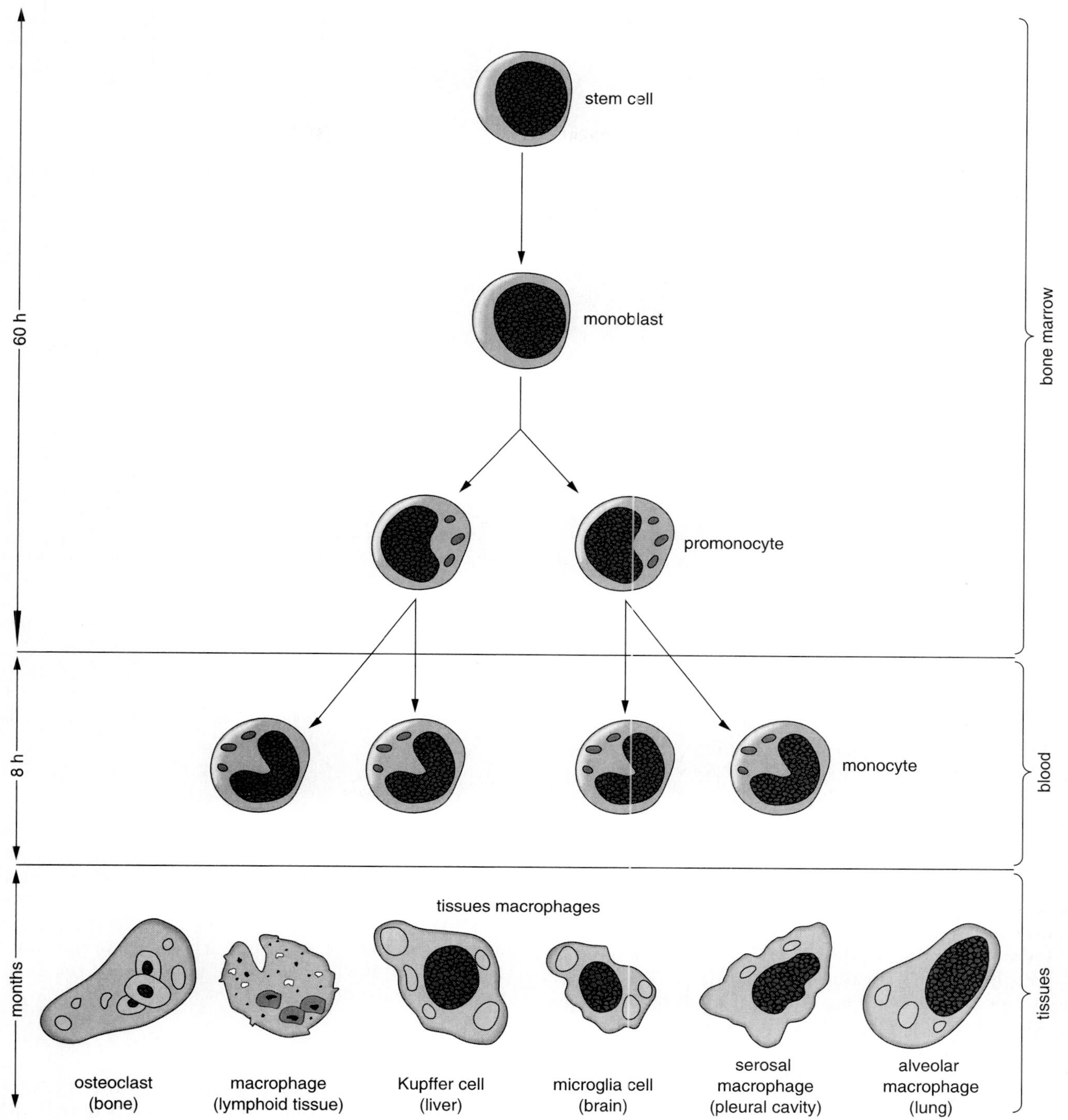


**figure 19.4**

The development and differentiation of monocytes. Note the monocytes can differentiate into a variety of different tissue macrophages.

granules, an extensive endoplasmic reticulum, a large Golgi complex, as well as many large mitochondria—all of which are indications that these cells are actively dividing(see figure 19.3). In the mature tissue macrophage, the endoplasmic reticulum becomes less prominent. The nucleus becomes more dense and nucleoli are no longer observed, indicating that cell division no longer occurs. The Golgi complex remains extensive, and granules are numerous throughout the cytoplasm. Apart from its size, another clear distinction between monocytes and tissue macrophages is the presence of numerous pseudopodia on the macrophage membrane. In contrast, the monocyte membrane is relatively smooth.

Histochemical analysis of the contents of the cytoplasmic granules reveals the presence of lysosomal and nonlysosomal enzymes. The lysosomal enzymes

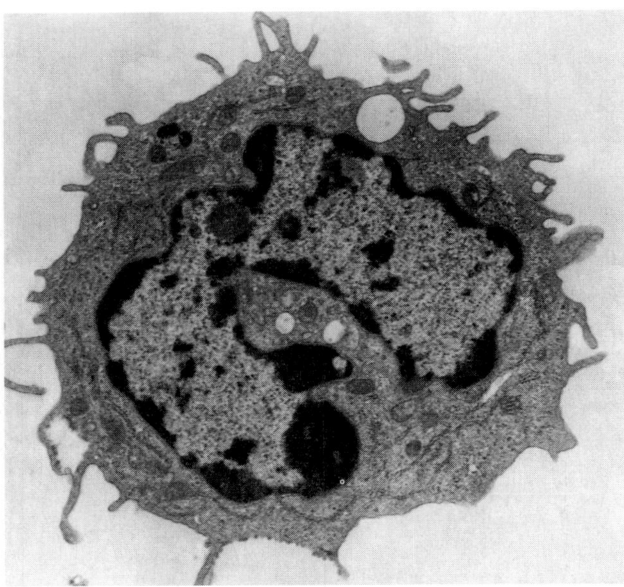

figure 19.5

A transmission electron micrograph of a monocyte. Note the large nucleus, endoplasmic reticulum, granules, pseudopods, and "lacy" appearance of the cytoplasm.

include β-glucuronidase, β-galactosidase, various esterases, acid phosphatase, arylsulfatase, and peroxidase. The nonlysosomal enzymes include aminopeptidase, succinic dehydrogenase, and cytochrome oxidase (table 19.1). All these enzymes function in killing and digesting foreign pathogens.

## METABOLISM OF MONOCYTES

A great deal of variation takes place in the metabolic activity of monocytes and macrophages. Less metabolic activity occurs in the monocyte than in the actively dividing immature tissue macrophage. Further, there is a significant difference in the rate of metabolism between activated mature macrophages and those not actively engaged in phagocytosis and digestion of foreign material. In many cases these changes in metabolism are mediated by secretions from other cell types, especially lymphocytes. The latter produce an array of antibodies and lymphokines that can stimulate macrophages to greater metabolic activity.

### Carbohydrate Metabolism

The principal source of energy for both monocytes and macrophages is through the breakdown of glucose in the Embden-Meyerhof (anaerobic glycolysis) pathway, even under aerobic conditions. However, we should note one exception: metabolism of alveolar macrophages is primarily aerobic. Alveolar macrophages will stop the process of phagocytosis when the environmental oxygen tension falls below 20 torr. The other macrophages in various body tissues do not appear to be dependent on the availability of oxygen.

**Table 19.1 Enzymes Present in Macrophages***

| |
|---|
| *Lysosomal Enzymes* |
| Beta-glucuronidase |
| Beta-galactosidase |
| Esterases |
| Acid phosphatase |
| Arylsulfatase |
| Peroxidase |
| *Nonlysosomal Enzymes* |
| Succinic dehydrogenase |
| Cytochrome oxidase |
| Aminopeptidase |

*Note that some enzymes are present only in lysosomal granules, and others are present only in nonlysosomal granules.

In general, macrophages use anaerobic glycolysis as their major source of glucose utilization, and a small amount (1 to 2%) is metabolized via the hexose monophosphate shunt. There is a direct correlation between the degree of maturity of the macrophage and its rate of glycolysis. The more mature the macrophage, the more it relies on the glycolytic pathway. The process of microbial killing is similar to that of the neutrophils. Macrophages are also dependent on the action of superoxide, hydrogen peroxide, hypohalides, singlet oxygen, and hydroxyl radicals as toxic agents.

### Lipid Metabolism

Both monocytes and tissue macrophages are capable of synthesizing fatty acids and cholesterol from acetate. Further, they are capable of using these products for intracellular esterification reactions. The rate of phospholipid synthesis is increased during phagocytosis, presumably to counteract the loss of phospholipid membrane to the newly formed phagosomes, just as in neutrophils.

### Protein Metabolism

Monocytes do not seem to be actively engaged in protein synthesis since their endoplasmic reticulum (ER) is rather limited. In contrast, tissue macrophages have a well-developed ER and are actively engaged in the production of many new proteins. Much of macrophage activity is geared toward production of lysosomal and nonlysosomal enzymes. Tissue macrophages also produce other biologically active compounds such as interleukin 1 (IL-1), interferon, and endogenous pyrogen.

Most lysosomal enzymes are packaged in the Golgi complex from which they bud off as lysosomal granules. In the phagocytic process these lysosomal granules fuse with phagosomes to produce phagolysosomes in which microbial killing and digestion take place. A direct correlation exists between the level of phagocytosis and the production of new lysosomal vesicles.

### Nucleic Acid Metabolism

High levels of DNA synthesis occur only in the early stages of development in the bone marrow and when the monocyte becomes an immature tissue macrophage. Monocytes and mature histiocytes do not appear to synthesize DNA. Apparently all stages of the monocyte-macrophage series are capable of RNA synthesis and will do so when properly stimulated.

## FUNCTIONS OF MONOCYTES AND MACROPHAGES

Blood monocytes and tissue macrophages are involved in a wide variety of defense functions in the body. These functions may be summarized as follows:

1. Monocytes and macrophages play a major role in defending the body against the invasion of pathogenic organisms (antigens) by *engulfment* and *phagocytosis.* In this they are much more efficient than neutrophils.
2. They play a major role in the *activation of acquired (specific) immune responses.*
3. They play a major role in *defending against host cells infected* by pathogenic organisms as well as against neoplastic cells.
4. Macrophages also function as *scavenger cells,* removing all dead cells and necrotic tissue. Whether these cells are destroyed because of mechanical damage or are killed as a result of immunological injury or simply have reached the end of their life span, they are all ingested and destroyed by various forms of tissue macrophages.
5. Finally, monocytes and macrophages *produce a large number of monokines,* including many growth factors that play a major role in the development and differentiation of other blood cells, as well as other cytokines involved in the activation of a large variety of cells.

Each of these functions is discussed in the following material.

### Engulfment of Foreign Particles

When an infection occurs, neutrophils form the first line of defense during the acute phase of inflammation. Later, during the subacute phase of inflammation, neutrophils are gradually replaced by macrophages and lymphocytes. Macrophages—with the help of products produced by lymphocytes—are much more effective in phagocytosing and destroying foreign microbes than are neutrophils, simply because they live longer, are larger, and are better equipped to phagocytose the pathogens than the polymorphonuclear neutrophils (PMNs). During the course of an infection, most pathogens become **opsonized** (= covered) with immunoglobulin molecules as well as with activated complement factors. Monocytes and macrophages possess many Fc receptors for immunoglobulin molecules as well as receptors for activated complement factors. Hence they are able to bind tightly to these foreign agents, which is the first step in the process of phagocytosis. Many microbes are slippery, and if they were not covered by opsonins, macrophages would have difficulty phagocytosing them. Since antibodies also act as **agglutinins** (cause clumping), these phagocytic cells can engulf many foreign particles at once. This will speed up the process of removing the pathogenic agents from the site of infection. Once ingested, these microbes are killed and digested in the phagolysosomes of the macrophages (figure 19.6). This phagocytic process is similar to the one described for neutrophils in chapter 17.

### Activation of the Acquired Immune System

Phagocytosis of foreign particles or **antigens** by macrophages is an absolute necessity for activation of the specific immune system. Pathogenic organisms first need to be digested before the cells of the acquired immune system (i.e., the lymphocytes) can interact with them. Digestion entails breaking up these large organic antigenic compounds into smaller particles. Most large proteins are degraded into small peptide chains of less than 10 amino acids, whereas complex polysaccharides are broken down into smaller chains of four to eight sugars. Most fats are broken down into simple hydrocarbon chains. Since practically all hydrocarbon chains are similar in structure, the body will not usually react against them as they resemble short-chain hydrocarbons frequently encountered in our own bodies. Fat molecules therefore lack the antigenic complexity needed to act as an **antigenic determinant** or **epitope** and are recognized by our bodies as belonging to the "self."

Many short peptide and carbohydrate chains produced from larger proteins and polysaccharides have a sufficient degree of complexity to function as antigenic determinants. These epitopes are then bound to major histocompatibility complex (MHC) molecules, and the complex is then incorporated into the macrophage cell membrane. Binding of epitopes to MHC molecules is a complex process accomplished in either of two ways, depending on the origin of the antigen. If the antigen is of **exogenous origin**—that is, it came from outside the body—it will bind to a **class II MHC** molecule, and this complex will then become incorporated into the macrophage cell membrane. This complex will now interact readily with **CD4$^+$ T cells** (= helper T cells). If the antigen is of **endogenous origin**—that is, it was produced inside the infected cell as a result of transcription and translation of either viral DNA or oncogenes (= cancer genes)—it will bind readily to **class I MHC** molecules which will then move to the surface and interact readily with **CD8$^+$ T cells** (= cytotoxic T cells).

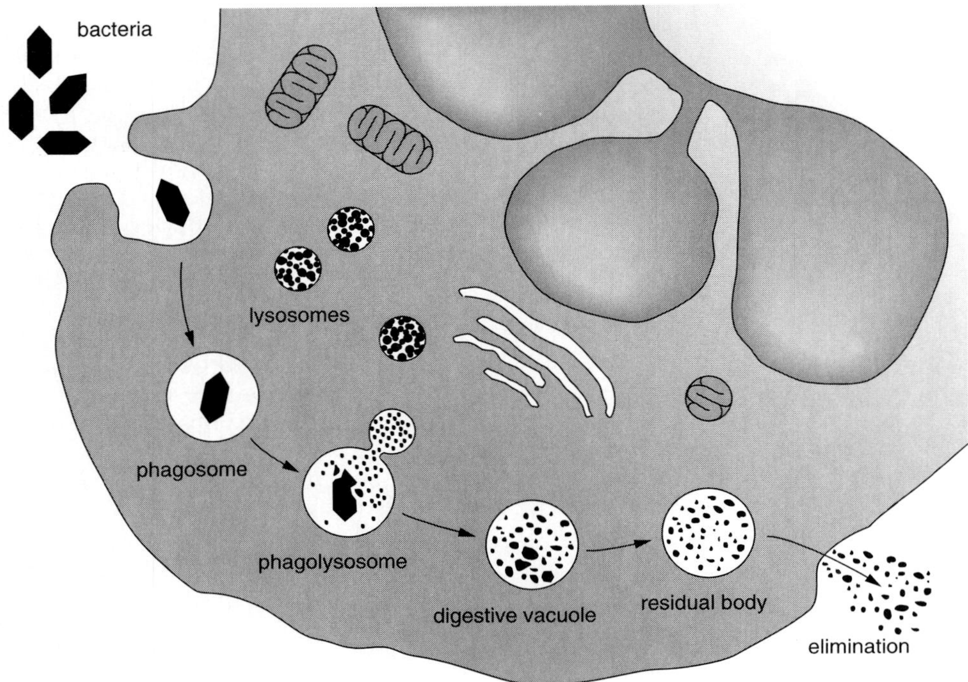


**figure 19.6**
Stages of phagocytosis.

The two processes of the binding of epitopes to the appropriate MHC molecule are described in the following material:

**1. Binding exogenous antigens to class II MHC molecules** This process is illustrated in figure 19.7. The first step involves binding the foreign antigen to the membranes of the monocyte or macrophage. As previously described, this process is facilitated considerably by opsonins. The second step involves the internalization of the antigenic particle by endocytosis, resulting in the production of a phagosome (also known as an endosome). The phagosome now fuses with a lysosomal granule, which forms into a phagolysosome. Inside the phagolysosome the large protein is degraded, producing one or more immunogenic peptides known as epitopes. Concurrently, in the nucleus, transcription of the messenger RNA (mRNA) for the class II MHC molecule has occurred. This mRNA is translated on the ribosomes into proteins that are then passed via the endoplasmic reticulum to the Golgi complex, where the proteins are glycosylated and given their final shape. In the cisternae of the Golgi complex they bud off, producing vesicles containing class II MHC molecules. These vesicles first fuse with the phagolysosomes followed by the binding of appropriate peptides to form the class II epitope-MHC molecules. The next step involves the fusion of the phagolysosomal vesicle that now contains epitope-MHC complexes with the outer cell membrane and the transfer of these complexes from the vesicular wall to the membrane wall. They are now incorporated into the cell membrane and ready to bind to $CD4^+$ cells.

**2. Binding endogenous antigens to class I MHC molecules** This process is fundamentally different from the preceding pathway. To begin with, in this process the macrophage must first produce an antigen from the foreign or cancerous DNA. The first step in this process involves ingestion of the foreign particle into the monocyte macrophage. The next step then involves transcription of the viral DNA or oncogenal DNA followed by protein synthesis from the viral or oncogenal mRNA at the ribosomes. These new proteins are now processed in cellular lysosomes, where they are broken down into a series of epitopes. In the meantime in the chromosomal MHC gene complex, class I MHC genes are transcribed into mRNA. This mRNA now moves to the ribosomes where it is translated into MHC protein molecules. These MHC proteins are then transported to the Golgi apparatus via the endoplasmic reticulum, where the final MHC molecule is formed. These class I MHC molecules are now concentrated in the cisternae of the Golgi apparatus from which they are budded off to form vesicles containing class I MHC molecules. These vesicles now fuse with lysosomal granules containing the newly formed viral or neoplastic proteins. Fusion allows the class I MHC molecules to bind to the viral or cancer proteins (epitopes). The newly formed class I MHC-antigen complex now fuses with and moves to the outer cell membrane where it is ready to bind with $CD8^+$ cells (figure 19.8).

Once the monocytes and macrophages have become activated either by foreign particles or foreign endogenous antigens, they move away from the area of infection to nearest lymphoid tissues and to the spleen, where they present the various epitopes coupled to MHC molecules to the appropriate $CD4^+$ and $CD8^+$ lymphocytes present in these lymphoid glands.

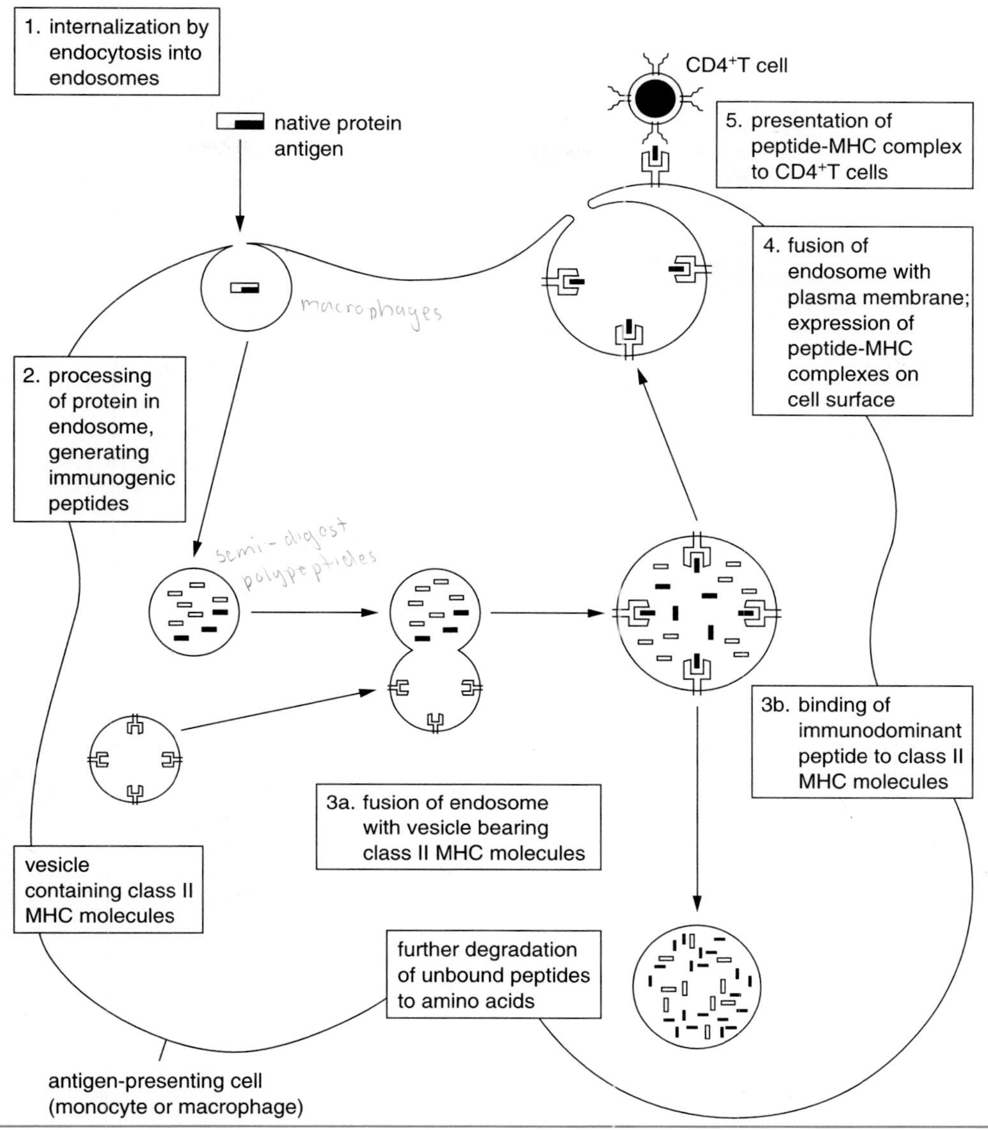


**figure 19.7**

The mechanism by which an extracellular protein antigen (epitope) binds to a class II MHC protein, and how this complex is incorporated into the phagocytic cell membrane.

Many different clones of lymphocytes are found in our lymphoid tissues, each of which is able to react with only one, or at the most a few, closely related epitopes. Stated another way, each clone of lymphocytes has a unique set of receptors able to bind to only one or at the most a few closely related epitopes. Once the macrophage has found a lymphocyte that is able to bind to one of the epitopes on its surface, it will bind to that receptor. This connection acts as a stimulus for the macrophage to release an activating and growth factor known as **interleukin 1 (IL-1),** which stimulates lymphocyte activation. This activation process involves the enlargement of the small lymphocytes (blast formation) and causes these cells to undergo multiple cell divisions. This results in the production of many similar active lymphocytes that now will be able to perform a number of immunological functions, depending on the type of cell activated. Eventually this will lead to the destruction of the pathogenic organism or to the death of the infected or cancerous cell. Because one role of the monocytes and macrophages is to present the processed antigen to the appropriate lymphocyte cells, they are also known as **antigen-presenting cells** or **APCs** (figure 19.9). Thus monocytes and macrophages play an essential role in the activation of the immune system. Unphagocytosed foreign agents are unable to induce an immune response, with the exceptions of some polysaccharides composed of long chains of similar sugars, or some synthetic polypeptides

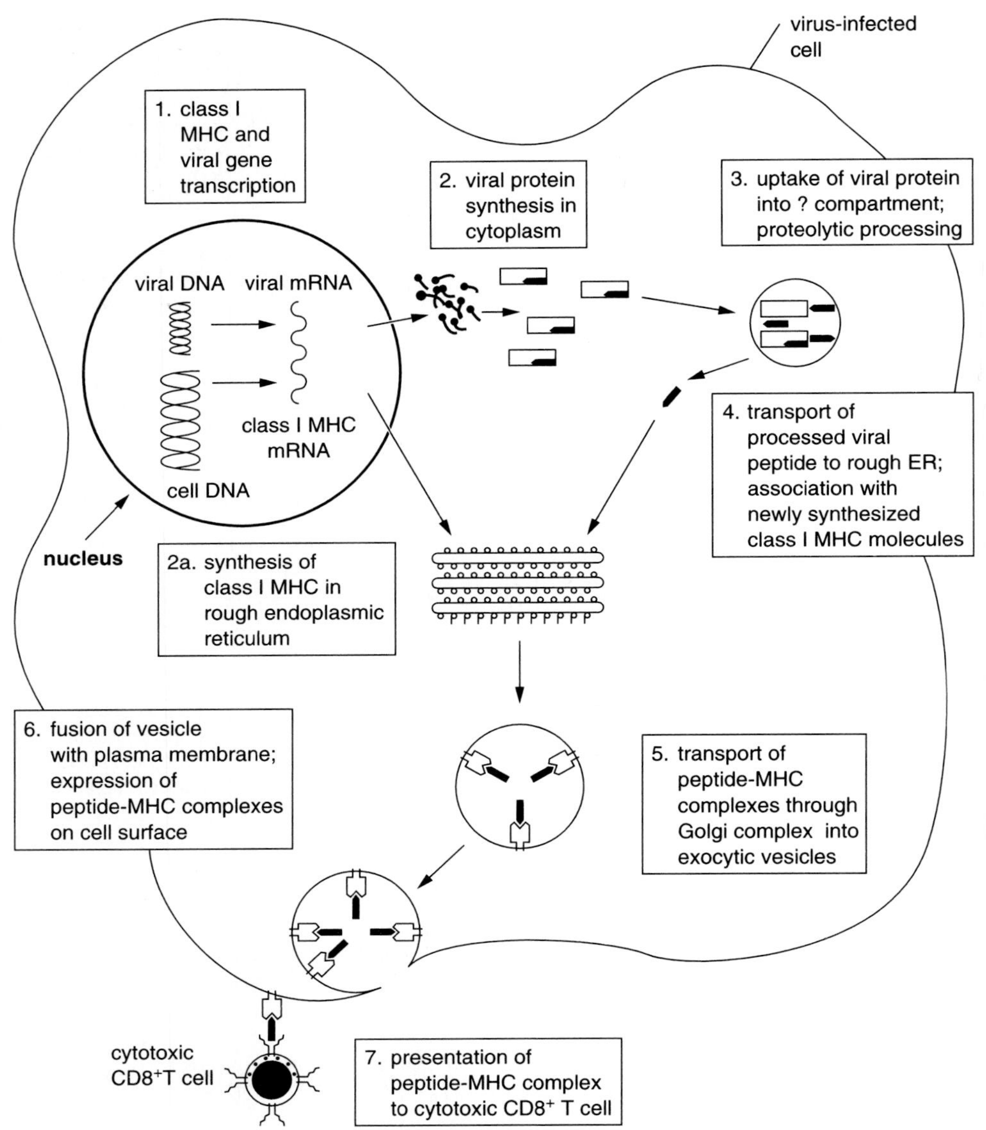


**figure 19.8**

The mechanism by which an intracellular viral antigen (epitope) binds to a class I MHC protein, and how this complex is incorporated into the phagocytic cell membrane.

composed of similar amino acids. The latter are referred to as **T-cell-independent antigens.** However most naturally occurring antigens must be processed first by the phagocytic cells before they are able to activate the immune system.

### Killing of Infected and Neoplastic Cells

Many studies have shown that monocytes and macrophages can eliminate infected body cells and tumor cells through both cytolysis and phagocytosis. Most neoplastic cells carry **tumor-specific antigens** on their surface. Similarly many cells infected by viruses, bacteria, or other microorganisms have microbial antigens incorporated in their cell membrane (figure 19.10). These infected cells are engulfed by macrophages that process the foreign antigens and present them to the appropriate B and T lymphocytes, such as cytotoxic T ($T_c$) cells and helper T ($T_H$) cells. B cells produce tumor-specific antibodies to cancer cells and microbe-specific antibodies to infected cells, which then bind to specific antigens on tumor cells or on infected host cells. Antibody-antigen complexes will then activate the complement system via the classical pathway. The role of lymphocytes, macrophages, antibody, and complement activity on the lysis of tumor cells is illustrated in figure 19.10.

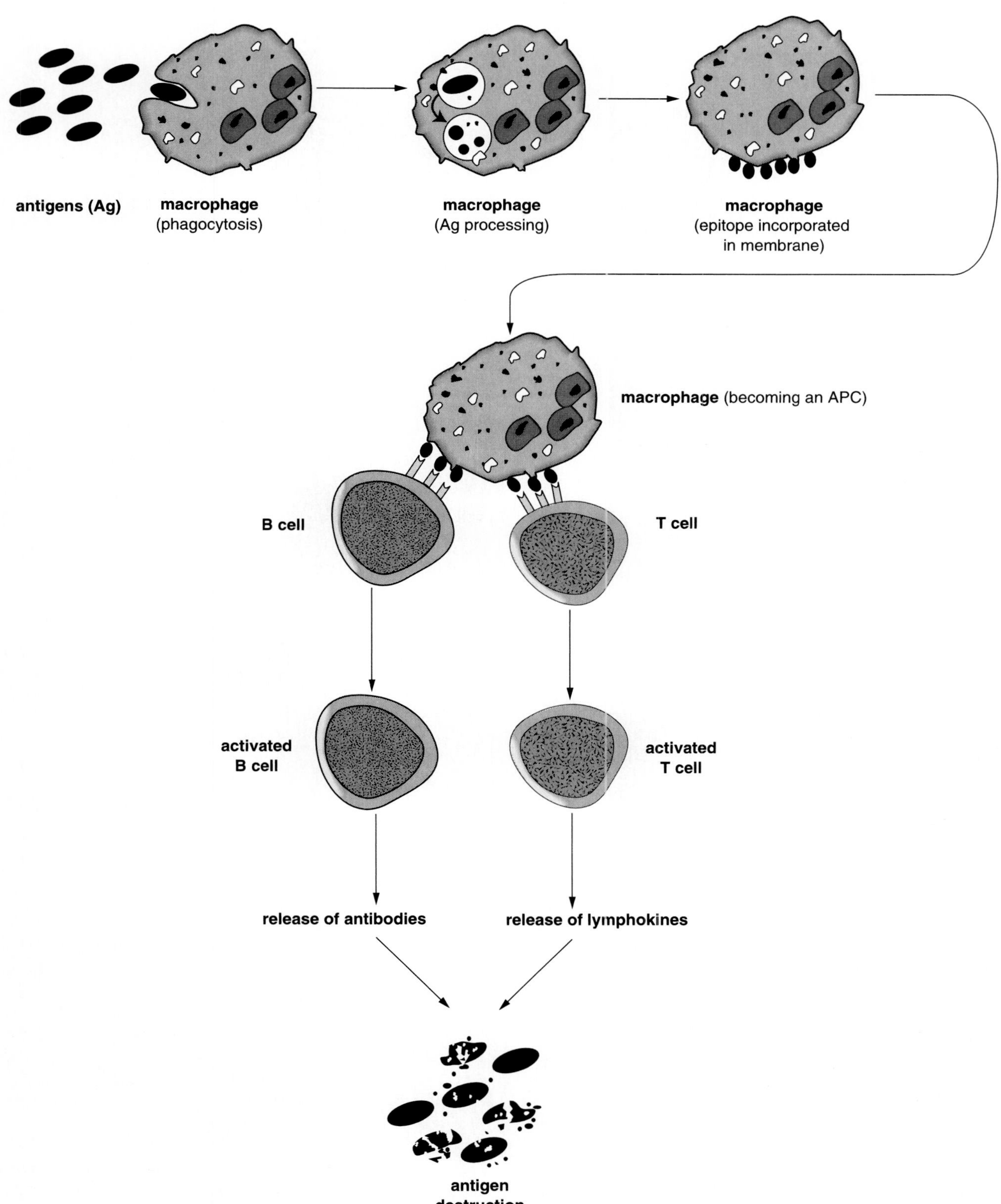


**figure 19.9**

Macrophage role in the activation of the specific (acquired) immune responses, i.e., their role as antigen-presenting cells (APCs).

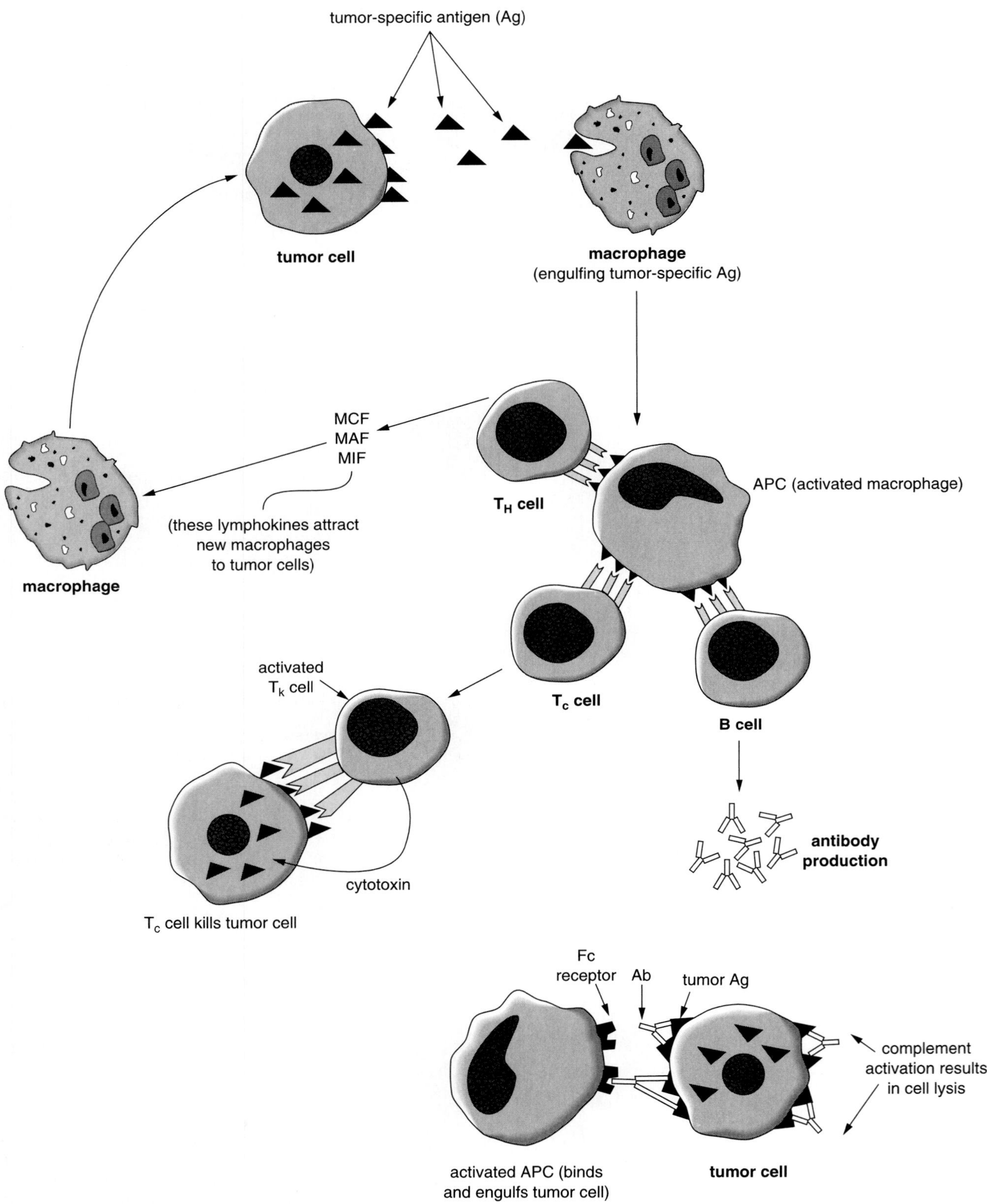


**figure 19.10**
The role of macrophages, antigen-presenting cells (APCs), in the killing of tumor cells. This process is enhanced by lymphokine, complement, and antibody production.

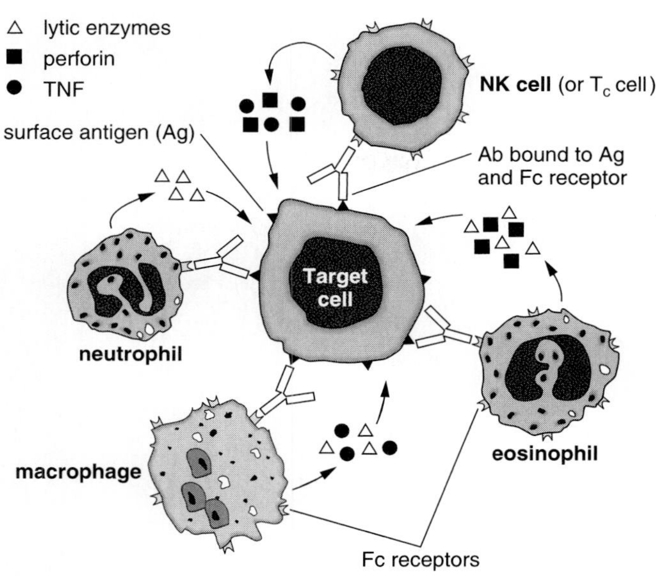


**figure 19.11**
The action of nonspecific cytotoxic cells in the destruction of target cells via antibody-dependent cell-mediated cytotoxicity (ADCC). Note that all these cells have Fc receptors for antibodies, and thus they can bind to the target cells by binding to the antibodies on the target cells. After they are bound, they release their cytotoxic chemicals, destroying the target cells.

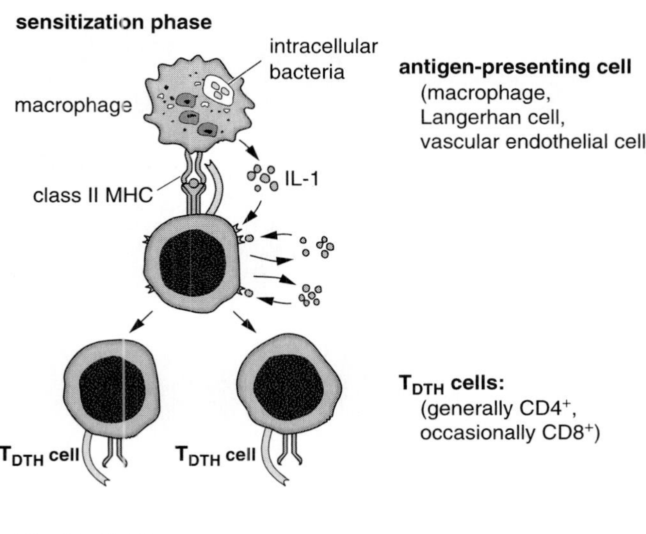


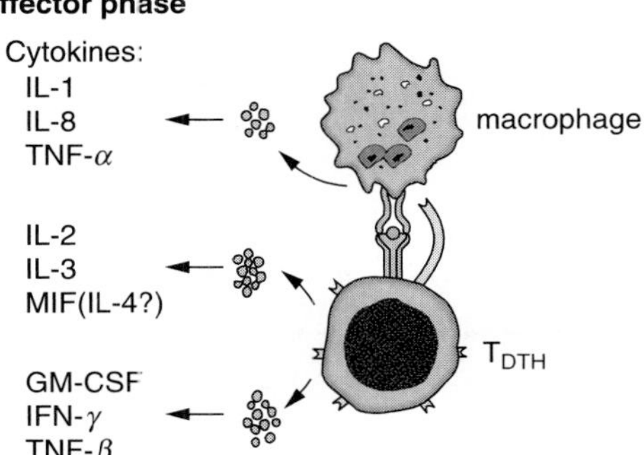


Cytokines attract and activate macrophages, resulting in increased lysosomal enzymes and increased killing; they induce the hematopoiesis of monocytes and neutrophils; and they stimulate extravasation of circulating neutrophils and monocytes.

**figure 19.12**
The stages associated with the delayed-type hypersensitivity (DTH) response. The sensitization phase takes at least a week, often more. Once a person is sensitized, the effector phase occurs within 1 or 2 days of the second contact with that antigen.

The receptors for the Fc fraction of the antibody molecules and the C3b complement factor on the macrophages enhance binding to these cancer cells (and infected cells), which initiates the process of phagocytosis. In addition, activated helper T lymphocytes will produce a number of lymphokines, which will further stimulate macrophages to greater cytolytic and phagocytic activity. These lymphokines include the following factors: (1) **macrophage chemotactic factor (MCF),** which attracts histiocytes to the tumor site; (2) **macrophage-activating factor (MAF),** which stimulates macrophages to greater phagocytic activity; and (3) **migration-inhibiting factor (MIF),** which retains activated macrophages in the tumor area. $T_c$ cells and natural killer (NK) cells will also bind to infected and neoplastic cells via their Fc receptors for antibody molecules. When they bind, they will also release cytotoxic chemicals that will cause the death of these cells. This type of killing involving interaction with antibody is known as **ADCC,** or **antibody-dependent cell-mediated cytotoxicity.** Other cells capable of ADCC are neutrophils and eosinophils. As illustrated in figure 19.11, each cell has its own cytolytic secretions.

Another way in which macrophages are involved in killing intracellular pathogens involves the **delayed-type hypersensitivity (DTH) response.** In this response the development or sensitization of the immune reaction is much slower than the ADCC response, hence the name delayed hypersensitivity. The term hypersensitivity may be somewhat misleading as it suggests that it is always detrimental. On the contrary, although the DTH response may cause inappropriate tissue damage in certain situations, its main role is to defend the body against intracellular exogenous pathogens including many intracellular bacteria, intracellular fungi, intracellular protozoans, and even certain intracellular viruses. A list of some common organisms affected by DTH is given in table 19.2. In the **sensitization phase** the intracellular pathogen is first processed by an APC, most commonly a macrophage. The APC activates a special type of CD4$^+$ T cell, known as a $T_{DTH}$ cell. In the **effector phase** activated $T_{DTH}$ cells secrete a series of lymphokines including IL-2, IL-3, MIF (IL-4), GM-CSF, interferon-$\gamma$ (IFN-$\gamma$), and tumor necrosis factor $\beta$ (TNF-$\beta$), which have the following three basic functions: (1) inducing development and differentiation of new monocytes

**Table 19.2** Antigens that can Induce a Delayed Hypersensitivity Response on T Cells

| Intracellular Antigens | Antigen-presenting Cells | $T_{DTH}$ Cells |
|---|---|---|
| *Bacteria* | Langerhans cells | $CD4^{+2}$ and |
| *Brucella abortus* | Macrophages | $T_H1$ subpopulation |
| *Listeria monocytogenes* | Vascular endothelial cells | Occasionally $CD8^+$ |
| *Mycobacterium leprae* | | in response to viral antigens |
| *Mycobacterium tuberculosis* | | |
| *Fungi* | | |
| *Candida albicans* | | |
| *Cryptococcus neoformans* | | |
| *Histoplasma capsulatum* | | |
| *Pneumocystis carinii* | | |
| *Parasites* | | |
| *Leishmania* sp. | | |
| *Schistosoma* sp. | | |
| *Viruses* | | |
| Herpes simplex | | |
| Measles | | |
| Variola (smallpox) | | |
| *Chemicals* | | |
| Contact dermatitis (poison ivy) | | |

and neutrophils, (2) attracting monocytes and neutrophils to the site of infection, and (3) activating these effector cells so they produce cytokines that will cause cytolysis of infected cells as well as an increase in phagocytosis (figure 19.12).

The influx and activation of these phagocytic cells in the DTH response provides an effective host defense against intracellular pathogens so they can be rapidly cleared from the system with relatively little tissue damage. In some special cases, however, when phagocytic cells have difficulty clearing the intracellular pathogens from the tissues, a prolonged DTH response may become destructive to the host. The macrophages respond by undergoing endomitosis and becoming very large giant cells that produce an intense inflammatory response. Further, these giant cells begin to adhere closely to one another and surround the infected cells, forming a hard cyst known as a **granuloma.** Because these granulomatous cells are active and release large quantities of lysosomal enzymes into the surrounding tissues, they will destroy normal tissue and replace it with hard nodules composed of giant cells. These granulomatous reactions may occur in a wide variety of tissues but are especially common in the lungs, heart, and kidneys.

### Removal of Old Cells and Necrotic Tissue

Macrophages also are the major agents of the body responsible for removal of damaged and senescent cells, including old and effete blood cells. The latter process occurs mainly in the spleen, when these cells percolate through the sinuses and cords of the red pulp area. Macrophages are also responsible for removal of injured and necrotic tissue cells, as well as for ingestion of inert particles such as coal dust and salt crystals. Finally, macrophages remove all debris at the end of an infection, including dead microorganisms, disintegrated neutrophils, and injured local tissue cells.

### Production of Monokines

A number of *in vitro* studies have shown that monocytes and macrophages produce several colony-stimulating factors that promote the growth and development of different white blood cells. These factors include GM-CSF, which promotes the development of granulocyte-monocyte precursor cells; G-CSF, which stimulates the production of new granulocytes; and M-CSF, which causes an increase in monocytes and phagocytes.

In addition, macrophages are also a major source of a number of interleukins, including IL-1. This monokine is a major source of activation of the helper T cells as well as B cells and natural killer (NK) cells. IL-6 is another important monokine that aids in the differentiation of myeloid stem cells and also increases the secretion of antibodies by activated B cells. IL-8 is a macrophage product that attracts neutrophils to an area of infection. It also induces adherence of neutrophils to vascular endothelial cells—that is, it promotes margination—and aids in their migration into the tissues.

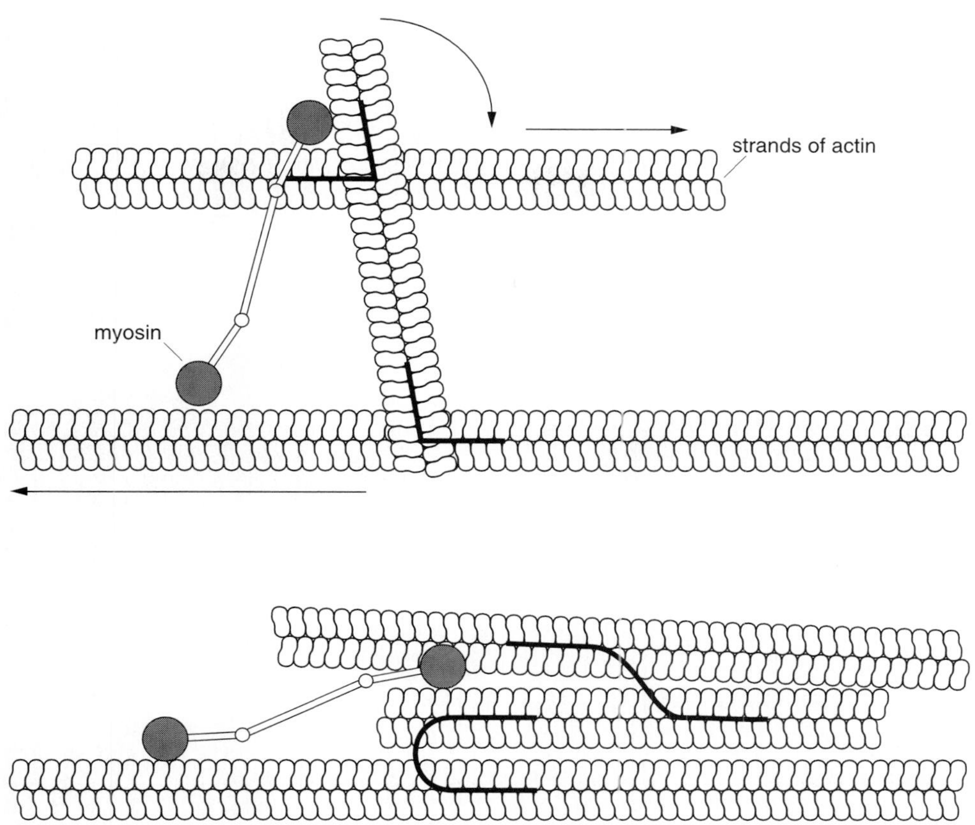


**figure 19.13**

Contractions in a pseudopod of a phagocytic cell may be induced by strands of myosin acting on a network of actin fibers.

Other factors produced by macrophages include tumor necrosis factor α (TNF-α) and transforming growth factor β (TGF-β). TNF-α has a cytotoxic effect on tumor cells but not on normal cells. It also induces numerous cell types to secrete various cytokines involved in the inflammatory response. TGF-β attracts monocytes and macrophages to an area of infection. It also induces increased IL-1 production in activated macrophages. Further, it inhibits the proliferation of hematopoietic stem cells, limits the inflammatory response, and promotes wound healing.

Indirectly monocytes and phagocytes also promote production of more erythrocytes since they have a great affinity for transferrin. Once the phagocytes have taken in this iron-transport compound, it is converted to ferritin. This storage form of iron is then delivered to the bone marrow tissues, where it is absorbed by the developing red blood cells, thus speeding the process of hemoglobin production. This in turn shortens the maturation time of new red blood cells.

## LOCOMOTION OF PHAGOCYTIC CELLS

All phagocytic cells must be able to move unidirectionally toward a site of infection. The process of locomotion involves the production and subsequent contractions of pseudopods by these phagocytic cells. This process of cellular movement is analogous to the contraction of muscle cells, since actin and myosin molecules are present in the pseudopodia of phagocytic cells. The process, however, is not exactly the same as in muscle cells. In phagocytic cells the actin polymers are not neatly arranged in parallel strands but form a scaffolding of cross-linked filaments connected to the cell membrane. The myosin molecules appear to bring these actin strands in apposition, thus rearranging the scaffolding (figure 19.13). Since actin strands are connected to the cell membrane, the membrane moves forward.

A more recent theory discounts the action of myosin on actin strands as an explanation for phagocytic locomotion. This theory suggests that development of a new pseudopod is associated with destruction of old actin formations, while new actin scaffolding is constructed in the newly developing pseudopod. Production of a new pseudopod usually occurs as a result of a chemical stimulus, such as the binding of an antibody or chemotactic lymphokine to the outer membrane of the cell. In response to this stimulus, the actin network in the immediate vicinity of this stimulatory compound begins to break down concentrating actin fragments and individual actin molecules in that area of the cell. Because of the high concentration of small molecules in that area, the

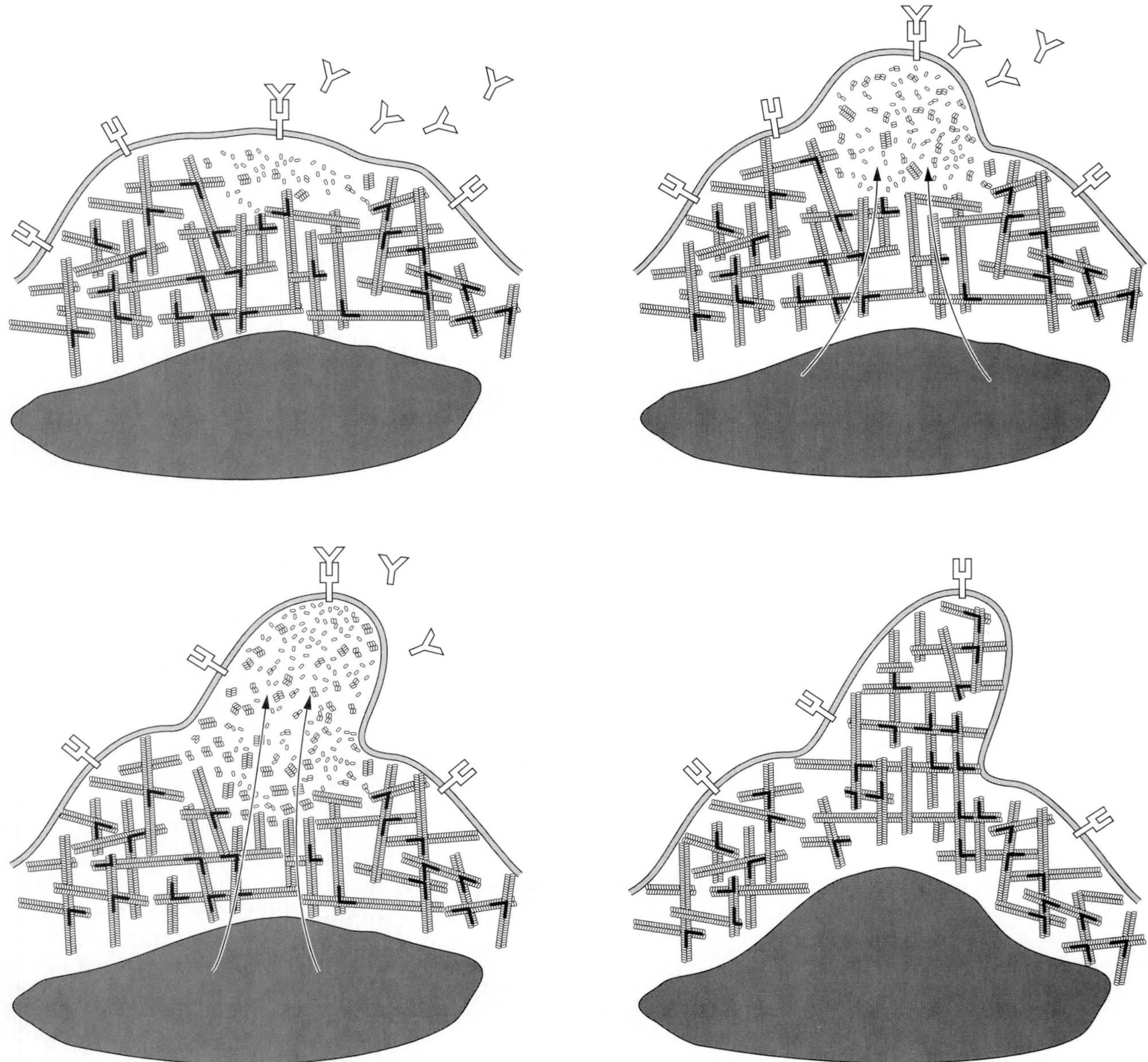

**figure 19.14**

The development of a new pseudopod is usually initiated by the binding of compound (e.g., an antibody) to a receptor. In the area of the occupied receptor(s), the actin network begins to break down, which results in the local accumulation of many individual actin molecules. These molecules produce an osmotic change in that area of the cell, and fluid is drawn to that area. The membrane bulges out, and a new pseudopod is formed. It is then stabilized by a new actin network that is quickly assembled.

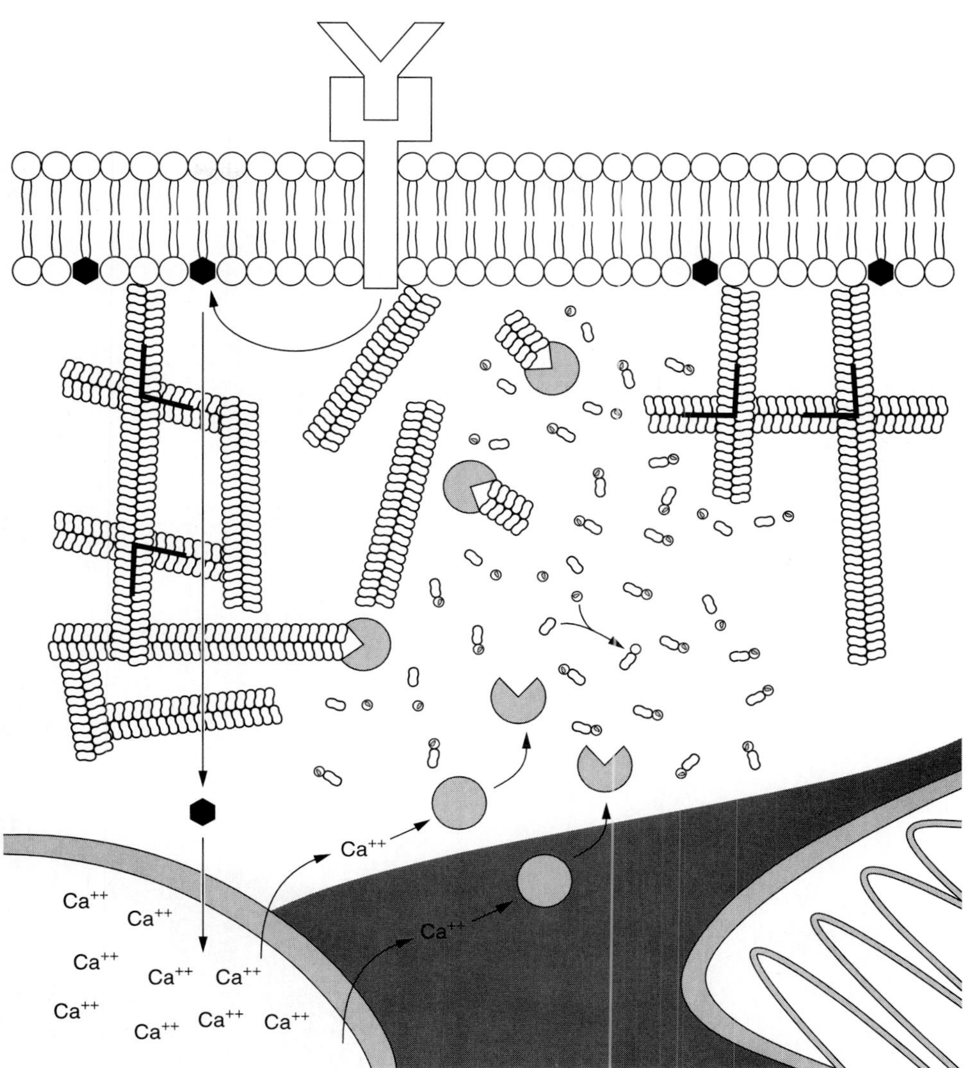


**figure 19.15**
The breakdown of the actin network. An internal signaling molecule is released from the cell membrane when a compound binds to a surface receptor. This internal signal, a six-carbon ring, stimulates the release of calcium ions from storage vesicles inside the cell. These calcium ions then bind to inactive gelsolin molecules, activating them. The activated gelsolin molecules then tear down the actin network.

osmotic balance in the cell is upset—and as a result fluid from the interior part of the cell moves to the area of the actin breakdown. This increased amount of fluid will cause distention of the outer membrane of the cell, thus forming a pseudopod. Once the pseudopod has reached a distinct shape, the actin scaffolding is reassembled in the newly formed pseudopod, thus giving it a more solid form that will not retract without further activity (figure 19.14).

The disassembly of actin polymers, and hence the breakdown of the structural network or scaffolding is controlled by two special compounds: gelsolin and profilin. Activation of **gelsolin** is controlled by calcium ions. Release of calcium ions from special cytoplasmic vesicles is controlled by an external signal, such as the binding of antibody or chemotactic lymphokines. The sequence of events is as follows according to this theory. First, the external stimulus binds to the receptor on the cell membrane. The activated receptor now is able to dispatch an internal signaling molecule, which is normally a six-carbon ring structure derived from phosphotidylinositol, a normal component of the cell membrane. This six-carbon ring structure detaches itself from the long fatty acid chains of phosphatidylinositol, which remain in the membrane and stimulate the release of calcium from storage vesicles (figure 19.15). The calcium ions then bind to inactive gelsolin and in doing so activate this compound, which is now able to disassemble the actin network. The resulting actin fragments are capped by gelsolin, and individual actin molecules are protected from further degradation by **profilin** molecules.

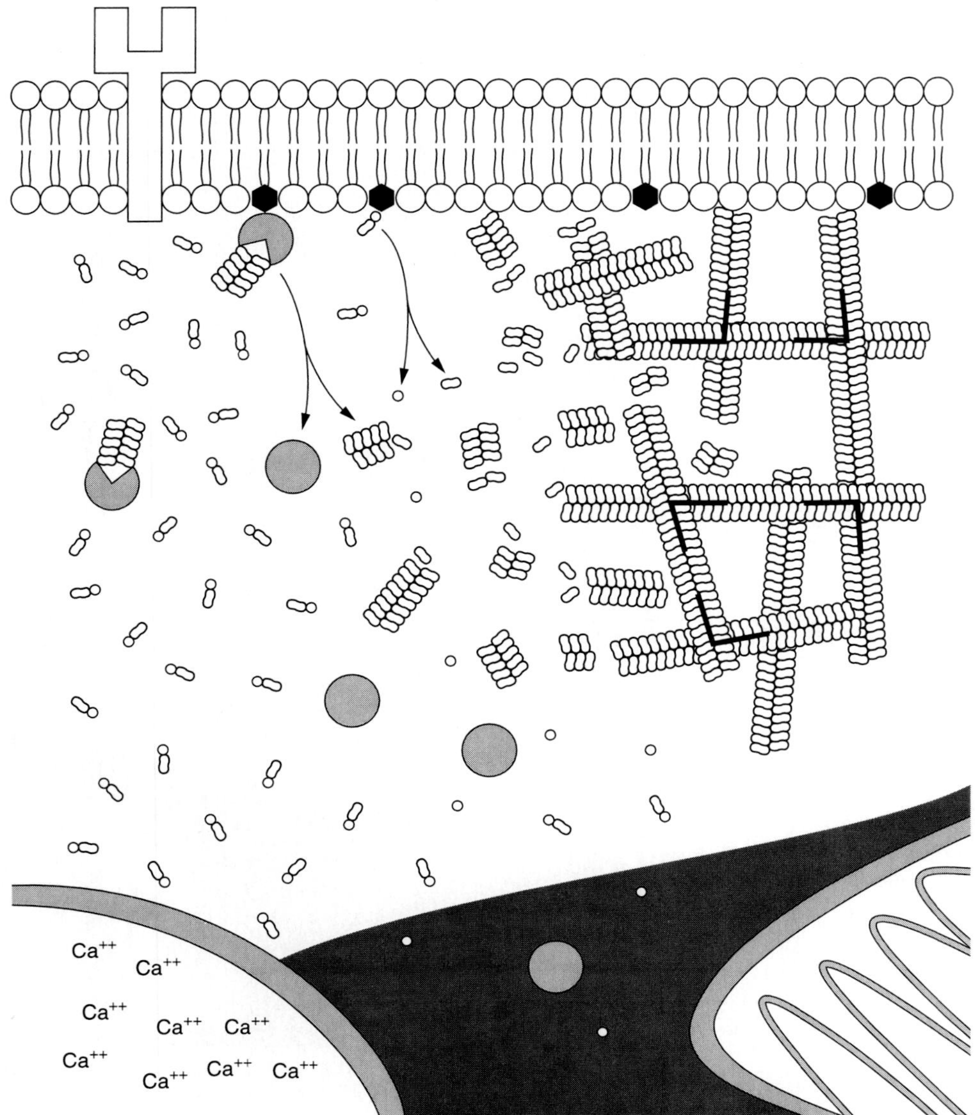


**figure 19.16**

The reassembly of the actin network. Gelsolin-actin complexes migrate to the cell membrane, where they interact with special membrane phospholipids (especially phosphatidylinositol). This interaction causes the gelsolin to dissociate from the actin, leaving free actin fragments. With the aid of profilin, individual actin molecules attach themselves to these actin fragments, thus lengthening the chains. Individual chains then attach to one another and to myosin fibers to form a new network.

Reassembly of the actin scaffolding requires removal of gelsolin and profilin from the actin molecules. The gelsolin-actin fragments may play a crucial role in initiating reassembly of the actin network. Gelsolin-actin complexes migrate to the cell membrane, where they interact with phosphatidylinositol. This interaction causes gelsolin to dissociate from the actin fragments. The actin-profilin complexes may act in a similar manner. These complexes also migrate to the cell membrane and interact with phosphatidylinositol molecules. This interaction results in dissociation of the complex and production of free actin molecules. The free actin fragments and free actin molecules are now able to join again to one another and thus regenerate the actin polymers, producing a new actin network (figure 19.16).

## Further Readings

Abbas, A. K., Lichtman, A. H., and Pober, J. S. Antigen Presentation and T Cell Antigen Recognition. Chapter 6 in *Cellular and Molecular Immunology*. W. B. Saunders Company, Philadelphia, 1991.

Beck, W. S. Reticuloendothelial (Mononuclear Phagocyte) System, Lymphatic System, and Spleen. Lecture 2 in *Hematology*, 4th ed., W. S. Beck, Ed., The MIT Press, Cambridge, Mass., 1985.

Cline, M. J. The Normal Macrophage. Part III A in *The White Cell*. Harvard University Press, Cambridge, Mass., 1975.

Kuby, J. Cell-Mediated Immunity. Chapter 13 in *Immunology*. W. H. Freeman and Company, New York, 1992.

Stossel, T. P. How Cells Crawl. *American Scientist*, Vol. 78 (Sept.–Oct. 1990).

## Review Questions

1. Name the five normal stages of development of monocytes and macrophages. Where does each stage occur?
2. Under what circumstances can a sixth stage occur in monocyte and macrophage development? Name the cell in the sixth stage and the reason for its name.
3. Where do monocytes originate? Which are the precursor cells? Which white blood cell is closely related to the monocytes?
4. What is the maturation time for monocytes? How long do monocytes stay in the bloodstream? How long do they live in the tissues?
5. How is the level of monocytes maintained in the blood? Name the major growth factors involved in the regulation of monocyte and macrophage production. Where are these cytokines produced?
6. A mature macrophage may develop into a number of different cell types, each with its own shape and function. List the major types of mononuclear macrophages in the body.
7. Distinguish between the size and shape of monocytes and tissue macrophages.
8. Name some of the major lysosomal and nonlysosomal enzymes found in monocytes and macrophages.
9. What is the major energy-providing pathway in monocytes and macrophages?
10. Why is phospholipid synthesis increased in macrophages during phagocytosis?
11. Why do mature tissue macrophages have a well-developed endoplasmic reticulum and Golgi apparatus?
12. List the five major functions of monocytes and macrophages.
13. Why are macrophages so efficient in the phagocytosis of foreign pathogens?
14. How do macrophages activate the acquired immune system? What are epitopes?
15. Distinguish between endogenous and exogenous antigens. How are endogenous and exogenous epitopes complexed to the major histocompatibility molecules? Which antigen binds to class I MHC molecules?
16. What is an APC? How do APCs bind to helper T cells? How do they activate these helper T cells?
17. Describe what happens to B cells, cytotoxic T cells, and NK cells that are activated by macrophages.
18. Monocytes and macrophages play a major role in cell-mediated immunity. Explain how they are able to kill tumor cells.
19. Explain how macrophages are able to kill virus-infected cells.
20. Describe the role of macrophages in delayed-type hypersensitivity reactions. What is the major function of this immune response? Mention some of the major pathogens that cause these reactions.
21. Explain what is meant by ADCC. List four different types of cells that may cause cytotoxic killing via ADCC. What is the nature of the cytotoxins produced by each of these cells?
22. Name some major monokines produced by monocytes and macrophages. What is the function of each of these monokines?
23. Explain how monocytes are able to move unidirectionally to an area of infection. Describe how pseudopods are formed.
24. Which compounds regulate the breakdown of actin polymers? How are these compounds removed from the actin fragments when a new actin scaffolding is formed?

# Chapter Twenty

## *Development and Structure of Lymphocytes*

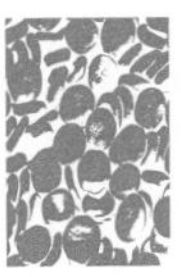

Lymphocytes are the smallest and second most common group of white blood cells in the bloodstream. About 20 to 40% of all leukocytes are lymphocytes in a normal healthy individual. Lymphocytes are easily recognized by their large round or oval nuclei that are deeply stained and the presence of very little cytoplasm. In most lymphocytes the total cytoplasm appears as a small crescent on one side of the nucleus. Only immature and recently activated lymphocytes will show larger amounts of pale cytoplasm when stained with Wright's stain.

## ONTOGENY OF LYMPHOCYTES

Lymphocytes are a heterogeneous collection of blood cells derived from lymphoid stem cells. As explained in chapter 3, pluripotent hematopoietic stem cells differentiate in either of two directions: they become either myeloid or lymphoid colony-forming units (CFUs). Except for the lymphocytes all blood cells are derived from myeloid stem cells (figure 20.1). The implication is that lymphocytes are not closely related to the other blood cells. Although all lymphocytes are ultimately derived from stem cells in the bone marrow, most lymphocytes are produced from earlier lymphocytes present in lymphoid tissues. As explained in chapter 3, lymphoid tissues become organs of hematopoiesis sometime after the fourth lunar month of fetal development. At that time lymphoid stem cells from the fetal liver begin to migrate into the developing lymphoid tissues and organs. Here they continue to divide and colonize these unique tissues. This process of migration and colonization continues throughout fetal life, first exclusively from the fetal liver and later on predominantly from the bone marrow. By the time a child is born, most of the lymphoid tissues will have filled with lymphoid cells. These lymphoid cells resemble small lymphocytes and have the unique ability to react with one, or at most a few, antigenic determinants. As explained in the last chapter, an antigenic determinant or epitope is a unique sequence of a few amino acids or a unique arrangement of short chained sugars able to bind to a given lymphocyte, causing it to become activated. Since it is plausible to hypothesize that there are several million different epitopes, the human body must produce several million different types of lymphocytes, each of which is able to react with one or a few epitopes. Most immunologists currently accept the idea that our lymphoid tissues contain several million different lymphocyte "clones," and each original clone may consist of several thousand cells.

Since our cells and tissues contain many thousands of different proteins and carbohydrates, it follows that we also have many different epitopes on our body cells and in our tissue fluids that are recognized as belonging to the **"self."** We can logically presume that the many different lymphocyte clones could react with these self-antigenic determinants. If our own lymphocytes could react with these self-epitopes, they would become activated and secrete products that would destroy our own tissues. This does not normally happen, but it does occur in autoimmune diseases. Why are the clones of lymphocytes against our self-antigenic determinants not activated? It has been postulated that **T-cell clones** against self-antigens are destroyed sometime during later prenatal or early postnatal life. This is known as **clonal deletion.** Most of the **B-cell clones,** on the other hand, are not deleted but are suppressed (i.e., made permanently inactive) during that same period of time. This is known as **clonal anergy.**

Normally lymphocytes in lymphoid tissues do not become active until after birth when children are exposed to many different foreign agents present in the air, in the food, or that enter their system via breaks in the skin. However, before birth fetuses accept foreign agents as part of the self, destroying or suppressing any lymphocyte clones that could be activated against these nonself particles. Thus most lymphocytes stay dormant during prenatal life and will become activated postnatally only when they are exposed to a specific foreign antigenic determinant with which they can interact. These epitopes will be presented to inactive lymphocytes in lymphoid tissues by activated macrophages that have processed them (antigen-presenting cells, or APCs), as was explained in the last chapter.

Possibly, and also likely, many lymphocyte clones in our bodies have never been activated because of lack of exposure to the specific epitope with which they can react. Many other clones, on the other hand, will have been activated over and over again.

## CLASSIFICATION OF LYMPHOCYTES

As explained in chapter 3, the two major categories of lymphocytes are B cells and T cells. B cells originate from the bone marrow and migrate to various lymphoid tissues, in which they settle in special areas. T cells also originate from the bone marrow, but first move to the thymus gland where they proliferate and mature before they settle in peripheral lymphoid tissues. T cells also have their own specific areas in the lymphoid tissues. There is little mixing of T cells and B cells in lymphoid organs.

From the morphological point of view, we cannot distinguish between mature inactive B cells and T cells in lymphoid organs or in the bloodstream. They all appear as small lymphocytes with a diameter of 6 to 10 μm. They can be separated, however, by using functional parameters. One of the major ways to distinguish between B cells and T cells in the peripheral blood is by combining them with sheep red blood cells (SRBCs). It was discovered accidentally many years ago that T cells have special receptors for SRBCs, whereas B cells lack such markers. By adding SRBCs to a mixture of lymphocytes, the SRBCs will form **rosettes** around T cells but not around B cells. By counting the number of lymphocytes surrounded with SRBCs and comparing them with the number

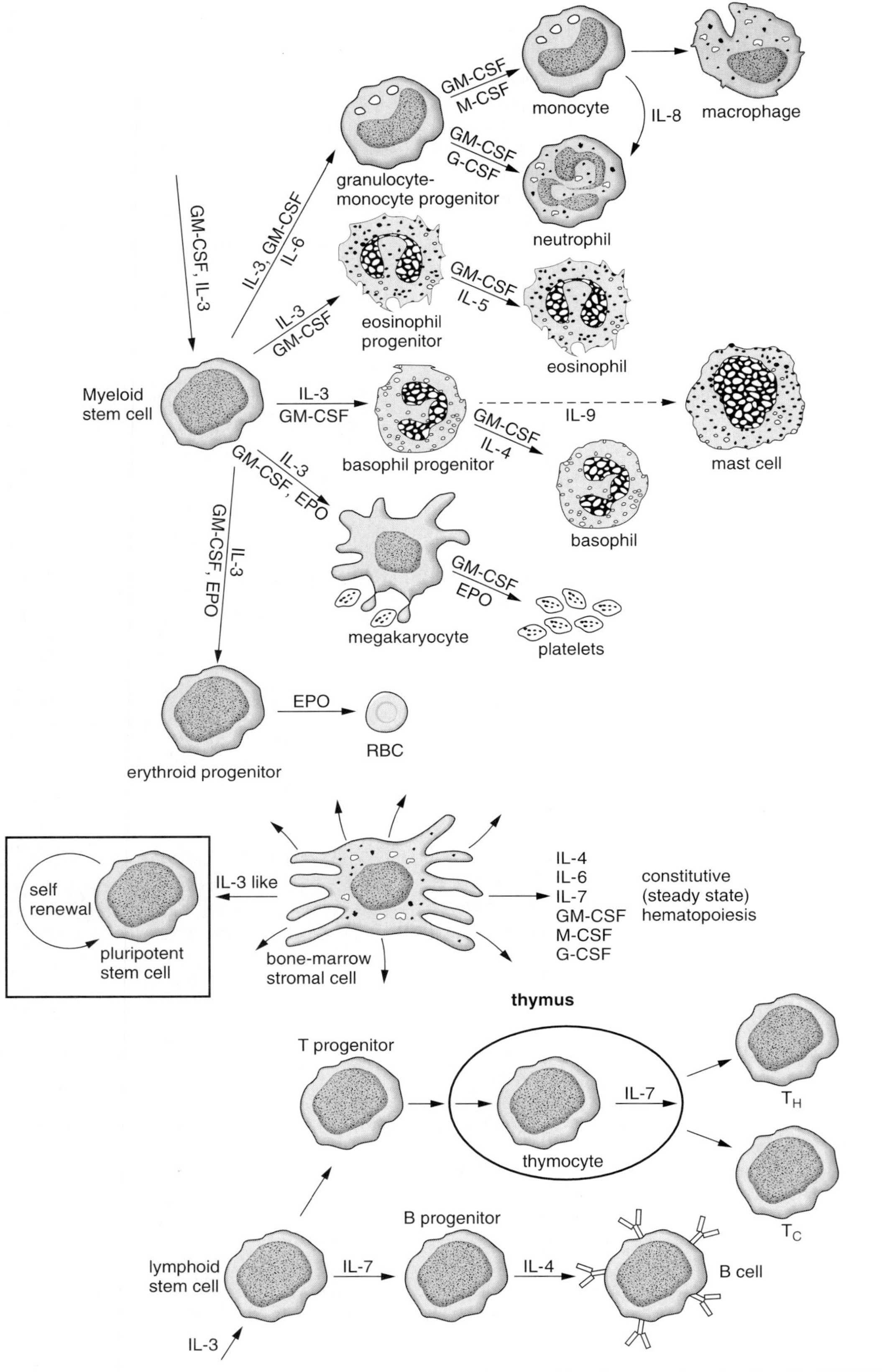


**figure 20.1**
An overview of hematopoiesis, including lymphopoiesis. Note the great variety of growth factors involved in the process.

that do not have such rosettes, we can determine the relative abundance of B and T cells in the blood.

To separate lymphocytes from other blood cells, we must first centrifuge a tube of anticoagulated blood. After centrifugation, a small whitish band will appear at the interface between the packed red blood cells and the plasma. This band is known as the **buffy coat** and contains the majority of the white blood cells. After removing the overlaying plasma, the buffy coat is carefully aspirated and placed in another tube containing a **lymphocyte separating medium (LSM).** After this mixture is centrifuged, all white blood cells will collect at the bottom of the tube except for the lymphocytes, which remain suspended in the medium.

The development and maturation processes of activated B and T cells follow different paths. Although both types of lymphocytes go through a series of maturational steps, these phases are quite divergent. Most activated B cells will develop into plasma cells or plasmacytes, with their unique morphological shape. Activated T cells, on the other hand, will still resemble small lymphocytes.

For reasons that are explained later, most lymphocytes in the bloodstream are T cells, whereas only 15 to 25% belong to the B-cell group.

We should note the existence of a third category of lymphocytes that does not express characteristics of either B or T cells. Since they do not possess any markers of B or T cells, they are commonly referred to as null cells. Even within this category we have different populations of cells. The great majority of these cells are classified as **natural killer cells** (or **NK cells**). They possess Fc receptors for IgG on their membranes. This allows them to bind to the Fc region of antibodies that are complexed with surface antigens on target cells. This binding acts as a trigger for the release of cytotoxic substances from these NK cells, and this results in the killing (cytolysis) of target cells (see figure 19.11). This type of cytolysis depends on the presence of antibody, and hence these cells are said to exhibit antibody-dependent cell-mediated cytotoxicity (ADCC). NK cells are also known as **large granular lymphocytes** (or **LGLs**) because they are usually much larger than mature T and B cells. Approximately 57% of the lymphocytes in the bloodstream are NK cells. They are also frequently present in peripheral lymphoid organs (i.e., the spleen and lymph nodes) and appear to be a heterogeneous group with respect to surface markers expressed. Apart from Fc receptors (which allow them to exhibit ADCC activity), many also have receptors for interferon (IFN) and interleukin 2 (IL-2), but they do not seem to possess receptors for specific antigenic determinants as do B and T cells. NK cells are spontaneously cytotoxic to a variety of targets, especially to virus-infected cells and cancer cells. Their primary task is to keep viral replication and neoplastic cell divisions in check until immune cells better suited for this task can be mobilized—that is, T cells and macrophages. In *in vitro* experiments certain viruses have been shown to quickly become lethal if NK cells are removed. The major lymphokines produced by these NK cells are tumor necrosis factor (TNF) and interferon-$\gamma$ (IFN-$\gamma$).

Finally, we should note that there are a small number of null cells in the circulation that are simply lymphoid stem cells or very immature B or T cells that have not yet expressed external markers.

## LYMPHOID TISSUES AND ORGANS

Lymphoid glands and tissues are part of the lymphatic system, which is primarily a draining system for extracellular fluids. The **lymphatic** or **immune system** is composed of numerous small and large ducts and lymph glands that filter these fluids (figure 20.2). Most lymph glands consist of well-defined peripheral lymph nodes. We should realize, however, that there are also many diffuse lymphoid patches scattered throughout the body, in addition to some larger lymphoid organs such as the thymus and the spleen.

Customarily lymphoid tissues and organs are divided into two major categories: the **primary** (or **central**) and the **secondary** (or **peripheral**) **lymphoid organs.** The primary lymphoid organs are those from which all lymphocytes are ultimately derived—that is, the **bone marrow** and the **thymus gland.** All other lymphoid organs are classified as secondary lymphoid glands and include the lymph nodes, the spleen, and other lymphoid tissues.

Each of these lymphoid tissues and organs is discussed in the following material, except for the bone marrow, which is described in chapter 3.

### Thymus Gland

The thymus gland is the first lymphoid organ to manufacture lymphocytes during embryonic development. The thymus is derived from the ventral side of the third and fourth pair of pharyngeal pouches, which are outpocketings of the embryonic gut (figure 20.3). During the sixth week of embryonic development, the third and fourth pair of pharyngeal pouches begin to grow rapidly, forming saclike protrusions that form the beginnings of the thymus gland and the parathyroid glands. The ventral portions of these pairs of pouches sever their connection with the pharynx and move under the breastbone into the chest cavity, where they meet, fuse, and settle above the heart as a bilobed organ known as the thymus gland (figure 20.4). The dorsal portions of these saclike protrusions will develop into the parathyroid glands. At the end of the third month of gestation, the first primitive lymphocytes from the liver arrive in the thymus tissue and start to colonize it. The gland becomes gradually surrounded by a capsule of connective tissue, from which trabeculae form, dividing it into irregular-shaped lobes. Each lobe has a light-staining center (medulla) containing mainly epithelial (secretory) cells, and an outer densely-staining cortex composed of lymphoid cells (figure 20.5). Eventually the

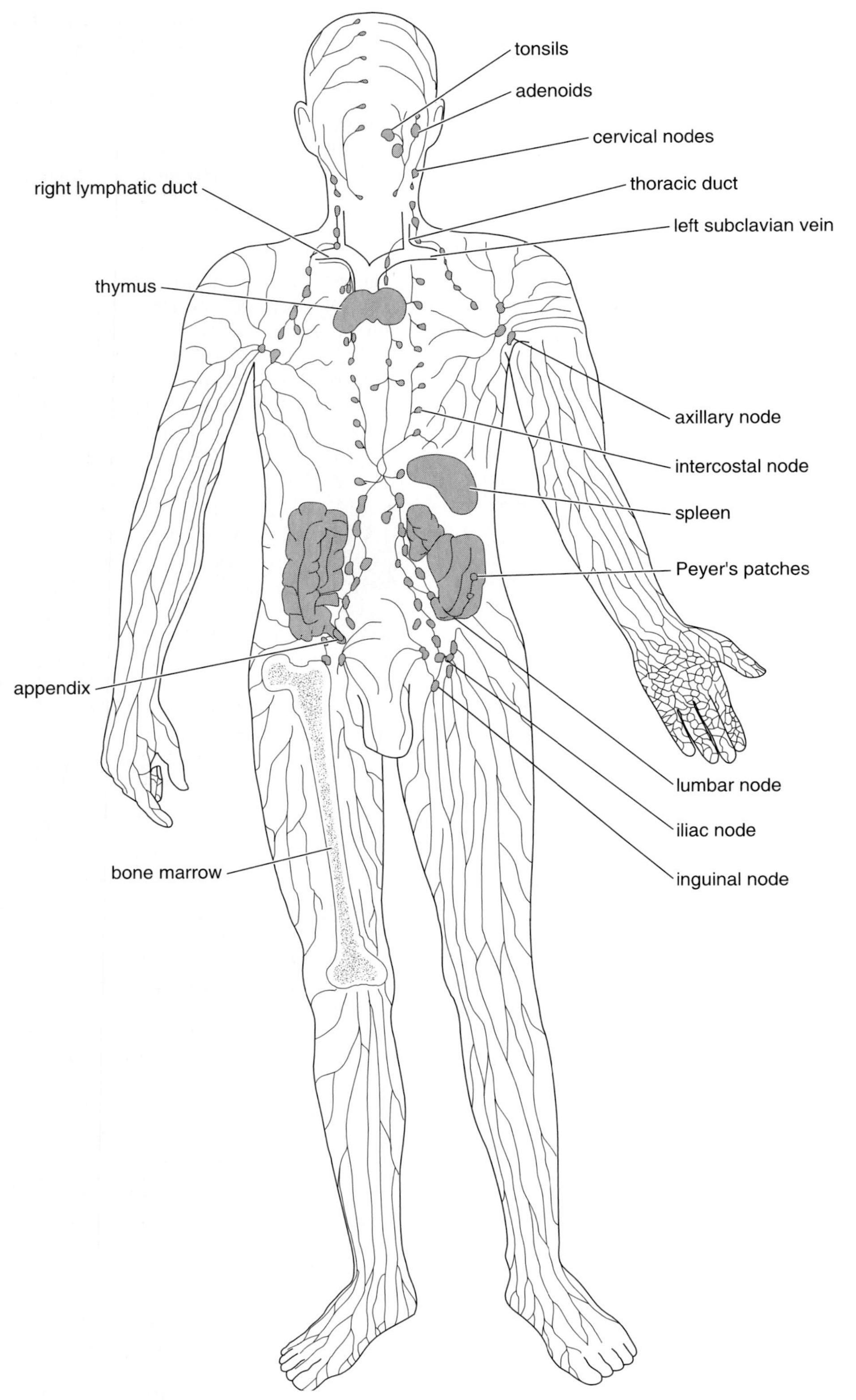


figure 20.2

The anatomic components of the lymphatic immune system.

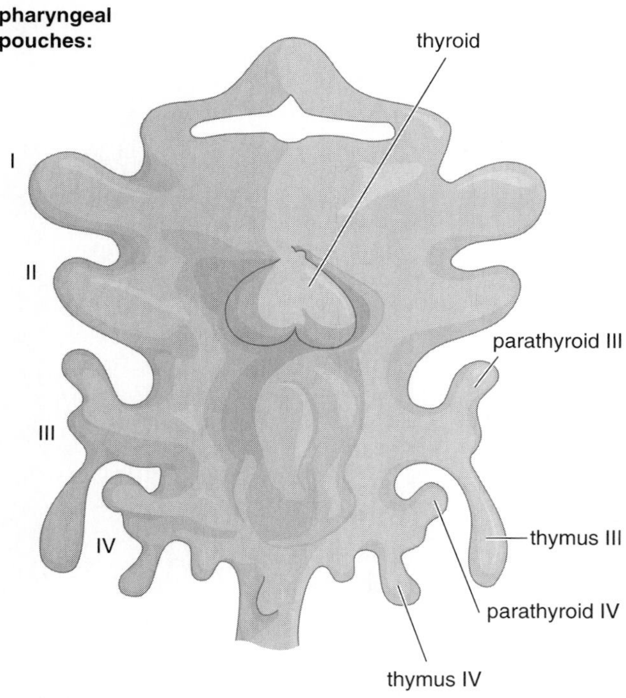


**figure 20.3**

A diagram of the embryological pharynx from which the thyroid, the parathyroids, and the thymus gland are derived. Note that the thymus is derived from the ventral portions of the third and fourth pairs of pharyngeal pouches.

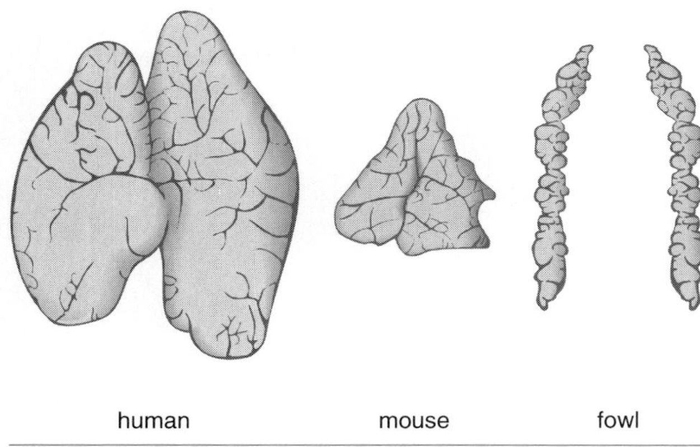


**figure 20.4**

The thymus glands of humans, mice, and fowl.



medullary cells start to secrete thymic hormones that function in the development and differentiation of T lymphocytes.

The thymus obtains its relative maximum size (i.e., as a percentage of total body weight) shortly after birth, when it weighs 10 to 15 g. The thymus continues to increase in size until it reaches an absolute maximum weight of 30 to 40 g at puberty. After adolescence the size of the gland begins to decline progressively until it eventually becomes less than 50% of its peak weight in later life. This gradual atrophy is associated with changes in its histology: the cortex especially becomes less prominent (figure 20.6).

Lymphocytes in the thymus are commonly referred to as **thymocytes.** More than 90% of all thymocytes are present in the cortical areas of the gland and appear as less developed lymphocytes. Most of the maturing thymocytes are found in the medullary area. We can easily distinguish between cortical and medullary thymocytes as they express different markers on their membranes. The mature T cells leave the thymus gland via the blood and lymph vessels in the medually areas to settle in the peripheral lymphoid tissues. The thymus gland has a higher rate of lymphocyte production than any other lymphoid tissue, at least in the early years of life, and many more thymocytes are produced than ever leave the gland. Remaining lymphocytes eventually die within the thymus. The reason for this phenomenon involves two processes. First there is a **positive selection** of thymocytes bearing receptors capable of binding to self-MHC molecules. Thus all immature thymocytes unable to interact with self-MHC molecules are programmed to die. Apparently the majority of thymocytes produced by the thymus belong to this category (figure 20.7). Second is a **negative selection** of thymocytes achieved by eliminating thymocytes bearing high-affinity receptors for self-antigens. This includes another large group of immature thymocytes. The only T cells allowed to mature and leave the thymus gland are both **self-MHC restricted**—that is, they are able to interact with self-MHC molecules—and are **self-tolerant**—that is, they will not become activated against self-antigens.

### Lymph Nodes

Lymph nodes are spherical or oval organs distributed throughout the body but appear to be especially numerous in certain discrete areas, such as the armpits (axillary lymph nodes), the groin area (inguinal lymph nodes), the neck (cervical lymph nodes), and the gut (mesenteric lymph nodes) (see figure 20.2). Lymph nodes vary greatly in size from a few millimeters in diameter to a width of a centimeter or more.

The histological appearance of all lymph nodes is rather similar. They are all covered with an outer **capsule** of connective tissue, and the tissue mass is divided into irregular lobes by **trabeculae** (or septa) of connective tissue (figure 20.8).

Blood vessels enter and leave the node in a depression known as the **hilus.** The lymph vessels enter the node at various places through the capsule **(afferent lymph vessels)** and leave the gland through a duct from the hilus **(efferent lymph vessels).** Just below the capsule is an open space lined by a network of reticular cells. The lymph percolates through these **subcapsular spaces** to the sinusoid vessels that permeate the entire lymph gland.

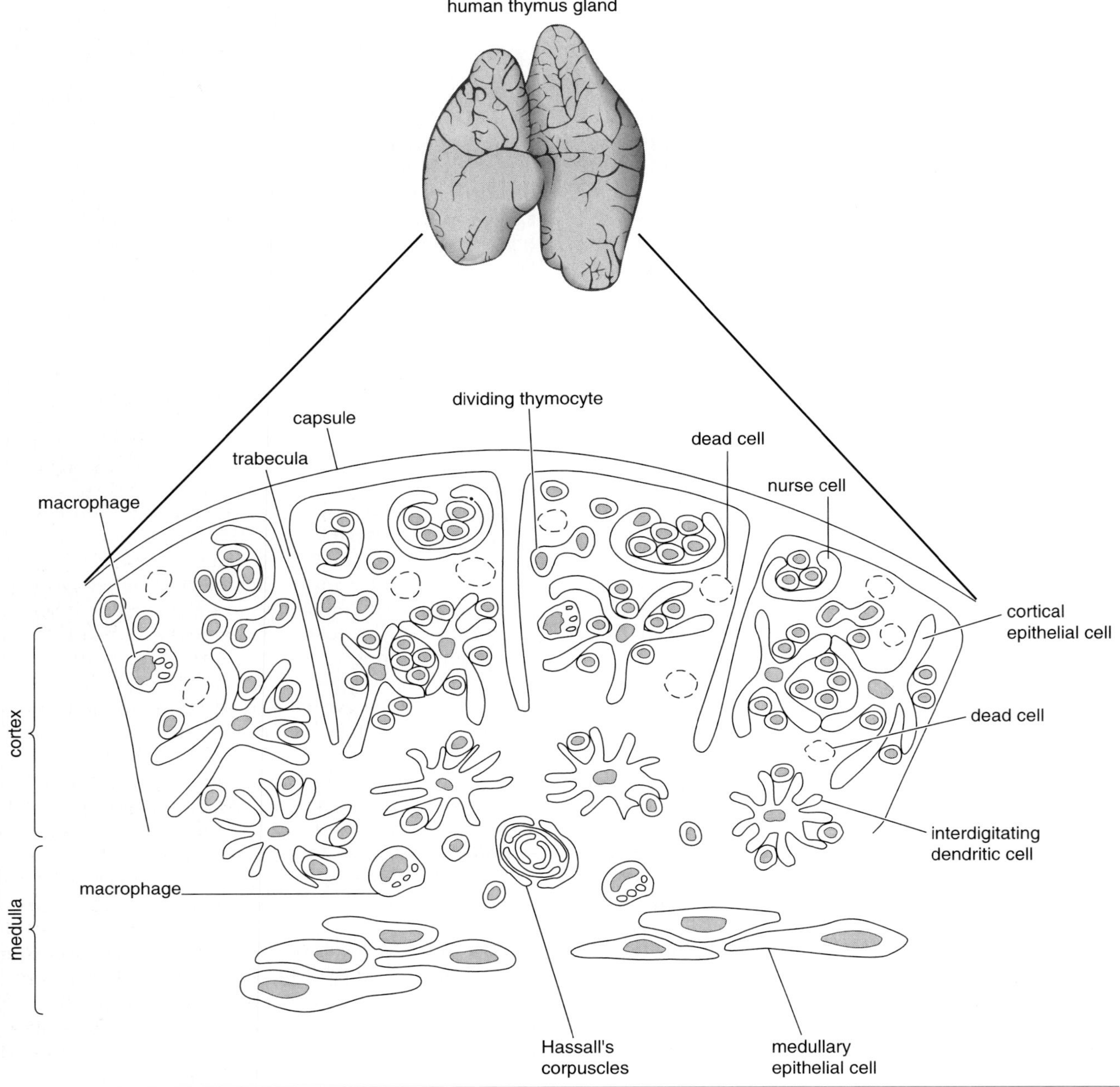


**figure 20.5**
Cross section through the thymus gland showing several lobules separated by connective tissue trabeculae.

The lymph node can be divided into the following three histological zones (figure 20.9):

1. **The outer cortex,** which contains many primary (i.e., nonactive) and secondary (i.e., active) follicles. This zone is the major source of B-cell production.
2. **The inner cortical area,** or **paracortical** zone, is the major area for T-cell production. This area also has primary and secondary follicles.
3. **The medulla,** or central area, is primarily composed of cords and sinuses. The cords separate the sinuses, and the spaces inside the cords are filled with many plasma cells, which are actively antibody-producing B cells. They secrete their antibodies directly into the sinuses, from which they are carried to all parts of the body.

The lymph nodes are dynamic structures constantly undergoing histological changes. We can easily distinguish between activated and nonactivated lymph nodes, as the former enlarge drastically due to increased production of lymphocytes as well as the accumulation of lymph fluid. Lymph nodes have many different functions. First, they are the principal residence of the lymphocytes. Second, they act as filters for nonself particles that escape into the bloodstream and body fluids. Such foreign agents are readily

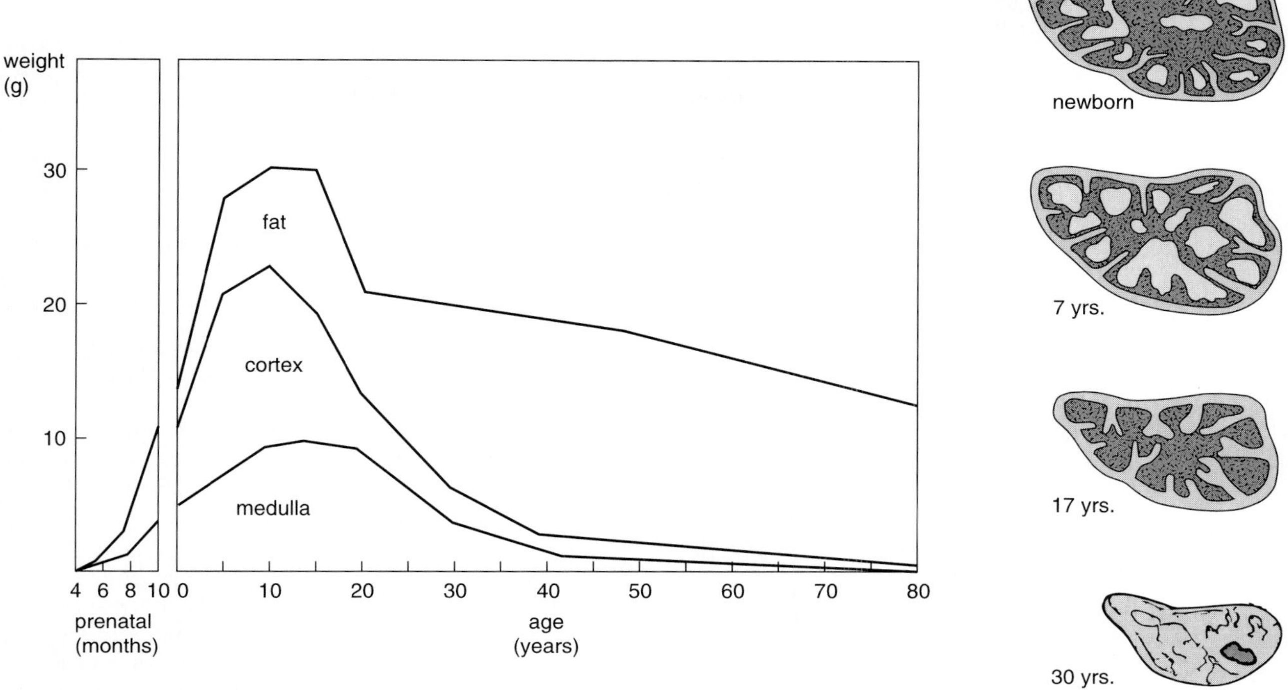


**figure 20.6**

The weight and content of the thymus gland during human life. Note that the lymphocytic tissue is gradually replaced by fat cells as the gland declines in size.

trapped in the lymph nodes, thus making it easier for macrophages to phagocytose them. Third, lymph nodes are the primary source for new lymphocytes whenever they are needed. Also, activated B cells will produce most of their antibodies in the lymph nodes.

Lymph nodes originate from the mesenchymal tissues found between developing blood vessels in the early embryo. The mesenchyme condenses to form roundish structures that become infiltrated with lymphocytes. Lymph vessels connect with these structures and form a sac surrounding these bulbous structures. Later, other mesenchymal cells that have developed into fibroblasts begin to form a capsule around the structure and form trabeculae on the inside, thus dividing the structure into lobes.

## Spleen

The spleen is a most interesting organ with a wide variety of functions. Located just below the diaphragm on the left side of the stomach, the spleen is held in place by various peritoneal folds (see figure 20.2). It is an oblong structure that is variable in size. Just like the thymus and the lymph nodes, the spleen is covered by a capsule of connective tissue from which trabeculae extend into the mass, dividing it into lobular areas. Both the capsule and the trabeculae contain a few smooth muscle fibers but not in a quantity sufficient to make the organ contractile. The spaces between the trabeculae are divided into three histologically distinct areas: the red pulp, the white pulp, and the marginal zones.

The **red pulp** gives the organ its purplish appearance and consists mainly of cords and sinuses, which are vascular spaces lined with mononuclear phagocytic cells. The cords may be considered as bands of reticular cells that separate the sinuses. Many arterioles terminate in these cord vessels. The sinuses are lined with endothelial cells and may be considered as venous vessels, in which the blood is collected after it moves through the cords. All blood cells must move between these mononuclear phagocytic cells from the cords to the sinuses (figure 20.10). During this process the old and imperfect cells are culled from the blood. They are phagocytosed by the phagocytic cells, which may be thought of as fixed macrophages. However not all imperfect cells are immediately destroyed. Many are only **pitted** at first—that is, parts of the blood cells containing such particles as Heinz bodies, Howell-Jolly bodies, and sideroblastic granules are removed, leaving a smaller and somewhat deformed cell. Apart from the culling and pitting functions of the red pulp, it also functions as a storage area for many blood cells and platelets. The red pulp of the spleen is also potentially a hematopoietic organ, as it retains a residual capacity for hematopoiesis in case of bone marrow failure.

The **white pulp** of the spleen contains mainly lymphoid tissue and is made up of scattered follicles with germinal centers. The white pulp is especially evident around many of the numerous smaller arteries of the spleen. These areas are known as **periarterial sheaths** and may be considered as sleeves of loose

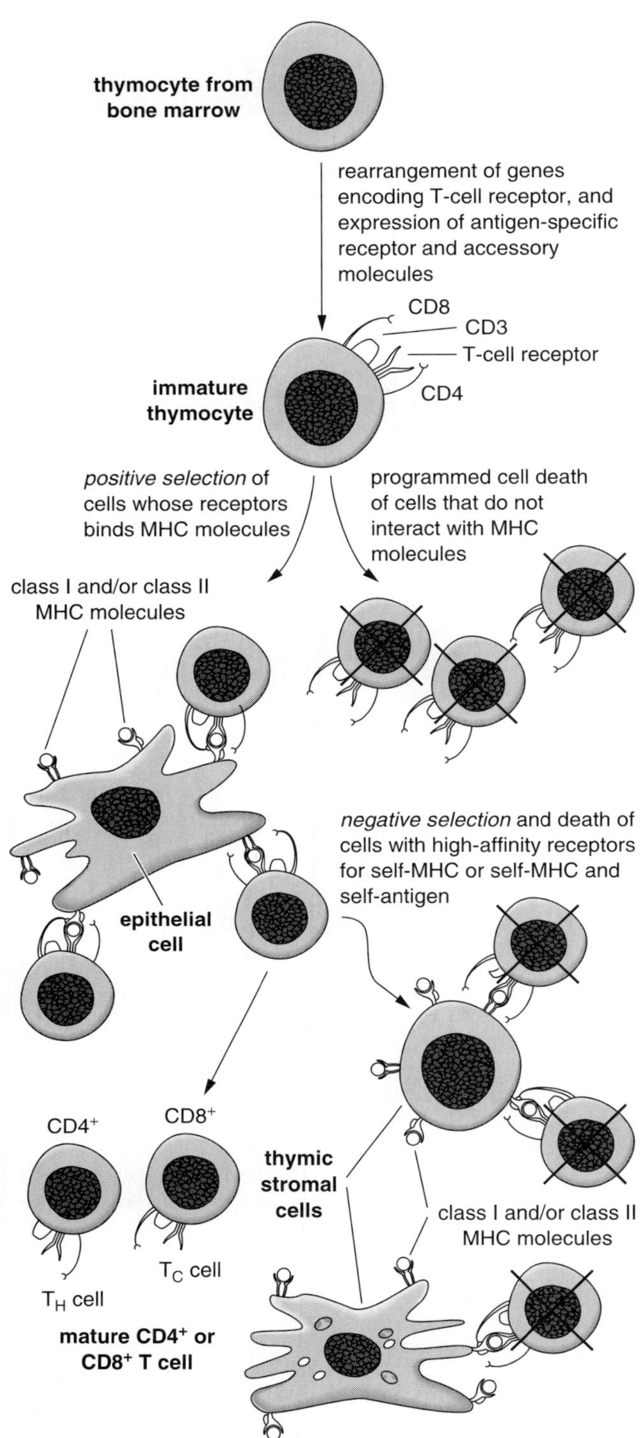


**figure 20.7**

Negative and positive selection of lymphocytes in the thymus gland.

reticular connective tissue packed with lymphocytes. Lymphocytes in the periarterial sheaths are mainly T cells, and lymphocytes in the peripheral **germinal follicles** are mainly B cells. The spleen also functions as a major filtering organ for foreign agents, especially early in life. The many macrophages in the spleen are always active in removing these nonself particles. The function of the white pulp is mainly immunological—that is, it plays a major role in the defense of the body. Finally, the spleen contains **marginal zones,** which are poorly defined regions between the red and white pulp and are made up of a network of reticular cells containing a variety of lymphocytes including B and T cells (figure 20.11).

### Other Lymphoid Tissues

The other lymphoid tissues may be grouped into the following three categories:

**1. Diffuse lymphoid tissues**

This category includes the lymphoid cells that are scattered along the lamina propria of the digestive and respiratory tracts. The lamina propria is the layer of connective tissue found immediately below the basement membrane. Sometimes these lymphoid cells appear clustered, but they are never organized in any recognizable pattern (figure 20.12).

**2. Solitary lymphoid follicles**

These are distinct clusters (or nodules) of lymphoid cells. They cannot be classified as glands, since they are not covered by a distinctive connective tissue layer (capsule). Moreover, they are not fixed entities and may come and go depending on the condition of the tissues in which they are found. Solitary lymphoid follicles are particularly common in the digestive and respiratory tracts of children (figure 20.13).

**3. Aggregated lymphoid follicles**

This category includes the **tonsils,** which are found in the back of the mouth and in the pharyngeal areas (figure 20.14); **Peyer's patches,** which are groups of domelike structures found between the villi of the small intestines (figure 20.15); and the **vermiform appendix,** which contains large numbers of aggregated follicles on its inside surface (figure 20.16).

Lymphoid follicles associated with the respiratory and intestinal tracts are commonly referred to as MALT, which is an abbreviation for mucosa-associated lymphoid tissue and lymphoid tissues associated with the intestinal tract only are often referred to as GALT (or gut-associated lymphoid tissue).

## B-LYMPHOCYTE DEVELOPMENT

From the functional point of view, the developmental series of the B cells includes at least five different cell types. The first cell in the B lymphocyte differentiation pathway is the **pre-B cell.** This is the earliest cell that can be recognized as belonging to the B-cell lineage. Pre-B cells are characterized by the presence of **cytoplasmic immunoglobulin (cIg),** but they lack distinct surface markers.

The second cell in this series is the **virgin B cell.** These cells not only express cytoplasmic IgM but also **membrane immunoglobulins (mIgs)**—that is, they have IgM antibody molecules on their cell surfaces.

The third cell is the **activated** or **stimulated B cell.** These cells not only possess cIg and mIg but also secrete immunoglobulin into the body fluids. These cells can produce different kinds of **secretory immunoglobulin (sIg),** depending on the type of stimulation or

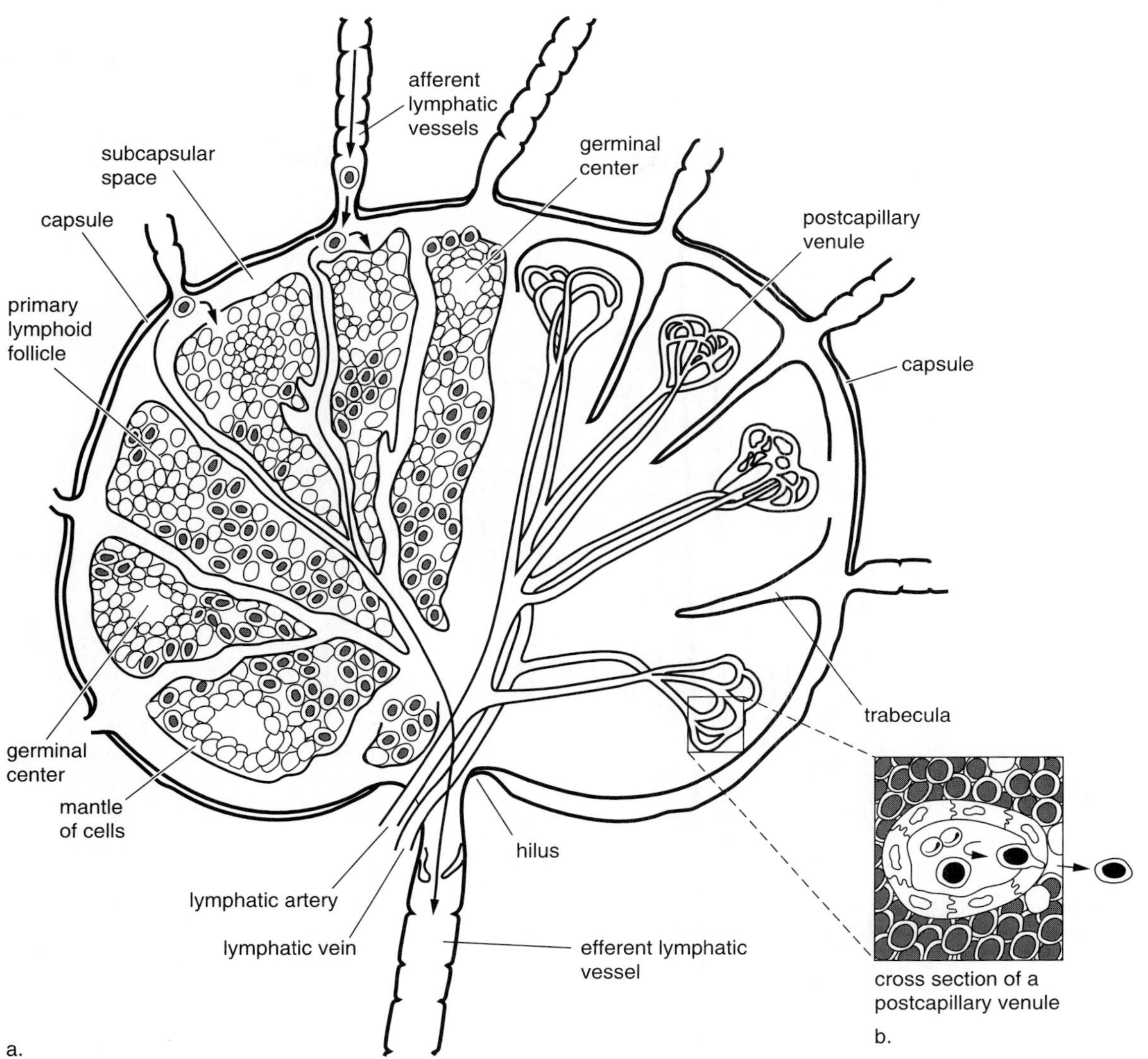


**figure 20.8**

(*a*) Cross section through a lymph gland. On the left is the arrangement of the various cells of the lymph gland. On the right is the division of the gland by trabeculae into a number of lobules, and the microcirculation of each lobule. Lymphocytes and macrophages (APCs) can enter the lymph node via the afferent lymphatic vessels and the lymphatic arteries (*b*), and they can leave the gland via the efferent lymphatic vessels and the lymphatic veins.

where they are found in the body. All activated B cells start by secreting IgM immunoglobulins. Later in the process of differentiation, they may switch to producing either IgG, IgA, or IgE. However an interim production of IgD may be needed for this switch to be effected (figure 20.17). For a B cell to reach this third stage of activation, it must first be stimulated by a foreign antigenic determinant (epitope) presented by an APC, such as a macrophage, helper T cell, or even another B cell.

The next cell is the **plasma cell** or **plasmacyte,** which is the mature antibody-secreting B cell found mainly in the cords of the lymph nodes and in activated lymphoid follicles of other lymphoid tissues. This cell has a unique shape and is larger than the normal lymphocyte. Plasma cells contain extensive endoplasmic reticulum with numerous ribosomes and an active Golgi apparatus—all indications of an actively secreting cell (figure 20.18). This plasma cell is a finite cell and is fairly short-lived; estimates are that an active plasma cell lives 10 to 14 days. If all cells of a given B-cell clone were stimulated and activated by a given epitope and all developed into plasmacytes, then this clone would be lost at the end of the infection. However this does not happen because a low percentage (about 5%) of the cells of an activated clone will develop into a fifth type of cell: the **memory cell.** These memory cells are produced by cell division from activated cell clones but do not differentiate into plasma cells; they remain as small lymphocytes. Memory cells form a reserve of epitope-sensitive clones to be called upon on subsequent exposure to an antigen. Thus each time the clone is stimulated by an antigenic determinant, the pool of memory cells will grow larger. This will continue until the number of memory cells is so great

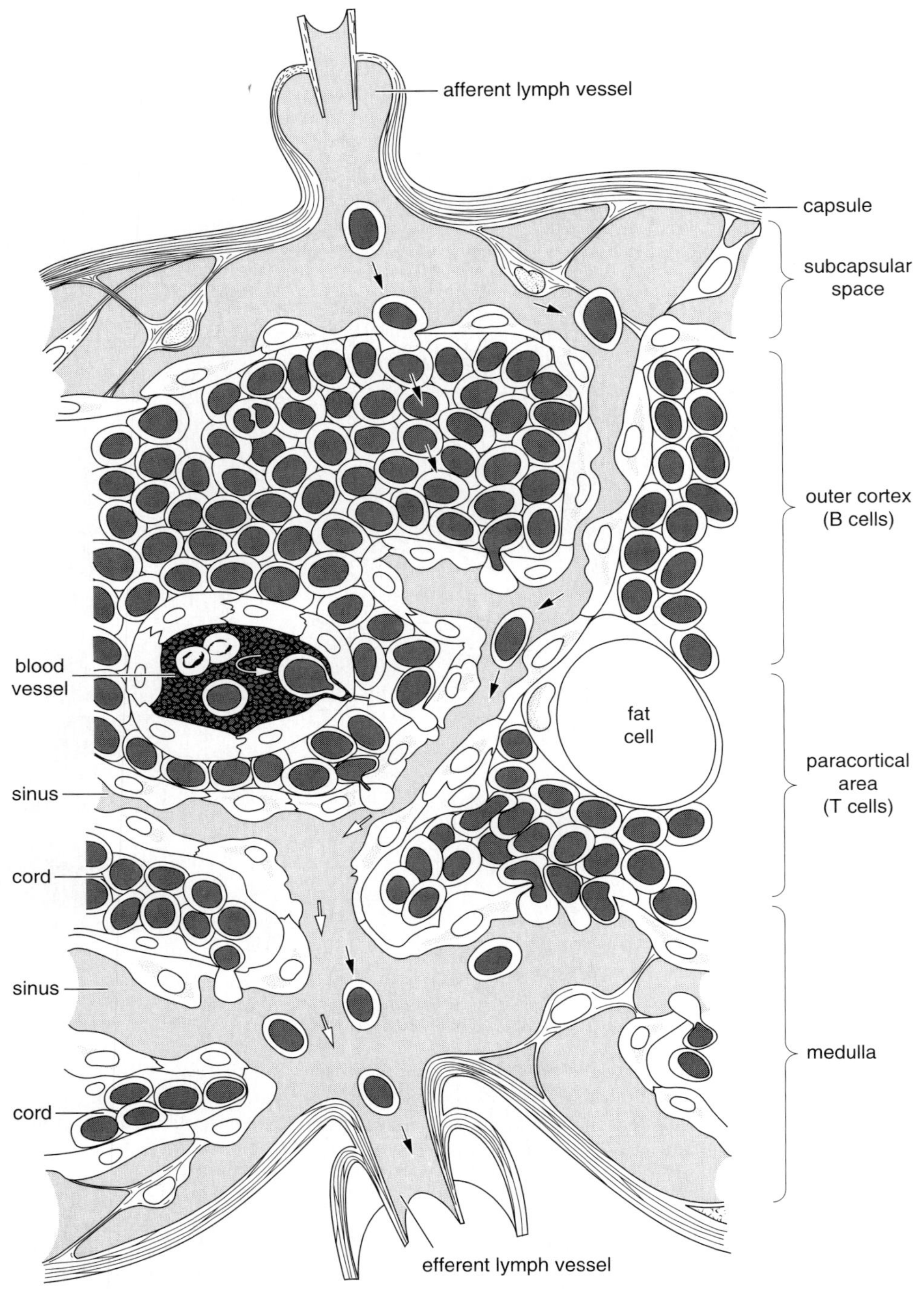


**figure 20.9**

Regions of a lymph node. Note the subcapsular space, the cords, the sinuses of the cords, and the sinuses of the medulla.

that the body will react immediately when stimulated again. When that happens, the body has achieved immunity against that particular antigenic determinant.

We can also look at the activation of B cells from the morphological point of view. As noted before many B-cell clones are produced during late prenatal life and the childhood years. They are stored in the outer cortex of the lymph nodes and in the germinal centers of the spleen and lymphoid follicles. At that stage they may be referred to as virgin B cells. When these B cells are presented by an APC with an epitope with which they can interact, they become activated and begin to enlarge. Such enlarged cells are frequently referred to as **plasmablasts.** They are similar to other early blast cells in that they have a large nucleus with many nucleoli. These activated B cells (blast cells) undergo several

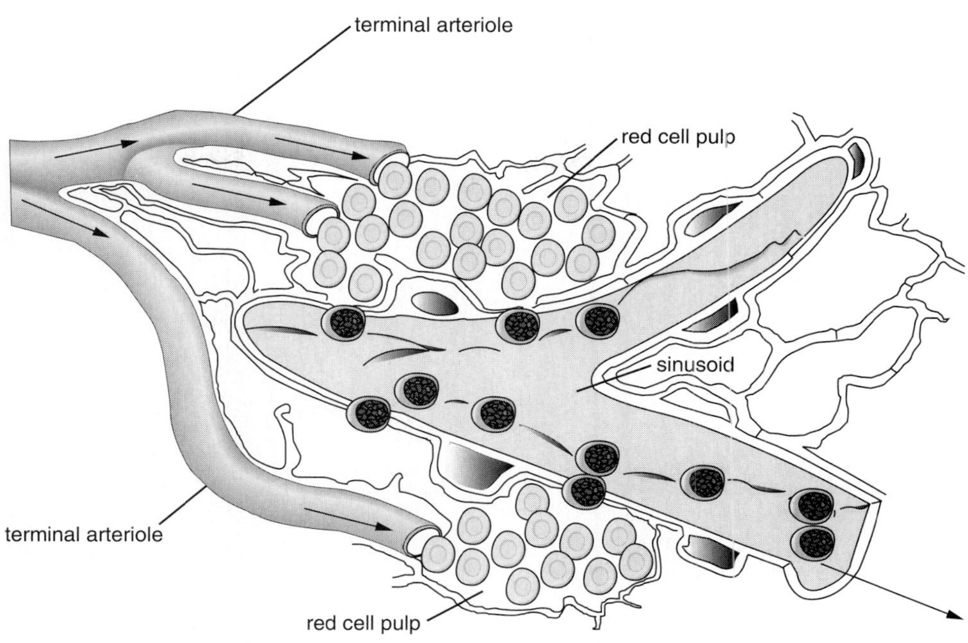


**figure 20.10**

The structure of the red pulp of the spleen. Note the terminal arteries and how the blood cells percolate through the cords to the venules. The reticular monophagocytic cells will remove senescent and abnormal cells, and will pit rigid granules from red blood cells.

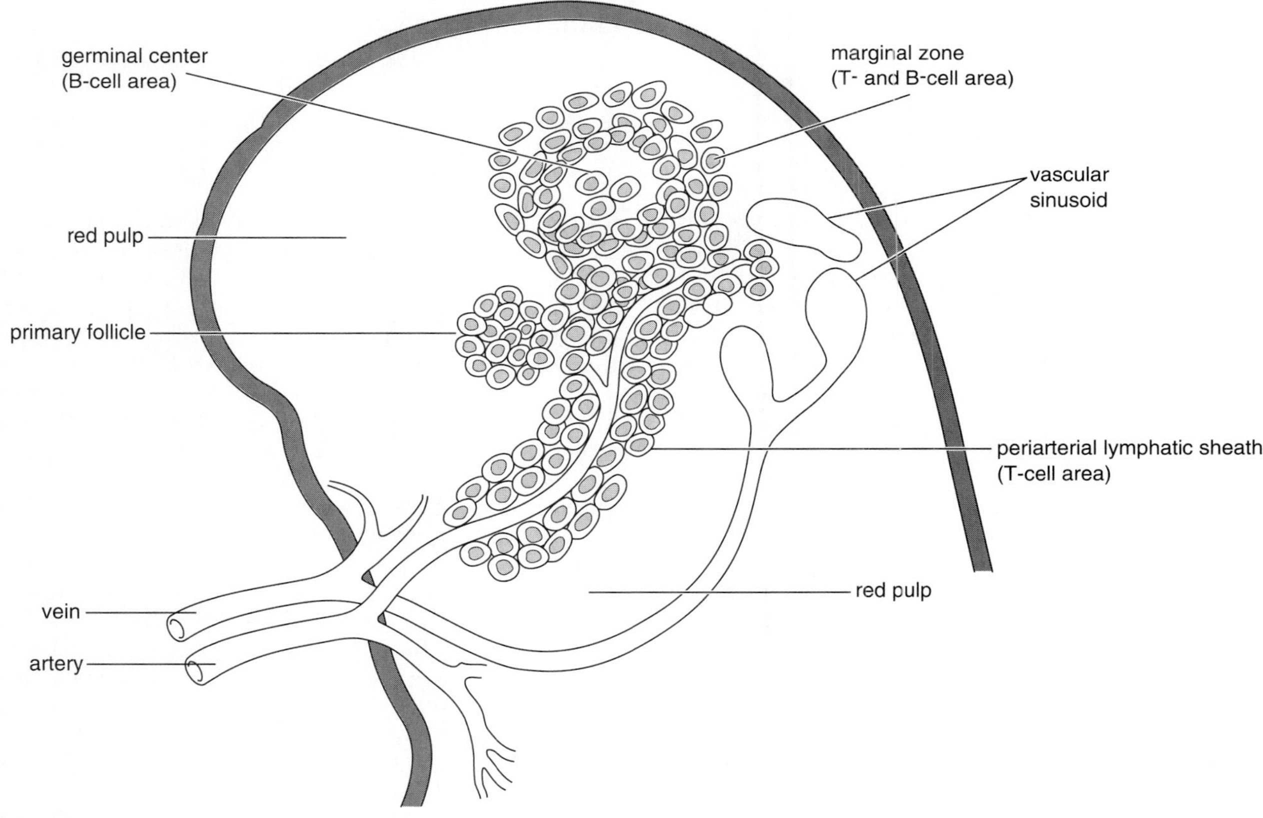


**figure 20.11**

A schematic diagram of the spleen showing the periarterial sheath (T-cell domain) and germinal centers (B-cell domain). These areas make up the white pulp of the spleen.

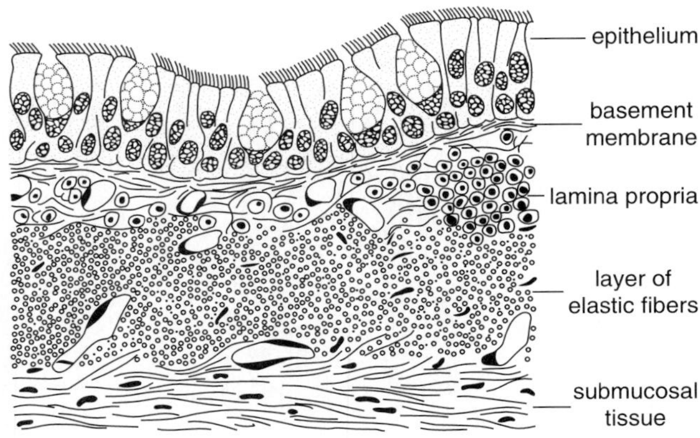


**figure 20.12**

Diffuse lymphoid tissue in the lamina propria of the mucous membrane of the respiratory system (trachea).

From O. F. Kampmeier, *Evolution and Comparative Morphology of the Lymphatic System*, 1969. Courtesy of Charles C Thomas, Publisher, Springfield, Illinois.

cell divisions and become either **proplasmacytes** or memory cells. The proplasmacytes undergo more divisions and maturation stages and become finite **plasmacytes.** The memory cells, like plasmacytes, do not have a characteristic morphology, but resemble other small lymphocytes (figure 20.19).

## T-LYMPHOCYTE DEVELOPMENT

As mentioned earlier, all T cells are ultimately derived from the bone marrow (and in the early fetus also from the embryonic liver and spleen). A number of pluripotent stem cells produced by these tissues become committed to follow the lymphoid line of development. These lymphoid stem cells may colonize the thymus gland, which predestines them to become T cells. Progenitor cells are attracted to the thymus by a chemotactic factor secreted by thymic epithelial cells. Most of the migration and colonization takes place during fetal life and early childhood. After adolescence the number of committed lymphoid stem cells entering the thymus gland declines with age.

In the cortical areas of the thymus gland, an active proliferation of lymphocytes takes place. As explained previously, many of these newly produced cells die in the thymus; some scientists have suggested this may involve as many as 90% of all thymocytes. Only thymocytes that survive the positive selection—that is, those that produce receptors that can bind to self-MHC molecules—and negative selection—that is, those that will not interact with self-antigens—will survive and move from the cortex to the medulla of the many thymic lobules. In the medulla they are affected by a number of thymic hormones that direct further development and differentiation. In contrast with B cells the development of T cells is not a single pathway but involves differentiation into several different functional T-cell lines.

Once a lymphoid precursor cell enters the thymus gland, it will express a unique membrane molecule known as **Thy-1.** These early thymocytes still lack other membrane molecules that are unique for T cells, such as T-cell receptors (TCRs) and the CD4 and CD8 markers. Double negative **$TCR^-$** and **$CD^{4-8-}$** thymocytes differentiate along one of two developmental pathways. A small group begins to develop a special type of TCR known as the γδ TCR but will not express any CD4 or CD8 molecules. They remain $CD^{4-8-}$. This small population of **$CD^{4-8-}$ γδ $TCR^+$ cells** now leave the thymus gland and move to the epithelial tissues of the skin, the digestive tract, and the respiratory tract. They never constitute more than 1% of the total T-cell population, and their role is still somewhat obscure. Possibly these γδ TCR cells represent the earliest and most primitive cell-mediated immune system, uniquely specialized to recognize changes in epithelial cells that have become infected by external pathogenic organisms. Such a system would not only protect the epithelial cells but also prevent the spread of infection into the interior milieu of the body.

The great majority of double-negative thymocytes develop along different pathways. One group expresses CD4 and CD8 membrane proteins but remains $TCR^-$. They are labeled **$CD^{4+8+}$ $TCR^-$ cells.** This step is followed by the expression of the normal αβ TCR in addition to the CD4 and CD8 receptors. These cells are labeled **$CD^{4+8+}$ $TCR^+$ cells.** These cells can develop

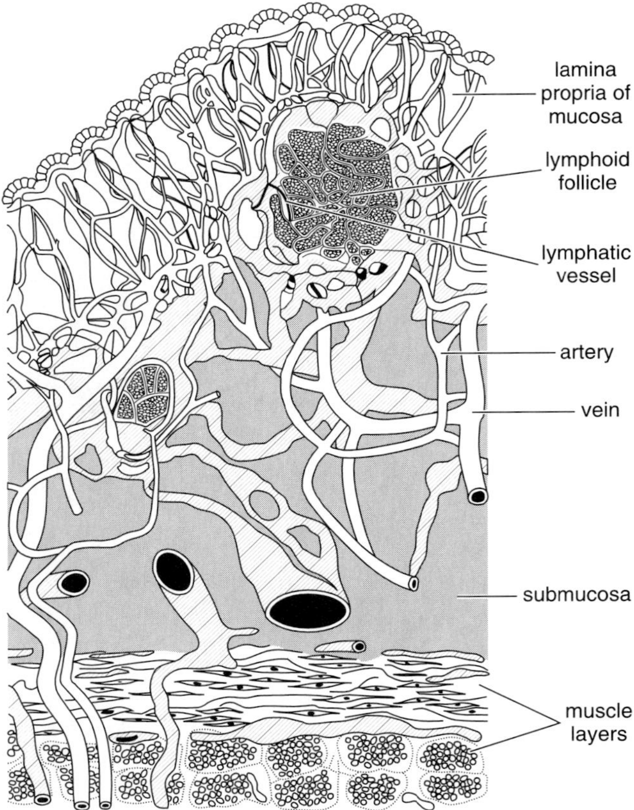


**figure 20.13**

A solitary lymphoid follicle in the wall of the colon of an infant.

From O. F. Kampmeier, *Evolution and Comparative Morphology of the Lymphatic System*, 1969. Courtesy of Charles C Thomas, Publisher, Springfield, Illinois.

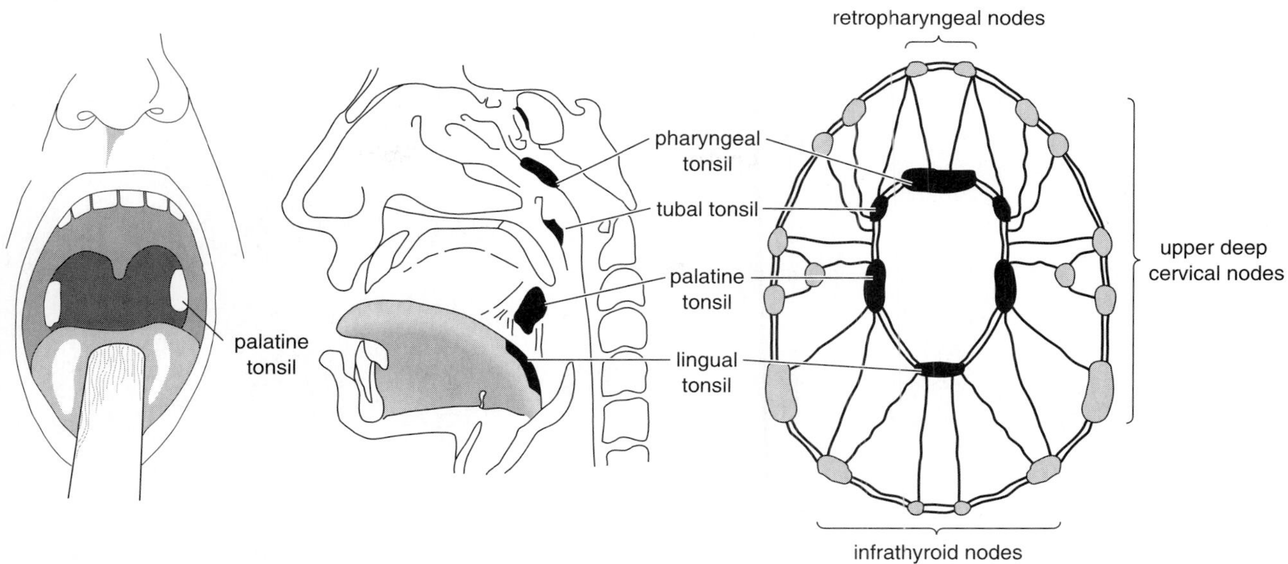


**figure 20.14**

The positions of the human tonsils.

From Jan Klein, *Immunology: The Science of Self-Nonself Discrimination.* Copyright © 1984 John Wiley & Sons, Inc., New York, NY. Reprinted by permission of John Wiley & Sons, Inc.

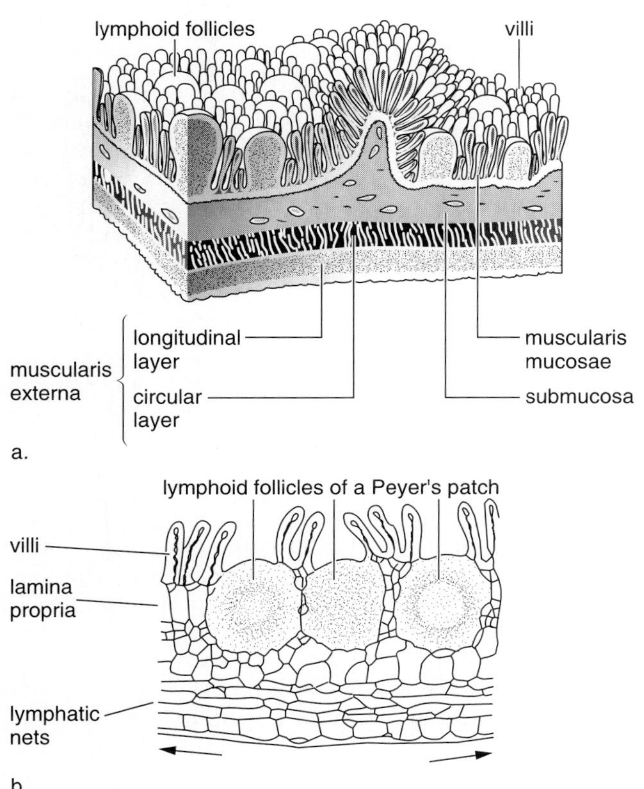


**figure 20.15**

Groups of lymphoid follicles known as Peyer's patches between the villi of the small intestine.

(*a*) From W. M. Copenhaver, R. B. Bunge, and M. B. Bunge, *Bailey's Textbook of Histology,* 1971, Williams and Wilkins Publishers, Baltimore, MD. (*b*) From O. F. Kampmeier, *Evolution and Comparative Morphology of the Lymphatic System,* 1969. Courtesy of Charles C Thomas, Publisher, Springfield, Illinois.

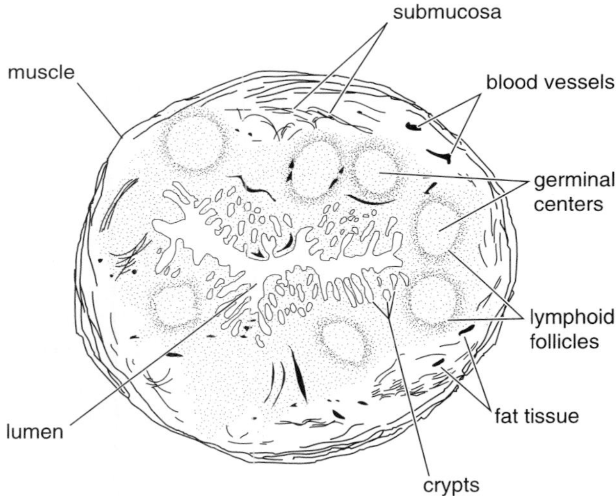


**figure 20.16**

Cross section of the appendix. Note lymphoid tissue and lymphoid follicles.

Source: Based on J. Sobatta: *Atlas und Lehrbuch der Histologie und Mikroskopischen Anatomie des Menschen,* (Lehmanns Verlag, Munchen, 1911) as appeared in Jan Klein, *Immunology: The Science of Self-Nonself Discrimination.* Copyright © 1984 John Wiley & Sons, Inc., New York, NY.

in one of three directions depending on thymic hormones and other chemical factors that act on a given cell (figure 20.20). One large group will develop into **$CD^{4+8-}$ αβ $TCR^{+}$ cells,** which will make up the majority of helper T lymphocytes (HTLs or $T_H$ cells). Another group develops into **$CD^{4+8+}$ αβ $TCR^{+}$ cells,** which will form the bulk of cytotoxic T lymphocytes (CTLs or $T_c$ cells) and suppressor T cells ($T_s$ cells). $T_c$ cells can be differentiated from $T_s$ cells by the expression of the

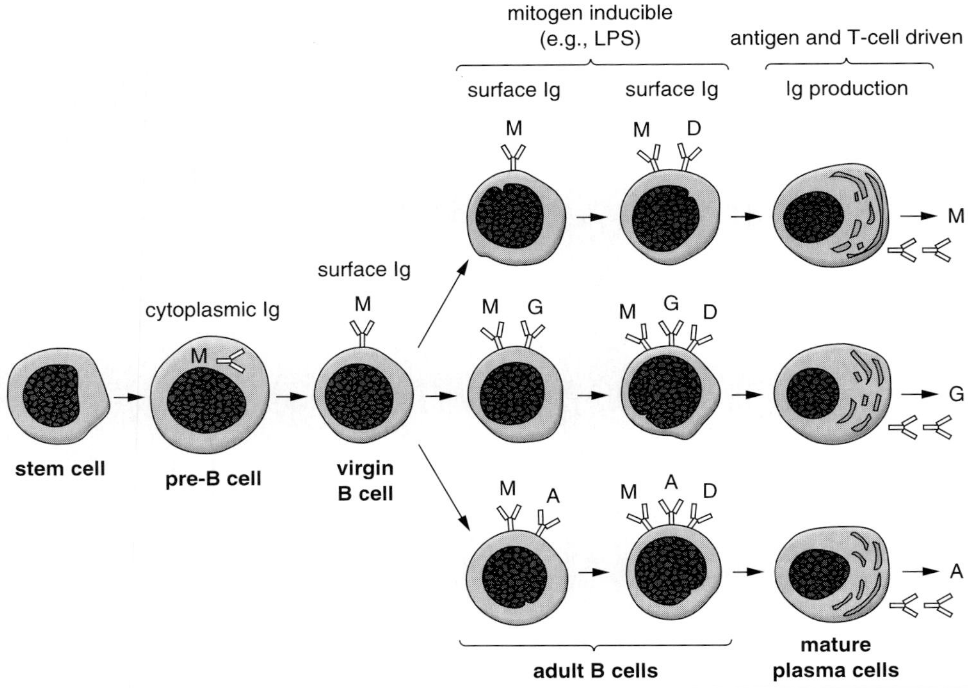


**figure 20.17**
The development of B lymphocytes.

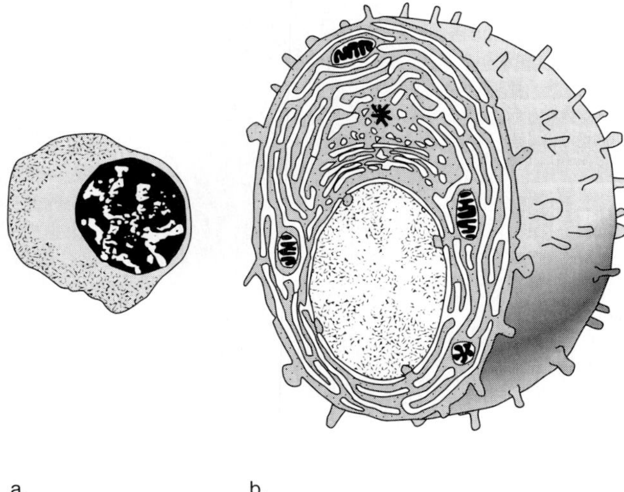


**figure 20.18**
The structure of a plasma cell (plasmacyte) as seen through (*a*) a light microscope; and (*b*) an electron microscope. Note the "cartwheel" appearance of the nucleus and the cytoplasmic structures indicative of an actively secreting cell.

From Jan Klein, *Immunology: The Science of Self-Nonself Discrimination.* 

CD11b receptors found on $T_s$ cell membranes. A third group will lose their CD4 and CD8 markers but will remain $TCR^+$. This group of **$CD^{4-8-}$ αβ $TCR^+$ cells** are known as null cells.

Once these cells have matured in the thymus gland, they migrate to the peripheral lymphoid organs. However they do not remain in one place forever, and neither do the B cells that have colonized the lymphoid tissue. Many lymphocytes are exchanged from one peripheral lymphoid gland to another. They may travel from the spleen to the lymph nodes and vice versa; they may travel from the GALT to the bloodstream and from there to other tissues and vice versa (figure 20.21).

In the peripheral lymphoid tissues, T cells wait until they are activated by an APC, just as B cells. When they become activated, the first step in development is also the formation of a large blast cell commonly known as **lymphoblast.** Just as plasmablasts and other activated hematopoietic cells, this lymphoblast can be recognized by its large nucleus containing many nucleoli (figure 20.22). This blast cell undergoes a number of cell divisions and becomes a **prolymphocyte,** which also has a large nucleus without nucleoli and has a considerable amount of cytoplasm. This stage is subject to cell division but once cell division stops, the prolymphocyte becomes an **immature lymphocyte** that is released into the bloodstream. This immature lymphocyte has a typical condensed round or oval nucleus and a large amount of cytoplasm. These cells then become **mature T cells,** with a small amount of cytoplasm surrounding the nucleus (figure 20.23). These are the active effector and regulator cells that we discussed previously. Apparently they are also finite cells, since memory cells for various clones of T cells have been identified (figure 20.24).

Most lymphocytes lead a rather monotonous life. They remain in a lymphoid tissue for some time, move

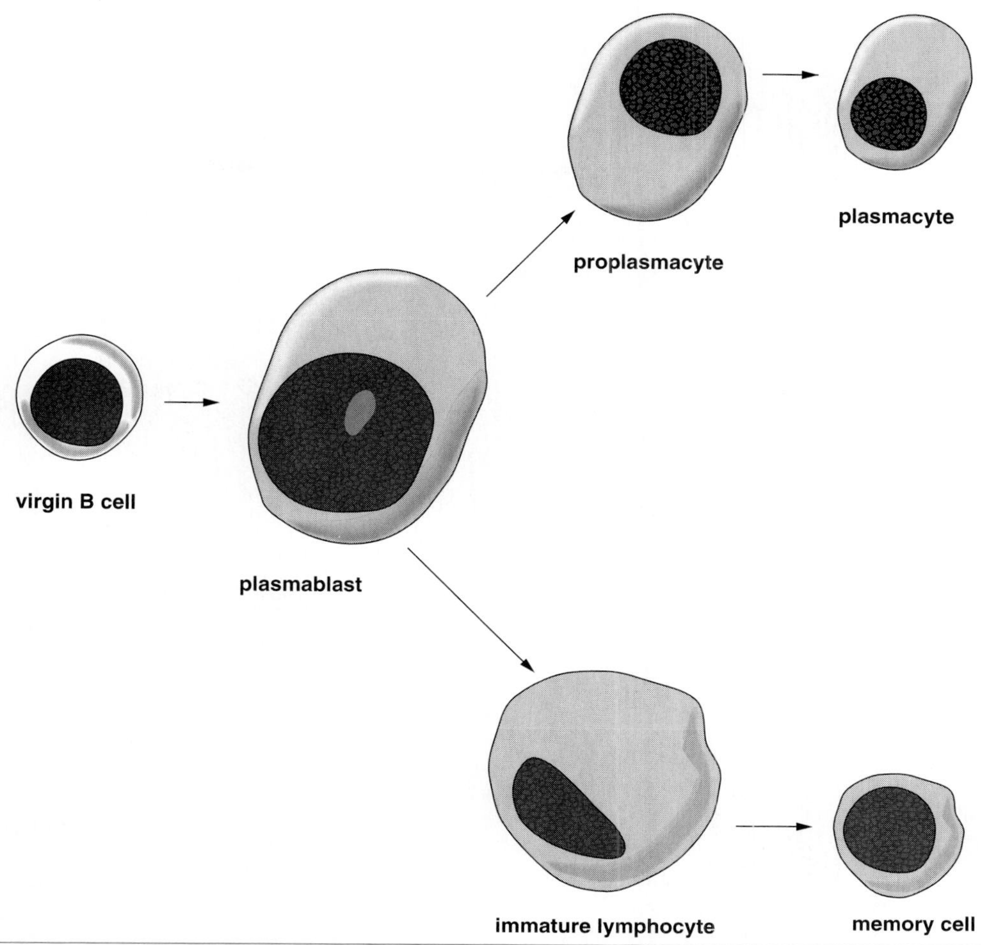


**figure 20.19**

B-cell activation.

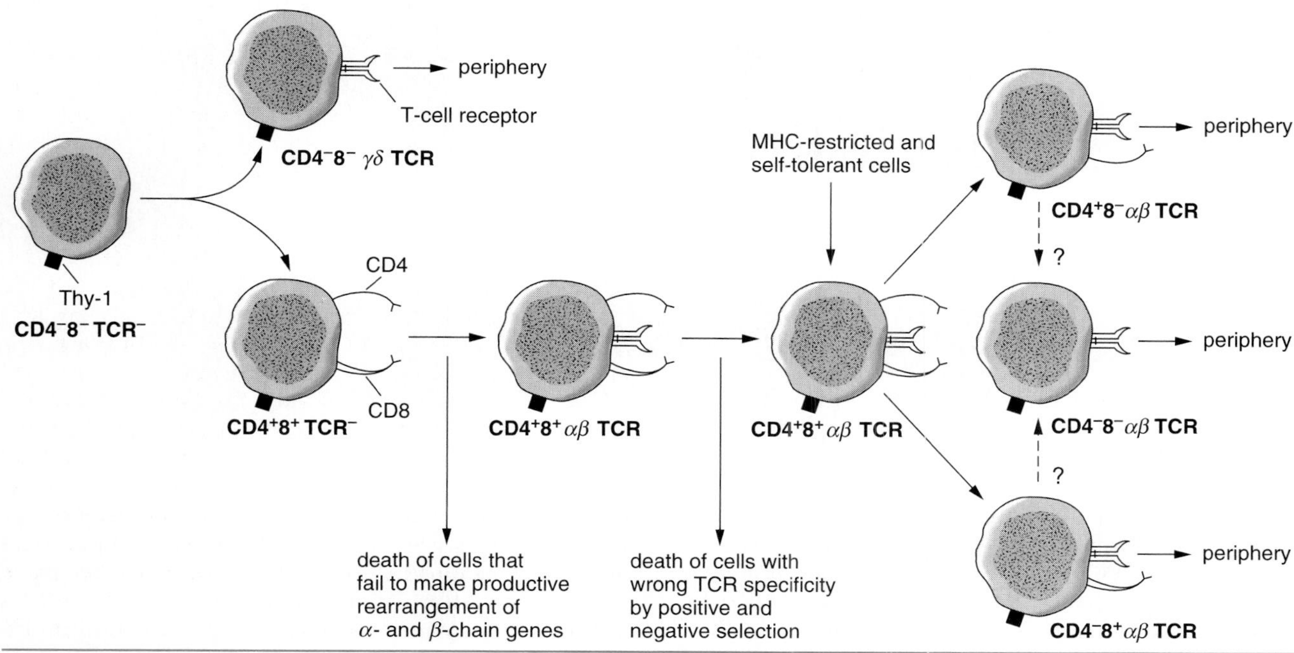


**figure 20.20**

The proposed pathways of T-cell development in the thymus gland.

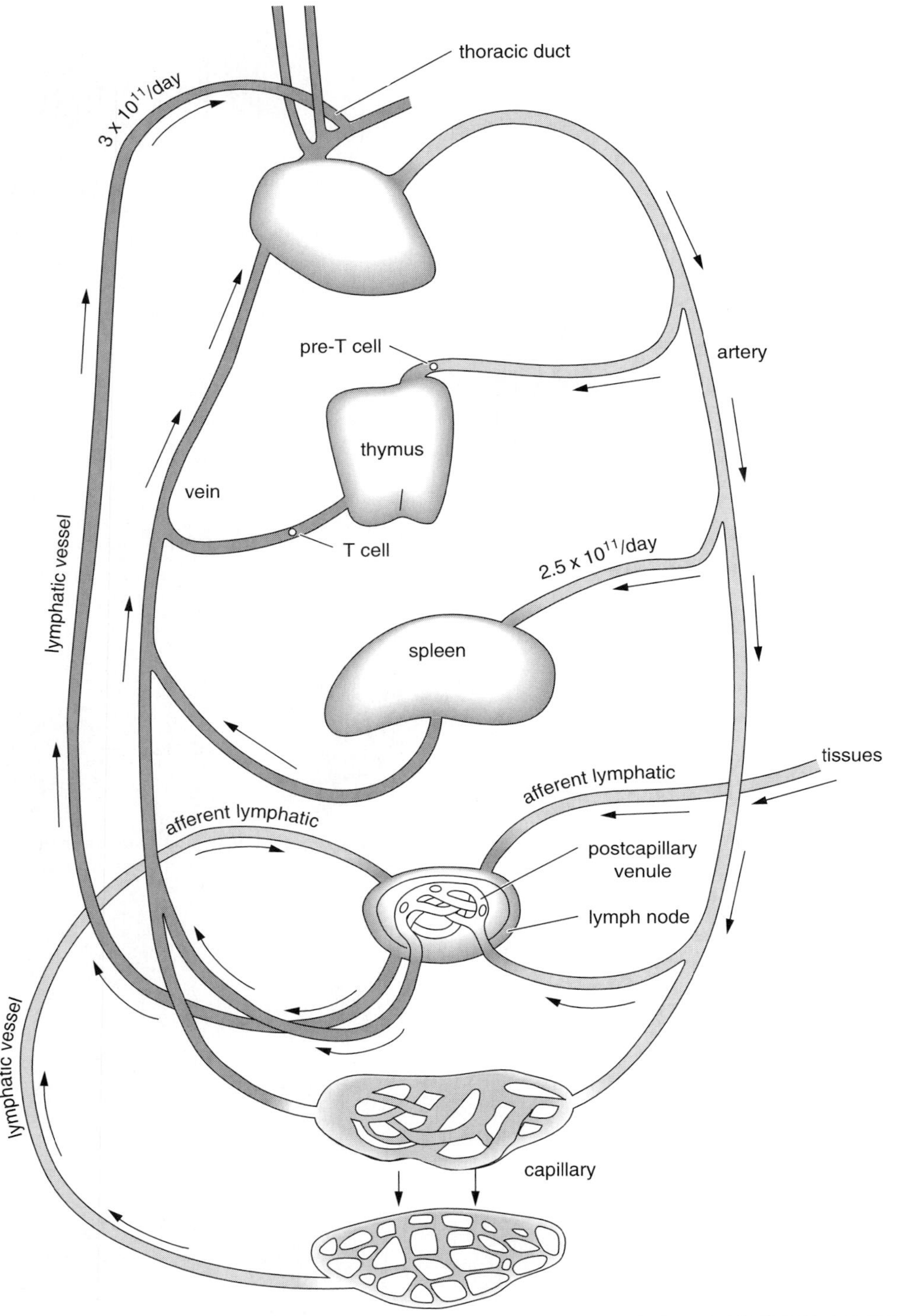


**figure 20.21**

Lymphocyte movement through the blood and lymph vessels and organs.

around a bit in the blood, lymph, or tissues, and return to the lymph glands again. Only when a lymphocyte gets stimulated by an antigenic determinant, or is stimulated otherwise by lymphokines, will it become active. This active phase is characterized by transformation (i.e., blast formation) and proliferation (i.e., cell division), and the production of regulator and effector cells, as well as memory cells. When the encounter takes place in the nonlymphoid tissues, the activated lymphocyte moves toward the nearest lymph node. When the encounter takes place in the blood, the activated lymphocyte usually settles in the spleen. When activated in a given lymphoid tissue, lymphocytes remain and proliferate, at least initially. Later the activated progeny may move to other lymphoid organs.

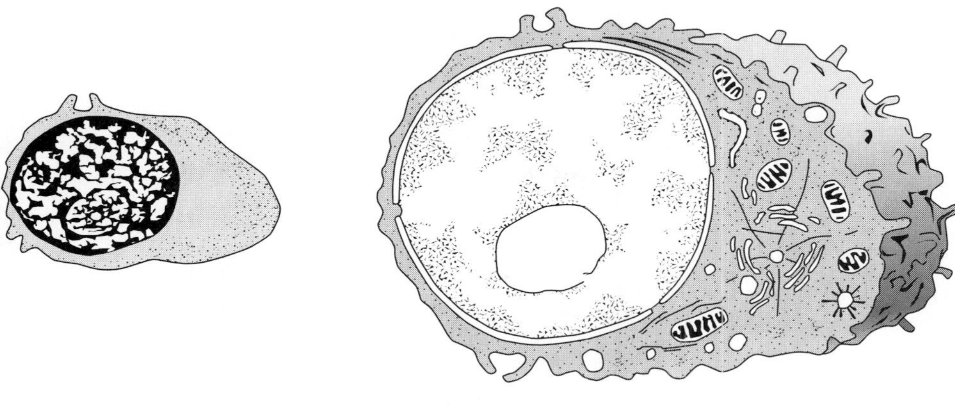


**figure 20.22**

The structure of an immature lymphocyte (blast cell) as seen through (*a*) a light microscope and (*b*) an electron microscope. Immature lymphocytes contain azurphilic granules, glycogen particles, vacuoles, and nucleoli.

Modified from M. Bessis: *Blood Smears Reinterpreted.* Springer International, 1977; as appeared in Jan Klein, *Immunology: The Science of Self-Nonself Discrimination.* Copyright © 1984 John Wiley & Sons, Inc., New York, NY. Reprinted by permission of John Wiley & Sons, Inc.

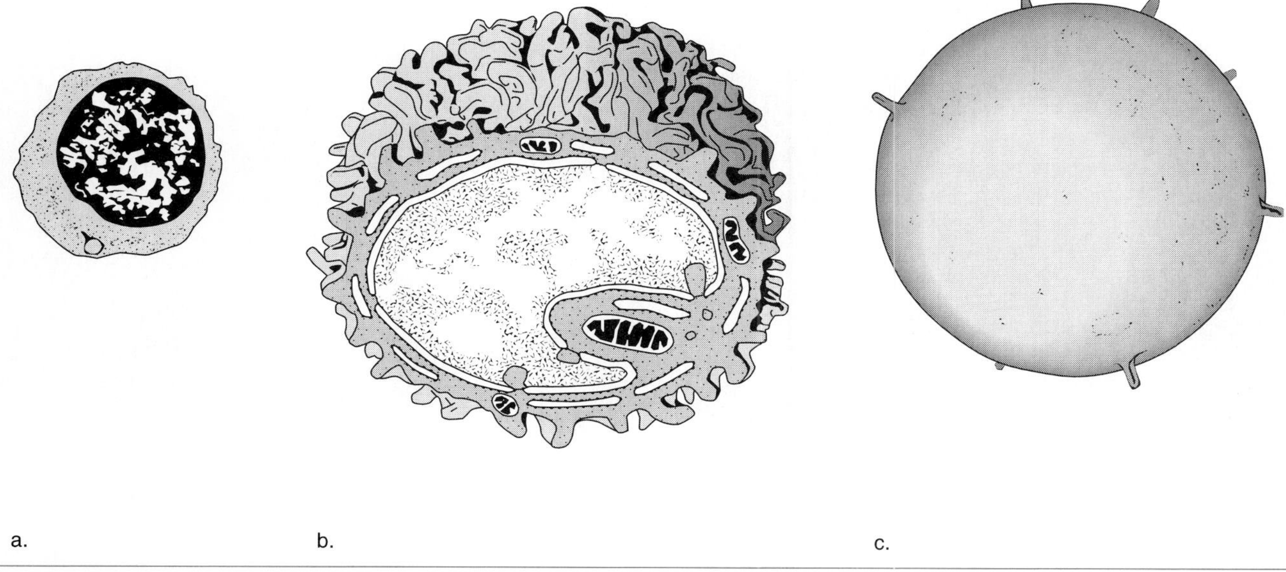


**figure 20.23**

The structure of a mature lymphocyte as seen through (*a*) a light microscope and (*b*) an electron microscope. Mature lymphocytes contain less cytoplasm and are smaller than their predecessors; *(b)* represents an activated lymphocyte and *(c)* an inactive lymphocyte.

From Jan Klein, *Immunology: The Science of Self-Nonself Discrimination.* Copyright © 1984 John Wiley & Sons, Inc., New York, NY. Reprinted by permission of John Wiley & Sons, Inc.

## LYMPHOCYTE METABOLISM

There appears to be quite a difference in metabolic activity of a resting and an activated lymphocyte. Unstimulated or resting lymphocytes are small cells with little cytoplasm containing only a few mitochondria, an underdeveloped endoplasmic reticulum, and small Golgi apparatus (see figure 20.23). Glycogen stores are also rather limited—all of which seem to indicate that these small lymphocytes have a rather low metabolic activity. Resting lymphocytes have a lower rate of oxygen consumption and glycolysis than activated lymphocytes or any other white blood cells.

The total lipid content of a small lymphocyte is about half that of a neutrophil; phospholipids are the major type of lipid present. Lymphocytes may possess a similar enzymatic machinery for lipid metabolism as do neutrophils.

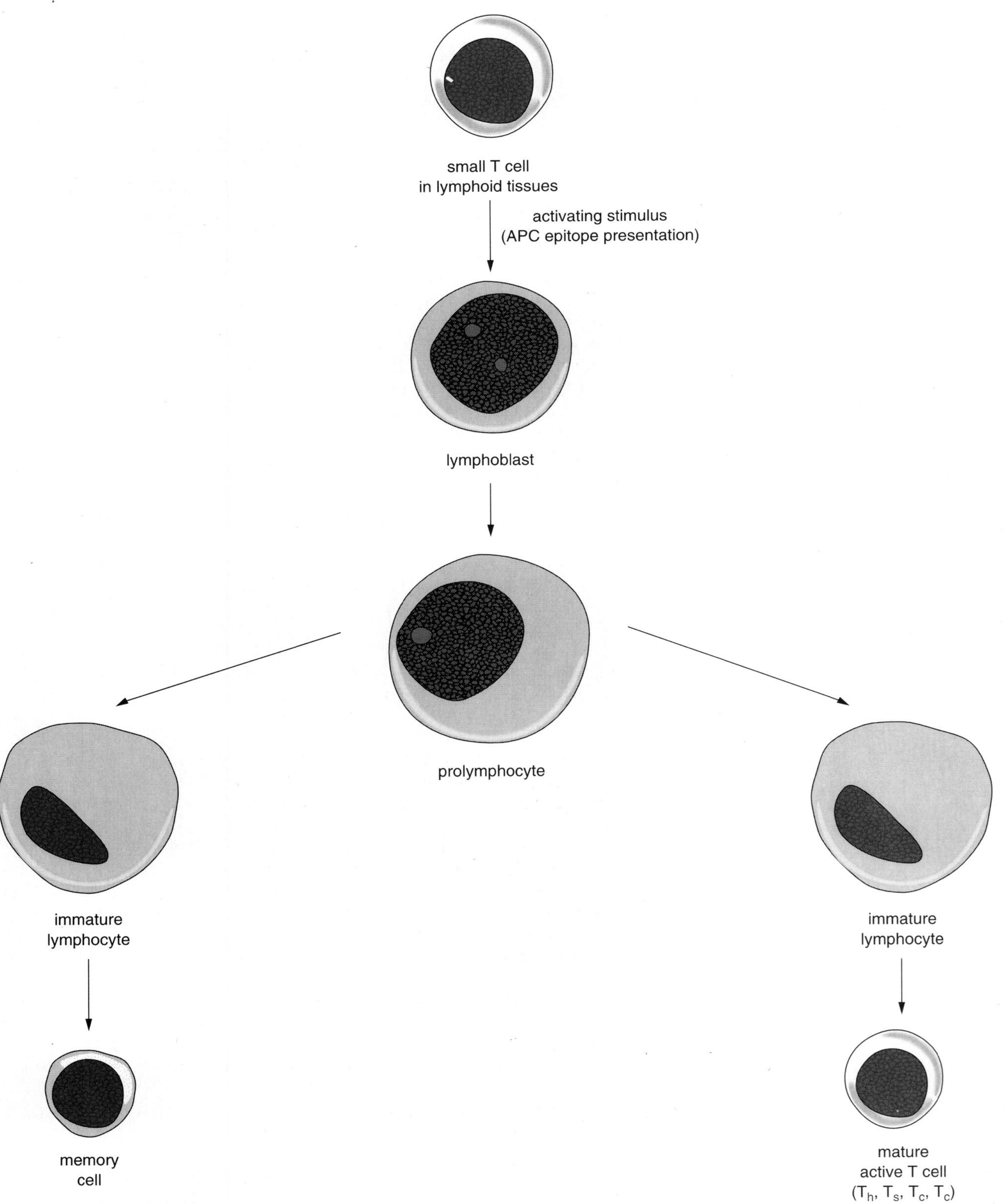


**figure 20.24**
T-cell development.

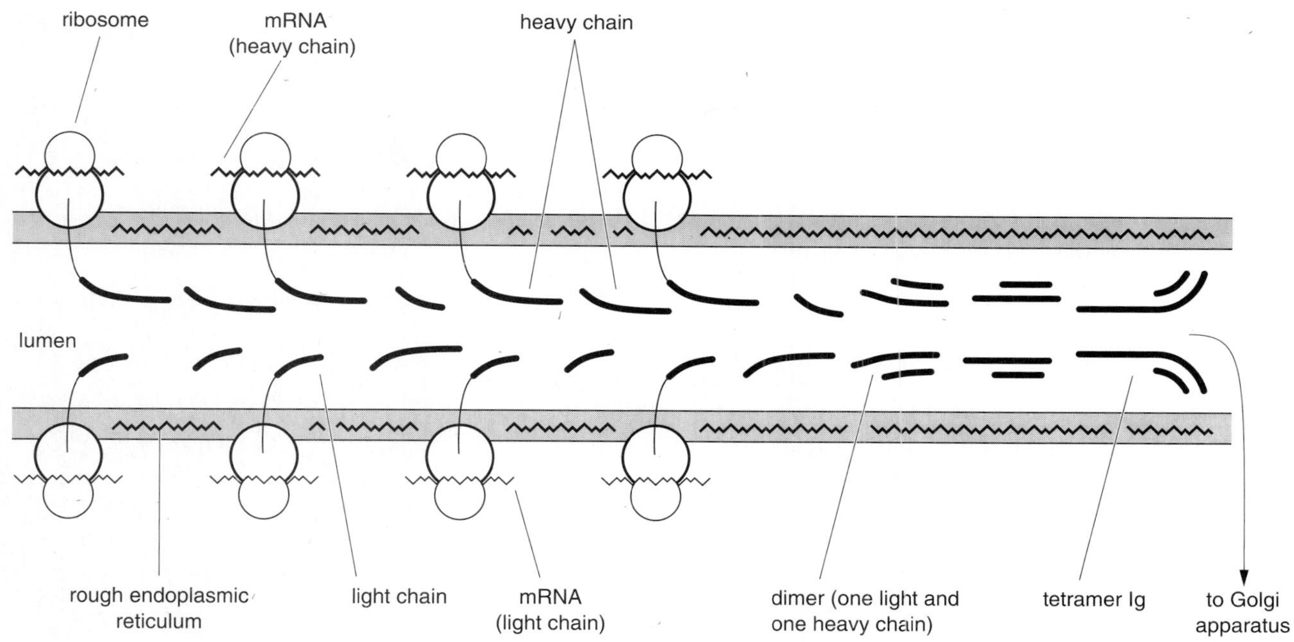


**figure 20.25**

The synthesis of immunoglobulin molecules on the ribosomes and their assembly in the endoplasmic reticulum. The carbohydrate moieties are attached in the Golgi apparatus.

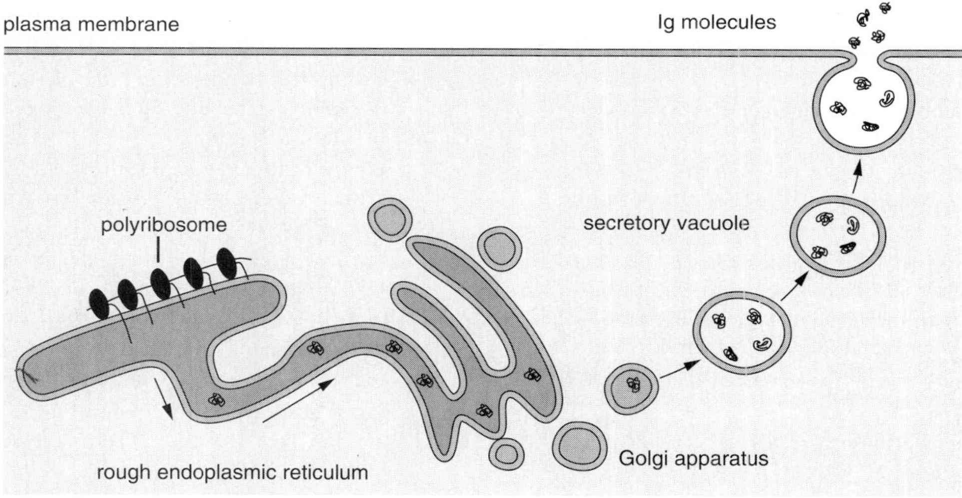


**figure 20.26**

Immunoglobulin passage from the endoplasmic reticulum, to the Golgi apparatus, to the secretory vacuoles, and to the surface of the cell.



Little is known about protein synthesis of the resting lymphocyte. On the other hand, a large body of knowledge exists regarding the production of immunoglobulins in an activated B cell. Heavy and light chain synthesis of the antibody molecule takes place on different polyribosomes (figure 20.25) and the immunoglobulin molecule receives its final configuration in the Golgi apparatus (figure 20.26). We should note that small lymphocytes are capable of limited antibody production despite the presence of limited ER and an ill-defined Golgi complex. These resting cells produce only cytoplasmic and membrane immunoglobulins.

Dramatic changes occur inside the lymphocyte once the cell becomes activated. The earliest changes involve a rapid increase in glycolysis and an exponential increase in RNA synthesis followed by a rapid rise in DNA and protein production. Within hours after the cell is activated, glycogen stores increase rapidly and glucose utilization as seen in lactate and pyruvate synthesis becomes quite evident.

## Further Readings

Boggs, D. R., and Winkelstein, A. Lymphocytes and the Immune System. Chapter 5 in *White Cell Manual,* 4th ed., F. A. Davis Company, Philadelphia, 1983.

Cline, M. J. The Normal Lymphocyte and Plasma Cell. Part II A in *The White Cell.* Harvard University Press, Cambridge, Mass., 1975.

Klein, J. Lymphoid Organs. Chapter 4 in *Immunology.* Blackwell Scientific Publications, Boston, 1990.

Kuby, J. T-Cell Receptor. Chapter 10 in *Immunology.* W. H. Freeman and Company, New York, 1992.

## Review Questions

1. Name the tissue from which lymphocytes are ultimately derived. Give the name of the lymphocyte precursor cell.
2. What is the normal percentage of lymphocytes in the blood?
3. Describe the appearance of the normal lymphocyte in a blood smear.
4. When do the first lymphocytes appear during fetal development? Which organs do they colonize?
5. List some unique characteristic of lymphocytes that are not present in other white blood cells.
6. Describe a lymphocyte clone. How many lymphocyte clones does an individual have?
7. Why do we not have T-cell and B-cell clones against self-antigens?
8. Why are lymphocytes usually not activated until after birth?
9. List the three major classes of lymphocytes. What are the relative percentages of each in the peripheral blood?
10. How would you separate white blood cells from the rest of the blood components? How would you separate lymphocytes from other white blood cells? How would you separate T cells from B cells?
11. Distinguish between the origin of B cells, T cells, and NK cells.
12. What is another name for NK cells? What is their major function? How do they act?
13. Distinguish between primary (or central) and secondary (or peripheral) lymphoid tissues. What are the major lymphoid organs in each group?
14. Name the embryonic tissue from which the thymus gland is derived.
15. Where in the body is the thymus gland found? Describe its normal shape.
16. When does the thymus gland reach its largest size? What happens to it after that time? Why is that understandable?
17. Describe the general structure of the thymus gland. Distinguish between the contents and functions of the cortical and medullary areas of the thymus.
18. Describe the general structure of a lymph node. Where are the lymph nodes found? Where are the cervical, axillary, mesenteric, and inguinal lymph nodes found?
19. Name the three histological zones of a lymph node. What is the major constituent of each zone?
20. Distinguish between cords and sinuses. Give the functions of each.
21. Describe the major functions of lymph nodes.
22. What is the embryonic origin of the lymph nodes?
23. Where is the spleen located? Describe its overall shape.
24. Distinguish between the red pulp and white pulp of the spleen regarding their histology and their functions.
25. Where in the spleen are most of the T cells found? Where are the B cells found? What is meant by the marginal zones?
26. Name the other three categories of lymphoid tissues (not including the thymus, spleen, and lymph nodes). Where are these lymphoid masses found?
27. Define the abbreviations MALT and GALT. What are the three categories of aggregated lymphoid follicles?
28. List the five stages in the development of B lymphocytes. How does each stage differ from the next?
29. Distinguish between plasma cells and memory cells. Why do memory cells develop in an activated B-cell clone?
30. Distinguish between cortical and medullary thymocytes.
31. List the stages in the development of T cells. What is characteristic of each stage? Name the final products of T-cell development. Which markers are present on each subpopulation?
32. Distinguish between a lymphoblast and a plasmablast.
33. Distinguish between the metabolism of a resting lymphocyte and an activated lymphocyte.

# Chapter Twenty-One

## *Functions of Lymphocytes*

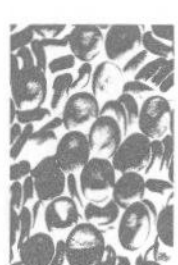

The functions of lymphocytes are many and varied, which is understandable since the many different classes of lymphocytes each have their own structure and functions. We can divide the functions of lymphocytes into the following five major categories:

**1. Nonspecific or innate immunity**
The body obtains this innate nonspecific immunity against viral-infected and neoplastic cells, especially from **natural killer (NK) cells,** which belong to the null cell group of lymphocytes.

**2. Regulation of the specific or acquired immune responses**
Certain lymphocytes determine whether the body will respond and how strongly, to foreign antigens and neoplastic cells. Lymphocytes involved in the regulatory process are mainly **helper T lymphocytes (HTLs** or **$T_H$ cells)** and **suppressor T lymphocytes ($T_S$ cells)**.

**3. Humoral immunity lymphocytes**
These lymphocytes function in defending the body against extracellular pathogens, and is commonly referred to as **humoral immunity.** Lymphocytes involved in this function belong to the **B-lymphocyte cell line** (resting B cells, activated B cells, plasma cells, and memory cells).

**4. Cellular immunity lymphocytes**
This group of lymphocytes defends the body against intracellular pathogens (especially viruses) and neoplastic cells. They consist primarily of **cytotoxic T lymphocytes (CTLs).** They are also known as **cytotoxic T cells ($T_C$ cells).** Another group of T cells plays a role in eliminating intracellular bacteria, fungi, and protozoans. They produce a slow-reacting immune response associated with inflammation. Originally it was thought that these cells were responsible only for slowly developing allergic reactions—as is the case with poison ivy allergies and contact dermatitis. Hence these T cells are known as **delayed-type hypersensitivity T cells** or **$T_{DTH}$ cells.** However their major function is in cellular immunity. These $T_{DTH}$ cells belong to the $CD^{4+8-}$ group of lymphocytes (just as $T_H$ cells).

**5. Production of hypersensitivity (allergy)**
Since there are many different types of hypersensitivity, it follows that different classes of lymphocytes are involved in each. Several groups of **B cells** are involved in a variety of allergic reactions. As mentioned in the previous paragraph, this is another type of allergy termed delayed hypersensitivity, because it is slow acting compared with other types of hypersensitivity reactions. This type is mediated by $T_{DTH}$ cells.

Each of these functions is discussed in more detail in the following material.

## LYMPHOCYTES INVOLVED IN INNATE IMMUNITY

As mentioned in chapter 20, another group of lymphocytes is labeled as null cells, since they cannot be classified as either B cells or T cells. The most common group of null cells is composed of natural killer cells (NK cells). They are large granular lymphocytes (LGLs) that possess a variety of markers but lack receptors for antigenic determinants. Hence they cannot function directly in the specific or acquired immune system. Most NK cells have Fc receptors for IgG antibodies and thus may play an indirect role in the attack against specific antigens (and hence acquired immunity) via ADCC (=antibody-dependent cell-mediated cytotoxicity) (*see figure 20.3*). What makes NK cells unique is their spontaneous ability to kill cancer cells and many virus-infected cells. They do so without being stimulated specifically by antigenic determinants. The mechanisms by which NK cells can recognize pathological target cells is still unknown and remains an active area of investigation. It is known, however, that γ-interferon (IFN-γ) and interleukin 2 (IL-2) promote the proliferation of both NK cells and lymphokine-activated killer (LAK) cells, which are especially effective in the destruction of tumor cells (figure 21.1).

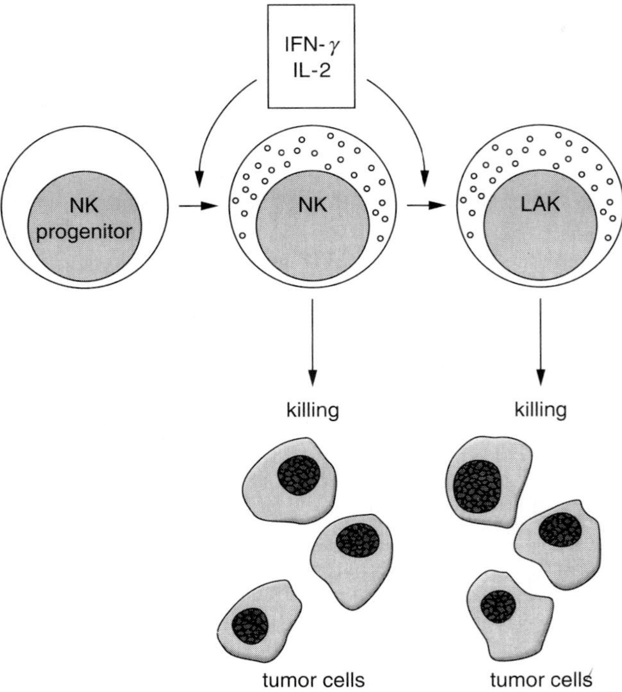


**figure 21.1**
The effect of γ-interferon (IFN-γ) and interleukin 2 (IL-2) on natural killer (NK) cells. These components promote the production of more NK cells and lymphokine-activated killer (LAK) cells, which are more effective in the destruction of tumor cells.

Virus-infected cells also possess receptors for both IFN-γ and IL-2 and produce large quantities of interferon. The interferon induces resistance in neighboring cells against viral infections, thus preventing the virus from spreading (figure 21.2). We can hypothesize that production of interferon by an infected cell will attract the NK cells and induce them to kill the cell once a sufficient number of their receptors are occupied. Such a

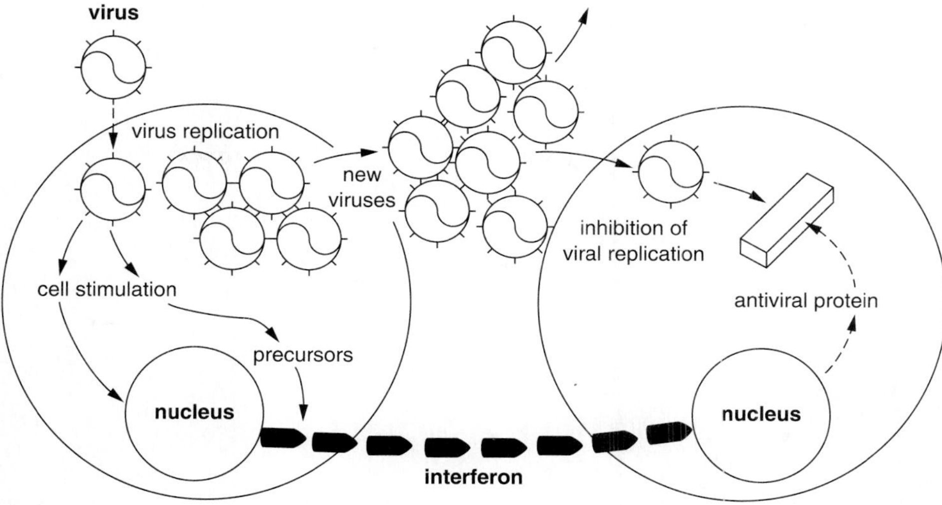


**figure 21.2**
The effect of interferon on viral replication.

condition could occur only when the NK cell is in close contact with the virus-infected cell. A similar scenario can be envisioned for cancer cells. Since cancer cells produce growth factors that may be similar to IL-2, binding these growth factors to IL-2 receptors may induce the NK cells to release cytotoxic substances and kill the neoplastic cell.

Thus NK cells form a first line of defense in the resistance against many viruses as well as against cancer cells. They keep viral replication and neoplastic growths in check until more effective cells of the specific cellular (or cell-mediated) immune system can be activated. Since interferons and interleukins are always produced in the largest quantities early in the infections, this forms another argument for the early nonspecific action of NK cells.

## REGULATORY FUNCTION OF LYMPHOCYTES

The major class of lymphocytes that plays an important role in regulating the specific (or acquired) immune responses of the body is the helper T cell ($T_H$ cell).

The overwhelming majority of foreign antigens cannot activate the immune system without the help of $T_H$ cells. These are known as T-cell-dependent antigens. The only T-cell-independent antigens are some simple carbohydrates composed of similar sugar molecules and a few artificial peptides composed of repeating amino acids. Practically all antigens that occur in nature require the assistance of helper T cells.

As mentioned in chapter 19, macrophages are also required, which function as antigen-presenting cells (APCs). APCs first process the antigens into antigenic determinants (or epitopes), which then bind to major histocompatibility complex (MHC) molecules, and the resulting complex is then incorporated int the membrane of the macrophages. These activate macrophages then move to the lymphoid tissues, where they present epitopes to various clones of T cells and B cells that can interact with these antigenic determinants. Once the APC binds to an appropriate $T_H$ cell, it releases interleukin 1 (IL-1), which stimulates the $T_H$ cell to undergo blast formation. This will result in cell replication and cell maturation. These activated $T_H$ cells are now able to bind to effector cells—that is, B cells and $T_C$ cells. Once this bond is produced, it acts as a stimulus for the $T_H$ cell to produce interleukin 2 (IL-2), which now causes these particular cells to become activated and undergo blast formation and maturation (figure 21.3). Clearly without the presence of $T_H$ cells, it is almost impossible to activate the acquired immune system. In acquired immunodeficiency syndrome (AIDS) the number of $T_H$ cells is drastically reduced, thus effectively wiping out the specific immunity of the individual despite the presence of many B cells and $T_C$ cells.

Suppressor T cells ($T_S$ cells) are able to suppress the specific immune response by inhibiting $T_H$ cell activity, effector T-cell activity, and even APC activity. An elevated $T_H$ count implies increased lymphocyte activity as helper cells predominate, whereas a high $T_S$ level implies depressed lymphocyte activity as a result of excessive suppressor activity.

Activated B cells are also capable of functioning as APCs and do so usually in repeated infections. Usually in the primary encounter with a given antigen, the processing of the antigen and the presenting of the epitopes to $T_H$ cells is done mainly by macrophages. In repeated encounters (anamnestic or booster responses), enlarged B-cell clones become the major APCs for the $T_H$ cells. They are also capable of engulfing and processing the primary antigens into epitopes, which are then bound to MHC molecules and incorporated into the B-cell membrane.

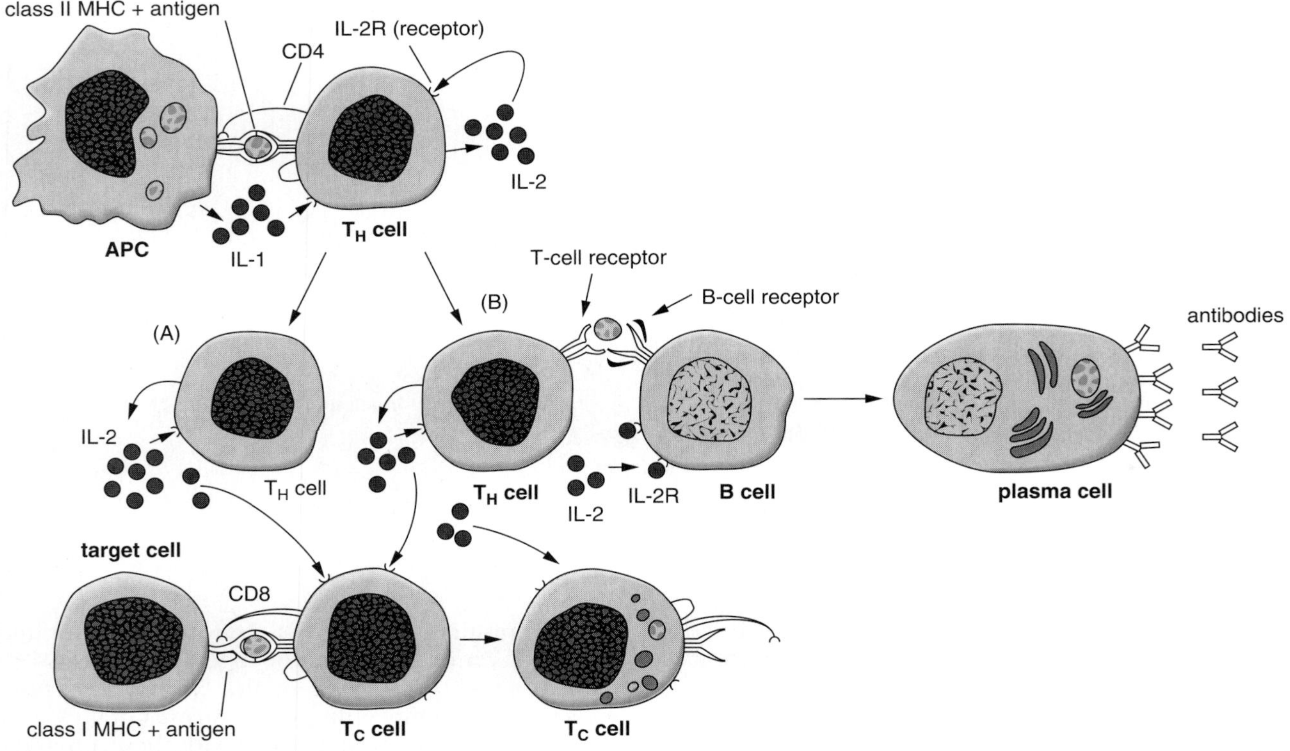


**figure 21.3**
The relationships between APCs, helper T cells ($T_H$), cytotoxic T lymphocytes ($T_c$), and B cells. The APCs stimulate the production of more $T_H$ cells. The $T_H$ cells, in turn, stimulate the production and activation of $T_c$ and B-cell clones.

## ROLE OF LYMPHOCYTES IN HUMORAL IMMUNITY

**B cells** form the major group of lymphocytes directly involved with the production of humoral immunity since they are the producers of immunoglobulins (antibodies). As explained in chapter 20, the major group of B cells responsible for antibody production are the plasma cells or plasmacytes. These are fairly large, short-lived cells found mainly in lymphoid tissues, from where they secrete large quantities of antibodies into the body fluids. Each plasma cell normally produces only one type of antibody: either IgM, or IgG, or IgA, or IgE. IgD is produced for only a short time when activated B cells switch from the production of IgM to the production of other antibody types (figure 21.4). The type of antibody produced is also highly specific—that is, it can bind to only one, or at the most a few, antigenic determinants. Antibodies confer immunity to the organism, and for that reason, plasma cells do not need to move to the area of infection. They can stay in the lymphoid tissues (especially in the cords of the lymph nodes and in the germinal centers of the spleen), since the effects are produced by highly specific immunoglobulin molecules.

Antibody molecules have many different functions. Some of the most important ones are described in the following material.

### Opsonization

One of the major characteristics of immunoglobulins is their ability to act as opsonins. An **opsonin** is a substance that can bind tightly to a receptor on a target cell or target molecule and makes it more susceptible to phagocytosis. The target cell or molecule displays epitopes on their outer surfaces allowing the antibody molecule to bind to it with its Fab fraction (figure 21.5). As a result the antibody's Fc fraction can bind to the Fc receptors present on most phagocytic cells. Binding the phagocytic cell to the Fc fractions of the opsonin allows the process of phagocytosis to occur quickly and efficiently. Without the presence of opsonins, many molecules and microbial cells would appear rather slippery to the phagocytic cell, and it would have considerable difficulty engulfing these particles. Opsonization of these particles, on the other hand, facilitates phagocytosis.

### Agglutination and Precipitation

The shape of the immunoglobulin molecule allows it to bind to two particles simultaneously. If this happens on a large scale with many antibody molecules and many particles, clumping or agglutination will occur (figure 21.6). Clumping prevents the pathogenic cell from moving away from the site of infection. It also allows the phagocytic cell to engulf many

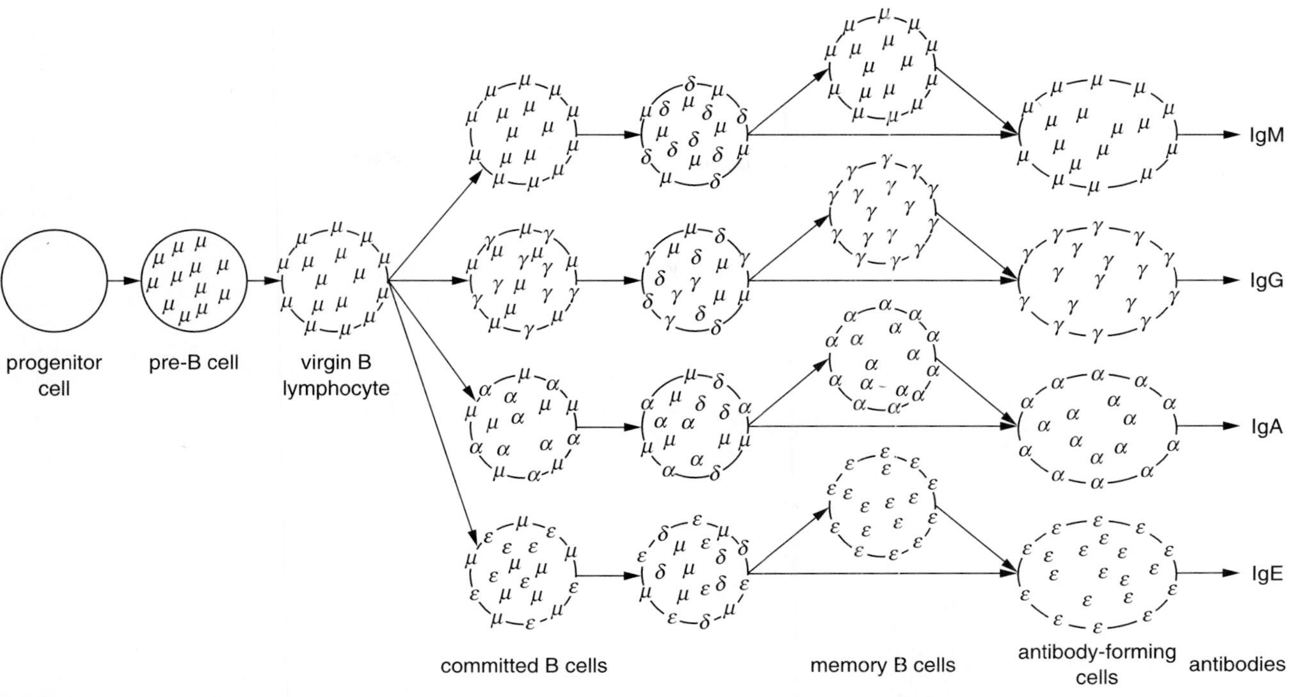


**figure 21.4**

Schematic representation of B-cell differentiation. Note that initially all B cells produce only IgM antibodies. IgD antibody production is necessary before the plasma cell can start producing only one class of antibodies.

From Jan Klein, *Immunology: The Science of Self-Nonself Discrimination.* Copyright © 1984 John Wiley & Sons, Inc., New York, NY. Reprinted by permission of John Wiley & Sons, Inc.

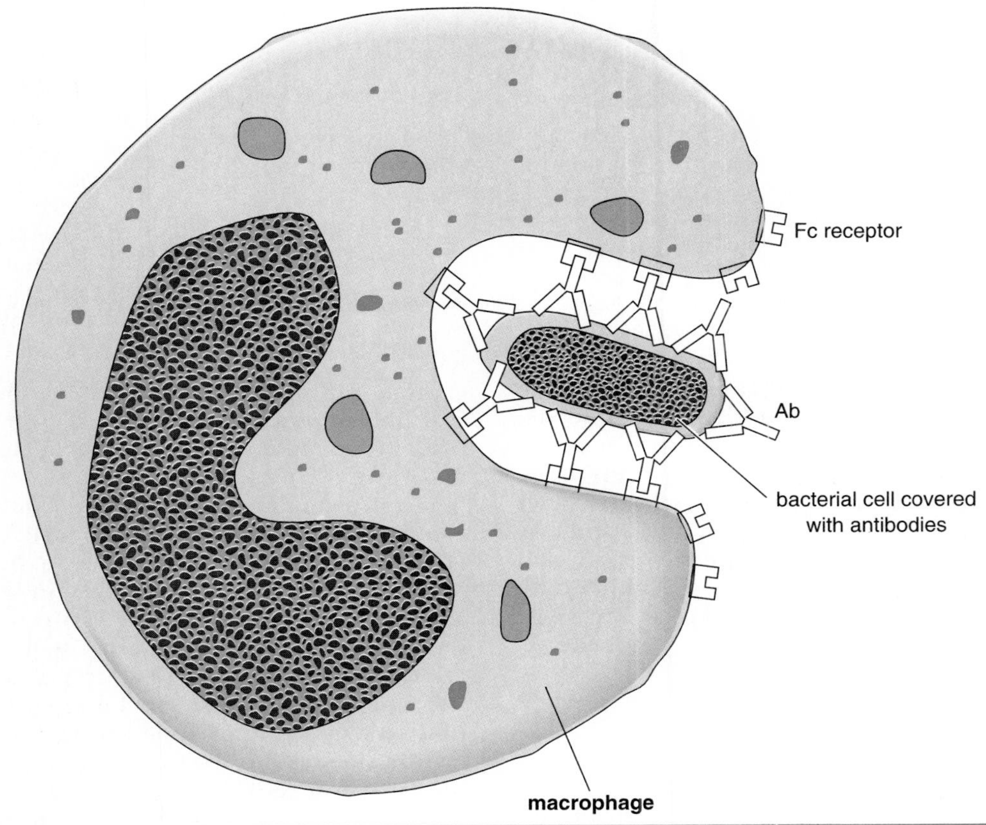


**figure 21.5**

The opsonization of bacterial cells by antibodies facilitates macrophage phagocytosis.

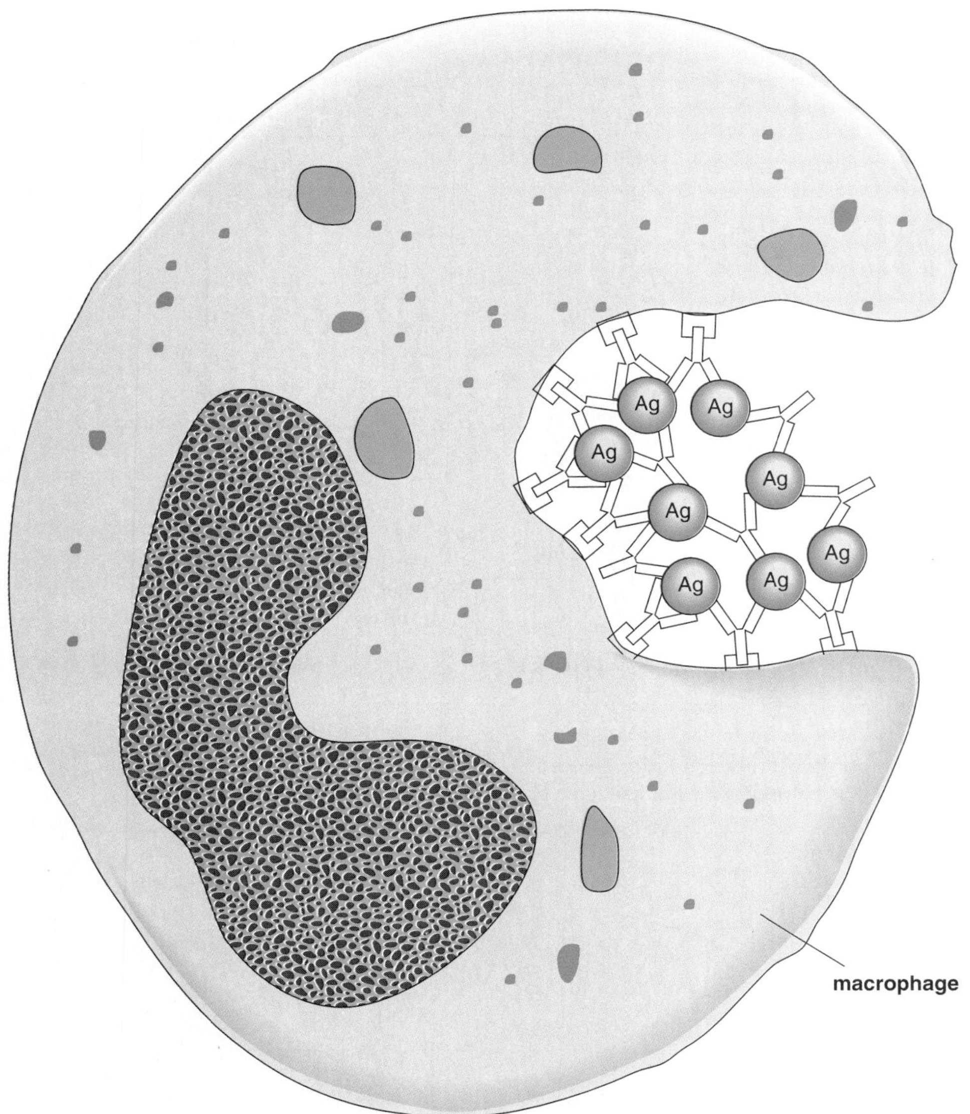


**figure 21.6**

Agglutination and precipitation also facilitate macrophage phagocytosis. Ag = antigen, which may be either a cell (agglutination), or a large protein (precipitation).

particles at once. Hence agglutination enhances the removal of pathogenic organisms from an infected area.

When antibody molecules bind to many large molecules in solution, the ensuing complex becomes so large that it can no longer stay in solution and precipitates out. A phagocytic cell can easily engulf and eliminate precipitated molecules, but it is rather difficult to attach to these molecules when they are singly in solution. Hence, precipitation also speeds the removal of foreign molecules.

### Neutralization

Many pathogenic organisms produce toxic compounds that can have deleterious effects on the organism. Most of these toxic compounds are enzymes that break down essential molecules of the cells and tissues, thus upsetting the metabolic activities of the body. Antibodies produced against these toxins will bind to them, and in doing so will render them unable to bind to a suitable substrate. Hence they will in effect inactivate or neutralize these toxic compounds.

Another aspect of neutralization involves the inactivation of viral particles. Viruses can replicate only inside a living cell. In order to get inside a cell, the virus must first bind to a receptor compound on the host cell membrane. This is usually accomplished by a viral envelope protein binding to a receptor site on the host cell. After attachment the phospholipid envelope of the viral particle fuses with the host cell membrane, and in this way it

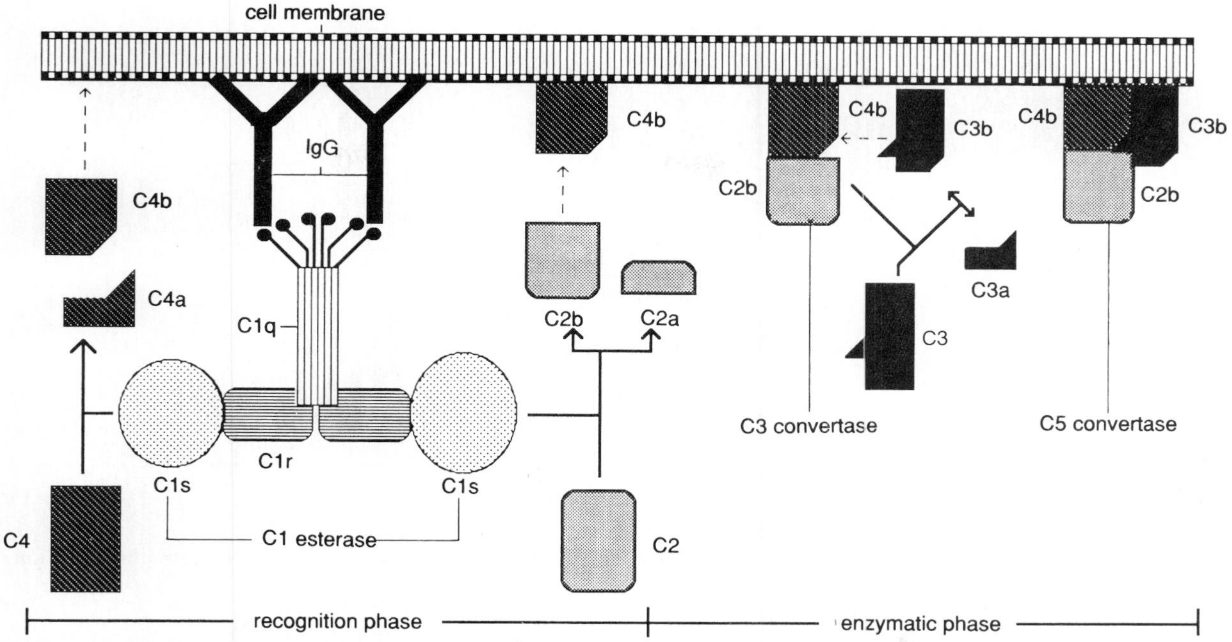


**figure 21.7**

Schematic representation of the initial chemical reactions, recognition and enzymatic phases, in complement activation via the classical pathway.

invades the cell. Antibody molecules will bind to the viral envelope proteins, thus preventing binding to the host cell receptors. In other words, binding the immunoglobulin molecule to the viral particle will effectively neutralize them—that is, prevent them from invading the host cell. The neutralizing antibody not only inactivates the virus but acts simultaneously as an opsonin. As previously discussed, opsonization will allow the phagocytic cells to engulf and destroy the viral particles before they harm the host.

## Complement Activation

The complement system consists of a series of at least 16 different proteins present in the plasma in an inactive form until they become activated. The activation of complement can occur in many ways—for instance, by snake venom or bacterial products or IgA aggregates. One other major mechanism of activation is via the formation of antibody-antigen complexes. When complement is activated via antibody-antigen complexes, it follows the **classical pathway.** Activation of complement proteins in the classical pathway is a sequential event that begins with activation of complement factor 1 (C1) and ends with the production of a hole in the antigenic cell membrane by a series of proteins that form the membrane attack complex (MAC). When many holes are made in a microbial cell wall, the organism will lyse. Thus the end result of all complement activation is cell lysis. However in the process of complement activation, many other biologically active products are generated, including C3a, C4a, and C5a, which are important chemotactic and anaphylactic factors.

The process of **complement activation** via the classical pathway consists of the following three phases:

**1. Recognition phase**

This occurs when the first complement factor binds to the antibody-antigen complexes (also known as **immune complexes**). To bind this first factor, two or more immune complexes in close proximity are needed; otherwise this first factor is unable to bind. Complement factor 1 (C1) is a complex factor composed of three different proteins. First is a huge protein, known as C1q, which consists of 18 polypeptide chains organized in six bundles of three chains. It looks like a cat-o-nine-tails. This C1q molecule binds to the *Fc* fractions of two bound antibody molecules (figure 21.7). Attached to this C1q structure are two molecules of C1r, and attached to each of these C1r compounds is a C1s molecule. The function of C1r is to bind C1s to the C1q structure. The function of C1s is to act as a catabolic enzyme that will cleave the next complement component (C4) into two parts.

**2. Enzymatic phase**

The second phase starts with the cleavage of C4 by C1s (also known as C1 esterase) into two components: C4a and C4b. Almost simultaneously C1s also cleaves C2 into C2a and C2b. Most of the C4b generated attaches itself to the cell surface (figure 21.7). The C2b produced then binds to C4b. C4a and C2a are released into the surrounding environment, where they are soon degraded. C4a has some anaphylatoxic activity, however, and can contribute to local inflammation. The C4b2b complex is enzymatically active and is now able to

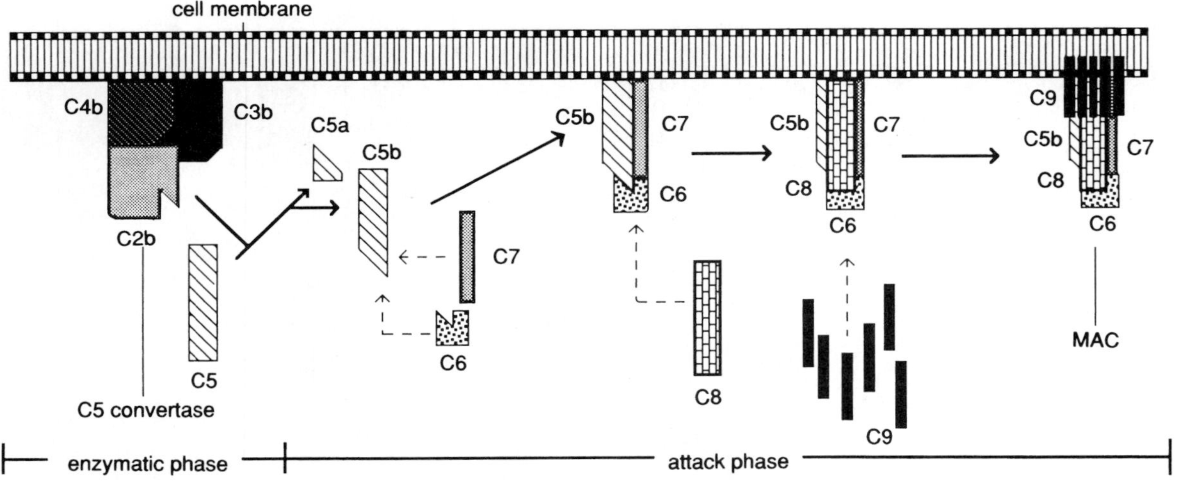


**figure 21.8**
Schematic representation of the final chemical reactions, enzymatic and attack phases, of complement system activation.

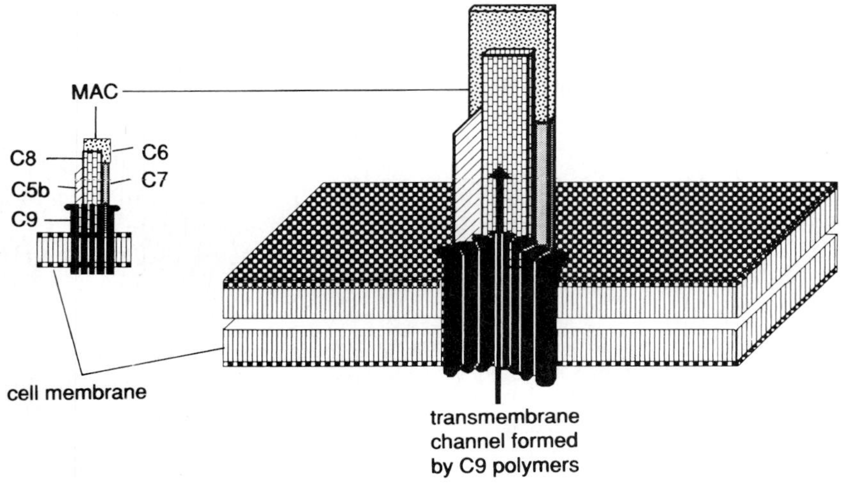


**figure 21.9**
The structure of a membrane attack complex (MAC).

interact with C3 and cleave it into C3a and C3b. Hence the C4b2b complex is also known as **C3 convertase.** C3b now joins the C4b2b complex to form a C4b2b3b complex, which is now able to bind to C5 and cleave it into C5a and C5b. Hence this complex is also known as **C5 convertase.** C3a and C5a are released in the body fluids, where they act as potent **chemotactic** and **anaphylatoxic factors** for neutrophils and monocytes. C5b attaches itself to the cell membrane away from the C4b2b3b complex.

**3. Attack phase**
The third and final phase starts with the attachment of C5b to the cell membrane (figure 21.8). Once this happens a steric change occurs in the molecule which allows it now to bind to C6 and C7. This produces a stable trimolecular complex known as C5b,6,7. This complex enables C8 to bind as well. Once C8 is incorporated into the complex, it allows several C9 molecules to bind, forming a **membrane attack complex (MAC)** (figure 21.9). C5b,6,7,8, and C9 molecules form a **transmembranal channel** or pore of approximately 100 Å in diameter. Formation of these pores will result in lysis of the cell (figure 21.10).

Antibody molecules play a major role in the killing of pathogenic agents once they are bound to antigenic determinants and have activated the complement cascade. Many phagocytic cells also have receptors for C3b molecules. Hence these C3b molecules on the membrane of the pathogenic cell will also act as opsonins.

## Production of Anaphylaxis

Antibody molecules may be involved in the production of local or systemic inflammation in several different ways. First, IgM and IgG antibodies can cause anaphylaxis in an indirect manner by activating complement. This will result in the production of C3a, C4a, and C5a—all of which are inflammatory agents.

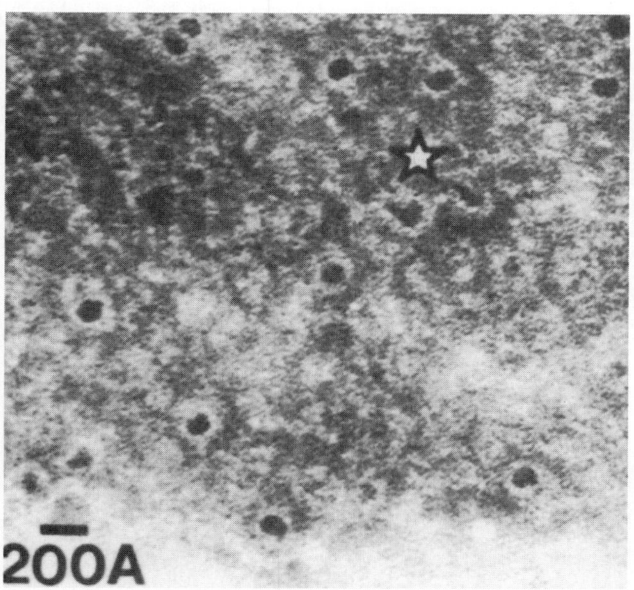


a.

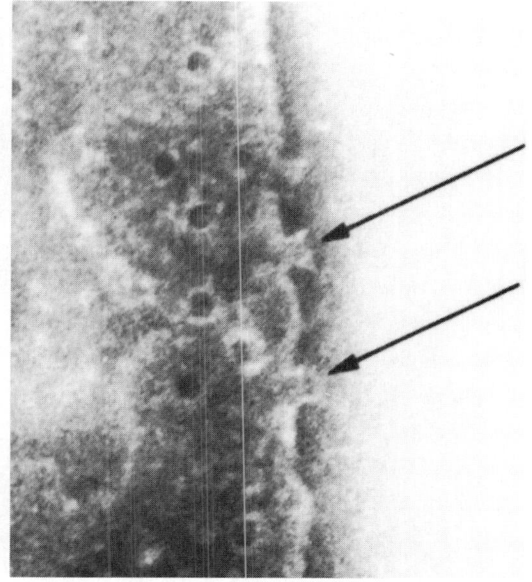

b.

figure 21.10

Scanning electron micrographs of transmembranal channels formed by MACs. (*a*) A view from above; and (*b*) a lateral view. Arrows indicate MAC insertion into the cell membrane.

Another way in which immunoglobulins can produce anaphylaxis is limited to IgE antibodies. These antibodies are produced only in those people who are allergic to certain substances such as pollen, food substances, dandruff, and dust. When such individuals come in contact with these allergenic substances, they will activate their humoral immune systems and produce large quantities of IgE antibodies. Certain individuals also exhibit a deficiency in suppressor T cells ($T_S$ cells) and as a result their IgE-producing B cells function at a higher rate than normal. As explained in chapter 18, these IgE antibodies have a great affinity for mast cells and basophils, which have Fc receptors for these immunoglobulins. As soon as they are produced, IgE molecules will bind to these mediator cells (figure 21.11). If a person is exposed again to these allergenic substances, they will now bind to the Fab fractions of the IgE antibody molecules. This binding acts as a signal for mast cells to release vasoactive amines, especially histamine, which is a powerful inflammatory agent. The release of these substances will result in an allergic reaction (*see chapter 18*).

## ROLE OF LYMPHOCYTES IN CELL-MEDIATED IMMUNITY

The major class of lymphocytes involved in specific cellular immunity are the $T_C$ cells, also known as cytotoxic T lymphocytes (CTLs). The $T_C$ cells produce a number of nonspecific substances **(lymphokines)** that result in the destruction of virus-infected and neoplastic cells. Activation of **cell-mediated immunity (CMI)** follows a similar pattern as activation of B cells. As with humoral immune responses, cellular immune responses are also evoked by an initial exposure to a particular antigenic determinant from a foreign substance. These viral or neoplastic antigens are first processed by the APCs and then presented to the appropriate clones of helper T lymphocytes (HTLs or $T_H$ cells) and CTLs ($T_C$ cells) in association with MHC molecules. The binding of APCs to HTLs will result in the production of IL-1. This compound stimulates the HTLs to undergo cell division and maturation. These activated HTLs in turn produce IL-2, which will activate the CTLs (figure 21.12). Development and maturation of these CTLs occurs mainly in the paracortical areas of the lymph nodes, and in the periarterial sheaths of the spleen. Large numbers of mature CTLs are released into the lymph and bloodstream. Once they make contact with virus-infected or neoplastic cells, they release cytotoxic secretions that will result in the death of these cells. Activated CTLs must seek out pathogenic cells individually, since their cytotoxins are nonspecific substances. If the cytotoxins were released into the general environment, they would destroy all cells they came in contact with, infected or not. Hence they cannot remain in the lymphoid tissues and release their secretions into the lymph and blood as activated B cells do. The exact nature of these lymphocytotoxic substances is still a matter of some controversy.

Membrane-bound granules of CTLs have been found to contain at least three different kinds of proteins: **perforins** (also known as cytolysins) and two types of **serine esterases.** The function of the esterases is not clear. The function of the perforins is to create pores in the target cells and thus produce cytolysis (figure 21.13).

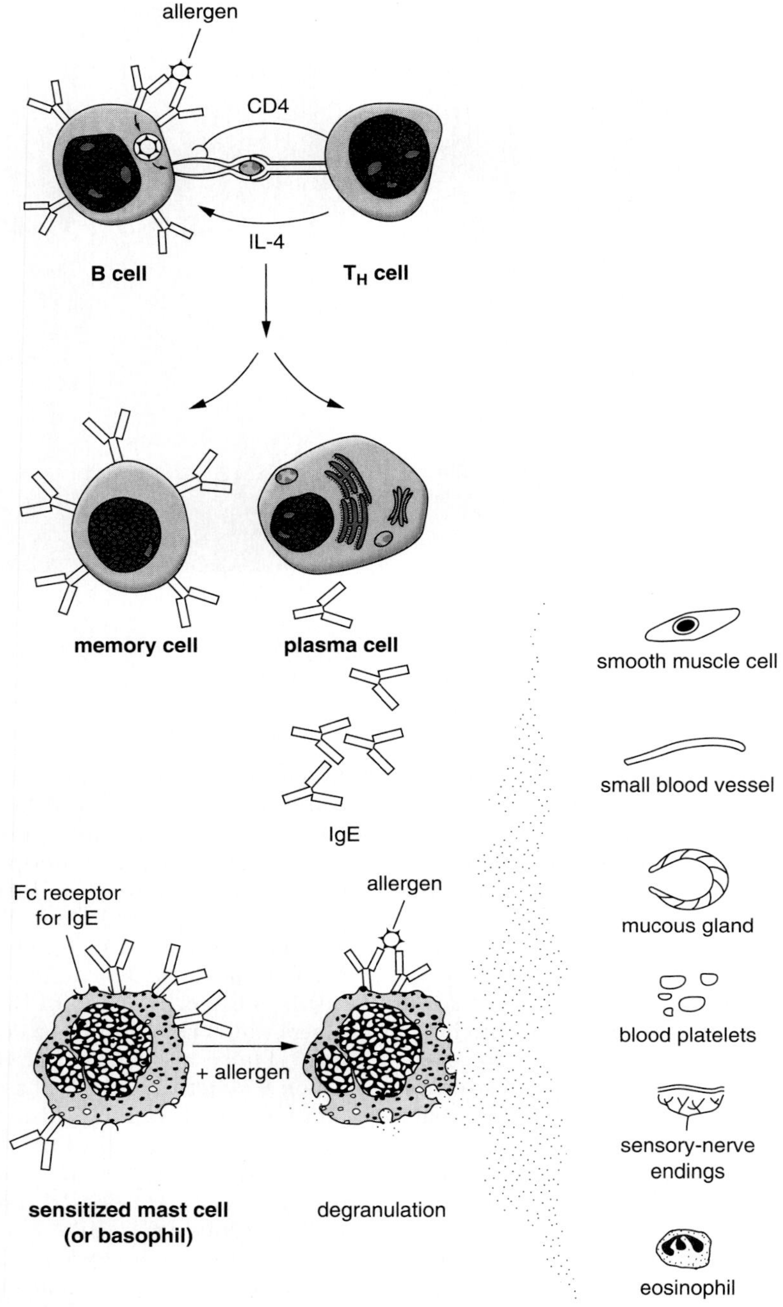


**figure 21.11**

The sensitization of mast cells and basophils by IgE antibodies. Vasoactive amines are released when allergens bind across IgE antibodies on these effector cells.

The second type of CMI involves $T_{DTH}$ cells, which play an important role in eliminating intracellular pathogens that are mainly nonviral in origin. Table 19.2 lists some important intracellular pathogens that cause activation of $T_{DTH}$ cells including bacteria, fungi, parasites, and some viruses affected by this type of CMI. In addition to microorganisms, chemicals such as formaldehyde or urushiol in the poison ivy plant may induce contact dermatitis caused by actions of the $T_{DTH}$ cells. This delayed-type hypersensitivity (DTH) response has two phases. The first is a **sensitization period** of about 1 to 2 weeks following the initial exposure to these antigens. During this period specialized $CD4^+$ cells (or $T_H$ cells) are activated by an APC, which presents the antigen to the $CD4^+$ cell together with a class II MHC molecule. Binding of the APC to the $CD4^+$ cell will result in the release of IL-1, and this will bring about the proliferation

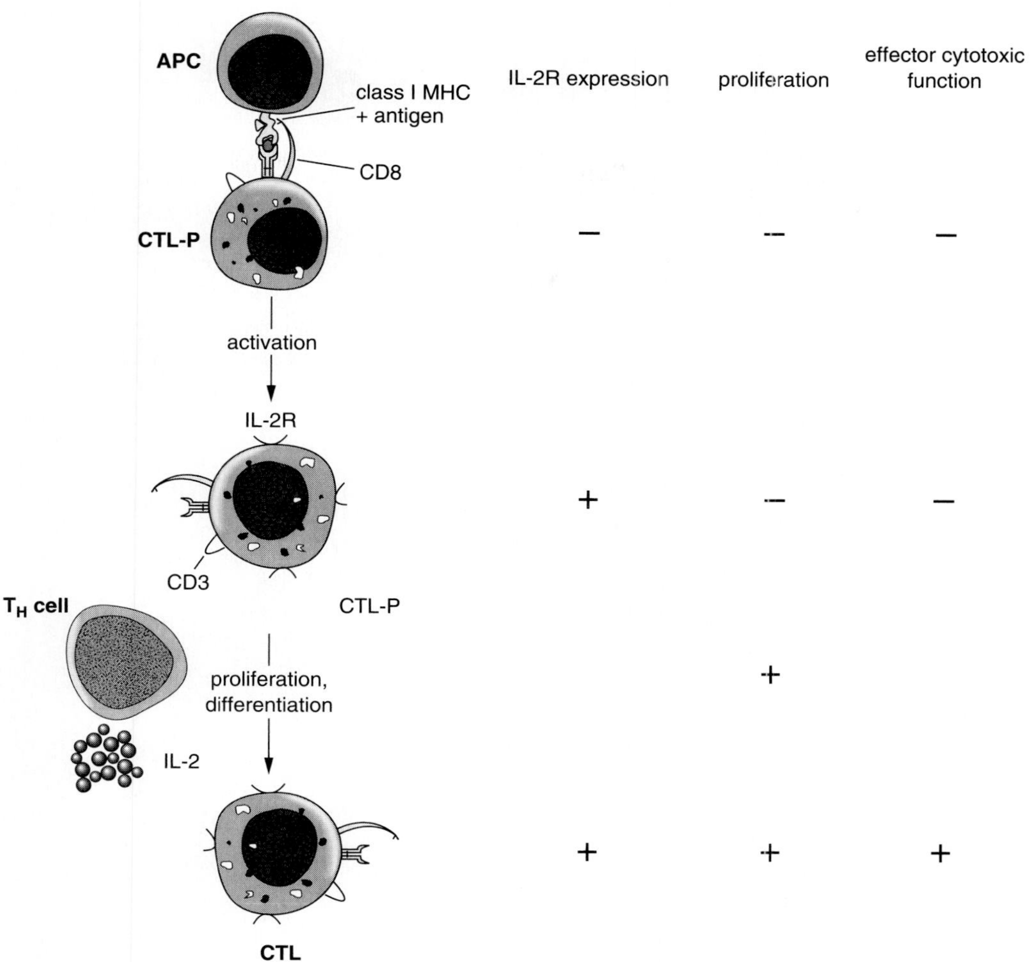


**figure 21.12**

The process by which CTL ($T_C$) precursor cells (CTL-P) become activated and differentiated with the help of APCs and HTLs ($T_H$ cells).

and expansion of a $T_{DTH}$ cell clone. This clonal expansion is aided by the autocrine action of the IL-2 produced by activated $T_{DTH}$ cells (see figure 19.12). A secondary contact with the same antigen will induce the second phase, which is known as the **effector phase.** During this period the activated $T_{DTH}$ cells will produce a number of powerful cytokines responsible for recruitment and activation of macrophages and other nonspecific inflammatory cells. It usually takes an average of 24 hours following the second contact with the antigen before the DTH response becomes apparent, and the activity usually does not peak until at least 48 to 72 hours later. This delayed onset of the immune response is a reflection of the time it takes for these cytokines to mobilize effector cells to the area of infection. Activated macrophages then produce a pronounced inflammation in the area of the infection. They also exhibit increased phagocytic activity and an increased ability to kill these microorganisms within these phagocytosed cells. Usually the pathogens are cleared with minimal long-term tissue damage. In certain cases in which the macrophages are not able to kill the pathogens quickly, there will be a prolonged DTH response. This will result in the production of multinucleated **giant macrophages,** which will form visible **granulomas** within normal tissues. The release of large quantities of lysosomal enzymes into the surrounding tissues by these giant cells may result in considerable damage to the surrounding tissues of the host.

## ROLE OF LYMPHOCYTES IN HYPERSENSITIVITY REACTIONS

As previously mentioned, lymphocytes have been implicated in several different types of hypersensitivity reactions: those caused by B cells and those caused by $T_{DTH}$ cells. **Hypersensitivity reactions** may be defined as inappropriate activations of the body's immune system, which may result in inflammation and/or tissue damage. For many years allergic reactions have

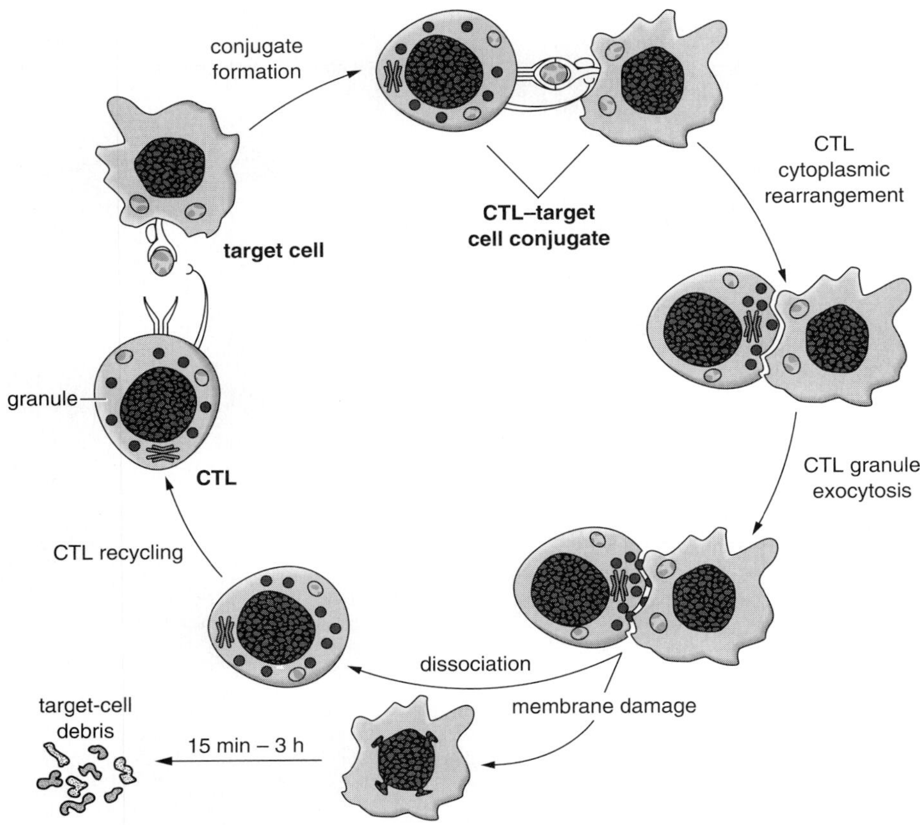


**figure 21.13**

Steps in the CTL-mediated killing of target cells.

been classified into the following four different categories or types. This classification was first proposed by the English scientists Gell and Coombs.

### Type I Hypersensitivity (Immediate Hypersensitivity)

Type I hypersensitivity reactions include such disorders as **allergic rhinitis** (hayfever), **urticaria** (hives), **bronchial asthma, atopic dermatitis,** and **food allergies.** These allergic reactions are frequently hereditary and are mediated by a special class of B lymphocytes that produces IgE antibodies. When a susceptible person is exposed to an allergen (i.e., an allergic substance), this allergen is first processed by the APCs, which then present appropriate antigenic determinants to appropriate B cells, which respond by producing special IgE antibodies. Nonallergic persons have extremely low levels of IgE in their blood compared to susceptible persons whose levels rise sharply. These IgE antibodies are **homocytotropic**—that is, they bind readily to mast cells and basophils, which have Fc receptors for these antibodies. When the individual is exposed to the same allergen for a second time, these allergens will bind to the IgE antibodies on the mediator cells (see figure 21.11). The binding acts as a trigger to produce high levels of cyclic guanosine monophosphate (cGMP) in the cell, which in turn causes these cells to degranulate and release vasoactive amines (histamine, serotonin) into the environment, resulting in inflammatory reactions.

### Type II Hypersensitivity (Cytotoxic Hypersensitivity)

These hypersensitivity reactions are mediated by B lymphocytes. Type II hypersensitivity reactions involve B cells that produce IgM and IgG. These antibodies react mainly with foreign red and white blood cells and platelets. When these antibodies bind to the blood particles, they activate the complement cascade, resulting in the lysis of these blood cells. Classic examples include **autoimmune hemolytic anemia (AIHA), hemolytic disease of the newborn (HDN)** and **transfusion reactions.** In AIHA these inappropriate immune reactions occur mainly as a result of the formation of **neoantigens** when drugs bind to the self-antigens on the blood cells or plasma proteins.

### Type III Hypersensitivity (Immune Complex Hypersensitivity)

This class of hypersensitivity reactions involves IgM and IgG antibodies produced by B cells and is also known as the **immune complex hypersensitivity reactions** because antibody-antigen complexes are always

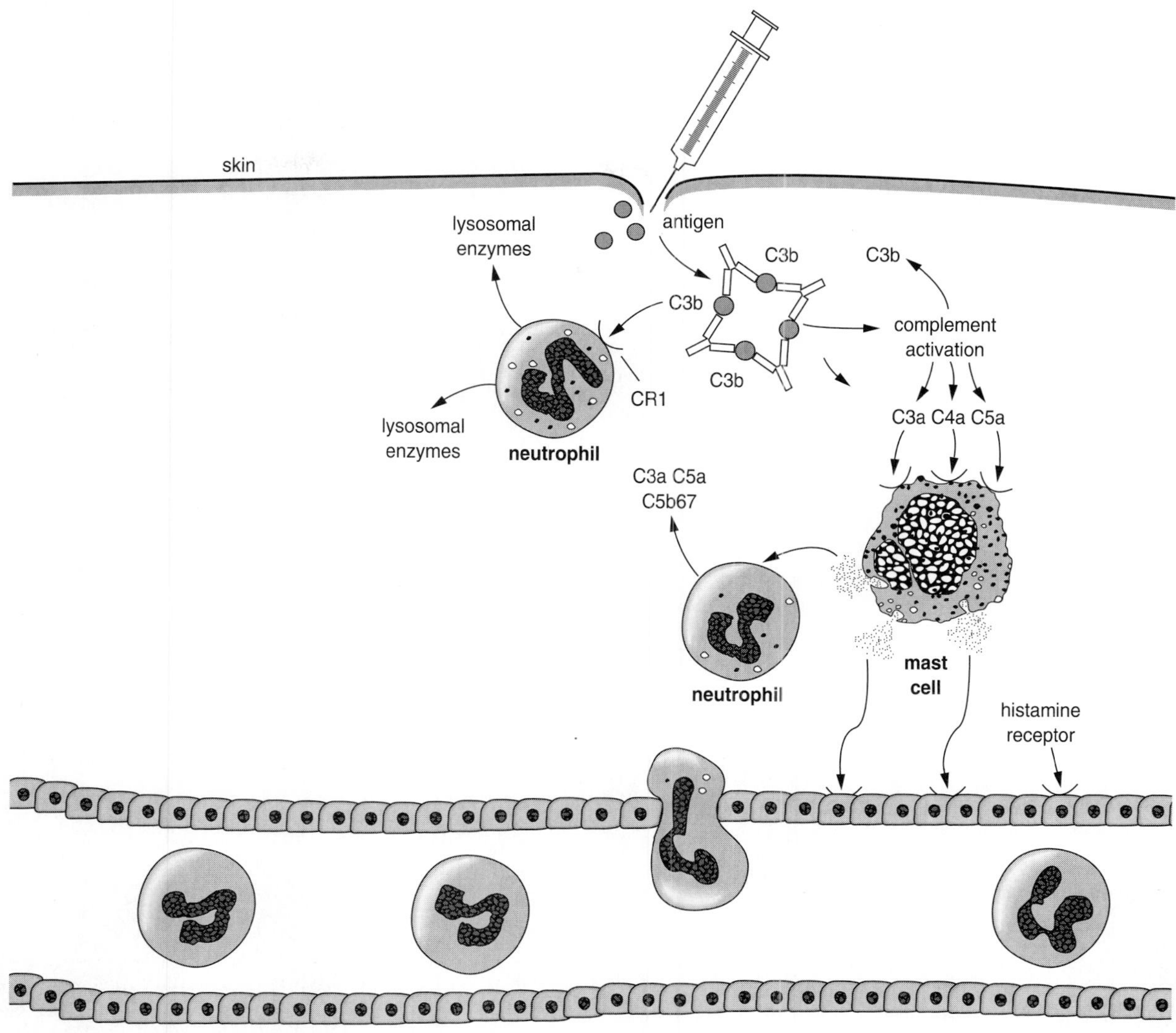


**figure 21.14**

Development of type III hypersensitivity reaction. Mast cells and neutrophils are activated by complement products. Mast cells release histamine, which causes localized anaphylaxis and the movement of neutrophils to the area of infection. Once there, the neutrophils release lysosomal enzymes that cause local tissue destruction.

the initiating stimuli. These soluble complexes frequently manage to avoid being engulfed by macrophages and are deposited in the glomeruli of the kidneys, in the lymph nodes, in the small blood vessels, in the joints, or in the skin. Once in these areas, these soluble complexes are able to activate the complement system, resulting in the production of large quantities of anaphylatoxins, C3a and C5a, which cause inflammation and chemotaxis—that is, they attract large numbers of PMNs to the area. These neutrophils are short-lived, and many disintegrate in these localized areas. They release their proteolytic and hydrolytic enzymes, which then cause local tissue destruction (necrosis) (figure 21.14). When this process occurs in the joints, it results in **rheumatoid arthritis;** in the glomeruli of the kidneys, it is called **glomerulonephritis;** in the lymph nodes it results in **lymphadenitis;** and in the blood vessels, it is called **vasculitis.** In all cases it results in the production of painful inflammatory lesions.

### Type IV Hypersensitivity (Delayed-Type Hypersensitivity)

This allergic reaction is caused by the $T_{DTH}$ cell, which causes inflammation through production of certain lymphokines. It is called delayed hypersensitivity because it usually occurs 24 to 48 hours after the presensitized person is exposed to a certain allergic substance. Type IV reactions are typically caused by antigens that are bound to cells. Initially, however, the antigen is first engulfed and processed by macrophages, then presented to $T_{DTH}$ cells. These $T_{DTH}$ cells now roam the tissues until they recognize a cell containing the epitope on its surface. Frequently these antigens are natural products such as

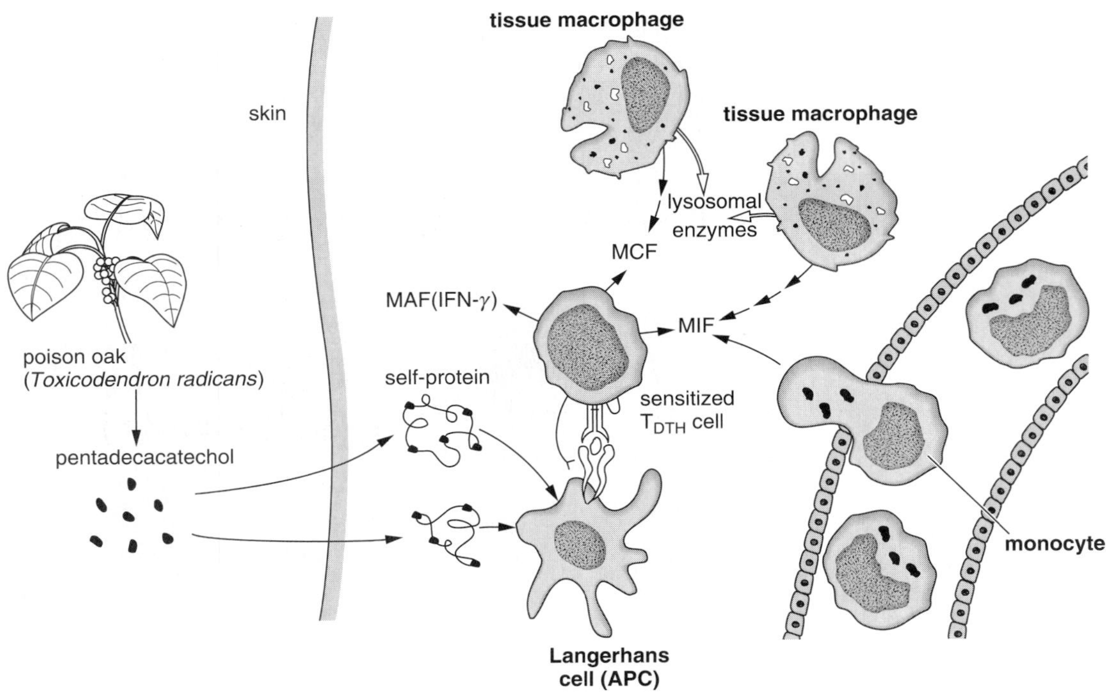


**figure 21.15**

The development of a type IV hypersensitivity reaction (delayed-type hypersensitivity) following exposure to poison oak. Tissue damage results from the release of lysosomal enzymes from tissue macrophages.

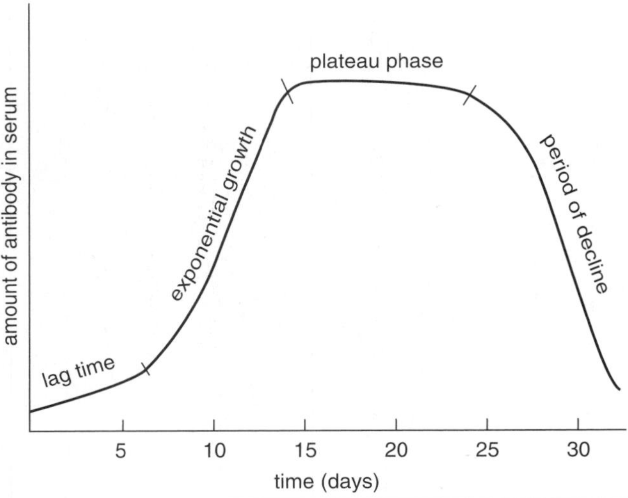


**figure 21.16**

The four phases of the humoral immune response.

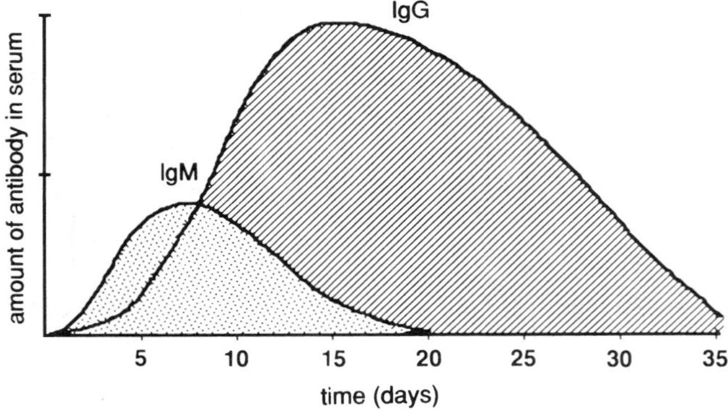


**figure 21.17**

The relative production of antibodies associated with the primary humoral response.

pentadecacatechols present in poison ivy and poison oak, or they may be industrial agents such as trinitrophenol chloride, dinitrochlorobenzene, nickel compounds, or dichromates used in the leather industry. When these compounds penetrate the skin, they bind to appropriate body cells, such as Langerhans cells. $T_{DTH}$ cells then bind to these cells, become activated, and produce a number of special lymphokines such as macrophage chemotactic factor (MCF), macrophage-activating factor (MAF), and migration-inhibiting factor (MIF). These macrophages interact with the antigen-coated cells and destroy them; in the process they may also cause local inflammation and tissue necrosis (figure 21.15). Examples of type IV hypersensitivity are **contact dermatitis** and **eczema.**

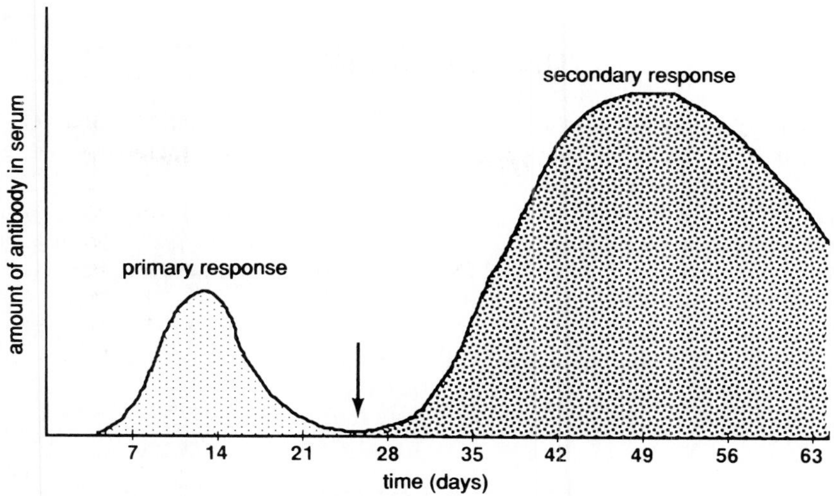


**figure 21.18**

Total antibody production during the primary and the secondary humoral immune responses. Reexposure to the antigen (marked by arrow) occurs at 25 days.

## STEPS IN B- AND T-CELL RESPONSES

The body goes through a series of steps in producing appropriate immune responses. When an antigen is first presented to an appropriate B- or T-cell clone by the APCs, the cells begin to divide and the clone enlarges and becomes active. However the total response is usually rather weak, as the clone is still small and few memory cells are present. In subsequent exposures to the antigen, however, the immune response becomes stronger and more specific as larger clones of activated cells develop.

### Humoral Immune Response

Following the initial activation of a B-cell clone, plasma cells and memory cells are produced. The production of antibodies by the plasma cells goes through four phases. First is the characteristic **lag period** of several days, when antibodies are not detectable. This is followed by a second phase, in which there is an **exponential increase** in immunoglobulins. The third phase is called the **plateau phase,** which occurs when a certain peak is reached and an increase in antibody levels can no longer be detected. The last phase is a **period of decline,** in which the antibody levels gradually diminish as they are cleared from the serum (figure 21.16).

This first exposure to a given antigenic determinant results in production of a biphasic response. Initially only IgM is produced, but after about a week the presence of IgG can be detected. Usually IgM disappears first from the serum, and the level of IgG lingers (figure 21.17). Further, the levels of IgG appear to be much greater than the initial levels of IgM. The actual levels of antibody production during this first exposure depend on a number of factors such as the amount of antigen, the condition of the host, the route of injection or infection, and the nature of the antigen.

If the individual is exposed a second time to the same antigenic determinant, the immune response is qualitatively and quantitatively different from the first response. Now a much larger clone is available because of the presence of many memory cells. These memory cells are long-lived and are able to produce a rapid secondary response. This **secondary** or **anamnestic response** has a shorter lag time and a greater exponential phase, resulting in greater antibody production, an extended plateau, and slower decline. The serum levels of IgG may be 10 to 100 times that of the primary response (figure 21.18).

### CMI Response

The cell-mediated immune response goes through a similar cycle. The initial clone of T cells is low, but once it is activated, it will produce large numbers of memory cells in addition to numerous activated HTLs ($T_H$ cells)and CTLs ($T_C$ cells). A second exposure to the antigen will result in a stronger anamnestic (or booster) response with shorter lag time, greater exponential growth, a longer plateau phase, and slower decline.

# Further Readings

Boggs, D. R. Lymphocytes and the Immune System. In *White Cell Manual,* 4th ed., F. A. Davis Company, Philadelphia, 1983.

Coleman, R. M., Lombard, M. F., and Sicard, R. E. Complement and Its Role in Immune Responses. Chapter 10 in *Fundamental Immunology,* 2nd ed., Wm. C. Brown Publishers, Dubuque, Iowa, 1992.

Goodman, J. W. Immunoglobulins I: Structure and Function. Chapter 9 in *Basic and Clinical Immunology,* 7th ed., Appleton & Lange, Norwalk, Conn., 1991.

Kuby, J. Cell-Mediated Immunity. Chapter 13 in *Immunology.* W. H. Freeman and Company, New York, 1992.

Terr, A. I. Allergic Diseases. Section IIIB in *Basic and Clinical Immunology,* 7th ed., Appleton & Lange, Norwalk, Conn., 1991.

# Review Questions

1. List the five major groups and functions of lymphocytes.
2. Name the lymphocytes involved in innate immunity.
3. How are NK cells activated? How do they function? What does the abbreviation ADCC stand for?
4. Which lymphocytes play a major role in immunoregulation?
5. How are helper T cells activated? How do helper T cells activate other cells of the immune system?
6. Name the group of lymphocytes primarily involved in humoral immunity.
7. List the five classes of immunoglobulins produced by activated B cells.
8. Give the five major functions of antibodies.
9. What is opsonization? Why is it important in the immune process?
10. Distinguish between agglutination and precipitation. Why are each of these processes important for the immune response?
11. Define neutralization. Name the two major forms. Why is the process of neutralization important in the immune response?
12. What is meant by the complement system? In what two ways can the system be activated? What is each pathway called?
13. List the complement components involved in the classical system of complement activation. Describe the process of complement activation via the classical pathway. What is the ultimate result of the activation of complement?
14. What is a MAC? Describe its makeup.
15. List the major functions of C3a and C5a.
16. Name the major groups of lymphocytes involved in cell-mediated immunity.
17. How and where are cytotoxic T lymphocytes activated? What are cytokines (or lymphokines)?
18. How do cytotoxic T lymphocytes kill their target cells?
19. Which groups of lymphocytes are involved in hypersensitivity reactions?
20. List the four types of hypersensitivity reactions according to Gell and Coombs. Provide alternate names for each of these classes of reactions.
21. Which cells and antibodies are involved in type I hypersensitivity reactions?
22. How are type I hypersensitivity reactions induced? Which are the effector cells?
23. Give examples of type I hypersensitivity reactions.
24. Which antibody classes are involved in type II hypersensitivity reactions and with what body cells do they react?
25. Give examples of type II and type III hypersensitivity reactions. Explain in each case how the hypersensitivity reaction is produced.
26. Which cells are involved in type IV hypersensitivity reactions? Describe the normal function of these cells.
27. Give examples of type IV hypersensitivity reactions. Explain how the allergic condition is produced.
28. List the four phases of the primary humoral immune response. What happens in each phase? How long does each phase last?
29. Compare and contrast the primary and secondary humoral immune responses. Why is the secondary immune response usually much bigger?
30. What is meant by an anamnestic or booster response? Why is it often needed?
31. Compare the humoral immune response to the CMI response.

# Chapter Twenty-Two

## *Qualitative Disorders of Granulocytes and Monocytes*

This chapter begins with a discussion of the qualitative disorders of granulocytes. This is followed by a description of the qualitative diseases of monocytes. All the major qualitative granulocytic disorders may be classified as **neutrophil** abnormalities. There are no major disorders associated with abnormal eosinophil or basophil structure.

## QUALITATIVE DISORDERS OF GRANULOCYTES

The major function of neutrophils is to execute a first line of defense to prevent the initial spread of infectious organisms. The job of the polymorphonuclear neutrophils (PMNs) is to prevent replication of the pathogenic organisms until the specific immune system is ready to assume the task of removing the foreign agents. To execute that role, neutrophils must perform several integrated functions: (1) unidirectional movement toward the pathogenic particles, (2) phagocytosing these foreign agents, and (3) killing these microorganisms or at least preventing them from replicating at will. The first two functions require the presence of both extrinsic and intrinsic factors. The extrinsic or external factors include the presence of chemotactic substances, activated complement factors, and antibodies. The intrinsic or internal factors include an intact metabolism, a functional microtubular cytoskeleton to ensure proper movement, and the presence of appropriate receptors for antibodies and activated complement factors. The killing function requires only the presence of intrinsic factors, such as the ability to produce hydrogen peroxide and other toxic substances. The absence of, or interference with, any of these essential extrinsic or intrinsic factors will result in an impaired function of the neutrophils, and hence an increased susceptibility to infectious diseases.

### Extrinsic Neutrophil Disorders

Neutrophils are unable to perform their functions properly if they are not activated by appropriate stimuli, nor are they able to phagocytose efficiently without the presence of opsonins covering the infectious agents. The most important of these extrinsic factors are discussed in the following material.

#### *Antibody Deficiencies*

As explained in chapter 21, opsonization of pathogenic organisms by specific antibodies is essential for effective phagocytosis. Neutrophils have receptors on their cell membranes for the Fc fractions of immunoglobulin molecules. By binding to the Fc fractions of antibody molecules coating the microbial particles, PMNs are able to make better contact with these pathogens. This is a necessary first step in efficient phagocytosis (figure 22.1). Impaired antibody synthesis, a condition that is commonly known as **hypogammaglobulinemia,** will result in deficient opsonization. A number of inherited and acquired disorders are associated with hypogammaglobulinemia, or sometimes even complete absence of antibody production known as **agammaglobulinemia** or **Bruton's disease** (*see chapter 25*).

#### *Complement Deficiencies*

Activated complement factors have a number of functions essential for proper neutrophil activity. Certain activated complement factors function as neutrophil **chemotactic agents,** such as C3a and C5a. These same activated complement factors also produce inflammation in the infected areas (i.e., **anaphylatoxins**), thus facilitating the passage of white blood cells from the circulation of the tissues. In addition, certain activated complement factors act as **opsonins,** for which neutrophils have membrane receptors such as C3b. Thus the presence of activated complement facilitates the process of phagocytosis by the PMNs (figure 22.1).

Clearly the absence or deficiency of complement factors in the blood will result in impaired anaphylactic and chemotactic responses as well as a reduction in the ability of the PMN to phagocytose. Absence of complement factors therefore will result in increased infections.

#### *Interference by Drugs and Toxins*

The role of neutrophils in preventing the spread of infectious diseases may also be impaired by a number of chemical compounds. For instance, certain steroid drugs not only block the chemotactic response of the neutrophils, but also inhibit the process of inflammation by tightening small blood vessels, thus preventing the PMNs from leaving the circulation at the appropriate places. Alcohol has been found to inhibit the unidirectional movement of neutrophils but not their phagocytic or microbicidal activity.

Antimetabolic drugs, such as the ones used in cancer chemotherapy, will prevent an increase in PMN production by the bone marrow at a time when greater numbers are needed. Hence people undergoing chemotherapy are more prone to infectious disease than others.

#### *Interference by High Osmolarity*

Increased osmolarity of the blood may affect the osmotic activity of neutrophils and cause them to become swollen. This will impair not only their movement but also their phagocytic ability. High osmolarity in PMNs is frequently observed in diabetic patients who show signs of ketoacidosis. As a result of this condition they are more prone to infections.

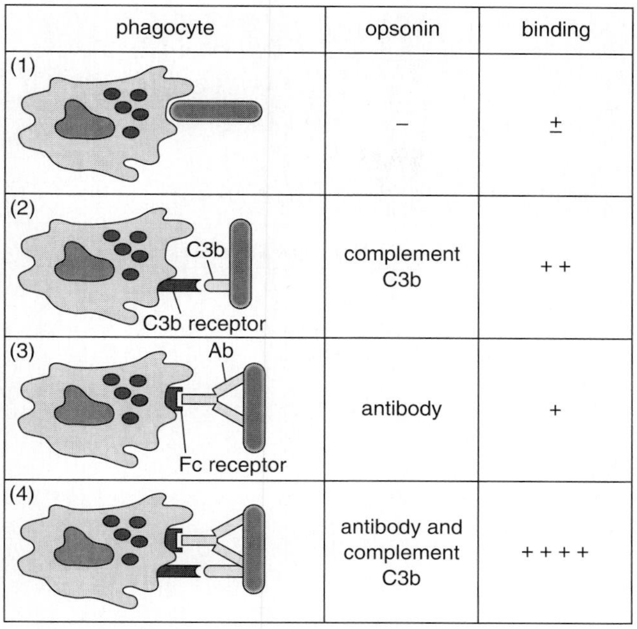

| phagocyte | opsonin | binding |
|---|---|---|
| (1) | – | ± |
| (2) C3b, C3b receptor | complement C3b | + + |
| (3) Ab, Fc receptor | antibody | + |
| (4) | antibody and complement C3b | + + + + |

**figure 22.1**

The effects of opsonization on the ability of macrophages to bind to foreign bacteria and other microorganisms. The stronger the binding, the quicker the process of phagocytosis.

*Effects of Bacterial Products*

A number of microorganisms have developed defense mechanisms that prevent them from being readily phagocytosed by neutrophils. For example, pneumococci have smooth polysaccharide coats, which protects them from opsonization. Therefore in the absence of specific opsonins, the PMNs are unable to make firm contact with these "slippery" bacteria, thus preventing their timely phagocytosis.

Other bacteria, such as *Staphylococcus aureus,* produce a certain protein, called **protein A,** that interferes with the Fc fraction of the immunoglobulin molecule, thus preventing the PMNs from binding to the antibodies that coat these bacteria. Other bacteria produce cytotoxins that cause degranulation of the neutrophils, thus rendering them ineffective as microbicidal agents.

## Intrinsic Neutrophil Disorders

Intrinsic neutrophil abnormalities involve disorders of locomotion and ingestion; others involve the inability of the PMNs to kill the ingested pathogens. Each of these abnormalities is associated with a distinct disorder. Some of the more well-known syndromes are discussed in the following material.

*Actin Dysfunction Syndrome*

Actin dysfunction syndrome is a disorder of locomotion and ingestion. It is a rare inherited disease in which the patient's PMNs are unable to move unidirectionally; in addition, the PMNs are unable to phagocytose properly because actin molecules in their cytoplasm cannot polymerize properly into long filaments.

*Chédiak-Higashi Syndrome*

Chédiak-Higashi syndrome is another rare congenital disorder that is transmitted as an autosomal recessive trait. This disease is associated with the presence of a few giant lysosomes in the PMNs. These huge lysosomes are nonfunctional—that is, they are unable to fuse normally with phagosomes containing ingested microorganisms. Hence they are unable to kill ingested pathogens. Further, these abnormal neutrophils also show defective chemotaxis—that is, they are unable to move properly to an infected area, which may be due to the presence of defective microtubular proteins. Children born with this disorder frequently die young as a result of repeated massive infections. Those that do survive into adulthood often develop a lymphoid malignancy later in life.

Finally, Chédiak-Higashi syndrome is usually associated with albinism—that is, hypopigmentation of the skin and the eyes, resulting in photophobia.

*Chronic Granulomatous Disease*

Chronic granulomatous disease (CGD) is another rare congenital disorder that may be inherited as either an X-linked or autosomal recessive trait. In this disease the phagocytic cell is unable to produce the respiratory burst necessary for the generation of hydrogen peroxide ($H_2O_2$) and superoxide ($O_2^-$), both of which are needed for microbial killing. This inability to metabolize molecular oxygen may be due to either the lack of oxidases or the inability of these enzymes to perform their function. As a result, these neutrophils are unable to kill catalase-positive bacteria—that is, bacteria able to break down the $H_2O_2$ they produce themselves. This group includes *Staphylococcus aureus,* most gram-negative enteric bacteria, and many fungi. In contrast, catalase-negative organisms produce $H_2O_2$ that is readily released into phagocytic vacuoles, where it interacts with granule-derived myeloperoxidase to produce toxic halides that are lethal to these microorganisms. In other words, the catalase-negative bacteria engineer their own death in these CGD phagocytes. These pathogens include many streptococci, pneumococci, and certain *Hemophilus* bacteria.

The onset of chronic granulomatous disease almost invariably starts during the first years of life. The major symptoms are (1) susceptibility to severe infections and

(2) formation of many tissue granulomas as a result of tissue macrophages developing into giant cells in an effort to contain infections. Hence the name chronic **granulomatous disease.**

The most direct way to diagnose the presence of CGD is to measure oxygen consumption before and after particle ingestion with either an oxygen electrode or a Warburg apparatus. Since these are comparatively complex tests, most laboratories use a simple dye test, the nitroblue tetrazolium (NBT) test. When NBT is added to normal neutrophils, the dye is taken up and reduced to an insoluble blue product that accumulates inside the cell, thus producing deep-blue stained cells. CGD neutrophils are also able to ingest the dye but are unable to reduce it, hence little blue color is produced in these cells.

Symptoms of CGD begin to appear during the first 2 years of life. The disease is seen primarily in males. Heterozygous females with the X-linked form exhibit diminished bacteriocidal ability, but not the severity exhibited by males.

### *Myeloperoxidase Deficiency Syndrome*

Myeloperoxidase deficiency syndrome is an autosomal recessive disorder in which the neutrophils are unable to produce the enzyme myeloperoxidase (MPO). As explained in chapter 17, myeloperoxidase is needed to oxidize $H_2O_2$ into toxic halides, which kill bacteria in the phagolysosomes of phagocytic cells. Lack of MPO reduces the microbicidal ability of the neutrophils. The defect is less severe than that produced by CGD phagocytes, and consequently bacterial infections tend to be less severe than in chronic granulomatous disease. In addition to the inherited form of MPO deficiency, there appears to be an acquired form of this disorder. In the latter condition both normal and MPO-deficient PMNs are found in the circulation.

### *Other Conditions*

The neutrophils of people with severe infections may have diminished microbicidal activity. The reasons for this phenomenon are not clear; however the increased demand for more granulocytes may temporarily overtax the bone marrow, resulting in impaired enzyme and granule formation. Patients with severe burns are deficient in granular enzymes in their PMNs and hence in their microbicidal activity. A summary of the various factors that may cause impairment in neutrophil function is given in table 22.1.

**Table 22.1 Extrinsic and Intrinsic Factors Associated with the Impairment of Neutrophil Functioning**

| **Extrinsic Neutrophil Defects** |
|---|
| *Antibody Deficiencies* |
| Congenital agammaglobulinemia |
| Acquired agammaglobulinemia |
| Immunological suppression |
| Lymphocytic leukemia |
| Multiple myeloma |
| *Complement Deficiencies* |
| Lack of factor C3 |
| Lack of factor C5 |
| *Drugs and Toxins* |
| Corticosteroids |
| Ethanol |
| Cytotoxic drugs |
| *Hyperosmolarity* |
| Diabetic ketoacidosis |
| *Bacterial Products* |
| *Staphylococcus aureus* (protein A) |
| Bacterial cytotoxins |
| **Intrinsic Neutrophil Defects** |
| *Abnormalities in Cytoplasmic Structures* |
| Actin dysfunction syndrome |
| Chédiak-Higashi syndrome |
| *Enzyme Deficiencies* |
| Chronic granulomatous disease |
| Myeloperoxidase deficiency syndrome |
| *Other Causes* |
| Overwhelming infections |
| Myelocytic leukemia |

## QUALITATIVE DISORDERS OF MONOCYTES

The functions of monocytes and macrophages are many and varied (*see chapter 19*), but ultimately all functions involve the ingestion and digestion of foreign materials as well as the phagocytosis of senescent and damaged body cells and cell products. Normal phagocytic cells will ingest only as much material as they can digest and dispose of by normal excretory processes. If more organic material accumulates within the cell than it can clear, the excess material will interfere with normal metabolic activities of the cell, which develops what is known as a "storage disease." Most qualitative monocytic disorders may be classified as **storage diseases.** Different causes for such storage disorders in monocytes and macrophages are listed in table 22.2.

One of the most common causes is the absence of, or a defect in, one of the enzymes needed for the breakdown of ingested materials. As a result half-catabolized products will accumulate which the cell is unable to dispose of. A second cause is the excess accumulation of products that are normally nontoxic to the cell when present in small quantities. For instance, normal monocytes and

**Table 22.2** Causes of Storage Disorders in Monocytes and Macrophages

*Lack of One or More Enzymes Necessary to Break Down Ingested Materials*
- Gaucher's disease
- Niemann-Pick disease
- Fabry's disease
- Wolman disease
- Tangier disease

*Excessive Accumulation of Normally Nontoxic Materials*
- Hemochromatosis
- Gout

*More Rapid Ingestion of Materials than Can be Cleared by the Cells*
- Chronic myelocytic leukemia
- Foamy alveolar macrophages

macrophages are able to accumulate small quantities of iron. When the normal body iron stores—that is, mucosal cells of the gut and mononuclear phagocytic cells of the bone marrow—become overloaded, the excess iron will be deposited in other body cells including the various types of macrophages where they may have a toxic effect on these cells. This condition is known as **hemochromatosis.** Finally, some storage disorders may be the result of a rapid accumulation of digested materials. Macrophages seem to accumulate materials faster than they can eliminate them. As a result there will be an excess storage of these materials, which may interfere with the normal functioning of these cells.

### Storage Disorders Caused by Enzyme Deficiencies

Most of the storage diseases in monocytes and macrophages are caused by deficiencies in enzymes involved with glycolipid degradation. Hence these storage disorders are collectively known as **lipid storage diseases** or **lipidoses.** As a result of the absence of these enzymes, various lipid substrates accumulate that cannot be acted upon by other enzymes. A specific clinical syndrome is associated with the presence of each of these abnormal substrates (table 22.3). Each of these disorders is discussed in the following material.

#### *Gaucher's Disease*

Gaucher's disease is an inherited syndrome characterized by the absence of the enzyme **β-glucocerebrosidase** and the accumulation of the substrate glucocerebroside in monocytes and macrophages. The disease is inherited as an autosomal recessive trait and appears to be especially common in Ashkenazi Jews (Jewish people of Eastern European descent). As a result of the accumulation of this lipid material in the cell, macrophages gradually lose their functional abilities.

Three different forms of the disease are recognized, commonly labeled type I, II, and III. **Type I disease** is chronic and first appears in older children and adults. It is also the most common form of Gaucher's disease and does not involve the nervous system. Estimates are that about one in 50 Ashkenazi Jews is a carrier for type I disease. **Type II** is an acute, progressive disease that appears in very young infants and is associated with severe neurological damage; **type III** is a subacute disease that appears in children between one and eight years of age and is most commonly associated with bone destruction. The prognosis for survival is very poor for types II and III: usually death occurs within a few years of the onset of the disease. Type I has a much better prognosis. Most people with this form of Gaucher's disease live fairly normal lives until adulthood or even old age. The clinical symptoms of this disease frequently include splenomegaly, liver abnormalities, bone pain and bone lesions, pulmonary problems, and skin hyperpigmentation. Sometimes bleeding and purpura is present as a result of thrombocytopenia. These clinical findings are usually the result of an accumulation of lipid-filled macrophages, called **Gaucher cells,** in these tissues.

Laboratory findings include first and foremost the presence of Gaucher cells in the bone marrow, lymphoid tissues, spleen, liver, and other organs. These cells are usually very large cells (between 20 and 80 μm in diameter) with a small eccentric nucleus. The cytoplasm appears to be filled with lipid inclusions that frequently look like crinkled tissue paper (figure 22.2). Sometimes the cytoplasm has a spongelike appearance, indicating the presence of many vacuoles. The cytoplasm will stain strongly positive with periodic acid-Schiff (PAS). Other laboratory findings may include a moderate normocytic-normochromic anemia and sometimes panleukopenia, especially when the splenomegaly is marked.

The same applies to the platelets—the larger the spleen, the fewer platelets in the circulation.

The most sensitive test for establishing Gaucher's disease, including the presence of the carrier (heterozygote) state, is the measurement of leukocyte or cultured fibroblast β-glucosidase activity. This test can also be done prenatally by measuring the same activity in cultured cells obtained by amniocentesis. Treatment for this disease is mainly supportive. Enzyme replacement therapy has been attempted but without much success. Management of the disease may include splenectomy and treatment to reduce bone pain and bone lesions.

#### *Niemann-Pick Disease*

This is another autosomal recessive disorder commonly seen in Ashkenazi Jews. This disease is caused by a deficiency of the enzyme **sphingomyelinase,** which results in an accumulation of the substrate sphingomyelin in the macrophages. Large monocytes and macrophages filled with sphingomyelin appear in the bone marrow and lymphoid organs as large cells with an eccentric nuclei and cytoplasm full of

**Table 22.3** The Major Lipid Storage Disorders

| *Disorder* | *Missing Enzyme* | *Accumulated Substrate* |
|---|---|---|
| Gaucher's disease | β-Glucocerebrosidase | Glucocerebroside |
| Niemann-Pick disease | Sphingomyelinase | Sphingomyelin |
| Fabry's disease | α-Galactosidase | Trihexosyl ceramide |
| Wolman disease | Acid esterase | Triglycerides and cholesterol esters |
| Tangier disease | High-density lipoprotein | Cholesterol esters |

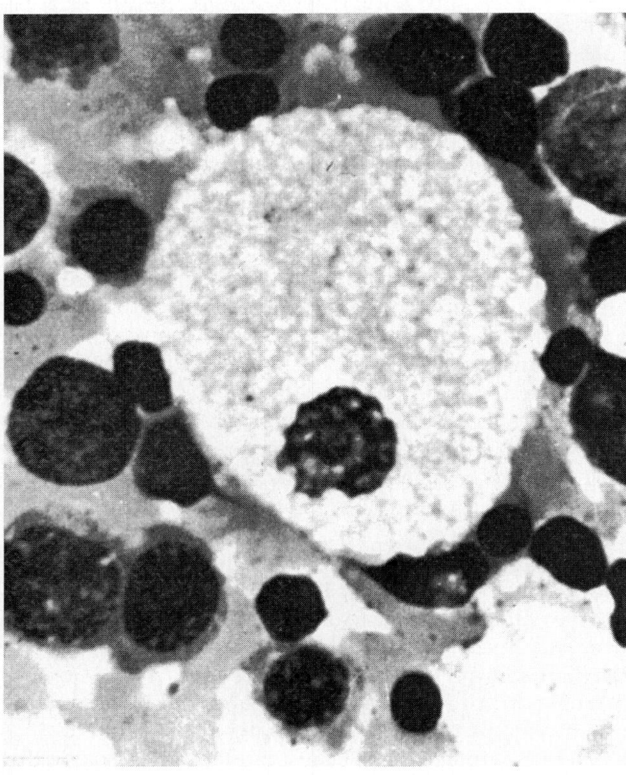

figure 22.2
Photomicrograph of a Gaucher cell.

globular inclusions, which give these cells a "foamy" look. Peripheral blood monocytes frequently show characteristic vacuoles, which are fat-filled liposomes. Usually the first signs of this rare disorder appear in young infants. These children show poor physical and mental development and soon exhibit considerable enlargement of the liver and spleen. Nervous system abnormalities and severe mental retardation become apparent during the second and third year of life. This is a progressive disease, and children rarely survive until their third birthday.

Niemann-Pick cells are similar in size (20 to 80 μm in diameter) to the Gaucher cells, but the cytoplasm has a globular (foamy) appearance rather than a crinkly one. Staining of these cells can be done with Oil Red O, or Luxor Fast Blue, or with Sudan Black B.

Because of the enlarged liver and spleen, moderate leukocytopenia and thrombocytopenia is frequently observed. Diagnosis of this disorder is done by demonstrating the absence of sphingomyelinase in the white blood cells.

*Fabry's Disease*

This is another lipid storage disorder. It is a sex-linked syndrome that also affects heterozygote female carriers to some extent. In this disease the macrophages accumulate the glycolipid trihexosyl ceramide because of a deficiency of **α-galactosidase.** People with this disorder usually live until adulthood, but many die in their thirties or forties due to renal failure.

*Wolman Disease*

This disorder is associated with a deficiency of **acid esterase.** As a result triglycerides and cholesterol accumulate in the monocytes and macrophages. Disease symptoms are similar to those of Niemann-Pick disease, and death usually occurs early in life.

*Tangier Disease*

This is a recessively inherited disorder in which the patient is unable to produce **high-density lipoprotein (HDL).** As a result cholesterol esters will accumulate in the macrophages. This disease is usually benign in childhood, but is frequently associated with splenomegaly, hepatomegaly, and enlarged lymph nodes. Later in life lipid deposits will occur in the pulmonary and coronary circulations, and this may result in blockage and infarcts.

### Storage Disorders Caused by a Toxic Accumulation of Materials

*Hemochromatosis*

This qualitative disorder of monocytes and macrophages is due to excess deposits of iron in these cells. This syndrome has several causes, one of which is multiple blood transfusions. As explained in chapter 11, the body has only a limited storage capacity for iron. It also is unable to eliminate excess iron at a rapid rate. The best the body can do is about 1 to 2 mg/day. As a result of frequent blood transfusions, excess iron will first be

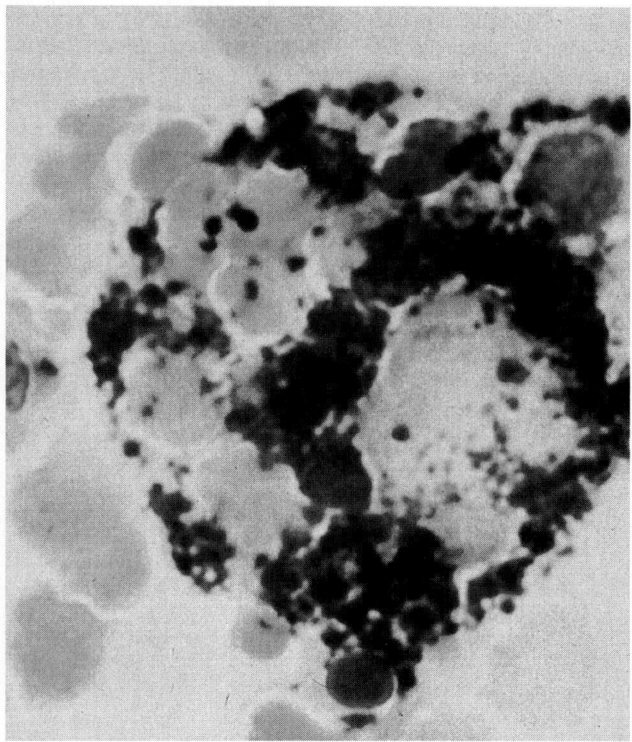

figure 22.3

A bone marrow aspirate shows a hemosiderin-laden macrophage from a patient with hemochromatosis.
From The American Society of Hematology Slide Bank, 3rd edition, 1990, used with permission.

stored in the mucosal cells of the gut and in the mononuclear phagocytic cells of the spleen, liver, and bone marrow. When these are saturated (figure 22.3) other cells become depositories of excess iron, including macrophages, β cells of the islets of Langerhans, cells of the gonads, and certain skin cells, which may cause bronzing of the skin, hence the term **hemochromatosis.** Hemochromatosis may also be the result of an inherited condition in which individuals absorb more iron from the intestine than normal people do. Usually people absorb only as much dietary iron as they lose during a day, so that iron levels stay constant. People with this inherited condition absorb more than they lose, and hence they accumulate iron in their tissues. In certain cases alcoholic cirrhosis may also result in an accumulation of iron in other than the normal storage cells. Finally, a number of cases of hemochromatosis are classified as idiopathic, as the cause is not known.

The clinical symptoms of hemochromatosis are many and varied. They may include anemia with its resulting fatigue and concentration problems, increased susceptibility to infections due to a lack of correctly functioning macrophages, and splenomegaly and hepatomegaly due to an accumulation of macrophages in these organs. Abdominal and joint pains have also been reported as a result of excess iron accumulation.

Laboratory investigations will show increased levels of serum iron, and increased total iron-binding capacity (TIBC) and ferritin. Treatment should include a diet low in iron. In certain situations periodic phlebotomies may be needed to reduce the accumulated iron.

#### *Gout*

This is a disorder in which there is an excessive production of **microcrystalline uric acid.** This is deposited in many tissues, such as the joints, where these substances cause inflammation. They can also be deposited in macrophages, and when they accumulate in these cells, they may interfere with the normal functioning of these cells. This may result in a greater risk of infection.

### Storage Disorders Caused by Clearing Problems

#### *Chronic Myelocytic Leukemia*

In chronic myelocytic leukemia macrophages will ingest and digest many of the abnormal granulocytes produced. They frequently will ingest more materials than they can digest and eliminate. As a result there is a tendency to accumulate excess phospholipid, and this accumulated material may interfere with the normal functioning of these macrophages.

#### *Foamy Alveolar Macrophages*

Pneumonitis is frequently associated with the presence of lipid-laden macrophages. These foamy alveolar macrophages may not be able to function properly in removing foreign agents from the lungs. The development of foamy alveolar macrophages may be due to the production of excessive surfactant. A major constituent of surfactant is the lipid dipalmityllecithin. Excessive surfactant is normally cleared away by the alveolar macrophages, which may account for the accumulation of lipid materials in these histiocytes.

## *Further Readings*

Cline, M. J. Abnormalities of Neutrophil Function. Chapter 8 in *The White Cell.* Harvard University Press, Cambridge, Mass., 1975.

Crocker, A. C. Histiocytoses and Lipidoses. Lecture 25 in *Hematology,* 4th ed., W. S. Beck, Ed., The MIT Press, Cambridge, Mass., 1985.

Powers, L. W. Metabolic Diseases, Deficiencies and Responses of Leukocytes. Chapter 21 in *Diagnostic Hematology*. The C. V. Mosby Company, St. Louis, 1989.

Stossel, T. P. Leukocytes II. Phagocytosis and its Disorders. Lecture 18 in *Hematology,* 4th ed., W. S. Beck, Ed., The MIT Press, Cambridge, Mass., 1985.

## *Review Questions*

1. In order to do their job properly, neutrophils must be able to perform which three important integrated functions?
2. List five important extrinsic factors that may interfere with the ability of leukocytes to phagocytose.
3. Explain how deficiencies in either antibodies or activated complement factors may interfere with proper phagocyte functioning.
4. How may drugs and toxins interfere with neutrophil functioning?
5. Explain how high osmolarity may interfere with proper neutrophil functioning.
6. Describe some of the methods developed by bacteria to escape phagocytosis and digestion by PMNs.
7. List three important disorders caused by intrinsic neutrophil deficiencies. In each case mention the nature of the deficiency.
8. What is the Chédiak-Higashi syndrome? What is its cause? Why is it usually a lethal disease?
9. What is CGD? What is its basic cause? What is the nature of the deficiency?
10. Explain why granulomas are formed in CGD.
11. How would you distinguish between X-linked and autosomal CGD?
12. Explain the function of myeloperoxidase in neutrophils. What happens when it is deficient?
13. Qualitative diseases of monocytes and macrophages are mainly storage disorders. Name the three major categories of storage disorders.
14. List five different lipid storage disorders.
15. What enzyme is missing in Gaucher's disease? What compound tends to accumulate?
16. In what group of people is there a high incidence of Gaucher's disease?
17. List the three types of Gaucher's disease. Which is the most common? Which one has the best prognosis?
18. What major clinical symptoms are associated with Gaucher's disease?
19. What are Gaucher cells? How are they stained?
20. What test is most sensitive for establishing the presence of Gaucher's disease?
21. What enzyme is missing in Niemann-Pick disease? What compound tends to accumulate?
22. Describe the appearance of the monocytes in Niemann-Pick disease.
23. Give the clinical symptoms and prognosis for Niemann-Pick disease. In which group of people is there a high incidence of this disorder?
24. What enzyme is missing in Fabry's disease? What compound tends to accumulate? What is the prognosis for this disorder?
25. Name the enzyme missing in Wolman disease. What compound tends to accumulate? Give the symptoms and prognosis.
26. What causes Tangier disease? What happens to the monocytes? Describe some of the major symptoms and effects.
27. What is hemochromatosis? Mention some of the major causes for this condition.
28. Name the normal storage areas for iron in the body. Which cells and tissues are affected in cases of iron overload?
29. How does hemochromatosis affect the functioning of monocytes and macrophages?
30. Name some important clinical symptoms of hemochromatosis.
31. What are the major laboratory findings of hemochromatosis?
32. What is a common treatment for hemochromatosis?
33. Define gout. How does it affect macrophages?
34. Explain how chronic myelocytic leukemia may affect the normal functioning of monocytes and macrophages.
35. What causes the development of foamy alveolar macrophages? Under what circumstances do these foamy alveolar macrophages usually develop?

# Chapter Twenty-Three

## *Quantitative Disorders of Granulocytes*

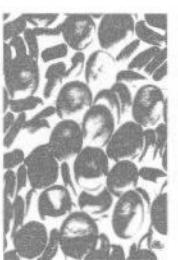

In this chapter the quantitative disorders of the neutrophils are discussed, followed by a discussion of eosinophilia and basophilia.

## QUANTITATIVE DISORDERS OF NEUTROPHILS: NEUTROPENIA AND NEUTROPHILIA

Abnormalities in the number of circulating polymorphonuclear neutrophils (PMNs) are the result of one or more of the following three factors: (1) rate of neutrophil production in the bone marrow, (2) rate of neutrophil utilization and/or destruction in the peripheral bloodstream and body tissues, and (3) distribution of the PMNs between the circulating and marginating pools.

A condition known as neutrocytopenia, or simply **neutropenia,** is produced when the number of circulating neutrophils falls below normal counts. Often the term **granulocytopenia** is used synonymously with neutropenia, since the overwhelming majority of granulocytes are neutrophils. When increased numbers of neutrophils are in the blood, the term **neutrophilia** is often used to denote that condition. It may be the result of an excessive production of PMNs by the bone marrow, or it may occur as a result of a shift in the various neutrophil compartments—for example, from the marginating pool to the circulating pool. Frequently the term **granulocytosis** is used as an equivalent for neutrophilia, as both are indications of increased numbers of neutrophils.

Whether or not clinical symptoms are associated with neutropenia depends on the magnitude of the reduction in the number of neutrophils in the blood. Overt symptoms are rarely noted as long as the count remains above 1,000 neutrophils per microliter. The risk of infection is moderately increased when the count is between 500 to 1,000 PMNs per microliter. Only when the count falls below 500 PMNs per microliter is the risk of infection greatly increased.

### Etiology of Neutropenia

From an etiological point of view, we can classify all neutropenias into three major categories: (1) those resulting from inadequate granulopoiesis, (2) those due to excessive destruction of neutrophils, and (3) those caused by abnormalities in distribution.

*Neutropenia Due to Inadequate Granulopoiesis*

Many different factors may result in inadequate granulopoiesis of the bone marrow. These include bone marrow injury, bone marrow replacement, nutritional deficiencies, abnormal control mechanisms, congenital abnormalities, or it may be secondary to other diseases or idiopathic.

**Bone Marrow Injury** Injury to the hematopoietic cells of the bone marrow may result from a variety of causes, some of which are easily identified, whereas others remain a mystery. The most common causes are as follows:

**Table 23.1 Cytotoxic Drugs that Cause Pronounced Neutropenia**

*Antimetabolites*
- Methotrexate
- 6-Mercaptopurine
- 6-Thioguanine
- Cytosine arabinoside
- Hydroxyurea

*Alkylating Agents*
- Cyclophosphamide
- Chlorambucil
- Busulfan (Myleran)

*DNA Binding*
- Daunorubicin
- Doxorubicin hydrochloride (Adriamycin)

*Mitotic Inhibitor*
- Vincristine (Oncovin)

**1. Irradiation**
Prolonged exposure to high doses of ionizing radiation will interfere with the normal DNA-replicating mechanisms and hence reduce the ability of the cells in the bone marrow to undergo mitosis.

**2. Cytotoxic drugs**
Many drugs used in cancer chemotherapy interfere with the normal mitotic mechanisms of the cells, including bone marrow cells. A list of commonly used drugs that are cytotoxic to the hematopoietic cells of the bone marrow is given in table 23.1.

**3. Nontoxic drugs**
Certain other drugs may interfere with the normal metabolic activities of the cells by reducing their activity without being cytotoxic. Nontoxic drugs may bind to neutrophilic stem cells, where they act as **neoantigens,** thus producing an autoimmune reaction in susceptible persons. These autoimmune reactions frequently result in lysis of affected stem cells. A list of the more common nontoxic drugs is given in table 23.2.

**4. Chemical compounds**
Long-term exposure to certain industrial chemicals may result in damage to the hematopoietic cells of the bone marrow. A list of the more dangerous industrial compounds is given in table 23.3.

**5. Idiopathic**
In many instances a cause for the sudden decrease in neutrophil production cannot be established. These cases are labeled as idiopathic.

**Bone Marrow Replacement** Neutropenia is also a common phenomenon in many malignant disorders

**Table 23.2 Nontoxic Drugs that Occasionally Cause Neutropenia**

*Antimicrobial Agents*
- Chloramphenicol
- Sulfonamides
- Ampicillin

*Phenothiazines*
- Mepazine
- Promazine

*Dibenzoxazepine Compounds*
- Imipramine

*Antihistamines*
- Pyribenzamine

*Anticonvulsants*
- Tridione

*Antithyroid Drugs*
- Methimazole
- Thiouracil

*Anti-inflammatory Agents*
- Indomethacin
- Phenylbutazone

*Diuretics*
- Thiazide
- Acetazolamide

*Type II Diabetes Drugs*
- Tolbutamide
- Chlorpropamide

*Miscellaneous*
- Propranolol
- Procaine amide
- Penicillamine

**Table 23.3 Industrial Chemicals that May Cause Neutropenia**

- Benzene
- Arsenic
- Nitrous oxide
- Carbon tetrachloride
- Thiocyanate
- Bismuth
- Dinitrophenol
- DDT

**Table 23.4 Congenital Disorders Associated with Neutropenia**

- Kostmann's syndrome (autosomal recessive gene)
- Familial benign neutropenia (autosomal dominant gene)
- Familial severe neutropenia (autosomal dominant gene)
- Agammaglobulinemia (uncertain cause; thought to be genetic)
- Chronic hypoplastic neutropenia (uncertain cause)

in which the bone marrow has become infiltrated by tumor cells. This invasion of cancer cells in the bone marrow is known as **myelophthisis** and involves the crowding out of healthy blood stem cells by tumor cells. It is commonly associated with such blood cell disorders as acute leukemia, lymphoma, and multiple myeloma, as well as with nonhematological diseases such as breast cancer, bone cancer, and lung cancer. Neutropenia may also be the result of **myelofibrosis,** a condition where normal hematopoietic elements are replaced by noncancerous fibrous tissue.

**Nutritional Deficiencies** Lack of folic acid, vitamin $B_{12}$, and other essential nutrients not only results in a decreased production of red blood cells but also affects the production of neutrophils, other white blood cells, and platelets.

**Abnormal Control Mechanisms** The level of neutrophil production in the bone marrow remains relatively steady in healthy individuals. A feedback mechanism keeps the number of neutrophils in each of the granulocyte compartments balanced, provided extrinsic or intrinsic factors do not upset the delicate balance. However some individuals may experience periodic oscillations of neutrophil counts, which originate from the bone marrow. These individuals are said to exhibit **periodic** or **cyclic neutropenia.** In this curious phenomenon the neutropenias occur every three weeks or so and last a few days each time. There is great variation in the cycles, some of which may be as short as 14 days, whereas others may occur every 30 to 40 days. The underlying cause is not well understood; it may be inherited or associated with other disorders. Clinical manifestations most often begin early in life; symptoms include fever, oral ulcerations, and susceptibility to skin infections. The disorder is usually benign and is fairly asymptomatic, as the polymorphonuclear neutrophil (PMN) count rarely falls below 500 PMNs per microliter. Many cases of cyclic neutropenia are associated with a corresponding increase in the number of monocytes (monocytosis).

**Congenital Abnormalities** Several inherited or congenital disorders of decreased neutrophil production have been identified. Luckily all of them are rare and most result only in mild neutropenia. However a few syndromes produce severe neutropenia. In the latter cases the children with the inherited trait usually die early due to overwhelming sepsis. A list of the better known congenital neutropenias is given in table 23.4.

**Neutropenia Secondary to Other Diseases** Disorders associated with neutropenia may also be secondary to other disorders. For instance, **paroxysmal nocturnal hemoglobinuria (PNH)** often shows granulocytopenia as a result of bone marrow hypoplasia. Little is known about the cause of this bone marrow depression. A similar scenario is associated with **systemic lupus erythematosus (SLE),** although in this disease both underproduction of neutrophils as well as premature destruction of granulocytes has been implicated. **Hepatic cirrhosis** is another clinical condition often associated with neutropenia.

**Idiopathic Neutropenia** Many cases of underproduction of neutrophils by the bone marrow do not fit into any of the previously discussed categories. No causative agent has been associated with them, and they do not appear to fit any familial patterns—hence they are called **idiopathic.** Clinical manifestations and infectious complications of these and other neutropenias vary, of course, with the severity of lack of PMNs.

*Neutropenia Due to Excessive Destruction*

Neutropenia may also occur when the peripheral depletion rate exceeds the bone marrow production rate for neutrophils. A number of causes may result in excessive neutrophil destruction. These include splenic sequestration, immunological injury, mechanical trauma, and inflammation.

**Splenic Sequestration** Splenomegaly, or splenic enlargement, is a condition that may result from a multitude of causes, as illustrated in table 23.5. Enlargement of the spleen is primarily due to an increase in the red pulp within its cords and sinuses. Blood cells, including neutrophils, readily enter the enlarged spleen, but do not exit it as quickly. The blood cells percolate within the enlarged cords and sinuses and are for all practical purposes no longer available for their normal tasks. Premature destruction of these blood cells may also occur in the enlarged spleen. This condition can be rectified only by the surgical removal of the spleen (splenectomy).

**Immunological Injury** The mechanisms that cause immunological neutropenia are the same that cause immunological hemolytic anemias, which occur as a result of actions by antibodies and complement on the blood cells. **Neoantigens** are produced on the PMN cell surface as a result of drugs or chemicals binding to self-proteins on the neutrophil membrane. The body no longer recognizes this new combination as "self," and starts to develop antibodies against this complex. The antibodies then bind to the neoantigens, thus activating the complement cascade, which will result in cell lysis. The more common drugs known to cause immunological injury to PMNs are chlorpromazine, phenylbutazone, and α-methyldopa.

**Table 23.5 Common Causes of Splenomegaly***

| |
|---|
| *Chronic Infection* |
| Tuberculosis |
| Malaria |
| *Malignant Myeloproliferative Disease* |
| Myelocytic leukemia |
| Myelofibrosis with myeloid metaplasia |
| Essential thrombocythemia |
| *Metabolic Storage Diseases* |
| Gaucher's disease |
| *Chronic Hemolytic Anemia* |
| Hereditary spherocytosis |
| Hereditary elliptocytosis |
| *Increased Pressure in the Portal Vein, Due to* |
| Thrombosis |
| Cirrhosis of the liver |

*Splenomegaly may result in neutropenia, as many PMNs may become sequestered in the enlarged spleen.

Immunological neutropenia may also be caused by maternal-fetal incompatibility. Each individual possesses **major histocompatibility (MHC) antigens** on their white blood cells, which are also known as **human leukocyte antigens (HLAs).** These HLAs are inherited from both parents and differ from individual to individual. A woman experiencing multiple pregnancies may have been exposed to the white blood cells of her fetus who carries HLAs derived from the father. Thus the mother's immune system recognizes the HLA types as foreign and will develop antibodies to these nonself antigens. Each time she is reexposed to a given antigen, the titer of her antibodies increases (each new exposure acts as a booster shot). These maternal antibodies can pass through the placenta into the fetal circulation, where they will attack and destroy the fetal granulocytes. As a result these infants are born with low levels of granulocytes, a condition known as **neonatal isoimmunization neutropenia.** This disorder may be considered analogous to hemolytic disease of the newborn (erythroblastosis fetalis). Neutropenia in newborns may be severe and may persist for the first two to three months of life. During that time these children are extremely susceptible to infection.

**Mechanical Trauma** Normal blood pumped through hemodialysis machines or oxygenators (used in heart-lung bypass procedures) often shows a transient drop in neutrophils. Apparently many of the neutrophils cannot survive the tubes and coils of these machines. They appear damaged and have a reduced phagocytic ability. This neutrophilic condition is usually of short duration and is rapidly rectified after the procedure has stopped, due to a new influx of granulocytes from the bone marrow storage compartment.

**Inflammation** In the early stages of a massive infection, there will be a temporary drop in circulating neutrophils as large numbers move to the site of inflammation.

*Neutropenia Due to Abnormal Distribution*

Neutropenia can also result from an abnormal shift of cells from the circulating pool to the marginating pool. This abnormal distribution is known as **shift neutropenia** (or pseudoneutropenia). This phenomenon is often observed in clinical disorders such as idiopathic hemolytic anemia, cirrhosis of the liver, macroglobulinemia, and multiple myeloma. Shift neutropenia is often mild to moderate since the numbers of circulating PMNs in the blood will remain above 1,000/μL.

Sometimes neutropenia may be caused by a combination of reduced production, increased destruction, and abnormal distribution. For example, an alcoholic patient with severe sepsis may have marrow hypoplasia, increased neutrophil utilization, and high levels of margination. Similarly, patients with Chédiak-Higashi syndrome may have abnormalities of production, destruction, and distribution.

### Etiology of Neutrophilia

**Neutrophilia** may be defined as a condition in which the individual has higher than normal numbers of neutrophils in the bloodstream. Generally speaking, neutrophilia occurs when the PMN counts are above 10,000 neutrophils per microliter. The neutrophil count may increase to above normal levels in the following four basic ways: (1) by increased PMN production in the bone marrow, (2) by increased release of neutrophils from the bone marrow storage compartment, (3) by a reduced rate of egress of neutrophils from the peripheral circulation, and (4) by a shift of cells from the marginating pool to the circulating pool of neutrophils.

Neutrophilia can be a temporary or a long-term event. Temporary neutrophilia is known as **acute granulocytosis,** and it occurs usually within hours after an appropriate stimulus occurs, such as production of endotoxins by bacteria or release of chemotactic substances by activated cells. This rapid onset of neutrophilia is usually the result of an increased release of PMNs from the bone marrow storage compartment and/or by a shift of marginating cells to the circulating pool.

Long-term neutrophilia, known as **chronic granulocytosis,** usually lasts several weeks to many months and reflects an increased production of PMNs by the bone marrow. This prolonged form of neutrophilia is usually initiated by lingering infections, prolonged administration of certain drugs (such as corticosteroids), or it may be an indication of malignant disease.

Most cases of neutrophilia involve both an increased production of neutrophils by the bone marrow as well as an increased release of these cells from the marrow storage compartment.

Thus, neutrophilia may be caused by a wide variety of physical, chemical, and biological factors. The following list includes seven of the major causes of neutrophilia:

1. **Infectious diseases**
Most infectious disorders result in an increased demand for PMNs, especially with infections caused by microorganisms.
2. **Inflammatory diseases**
Many inflammatory disorders such as rheumatoid arthritis, colitis, thyroiditis, and vasculitis, are associated with an increased production of neutrophils, which is reflected in increased numbers of PMNs in the blood.
3. **Drugs and chemicals**
Many drugs and chemical compounds will result in increased neutrophil production. These agents include corticosteroids, epinephrine, histamine, venoms, ethylene glycol, lithium carbonate, and mercury poisoning.
4. **Physical trauma**
Many forms of physical trauma result in neutrophilia. For instance, burns, electric shocks, anoxia, and exposure to extreme cold, are physical conditions that result in increased numbers of PMNs in the blood.
5. **Emotional stress**
Prolonged emotional stress also increases the number of PMNs. High stress levels produce an increase in epinephrine and corticosteroids, the latter of which stimulates the increase of PMNs in the bloodstream.
6. **Malignant diseases**
Many carcinomas, sarcomas, lymphomas, and leukemias are associated with dramatic increases in white blood cells, specifically increases in neutrophils and lymphocytes.
7. **Idiopathic**
Many cases of neutrophilia cannot be readily explained by any of the preceding causes. Hence they are classified as idiopathic.

*Reactive and Nonreactive Neutrophilias*

**Reactive neutrophilias** are initiated by known physical or chemical stimuli, which can be remedied by removing the underlying causes. Nonreactive neutrophilias, however, are caused by phenomena that are not known or understood. When the cause is not known, or if it is not possible to remove it (as is the case in a number of malignancies), then it is impossible to reduce the high levels of neutrophils—hence the term **nonreactive neutrophilias.** In most cases of reactive neutrophilias, the increased proliferation of neutrophils is beneficial to the individual and does not normally contribute to the symptoms of the underlying cause or disease. In nonreactive neutrophilias, on the other hand, the increase in PMNs seems to be purposeless. Further, the PMNs produced are usually abnormal in size and shape and are frequently nonfunctional. These nonreactive neutrophilias are practically always the result of an unregulated production of white blood cells in the bone marrow and hence are referred to as **myeloproliferative disorders.**

*Myeloproliferative Disorders*

The term myeloproliferative disorders refers to a group of diseases characterized by the purposeless proliferation of one or more bone marrow blood cell lines (table 23.6). Formerly these disorders were considered separate entities. However the idea that all blood cell lines are closely related, and the fact that a marrow blood cell line is almost never singly affected, has led to grouping these disorders into a single category. This concept has been reinforced because many of these disorders tend to evolve into other disorders. For instance, polycythemia vera may eventually develop into myelofibrosis with myeloid metaplasia or even into acute leukemia. This sequence of events is understandable because all myeloid cell lines are closely related and are derived from a common pluripotent stem cell of mesenchymal origin. The development of a proliferative abnormality can proceed in one direction or another or can go in several directions simultaneously, depending on the factor(s) causing the hyperproliferation.

The simultaneous or sequential hyperproliferation of the bone marrow blood cell lines that occurs in myeloproliferative disorders rarely includes lymphoid cell lines. The reverse is also true—the various lymphoproliferative disorders such as acute and chronic lymphocytic leukemia, various lymphomas, and infectious mononucleosis are rarely associated with excessive proliferation of myeloid elements. This is another indication that the commitment of stem cells to develop along either the myeloid or lymphoid pathway is made early in the process of blood cell differentiation.

*Characteristics of Myeloproliferative Disorders*

Although each of the myeloproliferative diseases has its own specific clinical and laboratory symptoms, they also share a number of common features. They all can exhibit either **hyperplasia, normoplasia,** or **dysplasia** of the bone marrow in various stages of the disease. **Metaplasia**—that is, conversion of tissue from one form to another—of the bone marrow and the extramedullary organs (liver and spleen) is frequently observed. Myeloproliferative diseases are often associated with signs of **hypermetabolism** such as an elevated basal metabolic rate (BMR), weight loss, tissue wasting, and night sweats. Also, there is a tendency toward **hyperuricemia**—that is, increased uric acid in the blood—due to increased nucleoprotein catabolism in the hyperactive hematological organs. In addition, myeloproliferative diseases usually have marked neutrophilia, a propensity for splenomegaly, and usually little or no tendency for spontaneous remission.

In the remainder of this chapter three of the myeloproliferative disorders are discussed, the other two (chronic and acute myelocytic leukemias) are discussed in chapter 24. This chapter ends with a short description of quantitative abnormalities of eosinophils and basophils.

**Table 23.6 Classification of the Major Myeloproliferative Disorders**

Polycythemia vera
Essential thrombocythemia
Myelofibrosis with myeloid metaplasia
Chronic myelocytic leukemia
Acute myelocytic leukemia

*Polycythemia Vera*

Polycythemia vera may be defined as a myeloproliferative disorder characterized by a neoplastic hyperproduction of erythrocytic, granulocytic, megakaryocytic, and fibrocytic cell lines—hence the name **polycythemia vera.** This disease is also called **polycythemia rubra vera,** since the most obvious symptom, and the one that causes the greatest clinical consequences, is the overproduction of red blood cells. However we should remember that this disease is also associated with an overproduction of all myeloid elements.

**Differential Diagnosis of Polycythemia Vera** Erythrocytosis—that is, an increase in the number of red blood cells in the blood—may be associated with many different diseases and conditions. In most instances hyperproliferation of erythrocytes is a **reactive phenomenon** and may be seen as a compensatory response to hypoxia, mediated by a normal erythropoietin feedback mechanism. Hypoxia may result from either a change in habitat (to high altitudes) or may result from internal abnormalities, such as cardiac or pulmonary disease. It may also be due to the presence of high levels of nonfunctional hemoglobin in the blood, such as methemoglobin or carboxyhemoglobin. Further, erythrocytosis may be caused by an inappropriate increase in erythropoietin production as a result of abnormal changes in the kidneys, such as kidney tumors. Finally, androgen levels may increase, either as the result of a tumor or due to administration of androgen. High levels of male sex hormones will also increase erythropoietin levels.

Polycythemia vera, on the other hand, is a **nonreactive** form of erythrocytosis. Since the causative agent for this disorder is not known, it is impossible to remove it. This form of erythrocytosis can be readily distinguished because the increase in red blood cells is always accompanied by an increase in white blood cells and platelets.

**Clinical Features of Polycythemia Vera** Polycythemia vera is a disease most common in middle-aged and elderly people. It is more common in men than in women, but this may reflect the fact that men are generally exposed to more environmental hazards in the workplace than women. The major symptoms include the following:

1. **Headache, pruritus, dyspnea,** and other ill-defined ailments. These conditions result from the fact that the increased amount of blood cells makes the blood more viscous, resulting in stagnation of the blood flow. The increased total number of blood cells also results in increased total blood volume, which may then produce blood vessel distention.
2. **Tendency to thrombosis.** This phenomenon is the result of three factors: (a) increased viscosity of the blood, (b) presence of a large number of platelets that are easily damaged due to the increased friction in the microcirculation, and (c) tendency to bleeding. Thrombosis may occur at any site; phlebitis is common as are myocardial infarcts, cerebrovascular accidents (=strokes), and thromboses in the hepatic-portal system.
3. **Tendency to bleeding.** Hemorrhage is promoted by the increase in the number of blood cells, which results in an increase in blood volume, which in turn causes distention of the blood vessel walls, especially in the microcirculation. Small leaks will produce platelet aggregation and clot formation, which may result in the formation of large thrombi (thrombosis). However the action of platelets may not sufficiently prevent leaks, since many platelets may have functional defects. Bleeding occurs in about one-third to one-half of the patients with polycythemia vera. The most common form is bleeding into the gastrointestinal tract.
4. **Splenomegaly.** The presence of an enlarged spleen may be explained by the fact that the spleen must work harder than normal to eliminate excess senescent and abnormal red blood cells.

**Laboratory Findings of Polycythemia Vera** Laboratory investigations of the blood and bone marrow of patients with polycythemia vera reveals the following features:

1. **Increased hematocrit value.** Due to the increased numbers of blood cells, especially red blood cells, the hematocrit value is always much higher than normal. Men usually have a hematocrit value of 55% or higher, whereas women have a hematocrit value of 47% and above.
2. **Red blood cell morphology is normal early** in the course of this disease unless iron deficiency is present. Iron deficiency is common in these patients due to increased red blood cell production and loss of iron in bleeding. In the later stages of polycythemia vera, myeloid metaplasia is present as well as abnormal blood cells in the peripheral circulation such as teardrop cells, elliptocytes, and nucleated red blood cells.
3. **White blood cell (WBC) counts are also increased.** Most of the WBC increase is in the number of granulocytes. Although there is a modest increase in the number of immature forms of granular leukocytes, the other granulocytes have a mature appearance.
4. **An increase in platelets** almost always occurs. However the platelet count is rarely above 1 million per microliter. Platelet function, on the other hand, is frequently impaired.
5. **Bone marrow is hypercellular** with large numbers of erythrocytic, granulocytic, megakaryocytic, and fibrocytic precursor cells. Later in the disease the fibroblasts become dominant, and the condition of **myelofibrosis** becomes prominent. Frequently this condition may lead to acute myelocytic leukemia.

**Treatment of Polycythemia Vera** The primary purpose of therapy in polycythemia vera is to reduce blood volume and viscosity. This may be achieved in the following three ways:

**1. Multiple phlebotomy**
Repeated bleeding is an effective way of reducing the blood volume, but it does not necessarily reduce its viscosity. This may be achieved by increasing the plasma volume after the bleeding has taken place. Reintroducing cell-free plasma from the patient's blood that was previously drawn or introducing other fluids will reduce blood viscosity. Bleeding is not without risks, however, since it may lead to hemodynamic changes that endanger the heart and blood vessels. It may also result in an increase in platelets, thus promoting the risk of thrombosis or iron deficiency. On the other hand, a lack of iron will slow the process of erythropoiesis and thus decrease the blood viscosity.

**2. Myelosuppressive drugs**
Myelosuppressive drugs are efficient in reducing hyperproliferation of various bone marrow cell lines. However these drugs also carry the risk of promoting myelofibrosis, which in turn may result in myeloid metaplasia.

**3. Radiation therapy**
Radiation is also effective in controlling the overproduction of myeloid cells in the bone marrow, but may increase the risk of producing acute leukemia. Therefore, using radiation or drugs should be limited to those patients for whom phlebotomy does not seem to be adequate.

Even without these treatments, myelofibrosis and myeloid metaplasia may occur. The latter condition may or may not develop into acute myelocytic leukemia.

### *Essential Thrombocythemia*

**Essential thrombocythemia** is a myeloproliferative disorder characterized by a predominant hyperproliferation of megakaryocytes in the bone marrow and large numbers of platelets in the peripheral circulation. This disease is also associated with a mild increase in red blood cells as well as myelocytic white blood cells. However the increase in platelets is the most dramatic feature of this disorder.

**Differential Diagnosis of Essential Thrombocytopenia** Essential thrombocytopenia is a nonreactive disease that must be distinguished from the more common reactive thrombocythemias associated with hemorrhage, infections, trauma, iron deficiency, splenectomy, and some solid tumors. Reactive thrombocytopenia can occur at any age. Nonreactive thrombocythemia, on the other hand, is of unknown etiology and is primarily a disorder of adults over 50 years of age. Further, the platelet counts in reactive thrombocythemias are usually not as high as in the nonreactive form. Moreover, many of the platelets in essential thrombocythemia are usually abnormal and large. Finally, essential thrombocythemia exhibits an increase in granulocytes and frequently erythrocytes, a feature not necessarily present in reactive thrombocythemias.

**Clinical Features of Essential Thrombocytopenia** Essential thrombocythemia is primarily a disease of middle-aged people and is often insidious in its beginnings. Its major symptoms are as follows:

1. **Hemorrhage**
The most common feature of this disorder is excessive bleeding. The most common site is the gastrointestinal tract. **Hematuria** and **easy bruising** are also common features.
2. **Thrombosis**
This symptom occurs less frequently than bleeding. When it occurs, it usually involves the splenic veins or the veins in the legs.
3. **Splenomegaly**
The spleen is enlarged in at least 80% of all cases, and many platelets are pooled in the spleen.

**Laboratory Findings of Essential Thrombocytopenia** Laboratory investigations of patients with essential thrombocytothemia normally reveal the following:

1. **High platelet count**
Platelet counts of over 1 million per microliter are quite common.
2. **Abnormal platelet morphology**
Many platelets are large and have bizarre shapes. Blood and marrow smears often show large aggregates of platelets.
3. **General hyperplasia of the bone marrow**
The most obvious feature is the presence of numerous megakaryocytes.
4. **Raised leukocyte count**
In most cases these counts, which are always raised, are over 10,000/μL.
5. **Low red blood cell counts**
These low counts are normal, despite an increase in erythropoiesis. This is due to the presence of a large spleen, which sequesters and destroys large numbers of erythrocytes. Poikilocytosis of red blood cells is also common.

**Treatment of Essential Thrombocytopenia** All therapy is aimed at reducing the number of platelets in the peripheral circulation. This can be achieved by reducing the hyperproliferation of megakaryocytes in the bone marrow by: (1) **whole-body radiation,** which is usually accomplished by intravenous injection of $^{32}P$ and (2) **chemotherapy.** Antimitotic drugs, such as busulfan, nitrogen mustards, and various anti-metabolites can be used. As explained earlier, both methods are inherently dangerous. However treatment must be given as the disease is potentially fatal, because of the possibility of thrombosis, embolisms, and serious bleeding.

*Myelofibrosis (MF) with Myeloid Metaplasia (MM)*

**Myelofibrosis (MF)** may be defined as a myeloproliferative disorder characterized by anemia, abnormal proliferation of all myelocytic hematopoietic tissues, varying degrees of fibrosis of the bone marrow, and myeloid metaplasia of the liver and spleen.

**Myeloid metaplasia (MM)** may be defined as the condition in which extramedullary tissue becomes hematopoietic and starts to resemble the histological pattern of bone marrow tissue.

This disorder is also known as myelosclerosis with myeloid metaplasia and is sometimes called **idiopathic myeloid metaplasia.**

A controversy that has not yet been solved satisfactorily deals with the significance of myeloid metaplasia in myelofibrosis. Some hematologists propose that myeloid metaplasia of the liver and spleen is the result of myelofibrosis of the bone marrow. They consider it a consequence of bone marrow failure brought on by the predominance of fibrous elements in the bone marrow. In other words, it is a compensatory phenomenon in which the extramedullary tissues take over the role of the failed bone marrow. Other hematologists, however, do not accept this viewpoint, since there have also been cases of MM without clear fibrosis of the bone marrow. They view myeloid metaplasia as an integral part of the general myeloproliferative process.

**Pathogenesis of MF with MM** Myelofibrosis with myeloid metaplasia is a disease that mainly occurs in middle-aged and elderly persons. The mean age of onset is 58 years. Both sexes are equally affected. Accurate statistics are not available regarding the incidence of this disorder, but it is almost as common as chronic myelocytic leukemia, which has an incidence of about 10 to 12 cases per 100,000 individuals.

As with other myeloproliferative disorders, the etiology of this disease is unknown. However radiation and benzene poisoning have been implicated as there have been many cases among survivors of Hiroshima and workers in the rubber industry. The onset of myelofibrosis with myeloid metaplasia may be the result of an inflammation of the bone marrow after it is stimulated by a toxic agent. This unknown stimulus causes hyperplasia of the bone marrow and also initiates formation of hematopoietic loci in the liver and spleen. Regarding fibrosis of the bone marrow, two theories have been proposed. One theory proposes that fibrosis is a response to excessive hematopoietic activity, since it occurs not only in the bone marrow but also around the metaplastic foci in extramedullary organs. The other theory suggests that fibrosis is the consequence of excessive proliferation of mononuclear phagocytic cells of the bone marrow, which occurs simultaneously with excess production of other bone marrow cell lines. There is a close connection between MF and other myeloproliferative disorders, since myelofibrosis frequently arises in the latter stages of polycythemia vera. Further, this disorder frequently evolves into acute myelocytic leukemia.

**Differential Diagnosis of MF with MM** Myeloid metaplasia in extramedullary tissues is known to occur in only the following two situations. First, it occurs when an increased demand for new blood cells cannot be met by the bone marrow. This frequently is seen in young children suffering from hemolytic disease of the newborn (HDN) or from β-thalassemia major. It may also occur in people as a result of prolonged severe bleeding. In these instances the myeloid metaplasia is reactive—that is, it will disappear once the underlying cause is removed. Second, it occurs when a myeloproliferative disorder is present. This kind of myeloid metaplasia is nonreactive and hence is also known as idiopathic myeloid metaplasia.

Many symptoms of MF are similar to those of chronic myelocytic leukemia; distinguishing it from the latter disease is easy because in myelofibrosis the Philadelphia chromosome is absent (see chapter 24). Further, in chronic myelocytic leukemia the level of leukocyte alkaline phosphatase is nearly always low, whereas in myelofibrosis it is usually fairly high.

**Clinical Features of MF with MM** Myelofibrosis with MM is an **insidious** disease in which the symptoms are expressed gradually. The average interval between the onset of symptoms and the diagnosis of the disorder is about 18 months. The major physical signs and symptoms are as follows:

**1. Anemia**
About two-thirds of all patients with MF show signs of anemia when they are first diagnosed.

**2. Splenomegaly**
An enlarged spleen is commonly present, and many patients first seek medical help to relieve discomfort in the upper abdominal area.

**3. Hepatomegaly**
An enlarged liver is present in at least three-fourths of all cases.

**4. Bone pain**
Pain in the bones, especially those of the legs, is rather common.

**5. Hypermetabolism**
An increased basal metabolic rate is normal and is expressed in fatigue, weight loss, weakness, and fever.

**6. Osteosclerosis**
X-ray examinations of the skeleton will reveal osteosclerosis in about half of all cases.

**Laboratory Findings of MF with MM** Analysis of the blood and bone marrow will reveal the following features:

**1. Red blood cells**
Moderate **erythropenia** is exhibited, depending on the size of the spleen. Increased splenic sequestration will result in premature destruction of many red blood cells. This process is accelerated by the abnormal shape of erythrocytes. Many exhibit a teardrop shape or show ovalocytosis. Anisocytosis and polychromatophilia are also common.

**2. White blood cells**
**Leukocytosis** is often present. The total leukocyte count is usually between 10,000 and 20,000 WBCs per microliter, although higher counts are not uncommon. Most of the extra white blood cells are fairly mature neutrophils, although many immature granulocytes are also present.

**3. Platelets**
In the early stages of this disease, **thrombocytosis** is quite common, and platelet counts of 1 million per microliter are not uncommon. Platelet morphology is often abnormal, with many giant forms present. In the later stages of this disease, however, thrombocytopenia and neutropenia become a frequent feature.

**4. Bone marrow**
Bone marrow in the classic case of myelofibrosis will show distinct hypocellularity. Early in the disease numerous megakaryocytes usually are present. Aspiration of the bone marrow will frequently exhibit numerous platelet aggregations and fibrin strands.

**Treatment of MF with MM** Presently effective therapy is unavailable for MF with MM. Radiotherapy and chemotherapy have been used extensively to control the increase in white blood cells and platelets both in the marrow and in extramedullary sites. However these therapies are much less effective in controlling hyperproliferation and splenomegaly than in chronic myelocytic leukemia. Busulfan may cause a reduction in the number of neutrophils and platelets, but its effects are unpredictable. Moreover, these therapies do not relieve the anemia and may necessitate the use of blood transfusions.

In certain situations splenectomy may be necessary, especially in cases in which traditional therapies do not result in a decrease in splenic size, and where an enlarged and infarcted spleen may cause great discomfort. Another reason to remove an enlarged spleen may be to relieve the anemia, since a large spleen sequesters large numbers of red blood cells which are needed in the peripheral circulation.

**Prognosis of MF with MM** The course of myelofibrosis with myeloid metaplasia is variable. The median survival time is five years. Some patients remain in fairly good health without therapy for long periods of time. In others there is a gradual decrease in health concurrent with an increase in the size of the spleen. Bone pain may become troublesome in the later stages. Bleeding due to thrombocythemia in the later stages may result in life-threatening episodes. In many cases the disease terminates in acute myelocytic leukemia; other causes of death include cardiovascular accidents, bleeding, and overwhelming infections.

## DISORDERS OF EOSINOPHILS

There is a lack of information concerning the clinical effects associated with qualitative and quantitative changes of eosinophils. The reason why so little is known about intrinsic and extrinsic factors that alter the structure and functions of eosinophils may be that most of the life cycle of eosinophils takes place outside the bloodstream, in bone marrow and body tissues. Further, low levels of eosinophils normally present in the circulation (1 to 3% of the total number of white blood cells) makes it difficult to observe any decrease in their number by routine blood analysis. On the other hand, an increase in the number of eosinophils is readily noted in a blood smear. Consequently the phenomenon of eosinophilia has been found to occur in a number of clinical conditions, including a number of well-defined diseases.

### Eosinophilia

Eosinophils play a major role in the destruction of helminth parasites. Hence eosinophilia is usually associated with **helminth infections,** especially with such parasites as blood flukes (schistosomiasis) and nematodes (trichinosis). A special form of eosinophilia, known as "tropical eosinophilia" is thought to represent a reaction of a *Filaria* worm. Further, **Löffler's syndrome**—a clinical condition characterized by blood eosinophilia, infiltration of eosinophils in the pulmonary system, and eosinophils in the sputum—is thought to be the result of helminths migrating through the lungs as part of their life cycle.

Second, eosinophils play a role in allergic reactions. Eosinophilia has been observed in **acute allergic reactions**—that is, type I or immediate hypersensitivity reactions due to the activation of mast cells and basophils by IgE antibodies and antigens (see chapter 21). Eosinophilia has also been observed in type III or immune complex hypersensitivity reactions.

Acute allergic reactions occur as a result of hypersensitivity to certain allergens—that is, compounds that induce an allergic reaction in susceptible individuals. Such allergens include food and drugs, which cause reactions such as hives, hay fever, and asthma. All are characterized by the production of inflammation due to the release of histamine by mast cells and basophils. Products of inflammation activate certain white blood cells to produce eosinophilic chemotactic factors, which attracts many eosinophils to the area of inflammation.

Eosinophilia associated with type III hypersensitivity is the result of inflammatory processes induced by immune complexes that activate complement factors, and attract neutrophils. Both neutrophils and activated complement further induce inflammation as well as other stimulating factors. In most of the areas of inflammation, eosinophils have an **antiphlogistic effect**—that is, an anti-inflammatory effect—although other investigators have argued for the opposite effect—that is, they may contribute to the process of inflammation.

Third, eosinophilia is also common in patients suffering from prolonged **infectious diseases** such as leprosy, tuberculosis, brucellosis, scarlet fever, and certain fungal infections.

Finally, many cases of eosinophilia cannot be etiologically defined and hence are labeled as **idiopathic.** Idiopathic eosinophilia is usually a benign condition. A summary of conditions associated with eosinophilia is presented in table 23.7.

**Table 23.7 Conditions Associated with Eosinophilia**

*Infections*
- Brucellosis
- Leprosy
- Mycoses
- Scarlet fever
- Trichinosis
- Tuberculosis
- Visceral larva migrans
- Löffler's syndrome

*Drug Sensitivities*
- Iodine
- Penicillin

*Malignancies and Leukemias*
- Carcinoma of the lung, ovaries, or stomach
- Hodgkin's disease
- Chronic myelocytic leukemia
- Eosinophilic leukemia

*Collagen Diseases*
- Periarteritis nodosa
- Rheumatoid arthritis

*Allergic Reactions*
- Asthma
- Hay fever
- Inflammatory processes

*Unknown Causes*
- Familial
- Idiopathic
- Sarcoidosis

## DISORDERS OF BASOPHILS

Disorders specifically related to qualitative and quantitative abnormalities of basophils are even more difficult to define than those of eosinophils. Basophils are so rare in the blood (0 to 1%) that it is practically impossible to determine whether there has been a decrease in their numbers or a change in their structure. The only feature that can be readily recognized in a routine analysis of a blood smear is a distinct increase in the number of basophils.

**Basophilia**

Increased numbers of basophils occur in certain conditions. High levels of basophils are commonly observed in myeloproliferative disorders such as chronic myelocytic leukemia and polycythemia vera. In cases of chronic myelocytic leukemia, the numbers of basophils may reach such high proportions that they suggest a diagnosis of basophilic leukemia. Basophilia has also been described in many instances involving type IV or delayed-hypersensitivity reactions. Finally, the endocrine disease, **myxedema** (low thyroid hormone production), has also been associated with high levels of basophils.

Low levels of basophils have been observed in type I or acute hypersensitivity reactions and in **hyperthyroidism**—that is, higher than normal levels of thyroid hormones in the blood. However in none of these cases has the presence or absence of basophils been directly linked to specific clinical features.

## *Further Readings*

Beck, W. S. Leukocytes III. Introduction to Pathology. Lecture 19 in *Hematology,* 4th ed., W. S. Beck, Ed., The MIT Press, Cambridge, Mass., 1985.

Castle, W. B. The Polycythemias. Lecture 20 in *Hematology,* 4th ed., W. S. Beck, Ed., The MIT Press, Cambridge, Mass., 1985.

Cline, M. J. The Abnormal Granulocyte. Part I B in *The White Cell.* Harvard University Press, Cambridge, Mass., 1975.

Conley, C. L. Polycythemia Vera, Diagnosis and Treatment. *Hospital Practice,* Mar. 1987.

## *Review Questions*

1. Define neutropenia and neutrophilia. Clinically speaking, at what WBC levels are these terms used?
2. Name the three major categories of causes that result in neutropenia.
3. List five important causes of neutropenia due to inadequate granulopoiesis.
4. List five different causes of bone marrow injury that may result in decreased granulopoiesis.
5. Name five different toxic compounds that may result in bone marrow injury and thus may produce neutropenia.
6. List five different groups of nontoxic compounds that may produce neutropenia. Give an example of each group.
7. What is myelophthisis? How may it produce neutropenia?
8. Name three different congenital disorders associated with neutropenia.
9. Name three different acquired disorders frequently associated with neutropenia.
10. List five different causes associated with neutropenia due to excessive granulocyte destruction.
11. Explain how splenomegaly may result in neutropenia.
12. List five common causes of splenomegaly.
13. Explain how neonatal isoimmunization neutropenia occurs.
14. Explain shift neutropenia. When is this phenomenon observed?
15. What are the four basic ways in which neutrophilia may be produced?
16. Distinguish between short-term and long-term neutrophilia. Under what circumstances is each form produced?
17. Mention a number of different biological, chemical, and physical factors that may produce neutrophilia.
18. Explain how emotional stress may result in neutrophilia.
19. Distinguish between reactive and nonreactive neutrophilias.
20. Define the term myeloproliferative disorder. List the five major myeloproliferative disorders.
21. What clinical features do all myeloproliferative disorders have in common?
22. Give five causes of reactive erythrocytosis. Define polycythemia vera. Why is it sometimes called polycythemia rubra vera?
23. Name the major clinical symptoms of polycythemia vera. Which age group is most commonly associated with this disease?
24. What are the major laboratory features of polycythemia vera? What is a unique laboratory feature of this disease?
25. List the three major methods of treatment for polycythemia vera.
26. Define essential thrombocythemia. Which age group is most commonly associated with this disease?
27. Describe the major clinical symptoms of essential thrombocythemia.
28. What are the major laboratory features of essential thrombocythemia? What is a unique laboratory feature of this disease?

29. Describe the major modes of treatment for essential thrombocythemia.
30. Define myelofibrosis with myeloid metaplasia. Which age group is mainly associated with this disease? What is the incidence of this disorder? Explain why radiation and toxic chemicals have been implicated as possible causes for this myelofibrosis.
31. Name at least three disorders associated with myeloid metaplasia.
32. What are the major laboratory features of myelofibrosis with myeloid metaplasia? What is a unique feature of this disorder?
33. List some major clinical features associated with myelofibrosis.
34. Describe some of the treatments used to combat myelofibrosis.
35. What is the common prognosis for myelofibrosis with myeloid metaplasia?
36. Explain why so little is known about the clinical effects associated with decreases in the numbers of eosinophils or basophils in the blood.
37. Mention a number of conditions associated with marked eosinophilia.
38. List the parasitic worm infections that are accompanied by increased levels of eosinophils.
39. Mention a few conditions associated with basophilia.

# Chapter Twenty-Four

## *Myelocytic Leukemias*

## INTRODUCTION TO LEUKEMIAS

**Leukemias** may be defined as malignant neoplasms characterized by a disorderly, purposeless proliferation of white blood cells with an overabundance of one cell type. It is a disease of the blood-forming tissues, predominantly the bone marrow. Leukemia is usually recognized in a blood smear by the presence of many abnormal white blood cells. The type of leukemia is established by the maturation level of the cells involved, special cytochemical stains, and the presence or absence of certain membrane markers. The latter may be readily detected by fluorescent antibodies. Further, many leukemias exhibit certain well-defined clinical symptoms as is explained in the following material. The existence of leukemias has been known for more than 150 years. In 1859 Rudolf Virchow, the father of the cell theory, found the presence of many white blood cells in people suffering from leukemia. He called such blood "weisses blut" (white blood).

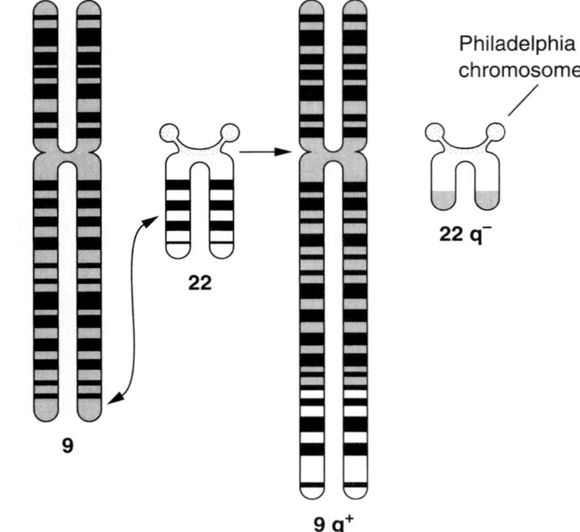


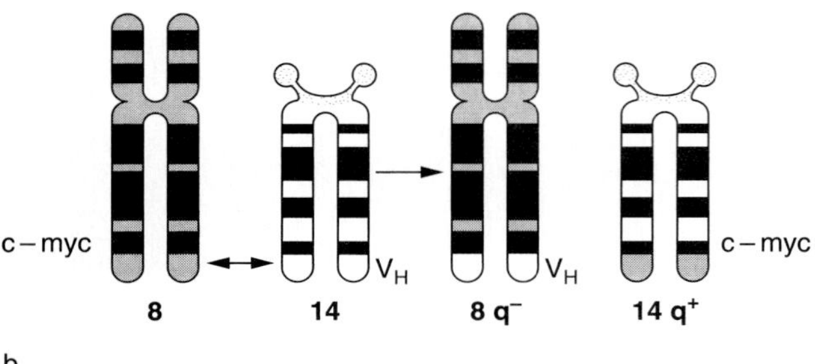


**figure 24.1**

Diagram of two common chromosomal translocations. (*a*) Chronic myelocytic (myelogenous) leukemia; (*b*) Burkitt's lymphoma.

### Etiology of Leukemia

Although the exact etiology of leukemia is unknown, it is generally agreed that the neoplasm is caused by a genetic mutation, frequently a chromosome translocation, or an activation of certain oncogenes. Most leukemic neoplasms result from the transformation of a single stem cell, which then undergoes uncontrolled clonal proliferation. Although the exact factors causing the genetic mutation or oncogene activation are not known, a number of physical, chemical, and biological agents have been implicated. These include the following:

**1. Heredity**
Some evidence points to a high **familial incidence** of certain tumors. In other words certain leukemias seem to run in families—for example, if one of a pair of monozygotic twins contracts leukemia, there is a high statistical chance that the other twin will also contract the same condition.

**2. Chromosomal abnormalities**
A number of leukemias are associated with distinct chromosomal abnormalities. For instance, the **Philadelphia chromosome,** a truncated form of chromosome 22 (figure 24.1), is found in all cases of chronic myelocytic leukemia (CML). Individuals with **Down's syndrome** (trisomy of chromosome 21), often develop acute myelocytic leukemia (AML). Cancer cells of patients with **Burkitt's lymphoma** exhibit a translocation that moves part of chromosome 8 to chromosome 14 (figure 24.1). The translocation of the proto-oncogene c-*myc,* which is normally involved in moving a resting lymphocyte into the cell cycle (i.e., prepares it for mitosis), may result in deregulation of that gene and make it an active oncogene that promotes continuous proliferation of a given cell. This is what happens in Burkitt's lymphoma, a B-cell neoplasm common in East Africa.

**3. Chemical agents**
Individuals in prolonged contact with certain chemical compounds have a greater incidence of certain leukemias than would be expected—that is, as compared with the general population. For instance, people working in the rubber industry who are in frequent contact with benzene experience an increased incidence of AML. Other **carcinogens** include chloramphenicol, phenylbutazone, carbon tetrachloride, and Ninhydrin.

**4. Ionizing radiation**
Persons exposed to prolonged large doses of ionizing radiation have a greater incidence of leukemia than the general population. The effects of radiation have been well documented in two instances. First, in the early days of radiology, the dangers of constant exposure to radiation were not yet appreciated—and before the introduction of protective clothing and better-shielded apparatus, the incidence of leukemia was much higher in radiologists than in other physicians. Second, as a result of exposure to radiation caused by atomic bombs dropped on Japan during

World War Two, the citizens of Hiroshima and Nagasaki had a much higher incidence of leukemia than the inhabitants of other Japanese cities. Moreover, the effects of radiation have been experimentally verified by numerous animal studies.

**5. Immunological defects**
Defects in the immune system—usually genetic in origin—have been associated with an increased incidence of leukemia, especially lymphocytic leukemias and lymphomas. This may be due to the breakdown of the immunosurveillance that normally keeps neoplastic growths in check—that is, most cancer cells are destroyed by the immune system before they can grow and metastasize.

**6. Viruses**
Neoplastic growths in both humans and animals have been associated with the presence of certain types of viruses. In 1910 Peyton Rous demonstrated the production of sarcomas in healthy chickens by injecting them with a certain virus (which is now known as the Rous sarcoma virus). In 1966, at the age of 85, Rous received the Nobel Prize for his work. Many other viruses have been implicated in neoplastic growths. For instance, the Epstein-Barr virus is considered a major contributing factor in Burkitt's lymphoma. Retroviruses such as the human T-cell lymphotropic virus, type I (HTLV-I) and type II (HTLV-II) have also been implicated in the production of cancer in infected persons.

Most cases of leukemia probably result from a combination of genetic and environmental factors resulting in the malignant transformation of a single susceptible cell type, which then will proliferate into a large clone of similar cells.

### Clinical Features of Leukemia

Leukemia is predominantly a neoplasm of the hematopoietic cells of the bone marrow. The cancerous cell clone will grow rapidly and at the expense of normal hematopoietic cells. The neoplastic cells will take nutrients from normal hematopoietic tissue and grow more rapidly than normal bone marrow cells. As a result the production of normal bone marrow elements will decrease and eventually will be crowded and displaced by the cancerous cell line. This will have the following three important consequences:

**1. Decreased red blood cell (RBC) production**
This will result in **anemia.**

**2. Decreased normal white blood cell (WBC) production**
Although there is usually a great increase in WBC production, all white blood cells belong to one clone of cells, which is almost always nonfunctional. As a result the patient will be much more susceptible to **infections.**

**3. Decreased platelet production**
This will result in a greater tendency to bleeding, as there are not enough platelets available to produce hemostatic plugs.

Other important general symptoms include the following:

1. **Systemic symptoms** include fatigue, headaches, dizziness, and other symptoms associated with anemia. Further, fever, increased sweating, and heat intolerance result from the activities of infectious agents. Clotting problems may result as a consequence of the production of defective platelets.
2. **Hypermetabolism** reflects the increased metabolic requirements brought on by the tremendous production of cancerous blood cells in bone marrow.

### Classification and Incidence of Leukemia

Since there are two main blood cell lines in the bone marrow—myeloid and lymphoid cell lines—there are two broad categories of leukemia: **myelocytic leukemia** and **lymphocytic leukemia.** The incidence of leukemia is about equally divided between myelocytic and lymphocytic leukemias. Both categories of leukemia may be further divided into two major forms: **acute leukemias,** which develop rapidly, and **chronic leukemias,** which develop slowly. About 50% of all leukemias may be classified as acute leukemias, and the other half are grouped as chronic forms. The four major types of leukemia are (1) acute myelocytic leukemia (AML), (2) chronic myelocytic leukemia (CML), (3) acute lymphocytic leukemia (ALL), and (4) chronic lymphocytic leukemia (CLL).

The myelocytic leukemias are discussed in this chapter, and the lymphocytic leukemias are described in chapter 25. Details about age, incidence, sex, death rates, and number of new cases are discussed under each type of leukemia. Overall, leukemias are more common in males than in females. This may reflect the fact that there are more men in the workplace than women, and thus men may be exposed to more biological, chemical, and physical factors that seem to play a role in the induction of leukemia.

## CHRONIC MYELOCYTIC LEUKEMIA (CML)

Chronic myelocytic leukemia (CML) is also known as chronic myelogenous leukemia and as **chronic granulocytic leukemia (CGL).** CML was the first leukemia to be described and is also the more well defined of all the myeloproliferative disorders.

**Chronic myelocytic leukemia** may be defined as a neoplastic disease with a unique chromosomal abnormality (the presence of the Philadelphia chromosome) and that is manifested by an excessive and apparently unrestrained production of granulocytes in the bone marrow and other hematopoietic organs. The peripheral blood smear exhibits large numbers of neutrophilic granulocytes, many of which are immature forms.

### Incidence of CML

Chronic myelocytic leukemia (CML) accounts for about 20% of all cases of leukemia, at least in the Western world. The overall incidence is about 10 persons per 100,000 individuals per year. Patients are typically between 20 and 60 years of age, with a peak incidence between 40 and 50 years. The disease is slightly more common in men than in women, but this may be more a reflection of occupation since more men tend to work in environments that predispose them to exposure to radiation and carcinogens.

### Etiology of CML

As with other myeloproliferative disorders, the causes of CML are not known, although statistical evidence links CML to exposure to large doses of ionizing radiation and to prolonged occupational exposure to such carcinogens as benzene. Although many chemical compounds have been suspected as being **leukemogens**—that is, leukemia-causing agents—few have been clearly identified as causative agents of CML. However benzene is generally agreed to be a definite cancer-causing agent.

### Pathogenesis of CML

Chronic myelocytic leukemia is associated with, and presumably results from, a somatic mutation in the bone marrow stem cells. In general, chromosome damage from excess radiation or mutagenic chemicals will result in damage to the chromatin so normal mitosis is no longer possible, resulting in the death of the transformed cells. In certain situations, however, these forces may produce clones of bone marrow stem cells that are not only able to survive but develop a propensity to proliferate excessively and independently of the normal regulatory forces. Analysis of bone marrow cells from individuals with CML have shown that they possess an abnormal chromosome, known as the **Philadelphia chromosome** (or **$Ph_1$ chromosome**). The Philadelphia chromosome is an abnormally small chromosome resulting from the translocation of a portion of the long arm of chromosome 22 to chromosome 9 (figure 24.1). This abnormality is not present in any other myeloproliferative disease. The relationship between this chromosomal abnormality and clinical manifestations of CML is not clear, although possession of the $Ph_1$ chromosome seems to promote additional chromosome abnormalities. These secondary chromosomal changes in $Ph_1$-positive cells may lead to the growth dominance of these cells, which progressively lose the capacity to differentiate properly and to function normally.

CML manifests itself by an increase in hematopoietic activity first in the bone marrow and later also in the liver and spleen, which may become greatly enlarged. Hematopoiesis usually expands fivefold to tenfold, especially the granulocytic cell lines, resulting in a tremendous increase in leukocytes in the peripheral bloodstream. Many of these cells do not readily leave the vascular environment as would be the case with normal healthy granulocytes, contributing to the increased numbers of white blood cells in the peripheral circulation.

### Clinical Features of CML

The signs and symptoms of CML develop gradually. CML is a typically insidious disorder that is usually discovered accidentally as a result of such symptoms as general malaise, fatigue, weight loss, and low-grade fever. Most of these symptoms are the result of the following:

**1. Anemia**
Anemia is a common feature of CML, especially in the later stages of the disease, when red blood cell production has decreased considerably as a result of the crowding out of erythrocytic blast cells by expanding granulocytic and fibrocytic tissues in the bone marrow.

**2. Bleeding**
Hemorrhage is common due to the production of many abnormal platelets and the pooling of many platelets in the enlarged spleen. Bruising (ecchymosis), retinal bleeding (which results in visual disturbances), and bleeding in the kidneys (resulting in hematuria) are not uncommon. Bleeding may further exacerbate the already existing anemia.

**3. Splenomegaly and hepatomegaly**
An enlarged spleen is almost always present, and an enlarged liver is fairly common as well. This enlargement is the result of the development of extramedullary hematopoiesis. Moreover, the spleen must work harder to eliminate the many abnormal white blood cells.

**4. Bone pain**
Because of the increased hematopoietic activity in the bone marrow, many patients will complain about a tenderness of the bones, especially the sternum and ribs, which are the major sites of blood cell production.

**5. Hypermetabolism**
An increase in the basal metabolic rate is also a common feature of CML, resulting in weight loss, night sweats, and low-grade fever.

**6. Prolonged infections**
It is an apparent paradox that CML is associated with increased susceptibility to infections and with the prolonged presence of infectious diseases since the blood contains so many white blood cells. Unfortunately most of these cells are abnormal and apparently have lost the ability to respond to chemotaxis. Their phagocytic abilities also seem impaired.

### Laboratory Findings of CML

Laboratory investigations of CML will show a number of different features, depending on the phase of the disease. These findings include the following:

**1. White blood cells**
Distinct leukocytosis is the most obvious sign. The number of white blood cells is usually between 50,000 and 300,000 WBCs per microliter. The great majority of these cells belong to the neutrophilic cell line. The total number of lymphocytes and monocytes is always less than 10%. A complete spectrum of neutrophilic cells can be seen, although the great majority are myelocytes, metamyelocytes, band cells, and mature

polymorphonuclear neutrophils (PMNs) (figure 24.2). Relatively few myeloblasts and promyelocytes are encountered in the peripheral blood. Usually a distinct increase in the number of eosinophils and basophils occurs, especially in the later phases of the disease.

**2. Red blood cells**

Chronic myelocytic leukemia is always associated with a decrease in the number of red blood cells in the peripheral circulation. The erythrocyte count seems to decrease as the disease progresses. In the early stages of this form of leukemia, anemia is usually normocytic and normochromic, and there is little evidence of iron deficiency (except in cases of excessive bleeding). As the myeloid overgrowth progresses, the number of erythrocyte precursor cells in the bone marrow decreases, and the circulating red blood cells begin to show variations in size and shape. For instance, many cells become teardrop shaped. These abnormal erythrocytes have a reduced life span, a fact further accentuated by an enlarged spleen.

**3. Platelets**

Thrombocytosis is common in the early stages of CML. There may be 1 million or more platelets per microliter in the peripheral circulation. Frequently these platelets are larger than normal in size, but their function is below normal. Later in the disease platelet numbers may decrease drastically due to sequestration in the spleen and the crowding out of megakaryocytes by granulocytic precursor cells.

**4. Bone marrow**

Aspiration of bone marrow tissue will show hypercellularity with predominant granulocytic cells. The majority of these granulocytic cells are myelocytes and metamyelocytes. This is in contrast with acute myelocytic leukemia, in which the predominant bone marrow cells are myeloblasts and promyelocytes. As previously mentioned, the erythrocytic precursor cells are usually diminished and the number of megakaryocytes is normally increased, especially in the early stages. Bone marrow biopsy will reveal a moderate increase in myelofibrosis, which may become a marked phenomenon in the later stages of this disease.

**5. Philadelphia chromosome**

Chromosome analysis of the bone marrow precursor cells will reveal a $Ph_1$ chromosome. This is considered a definitive sign of CML.

**6. Biochemical abnormalities**

Most cases of CML have a distinct increase in lactate dehydrogenase (LDH) production and a marked decrease in leukocyte alkaline phosphatase (LAP) levels. Further, serum vitamin $B_{12}$ levels are usually much higher, up to 15 times the normal level. Uric acid secretion is also above normal, which frequently results in the production of kidney stones and attacks of gout.

## Treatment of CML

Treatment of chronic myelocytic leukemia is designed to achieve the following three goals: (1) lowering the level of circulating white blood cells to 10,000 to 20,000/μL, (2) reducing thrombocytosis, and (3) reducing the size of the spleen.

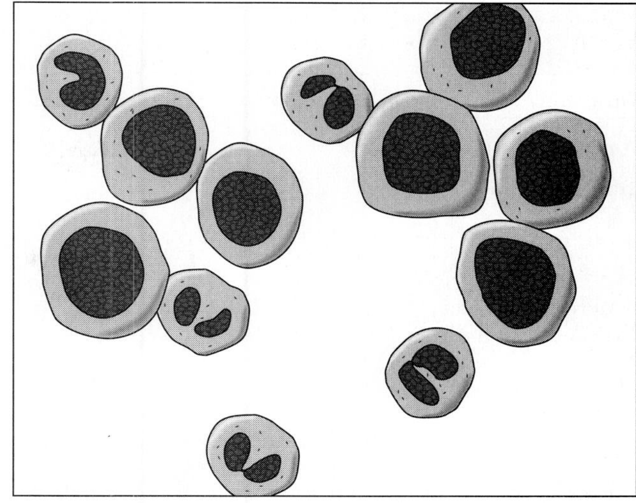

**figure 24.2**

Typical blood smear of a patient with chronic myelocytic leukemia (CML).

To achieve these aims, two major forms of therapy are used: chemotherapy and radiation of the spleen. Chemotherapy uses a number of cytotoxic drugs that act on the hematopoietic cells of the bone marrow and extramedullary sites, thus preventing these cells from proliferating. This treatment aims to achieve all three of the preceding three goals. A list of chemotherapeutic agents with significant activity in CML is given in table 24.1. Many hematologists believe that the alkylating agent busulfan is the treatment of choice. Some regimens employ 6-mercaptopurine or 6-thioguanine in combination with busulfan. Regular blood counts allow the dose to be titrated in individual patients. In patients resistant to (or experiencing side-effects from) busulfan, hydroxyurea and dibromomannitol are useful alternatives.

Splenic radiation is usually reserved for patients whose splenomegaly does not respond to chemotherapy.

## Prognosis of CML

Despite modern diagnostic methods, better medical care, development of new drugs, and improved radiologic equipment, the survival time of patients with chronic myelocytic leukemia has not changed significantly over the last few decades. The median survival time remains 3 to 4 years with or without treatment. Five to ten percent of all patients die during the first year of the disease, and about twenty percent survive 10 years or more. Although survival time has not markedly improved, the quality of life of patients with CML has improved dramatically with modern treatment. Ultimately, however, the disease becomes refractory to treatment. The large and continued doses of drugs and/or radiation will cause a reduction of white blood cells and a reduction of red blood cells and platelets, resulting in distinct anemia, a greater tendency to bleeding, and increased

**Table 24.1 Drugs Commonly Used in the Treatment of CML**

| Drug | Mode of Action | Common Side-effects* |
|---|---|---|
| *Antimetabolites* | | |
| Methotrexate | Inhibits pyrimidine or purine synthesis, incorporation into DNA | Mouth ulcers, gut toxicity |
| 6-Mercaptopurine | | Jaundice |
| Cytosine arabinoside | | Intestinal toxicity, hemolytic anemia |
| Hydroxyurea | | Gut toxicity, skin atrophy |
| *Alkylating Agents* | | |
| Cyclophosphamide | Cross-links DNA, impedes RNA formation | Hemorrhagic cystitis, cardiomyopathy, loss of hair |
| Chlorambucil | | Marrow aplasia, |
| Busulfan | | pulmonary fibrosis, hyperpigmentation |
| *DNA Binding* | | |
| Daunorubicin Doxorubicin hydrochloride | Binds to DNA and interferes with mitosis | Cardiac toxicity, hair loss |
| *Mitotic Inhibitor* | | |
| Vincristine | Spindle damage, absent metaphase | Neuropathy (peripheral, bladder, gut) hair loss |
| *Miscellaneous* | | |
| Corticosteroids | Not identified | Peptic ulcer, obesity, diabetes, osteoporosis |
| L-Asparaginase | Deprives cells of asparagine | Hypersensitivity, low albumin and coagulation factors, pancreatitis |

*Most chemotherapeutic drugs cause nausea and vomiting.

infection. Finally, this chronic condition will change to an acute form of leukemia with the production of large numbers of immature myeloblasts in the bone marrow and the bloodstream. This new blastic phase is fatal, and survival is rarely more than 12 months after the transformation to the acute phase has begun. Treatment with agents used in acute leukemia has not been successful.

## ACUTE MYELOCYTIC LEUKEMIA (AML)

Acute myelocytic leukemia (AML) is also known as acute myelogenous leukemia, acute myeloblastic leukemia, and **acute granulocytic leukemia (AGL). Acute myelocytic leukemia** may be defined as a myeloproliferative disease in which there is a preponderance of primitive blast cells in the bone marrow

**Table 24.2 FAB Classification of Various Forms of Acute Myelocytic Leukemia (AML)**

M1 Myeloblastic leukemia without maturation
M2 Myeloblastic leukemia with maturation
M3 Promyelocytic leukemia
M4 Myelomonocytic leukemia
M5 Monocytic leukemia
M6 Erythroleukemia
M7 Megakaryoblastic leukemia

and in the circulation. Analysis of bone marrow and blood smears of patients with AML show different types of blast cells in different individuals, indicating the presence of a heterogeneous disease.

### Incidence of AML

Acute myelocytic leukemia is predominantly a disease of adults, although it does occur with some frequency in older adolescents. There appear to be two peaks in this disorder: one at 15 to 20 years of age and another peak after 50 years of age. This form of leukemia occurs with equal frequency in all races and in all geographical areas. The incidence of AML is about five cases per 100,000 individuals per year, and it accounts for about 60 to 65% of all cases of acute leukemias.

### Classification of AML

Acute myelocytic leukemias show a wide variety of morphological characteristics. Over the years many attempts have been made to classify the various forms of AML. Some classifications remain in use for some time but are later replaced by more up-to-date versions. The most recent effort to classify acute leukemias was an international cooperative endeavor that resulted in formation of a semipermanent committee known as the **French-American-British (FAB) Cooperative Group,** which publishes and periodically updates a scheme for classifying AMLs based on morphological characteristics. Currently the FAB group recognizes seven different types of acute myelocytic leukemias and three different types of acute lymphoblastic leukemias. In this chapter only the seven forms of AML, labeled M1 to M7, are discussed (table 24.2). Since the introduction of this classification, a great deal of new information has become available to researchers using immunological markers to identify specific stages of myelocytic precursor cells. However much of that information has not yet been incorporated into the FAB classification. The FAB system will most probably be modified sometime in the future when more data regarding cell types becomes available.

Currently the seven major types of AML, also known as acute nonlymphocytic leukemia (ANLL) are as follows:

**M1: Acute myeloblastic leukemia without maturation** This is the most primitive type of AML and consists almost entirely of myeloblasts—that is, the first stage of granulocytic development. These myeloblasts are found in large numbers both in the bone marrow and in the peripheral bloodstream. These cells appear not to be functionally activated as they do not exhibit any myeloperoxidase activity (figure 24.3).

**M2: Acute myeloblastic leukemia with maturation** These blast cells are not as uniform in shape as the M1 myeloblasts; many have irregular outlines, and some promyelocytes and myelocytes may also be present in the bloodstream. These cells are clearly a further stage in the development of this cell line since they show myeloperoxidase activity. In about 50% of all cases of M2-type AML, azurophilic rods called **Auer rods** are observed in the cytoplasm of these myeloblastic cells. Auer rods are composed of lysosomal material and are essentially made up of fused primary granules (figure 24.3).

**M3: Promyelocytic leukemia** This rare type of AML is characterized by a predominance of promyelocytes. The promyelocytes contain many large lysosomal granules, many fused to form Auer rods. The nuclei are often more immature than those of normal promyelocytes, and they are often irregular in shape (figure 24.3). This form of AML is often associated with disseminated intravascular coagulation (DIC) as a result of the lysis of these promyelocytes in the bloodstream. The resulting release of the lysosomal contents into the blood causes local coagulation activity. This may result in bleeding problems, as many coagulation factors and platelets are consumed in DIC activity. Therapeutic administration of anticoagulants, especially heparin, may be required to control DIC and to prevent hemorrhage.

**M4: Myelomonocytic leukemia** This form of AML is characterized by the presence of monocytic blast cells that have lacy or reticular nuclei with many nucleoli. They are classified as monocytic cells as they show nonspecific esterase activity. The abundant cytoplasm is vacuolated and contains many Auer rods (figure 24.4). Tissue infiltration by these monocytic blast cells is a common occurrence. They are especially common in the gingivae, mucosal surfaces, and meningeal linings of the central nervous system and the lymph nodes. This M4 type of AML was formerly known as the **Naegeli type** of monocytic leukemia.

**M5: Monocytic leukemia** This type of AML was formerly known as the **Schilling type** of monocytic leukemia. An individual is diagnosed with the M5 type of AML when more than 80% of the nonerythroid bone marrow cells are classified as belonging to the monocytic series. If most of the monocytic cells are early monoblasts, this type is

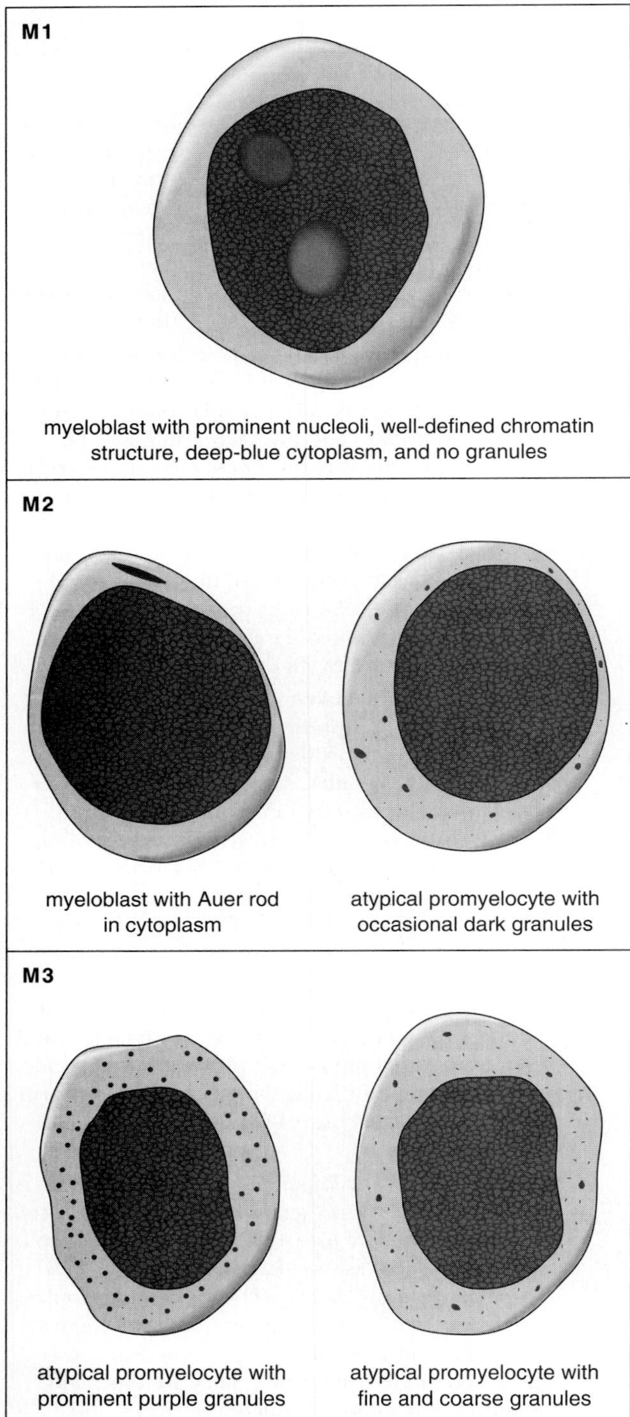


**figure 24.3**
Cells commonly found in patients with myeloblastic leukemias. M1: acute myeloblastic leukemia without maturation; M2: acute myeloblastic leukemia with maturation; M3: hypergranular promyelocytic leukemia.

often referred to as **M5a;** if most belong to the more developed monocyte type, it is classified as **M5b** (figure 24.4).

**M6: Erythroleukemia** This form of AML is also known as **Di Guglielmo's disease.** In this condition all myeloid blood cell lines exhibit some abnormalities in proliferation, but normoblasts account for more than 50% of all bone marrow cells. Many of these cells have bizarre morphological features, such as clover-shaped nuclei or are multinucleated (figure 24.5). Megaloblastic forms are common (most probably due to their multinucleated appearance) and ring sideroblasts and Auer rods frequently occur. Gigantic megakaryocytes are also common. These cells react positively with periodic acid-Schiff (PAS) stain, whereas normal erythroblasts would react negatively. The prognosis for this type of AML is always poor, and the response to chemotherapy is usually minimal.

**M7: Megakaryoblastic leukemia** This rare form of AML is characterized by a proliferation of atypical megakaryocytes in the bone marrow and the appearance of many large poorly differentiated blast cells in the peripheral blood. Leukemic megakaryocytes are typically smaller than normal megakaryocytes with an increased nucleus to cytoplasm ratio (figure 24.6). Vacuoles are often present in the cytoplasm. The megakaryoblasts typically stain negatively with Sudan black B or myeloperoxidase but often stain positively for nonspecific esterase activity. The bone marrow often shows an increase in fibrous tissue. The peripheral blood smear is characteristically pancytopenic. The lack of platelets is usually associated with increased bleeding tendencies.

## Pathogenesis of AML

The leukemic cell populations of persons with AML are probably the result of a continuous proliferation of one or more clones of abnormal blast cells. These cells fail to differentiate properly and continue to proliferate without maturing to the normal nonproliferating stages. The accumulation of these immature, continually dividing cells results in the replacement of the normal hematopoietic precursor cells by these neoplastic cells. Ultimately this will cause complete bone marrow failure. The leukemia cells do not divide any faster than precursor cells, but they differ in failing to stop after three to four mitotic divisions. The presence of this disease may be recognized when the number of blast cells increases to more than 5% of the total cell mass of the bone marrow.

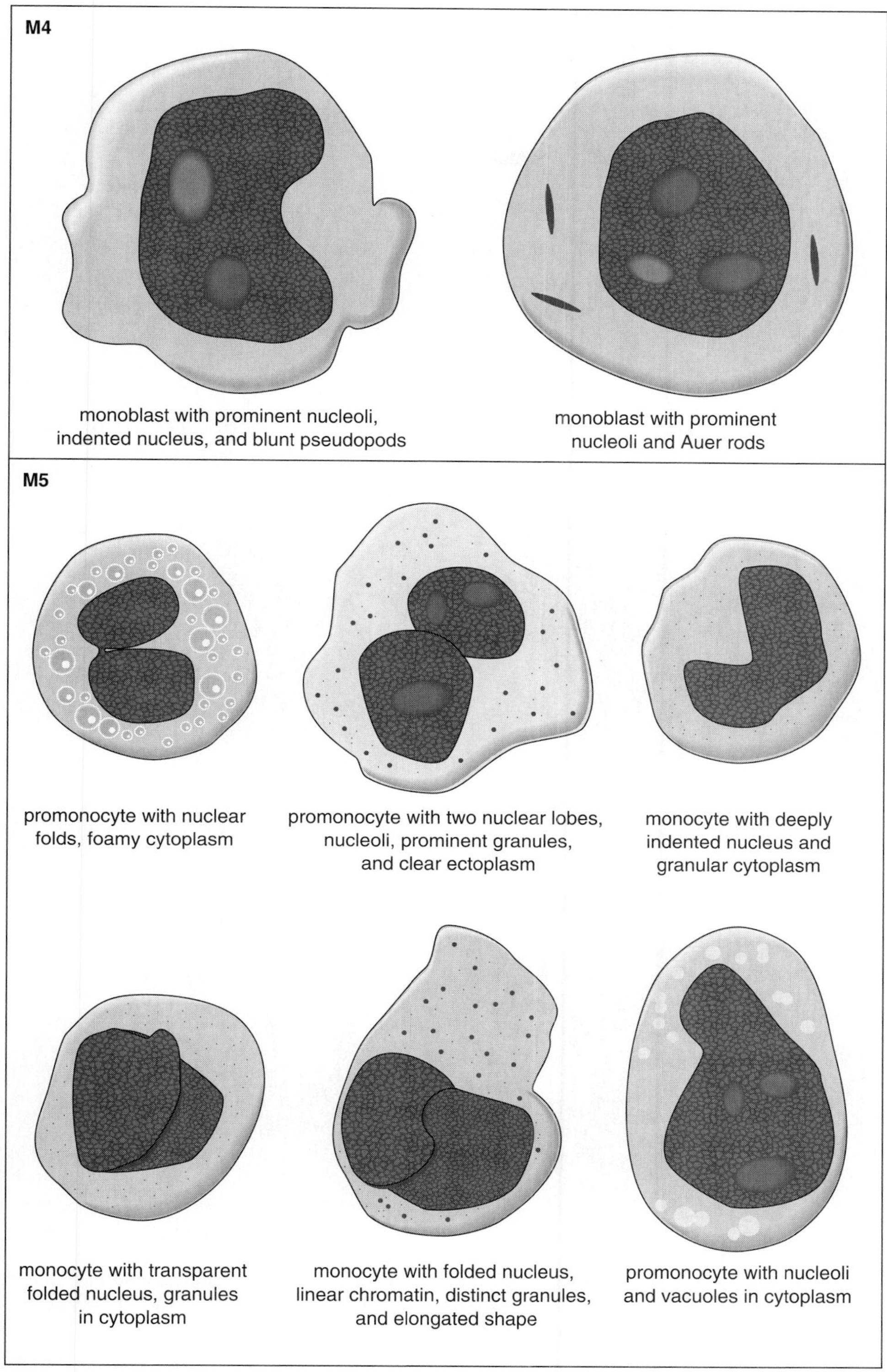


**figure 24.4**

Cells commonly found in patients with the following forms of acute myeloblastic leukemia. M4: myelomonocytic leukemia; M5: monocytic leukemia.

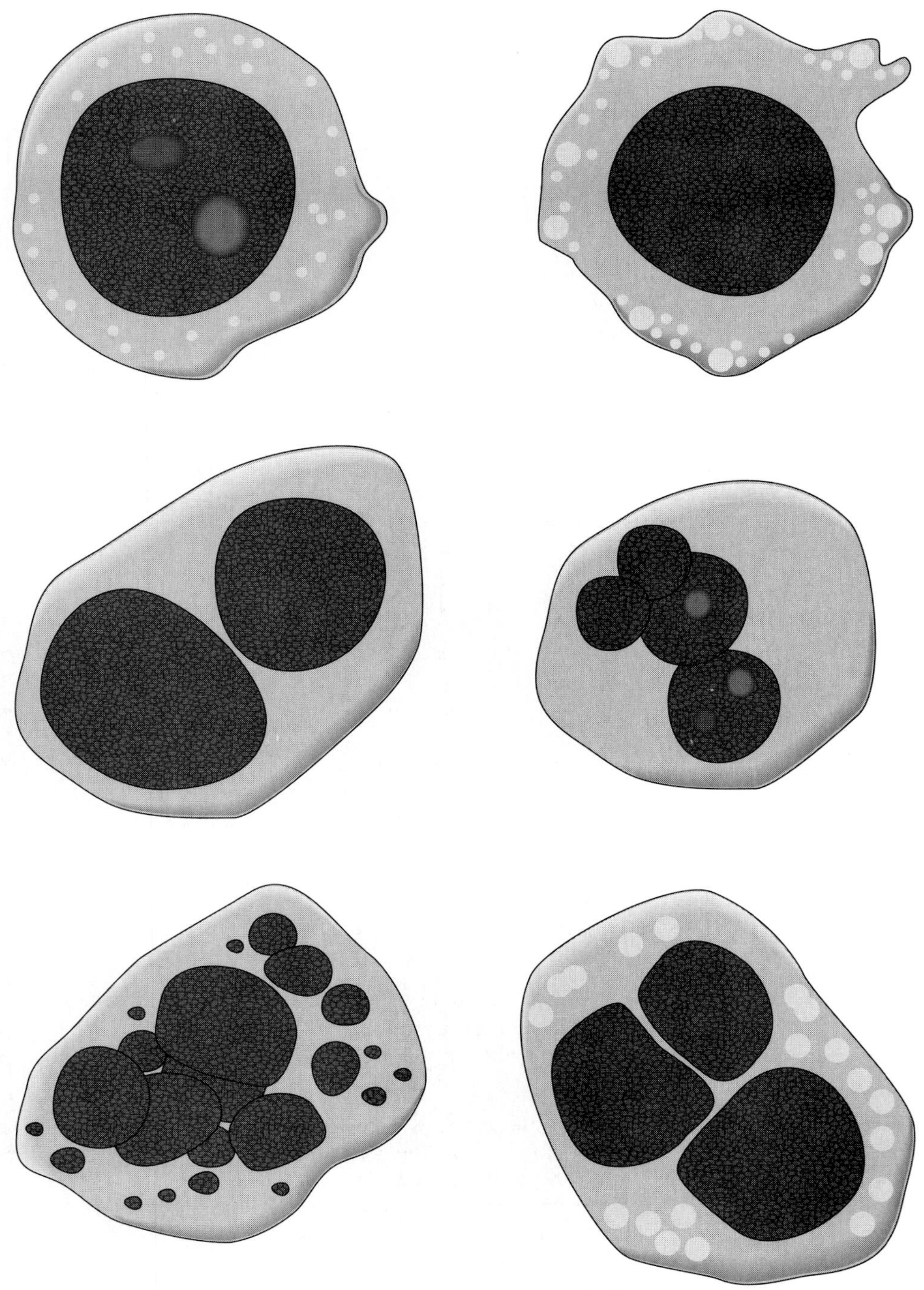

**figure 24.5**

Nucleated red blood cells found in the bone marrow of patients with erythroleukemia (Di Guglielmo's disease), a malignant blood dyscrasia.

## Etiology of AML

As with other myeloproliferative disorders, the etiology of acute myelocytic leukemia is unknown. A number of distinct factors have been implicated, however the relative importance of each of these factors as well as their interaction presently is not clear. The major factors implicated are as follows:

**1. Ionizing radiation**

Frequent exposures to large amounts of radiation result in chronic myelocytic leukemia, which then changes into acute myeloblastic leukemia.

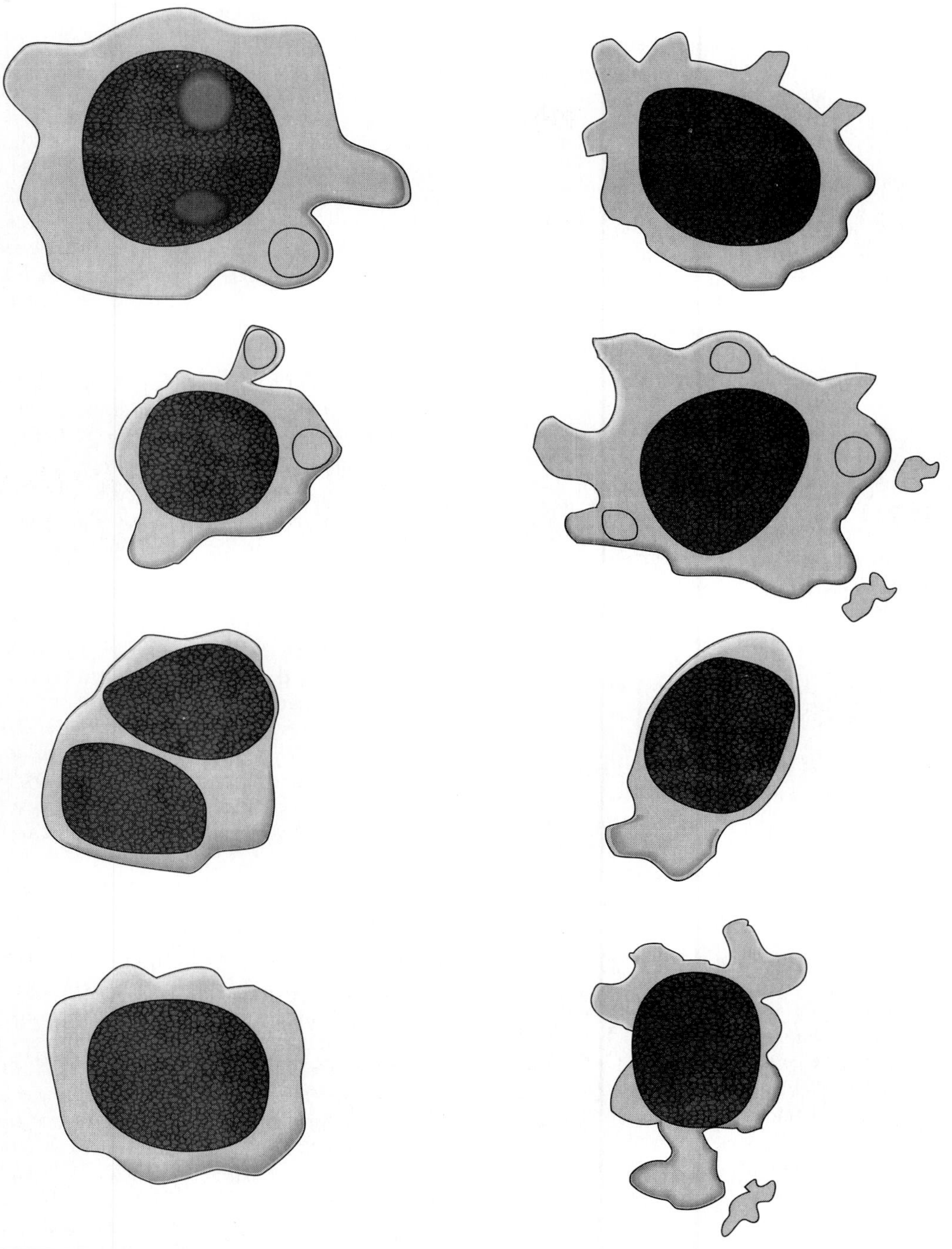

**figure 24.6**

Various forms of micromegakaryocytes commonly found in patients with acute myeloblastic leukemia. Notice that usually there is a single nucleus and that platelet formation is taking place.

**2. Leukemogens**

Certain cancer-causing chemicals have been implicated as agents that may produce AML. As with radiation, these mutagenic chemicals usually produce chronic myelocytic leukemia, which later may change to the acute form of the disease. Further, AML has been linked with certain anticancer drugs used in the treatment of solid tumors.

**3. Congenital factors**

If leukemia occurs in newborns, it is almost always AML. Hereditary factors may be involved, since if one of a pair of identical twins develops AML, the other has a high probability of contracting the disease. Also, a number of reports have mentioned high familial incidences of AML.

**4. Viruses**

Although no definite evidence implicates viruses in human AML, many animal studies point in that direction. DNA and RNA viruses have been shown to be causative agents of both myelocytic and lymphocytic leukemias in birds and mammals.

**5. Neoplasia**

Other neoplastic diseases of the bone marrow, such as polycythemia vera, essential thrombocytosis, and myelosclerosis may eventually transform into acute myelocytic leukemia.

**6. Chromosomal changes**
Various chromosomal abnormalities have been found in more than half of all patients with AML. Individuals with Down's syndrome (trisomy at chromosome 21) have up to a thirtyfold greater risk of AML than the general population.

## Clinical Features of AML

The symptoms of acute myelocytic leukemia are all associated with decreased production of normal hematopoietic tissues as well as with the structural and functional consequences of infiltration of tissues and organs with abnormal white blood cells. The most common symptoms are as follows:

**1. Anemia**
Acute myelocytic leukemia is almost invariably associated with anemia. In many cases the symptoms of anemia prompt the patient to seek medical help, and the existence of AML is often diagnosed as a result of routine blood tests. The anemia is normally caused by a combination of three factors: (1) depressed erythropoiesis due to the increased production of neoplastic myeloblasts, (2) increased bleeding due to insufficient numbers of platelets, and (3) accelerated destruction of red blood cells due to an enlarged liver. The degree of anemia depends on the speed of the development of AML as well as on iron and folic acid stores. Most patients with AML have hemoglobin levels of less than 10 g/dL.

**2. Infections**
Many patients with AML seek medical advice because of a general feeling of malaise accompanied by low-grade fever and complaints of infections of the skin, mouth, throat, and respiratory and urinary tract. These infections are usually not associated with inflammation due to the absence of functional neutrophils but rather run the gamut of pathogenic agents from bacterial to fungal to viral infections and are usually severe.

**3. Bleeding**
More than 60% of all patients with AML show some sign of bleeding at the time of diagnosis. This is usually in the form of petechiae, ecchymoses, or bleeding gums. Occasionally serious internal bleeding occurs. The presence of abnormal platelets, or the lack of platelets, causes the bleeding to continue. Severe bleeding rarely occurs as long as the platelet count is above 25,000/μL.

**4. Coagulation**
Disseminated intravascular coagulation (DIC) is commonly observed in AML, especially in the promyelocytic form (M3). The coagulation process may be initiated by factors produced by infectious agents present in the body—for example, bacterial endotoxins or viral compounds.

## Laboratory Findings of AML

Laboratory investigations of blood and bone marrow of patients with acute myelocytic leukemia will reveal the following features:

**1. Red blood cells (RBCs)**
The red blood cell counts are low. Blood smears normally will reveal few abnormal RBCs, although a number of nucleated erythrocytes may be present.

**2. White blood cells (WBCs)**
There is tremendous variation in the white blood cell count, which may be increased, normal, or even decreased. One-quarter of the patients will have WBC counts below 5,000/μL, another quarter will have counts above 50,000 WBCs per microliter. Half the patients will have WBC counts of 10,000 to 20,000/μL. High WBC counts—that is, more than 100,000 WBCs per microliter are relatively rare and occur in less than 10% of all cases. When the WBC count is elevated, the circulating cells are predominantly granulocytic blast cells. In neutropenic patients the circulating cells are mainly lymphocytes. Many of the myeloblasts in the peripheral bloodstream have Auer rods.

**3. Platelets**
In most cases of AML, the platelet count is low. However major bleeding as a result of thrombocytopenia occurs only when the platelet count goes below 25,000/μL of blood.

**4. Bone marrow**
The bone marrow is hypercellular with a marked proliferation of early granulocytic blast cells, which comprise more than 75% of all bone marrow cells.

## Treatment of AML

The treatment of acute myelocytic leukemia has two major goals, one of which is to eradicate leukemic cells without destroying normal bone marrow cells. This is attempted with therapeutic chemicals (chemotherapy). The other goal is to reduce or eliminate the side-effects of AML such as anemia, infections, bleeding, and the danger of coagulation. This is achieved by aggressive supportive care.

### *Chemotherapy*

Chemotherapy attempts to eliminate large numbers of abnormal blast cells without destroying normal hematopoietic tissues. In other words, it tries to eliminate neoplastic cells and achieve a complete remission of the disease. **Complete remission** may be defined as the presence of less than 5% of blast cells in the bone marrow together with normal erythropoiesis, granulopoiesis, and thrombopoiesis. A course of treatment for acute myelocytic leukemia typically involves three different phases: (1) an initial induction phase, (2) a consolidation phase, and (3) a prolonged maintenance program. A list of drugs commonly used in the treatment of AML is given in table 24.3, together with their proposed mechanism of action and their side-effects. None of these drugs are used singly; all modern chemotherapy involves a combination of drugs.

A typical treatment would start with a short intensive drug regimen lasting 5 to 7 days. After a rest period of about 30 days to permit the bone marrow to regenerate, this course is repeated once or twice to achieve a maximum degree of remission. This is followed by a consolidation phase of similar length, followed by a maintenance program for which similar or different drugs may be used.

**Table 24.3** Chemotherapeutic Agents Commonly Used in the Treatment of Acute Myeloblastic (Myelocytic) Leukemia

| Drug | Mode of Action |
|---|---|
| Cytosine arabinoside | Inhibits DNA synthesis |
| Doxorubicin | Inhibits DNA and RNA synthesis |
| 5-Azacytidine | Inhibits DNA and RNA synthesis |
| 6-Thioguanine | Inhibits purine synthesis |
| Methylglyoxal Bis (guanylhydrazone) | Unknown |
| acridinylamine methanesulfon-m-anisidide (AMSA) | Binds to DNA |
| Prednisone | Lyses lymphoblasts |
| Vincristine | Inhibits RNA synthesis and assembly of mitotic spindles |
| Asparaginase | Depletes endogenous asparagine |
| Daunorubicin | Inhibits DNA and RNA synthesis |
| Doxorubicin | Inhibits DNA and RNA synthesis |
| Methotrexate | Inhibits pyrimidine synthesis |
| 6-Mercaptopurine | Inhibits pyrimidine synthesis |
| Cyclophosphamide | Cross-links DNA strands |
| Cytosine arabinoside | Inhibits DNA synthesis |

A commonly used drug combination includes cytosine arabinoside, cyclophosphamide, prednisone, and vincristine. Such drug regimens have a remission rate of better than 50%. However the duration of the remission is usually rather short—typically about 6 months—after which time retreatment is needed, since maintenance programs currently have not been very successful. The median survival period of patients with at least one remission is about 12 months, as compared with 2.8 months for those who do not respond to treatment. Success in treatment depends largely upon the regenerating capacities of the normal bone marrow elements after chemotherapy is stopped. Although only about half of all AML patients achieve a satisfactory remission with the present therapeutic drugs, continuing these procedures is important despite the initial side-effects, since the improvement in the quality of life during the period of remission is such that it makes it worthwhile. Further, some patients with AML have survived for long periods of time, which suggests the possibility of a cure in some cases. About 25% survive for more than 16 months.

### *Supportive Management*

Successful remission is intimately connected with aggressive supportive therapy, especially during the early phases of treatment. During the early stages of chemotherapy, severe hypoplasia of the bone marrow is common and can be potentially fatal. Anemia produced as a result of the therapy must be relieved by repeated blood transfusions. Similarly, since bleeding is almost always present, infusions of platelet concentrates are necessary. The infections should be attacked by antibiotic therapy. Since many of the infections arise from the patient's own commensal bacteria, especially gram-negative gut bacteria, bowel sterilization with nonabsorbed antibiotics and antifungal agents is commonly used.

Ascertaining the presence of infections often is difficult since the usual signs of inflammation are normally absent. For that reason, patients should be constantly monitored for presence of pathogens by taking regular mouth, throat, and perianal samples, and by blood cultures and urine analysis. Further, if it is feasible, granulocyte transfusions should be given, especially to those who do not respond well to antibiotic treatment, and also to others since neutrophils will induce signs of inflammation if infections are present. However harvesting white blood cells in sufficient numbers is an expensive process that is not available at most medical centers.

### *Bone Marrow Transplantation*

Bone marrow transplantation is now used in many major hospitals with some success. It is especially successful in patients who are one of two identical twins, where the healthy donor twin can provide bone marrow accepted as self by the recipient twin. In other instances marrow elements can be harvested from compatible siblings, and although the success rate is not as high, it has been encouraging, especially in younger people, who seem to respond better than older persons.

# Further Readings


Foucar, K. Acute Leukemias: Part 2. Acute Nonlymphocytic Leukemia. *Laboratory Medicine* 12 (8), 1981.

Harmon, D. C. The Leukemias. Lecture 21 in *Hematology*, 4th ed., W. S. Beck, Ed., The MIT Press, Cambridge, Mass., 1985.

Hoffbrand, A. V., and Pettit, J. E. The Leukaemias. Chapter 7 in *Essential Haematology*, 2nd ed., Blackwell Scientific Publications, Oxford, 1984.

Powers, L. W. Primary Disorders of the White Blood Cells. Chapter 20 in *Diagnostic Hematology, Clinical and Technical Principles*. The C. V. Mosby Company, St. Louis, 1989.

Thomas, E. D. Bone Marrow Transplantation in Hematologic Malignancies. *Hospital Practice*, Feb. 1987.


# Review Questions

1. Define leukemia. List the four major classes of leukemia. What are the major parameters used in the classification of leukemias?
2. List the major causes of leukemia. Mention five important factors implicated as major causative agents of leukemia.
3. What evidence exists that chromosomal abnormalities are involved in the induction of anemia?
4. Define carcinogens. Mention some important carcinogens. What evidence implicates chemical agents as causative factors of leukemia?
5. What evidence implicates ionizing radiation in the production of leukemia?
6. Describe the evidence implicating viruses as a causative factor in the onset of leukemia.
7. List five major clinical symptoms associated with leukemia. Explain the cause of each of these symptoms.
8. Distinguish between the incidence of myelocytic and lymphocytic leukemias. Distinguish between the incidence of acute and chronic leukemias.
9. Give a possible explanation for leukemias being slightly more common in men than in women.
10. What does the abbreviation CML stand for? What is another name for CML? What is the incidence of CML? Which age group is predominantly affected by CML?
11. List the major causes of CML.
12. What is the Philadelphia chromosome? Why is it an important diagnostic tool?
13. List the major clinical symptoms of CML.
14. What are the major laboratory features of CML? What is a unique feature of CML?
15. Describe the treatment of choice for CML.
16. List some of the drugs of choice used to treat CML. How does each of these drugs act (i.e., what effect does each have on the body)?
17. What is the prognosis of CML?
18. What does the abbreviation AML stand for? What other names are used for AML?
19. What is the incidence of AML? Which age groups are particularly affected?
20. List the seven major forms of AML according to the FAB classification. What are important features of each?
21. What are the Schilling and *Naegeli* types of AML?
22. Give the major causes of AML.
23. List the major clinical symptoms of AML. Explain why these symptoms occur.
24. What are the major laboratory features of AML?
25. What is the major treatment of AML? Define complete remission.
26. Describe a typical chemotherapy treatment for AML. Explain why each of these treatments lasts for only a short time.
27. Describe the prognosis for AML with and without chemotherapy.
28. What is meant by supportive management for AML?
29. Describe the problem with bone marrow transplantations in the cure of AML.

# Chapter Twenty-Five

## *Lymphoproliferative Disorders I: Infectious Mononucleosis and Lymphocytic Leukemias*

**Lymphoproliferative disorders** are complex and varied diseases associated with an increased production of various types of lymphoid cells. This group runs the gamut from a benign self-limiting proliferation of lymphocytes (e.g., infectious mononucleosis) to a rapid-growing, malignant neoplastic disease (e.g., acute lymphocytic leukemia) to chronic lymphoproliferative disorders that may or may not be fatal (e.g., Hodgkin's disease).

## REACTIVE AND NONREACTIVE LYMPHOCYTOSIS

This chapter describes forms of lymphocytosis known as **reactive** lymphocyte proliferations. Generally self-limiting, these types of lymphocytosis are exemplified by a disease known as **infectious mononucleosis.** This is followed by a discussion of two major types of **nonreactive** lymphoproliferative disorders: **acute** and **chronic lymphocytic leukemia.** In chapter 26 the other forms of nonreactive lymphocytosis are discussed. For a summary of all the lymphoproliferative disorders see table 25.1.

**Table 25.1 A Classification of the Major Lymphoproliferative Disorders**

*Reactive Lymphocytosis*
- Infectious mononucleosis
- CMV mononucleosis
- HIV-induced lymphocytosis
- Lymphocytosis associated with viral hepatitis, *Toxoplasma,* drug hypersensitivity, bacterial infections, massive blood transfusions

*Nonreactive Lymphocytosis*
- Lymphocytic leukemia
    - Acute lymphocytic leukemia (ALL)
    - Chronic lymphocytic leukemia (CLL)
- Lymphomas
    - Hodgkin's disease
    - Non-Hodgkin's lymphoma
- Plasma cell leukemias
    - Multiple myeloma
    - Waldenström's macroglobulinemia
    - Heavy chain disease

## DISORDERS OF REACTIVE LYMPHOCYTOSIS

In **reactive lymphocytosis** increased numbers of atypical lymphocytes are produced as a result of exposure to certain antigenic stimuli. Frequently these stimuli are viruses, but they can also be drugs or pathogenic organisms. The resulting lymphocytes are large, immature appearing, and are referred to as stimulated or "reactive." They frequently appear abnormal and may have bizarre shapes but are not malignant. These atypical cells are usually larger than normal and may reach a diameter of 30 μm or more. In most cases both the nuclear and cytoplasmic mass are increased. The shape of the nucleus of the reactive lymphocyte is frequently irregular, and its cytoplasm often appears indented due to the presence of nearby red blood cells (figure 25.1).

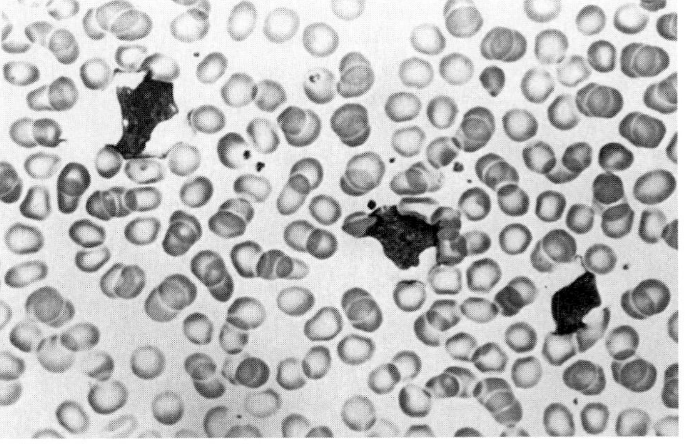

figure 25.1

A blood smear of a patient with infectious mononucleosis. Note the presence of atypical lymphocytes.

### Etiology of Reactive Lymphocytosis

Reactive lymphocytosis is most commonly associated with viral infections. Some of the most important viral disorders associated with reactive lymphocytosis are infectious mononucleosis, cytomegalovirus (CMV) infections, viral hepatitis, and human immunodeficiency virus (HIV) infections, which may cause acquired immunodeficiency syndrome (AIDS). Other causes include toxoplasmosis, certain bacterial infections, drug hypersensitivity, and the infusion of large amounts of blood (e.g., during open heart surgery). The most common of all these causes is infectious mononucleosis, which is discussed in some detail in the following material.

### Infectious Mononucleosis

This clinical syndrome was originally described as **glandular fever,** and it is still known under that name in some English-speaking countries such as Australia. In certain European countries, it is also known as **Pfeiffer's disease,** so-named after the German physician Emil Pfeiffer, who first described this syndrome. The term *infectious mononucleosis* was coined in 1920 by the American physicians Sprunt and Evans. The major overt symptoms include sore throat, enlarged lymph nodes in the neck (cervical lymphadenopathy), fever, and tiredness.

*Etiology of Infectious Mononucleosis*

For many years the cause of infectious mononucleosis was unknown, but for about the last 25 years the causative agent is generally acknowledged to be the **Epstein-Barr virus (EBV),** which is a DNA virus that belongs to the group of **herpesviruses.** Nearly all people with infectious mononucleosis have high titers of anti-EBV antibodies in their blood. In addition, persons that have had infectious mononucleosis have continuous lymphoid cell lines containing EBV in their blood. Apart from lymphocytes (B cells), EBV also readily infects the pharyngeal epithelium. The virus is transmitted via exchange of oral secretions. Hence it is also called the **kissing disease.** It affects mainly teenagers and college students between 15 and 20 years of age. Socioeconomic status may also be a factor in acquiring the disease. In most lower socioeconomic groups, 80% of children have immunity against EBV by the age of 4 years and thus are less prone to acquire it at a later age. In early childhood infection with EBV is most often asymptomatic. In more affluent environments, only 10 to 15% of children are immune to EBV at age 4 years. Hence they are much more susceptible to this disease later in life.

*Clinical Features of Infectious Mononucleosis*

Teenagers or young adults with infectious mononucleosis typically will exhibit a sore throat, fatigue, fever, and swollen cervical lymph nodes. Splenomegaly and hepatomegaly are also quite common, and less frequently jaundice and facial edema are present.

The usual case of infectious mononucleosis is mild and uncomplicated. The worst of the symptoms disappear within 3 to 5 weeks, although it may take several months before the individual feels completely well again.

*Laboratory Findings of Infectious Mononucleosis*

Almost all individuals with infectious mononucleosis have an increase in the number of lymphocytes, with at least 10% showing reactive features—that is, their lymphocytes appear large and immature with bizarre nuclei. The disease is usually diagnosed by a **positive heterophile antibody test** result. A heterophile antibody is an antibody that reacts with an antigen of an unrelated animal species. In other words, immunologically speaking, it reacts across species lines. In 1931 Paul and Bunnell discovered heterophile antibodies in the serum of patients with infectious mononucleosis. They found that sheep and horse red blood cells could be agglutinated by the serum of a person infected with mononucleosis. This characteristic is the basic principle of the various rapid slide tests now used to diagnose the presence of infectious mononucleosis.

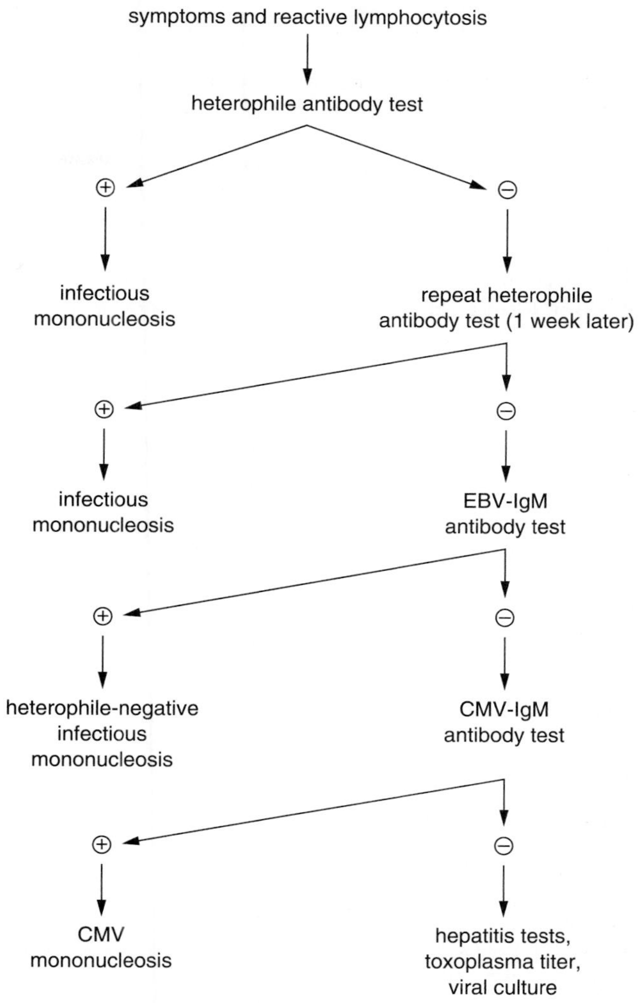


figure 25.2

Flow chart of laboratory tests in the differential diagnosis of infectious mononucleosis and similar diseases.

People who are heterophile-negative can still show mononucleosislike symptoms, which may be caused by either cytomegalovirus (CMV), the hepatitis virus, or HIV. This may be diagnosed by looking for the presence of antibodies against these viruses in the blood of the patients.

Normally it takes 2 to 3 weeks before a strong antibody titer against EBV is present. Thus early negative test results should be repeated 7 to 10 days later (figure 25.2).

*Treatment of Infectious Mononucleosis*

The therapy for infectious mononucleosis depends on the severity of the symptoms. In most routine cases no specific therapy is needed other than bed rest and plenty of fluids. In cases of severe neck inflammation, steroid therapy may be needed to reduce the local swelling.

## ACUTE LYMPHOCYTIC LEUKEMIA (ALL)

Acute lymphocytic leukemia is also known as **acute lymphoblastic leukemia,** since it is characterized by a great unrestrained proliferation of poorly differentiated lymphoid blast cells (primitive lymphoblasts). ALL is associated with a large reduction of normal hematopoietic cells and thus is always accompanied by anemia, granulocytopenia, and thrombocytopenia. If not treated promptly, ALL may be fatal due to a combination of anemia, infection, and bleeding.

### Incidence of ALL

Acute lymphocytic leukemia (ALL) is the major neoplastic disease of childhood; 60 to 70% of all cases of ALL occur in young children, with a peak incidence around 4 years of age. Only 20% of all cases of ALL occur in adults. This disease affects males and females in equal numbers.

Acute lymphocytic leukemia accounts for 20 to 25% of all reported cases of leukemia. It appears to be much more frequent in whites than in blacks or Asians.

### Etiology of ALL

Although the exact etiology of acute lymphocytic leukemia is unknown, genetic, hereditary, and environmental factors have been implicated, as follows:

**1. Genetic factors**
Several congenital abnormalities are associated with an increased incidence of ALL. These disorders are characterized by chromosome abnormalities. For instance, a much higher incidence of ALL occurs in persons with Down's syndrome, Bloom's syndrome, and those with ataxia-telangiectasia.

**2. Hereditary factors**
Some striking familial clusters of ALL have been described, although they are uncommon. We should keep in mind, however, that some of these clusters could be the result of unknown environmental factors.

**3. Environmental factors**
There is a high correlation between exposure to ionizing radiation, mutagenic chemicals, certain viruses, and the incidence of acute leukemia. In most instances, however, the leukemia has been myelocytic rather than lymphocytic in nature. Currently we cannot rule out the previously mentioned environmental factors as possible causes of ALL.

Certain DNA viruses may be involved in the transmission of ALL. DNA viruses are usually transmitted to the offspring with the genome. As a result of this transmission, the body becomes immunologically tolerant of these viral antigens. Later in life this viral genetic material may become activated by some external event—such as exposure to certain chemicals or radiation—thus causing these cells to become malignant. However, this oncogene theory has not currently been proven as a cause of ALL.

### Classification of ALL

Originally all cases of ALL were considered a single form of leukemia. But with the development of more accurate staining techniques and improved microscopy, it became clear that there were several subtypes of ALL—a conclusion later reinforced by the fact that each subtype had its own specific surface markers. We now divide all forms of ALL into the three major categories established by the French-American-British (FAB) Cooperative Group. However, with the development of new immunological techniques it became possible to divide ALL even further. Hence both the FAB classification and the immunological classification for ALL are described.

#### *FAB Classification of ALL*

The FAB classification of ALL is based on morphological features and not on surface markers, and includes the following subtypes:

**L1: ALL dominated by small blast cells** These small cells exhibit small amounts of cytoplasm that does not stain deeply. The nuclei are round or oval and may show cleftlike features. The nuclei stain dark and do not usually contain nucleoli (figure 25.3). This is the most common type of ALL (about 85% of all cases in children) and also has the most favorable prognosis.

Immunological markers indicate that most of these lymphoblasts belong to the non-B-cell groups of lymphocytes.

**L2: Large pleiomorphic lymphoblasts** This form is characterized by the presence of large and small lymphocytes. The heterogenicity in size is a distinct feature of this type of ALL. Because of this variation in size, the amount of cytoplasm varies as does the staining of the cytoplasm. Sometimes the cytoplasm stains deeply, in other cases only lightly. The nuclei are frequently irregular in shape, and clefts and indentations are common (figure 25.3). About 14% of all cases of ALL are classified as L2. Immunological markers indicate that this type of ALL also involves only non-B lymphocytes—that is, T cells and natural killer (NK) cells.

**L3: Burkitt's type of ALL** This is the rarest form of ALL—less than 1% of all cases belong to this category. It is also the most distinct type of ALL in that all cells are large and morphologically homogeneous. The nuclei are large, round or oval and have finely stippled chromatin. The cytoplasm is abundant, deeply basophilic, and contains many vacuoles (figure 25.3). Immunological markers are indicative of lymphoblasts that belong to the B-cell group. A summary of the morphological classification of ALL is given in table 25.2.

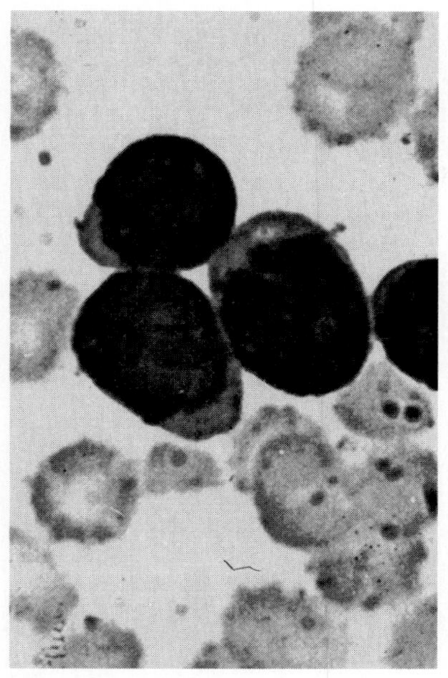
L1

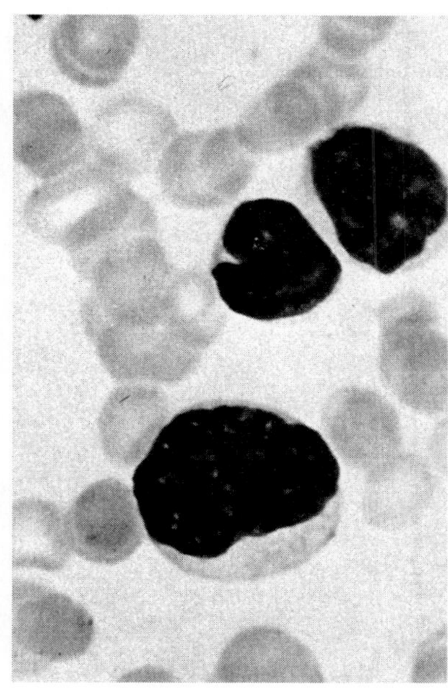
L2

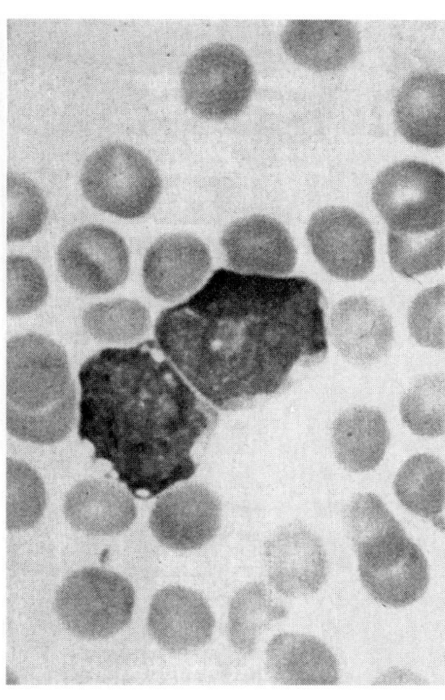
L3

figure 25.3

Blood morphology in acute lymphoblastic leukemia (ALL), according to the FAB classification.

**Table 25.2** Diagnostic Features of the Three Forms of Acute Lymphoblastic (Lymphocytic) Leukemia (FAB system)

| Feature | L1 | L2 | L3 |
|---|---|---|---|
| *Cell Morphology* | | | |
| Cell size | Small | Variable | Large |
| Nucleus: | | | |
| Shape | Round | Irregular, clefted | Round, oval |
| Chromatin | Homogenous | Heterogenous | Homogenous |
| Nucleoli | Indistinct | One or more | One or more |
| Cytoplasm: | | | |
| Amount | Scant | Moderate | Moderate |
| Color | Moderate basophilic | Variable basophilic | Deep basophilic |
| Vacuoles | Variable | Variable | Prominent |

*Immunological Classification of ALL*

The new immunological classification is based on the presence or absence of certain surface markers of the neoplastic lymphocytes. These surface molecules indicate not only the type of lymphocytes involved—that is, such as B cells, T cells, and NK cells—but also the stage of lymphocyte development (e.g., pre-B cell, early B cell). Based on the surface markers, ALL can be divided into the following five different categories:

**1. Undifferentiated lymphocyte ALL**

This type of ALL consists of lymphocytes that cannot be classified as either B or T lymphocytes. They all possess a marker known as CALLA (=common ALL antigen) and also have the enzyme terminal deoxynucleotidyl transferase (TdT) and class II MHC molecules (i.e., HLA-DR).

**2. Pre-B-cell ALL**

These lymphoblasts exhibit all the preceding cell membrane markers as well as cytoplasmic immunoglobulin (cIg), which classifies them as belonging to a B-cell category. Since they do not show specific B-cell membrane markers, they were previously classified as non-B cells belonging to either the L1 or L2 morphological category of ALL.

**3. B-cell ALL**

This category may or may not show the CALLA marker; it exhibits HLA-DR but not TdT. What distinguishes this cell is the presence of B-cell receptors on its surface—that is, membrane immunoglobulins (mIgs). In the FAB classification this group is classified as L3.

**Table 25.3** Immunological Markers that Help Determine which Type of Lymphocyte is the Cause of the Acute Lymphocytic Leukemia

| Immunological Marker | Predominant Lymphocyte Present | | | | |
|---|---|---|---|---|---|
| | Undifferentiated Lymphocyte | Pre-B Cell | B Cell | T Cell | Null Cell |
| CALLA | + | + | + or – | + or – | – |
| TdT | + | + | – | + | + |
| HLA-DR | + | + | + | – | + |
| cIg | – | + | – | – | – |
| mIg | – | – | + | – | – |
| CD2 | – | – | – | + | – |
| CD7 | – | – | – | + | – |

**4. T-cell ALL**

This group of ALL is characterized by the absence of either cIg or mIg as well as HLA-DR. It exhibits a unique T-cell marker known as CD2, which allows these cells to form rosettes with sheep red blood cells (SRBCs). They also show the presence of TdT and CD7 (another unique T-cell marker).

**5. Null cell ALL**

This form of ALL is characterized by cells that do not possess markers unique to either B cells or T cells, but they do exhibit TdT and HLA-DR. In the FAB classification they are classified as either L1 or L2. A summary of the immunological classification of ALL is given in table 25.3.

## Laboratory Findings of ALL

Laboratory investigations usually reveal a marked leukocytosis with the presence of many primitive blast cells (the type of blast cell depending on the form of ALL), anemia, granulocytopenia, thrombocytopenia, and bone marrow packed with poorly differentiated lymphoblasts.

**1. White blood cells**

The peripheral leukocyte count may vary from low levels (less than 1,000 WBCs/µL) to high levels (more than 500,000 WBCs/µL). Roughly 50% of all children with ALL have an aleukemic picture—that is, they have normal peripheral white blood cell counts and few blast cells in the circulation. Generally speaking, patients with low white blood cell counts usually have a better survival rate than those with very high numbers of leukocytes, possibly because very high numbers may cause greater physiological complications such as intravascular cell clumping. This may lead to vascular occlusions in the small blood vessels, especially in the brain, resulting in poor perfusion of the brain, stroke, or cerebral hemorrhage.

**2. Red blood cells**

Anemia is very common in ALL, affecting more than 90% of all patients with this disorder. The red blood cells are usually normocytic and normochromic, except in cases of excessive bleeding, when they may become hypochromic. The low numbers of erythrocytes in the peripheral circulation may be the result of crowding out of erythroid precursor cells by cancerous lymphoblastic cells in the bone marrow. In many cases anemia may be caused by the antimitotic and antimetabolite drugs used in the treatment against ALL. Most likely, it is a combination of these factors.

**3. Platelets**

Thrombocytopenia is also observed in more than 90% of all patients with ALL. In many instances, the few platelets present are large and bizarre in shape. Reduced platelet numbers in the peripheral bloodstream may also be explained by the crowding out of megakaryocytes by malignant lymphoblasts or by drug therapy (just as the reduction in red blood cells).

**4. Bone marrow**

Bone marrow biopsies of patients with ALL generally exhibit a marrow crowded with primitive blast cells (figure 25.4). The marrow fat spaces are greatly reduced or absent. Normal erythropoietic, granulopoietic, and megakaryocytic elements are also greatly reduced.

The blast cells are, in most cases, easily recognized as lymphoblasts rather than erythroblasts or myeloblasts. The lymphoblasts demonstrate nuclear arylsulfatase activity, a unique characteristic not present in either erythroblasts or myeloblasts.

We should remember that cancerous white blood cells do not proliferate faster than normal cells but rather divide continuously—that is, they do not stop after three or four cell divisions, as is the case with normal white blood cells.

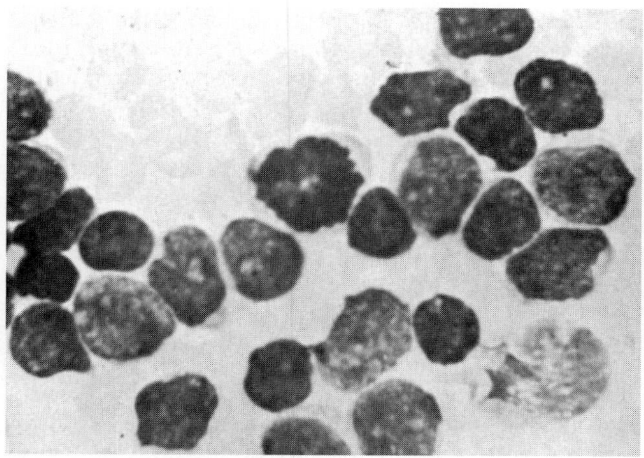

a.

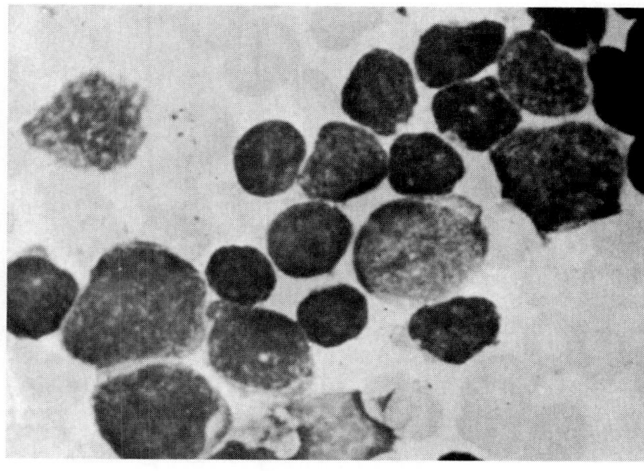

b.

**figure 25.4**

(*a*) Bone marrow smear and (*b*) peripheral blood smear of acute myelocytic leukemia (AML). Note the many similarities.

## Clinical Features of ALL

The clinical symptoms of acute lymphocytic leukemia are many and varied, and all are potentially able to cause the death of the patient. Clinical symptoms include the following:

**1. Anemia**
In many cases anemia is due to (1) a reduction in erythropoiesis; (2) an increased tendency to bleeding, as a result of thrombocytopenia; and (3) a shorter life span of red blood cells, as a result of the presence of abnormal hemoglobins such as HbF. The clinical symptoms depend on the severity of the anemia. The hemoglobin level rarely falls below 7 g/dL. In general the effects of anemia tend to be more severe in adults with ALL than in children.

**2. Infection**
A decrease in the number of functional white blood cells (including functional B and T cells) will result in an increased susceptibility to viral, bacterial, fungal, and protozoan infections. Overwhelming infection is a major cause of death in ALL.

**3. Bleeding**
Virtually all patients with ALL will experience bleeding at some time. The bleeding usually becomes more serious as the disease progresses. Intracerebral hemorrhage (as a result of clumping of lymphoblasts and vessel rupture) is a common cause of death of ALL patients.

**4. Organ infiltration**
Acute lymphocytic leukemia is not just a blood disease but may also be viewed as a systemic disorder, since the malignant lymphoblasts are capable of infiltrating almost any body organ. Consequently the clinical manifestations of ALL are not limited to anemia, infection, and bleeding, but may include many other symptoms, depending on the organ(s) affected. For instance, approximately 75% of ALL cases exhibit lymph node enlargement (lymphadenopathy). This is frequently an early warning sign of ALL. Similarly, the liver and the spleen are also frequently enlarged, without unduly interfering with the functions of these organs. Infiltration of the skin by malignant lymphoblasts may result in local lesions, petechiae, and ecchymoses. Infiltration of the bones and the joints may cause localized lesions resulting in bone and joint pain. Swollen arthritic joints occur in half the children with ALL.

Leukemic lymphoblasts may also infiltrate the brain and the membranes surrounding the brain. This may result in a special type of acute lymphocytic leukemia known as **meningeal leukemia.** Symptoms of this disease initially include severe headaches and nausea, which may later be followed by convulsions and palsylike symptoms due to the increased intracranial pressure.

The lung is another common site of infiltration by cancerous lymphoblasts, which may result in pulmonary lesions. Lesions in the upper airways contribute to bleeding and infections.

The gastrointestinal (GI) tract is another favorite target of the leukemic lymphoblasts in ALL. However, most of the damage that occurs in the GI system is caused by cytotoxic drugs used to control the acute leukemia. These drugs damage the rapidly dividing epithelial cells of the intestinal tract and thus may cause local ulceration, infection, and hemorrhage.

Infiltration of the kidneys is also quite common in ALL. However, impairment of renal function as a result of the invading lymphoblasts is quite rare. Major kidney damage is usually caused by high levels of uric acid in the blood and urine due to increased purine metabolism in the continuously dividing leukemic cells. High levels of uric acid tend to crystallize out, especially at low pHs, which may lead to malfunctioning of the nephrons, and vena anuria. This will lead to rapid death if not treated promptly.

Finally, testicular infiltration by malignant lymphoblasts is also quite common in male patients with ALL.

### Treatment of ALL

The therapy of acute lymphocytic leukemia is designed to resolve the dual problem of eradicating the cancerous lymphoblast cell population while removing ALL's clinical symptoms. The former is normally achieved by chemotherapy, and the latter is accomplished by supportive treatment. Another promising form of treatment, especially in young children, has been the introduction of bone marrow transplants.

*Chemotherapy*

The purpose of chemotherapy is to remove the cancerous cells while sparing the normal marrow cell populations. This is achieved by exploiting the subtle differences that exist between normal white blood cells and leukemic cells regarding metabolism, kinetics, and regeneration potential. Antileukemic drugs function by inhibiting the mitotic activity of these cancer cells. Some of these drugs prevent the replication and transcription of DNA. Others block the availability of critical enzymes and vitamins needed to replicate new DNA, and still others inhibit the mitotic apparatus.

Anticancer drugs may be divided into two categories according to their mode of action: (1) **cell-cycle-specific drugs,** which are reactive during a specific part of the cell cycle and (2) **cell-cycle-nonspecific drugs,** which have cytotoxic action on cells in any phase of the cell cycle. This group includes alkylating agents, antibiotics, and glucocorticoids.

The ultimate purpose of drug therapy is to produce complete remission of the disease. This is achieved when less than 5% of all bone marrow cells are blast cells, and normal erythropoiesis, granulopoiesis, and thrombopoiesis can again take place. However, even when the disease is in remission, and all other effects of acute lymphocytic leukemia have been eradicated, aggressive treatment must continue to ensure that any residual intact leukemic cells are eliminated. Only by long-term treatment can this disease be controlled and the patient considered cured.

*Supportive Therapy*

Supportive treatment works to remove the side-effects of acute lymphocytic leukemia—especially anemia, infection, and hemorrhage. The effects of anemia may be reduced by blood transfusions with packed red blood cells or whole blood. Bleeding may be controlled by the infusion of platelet-rich concentrates. Infections can be treated with aggressive antibiotic therapy. Special isolating techniques, such as maintaining the patient in laminar-flow rooms, have been used to reduce the chance of infection during therapy.

*Bone Marrow Transplantation*

Bone marrow transplantation following intensive chemotherapy and/or radiotherapy has also been used with considerable success in recent years. This treatment is especially necessary for patients who have shown signs of drug resistance. Bone marrow grafts from compatible donors are now commonly performed, and in most cases have produced prolonged disease-free survival.

### Prognosis of ALL

Before chemotherapy was used, the survival time of children with acute lymphocytic leukemia was less than 3 months after the initial diagnosis was made. Currently the median survival time is 6 years or more, and prolonged survival and perhaps even cure has been achieved in more than one-third of all patients.

The prognosis for adults with ALL becomes progressively poorer with age. However, the outlook of patients with ALL is generally better than for those with AML.

## CHRONIC LYMPHOCYTIC LEUKEMIA (CLL)

Chronic lymphocytic leukemia is another lymphoproliferative disorder. However, in contrast with ALL and AML, this disease is characterized by the presence of increased numbers of small, fairly normal-looking lymphocytes in the blood, bone marrow, and lymphoid tissues. Moreover, in most instances, the increase in small lymphocytes in these tissues is not as dramatic as in most acute leukemias. This lymphocytic condition is primarily due to an increase in B cells. Over 90% of all cases of CLL are caused by B-cell tumors, the remaining are T-cell cancers. The B-cell type of CLL is usually more benign than the T-cell type. T-cell-type CLLs tend to progress more rapidly and do not respond as well to traditional therapies.

### Incidence of CLL

Chronic lymphocytic leukemia (CLL) is primarily a disease of middle-aged and elderly persons. It rarely begins before 40 years of age, and there is a peak occurrence between 50 and 60 years of age. CLL is a rather common type of leukemia, as it accounts for one-quarter (25%) of all leukemias.

### Laboratory Findings of CLL

One of the most obvious features of CLL is the moderate to high increase in the white blood cell (WBC) count. In most patients the total WBC count is 10,000 to 30,000/μL. However in about 20% of patients with CLL, there is a marked increase in the total number of leukocytes of more than 100,000/μL. The differential white blood cell count will reveal that 60 to 80% of the leukocytes in the blood smear are lymphocytes. Most of these lymphocytes appear as small lymphocytes, although a small percentage of large immature-looking lymphocytes will be present. Many of these leukemic lymphoblasts are rather fragile; and consequently become ruptured or broken in the process of smear formation. This will result in the presence of

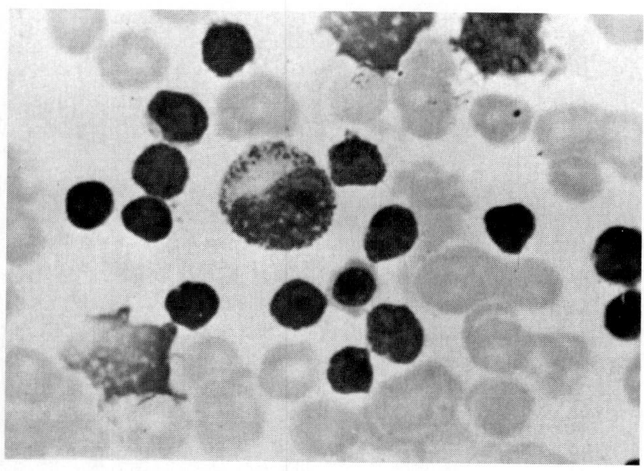
a.

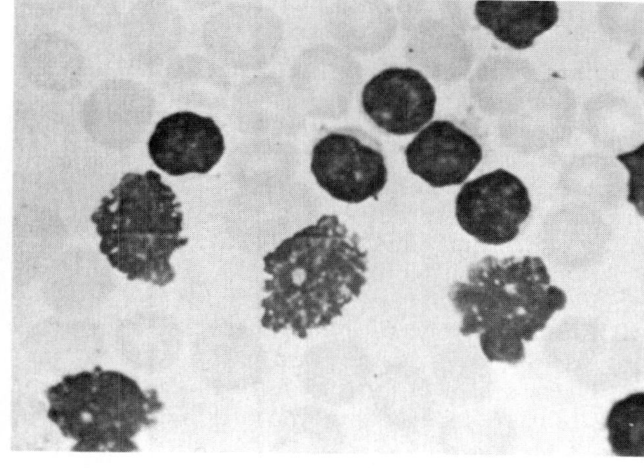
b.

**figure 25.5**

(*a*) Bone marrow smear and (*b*) peripheral blood smear of chronic lymphocytic leukemia (CLL). Note that there are more lymphocytes in the bone marrow than in the peripheral blood.

**smudge cells** on the blood smear of patients with CLL. A smudge cell is a large spread-out nucleus of a broken white blood cell (figure 25.5*b*).

Since most cases of CLL involve abnormal B-cell proliferation, these patients will have a shortage of normal B cells. Hence CLL is commonly associated with a decrease in antibody production. This condition of **hypogammaglobulinemia** can be easily diagnosed by serum electrophoresis.

Bone marrow smears will also show increased numbers of small lymphocytes (figure 25.5*a*). However the increase is not as dramatic as in acute leukemias and chronic myelocytic leukemias. Although there is some displacement and crowding out of normal hematopoietic elements in the bone marrow, the effect is not as dramatic as in acute leukemias and CML. In most cases red blood cell production is not greatly reduced. Although we do not know why, certain patients with CLL (10 to 15%) will also develop **autoimmune hemolytic anemia (AIHA).** This will result in premature destruction of erythrocytes and produce low RBC counts and a marked increase in reticulocytosis. A positive direct Coombs' test will indicate which patients with CLL also have AIHA.

In most cases of CLL, platelet production is not seriously impaired, especially in the early stages; however some thrombocytopenia is commonly observed later in the disease. This is due to the presence of splenomegaly in many patients with CLL, and most of the platelets will be located in the enlarged spleen.

### Clinical Features of CLL

Anemia is not an obvious symptom in most cases of CLL, certainly not in the early stages. Later in the disease red blood cell production may decrease because of increased myelophthisis (crowding out). In 10 to 15% of patients with CLL, the anemia is distinct due to an increased destruction of red blood cells by autoantibodies (AIHA).

Splenomegaly and hepatomegaly are common since these organs attempt to eliminate excess lymphocytes. The superficial lymph nodes may also be enlarged because of the increased numbers of small lymphocytes.

The frequently observed condition of thrombocytopenia in patients with CLL contributes to easy bruising and purpura.

Pruritus is another common feature of CLL. Itching is caused by herpes zoster infections of the skin. In some patients, herpes zoster infection may be the first indication that the patient has chronic lymphocytic leukemia, as it is generally an insidious disease. In many older patients with laboratory features indicating the presence of CLL, the disease is fairly benign and may remain stable and asymptomatic for many years.

As the disease progresses, there may be an increased incidence of bacterial, viral, or fungal diseases due to a lack of functional white blood cells. However this is usually not an obvious symptom in the early stages of disease.

In general patients below 50 years of age and patients with a T-cell type of CLL usually have a more aggressive type of disease, characterized by fatigue, weight loss, night sweats, lymphadenopathy, enlargement of the liver and the spleen, and bone marrow failure.

### Treatment of CLL

Benign forms of CLL usually do not require any special treatment. However, if CLL is associated with high levels of small lymphocytes in the blood, marrow, and lymphoid tissues, chemotherapy may be needed to reduce the numbers of these white blood cells. The usual treatment is with drugs that prevent

DNA replication or RNA formation, such as busulfan, cyclophosphamide, or chlorambucil. A decrease in lymphocyte production in the bone marrow will prevent bone marrow failure. In most cases in which bone marrow failure has taken place, the patient should be treated with corticosteroids, such as prednisolone, until there is a significant recovery of hematopoietic tissue, and an increase in erythrocytes, granulocytes, and platelets in the blood.

Radiotherapy may be needed to reduce the lymph nodes and spleen to normal size. In case of AIHA, splenectomy may be needed if the spleen does not respond to corticosteroids.

Overt symptoms associated with bacterial and fungal infections may be reduced with aggressive antibiotic therapies.

Since some patients show low concentrations of immunoglobulins (antibodies) in the blood, regular injections with gamma-globulin may also help in reducing the incidence of pathogenic infections. In case of temporary bone marrow failure, transfusions of packed red blood cells and platelets may also be used as a means of supportive therapy.

**Prognosis of CLL**

The prognosis of patients with chronic lymphocytic leukemia is usually quite good. CLL frequently is quite benign and slow to progress, which is remarkable since it is primarily a disease of older individuals.

People diagnosed with benign forms of CLL routinely live 10 or more years after the initial diagnosis. In more severe cases requiring more aggressive treatment, survival times of 3 to 5 years are also common.

Unlike chronic myelocytic leukemia (CML), CLL does not normally transform into acute leukemia. If death does occur in patients with this disease, it is usually from complications of infections, anemia, or bleeding, rather than from this cancerous condition.

## HAIRY CELL LEUKEMIA (HCL)

Hairy cell leukemia is considered a special form of chronic lymphocytic leukemia. This disease is characterized by the presence of small lymphocytes with many fine cytoplasmic extensions radiating from a small cytoplasm that surrounds the nucleus. Hence the term **hairy cell** (figure 25.6). This condition is a monoclonal proliferation of B lymphocytes. The hairy cells are not only found in the blood but also in the bone marrow, the spleen, and other lymphoid organs.

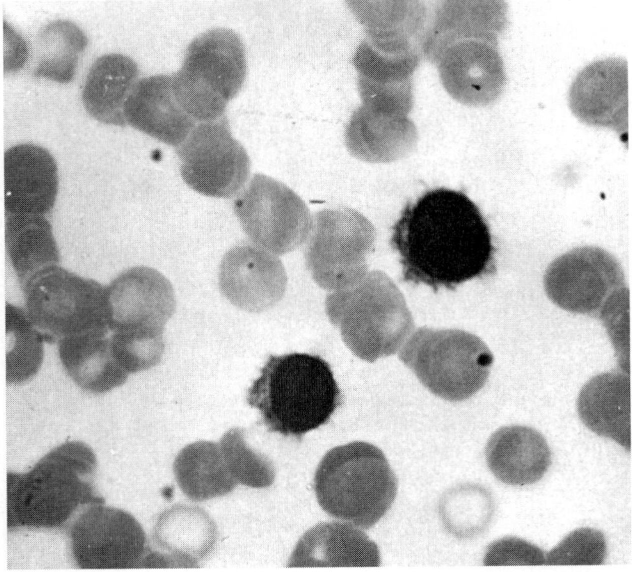

figure 25.6

Peripheral blood smear showing hairy cell lymphocytes, an indication of hairy cell leukemia.

Splenomegaly is a common condition associated with HCL, and the bone marrow often exhibits increased fibrosis. This disorder is often associated with considerable pancytopenia of the peripheral bloodstream. Consequently anemia, infection, and bleeding are common symptoms of this disease, and supportive therapy for these conditions is often necessary. Hairy cell leukemia is also a disease of middle-aged and elderly persons, with a peak occurrence between 40 and 60 years of age. This disorder is typically more common in males, who are four times as likely to contract it as females.

In the past, HCL was believed to be a rare disease but now is considered fairly common and to comprise 3% of all leukemias.

The prognosis of people with HCL is usually quite good, since it develops slowly and is relatively benign. More than half of the people diagnosed with HCL live for 10 years or more after the initial diagnosis. Similar to other forms of chronic lymphocytic leukemia, death is rarely due to the disorder, but rather results from secondary conditions such as an overwhelming bacterial infection or senescence.

## Further Readings


Cohen, D. L., Duval-Arnould, B., and Olson, T. A. Acute Lymphoblastic Leukemia of Childhood. *American Family Physician* 30(4), Oct. 1984.

Harmon, D. C. The Leukemias. Lecture 21 in *Hematology,* 4th ed., W. S. Beck, Ed., The MIT Press, Cambridge, Mass., 1985.

Powers, L. W. Primary Disorders of White Blood Cells. Chapter 20 in *Diagnostic Hematology, Clinical and Technical Principles.* The C. V. Mosby Company, St. Louis, 1989.


## Review Questions

1. Name the two major categories of lymphoproliferative disorders. Give examples of each.
2. What is the cause of infectious mononucleosis? How is it normally transmitted?
3. What are the other names for infectious mononucleosis?
4. Name a unique laboratory feature of infectious mononucleosis.
5. Name a common screening test for infectious mononucleosis. What is a heterophile antibody?
6. How can we distinguish infectious mononucleosis from other forms of reactive lymphocytosis?
7. List the clinical manifestations of infectious mononucleosis.
8. What is the common treatment of infectious mononucleosis? What is the prognosis for infectious mononucleosis?
9. What is the incidence of ALL? What age group is most commonly affected by this leukemia?
10. List the possible causes of ALL.
11. Describe the basis for the FAB classification of ALL. Describe the basis for the immunological classification of ALL.
12. Name the major categories of ALL according to the FAB classification. What are the major features of each category? How do they differ from one another?
13. List the major categories of ALL based on immunological characteristics. Describe the unique features of each category.
14. Give the average RBC, WBC, and platelet counts in ALL.
15. Describe the bone marrow picture in ALL.
16. List the major clinical features of ALL. Explain why these symptoms occur.
17. What are the major forms of treatment for ALL?
18. List some of the major side-effects of chemotherapy used in ALL.
19. Describe supportive treatment for ALL.
20. List some of the problems involved with bone marrow transplantation.
21. What is the prognosis for ALL? Which forms of ALL have the best prognosis? Which have the worst prognosis?
22. What is the incidence of CLL? In what age group is it most common?
23. What are the causes of CLL?
24. What are some common laboratory features associated with ALL?
25. Explain why CLL is usually an insidious disease. How is it commonly detected?
26. What are some of the major clinical features of CLL? Explain why these features may occur.
27. List some common modes of treatment for CLL.
28. What is the prognosis for CLL?
29. Describe a unique feature of hairy cell leukemia. In what group is HCL commonly encountered?
30. How would you classify hairy cell leukemia? Which blood cell is involved?
31. What is the incidence of hairy cell leukemia? In what age group does it occur most frequently?
32. What is the prognosis for people with hairy cell leukemia? What is the common cause of death in hairy cell leukemia?

# Chapter Twenty-Six

## *Lymphoproliferative Disorders II: Malignant Lymphomas and Plasma Cell Leukemias*

**Malignant lymphomas** may be described as a group of diseases that are cancers of the lymph nodes. They are usually divided into two separate categories—**Hodgkin's disease** and **non-Hodgkin's lymphomas**—on the basis of the presence or absence of a unique type of cell known as the Reed-Sternberg (RS) cell. These RS cells are present in the various forms of Hodgkin's disease, but are not found in the other malignant lymphomas.

**Plasma cell leukemias** may be defined as neoplastic diseases of plasma cells in the bone marrow. Hence they are also known as plasma cell myelomas, or multiple myelomas. Each of these major categories is discussed in the following material.

## HODGKIN'S DISEASE

Hodgkin's disease is a form of malignant lymphoma that is closely related to other malignant lymph node tumors. The name is restricted, however, to lymphoid tumors showing the presence of Reed-Sternberg cells. This disorder was first described in 1832 by Thomas Hodgkin, hence the name of the disease.

### Incidence of Hodgkin's Disease

Hodgkin's disease can occur at any age, but seems to manifest itself mainly during either of two periods in the human life cycle. It is the most common malignancy of young adults, at least in the Western world, with a peak occurrence between 20 and 30 years of age. A second group is also more prone to this disease: middle-aged people between the ages of 50 and 60. The disorder is rare in children. Finally, there is greater incidence of this disease in men versus women (ratio of 2:1).

### Clinical Features of Hodgkin's Disease

Patients with Hodgkin's disease display a wide range of features from the presence of local enlarged lymph nodes with no other obvious symptoms to widespread lymphadenopathy with many other overt signs of cancer such as fever, fatigue, weight loss, night sweats, and general malaise. Usually Hodgkin's disease originates in a single peripheral lymph node region, from which it later spreads to other lymphoid areas. In most cases this disorder is first diagnosed by the presence of a painless, asymmetrical enlargement of the superficial lymph nodes of the neck (in 60 to 70% of all cases), or the axillary region (10 to 15%), or the inguinal area (5 to 15%).

Later in the disease the malignancy spreads to other nodal areas and to the liver, the spleen, and the bone marrow. The latter stages can be determined by the use of computed axial tomography (or CAT scans). Splenomegaly and hepatomegaly are also common features of this disease. Other typical symptoms include the presence of cyclical fever (high one week, low the next), pruritus, and pain in the enlarged lymph nodes after alcohol consumption.

Finally, in the late stages when the malignancy has spread to the bone marrow and is crowding out the normal hematopoietic tissue **(myelophthisis),** anemia, infection, and bleeding may become overt symptoms.

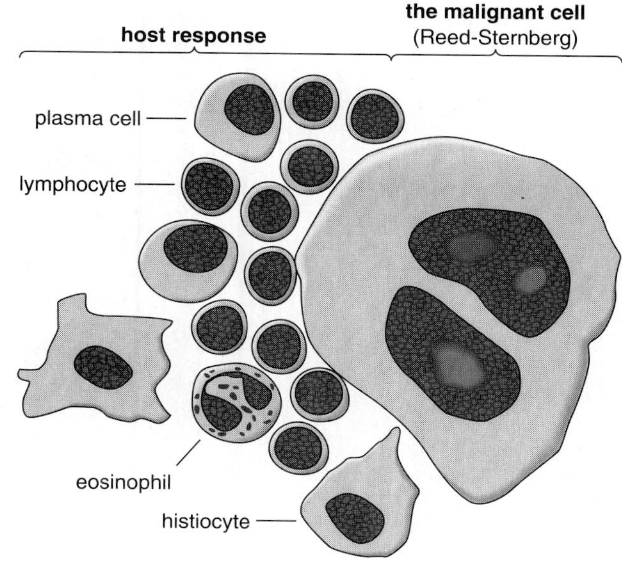


figure 26.1

Blood picture commonly seen in Hodgkin's disease. Note the giant multinuclear Reed-Sternberg cell.

### Laboratory Findings of Hodgkin's Disease

The single most unique feature of Hodgkin's disease is the presence of **Reed-Sternberg (RS) cells** in the lymph nodes. RS cells are large malignant cells (usually with diameters of 50 to 100 μm) that frequently exhibit more than one nucleus in a common cytoplasm (figure 26.1). To diagnose this disease, a **lymph node biopsy** is required. On the basis of this biopsy, we can characterize Hodgkin's disease into four different subtypes as follows:

1. **Lymphocyte-predominant type**
In this category few RS cells are encountered. The major types of cells are *mature (small) lymphocytes and macrophages.* This form is not common and is seen mainly in young adult males with Hodgkin's disease.
2. **Lymphocyte-depleted type**
This type of Hodgkin's disease has a pattern consisting of *many RS cells and few small lymphocytes.* The histological sections appear disorganized and contain connective tissue elements. This is the rarest form of Hodgkin's disease (1 to 2%) and is also the most malignant. It is mainly found in older adults.

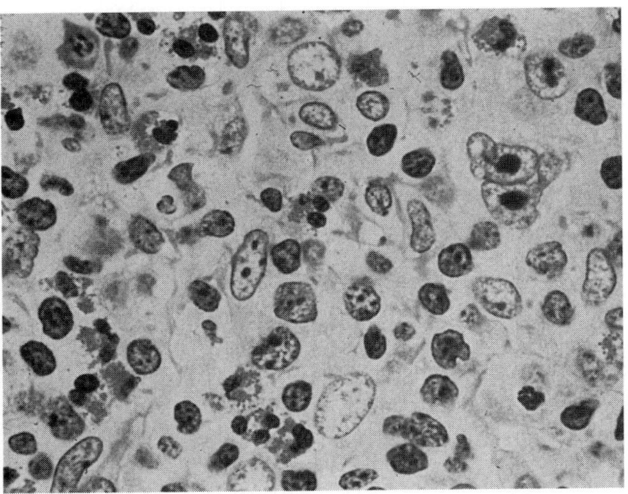

figure 26.2
Histological section of lymph node in patient with Hodgkin's disease with mixed cellular type. A Reed-Sternberg cell with two nuclei is located near the center of the section.
From The American Society of Hematology Slide Bank, 3rd edition, 1990, used with permission.

**3. Mixed cellular type**
This form exhibits a *mixture of RS cells and small lymphocytes.* It is the second most common type of Hodgkin's disease and is mainly found in young adults (figure 26.2).

**4. Nodular sclerosis**
This type is associated with the presence of *abnormal RS cells* (known as **lacunar cells**), abnormal macrophages, and the appearance of collagen bands that encapsulate the nodules of abnormal tissue. This is the most common form of Hodgkin's disease and is especially typical in young adult women.

Other clinical findings typically include the following (especially in later stages of the disease):

**Red blood cells** Low numbers of normocytic-normochromic erythrocytes occur.

**White blood cells** Many cases of Hodgkin's disease have increased granulocytes in the blood, especially neutrophils and eosinophils. This is probably because patients with Hodgkin's disease frequently suffer from infections, especially viral and fungal infections. In advanced stages of the disease, the number of peripheral lymphocytes usually drops significantly.

**Platelets** The platelet count is normal during the early stages but will become progressively lower as the disease advances.

**Bone marrow** The bone marrow usually appears normal early in the disease but may show abnormal features as the disease spreads to different sites. Patients in whom the bone marrow is involved often exhibit hypercalcemia, hypophosphatemia, and an increased serum alkaline phosphatase level. The malignant lymphoid tissue may also spread to the liver, the lungs, and the skin.

### Treatment and Prognosis of Hodgkin's Disease

The treatment patterns of Hodgkin's disease will depend on two major factors: (1) the subtype of disease involved and (2) the spread of the disease at the time of diagnosis. Traditionally the spread of the disease is divided into the following four different stages (figure 26.3):

**Stage I:** Only one lymph node area is involved.

**Stage II:** Two or more lymph node areas are involved, but all are confined to one side of the diaphragm.

**Stage III:** Lymph nodes above and below the diaphragm are involved.

**Stage IV:** Tissues and organs outside the lymph node areas are involved such as bone marrow, liver, and lungs.

Generally speaking, the prognosis is usually quite good as long as the disease is confined to one side of the diaphragm. More than 85% of all patients with stage I and II disease survive this malignancy. Patients with stage III disease survive 70% of the time, whereas only 50% survive stage IV disease. Survival also favors young adults above older people. Moreover, disease conditions with few RS cells have a better survival rate than those with many RS cells.

Radiation therapy is the treatment of choice for patients with stage I and II disease. Many patients can be cured with radiotherapy alone. Patients with stage III and IV disease usually need a mixture of chemotherapy and radiotherapy. Chemotherapy usually involves a mixture of anticancer drugs—for example, a commonly used combination involves mechlorethamine, vincristine, procarbazine, and prednisone.

## NON-HODGKIN'S LYMPHOMAS

### Classification of Non-Hodgkin's Lymphomas

Classification of non-Hodgkin's lymphomas is currently in a state of flux. Many scientific conferences have attempted to organize the multitude of histopathological groupings. However the simplest way has been to classify the non-Hodgkin's lymphomas into those **derived from B cells,** those **derived from T cells,** and those that are **non-B, non-T cell** in origin (these lymphocytes are referred to as null cells). The majority of non-Hodgkin's lymphomas are of B-cell origin (about 80%); the other 20% are equally divided belonging either to the T-cell group (10%) or to the null cell group (10%).

### B-Cell Lymphomas

*Classification of B-Cell Lymphomas*

B-cell-derived non-Hodgkin's lymphomas may be further categorized into those consisting of small

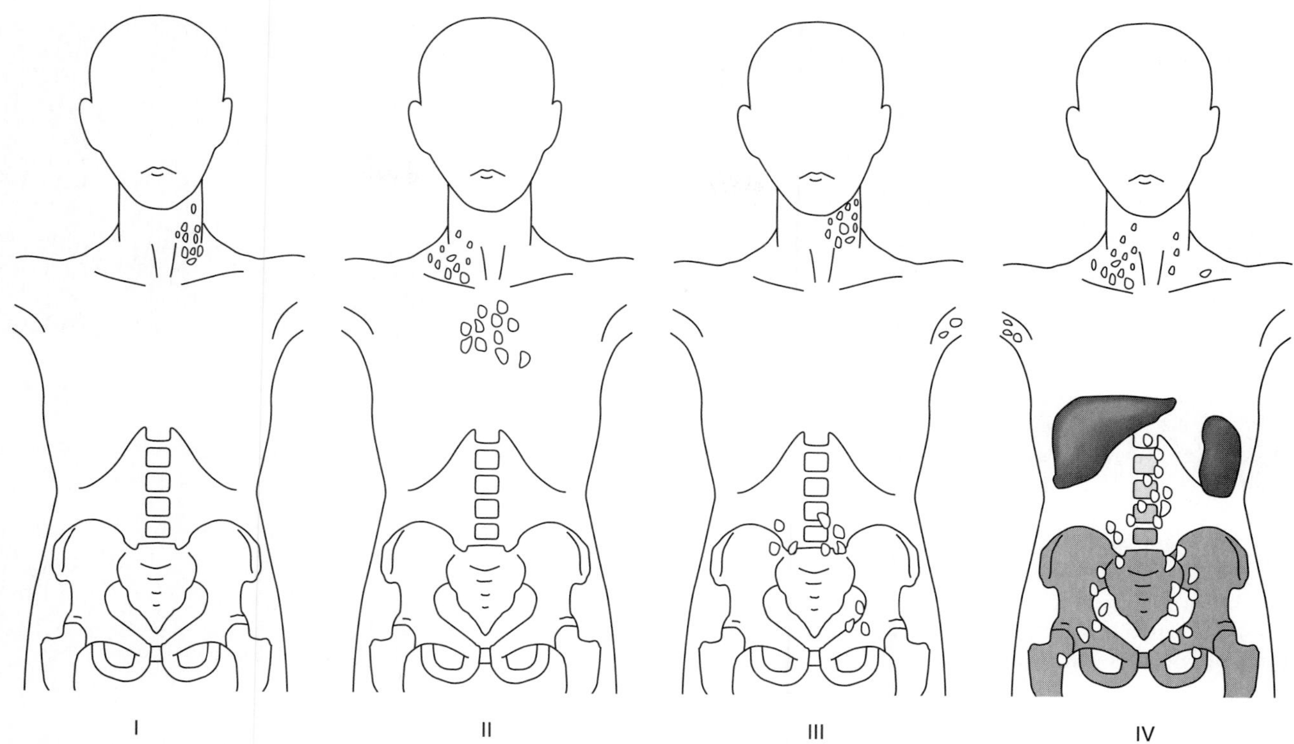


**figure 26.3**
The four stages in Hodgkin's disease *(see text for explanation)*.

lymphocytes, those consisting mainly of large lymphocytes, and those of mixed lymphocyte origin.

**1. Small-cell type (B-cell) lymphoma**
This type of lymphoma is typically found in middle-aged and elderly populations. Patients frequently have enlarged lymph nodes (lymphadenopathy) and an enlarged liver (hepatomegaly). Biopsies of the enlarged lymph nodes exhibit diffuse patterns of many small lymphocytes (figure 26.4*a*), and the blood smear may resemble the pattern commonly encountered in chronic lymphocytic leukemia (CLL), as it also contains many small (mature) lymphocytes. Knowing whether such blood smear patterns are caused by CLL or by small-type B-cell lymphoma is not crucial as both have the same prognosis and treatment patterns. Indeed, many pathologists regard small-type B-cell lymphoma as the tissue phase of CLL. Patients with this type of lymphoma usually have a long-term survival rate.

**2. Large-cell type (B-cell) lymphoma**
Lymph node biopsies of patients with this condition have many large lymphocytes with diameters between 20 and 40 μm. The cells also have large nuclei, frequently with one or more nucleoli (figure 26.4*b*). Usually this type of lymphoma originates as a single node enlargement and progresses to invade other lymphoid tissues and even the bone marrow. If treated in time, this type of non-Hodgkin's lymphoma also has an excellent prognosis.

**3. Mixed-cell type (B-cell) lymphoma**
In this type of lymphoma, there is no predominant large or small lymphocyte cell type present, but a mixture of both. This is mainly a disease of older people. It also originates in a localized area from which it spreads to many other lymph nodes and to the bone marrow. If treated in time, the prognosis is quite good.

### *Clinical Features of B-Cell Lymphomas*

In most cases patients will have an asymmetrical swelling in one or more of the superficial lymph node regions. The enlarged area is usually not painful. In the later stages of the disease, when the malignancy has metastasized, symptoms common in many cancers are usually present including fatigue, weight loss, and night sweats.

In cases where the malignancy has spread to the bone marrow, anemia, infection, and bleeding may occur as a result of the crowding out of normal hematopoietic tissue. The liver and spleen are frequently enlarged. The gastrointestinal lymph nodes may also be involved, which may lead to acute abdominal pain.

In advanced cases the tumor cells may spread to other organs such as the brain, skin, testes, and thyroid gland.

### *Laboratory Findings of B-Cell Lymphomas*

In instances where malignant lymphocytes have invaded the bone marrow, low red blood cell, granulocyte, and platelet counts are commonly encountered. The erythrocytes are usually normochromic and normocytic.

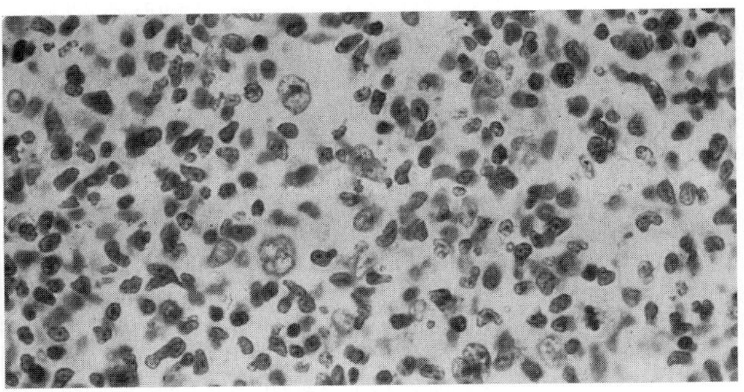

a.

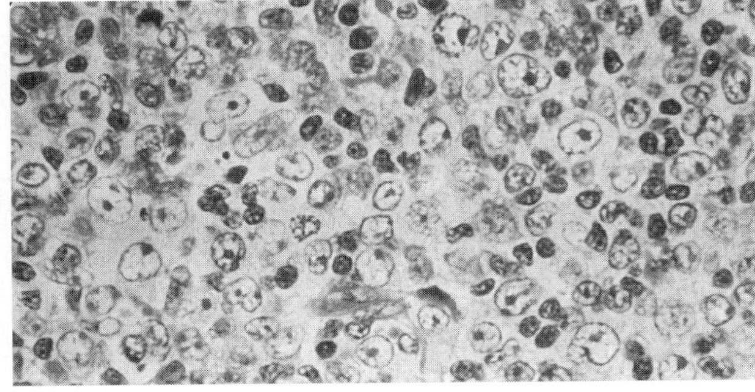

b.

figure 26.4

Histological lymph node sections in non-Hodgkin's lymphoma. (*a*) Small-cell type lymphoma and (*b*) large-cell type lymphoma.

From The American Society of Hematology Slide Bank, 3rd edition, 1990, used with permission.

Increased numbers of lymphocytes may or may not be present in the peripheral bloodstream. If lymphocytosis is observed in peripheral blood smears, the picture will show an increase of small lymphocytes or an increase in large lymphocytes or a mixture of both.

Lymph node biopsies reveal the presence of increased numbers of unorganized lymphocytes that are either large, small, or a mixture of both. Bone marrow biopsies may also have large numbers of abnormal lymphocytes as well as considerable fibrosis of the marrow tissue. Since most malignant lymphomas are the result of monoclonal B-cell tumors, there may be an occasional increase in antibody production, either IgG or IgM, which may be observed in an electrophoretogram. Frequently the level of uric acid in the serum increases. In cases where the liver is involved, abnormal liver function tests are also common.

#### *Treatment of B-Cell Lymphomas*

Treatment for non-Hodgkin's lymphoma is based on the type and the stage of the disease. In patients with low-grade malignancy—that is, where the tumor is limited to a certain group of peripheral lymph nodes with predominantly small lymphocytes, minimal treatment is required beyond local radiotherapy. In cases where the malignancy has spread, radiotherapy must be combined with chemotherapy for best results. Intensive chemotherapy programs, similar to those used in acute lymphocytic leukemia, may be required where high-grade malignancy is present.

#### *Prognosis of B-Cell Lymphomas*

In patients in whom the lymphoma is limited to a few localized lymph nodes (low-grade disease) and treated in time, the prognosis is excellent, especially if the lymphocytes are small. Most patients survive 10 or more years after the initial diagnosis.

The outlook for patients with high-grade disease—that is, where the malignancy is actively spreading to many tissues at the time of diagnosis—is not as good. However with intensive radiotherapy and chemotherapy, many patients have had survival rates of 2 to 5 years, and some have been completely cured.

### Burkitt's Lymphoma

**Burkitt's lymphoma** is a unique type of malignant non-Hodgkin's lymphoma of the B cells that is primarily a disease of childhood. It was originally described in African children by the British missionary-physician Denis Burkitt and involved lesions and growth of the jaw and other facial bones (figure 26.5). Later, similar cases were described in the United States. However in the American cases, jaw tumors are uncommon. Over half the cases in the United States involve tumors of the abdomen usually associated with the ileum, the ovaries, or the kidneys. Burkitt's lymphoma and Burkitt-like lymphomas account for about one-third of childhood non-Hodgkin's lymphomas in the United States. The male to female ratio is about 3:1.

The malignant lymphocytes usually belong to the small lymphocyte type (less than 20 μm in diameter). What contributes to the uniqueness of this type of non-Hodgkin's lymphoma is that it was the first lymphoma described as being associated with chromosomal abnormalities. A translocation involving chromosomes 8 and 14 occurs in which a piece of DNA of chromosome 8 is transposed to a region of chromosome 14 responsible for immunoglobulin heavy chain production (see figure 24.1). Another interesting facet of this disease is that all the neoplastic B cells are infected with **Epstein-Barr virus (EBV),** a type of DNA virus that belongs to the herpes class of viruses. Thus EBV may be involved in the onset of this malignancy. The prognosis of young patients with Burkitt's lymphoma is poor. Although radiotherapy and chemotherapy may frequently result in an initial remission, relapses are common, and more than two-thirds of the patients with this disease will eventually die from it. Burkitt-like lymphomas are common in

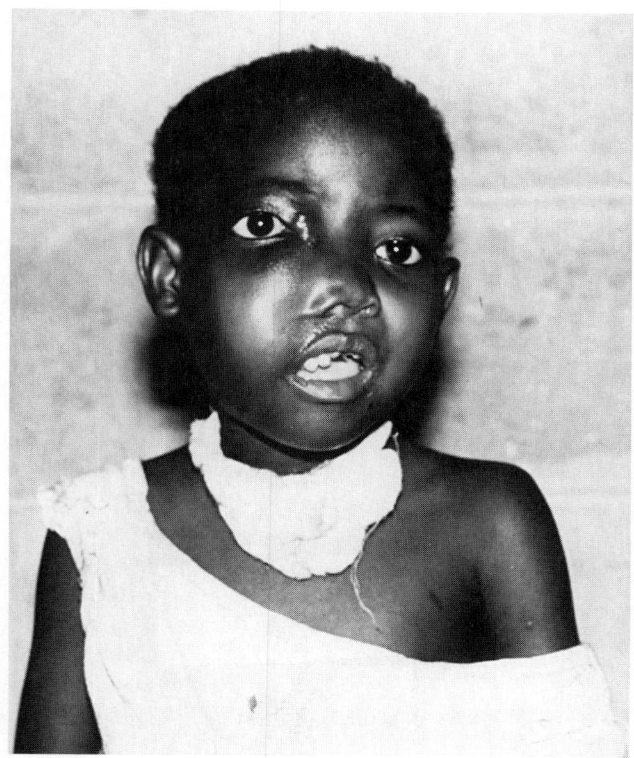

figure 26.5
Child with Burkitt's lymphoma.

patients with acquired immunodeficiency syndrome (AIDS), presumably because they are also infected with EBV.

### T-Cell and Null Cell Lymphomas

T-cell and null cell lymphomas form a small percentage (from less than 10% to close to 20%) of the non-Hodgkin's lymphomas. These non-B-cell lymphomas can also occur in three forms, just as B-cell lymphomas. They can occur as a large-cell type, a small-cell type, and a mixed-cell type. Most belong to either the small- or the large-cell type. Two forms in particular should be noted. First is a type of T-cell lymphoma that occurs mainly in young children and adolescents and is always associated with mediastinal masses and the presence of many large lymphoblasts in the peripheral bloodstream. These growths are of thymic origin and thus can be definitely classified as **T-cell lymphomas.** This disorder is also known as **T-cell lymphoblastic lymphoma (TLBL)** and accounts for 40% of all non-Hodgkin's lymphomas in children. If untreated, it is rapidly fatal, usually terminating in acute lymphoblastic leukemia. Chemotherapy may be beneficial, but the prognosis is not as good as for acute lymphocytic leukemia (ALL).

Second, is **Sézary syndrome,** which is also associated with nodular masses and the presence of many mature T lymphocytes in the blood. These lymphocytes frequently show a typical cerebriform nucleus (figure 26.6). Lymphocytes exhibiting cerebriform nuclei are known as **Sézary cells.**

In general, the prognosis for individuals with T-cell and null cell non-Hodgkin's lymphomas is similar to that for B-cell lymphomas. Patients with low-grade malignancy (usually associated with the presence of small lymphocytes) have a survival rate of 5 years or more with radiotherapy and chemotherapy. Patients with high-grade malignancy—as exemplified by the presence of many immature blast cells—have a less favorable prognosis and will usually die within 2 years after initial diagnosis. Approximately 50%, however, may survive 5 or more years with radiotherapy and intensive chemotherapy. Some may even be considered completely cured.

## PLASMA CELL LEUKEMIAS

The most common type of plasma cell leukemia is frequently called **multiple myeloma,** which is a neoplastic disease of the bone marrow plasma cells and occurs as a result of a monoclonal proliferation of one clone of activated B cells. Variations of this disease occur, such as **Waldenström's macroglobulinemia** and **heavy chain disease,** which are discussed at the end of this chapter. This text focuses primarily on multiple myeloma as it is the principal form of plasma cell leukemia.

### Multiple Myeloma

Multiple myeloma is characterized by the accumulation of large numbers of plasma cells in the bone marrow, the presence of high levels of monoclonal antibody in the blood and urine, and the frequent occurrence of bone lesions.

Multiple myeloma is typically a disease of middle-aged and elderly persons. Over 80% of all cases occur in individuals over 40 years of age.

#### *Clinical Features of Multiple Myeloma*

One of the most characteristic overt features of multiple myeloma is the presence of bone pain as a result of osteolytic bone lesions. Because normal hematopoietic blood cells are crowded out by malignant plasma cells, most patients usually also exhibit anemia, frequent infections, and a tendency to bleeding. Infections are caused by a decrease in granulocytes and monocytes and also the absence of normal B cells, which results in a deficiency of appropriate antibody production.

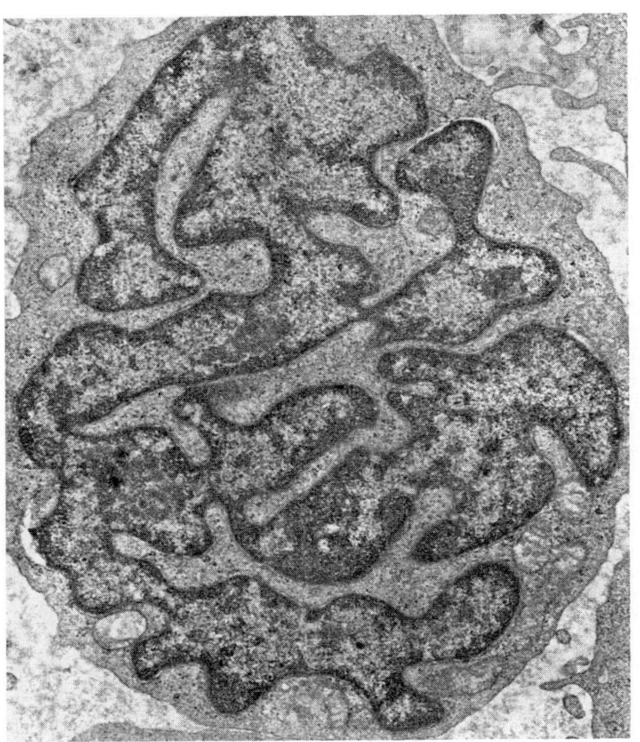
a.

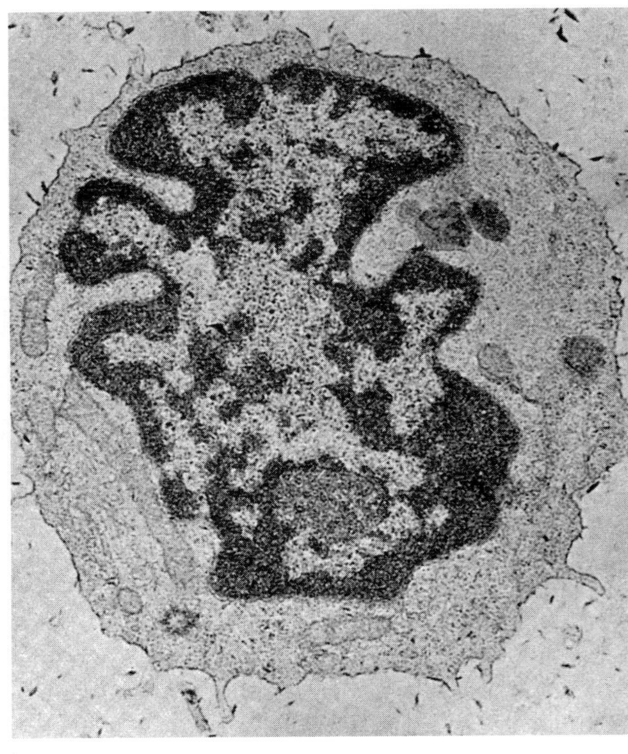
b.

figure 26.6

Electron micrograph of Sézary cells. (*a*) Typical Sézary cell with cerebriform nucleus; and (*b*) small-cell variant.

Bleeding results from a lack of platelets and from the interference of abnormal myeloma proteins (abnormal monoclonal antibody) with normal platelet function.

Another common complication in multiple myeloma is renal failure caused by large amounts of myeloma proteins clogging the nephrons.

In patients with pronounced lytic bone disease, the blood serum will contain high levels of calcium (hypercalcemia). This may result in polyuria, polydipsia, anorexia, vomiting, constipation, and mental disturbances.

### *Laboratory Findings of Multiple Myeloma*

The single most important diagnostic test for multiple myeloma is the electrophoresis of serum proteins. A typical electrophoretogram of this disease exhibits a sharp peak in the gamma-globulin region, with reduced beta-globulins and other gamma-globulins (figure 26.7). In two-thirds of all cases, the serum immunoglobulin belongs to the IgG class, the other one-third is usually IgA. In rare cases IgE or IgD—or a mixture of the two—occurs.

Another important diagnostic feature is the presence of **Bence-Jones proteins** in the urine. Bence-Jones proteins are free light chains of the immunoglobulin molecule. In multiple myeloma there is an excess production of light chains classified as either kappa (κ) or lambda (λ), but never both.

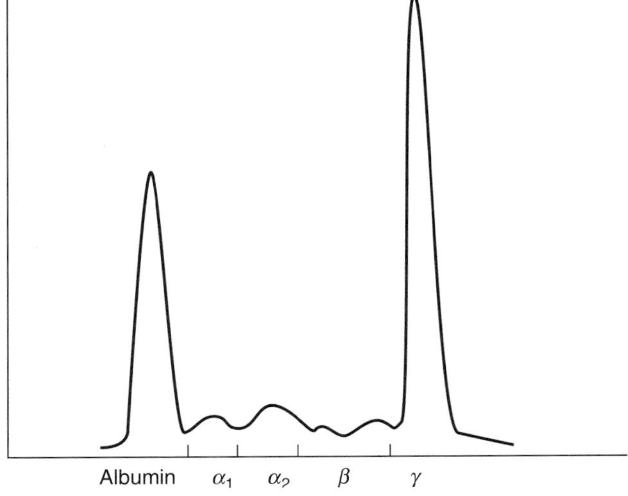


figure 26.7

Typical serum protein electrophoresis densitometer pattern of multiple myeloma, showing the abnormally peaked gamma-globulin fraction.

Bone marrow analysis reveals a sharp increase in plasma cells, many with abnormal shapes. These cells are known as **myeloma cells** (figure 26.8).

Further, skeletal analysis will frequently show **osteoblastic sclerosis**—that is, X rays of the skull and other flat bones will reveal many dark spots, indicating

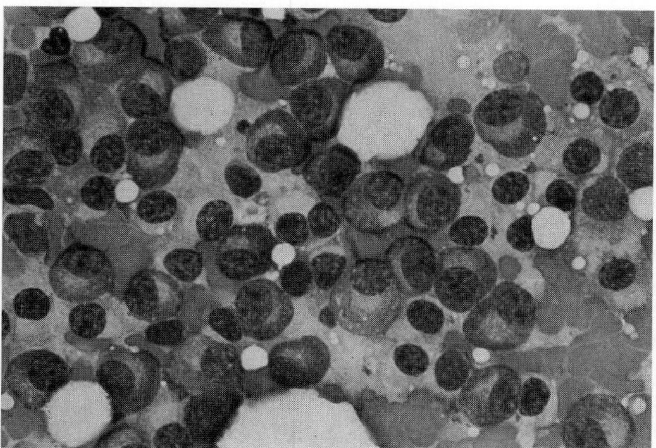

figure 26.8

Typical bone marrow smear of multiple myeloma, showing many large plasma cells.

From the American Society of Hematology Slide Bank, 3rd edition, 1990, used with permission.

bone resorption (figure 26.9). Practically all multiple myeloma patients will have at least two of the foregoing symptoms.

Because of the obvious myelophthisis (crowding out of normal hematopoietic tissue in bone marrow), the red blood cell count usually is low. Blood smear analysis, however, will show that the red blood cells have a normocytic and normochromic appearance. Rouleau formation is often present because of the increase in abnormal myeloma proteins in the blood. The level of white blood cells is also decreased in most cases. Moreover, 20% of all patients will have high numbers of abnormal plasma cells in the blood smear. The level of platelets is also reduced, especially in the later stages of the disease. About half of patients will have a marked increase in serum calcium levels, and about 20% will have a higher level of blood urea.

### *Treatment of Multiple Myeloma*

Chemotherapy is the treatment of choice. Antimetabolite and alkylating agents are used to reduce plasma cell proliferation. The objective of treatment is to reduce the plasma cell numbers while increasing normal bone marrow cell production.

Other forms of treatment may also be necessary to combat the symptoms of multiple myeloma. For instance, the blood viscosity produced by myeloma proteins may be reduced by **plasmapheresis,** the infusion of normal plasma. This process will also reduce the danger of bleeding. Anemia may be counteracted by packed red blood cell transfusions. Hypercalcemia may be treated with phosphates, which bind with calcium molecules; the resulting calcium phosphate salts will then be deposited into the bone, thus reducing the blood calcium levels. Steroid therapy is another treatment method. Steroids inhibit activity of bone osteoclasts and thus prevent breakdown of bone salts. In this way calcium levels are also decreased.

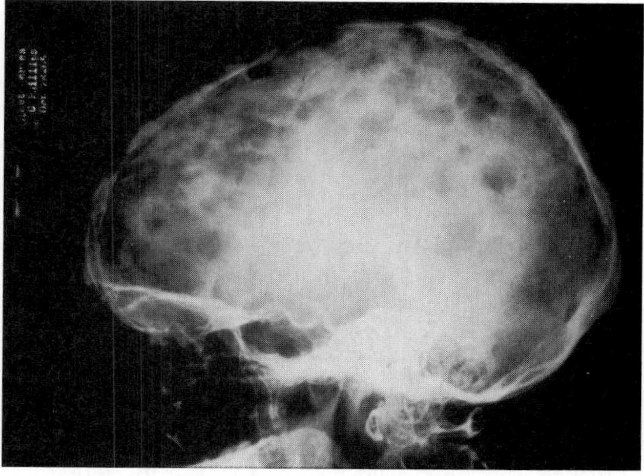

figure 26.9

Skull X ray of patient with multiple myeloma. Note characteristic "punched out" lesions.

From the American Society of Hematology Slide Bank, 3rd edition, 1990, used with permission.

### *Prognosis of Multiple Myeloma*

The prognosis for multiple myeloma is rather poor, with a median survival rate of about two years. Usually the level of Bence-Jones proteins in the urine is a good prognostic indicator. The higher the level of Bence-Jones proteins in the urine, the shorter the survival time. The same applies to blood urea levels—the higher the blood urea levels, the shorter the survival time.

## Waldenström's Macroglobulinemia

This is a special form of plasma cell leukemia. Waldenström's macroglobulinemia is a rather rare condition seen mainly in males over 50 years of age. A unique feature of this disease is that plasma cells produce high levels of **monoclonal IgM.** IgM is a very large protein with a molecular weight of about 900,000 daltons, which is about five times that of IgG. Hence the term **macroglobulinemia.**

### *Clinical Features of Waldenström's Macroglobulinemia*

This is a slowly developing disease that is insidious in its onset. Usually the patient goes to the physician with such complaints as visual disturbances, lethargy, and confusion. All these symptoms may occur as a result of the **hyperviscosity** of the blood. Other patients complain of weakness and muscular fatigue (due to anemia). Still others notice prolonged bleeding symptoms (due to the interference of IgM immunoglobulins with the proper functioning of the platelets).

The visual disturbances may be the result of swollen veins and breakage of the small delicate blood vessels in the retina due to increased blood viscosity. The high concentration of IgM also results in cardiovascular complications. Frequently the size of the liver, spleen, and lymph nodes is moderately increased as a result of the body's effort to reduce the level of macroglobulins in the blood.

*Laboratory Findings of Waldenström's Macroglobulinemia*

The single most important diagnostic feature is the presence of very high levels of IgM (usually greater than 15 g/L), which may be shown with radial immunodiffusion (RID) or immunoelectrophoresis.

The bone marrow exhibits increased numbers of plasma cells and other lymphocytes, and the blood serum will also show increased lymphocyte counts, especially plasmacytoid lymphocytes. The lymph nodes will have a pattern similar to that found in bone marrow. In advanced disease there will be low counts of erythrocytes, granulocytes, and platelets.

*Treatment of Waldenström's Macroglobulinemia*

The treatment of choice is chemotherapy with antimetabolite drugs and prednisolone, as in multiple myeloma. Hyperviscosity of the blood should be treated with repeated plasmapheresis. Anemia may be alleviated by blood transfusions and infections controlled with aggressive antibiotic treatment.

**Heavy Chain Disease**

This is another rare syndrome characterized by the excess production of either alpha-, gamma-, or mu-chains of the immunoglobulin molecule. Heavy chain disease is a special type of plasma cell leukemia exhibiting many of the symptoms of multiple myeloma and Waldenström's macroglobulinemia, and hence requires a similar type of treatment.

Gamma ($\gamma$)-chain disease is also known as **Franklin's disease.** Patients with this disorder have excess gamma-chains or more commonly fragments of gamma-chains in their blood and urine. Normal immunoglobulin levels are always low; hence these patients are prone to infections.

Alpha ($\alpha$)-chain disease is the most common form of heavy chain disease. This disorder is mainly observed in persons of northern African or eastern Mediterranean descent. Typical patients will also have abdominal lymphoma. The disorder is more common in males than in females.

Mu ($\mu$)-chain disease is the rarest form of heavy chain disease. Unlike the other forms discussed here, excess mu-chains are found only in plasma but not in urine.

# *Further Readings*

Boggs, D. R., and Winkelstein, A. Lymphocytes and the Immune System. Chapter 5 in *White Cell Manual,* 4th ed., F. A. Davis Company, Philadelphia, 1983.

Churchill, W. H. Plasma Cell Disorders and Dysproteinemias. Lecture 24 in *Hematology,* 4th ed., W. S. Beck, Ed., The MIT Press, Cambridge, Mass., 1985.

Harris, N. L. The Malignant Lymphomas II. Pathology. Lecture 23 in *Hematology,* 4th ed., W. S. Beck, Ed., The MIT Press, Cambridge, Mass., 1985.

Hoffbrand, A. V., and Pettit, J. E. Malignant Lymphomas. Chapter 8 in *Essential Haematology,* 2nd ed., Blackwell Scientific Publications, Oxford, 1984.

Rosenthal, D. S. The Malignant Lymphomas I. Clinical Aspects. Lecture 22 in *Hematology,* 4th ed., W. S. Beck, Ed., The MIT Press, Cambridge, Mass., 1985.

# *Review Questions*

1. List some of the major distinguishing features of Hodgkin's disease and non-Hodgkin's lymphomas.
2. What is the incidence of Hodgkin's disease? What age group does it affect primarily?
3. What is the cause of Hodgkin's disease?
4. List the major clinical features of Hodgkin's disease.
5. Give the major laboratory findings associated with Hodgkin's disease. Describe a unique laboratory finding of Hodgkin's disease.
6. Name the four different subtypes of Hodgkin's disease.
7. List the four different stages of Hodgkin's disease.
8. What is the preferred treatment for Hodgkin's disease?
9. What is the prognosis for patients with Hodgkin's disease? On what features does the prognosis depend?
10. Name three types of non-Hodgkin's lymphoma. Which of these types is the most common?
11. List the three major types of B-cell lymphoma. Describe the distinguishing features.
12. Describe the major clinical features of non-Hodgkin's malignant lymphoma.
13. List the major laboratory findings associated with malignant lymphoma.
14. What is the preferred treatment for non-Hodgkin's malignant lymphoma?

15. What is the prognosis for B-cell malignant lymphoma? In what situation is the prognosis more favorable? When is it less favorable?
16. Name the two major forms of T-cell malignant lymphoma.
17. Distinguish between non-Hodgkin's malignant lymphoma and multiple myeloma.
18. Define Burkitt's lymphoma. How is it classified? Name a unique feature of this lymphoma. Where is it found and in what age group? Describe the cause of Burkitt's lymphoma.
19. What is multiple myeloma?
20. How is multiple myeloma diagnosed?
21. List some of the major laboratory findings of multiple myeloma.
22. Describe the major clinical features of multiple myeloma.
23. What is the preferred treatment for multiple myeloma?
24. What is the prognosis of patients with multiple myeloma?
25. List the major clinical features of Waldenström's macroglobulinemia. Give an explanation of these features.
26. Name a unique laboratory finding of Waldenström's macroglobulinemia. How would you diagnose this disease?
27. What is the treatment for patients with Waldenström's macroglobulinemia?
28. What is heavy chain disease? List the three forms of heavy chain disease. Which is the most common form? Which is the least common form? What is Franklin's disease?
29. Describe the symptoms and treatment for heavy chain disease.

# Chapter Twenty-Seven

## *Lymphocyte Deficiency Disorders*

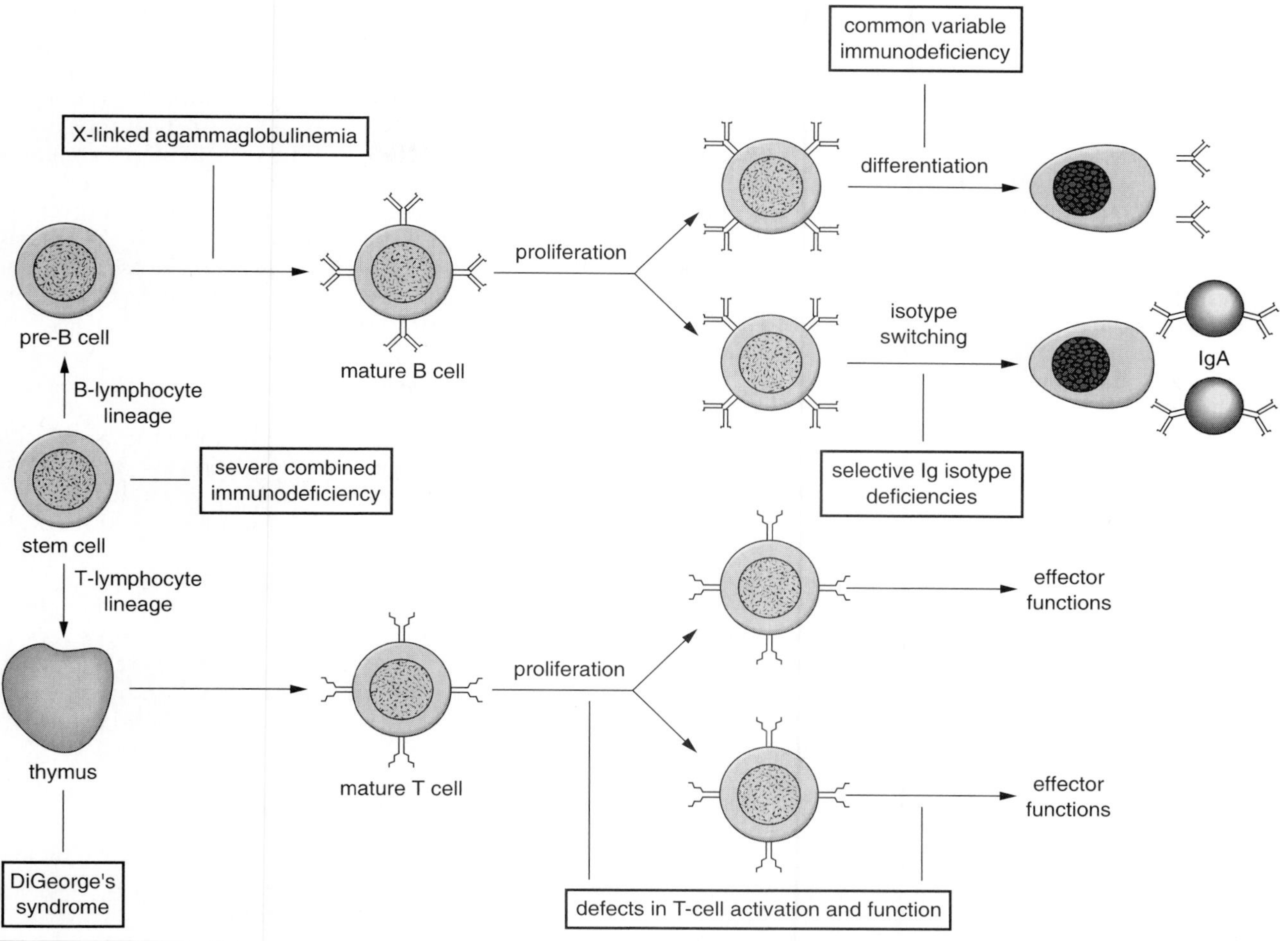


**figure 27.1**
The major sites of cellular abnormalities in immunodeficiency syndromes.

Chapters 25 and 26 dealt with disorders associated with an excess production of various types of lymphocytes. Hence they were called lymphoproliferative disorders. This chapter discusses disorders that occur as a result of a decreased production of the major classes of lymphocytes. This group of disorders is often referred to as **immunodeficiency diseases.** We should realize, however, that deficiencies in immune responses are not just the domain of lymphocytes. Deficiencies in nonspecific immune responses are mainly the result of abnormalities of or lack of **granulocytes, monocytes,** and **macrophages.** The deficiencies caused by these white blood cells have been dealt with in previous chapters. **Lymphocyte deficiency disorders** result in defective specific immunity. This type of immunodeficiency is caused by the abnormal development, activation, or functions of B and/or T lymphocytes. Hence all lymphocyte deficiency syndromes can be divided into three different categories: (1) those involving **B-cell deficiencies,** (2) those involving **T-cell deficiencies,** and (3) those involving *both* **B-** and **T-cell deficiencies** (figure 27.1). Each of these categories is discussed in this chapter.

## B-CELL DEFICIENCY DISORDERS

Abnormalities in B-lymphocyte development, activation, and function will result in deficiencies in antibody production. Since antibodies (= immunoglobulins) are the principal agents that maintain **humoral immunity,** that lack of immunoglobulins will result in a greater susceptibility to **extracellular pathogens** such as many bacteria (e.g., streptococci and pneumococci), as well as certain viruses (e.g., poliovirus), and protozoans (e.g., *Giardia*).

Several well-defined B-cell deficiencies have been described through the advancements made in laboratory technology. The principal ones are (1) X-linked agammaglobulinemia, (2) selective dysgammaglobulinemia, (3) transient hypogammaglobu inemia of infancy, and (4) acquireda gammaglobulinemia. Each of these four disorders is discussed in the following material.

### X-Linked Agammaglobulinemia (Bruton's Disease)

**X-linked agammaglobulinemia** is also known as **Bruton's disease** and is named after Ogden Bruton, who first described this condition in 1952. This immunodeficiency disorder is characterized by a total lack of plasma cells in the lymphoid tissues and an absence of B cells in the peripheral circulation. Lack of functional B cells will result in a total deficiency of **immunoglobulin production** and hence a complete absence of humoral immunity. Individuals with this disease are extremely prone to pyogenic (= pus-forming) bacterial infections, as well as to infectious diseases caused by other extracellular parasites.

Bruton's disease is one of the most common congenital B-cell deficiency syndromes. It is a **sex-linked disease,** which means it is inherited on the X chromosome as an abnormal gene. As a result females usually are the carriers of this disease, whereas the males will usually express this syndrome. The mothers of these males are heterozygous, having one normal X chromosome gene and one abnormal X chromosome gene that is not expressed. In males there is no counterpart to this defective gene—and hence, if present, is always expressed on the X chromosome.

#### *Clinical Features of X-Linked Agammaglobulinemia*

Affected boys are highly susceptible to many bacterial infections and to certain viral and protozoan infections. Since T-cell production is not affected, the cell-mediated immunity of the individual is not impaired. Therefore the ability to fight intracellular pathogens and neoplastic cells is not affected. Children with Bruton's disease suffer from many different recurrent infections caused mainly by pyogenic bacteria that attack the throat, the conjunctiva of the eyes, the middle ears, the skin, the respiratory system, as well as the gastrointestinal tract.

This immunodeficiency disease usually does not become apparent until the end of the first year of life, when the passively transferred maternal antibodies have disappeared from the infant's circulation. The end of the first year of life is also the time that most children are increasingly exposed to environmental pathogens. Hence at this time these male infants begin to exhibit such problems as frequent middle ear infections, bronchitis, pneumonia, dermatitis, meningitis, and diarrhea.

#### *Laboratory Findings of X-Linked Agammaglobulinemia*

Unique features of X-linked agammaglobulinemia include (1) the absence of serum immunoglobulin, (2) the absence of B cells in the peripheral bloodstream, (3) the absence of germinal centers in the lymph nodes, (4) the absence of plasma cells in the medullary spaces of the lymphoid tissues, and (5) the presence of normal numbers of pre-B cells in the bone marrow. The latter feature indicates that this disorder is caused by an abnormality in the maturation process of B cells. Thus a block occurs that prevents pre-B cells from developing into virgin B cells—that is, B cells able to express both cytoplasmic and membrane immunoglobulins. The numbers and types of T lymphocytes are usually within the normal range, although in some patients T-cell numbers are slightly depressed also. Possibly this may be explained by the fact that B cells are normally involved in the activation of T cells (because of their ability to present processed antigens to the inactive T cells). Diagnosis of Bruton's disease involves serum electrophoresis, which will exhibit an absence of **gamma globulin** in the serum electrophoretogram (figure 27.2).

#### *Treatment and Prognosis of X-Linked Agammaglobulinemia*

If treatment is not administered, most infants will succumb early in life due to complications of overwhelming infections. However the lives of such children may be saved by repeated (monthly) injections with gamma-globulin pooled from several donors. Such preparations contain pre-formed antibodies against most common pathogens, and thus will provide effective (passive) immunity for the child. Patients receiving this treatment have survived for several decades, although many of these individuals may develop chronic lung disease, lymphocytic leukemia, or lymphoma later in life.

### Selective Dysgammaglobulinemia

Immunodeficiency syndromes may selectively involve one or more classes of immunoglobulin molecules. The most common type is **selective IgA deficiency.** Approximately 1 in 500 people is estimated to be affected by this syndrome, which appears to occur in various forms. Some forms of IgA deficiency are inherited as autosomal dominant or recessive traits, whereas others are acquired as a result of a fetal *Rubella* infection or drug exposure. The clinical symptoms of this condition are also varied. Many individuals are completely asymptomatic despite low IgA levels; others have occasional gastrointestinal or respiratory problems (IgA is primarily a secretory antibody released in the mucous layers of the digestive and respiratory tracts). However, severe infections that result in damage to the gut or airways are rare in patients with IgA deficiency. The specific mechanism that inhibits IgA production is not known. Since normal levels of IgM and IgG are produced, it is possible that one or more cytokines needed to switch B cells from producing IgM to IgA are lacking.

**Selective IgM deficiency** is a rare autosomal recessive disease usually associated with severe bacterial infections. Patients with IgM deficiency produce normal virgin B cells with membrane IgM and IgD, but for some reason these cells are unable to differentiate into IgM-secreting plasma cells. The defect may result from a lack of helper T cell availability or from the inability of certain helper T cells to produce required growth factors. **Selective IgG deficiencies** are also very rare and are also the result of blocked B-cell differentiation, presumably due to helper T-cell inefficiency.

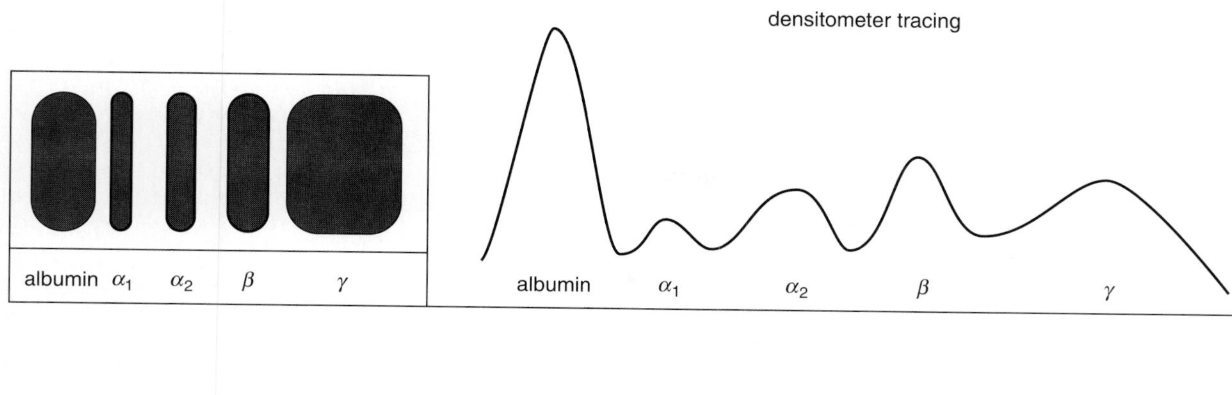


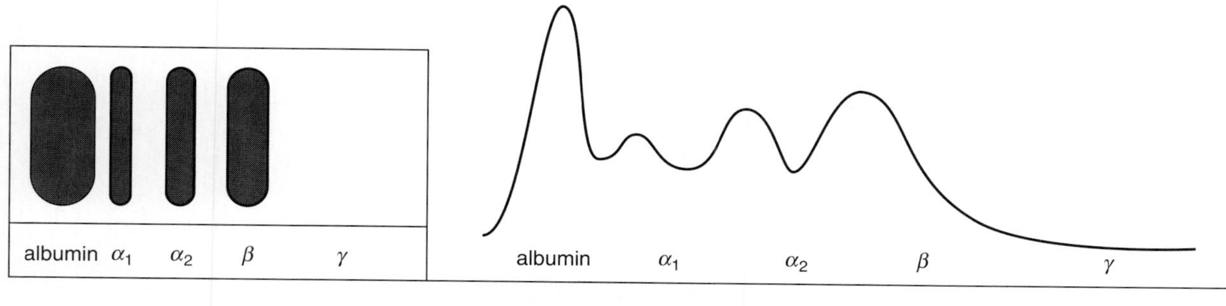


**figure 27.2**

Serum electrophoresis patterns and corresponding densitometer patterns of *(a)* normal serum and *(b)* serum from a patient with Bruton's disease (hypogammaglobulinemia).

Finally, there is a selective deficiency involving a combined lack of **IgA** and **IgG.** Patients with this condition are able to produce IgM antibodies. However, since the concentration of IgM is normally lower than IgG and IgA, these individuals are also prone to severe pyogenic bacterial infections. To complicate things further, many of the IgM antibodies produced by these individuals also act as **autoantibodies** and attack their own blood cells. This may lead to anemia, bleeding, and further infections.

Diagnosis of these selective dysgammaglobulin emias is usually performed by **immunoelectrophoresis**—that is, the initial serum electrophoresis is followed by immunodiffusion with selected antibodies against the various immunoglobulin classes (figure 27.3).

Treatment for deficiency of IgG and IgM may involve intravenous or intramuscular injections with pooled IgG and IgM antibodies as well as antibiotic therapy. Treatment for IgA deficiency does not involve gamma-globulin injections but only antibiotic therapy.

### Transient Hypogammaglobulinemia of Infancy

Normally all infants experience a short period of hypogammaglobulinemia at approximately 5 to 6 months of age. During that period the level of maternal antibodies in the child's blood gradually declines. This decline is normally matched by a steady increase of antibodies, which are produced by the infant. However, immunoglobulin production by the child will not reach adequate levels until the end of the first year or even later (figure 27.4).

Sometimes the decrease of maternal immunoglobulins is not followed by a corresponding increase in endogenous antibodies. Such children become susceptible to pyogenic infections at about 5 to 6 months of age. However such infections are transient and gradually disappear when the child gets older and antibody production reaches normal levels (figure 27.4). In other words this syndrome in infancy reflects a delayed maturation of the humoral component of the immune system.

Thus during the lag time the child is susceptible to many infections, especially respiratory infections, which may be fatal if not treated promptly. Here again, as with other B-cell immunodeficiency diseases, replacement with pooled immunoglobulins is necessary, usually combined with antibiotic treatment.

### Acquired Agammaglobulinemia

Patients with acquired agammaglobulinemia have clinical symptoms similar to those found in boys with Bruton's disease. The difference is that this form of agammaglobulinemia is not inherited but is acquired later in life. This disorder rarely occurs before puberty and is most common in individuals between 15 and 35 years of age.

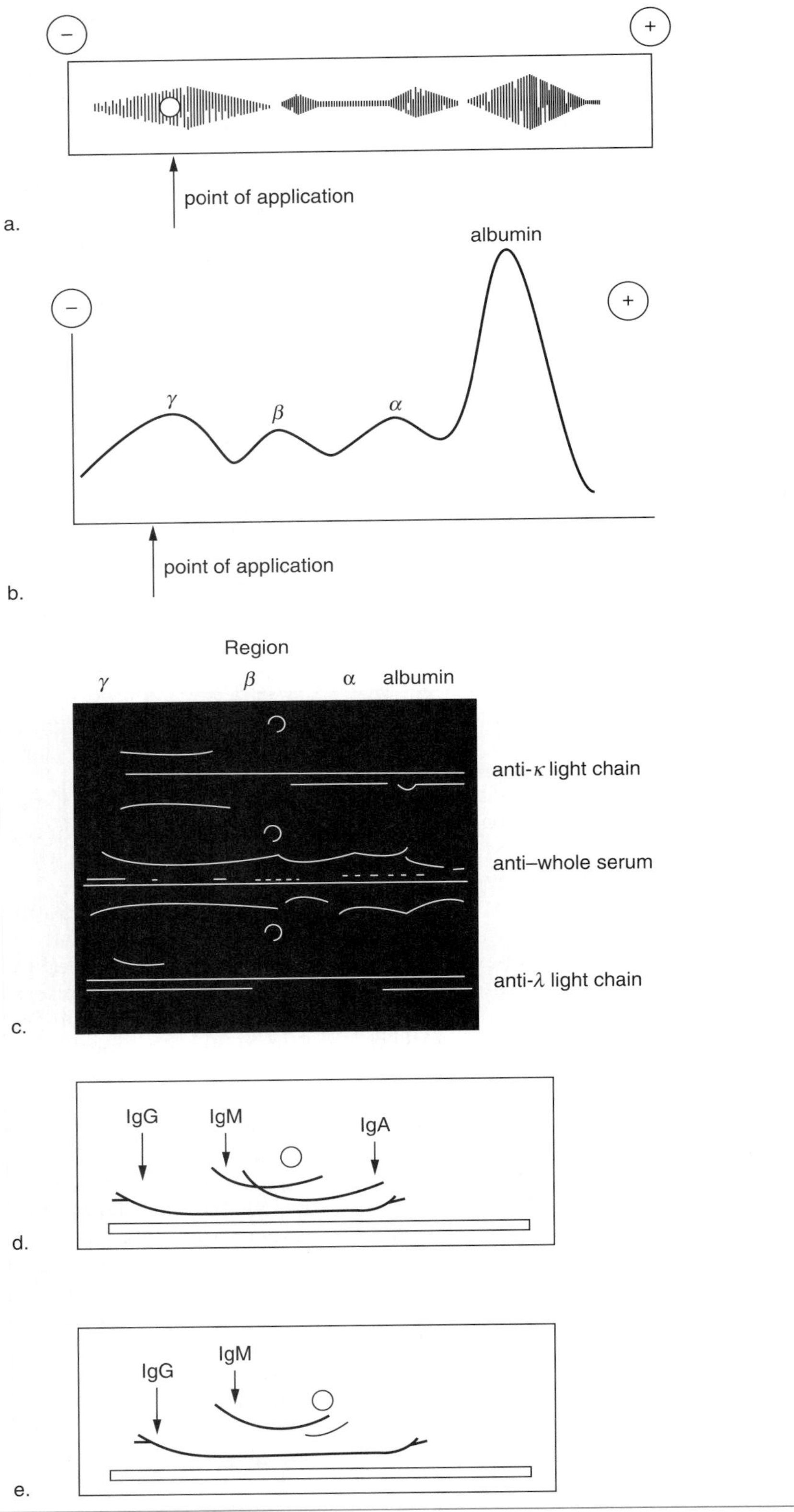


**figure 27.3**

Serum electrophoresis and immunoelectrophoresis patterns. (*a*) Normal serum electrophoresis pattern; (*b*) corresponding densitometer pattern; (*c*) actual immunoelectrophoresis patterns when serum is reacted with anti-κ antibodies, whole serum antibodies, and anti-λ antibodies. (*d*) Diagram showing the relative positions of IgG, IgM, and IgA when reacted with anti-whole serum. (*e*) Diagram showing a selective IgA deficiency.

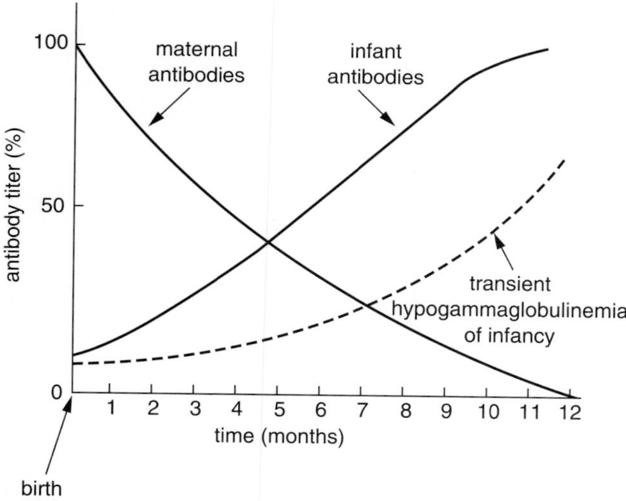


**figure 27.4**
Normal levels of maternal and infant's immunoglobulins during the first year of life (solid lines), and immunoglobulin levels in transient hypogammaglobulinemia of infancy (dotted line).

The cause of acquired agammaglobulinemia is unknown, although several explanations have been proposed. One theory postulates that patients develop a class of suppressor T cells that prevents activation of B-cell clones and thus the formation of new plasma cells and antibody production. Other theories point to the development of intrinsic B-cell defects or to deficient T-cell help. Most likely, however, the cause in the majority of cases may be a defect in the terminal differentiation of B cells to active antibody-producing plasma cells. This conclusion is based on the fact that the lymph nodes contain an adequate number of B cells but no plasma cells.

In addition to an increased susceptibility to bacterial infections, these patients also have a high incidence of autoimmune diseases, such as rheumatoid arthritis and autoimmune hemolytic anemia (AIHA).

Laboratory investigations of the blood of these patients exhibit normal numbers of B lymphocytes but low levels of immunoglobulins in the serum.

The treatment for acquired agammaglobulinemia is similar to that of Bruton's disease: monthly injections with pooled gamma-globulin together with antibiotic therapy. This regimen will control most infections and allow these patients to live a normal life. This disease affects both males and females in equal numbers.

## T-CELL DEFICIENCY DISORDERS

Just as B-cell deficiency syndromes result in impaired humoral immunity, T-cell deficiency diseases produce a decrease in **cellular** (or cell-mediated) **immunity.** People with T-cell deficiencies also show deficiencies in humoral immunity because the maturation and activation of functional B cells requires the presence of helper T cells. Because T cells are primarily involved in cellular immunity, people who lack T cells are especially susceptible to viral, fungal, and protozoan infections as well as to neoplastic growths.

A number of immunodeficiency diseases resulting from a decreased production, maturation, or activation of T cells have been known for many years. The most important of these is **congenital thymic aplasia** (also known as **DiGeorge's syndrome**). A more recent T-cell deficiency condition is **acquired immunodeficiency syndrome** or **AIDS.** In contrast with DiGeorge's syndrome, AIDS is not an inherited disease but is acquired as a result of an infection with the human immunodeficiency virus (HIV).

Apart from these two well-known disorders, there have been numerous reports of other T-cell deficiencies. With continuing advances in the clinical laboratory, pinpointing the causes of these abnormalities has become easier. In some cases the T-cell deficiency was attributed to the inability of the T cell to express the appropriate receptors on its membrane (e.g., CD3); in others the activated T cells did not produce the appropriate cytokines (e.g., IL-2). Specific clinical syndromes associated with each of these causes will most likely be described in the coming years as more cases are recognized. In this text the T-cell deficiency syndromes are limited to descriptions of DiGeorge's syndrome and AIDS.

### Congenital Thymic Aplasia (DiGeorge's Syndrome)

DiGeorge's syndrome is a congenital disease characterized by aplasia or hypoplasia of the thymus and parathyroid glands. It is associated with lymphocytopenia in the peripheral bloodstream and an absence of cellular (cell-mediated) immunity. The decrease in the number of lymphocytes in the blood is a reflection of decreased T-cell production, since 80 to 90% of all lymphocytes in the circulation of a normal individual may be classified as T cells. The lack of T lymphocytes is associated with the absence of a functional thymus gland.

Interference in normal embryonic development of the third and fourth pair of **pharyngeal pouches** of the primitive gut results in the failure of the thymus gland to form (see figure 20.3). Normally around the twelfth week of gestation, the **ventral** portions of these pouches begin to differentiate into **thymic tissue** and move to an area above the heart. The **dorsal** portions of these same pouches develop into the **parathyroid glands,** which then separate and migrate dorsally until they become embedded into the thyroid tissue. During that same period the ears and lips are also formed. Thus many patients with DiGeorge's syndrome not only lack a thymus gland and parathyroids, but are also born with low-set ears and a fish-shaped mouth, as well as abnormally slanted eyes.

#### *Clinical Features of Congenital Thymic Aplasia*

The first symptoms of this disease are not associated with a lack of T cells because of the absence of a functional thymus gland but occur as a result of a lack of **parathyroid hormone (PTH).** Most infants with DiGeorge's syndrome will have involuntary muscle spasms and twitchings within 24 hours after birth, due to a lack of calcium in the blood. The calcium level is primarily regulated by PTH. Many infants also exhibit renal and heart abnormalities, which frequently require immediate surgery. Patients that survive the early neonatal period usually develop severe viral, bacterial, and protozoan infections by the end of the first year of life, which are fatal if not treated immediately.

*Treatment of Congenital Thymic Aplasia*

Treatment usually involves the transplantation of fetal thymic tissue into the deficient infant. The earlier this can be done, the greater the chance that the graft will be accepted, resulting in a restoration of T-cell production and cellular immunity. The hypocalcemia must also be controlled; this can be accomplished by oral administration of calcium together with vitamin $D_3$ (cholecalciferol) or parathyroid injections.

In some cases, transplantation may not be necessary because T-cell function tends to improve with age and is almost normal by the time the child is 5 years of age. This is probably because some thymic tissue has been present from the beginning but is located at an abnormal site. It is also possible that there are extrathymic sites for T-cell development. Although such sites are likely, none have currently been identified.

There is an animal model that mimics DiGeorge's syndrome. Scientists have been able to produce a strain of mice with an inherited defect of the epithelial cells of the skin, leading to hairlessness. This same epithelial defect also occurs in the pharyngeal pouches, thus causing thymic aplasia. Hence these mice are known as **nude** or **athymic** mice (figure 27.5). The presence of this animal model allows scientists to experiment with therapies for patients with DiGeorge's syndrome.

## Acquired Immunodeficiency Syndrome (AIDS)

Acquired immunodeficiency syndrome is a disease that was first described in the early 1980s. AIDS is associated with a pronounced lack of specific (or acquired) immunity. As a result all patients with AIDS are highly susceptible to many opportunistic pathogenic microorganisms as well as to malignant diseases. AIDS is caused by a retrovirus known as the **human immunodeficiency virus** or **HIV.** HIV primarily infects helper T cells and to a lesser extent also monocytes, macrophages, and certain other mononuclear cells (table 27.1), thus severely impairing the immune system.

AIDS is spreading rapidly and globally. This syndrome is thought to have originated in sub-Saharan Africa, where large numbers of people are now infected with this virus. In the United States approximately 2 million people have been positively identified as being HIV-positive, and more than 50,000 individuals have died as a result of this disease. Since there is currently no known cure for AIDS, most of the infected people will likely die from the complications of this infection.

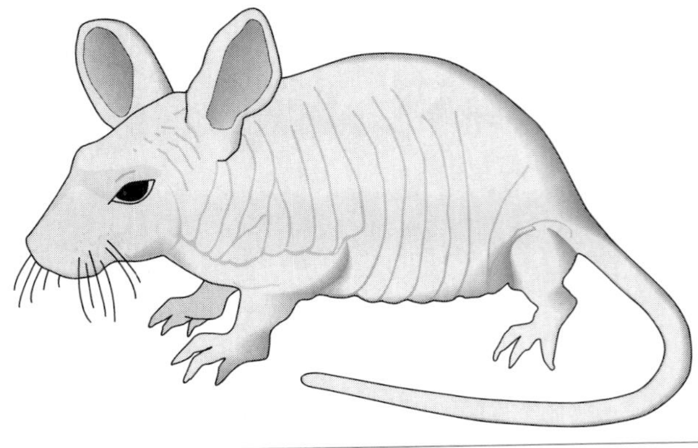

**figure 27.5**

The nude (athymic) mouse.

*Etiology of AIDS*

Acquired immunodeficiency syndrome is the result of an infection with HIV, which is a member of the lentivirus family of **retroviruses.** HIV together with other lentiviruses such as simian immunodeficiency virus (SIV), bovine immunodeficiency virus (BIV), and feline immunodeficiency virus (FIV) may be described as infectious agents causing cellular infections resulting in potentially fatal diseases. Two forms of the human immunodeficiency virus have been identified on the basis of their genomic structure, and they have been labeled **HIV-1** and **HIV-2.** Although both forms can cause the same symptoms, HIV-1 is known to spread more rapidly and appears to be more pathogenic than HIV-2. HIV-2 was originally isolated in West Africa and has not spread much beyond that region. In the United States, HIV-1 is the more common causative agent of AIDS. HIV infections occur when viral particles or HIV-infected cells in the blood, semen, or other body fluids are introduced into noninfected individuals. At the end of 1993, there were approximately 362,000 cases of AIDS in the United States. Epidemiologists at the U.S. Centers for Disease Control (CDC) in Atlanta have identified the following eight major groups that are greatly at risk for developing AIDS:

1. Men who have sex with other men (homosexual or bisexual males) who have unprotected anal intercourse. This group consists of approximately 54% of all cases of AIDS.
2. Intravenous drug abusers who share needles. This group makes up about 24% of all cases of AIDS.
3. Men who have sex with other men and are *also* IV drug users comprise 6.5% of all cases of AIDS.
4. Recipients of unscreened (or improperly screened) blood, plasma, or blood cell group consists of 2 to 3% of all cases of AIDS.
5. Hemophiliacs who were transfused prior to March 1985 with pooled concentrated coagulation factors compose less than 1% of AIDS cases.
6. Heterosexual partners of members of the foregoing high-risk groups compose approximately 6.5% of AIDS cases.

**Table 27.1 Cell Types That Can Be Infected with HIV**

| Blood/Immune Cells | Neural Cells | Others |
|---|---|---|
| T lymphocytes | Astrocytes and oligodendrocytes | Fibroblasts |
| B lymphocytes | Microglia | Sperm |
| Monocytes/macrophages | Glial cell lines | Liver sinusoid epithelium |
| Bone marrow precursor cells | Fetal neural cells | Bowel epithelium |
| Dendritic cells | Brain capillary endothelium | Colon carcinoma cells |
| Langerhans cells | | Osteosarcoma cells |
| | | Rhabdomyosarcoma cells |
| | | Fetal chorionic villi |

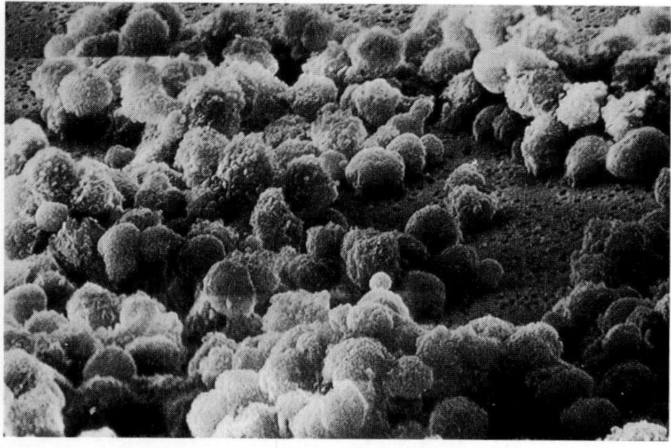
a.

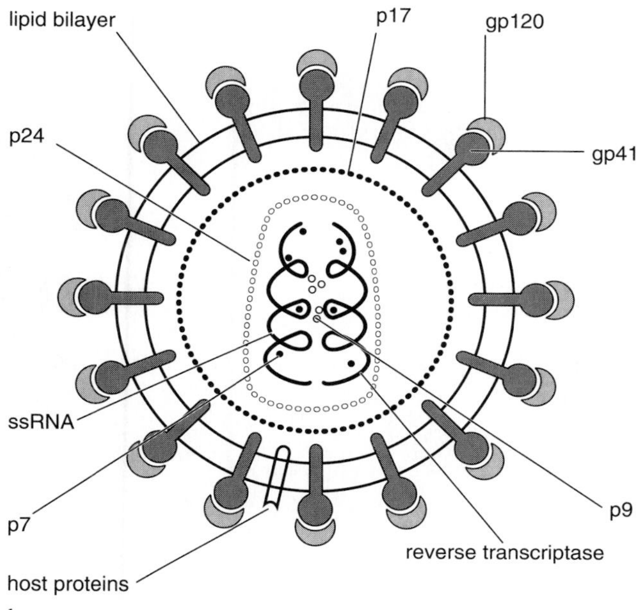

b.

**figure 27.6**
The human immunodeficiency virus (HIV). (*a*) A scanning electron micrograph of a HIV-infected lymphocyte; (*b*) diagram of the HIV.

7. Incidence of babies born from HIV-infected mothers is approximately 1% of AIDS cases.
8. The final category, the group in which the risk is unknown or has not been identified, is approximately 6.5%.

Once the virus is within the physiological milieu of the body, it attempts to enter the white blood cells, specifically the helper T cells. Certain proteins (gp120 and gp41) of the viral envelope are able to bind to the **CD4 receptor proteins** on the helper T cells (figure 27.6). $CD4^+$ is a unique receptor protein of the helper T lymphocytes. Once the virus has absorbed to $CD4^+$ cells, it enters by fusing with the host cell membrane or by **receptor-mediated endocytosis.** Once intracellular, the HIV virion begins to replicate as follows (figure 27.7).

First, the RNA genome of the HIV is transcribed into double-stranded DNA, which is then integrated into the host cell genome. The integrated DNA form of HIV is known as the **provirus,** which may remain inactive for months or years. As long as it is inactive, no side-effects will occur, but it may be passed on to daughter cells upon mitosis. This is known as a **latent infection.** Eventually the provirus will become activated by white blood cell (WBC) cytokines, such as IL-6 or tumor necrosis factor (TNF) or by the presence of certain antigens. Once activated, the provirus transcribes new viral RNA material, some of which will be translated into proteins that form new viral envelopes around the viral RNA. After their release from the host cells, these new virions continue the cycle of infecting other helper T cells. In some host cells the infection process occurs at low levels, allowing the host cells to survive. This is known as **controlled growth.** However, in other cells the activation of the provirus results in rapid viral assembly, leading to massive membrane damage upon viral budding and release, so that the host cell is lysed by the process. This is known as a **lytic infection** (figure 27.7).

Thus the process involving the formation of new HIV particles results in either inactivation or destruction of helper T cells. Since helper T cells are essential for proper development, maturation, and

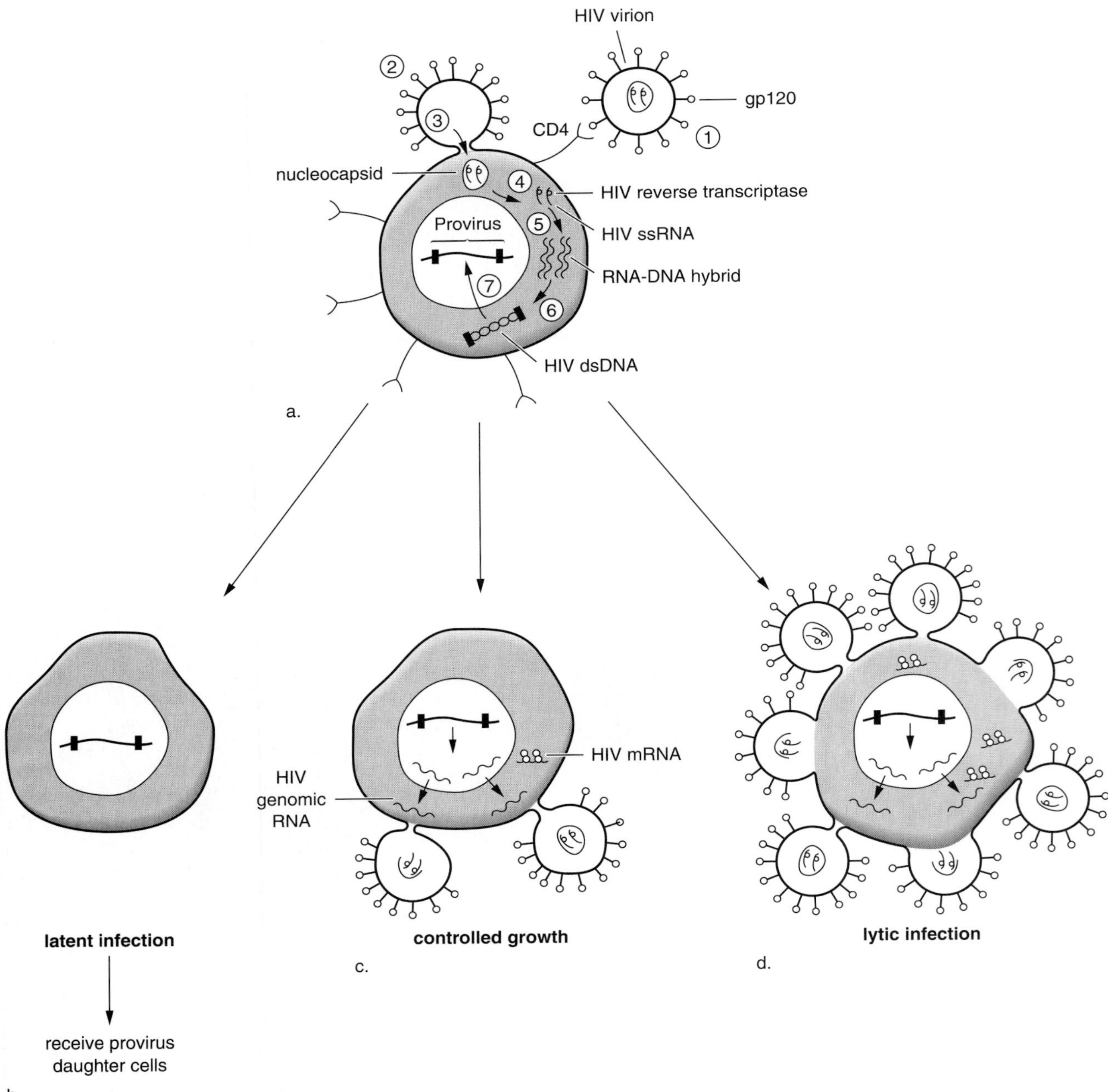


**figure 27.7**

(*a*) The seven stages of infection of a susceptible cell with the HIV virus: (1) the binding of gp120 to the CD4 receptor; (2) the fusion of the nuclear envelope with the cell membrane; (3) the entrance of the viral core into the target cell; (4) the removal of the outer core proteins, resulting in the release of the viral RNA and the enzyme reverse transcriptase; (5) the production of RNA-DNA hybrids; (6) the production of double-stranded DNA after removal of the original RNA; (7) the incorporation of the viral DNA into the host's chromosomal DNA. The incorporated DNA is now known as a provirus. The HIV provirus can now follow one of three pathways: (*b*) it can remain latent; (*c*) it can undergo controlled growth, without damaging the host cell; (*d*) it can undergo rapid viral replication leading to massive damage and eventual lysis of the host cell.

activation of B lymphocytes and cytotoxic T lymphocytes (CTLs), both humoral and cellular (cell-mediated) immunity will be greatly impaired by the absence of these helper T cells. The specific mechanism by which helper T cells are eliminated or inactivated by HIV virions is not known, but several mechanisms have been implicated. For example, large amounts of replicated viral RNA may interfere with normal cellular functions, and hence these products may become toxic to cells. In addition, some of the viral proteins produced by the HIV genes may interfere with the metabolic activities of the helper T cells, and hence may be lethal for these cells. Viral proteins may also block the action of the cytokines necessary for development and maturation of new helper T cells. Finally, viral particles may bind to the CD4 receptors on the T cells and thus produce neoantigens to which the body may develop autoantibodies, which will eventually destroy the affected helper T cells via the action of complement.

### *Clinical Features of AIDS*

Because of the complex biology of the HIV virion, the clinical symptoms associated with AIDS also vary greatly. The infection may run the gamut of features, from being completely asymptomatic, to mild flulike symptoms (e.g., fever, sore throat, headache, muscle ache), to severe incapacitating conditions caused by the presence of overwhelming opportunistic infections, rapidly growing malignant tumors, swollen glands, multiple open skin sores, and brain infections that produce progressive dementia.

The flulike symptoms usually occur when the virus is actively replicating and spreading via the blood and cerebrospinal fluid. After this initial phase the virus usually disappears from the body fluids, but by this time sufficient anti-HIV antibodies have been produced, so that the patient can easily be identified as being **seropositive** for HIV. High levels of serum antibody against HIV are usually present within 3 to 20 weeks after the initial exposure to the virus. After the virus disappears from the circulation, it may remain latent anywhere from 2 to 10 years, and the patient may not exhibit symptoms for most of that time.

The CDC recognizes four different stages of HIV disease. As previously mentioned, the initial HIV infection is usually associated with an acute mononucleosislike illness. The glandular feverlike symptoms last for a few weeks, after which the patient becomes asymptomatic. This is the second stage. The third phase begins when the patient exhibits persistent generalized lymphadenopathy, characterized by the enlargement of multiple lymph nodes but no concurrent illness. In all three phases the patient will be seropositive for HIV. Eventually the patient will develop AIDS. In the last phase, individuals may be described as having AIDS if they test positive for the HIV antibody and exhibit combinations of symptoms from one or more of the categories of diseases given in table 27.2. Most AIDS patients exhibit symptoms of constitutional disease, including persistent fever, weight loss, chronic diarrhea, and general weakness. About 60% of all AIDS patients also show symptoms of neurological disease such as dementia, myelopathy, and neuropathy. Many patients suffer from opportunistic infections caused by intracellular pathogens, indicating a general deficiency of cell-mediated immunity (CMI). The most common of these is *Pneumocystis carinii* infection, which causes pneumonia. This infectious disorder occurs in about 60% of all cases of AIDS. Other infectious agents include *Cryptosporidium, Toxoplasma, Mycobacterium avium-intracellulare, M. tuberculosis, Salmonella, Candida, Cryptococcus neoformans, Histoplasma capsulatum,* cytomegalovirus (CMV), herpes simplex virus (HSV), and varicella-zoster (VZ) virus.

Finally, malignant tumors are also common in AIDS patients, including Kaposi's sarcoma, which may invade the skin, the mucosa and many visceral organs, malignant lymphoma including Burkitt's lymphoma, and tumors of the central nervous system (CNS).

### *Laboratory Findings of AIDS*

Since AIDS has so many different aspects and can express itself in so many different ways, few laboratory features are unique to this disease. Thus laboratory findings depend on the type of infection and the stage of the disease. However some common laboratory test results include low numbers of helper T cells and the presence of HIV antibodies in the serum. Usually the ratio of helper T cells ($CD4^+$ cells) and cytotoxic T cells ($CD8^+$ cells) is 2:1. In AIDS this ratio is reversed to 1:2. Once the $CD4^+$ T-cell count falls to 200/μL of blood, the individual becomes more susceptible to opportunistic infections and neoplastic disease. Although monocytes and macrophages are frequently infected with the HIV virion, they are usually not killed by this virus, and thus their numbers are not depleted as is the case with the helper T cells. Second, HIV patients usually exhibit elevated levels of immunoglobulins and abnormalities in B-cell function. This may be the result of the polyclonal activation of B cells. For example, AIDS patients are unable to mount an adequate IgM response.

The screening test for HIV infection is the **ELISA (enzyme-linked immunosorbent assay)** test for serum antibodies to HIV antigens. In this test the viral antigens are adsorbed to a solid phase. The patient's serum is added, and after incubation the unbound antibody is washed away. An enzyme-conjugated goat antihuman immunoglobulin reagent is then added. After an incubation period the excess reagent is washed away, and a color substrate for the enzyme is added. Since the enzyme is linked to HIV antibodies, a colored reaction product indicates that the patient has antibodies to HIV antigens and must therefore have been exposed to the virus. This process is summarized in figure 27.8. A lag period occurs between the actual exposure to HIV and the appearance of HIV antibody. Generally speaking, it takes 4 to 7 weeks after the initial infection before the antibody titer is high enough for a positive identification to be made.

The initial diagnosis from an ELISA positive patient is customarily confirmed with more sensitive tests such as the **Western blot assay.** In this test HIV proteins are first separated by electrophoresis and then transferred to a nitrocellulose membrane. The patient's serum is then added to the nitrocellulose membrane and allowed to react. Then radioactive labeled goat antihuman immunoglobulin reagent is added. If the patient's antibodies bind to the HIV proteins, then the radioactive goat antibodies will bind to the human antibodies. The nitrocellulose

**Table 27.2** Classification of the Four Stages of HIV Infection as Proposed by Centers for Disease Control (CDC) and the Clinical Symptoms Normally Associated with Each Stage

1. Acute infection: glandular feverlike illness lasting a few weeks
2. Asymptomatic: no symptoms at the time of infection
3. Persistent generalized lymphadenopathy (PGL): lymph node enlargement persisting for 3 or more months with no evidence of infection
4. AIDS
   - *Constitutional disease, including 1 or more of the following:*
     - Fever persisting for more than 14 days without infectious cause
     - Weight loss greater than 10% of body weight with no apparent cause
     - Diarrhea persisting for more than 30 days without definable cause
   - *Neurological disease, including 1 or more of the following:*
     - Dementia
     - Myelopathy
     - Peripheral neuropathy
   - *Opportunistic infections, including 1 or more of the following:*
     - *Pneumocystis carinii* pneumonia
     - Disseminated atypical mycobacterial disease
     - Cerebral toxoplasmosis
     - Esophageal candidiasis
     - Cryptosporidiosis
     - Extrapulmonary cryptococcosis
     - Cytomegalovirus retinitis
     - Disseminated coccidioidomycosis
     - Disseminated histoplasmosis
     - Tuberculosis involving at least 1 extrapulmonary site
     - Recurrent nontyphoid septicemia
     - Extraintestinal strongyloidosis
     - Progressive multifocal leucoencephalopathy
   - *Neoplasms:*
     - Kaposi's sarcoma
     - Primary lymphoma of the brain
     - Non-Hodgkin's lymphoma
   - *Other:* clinical findings or diseases not classified above but which are attributed to HIV infection

membrane is then overlaid with a radiosensitive film, forming a dark spot, thus indicating the presence of HIV antibodies to specific HIV proteins (figure 27.9).

### *Treatment and Prognosis of AIDS*

The treatment or cure for HIV infections and for AIDS remains unresolved. Experimental drugs have been developed that seem to help in some cases. Currently the most successful of these drugs is **AZT (3′-azido-3′-deoxythymidine triphosphate),** which is an inhibitor of reverse transcriptase and hence can stop viral replication. Unfortunately this and other drugs do not eliminate the virus and frequently have unpleasant side-effects. Presently an effective vaccine for HIV has not been developed. Recently the FDA approved a limited trial using an experimental vaccine composed of gp120. If the vaccine is successful, individuals would develop antibodies to the envelope protein gp120. High titers of anti-gp120 antibodies would presumably eliminate the HIV before it could infect $CD4^+$ lymphocytes and other cells. Thus far the best treatment involves the use of AZT and the aggressive use of antibiotic drugs, but they are not effective against all infectious pathogens.

Since effective treatment currently does not exist, most people with HIV will likely succumb within 10 years to the complications of this infection.

## COMBINED B- AND T-CELL DEFICIENCY DISORDERS

Combined immunodeficiency diseases that involve the absence, decrease, or malfunction of both B and T lymphocytes are the result of **primary lymphoid abnormalities**—that is, the inability of early stem cells to develop along lymphocytic pathways. These abnormalities result from genetic variations in the normal genome. The most important of the combined

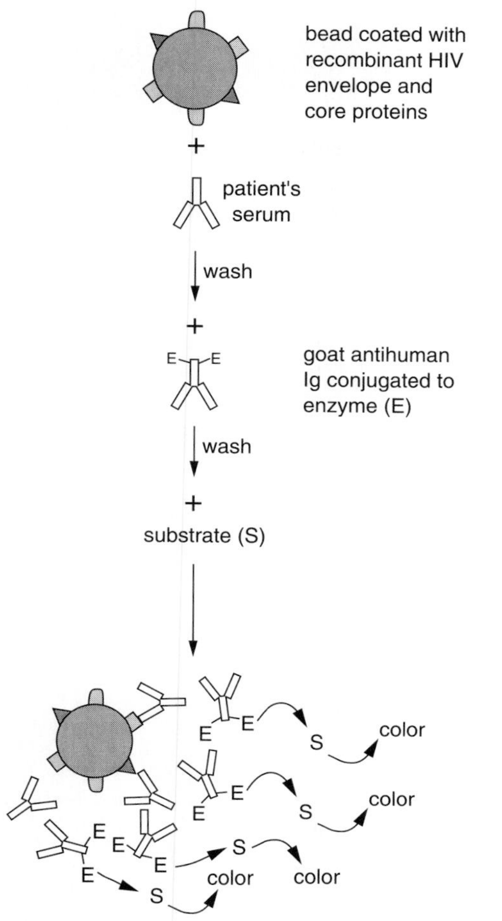


figure 27.8

The screening test for HIV infection is an enzyme immunoassay to detect the presence of anti-HIV antibodies in the blood.

immunodeficiency syndromes is known as **severe combined immunodeficiency disease (SCID).** Five different types of SCID have been identified on the basis of the genetic abnormality involved. In addition to the various forms of SCID, other immunodeficiency diseases have been associated with T- and B-cell impairment. The most important disorders of the second category are **Wiskott-Aldrich syndrome** and **ataxia-telangiectasia,** which are discussed following the description of SCID.

### Severe Combined Immunodeficiency Disease (SCID)

The term severe combined immunodeficiency disease is applied to a heterogeneous group of disorders characterized by defective development of both B and T lymphocytes. As a consequence there will be a profound lymphocytopenia occurring in both the bloodstream and in the lymphoid tissues. The lack of functional lymphocytes results in deficient humoral and cellular (cell-mediated) immunity.

#### *Etiology and Classification of SCID*

Although all cases of SCID are presumed to be the result of genetic abnormalities, in many cases the exact genetic deficiency is unknown.

Originally SCID was thought to be the result of an autosomal recessive trait. Later certain cases of SCID were associated with X-linked recessive genes. To separate the two, the original autosomal recessive condition was named **Swiss-type lymphopenic agammaglobulinemia,** as it was first described in Switzerland in the 1950s. Later autosomal recessive causes for SCID were discovered, each resulting in the

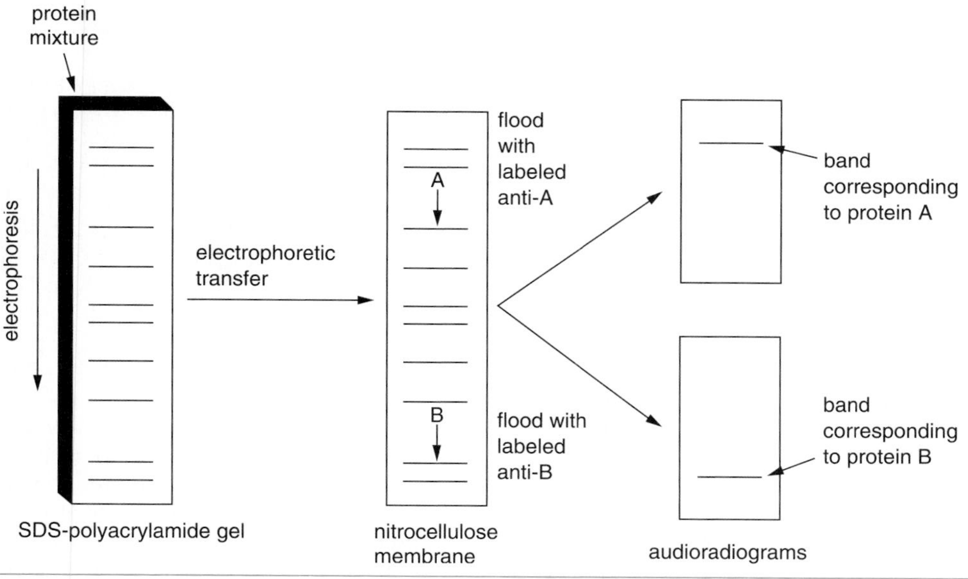


figure 27.9

The Western blotting procedure is used to detect the presence of anti-HIV antibody protein in a person's serum. Serum proteins are separated by electrophoresis, transferred to a polymer membrane, and labeled with radioactive antibodies. A dark band indicates the presence of HIV antibodies specific to HIV proteins.

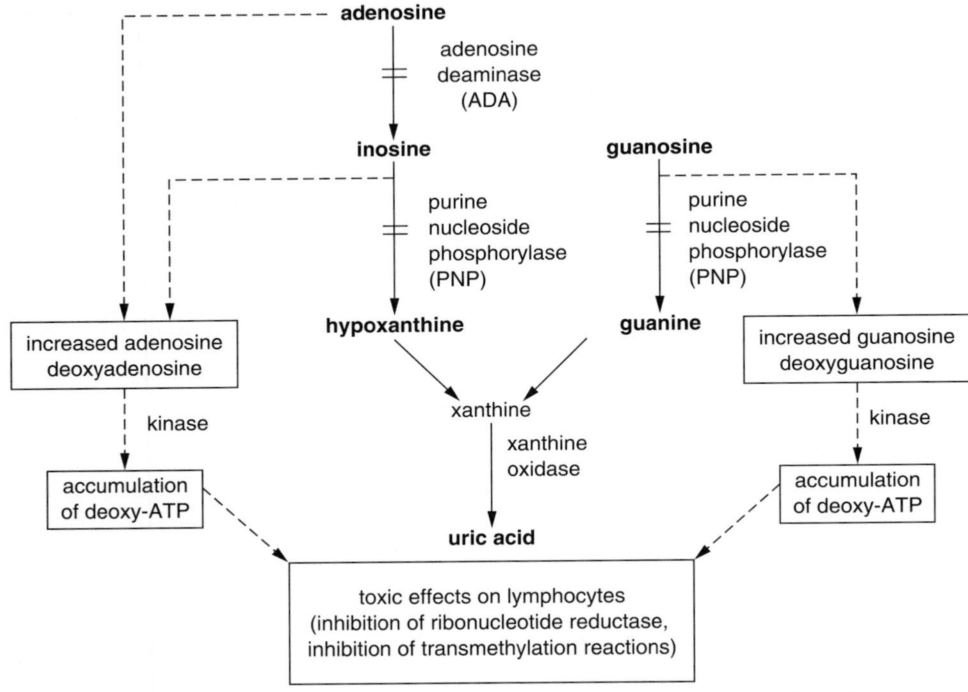


**figure 27.10**

Abnormalities in purine (adenosine and guanosine) metabolism are caused by the lack of the enzymes adenosine deaminase (ADA) and purine nucleoside phosphorylase (PNP). In the absence of these enzymes, different metabolic pathways are utilized (dashed lines), and compounds that are toxic to lymphocytes accumulate.

absence of some essential enzyme. Currently SCID has the following five known causes:

**1. X-linked SCID**
This type of SCID is associated with a marked decrease in T cells in the lymph nodes and in the circulation. The number of B cells appears normal or may be slightly increased. However these B cells are functionally impaired since serum immunoglobulin concentrations are usually low, which can be explained by the absence of the helper T cells necessary for B-cell development, maturation, and activation.

**2. Autosomal recessive SCID**
This is the original Swiss-type lymphopenic agammaglobulinemia. In this form of SCID, both T- and B-cell numbers are very low. The maturation process of both types of lymphocytes appears severely impaired at the stem cell stage. Consequently patients not only have low B- and T-cell counts in the blood and lymphoid tissues but also low serum immunoglobulin concentrations.

**3. ADA deficiency SCID**
This form is caused by a deficiency of the enzyme **adenosine deaminase (ADA).** This condition is also inherited as an autosomal recessive deficit. Approximately 50% of all autosomal recessive SCID cases are in this category. ADA is an enzyme that deaminates adenosine to inosine. The ADA gene is located on chromosome 2. A lack of this enzyme leads to accumulation of deoxyadenosine and deoxy-ATP in developing lymphocytes (figure 27.10). These metabolites are highly toxic to developing cells because they block DNA synthesis. As a result few lymphocytes are produced, and those that are produced fail to mature properly, resulting in severe immunodeficiency. The gene for ADA has been identified and cloned, providing an ideal opportunity for gene transfer technology in the future.

**4. PNP deficiency SCID**
This is a rarer form of an autosomal recessive disease in which the enzyme **purine nucleoside phosphorylase (PNP)** is absent. This enzyme is required in the conversion of hypoxanthine and guanosine to guanine (figure 27.10). Deficiency of PNP leads to the accumulation of deoxyguanosine and deoxyguanosine triphosphate (deoxy-GTP) in T cells, which causes inhibition of DNA replication. Thus the lack of the functional PNP gene located on chromosome 14 results in decreased T-cell production. B cell levels do not appear to be affected. However since helper T cells are necessary for B-cell maturation and activation, B-cell function and antibody production will also be impaired.

**5. Bare lymphocyte syndrome**
This is a form of SCID that has only recently been described. This form of combined immunodeficiency disease is caused by a deficiency of **class II MHC protein** expression on the membranes of white blood cells, especially the lymphocytes. The presence of class II MHC proteins is important for the helper T cells, as they are essential membrane receptors for activation of specific immune responses. Thus functional defects of helper T cells results in lack of activated B cells and cytotoxic T cells. Consequently this results in impaired humoral and cellular (cell-mediated) immunity.

The actual incidence of SCID is not known, since most young patients with this condition die before a correct diagnosis is made. Evidence that most cases of SCID are caused by abnormalities in the early lymphocyte precursor cells is the fact that immunocompetence can be successfully restored with bone marrow transplantation.

*Clinical Features of SCID*

The clinical symptoms of SCID include a failure to thrive, chronic diarrhea, persistent thrush (oral candidiasis), middle ear infections, pneumonia, dermatitis, and a high susceptibility to infectious diseases. Symptoms usually appear at about 6 months of age, when the level of maternal antibodies become lower and are no longer able to offer protection.

*Laboratory Findings of SCID*

Laboratory investigations exhibit low numbers of T lymphocytes and frequently a diminished number of B cells. Immunoglobulin concentrations are also decreased, indicating a lack of functional B lymphocytes.

Lymph node biopsies reveal a severe depletion of lymphocytes and an absence of proper tissue organization—that is, a lack of germinal centers. Plasma cells are also absent in the cords.

*Treatment and Prognosis of SCID*

SCID may be considered a fatal disease unless early detection of this condition is followed by successful bone marrow transplantation. Concurrent with the time period that the bone marrow cells are becoming established, aggressive antibiotic treatment and gamma-globulin injections are necessary in order to prevent overwhelming infections. Successful marrow transplantations will result in complete remission of this disease and may give the patient a chance for a normal life span.

**Wiskott-Aldrich Syndrome**

Wiskott-Aldrich syndrome is a combined immunodeficiency disease that is also characterized by lack of platelets (thrombocytopenia), eczema, and recurrent infections. It is an inherited X-linked recessive disorder. At birth patients are normal but by 1 year of age, they begin to develop bleeding problems, experience severe eczema, and thereafter soon become more susceptible to bacterial and viral infections, indicating a progressive deterioration of the immune system.

*Laboratory Findings of Wiskott-Aldrich Syndrome*

Initially patients with Wiskott-Aldrich syndrome will have normal lymphocyte counts, although these cells tend to be smaller than normal. With increased age, both B-cell and T-cell numbers begin to decrease, and immunoglobulin levels remain low. Both platelet counts and platelet size are decreased from the very beginning. This genetic disease has been associated with an enzyme deficiency. An enzyme may be missing that catalyzes the **glycolysation** of certain membrane proteins. This results in reduced expression and function of glycoproteins in the cell membranes of lymphocytes and platelets.

*Treatment and Prognosis of Wiskott-Aldrich Syndrome*

Treatment for the infections is usually limited to antibiotic therapy; injection with gamma-globulin may be risky as it may result in bleeding problems. However with aggressive antibiotic regimens most pyogenic infections can be controlled, and the long-term prognosis is usually rather good. Many young patients with Wiskott-Aldrich syndrome reach sexual maturity; however some eventually develop either myelocytic or lymphocytic leukemia later in life.

**Ataxia-Telangiectasia**

This syndrome is a combined immunodeficiency disease characterized by **ataxia** (improper balance and movement) and **telangiectasia** (pathological dilations of small veins).

Ataxia-telangiectasia is also an inherited disorder transmitted by autosomal recessive genes. Similar to the forms of SCID, T-cell function is more impaired than B-cell function. B-cell activity is also decreased since they require helper T cells for maturation and activation.

*Clinical Features of Ataxia-Telangiectasia*

The onset of clinical symptoms normally begins around 2 years of age with pathological alterations of the small veins, which become highly dilated. The first organ to be affected by these abnormal venous networks is the cerebellum, resulting in disturbances of movement and balance. As the patient grows older, additional neurological symptoms begin to appear. Further, the signs of telangiectasia become visible in the skin and the whites of the eyes. Around the fifth year of age, the immune system begins to deteriorate rapidly. Cellular (cell-mediated) immunity becomes impaired as the number of peripheral T cells decreases, and those that are present seem unable to function properly. Humoral immunity is also impaired as the antibody response to specific antigens becomes depressed and the immunoglobulin level in the serum falls, although the level of B cells may remain close to normal. As a result the young patients become highly susceptible to viral, fungal, and bacterial diseases. They seem to be especially vulnerable to infections of the respiratory system (**sinopulmonary infections**). When these children reach puberty, they become progressively mentally retarded and seldom develop secondary sex characteristics. Finally, they develop an inability to repair any damage that occurs to their DNA. As a result their neurological and immunological functions progressively deteriorate, and many succumb before they reach adulthood because of complications from infections or because they develop leukemia or lymphoma.

Presently treatment is unavailable for this disease, although early and continuous therapy with antibiotics may limit the effects of many of the sinopulmonary infections early in life.

## Further Readings


Abbas, A. K., Lichtman, A. H., and Pober, J. S. Congenital and Acquired Immunodeficiencies. Chapter 19 in *Cellular and Molecular Immunology*. W. B. Saunders Company, Philadelphia, 1991.

Kuby, J. Immunodeficiency Diseases. Chapter 20 in *Immunology*. W. H. Freeman and Company, New York, 1992.

Kuby, J. The Immune System in AIDS. Chapter 21 in *Immunology*. W. H. Freeman and Company, New York, 1992.


## Review Questions

1. List the three major categories of lymphocyte immunodeficiency disorders.
2. Name the major categories of B-cell deficiency syndromes.
3. Provide another name for Bruton's disease. When does this disorder become apparent and why?
4. List the major clinical features of X-linked agammaglobulinemia.
5. Name the major laboratory findings of X-linked agammaglobulinemia.
6. Describe the treatment for X-linked agammaglobulinemia. What is the prognosis?
7. Define selective dysgammaglobulinemia. List the major types of this disease. What is the most common form? The least common type?
8. Define transient hypogammaglobulinemia of infancy. What is its cause?
9. Define acquired agammaglobulinemia. When does it normally appear? What disorder does it resemble?
10. List the major T-cell deficiency syndromes.
11. Provide another name for DiGeorge's syndrome.
12. What are the usual indications that a newborn infant may be suffering from congenital thymic aplasia? Explain this phenomenon.
13. List the clinical features of congenital thymic aplasia. What are the major laboratory findings?
14. What is the treatment of choice for DiGeorge's syndrome?
15. What is the cause of AIDS?
16. Distinguish between the origin and pathogenicity of HIV-1 and HIV-2.
17. Describe the structure of HIV.
18. Explain how HIV enters the interior milieu of the body.
19. Explain why $CD4^+$ cells are the favorite target for HIV. Mention some other cells that may be infected.
20. Explain how HIV enters the $CD4^+$ cells.
21. Distinguish between (a) latent infections, (b) lytic infections, and (c) controlled growths of HIV.
22. What is meant by the HIV provirus? What causes activation of the provirus?
23. List the four stages of AIDS.
24. When is someone said to be suffering from AIDS?
25. What is the normal screening test to find out if someone is infected with HIV? How is that test performed?
26. Name the test used to confirm a positive HIV infection. How is this test performed?
27. List the current modes of treatment for AIDS.
28. List the eight major groups of people who are currently at risk for developing AIDS.
29. Name the major combined T- and B-cell deficiency syndromes.
30. What does the abbreviation SCID stand for? Name the five forms of SCID.
31. List the major clinical features of SCID.
32. What common laboratory features are associated with SCID?
33. What is the treatment and prognosis for SCID?
34. Define the Wiskott-Aldrich syndrome.
35. What major laboratory features are associated with the Wiskott-Aldrich syndrome?
36. Describe the treatment and prognosis for the Wiskott-Aldrich syndrome.
37. Define ataxia-telangiectasia. What are its major clinical features?

# SECTION FOUR

## *Platelets and Hemostasis*

# Chapter Twenty-Eight

## *Primary Hemostasis*

## CONCEPT OF HEMOSTASIS

**Hemostasis** is the term used to describe the normal functioning of the circulatory system. Its components include the mechanisms that allow the smooth movement of blood through the blood vessels. Second, it includes the factors involved in the rapid repair of damaged blood vessels, thus minimizing loss of blood. Further, it includes the chemical compounds that prevent the inappropriate formation of blood clots in intact blood vessels. Finally, included are the compounds that can rapidly degrade such clots should they occur.

### Components of Hemostasis

A large number of components are involved in the maintenance of hemostasis and traditionally are classified into three different categories. First, are the **blood vessels** themselves: their shapes and structures are designed for smooth movement of blood from the heart to the tissues and back to the heart. Second, the blood **platelets** help maintain the integrity of the endothelial lining of the various blood vessels, as well as plug any ruptures in the circulatory vessels. Third, soluble **plasma proteins** and other chemical compounds function in the maintenance of hemostasis. This chapter's discussion is confined to a description of the structure and functions of the blood vessels and platelets and how they interact to form a temporary plug when damage occurs in the circulatory vessels. This is known as **primary hemostasis.** Chapter 29 discusses the structure and functions of soluble plasma proteins involved in the complex process of converting fibrinogen into insoluble fibrin, which is needed to strengthen the hemostatic plug. This process of coagulation is also known as **secondary hemostasis.** Chapter 29 also discusses the compounds involved in degrading the fibrin plug after the break in the vessel is repaired **(fibrinolysis).**

## BLOOD VESSELS

The circulatory system is composed of several types of blood vessels. These vessels may be classified into two categories. First are a relatively few number of large blood vessels that compose the **macrocirculation.** The large arteries and veins belong to this group, whose major function is to relay blood from the heart to the tissues and back to the heart. The largest number of blood vessels, however, belong to the **microcirculation.** This category includes all the arterioles (the smallest arteries), the venules (the smallest veins), and the capillaries (the smallest blood vessels, which connect arterioles with venules). The human body contains approximately 1,000 large veins and arteries compared to billions of arterioles, venules, and capillaries (table 28.1).

Several basic differences exist between the macrocirculation and microcirculation. The macrocirculation transports the blood rapidly to and from the heart in contrast to the slower-moving microcirculation, which is composed of smaller blood vessels. The major functions of the microcirculation are to exchange blood gases, electrolytes, food substances, and waste products with the surrounding tissue cells. This takes time—hence the need for slower movement, which is ensured by the small diameters and lower hydrostatic pressure of these tiny blood vessels. This difference in movement also produces a difference in the distribution of blood components in the large blood vessels as compared with those of the microcirculation. In the smaller blood vessels the various blood components are uniformly distributed. This is not the case in the large blood vessels. In the macrocirculation blood cells are found in the center of the blood vessels, whereas plasma is concentrated near the edge of vessel lumen (figure 28.1*a*). This is caused by the hydrodynamic forces generated by the rapid flow of blood.

**Bernoulli's principle** states that the pressure within a moving fluid is inversely proportional to the velocity of the flow. When a liquid moves rapidly through a tube, the linear velocity is always highest in the center and lowest at the periphery. This is caused by the friction the wall exerts on the fluid. Since the pressure is lowest in the center of the tube, the solid particles tend to be pushed away from the higher pressure of the edge and become concentrated in the center of the tube. This is known as **laminar flow,** with rapidly moving blood cells in the center and slower-moving plasma on the periphery. In the microcirculation these principles do not hold, as the pressure and velocity are too low in the small blood vessels. Moreover, the diameters of the arterioles, venules, and capillaries are such that blood cells touch the endothelial lining of these tiny blood vessels. In fact, many capillaries are so small that red blood cells must be folded to press through and white blood cells also must adjust their shape (figure 28.1*b*).

### Structure of Blood Vessels

Although blood vessels vary greatly in size and shape, they all have the same basic structure. Blood vessels are hollow tubes lined with a single continuous layer of tightly interconnected **endothelial cells,** which prevents minimal fluid loss from a vessel. The endothelial surface extending into the lumen of the blood vessels is covered by a thin layer of mucopolysaccharide called the **glycocalyx.** The outside surfaces of the endothelial cells are attached to a **basement membrane** that consists of **collagen** fibers in a protein matrix that are wrapped tightly around the cellular

**Table 28.1** Comparison of Various Characteristics of Capillaries, Arterioles, Large Arteries, Aorta, Large Veins, and Venae Cavae

| Characteristic | Blood Vessel Type | | | | | |
|---|---|---|---|---|---|---|
| | **Capillaries** | **Arterioles** | **Large Arteries** | **Aorta** | **Large Veins** | **Venae Cavae** |
| Total number | 10 billion | ~500,000 | Several 100 | 1 | Several 100 | 2 |
| Wall thickness | 1 μm | 20 μm | 1 mm | 2 mm | 0.5 μm | 1.5 mm |
| Internal radius | 3.5 μm | 30 μm | 0.2 cm | 1.25 cm | 0.5 cm | 3 cm |
| Total cross-sectional area | 6,000 $cm^2$ | 400 $cm^2$ | 20 $cm^2$ | 4.5 $cm^2$ | 40 $cm^2$ | 18 $cm^2$ |
| Functions | Major site of exchange; distribution of extracellular fluid between plasma and interstitial fluid | Resistance vessels; distribution of cardiac output | Circulation from heart to tissues; pressure reservoir | | Circulation to heart from tissues; blood reservoir | |

layer. The collagen fibers and the protein matrix are secreted by connective tissue cells that are also found in the basement membrane. This is the basic structure of the capillaries (figure 28.2).

The arterioles and venules have one additional layer of **elastic connective tissue,** in which smooth muscle cells are embedded (figure 28.2). The muscle cells allow these small vessels to contract and expand when necessary.

The large arteries and veins of the macrocirculation have two additional layers of muscle and elastic exterior to the tissue found in the arterioles and venules. The **muscle layer** is usually the thickest of all the blood vessel layers. It allows the blood vessels to contract and thus propel the blood forward (figure 28.2). Since arteries propel blood under much greater pressure than veins, the muscular walls of arteries are much thicker than that of veins. Further, the lumen of the arteries is always round and taut, whereas that of the veins has a semicollapsed appearance. The outer layer of the arteries and veins is composed of a **fibrous connective tissue coat.**

Table 28.1 provides a summary of some important physical characteristics of the major types of blood vessels. Not only are capillaries the most common type of blood vessel, but they also have the greatest cross-sectional area. Thus most biological exchanges occur in these smaller blood vessels as well as most hemostatic problems, such as clot formation and vessel breakage. The components of the hemostatic mechanism are well adapted to dissolve such clots and repair tears in the microcirculation. These hemostatic factors, however, are less efficient in removing large clots and repairing the vessels of the macrocirculation. Ruptures in large blood vessels are dependent on the ability of these blood vessels to contract for prolonged periods of time, thus slowing the movement of blood. Only when the blood flow is slowed sufficiently can the platelets and clotting factors plug the break and the fibroblasts produce new basement membrane and connective tissue layers around the newly formed hemostatic plug.

### Special Functions of Vascular Endothelium

The vascular endothelium also has special functions in the process of hemostasis. First, the intact endothelium provides a site for many of the plasma proteins involved in clotting and plug-formation processes.

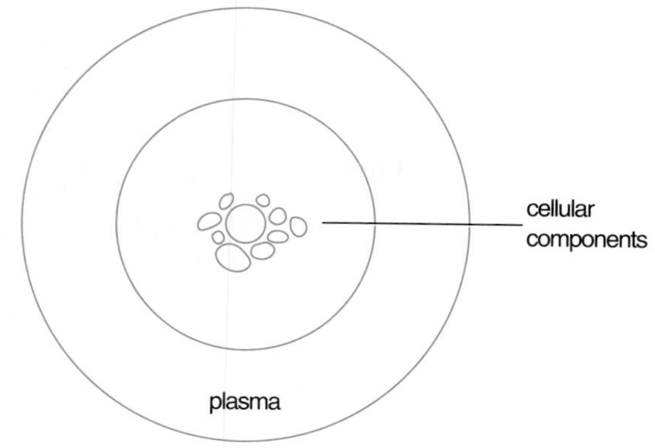


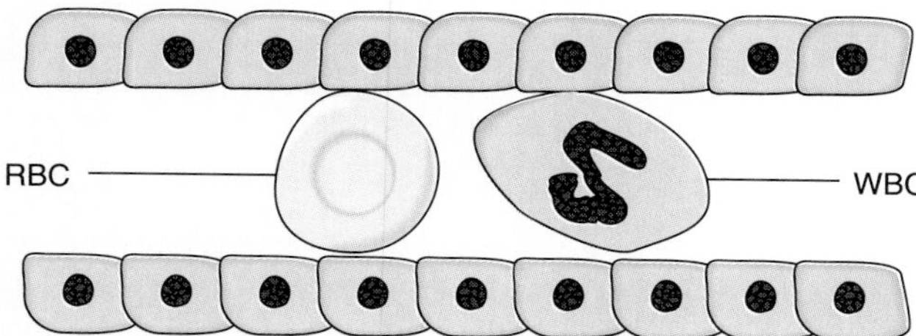


**figure 28.1**

(*a*) Cross-sectional diagram of a large blood vessel showing the laminar flow of the various components of blood (e.g., leukocytes, erythrocytes, and platelets). (*b*) Diagram of blood cells moving through a capillary.

These factors remain inert as long as the endothelium is intact. Injury, however, to the vascular endothelium exposes the collagen fibers of the basement membrane, which activates the plasma-borne inactive coagulation factors. This leads to the formation of blood clots.

When blood vessels are injured, they tend to constrict to reduce the blood flow and thus the blood loss. Vasoconstriction occurs almost immediately after injury takes place, and products from activated platelets, such as **thromboxane $A_2$ ($TXA_2$),** and **serotonin,** activate the muscle cells surrounding the blood vessels causing them to constrict. After the damaged vessel is plugged, the constriction needs to be relaxed, which is another function of endothelial cells. They secrete **prostacyclins** (e.g., $PGI_2$), which cause vasodilation. $PGI_2$ can also inhibit the activation of platelets.

The endothelial cells also produce the basement membrane surrounding it. Endothelial cells synthesize collagen fibers and elastin, as well as fibronectin, a glycoprotein that is necessary for cell adhesion. Certain essential coagulation factors, such as **von Willebrand factor,** are also synthesized in the vascular endothelium. Further, these same endothelial cells produce **cell adhesion factors** important in inflammation and in activation of the immune system. For instance, they synthesize and secrete **endothelial leukocyte adhesion molecule type 1 (ELAM-1), intercellular adhesion molecule type 1 (ICAM-1),** and **vascular cell adhesion molecule type 1 (VCAM-1)**—all of which are receptor molecules for various blood cells.

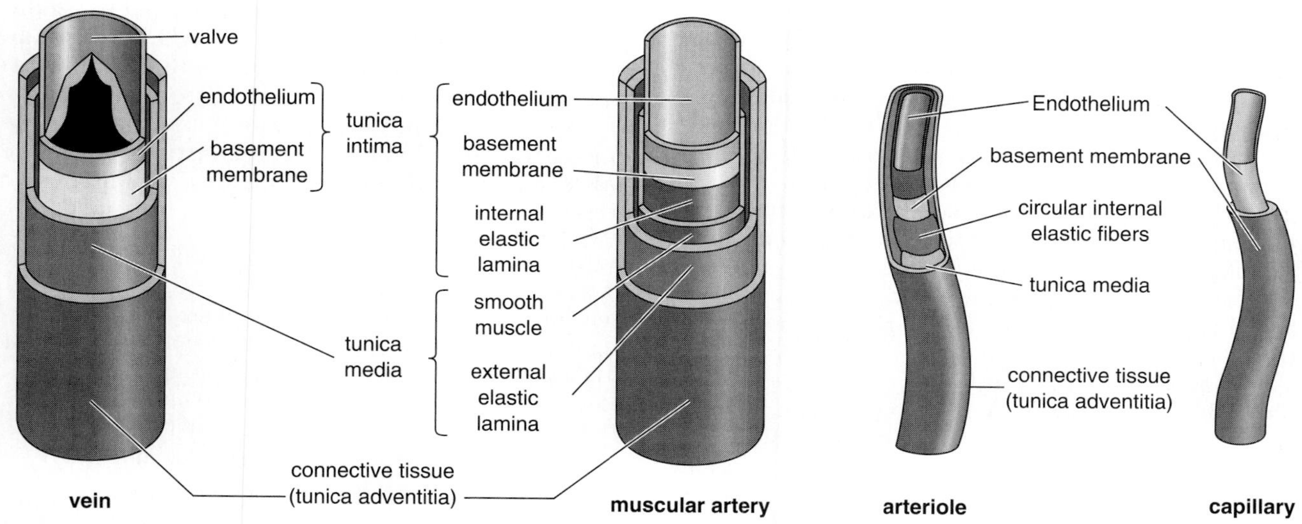


**figure 28.2**

Wall structures of veins, arteries, arterioles, and capillaries.

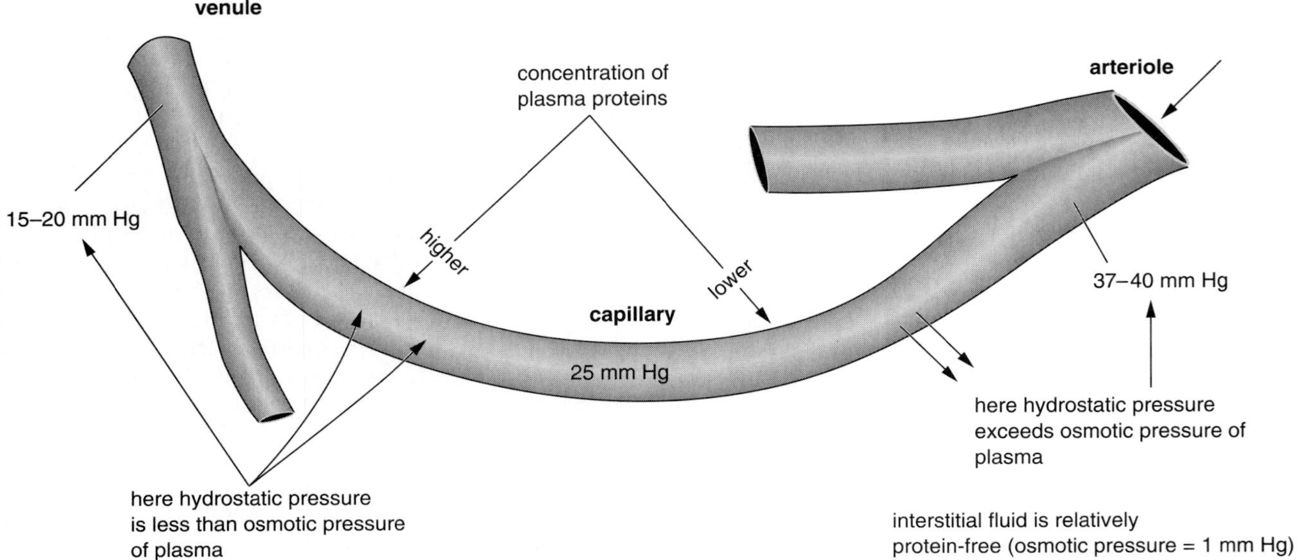


figure 28.3
The movement of water and electrolytes in and out of the microcirculation.

Endothelial cells also assist in the processing of blood-borne antigens that activate the cellular (cell-mediated) immune system.

Finally the endothelial layer forms a semipermeable membrane that prevents blood cells, as well as many macromolecules, from leaving the bloodstream.

### Fluid Exchange Between Blood and Tissues

No direct contact occurs between blood cells and tissue cells. However, direct contact does occur between blood plasma and the **interstitial fluid.** The exchange of water and solutes between the blood and interstitial (intercellular) fluid is controlled by a mixture of **hydrostatic** and **osmotic forces.** Hydrostatic forces on the blood are regulated by the pumping action of the heart and the pressure of the blood vessel walls. Osmotic pressure of the blood is regulated by the blood level of plasma proteins. Approximately two-thirds of the plasma proteins are albumins, and the remaining are the globulins. Hence the osmotic pressure of the blood is mainly a function of the **albumin content** of the plasma.

**Osmosis** may be described as the movement of water through a semipermeable membrane from an area of high water concentration (i.e., solution with few solutes) to an area of low water concentration (i.e., a solution with many solutes). **Osmotic pressure** is defined as the pressure that needs to be applied to a given system so that the influx of water equals its efflux. In the small blood vessels, the average osmotic pressure is approximately 25 mm Hg. In the arterioles the hydrostatic pressure is about 40 mm Hg. Hence at the arteriolar ends of each capillary, hydrostatic pressure exceeds osmotic pressure, thus forcing water and small electrolytes out of the circulation (figure 28.3). At the venous end of the capillaries is a reverse situation. Here the osmotic pressure exceeds the hydrostatic pressure found in small venules, thus forcing water back into the blood vessels. As a result, the overall level of water in the blood vessels remains fairly constant.

In cases of hypoalbuminemia the osmotic pressure of plasma is decreased, which means that water will leave the capillaries more quickly at the arteriolar end but will not return as readily at the venular end. Hence the interstitial fluid receives a net gain of water. This produces tissue swelling known as **edema.**

However diminished plasma protein is not the only cause of edema. On the contrary, there are other important causes including the following:

**1. Vessel inflammation**
Inflammation of the small blood vessels will cause an expansion of their lumen. As the vessel expands, the tight junctions between the endothelial cells are loosened, and plasma proteins diffuse out of the blood into the interstitial fluid, which is relatively protein-free. This upsets the osmotic balance between the plasma and intercellular fluid. The increase in proteins in the interstitial fluid inhibits the return of water to the blood at the venular end of the capillaries—thus increasing the water level in the tissues and producing edema.

**2. Lack of platelets**
Low platelet counts affect the integrity of the endothelial cells of the blood vessel walls. Platelets perform a major nurturing role in keeping endothelial cells in good shape. Lack of platelets tends to result in contraction of individual endothelial cells so that they no longer maintain tight intracellular junctions. Once the spaces between the

endothelial cells are increased, plasma proteins can leak into the interstitial fluid much more easily, which will upset the osmotic balance and cause edema.

**3. Venous blockage**

An increase in hydrostatic pressure on the venous side of the capillary vessel will slow the return of water into the capillary. More water will leave the microcirculation than will return, increasing the water level in the interstitium, thus producing edema. Venous blockage may be caused by venous thrombosis—that is, blood clots in the veins—or tight garments, which can cause venous constriction.

**4. Impaired lymphatic drainage**

Fluid forced out of the circulation at the arteriolar end of the capillary is never completely protein-free. There is a continuous small amount of leaking of proteins into the interstitium. These proteins cannot be returned directly to the venous blood. Normally these proteins enter the lymphatic circulation and are eventually returned to the blood circulation via the thoracic or subclavian ducts. However if the lymphatics are impaired or blocked, protein will accumulate in the interstitium, thus causing local edema.

## PLATELETS

Platelets form the second major component of the hemostatic system. Platelets are frequently referred to as **thrombocytes** because of their sealing function—that is, they form a **thrombus,** or plug, when there is a break in the circulatory system. The term is really a misnomer, since platelets are not cells, but are small packages of cytoplasm that are nipped off from the cytoplasm of large mother cells in the bone marrow known as megakaryocytes. On a peripheral blood smear, platelets appear as small bluish, granular structures, discoid in shape, that are between 2 and 3 μm in diameter. The normal concentration of platelets in the blood is between 150,000 and 400,000/μL.

### Platelet Development

Platelet development is frequently referred to as **thrombopoiesis.** All platelets are derived from large polyploid cells in the bone marrow called **megakaryocytes.** These platelet mother cells are derived from the same pluripotent stem cells as are other blood cells. The platelet-producing cells follow the myeloid pathway of development, just as erythrocytic and granulocytic-monocytic blood cells.

Under the influence of certain hormonal growth factors, myeloid stem cells differentiate into megakaryocyte precursor cells known as **megakaryocyte colony-forming units (MEG-CFUs).** The development and differentiation of MEG-CFUs into megakaryoblasts and megakaryocytes is controlled by certain regulatory factors. The most important ones are (1) granulocyte-monocyte colony-stimulating factor (GM-CSF), (2) interleukin 3 (IL-3), (3) erythropoietin (EPO), and (4) thrombopoietin. The stimulation of MEG-CFUs by GM-CSF and IL-3 results in the development of megakaryoblasts, the first stage in the formation of megakaryocytes. As with other myeloid cells, the developmental series of the megakaryocytes can also be divided into six different stages (figure 28.4).

The first stage is the **early megakaryoblast,** which is a medium-sized cell (with a diameter of 10 to 24 μm) and a single large nucleus containing many nucleoli and a scant basophilic cytoplasm without granules. This primitive blast cell is difficult to distinguish from other primitive myeloid blasts cells such as the erythroblasts and myeloblasts.

The second stage is the **late megakaryoblast.** This cell can be recognized as a larger cell (with a diameter of 15 to 30 μm) with multiple nuclei (usually 2 or 4), each still showing nucleoli. The cytoplasm is larger than in the previous stage and stains a deep blue basically devoid of any granules.

Stage three is the **promegakaryocyte,** a still larger cell containing more nuclei (usually 8, 16, or 32). These multiple nuclei usually appear as a large lobed mass. The nucleoli are no longer obvious at this stage. The abundant cytoplasm is beginning to show numerous **granules.** The cytoplasm appears pinkish, and the granular material looks deep blue.

Stage four is the **immature megakaryocyte.** Usually this cell is larger than the previous cells (with diameters of 20 to 50 μm). The nuclear mass is somewhat compacted and appears as a single large structure, although individual nuclei are still recognizable. At this stage the cytoplasm develops a cytoplasmic membrane system known as the **demarcation membrane system (DMS)** by invaginations of the outer cell membrane. As it forms, small areas of megakaryocytic cytoplasm become separated from the remaining cytoplasm. Each of these areas will eventually become a platelet.

Stage five is the **mature megakaryocyte,** which actually releases platelets from its cytoplasm. The mechanism of **platelet shedding** is not thoroughly understood, but the platelets may be shed in large groups called **proplatelets,** which then move from the extravascular spaces, where they are produced, into the vascular sinuses of the bone marrow. Here the proplatelet mass is further subdivided into individual platelets. The final stage is the **old megakaryocyte.** In this phase, when the cytoplasm has been removed in the form of platelets, basically nothing remains except a **bare nuclear mass.** The nucleus then deteriorates and is removed by macrophages.

Although there are six separate stages in the life cycle of the megakaryocyte, the process of **thrombopoiesis** (platelet development) is usually divided into the following three different phases (figure 28.4):

**1. Proliferation**

This is the process of **endomitosis**—that is, nuclear proliferation in a common cytoplasm. With each mitotic division the cytoplasm enlarges greatly. A direct correlation exists between the number of nuclei and the size of the cytoplasm. The higher the number of nuclei, the larger the

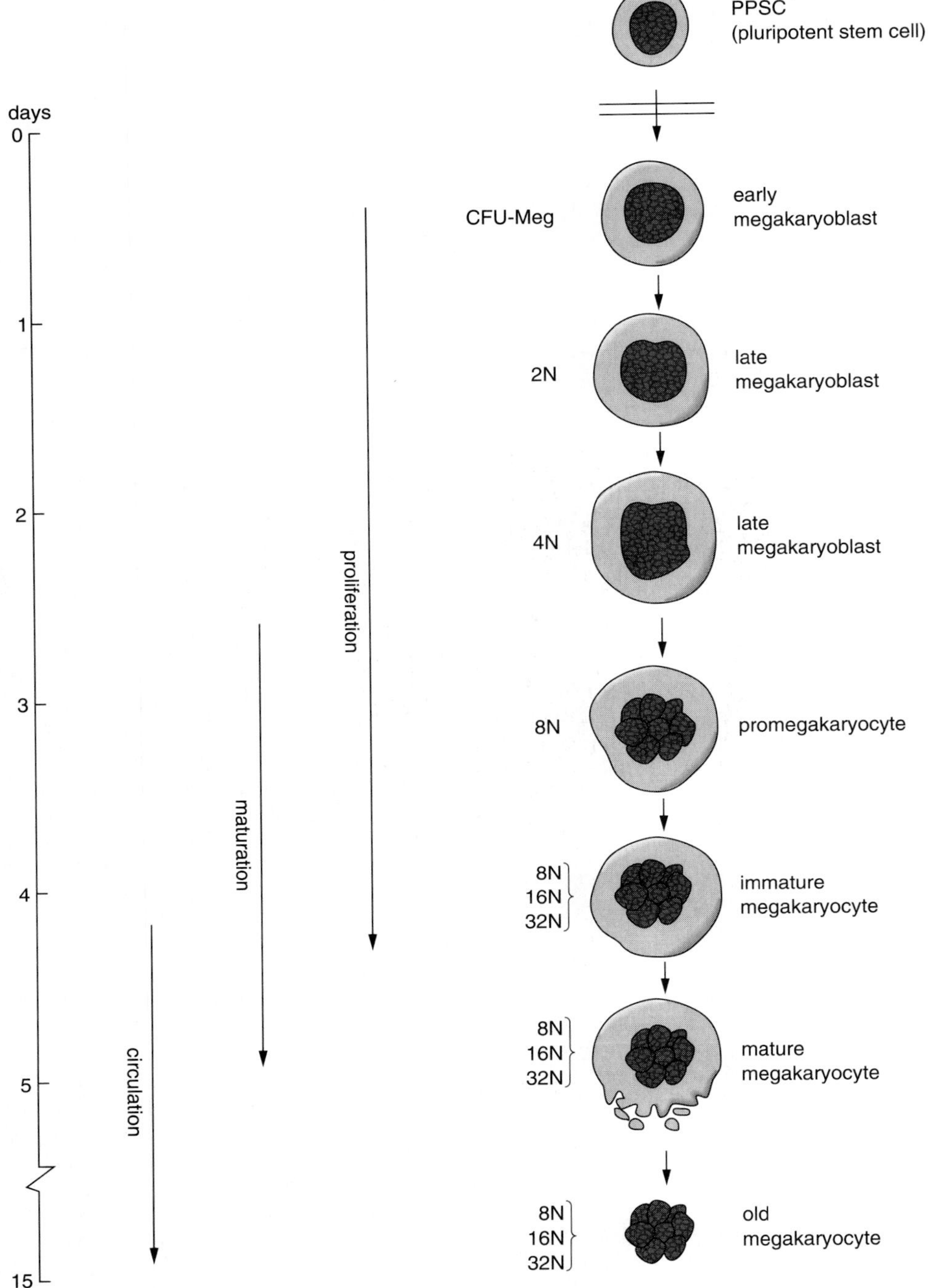


**figure 28.4**

The development and maturation of a megakaryocyte. N = nuclei.

cytoplasm. If we compare the various multinucleated megakaryocytes in normal bone marrow, we may observe that approximately 65% of all megakaryocytes have 8 nuclei; 25% have 16 or more nuclei; and about 10% have 2 to 4 nuclei.

**2. Maturation**

The maturation process includes the development of cytoplasmic granules and formation of the demarcation membrane system (DMS). Maturation begins early, but generally follows the proliferation phase. In normal bone marrow about 25% of the developing megakaryocytes have no granules, 25% are partially granulated, and 50% appear to be fully granulated.

**3. Circulation**

In this phase the mature megakaryocytes release proplatelet packages into the bloodstream. Mature megakaryocytes are located in close proximity to bone marrow sinuses, thus allowing them to shed their platelets directly into the sinuses of the circulatory system.

In the normal healthy individual, it takes 5 to 6 days for a megakaryoblast to develop into a mature platelet-producing megakaryocyte.

**Platelet Kinetics**

Since the number of platelets produced by a megakaryocyte is directly proportional to the amount of cytoplasm, the total platelet-producing capacity of the bone marrow is the product of the average cytoplasmic volume and the number of megakaryocytes. A typical megakaryocyte produces 40 to 60 platelets per day until its cytoplasm is depleted. A 50% or more decrease of platelets released may be caused by defective platelet formation or ineffective platelet delivery to the circulation. It may also be caused by intramedullary destruction or by interference caused by bone marrow tumors. Determining the total number of platelets produced by each megakaryocyte is difficult because megakaryocytes differ enormously in size. Since size is directly dependent on the number of nuclei in the nuclear mass, a megakaryocyte with 32 nuclei will produce more platelets than a megakaryocyte with only 4 or 8 nuclei. Each megakaryocyte is estimated to produce 150 to 200 platelets per nucleus.

The maintenance of the megakaryocyte mass is regulated by a special growth factor known as **thrombopoietin,** which regulates the rate of cytoplasmic maturation and platelet release. It is also important in regulating the amount of cytoplasm present in each megakaryocyte by virtue of its ability to increase intracellular mitoses.

The normal count of platelets in the circulatory system is between 150,000 and 400,000/μL. Approximately two-thirds of all platelets released into the circulation are found in the bloodstream; the other one-third is sequestered in the spleen. A level of equilibrium exists between the platelets in the blood and those in the spleen. Splenomegaly is associated with a marked increase in splenic pooling. A greatly enlarged spleen can sequester 80 to 90% of all platelets in the circulation. The distribution of platelets between the systemic and splenic compartments can be estimated by determining the proportion of $^{51}$Cr-labeled platelets remaining in the circulation after they are infused into the blood. This is an important test used to determine the extent of hypersplenism.

The average life span of platelets is approximately 10 days. The average volume of a platelet is about 5 μm, with a gradual dimunition later in their life span. In cases in which the need for more platelets is greatly increased, there may be a premature release of platelets from the bone marrow. Such platelets are usually larger and more competent than normal platelets. These larger **shift platelets** are characteristic in cases of severe bleeding. In contrast, smaller than normal platelets are frequently observed in inflammation and in iron deficiency anemia.

**Structure of Platelets**

Platelets are small discoid-shaped packages of cytoplasm surrounded by a membrane similar in structure to the normal cell membrane. In contrast with red and white blood cells, which have relatively smooth surfaces, the platelet surface has many openings that extend deeply into the interior of the structure (figure 28.5). These channels, known as **canaliculi,** give the platelets a spongelike appearance. The surfaces of these channels are continuous with the platelet membrane and increase the surface area of the platelets. This is important as platelets need numerous receptors to selectively adsorb the many coagulation factors necessary for the clotting process.

As the platelet membrane is similar to the cytoplasmic membrane of other cells, it also consists of a double layer of phospholipid in which a large number of integral proteins are embedded. Most of these integral proteins are glycoproteins that serve as receptors for the many factors involved in the clotting process. The majority of these receptor proteins are occupied by coagulation factors selectively adsorbed from the plasma. Some of the more important coagulation proteins that are adsorbed to the platelet membrane are **factor V, factor VIII,** and **factor I** (the latter is also known as fibrinogen). These adsorbed clotting factors form a surface coat of proteins collectively known as the **glycocalyx.** These clotting factors are found not only on the exterior of the platelet surface but also on the membranes lining the canaliculi in the interior of the platelets.

A filament-rich microtubular layer is found just below the platelet surface. This narrow layer consists of microtubules and microfilaments that form a submembranal scaffolding. This layer not only maintains the discoid shape of the platelets but also acts as a contractile system that changes their shape once they are activated. Microtubules are composed of bundles of tubulin, whereas microfilaments are composed of actin, which readily binds to myosin molecules. All of these fibrous proteins are responsible for contractile activity.

Below the microtubular layer is the **organelle zone.** This inner layer of platelets consists of mitochondria, glycogen particles, and at least three different types of granules: dense bodies, alpha granules, and lysosomal granules. The contents of the granules are also important for platelet functioning. The **dense bodies,** so-named because they appear more dense than other granules in electron microscope preparations, contain large quantities of ADP, ATP, and other nucleotides, as well as serotonin, phosphate, and calcium ions. **Alpha granules** are the most common of the three types of intracellular particles and consist of proteins that assist with coagulation and other platelet functions. Finally, **lysosomal granules** contain various hydrolytic enzymes required in the degradation of proteins.

The inner zone of the platelets is crisscrossed by surface-connected canaliculi. Some of these channels are formed by invagination of the outer membrane

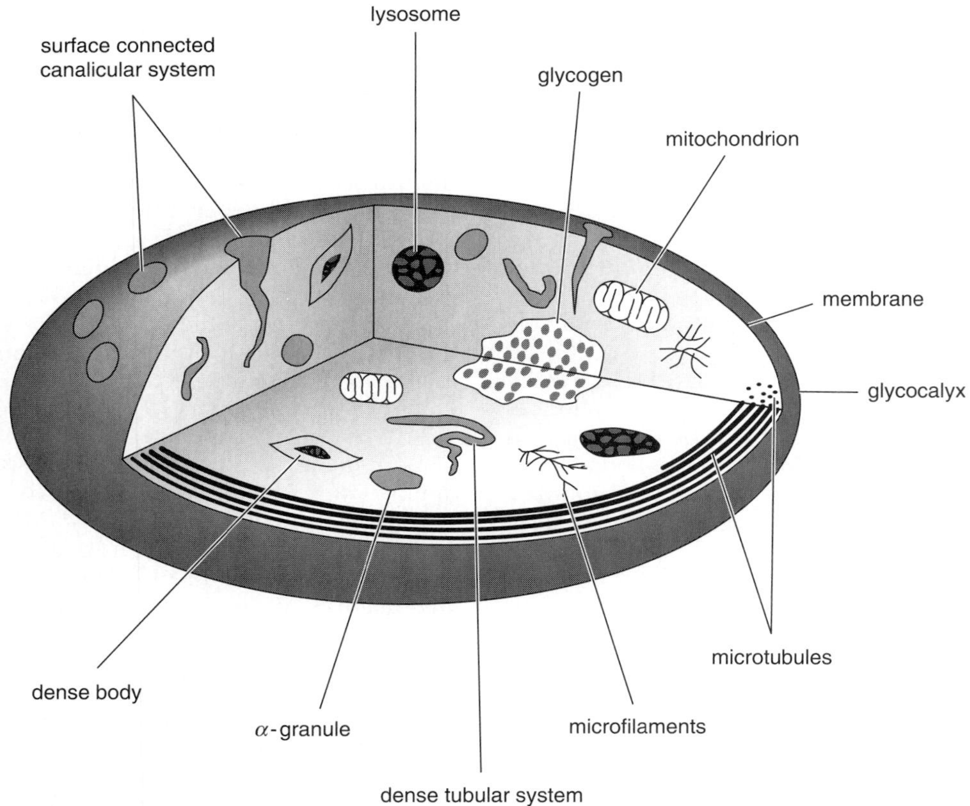


**figure 28.5**
Structure of a platelet.

during the maturation process. Other parts of the canaliculi system are remnants of the rough endoplasmic reticulum that were present in the megakaryocytes. These two types of membranes fuse at various sites within the platelet cytoplasm. In addition to adsorbing certain clotting factors, these membrane complexes are important in regulating intraplatelet calcium levels. Calcium levels in turn are important in the regulation of platelet metabolic activity and activation.

### Functions of Platelets

Platelets have several functions, all of which are related to hemostasis. Blood loss from intact blood vessels is prevented by the structure of the endothelial cells that form the vessel walls. Endothelial cells form tight junctions, which prevent escape of blood cells and plasma proteins from the circulation. To maintain the structural integrity of the vessel walls, platelets are required. When endothelial cells accidentally separate in the intact blood vessels, platelets fill any small gaps that may have occurred. They bind readily to the basement membrane and thus prevent the escape of plasma and blood cells from the circulation (figure 28.6). When the number of platelets is reduced, nurturing does not occur and endothelial cells tend to become thin and lose their tight connections. Thus gaps appear between the cells, and blood leaks out of the blood vessels into the tissues.

Finally, platelets form plugs in damaged blood vessel walls that occur as a result of vessel injury. This process is known as the **primary hemostatic plug formation,** and it greatly limits the loss of blood from the circulation. When a vessel is cut, two things happen almost immediately. One, the blood vessel will contract and constrict, thus reducing the flow of blood. Two, the platelets congregate at the edges of the cut, thus initiating the process that will result in plug formation. Vascular constriction is a transient process, usually lasting less than a minute. However this amount of time is sufficient for plug formation to start. Vessel constriction is caused by powerful vasoconstrictor

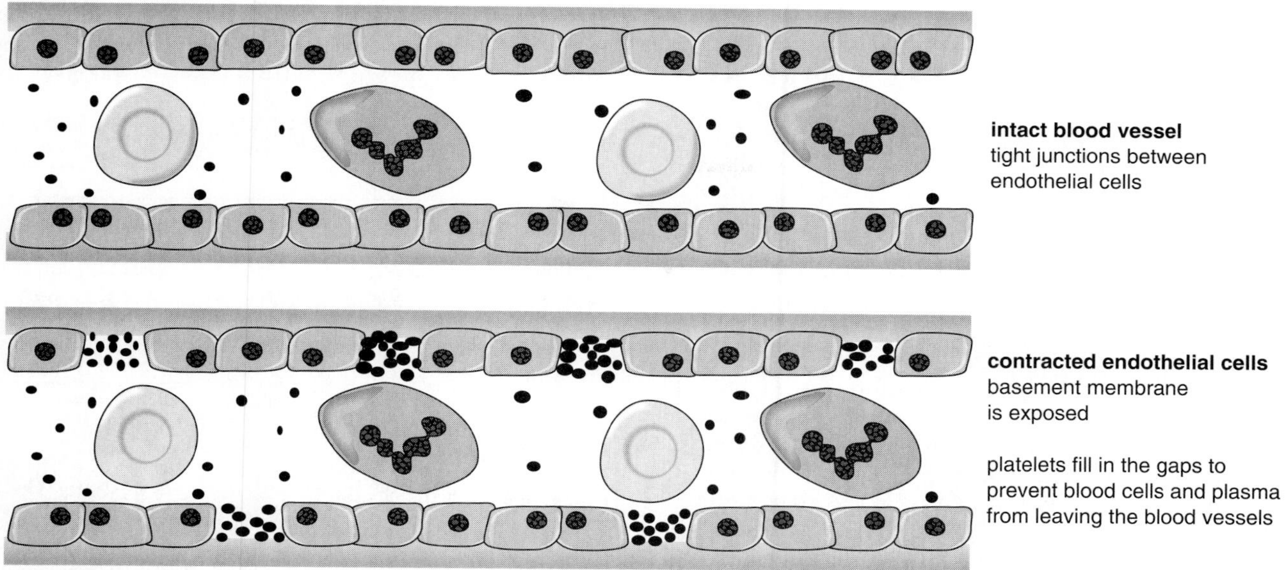


**figure 28.6**

The role of the platelets when the endothelial cells contract.

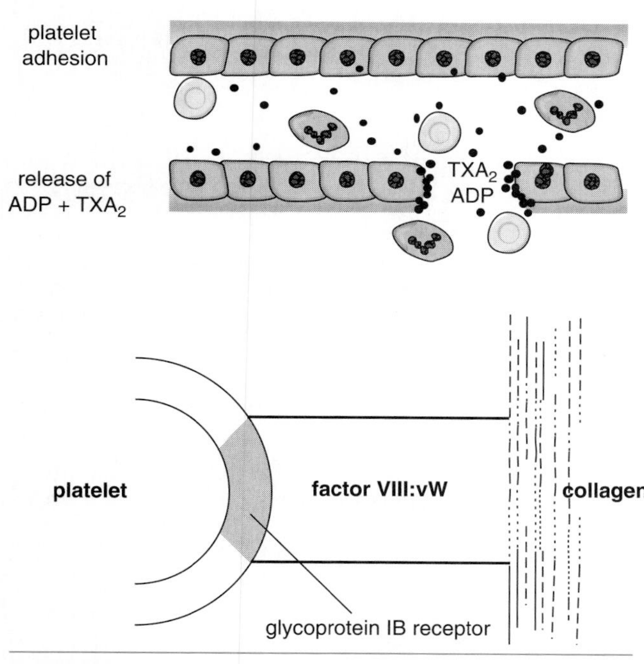


**figure 28.7**

Platelets adhere to the collagen of the exposed basement membrane via the von Willebrand factor (VIII:vW).

molecules released from the injured cell walls and by the platelets at the site of injury. These vasoconstrictor molecules are known as **thromboxanes,** the most important of which is thromboxane $A_2$ ($TXA_2$).

### Formation of the Primary Hemostatic Plug

The process of primary hemostatic plug formation follows a specific sequence of events that can be summarized into the following five steps:

**Step I** involves the initial **adhesion of platelets** to the exposed basement membrane at the site of injury. Platelets adhere to the exposed collagen fibers of the subendothelial vessel wall. This adhesion requires the presence of a compound known as the **von Willebrand factor** (figure 28.7), which is part of a complex clotting factor known as **factor VIII** (see chapter 29). The normal platelet has receptors on its membrane, commonly known as **glycoprotein 1B.** Thus von Willebrand factor acts as a bridge between the platelet and exposed collagen fibers of the basement membrane.

**Step II** occurs immediately following the initial platelet adherence to the basement membrane. Adhesion triggers both morphological and functional changes in the attached platelets. The platelets change shape from a discoid to a sphere from which many spiny projections radiate. These pseudopods increase the chance of making contact with other platelets.

The more contact they have, the better they will stick together (figure 28.8). In addition to these obvious morphological changes, the platelet surface also begins to show an increase in the number of receptors, especially for thrombin, collagen, ADP, and epinephrine. When these receptors are occupied, signals are sent to the interior of the platelet, which responds by releasing the contents of its dense granules (especially ADP) and by the production of eicosanoids from its plasma membrane. **Eicosanoids** are chemical compounds derived from arachidonic acid in the membrane, including thromboxanes (e.g., $TXA_2$) and prostaglandins (e.g., $PGE_2$ and $PGF_2$).

**Step III** involves **platelet aggregation.** Both ADP and $TXA_2$ act as chemotactic factors that cause platelets to aggregate and form an unstable plug at the site of injury. Once enough platelets have aggregated to form a weak plug, they begin to discharge the chemical compounds from their granules. These secreted substances help to strengthen the platelet plug by causing additional platelets to adhere and aggregate. Eventually a mechanical plug is formed that is large enough and strong enough to seal the injury and prevent further blood loss. The time required to stop the bleeding depends on the depth of the injury and the size of the vessel involved. In minor skin wounds bleeding usually stops within a few minutes.

**Step IV** involves stabilization and anchoring of the plug to the vessel wall, which requires fibrin, the production of which starts during this phase. This complex process involves many different clotting factors, and it is known as **secondary hemostasis,** which is discussed in detail in chapter 29. Platelets also play a major role in the process of fibrin formation. The change in morphology during platelet aggregation leads to exposure of receptors for many of the coagulation proteins. Coagulation is a cascade in which individual coagulation factors are sequentially activated by a specific enzyme. Binding of these inactive coagulation factors to their platelet receptors provides the proper orientation so the activating enzyme can bind and modify the factor. The strands of fibrin produced weave through the platelet plug and connect with the blood vessel walls.

**Step V** is the final step of the platelet plug formation. In this phase the long strands of fibrin that have woven through the clot and anchored it to the vessel wall begin to contract. This tightens and strengthens the primary plug and anchors it more firmly against the blood vessel wall. The enzymes performing these contractions are energy dependent, and this energy is derived from glycogen stored in the platelets.

This stable plug is only temporary. It plugs the hole in the blood vessel until the endothelial wall is repaired. At that time the clot is gradually degraded by a series of enzymes also present in the blood.

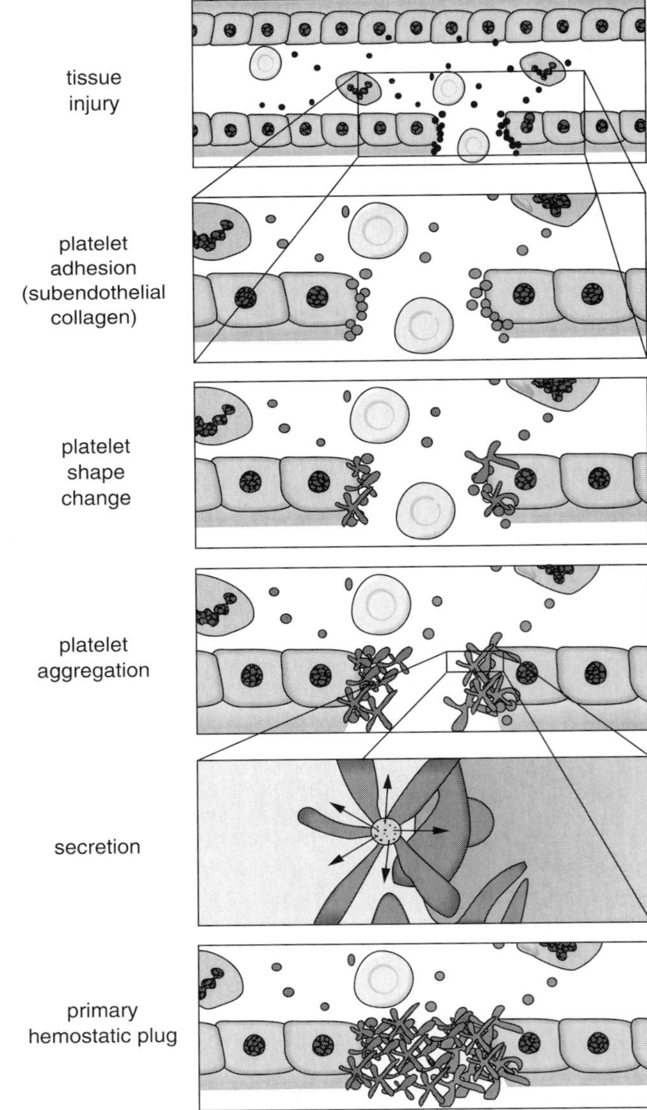


**figure 28.8**

Platelet plug formation.

## Further Readings

Handin, R. I. Hemorrhagic Disorders II. Platelets and Purpura. Lecture 27 in *Hematology*, 4th ed., William S. Beck, Ed. MIT Press, Cambridge, Mass., 1985.

Hirsch, J., and Brian, E. A. An Introduction to Normal Hemostatic Mechanisms. Chapter 1 in *Hemostasis and Thrombosis, A Conceptual Approach*, 2nd ed. Churchill Livingstone, New York, 1983.

Hoffbrand, A. V., and Pettit, J. E. Platelets, Blood Coagulation and Haemostasis. Chapter 11 in *Essential Haematology*, 2nd ed. Blackwell Scientific Publications, Oxford, 1984.

Larson, L. Primary Hemostasis. Chapter 20 in *Textbook of Hematology* by Shirlyn B. McKenzie. Lea & Febiger, Philadelphia, 1988.

Ogston, D. Platelet Structure and Function. Chapter 2 in *The Physiology of Hemostasis*. Harvard University Press, Cambridge, Mass., 1983.

## Review Questions

1. Define hemostasis.
2. List the three major components of the hemostatic mechanism.
3. Distinguish between blood vessels of the macrocirculation and microcirculation. What is the function of each?
4. Distinguish between the structure and function of arteries and veins.
5. Name the three types of blood vessels that make up the microcirculation.
6. Distinguish between the structure of a capillary and an arteriole.
7. Distinguish between the numbers of blood vessels belonging to the macrocirculation and microcirculation.
8. Which blood vessels are the major sites for blood vessel injury and why?
9. List five important functions of the vascular endothelium.
10. What are some important cell adhesion factors produced by the vascular endothelium?
11. Which forces regulate the exchange of fluids between the blood and the interstitial spaces? What is the origin of these forces?
12. Explain why the albumin levels of the blood are primarily responsible for the osmotic pressure of the blood.
13. Explain why water leaves the circulation at the arteriolar end of the microcirculation and why it returns at the venular end.
14. What is edema? List five major causes of edema.
15. Explain why vessel inflammation may result in edema.
16. Why may a lack of platelets result in edema?
17. Explain why venous blockage may result in edema.
18. Explain why impaired lymphatic drainage may result in edema.
19. Why are platelets sometimes called thrombocytes? Why is this term a misnomer?
20. Where are platelets produced? From which cells are they produced?
21. What is the normal level of platelets in the blood?
22. Name some important growth factors that stimulate the production of platelets.
23. List the six stages in the development of platelets. How do we distinguish between each stage?
24. Name the three phases in the process of thrombopoiesis. What is the length of the maturation process of a platelet?
25. What is the normal life span of a platelet?
26. What percentage of the platelets is normally sequestered in the spleen? Describe what happens when splenomegaly occurs.
27. How many platelets does an average megakaryocyte produce per day? How can we estimate the total platelet-producing capacity of a given megakaryocyte?
28. Describe the general structure of a platelet.
29. Explain the function of the platelet canaliculi.
30. How does a platelet maintain its shape?
31. Name the three different types of granules found in platelets. What is the function of each of these granules?
32. List the major functions of platelets.
33. List the five steps of hemostatic plug formation.
34. What allows platelets to adhere to the exposed basement membrane?
35. Why do platelets change shape after adhesion to the basement membrane?
36. Describe the function of fibrin strands in hemostatic plug formation.

# Chapter Twenty-Nine

## *Secondary Hemostasis*

In **secondary hemostasis** soluble plasma proteins interact in a series of complex enzymatic reactions to convert fibrinogen into long strands of fibrin. This process is also known as **coagulation** and occurs in a series of well-defined sequential steps. This reaction sequence, frequently referred to as the **coagulation cascade,** involves the serial activation of inactive plasma proteins to active enzymes and cofactors. Each activated enzyme reacts with the next inactive plasma protein to convert it into an active enzyme. Fibrinogen, the final substrate, is converted to fibrin by the activated enzyme, thrombin. All enzymes in the cascade are **serine proteases**—that is, they act as cleaving enzymes at the site of serine residues. The coagulation cascade is assisted and amplified by **cofactors,** which greatly increase the rate and magnitude of the chemical reactions.

All enzymatic reactions (except the conversion of fibrinogen to fibrin) require a **phospholipid surface,** which is provided by the platelet membranes and by the injured vessel walls. The phospholipid surface limits the reactions to a localized area, and thus fibrin formation is confined to the site of injury. As mentioned in chapter 28, both primary and secondary hemostasis are needed for hemostatic plug formation. Disorders in either type of hemostasis will result in prolonged bleeding. Lack of platelets is usually associated with small pinpoint hemorrhages beneath the skin known as **petechiae.** Lack of coagulation proteins usually results in large bruises beneath the skin and in other less visible parts of the body. These bruises are known as **ecchymoses.**

Fibrin formation is a well-regulated process usually limited to the area of injury. The activated clotting factors must remain localized to the area around the hemostatic plug because sufficient thrombin could be generated in 10 mL of plasma to clot the entire blood volume of the body in less than 1 minute. However this does not occur because each enzymatic step in the coagulation cascade is limited by one or more **inhibitory** or **regulatory factors.** Further, the process of coagulation also initiates **fibrinolysis,** which proceeds slowly at first but gradually accelerates after coagulation is completed. Plasma contains an inactive protein called **plasminogen,** which is activated by serine proteases of the coagulation cascade and other factors. These enzymes convert plasminogen into **plasmin,** which is an active enzyme capable of degrading fibrin into smaller nonreactive fragments.

Thus secondary hemostasis consists of a series of biochemical events resulting in the formation of a fibrin clot. The process involves many coagulation factors and inhibitors that limit the clot to the area of vessel injury, and that will dissolve the hemostatic plug as soon as the vessel damage is repaired.

**Table 29.1 The Nomenclature of the Original Blood Coagulation Factors**

| Factor | Synonyms |
|---|---|
| I | Fibrinogen |
| II | Prothrombin |
| III | Tissue thromboplastin, tissue factor |
| IV | Calcium |
| V | Labile factor, proaccelerin, prothrombin accelerator |
| VI | Deleted (actually activated factor V) |
| VII | Stable factor, proconvertin, serum prothrombin conversion accelerator (SPCA) |
| VIII | Antihemophilic factor (AHF), antihemophilic globulin (AHG), platelet cofactor I |
| IX | Antihemophilic factor B, Christmas factor, plasma thromboplastin component (PTC), platelet cofactor II |
| X | Stuart-Prower factor, thrombokinase |
| XI | Plasma thromboplastin antecedent (PTA), antihemophilic factor C |
| XII | Hageman factor, contact factor |
| XIII | Fibrin stabilizing factor, fibrinase, Laki-Lorand factor |

## COAGULATION FACTORS

Many different factors play a role in the coagulation process. To clarify coagulation factor terminology, an international system of nomenclature for clotting factors was developed by an international committee. A series of **12 factors** was designated with Roman numerals I through XII, according to the order of their discovery (not their reaction sequence). The committee accepted the 12 coagulation factors for three major reasons: (1) the availability of reliable data regarding molecular weight, plasma concentration, stability, absorbability, and half-life; (2) a clinically identifiable disease state could be associated with the deficiency of each of these factors; and (3) availability of reliable assay methods to measure each of these factors. Later factor VI was found to be only an activated form of factor V and is no longer included in the list.

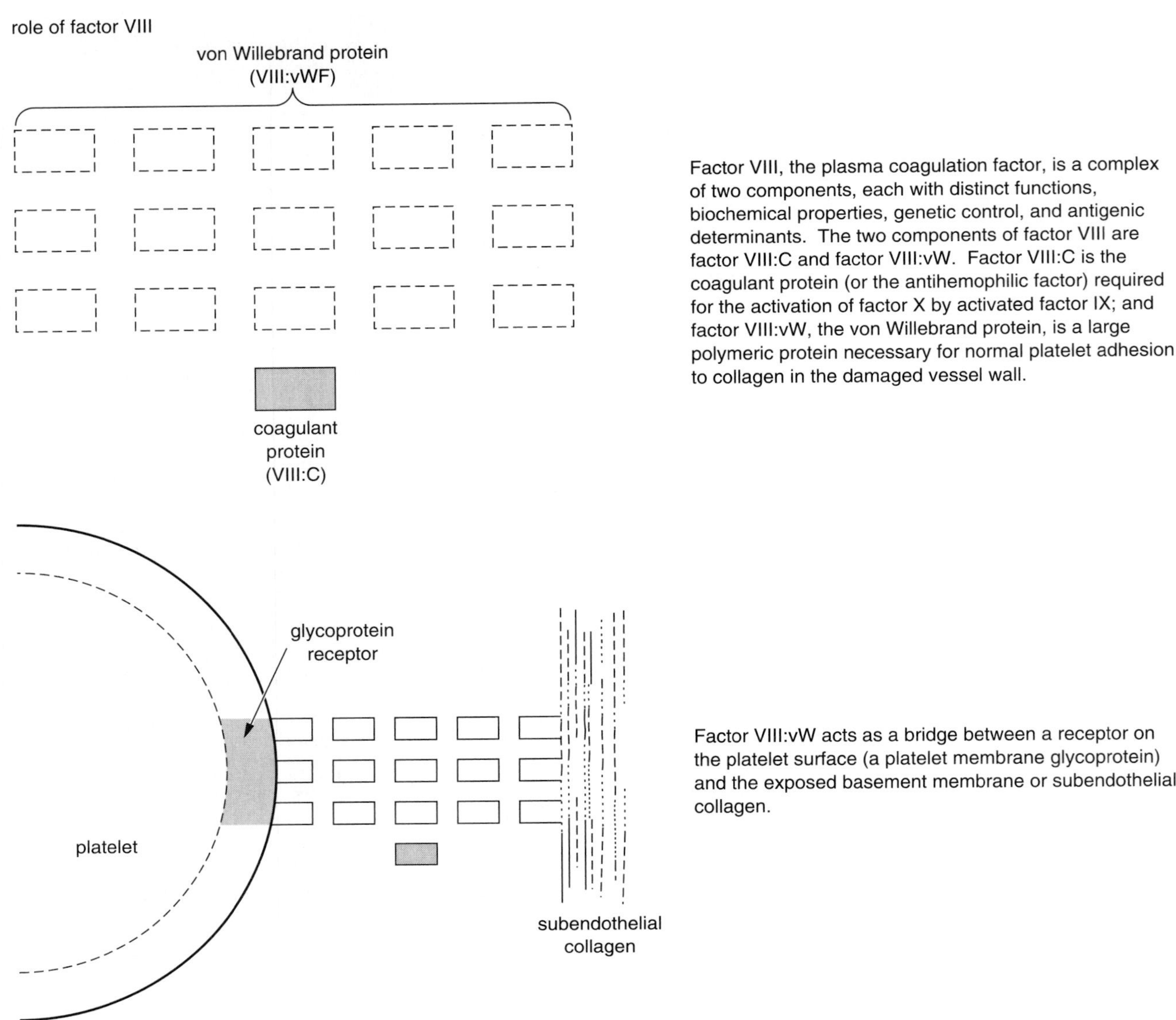


**figure 29.1**
The two components of factor VIII: the von Willebrand protein (VIII: vWF) and the coagulant protein (F VIII:C).

Sometime later another essential factor was discovered that stabilizes the fibrin clot and was designated factor XIII (table 29.1).

Two other clotting factors—**factor III (tissue thromboplastin)** and **factor IV (calcium)**—do not meet these criteria exactly. Tissue thromboplastin, also known as **tissue factor,** is not a blood component. It is found in most body tissues and is present in large amounts in the brain and lungs. Since tissue factor is universally present in the body, it is now preferentially known by its synonym rather than by its Roman numeral. The same is true for calcium, which is also found in all cells and fluids of the body. Some hematologists refer to these two factors as the **universal factors.**

All other coagulation factors are present mainly in the blood. Most of them occur in exceedingly small quantities. The exception is **fibrinogen** (also known as **factor I**), which occurs at levels of 200 to 400 mg/dL of blood.

With the exceptions of factors III and IV and segments of factor VIII, the coagulation factors are synthesized in the liver, as is plasminogen and other factors that inhibit coagulation. In severe liver disease these plasma protein levels are insufficient. Hence people with liver disorders are prone to excessive bleeding.

Factor VIII is the most interesting and most complex clotting factor and is composed of two distinct portions. A large polymeric portion binds platelets to collagen fibers; this fraction is known as the **von Willebrand factor** and is designated **factor VIII:vW.** A smaller protein portion functions as a cofactor in the coagulation cascade; this fraction is designated

**Table 29.2** The Twelve Coagulation Factors Currently Recognized, with Some of Their Chemical and Biological Properties

| Clotting Factor (synonym) | Molecular Weight (daltons) | Concentration (μg/mL) | Half-life (hours) | Type of Factor |
|---|---|---|---|---|
| *Intrinsic System* | | | | |
| **1.** *Factor XII* (Hageman factor) | 80,000 | 24–45 | 40–60 | Protease |
| **2.** *Prekallikrein* (Fletcher factor) | 80,000 | 30–50 | unknown | Protease and cofactor |
| **3.** *High-molecular-weight kininogen* (Fitzgerald factor) | 150,000 | 70–100 | 5–6 days | Cofactor |
| **4.** *Factor XI* (plasma thromboplastin antecedent) | 160,000 | 2–7 | 60–70 | Protease |
| **5.** *Factor IX* (Christmas factor) | 57,000 | 3–4 | 20–24 | Protease |
| **6.** *Factor VIII* | 1–2,000,000 | | | Cofactor, protease, and substrate |
| antihemophilic factor | | 5–10 | 10–16 | |
| von Willebrand factor | | 16 | 24–40 | |
| *Extrinsic System* | | | | |
| **7.** *Factor VII* (proconvertin) | 55,000 | 0.5–1.0 | 1–5 | Protease |
| *Tissue factor* (tissue thromboplastin) | 45,000 | 0 | | Cofactor |
| *Common Pathway* | | | | |
| **8.** *Factor X* (Stuart-Prower factor) | 59,000 | 3–10 | 24–65 | Protease |
| **9.** *Factor V* (proaccelerin) | 330,000 | 4–14 | 15–25 | Cofactor |
| **10.** *Prothrombin* (factor II) | 70,000 | 80–150 | 70–100 | Protease |
| **11.** *Fibrinogen* (factor I) | 340,000 | 2,000–4,000 | 70–120 | Substrate |
| **12.** *Factor XIII* (fibrin stabilizing factor) | 300,000 | 10 | 100–175 | Transaminase |

**factor VIII:C** (figure 29.1). The factor VIII:C portion of the molecule is synthesized in the liver, and the factor VIII:vW is synthesized by megakaryocytes and epithelial cells.

Since the original selection of the 12 major coagulation factors, several new factors have been identified that are intrinsically associated with the coagulation process. The two most important are (1) **prekallikrein** (also known as the **Fletcher factor**) and (2) **high-molecular-weight kininogen (HMWK),** which is also known as the **Fitzgerald factor.** Prekallikrein and HMWK have now replaced the original factors III and IV, so that the total number of plasma coagulation factors remains at 12. However, these two new factors have never received a Roman numeral designation (table 29.2).

## PROPERTIES OF COAGULATION FACTORS

The coagulation factors can be divided into three different groups depending on their biochemical and physiological properties. These groups are the fibrinogen group, the prothrombin group, and the contact group (table 29.3).

### Fibrinogen Group (Factors I, V, VIII, XIII)

The fibrinogen coagulation factors are large molecules that are consumed during the coagulation process. They are present in the plasma but absent in the serum. Hence **serum** may be defined as plasma minus the consumable clotting factors. This fibrinogen group

**Table 29.3** The Three Groups of Coagulation Factors and Some of Their Basic Properties*

| Groups* | Coagulation Factors | Basic Properties |
|---|---|---|
| Contact group | XII, XI, prekallikrein, HMW kininogen | Require contact with a surface for activation |
| Prothrombin group | II, VII, IX, X | Require vitamin K for synthesis; absorbed from plasma by $BaSO_4$ |
| Fibrinogen group | I, V, VIII, XIII | Large molecules; absent from serum |

*The coagulation factors can be divided into three groups based on their physiologic properties and biochemical functions in the coagulation cascade.

of coagulation factors is also susceptible to denaturation (especially factors V and VIII). Hence their activities are often reduced in stored plasma. Further, the fibrinogen group of factors is not dependent on vitamin K for synthesis, nor are these factors adsorbed by barium sulfate. These factors tend to increase during inflammation as well as during pregnancy and are also increased in women using the contraceptive pill.

### Prothrombin Group (Factors II, VII, IX, X)

These coagulation factors are dependent on vitamin K for their synthesis in the liver. Vitamin K is a fat-soluble compound absorbed only in the gut in the presence of bile salts. Vitamin K is present in some vegetable oils and leafy plants and is also synthesized in the gastrointestinal tract by various indigenous bacteria. These factors are decreased in patients who have difficulty with gastrointestinal absorption (e.g., sprue, biliary obstruction, or pancreatic disease).

The production of the prothrombin group of coagulation factors in the liver is inhibited by coumarin drugs. The coumarins are used as anticoagulants as they interfere with the synthesis of certain coagulation factors and thus interfere with the process of coagulation.

The prothrombin factors are not consumed during coagulation, are readily adsorbed from the plasma by barium sulfate ($BaSO_4$), and are stable compounds that are well preserved in stored plasma. They have molecular weights between 50,000 and 100,000 daltons and require calcium as a cofactor for binding to phospholipid surfaces.

### Contact Group (Factors XI, XII, HMWK, Prekallikrein)

The contact coagulation factors are necessary in the initial phases of the coagulation process. They require contact with a negatively charged surface for activation. The contact group is fairly stable, is not consumed during coagulation, is not adsorbed by $BaSO_4$, and is not dependent on vitamin K for synthesis. These factors are also closely related to fibrinolytic, kinin, and complement systems.

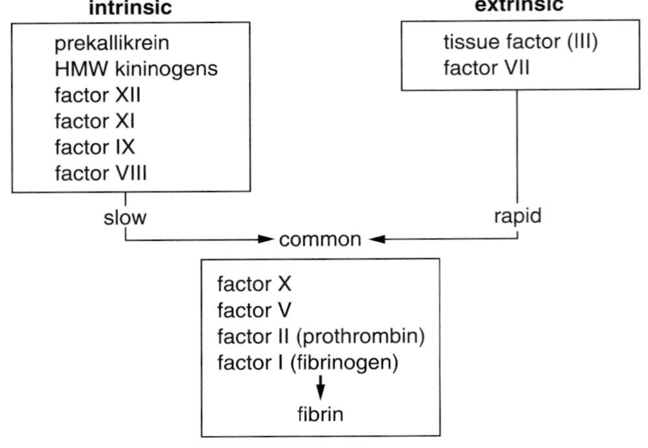


**figure 29.2**

The three components of the coagulation cascade: the intrinsic pathway, the extrinsic pathway, and the common pathway.

## COAGULATION CASCADE

The coagulation cascade is traditionally divided into three interacting pathways: (1) the intrinsic pathway, (2) the extrinsic pathway, and (3) the common pathway (figure 29.2). Each pathway involves reactions with a specific group of factors. Both the intrinsic and extrinsic systems eventually activate the common pathway. (All three pathways influence one another; division of the coagulation cascade into extrinsic and extrinsic systems is theoretical.) The clotting process that occurs *in vitro* in a test tube in the absence of a damaged blood vessel involves factors belonging to the intrinsic pathway. Factors belonging to the extrinsic pathway are activated as a result of damage to endothelial cells lining the blood vessel walls. However in *in vivo* situations, both pathways are activated simultaneously, and products of the common pathway may amplify both the intrinsic and extrinsic systems (figure 29.3).

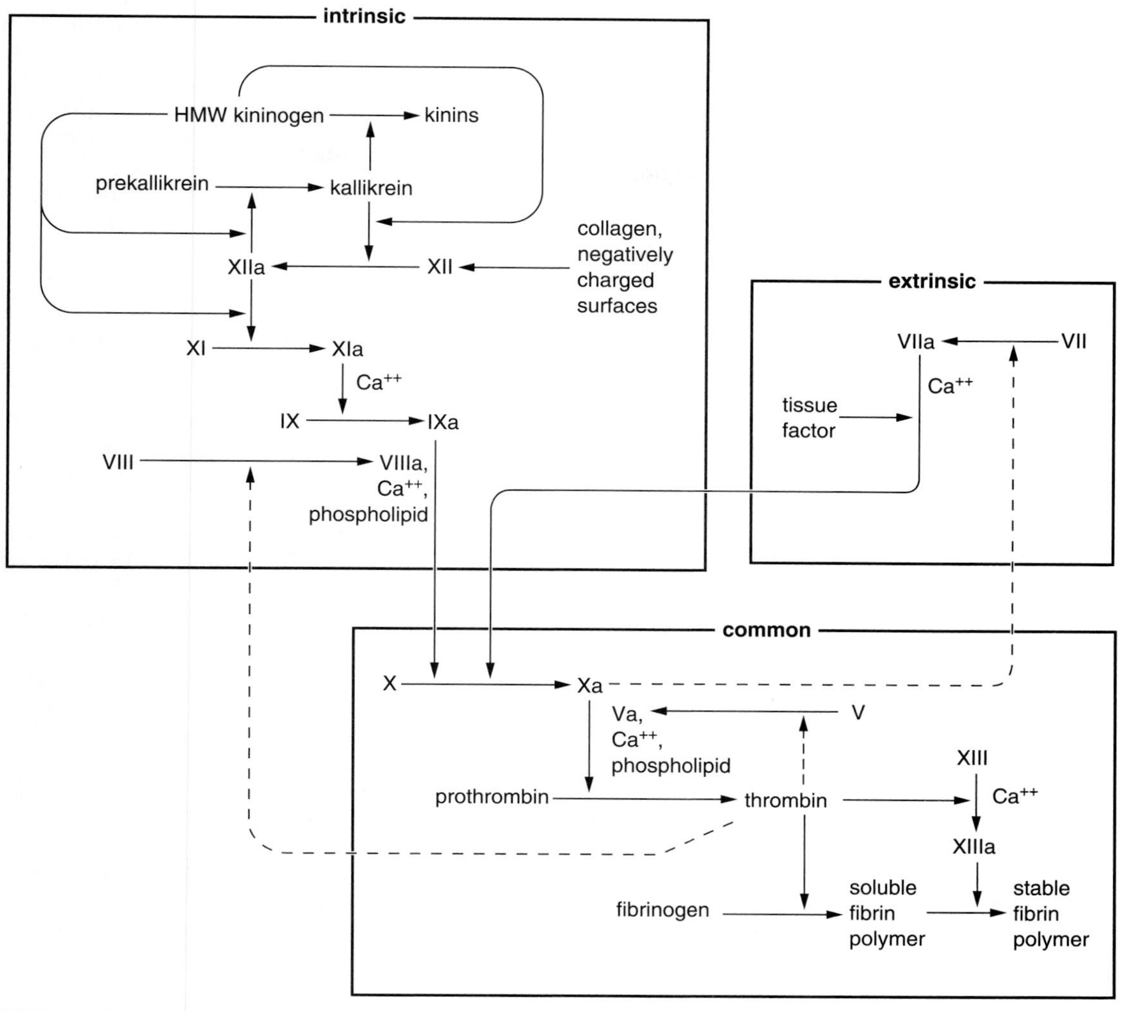


**figure 29.3**
The coagulation cascade. The factors associated with each of the three coagulation pathways.

## Intrinsic Pathway

The components of the intrinsic pathway are present in the bloodstream, hence the name **intrinsic.** Although the intrinsic pathway requires more time to become activated than the extrinsic pathway, it is generally thought to be the more important of the two.

For the intrinsic pathway to be activated, the blood must have direct contact with a foreign surface, such as a glass test tube wall or a damaged blood vessel wall. If blood is drawn into a glass test tube, it will clot within 10 minutes. If, on the other hand, it is drawn into a nonwettable plastic test tube, it will remain fluid for a period of time (frequently for more than 1 hour) (figure 29.4).

A glass surface will activate factor XII, hence it is also known as the **glass factor,** or the **contact factor.** Activation also occurs when factor XII is exposed to collagen beneath the damaged endothelium. There are four contact factors: factor XII, factor XI, prekallikrein, and HMWK. Factors XII, XI, and prekallikrein act as enzymes; HMWK is the cofactor necessary for their activation.

Initial exposure of factor XII to a negatively charged surface (glass, koalin, bacterial cell walls, exposed collagen) may result in a weak nonproteolytic activation of factor XII—that is, the molecule is not cleaved but is partially activated. Factor XII then reacts weakly with factor XI, prekallikrein, and plasminogen, which are then converted into activated factor XI (= factor XIa), kallikrein, and plasmin (figure 29.5). The products of these reactions, particularly kallikrein, feed back to factor XII and cause enzymatic cleavage, thus producing activated factor XII (= factor XIIa). Factor XIIa is more reactive than weakly activated factor XII and is better able to convert factor XI into factor XIa, prekallikrein into kallikrein, and plasminogen into plasmin. These three activated enzymes activate even more factor XII, thus amplifying the process. Kallikrein also converts kininogen into bradykinin, which is an important compound in inflammation.

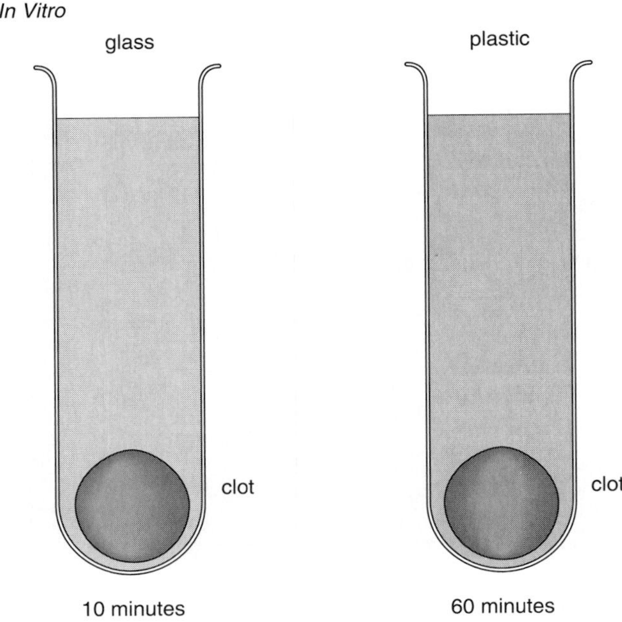


**figure 29.4**
Comparison of the blood clotting times in glass and plastic test tubes.

Plasmin, as previously mentioned, is important in fibrinolysis. Both plasmin and kallikrein are also capable of activating complement.

Factor XII is a single polypeptide chain cleaved by kallikrein (and to a lesser extent also by plasmin and factor XIa) into two polypeptide chains joined by a disulfide bridge (figure 29.6). Activated factor XII is known as factor XIIa, which can be further degraded by the same three enzymes into **factor XII fragments.** These factor XII fragments are powerful compounds that can convert prekallikrein to activated kallikrein and thus are able to amplify the process of factor XII activation.

Inactive factor XI consists of two polypeptide chains joined by a disulfide bridge (figure 29.7). Activated factor XII (factor XIIa) cleaves each factor XI polypeptide chain into two fragments, but each fragment remains bound to the other by an S-S bond. Hence activated factor XI (factor XIa) consists of four polypeptide chains. The smaller of the fragments (the light chains) contains the active enzyme sites. The function of factor XIa is to activate factor IX.

Factor IX is a single polypeptide chain containing a disulfide bond (figure 29.8). Activation of factor IX is a two-step process involving two cleaving activities. First is an intermediate step in which factor IX is degraded by factor XIa into two polypeptide chains joined by an S-S bond. The next step involves a further cleaving of the heavy chain and the removal of the intermediate piece. The remaining two polypeptide fragments are still joined by the disulfide bridge (figure 29.8). To activate factor IX, calcium is required as a cofactor. Calcium assists in binding factor IX to the phospholipid surface of the platelet (figure 29.9). Factor IX can also be activated via the activated components of the extrinsic pathway—that is, the complex of factor VIIa with tissue factor (factor III) and calcium. This complex can act directly on factor IX and in the coagulation amplification process (figure 29.10). The main function of factor IXa is to activate factor X, which is the first factor activated in the common pathway. The activation of factor X is a complex process involving at least three other factors. As soon as factor IX is activated, it forms a complex with factor VIII (which is not an enzyme, but a cofactor) and calcium on the platelet phospholipid surface to activate factor X. Thus platelet phospholipid is often referred to as **platelet factor 3 (PF3).**

As previously mentioned, factor VIII consists of two portions: factor VIII:C, which is the unit associated with the coagulation process, and factor VIII:vW, which anchors the platelet to the exposed basement membrane. The coagulant portion of this complex molecule is also known as the **antihemolytic factor A,** which serves as a cofactor in the activation of factor X. Although the precise biochemical mechanisms in factor X activation are not precisely known, it is thought that factor VIII:C adheres first to the platelet membrane followed by the activating factor IX (factor IXa), which then binds to factor X and in doing so activates factor X to factor Xa (figure 29.9). Activated factor II (factor IIa = thrombin) may increase the activity of factor VIII:C, thus greatly accelerating the binding of factor IXa and the activation of factor X (see figure 29.3).

### Extrinsic Pathway

The end result of both the intrinsic and the extrinsic pathways is to activate factor X of the common pathway. The name **extrinsic** is derived from the fact that activation of this pathway requires a factor not normally present in the blood. It is present, however, in most cells of the body. This factor is known as the **tissue factor** (or **tissue thromboplastin,** or **factor III**) and usually is released only upon cell injury. When a vessel is injured, tissue factor is released into the bloodstream where it binds to activated factor VII (factor VIIa) and calcium. This complex can now activate factor X (figure 29.10).

Factor VII is a single chain glycoprotein with a molecular weight of about 55,000 daltons. Specifically how factor VII is activated is unknown. Activated factor X (factor Xa) may be the principal agent in the activation of factor VII, although factors XIIa and XIa are also capable of activating factor VII. Whether factor VII needs to be cleaved to become activated is unknown; however it is likely as most other coagulation factors share similar cleavage mechanisms.

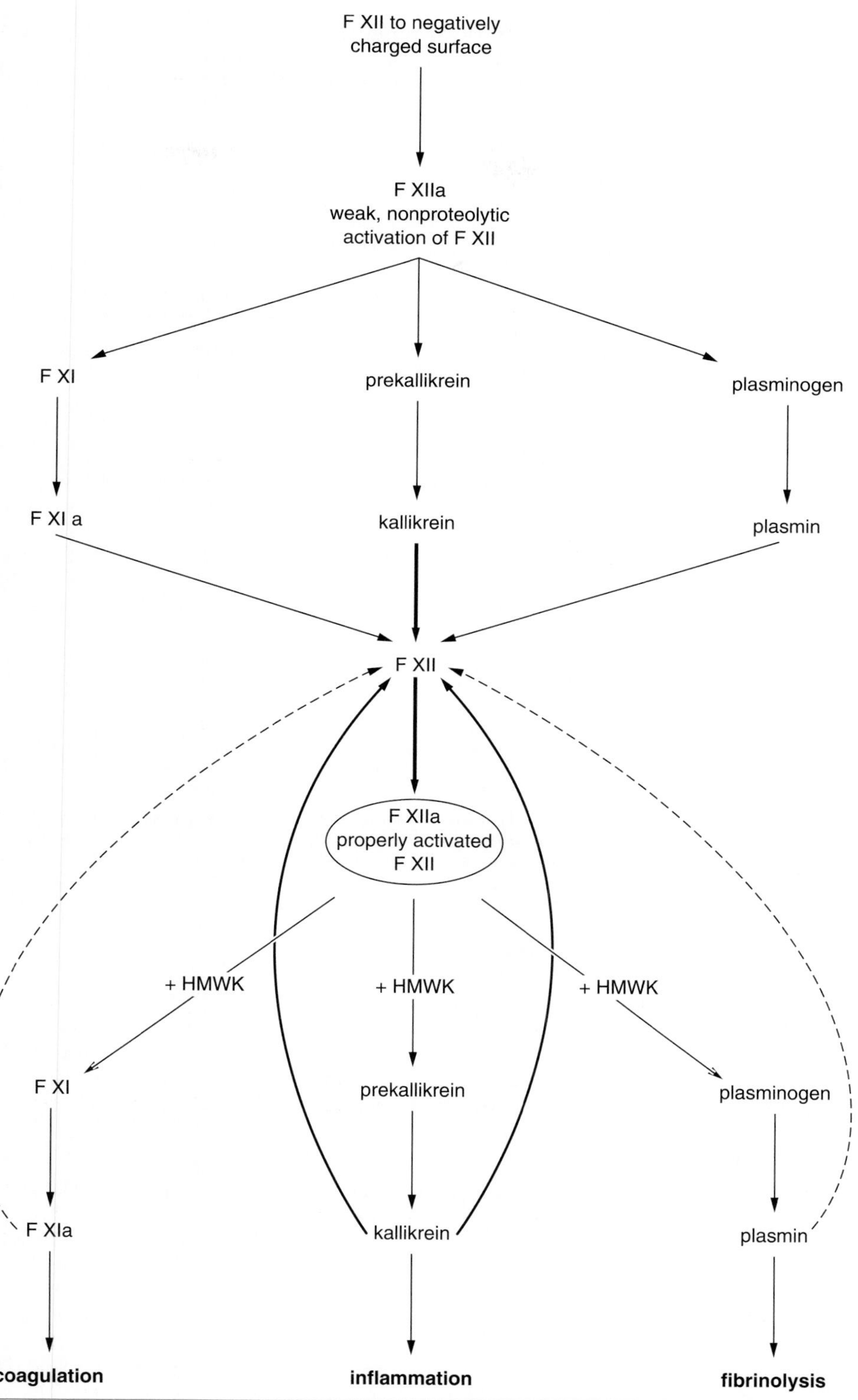


figure 29.5

The activation of factor XII.

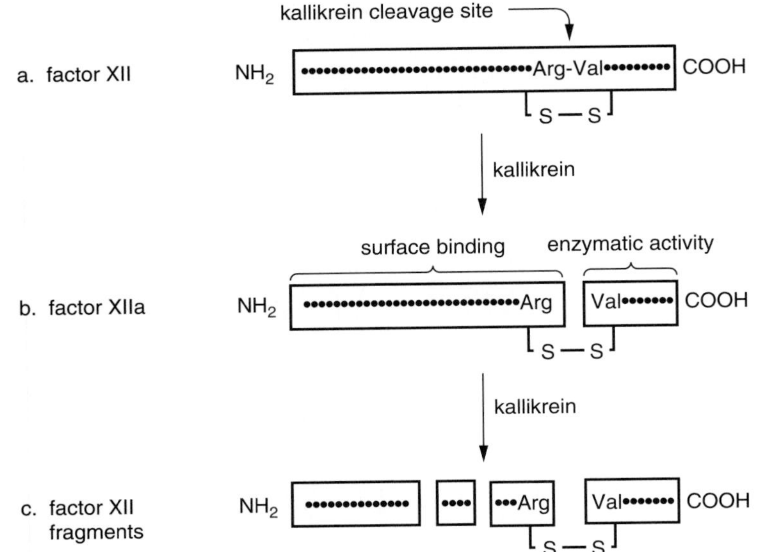


**figure 29.6**

The various forms of factor XII: (*a*) inactive (XII), (*b*) active (XIIa), and (*c*) factor XII fragments.

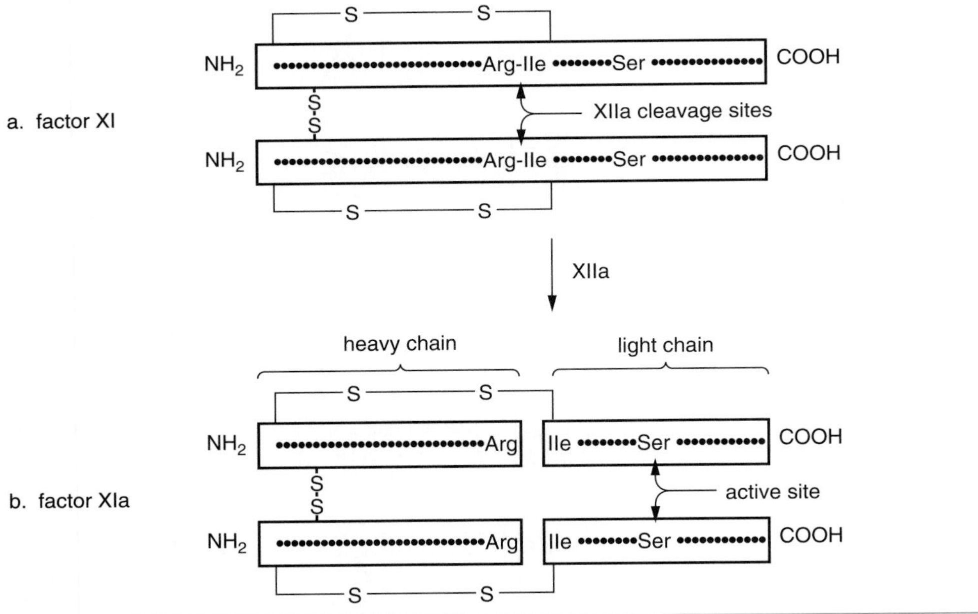


**figure 29.7**

The difference between (*a*) inactive (XI) and (*b*) active factor (XIa).

## Common Pathway

The common pathway consists of five different steps: (1) activation of factor X, (2) conversion of prothrombin to thrombin, (3) cleavage of fibrinogen to fibrin, (4) polymerization of fibrin, and (5) stabilization of fibrin polymers by factor XIII.

### *Activation of Factor X*

Factor X is composed of two polypeptide chains joined by a disulfide bond (figure 29.11). As previously explained, factor X can be activated in two ways: (1) by a complex derived from activation of the intrinsic pathway and (2) by a complex derived from

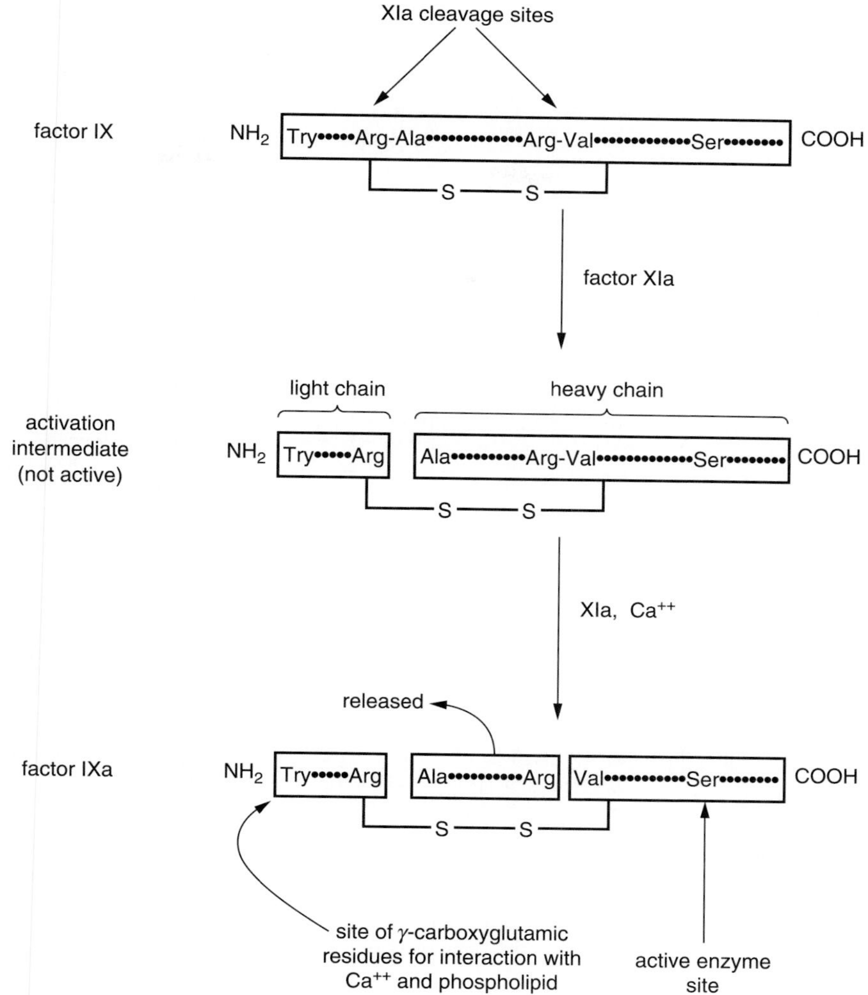


**figure 29.8**

The activation of factor IX is a two-step process involving two cleavages by factor XIa and $Ca^{2+}$ (a cofactor).

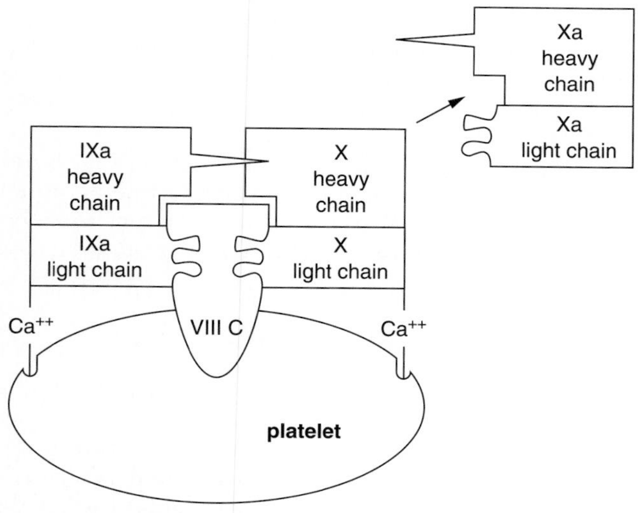


**figure 29.9**

Schematic representation of the activation of factor X in the intrinsic pathway. This process involves four different components: factor IXa, factor VIII:C, $Ca^{2+}$, and phospholipid (PF3), which is provided by the platelet membrane.

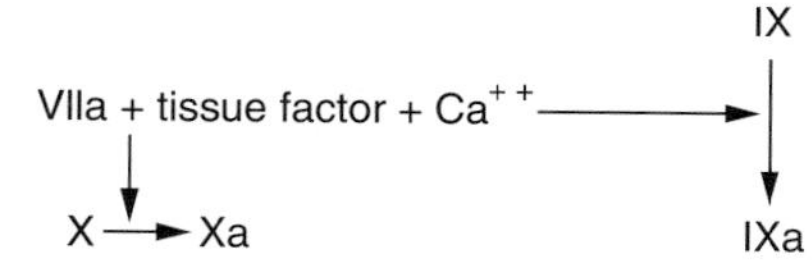


**figure 29.10**

The activation of factor X in the extrinsic pathway involves factor VIIa, tissue factor (F III), and $Ca^{2+}$. This set of factors is also capable of activating factor IX.

activation of the extrinsic pathway (figure 29.12). The previously mentioned complexes cleave a portion of the heavier of the two chains to form activated factor X (factor Xa). The two chains remain joined by an S-S bridge. The active site of the molecule is present on the remaining portion of the heavy chain, which is known as the alpha-form of factor Xa because alpha-factor Xa is often further cleaved by autocatalysis to an even smaller active enzyme known as beta-factor Xa (figure 29.11). Both alpha-factor Xa and beta-factor Xa seem to have similar precoagulant activity.

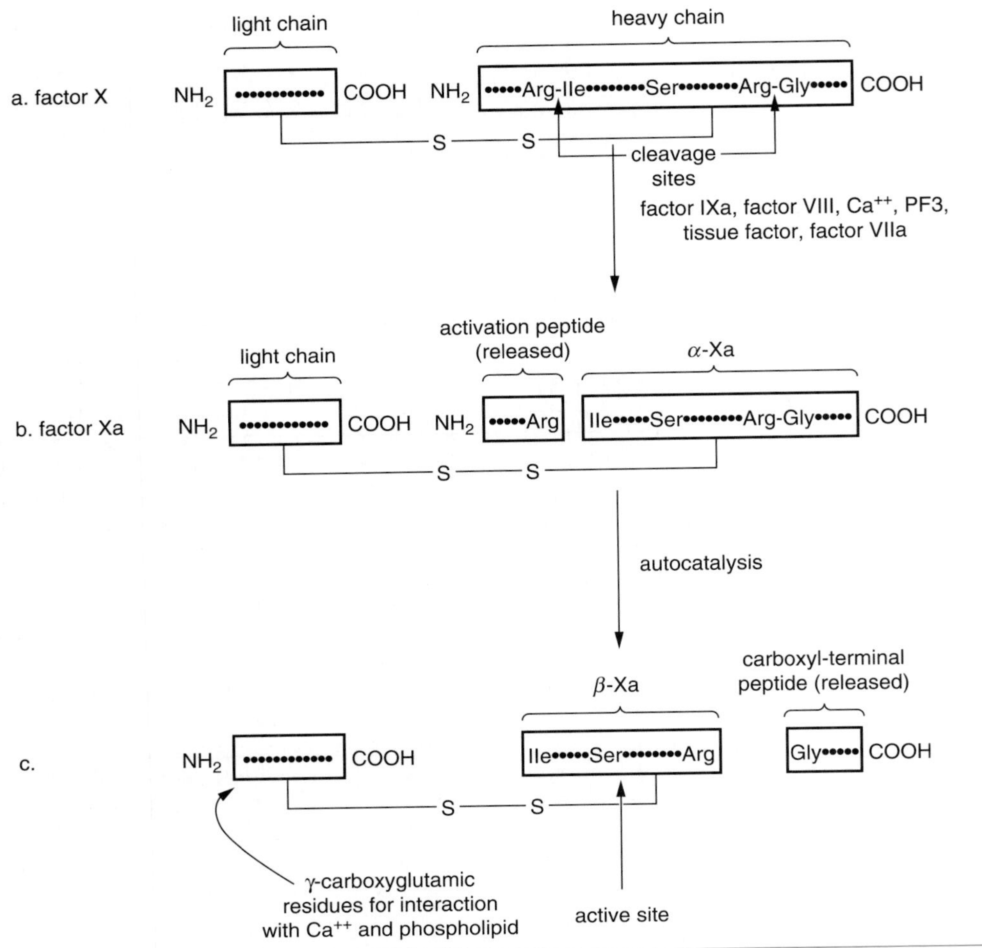


**figure 29.11**

The difference between (*a*) inactive and (*b*),(*c*) active factor X. Note that factor Xa can occur in two forms: α-Xa and β-Xa. Both have similar anticoagulant activity.

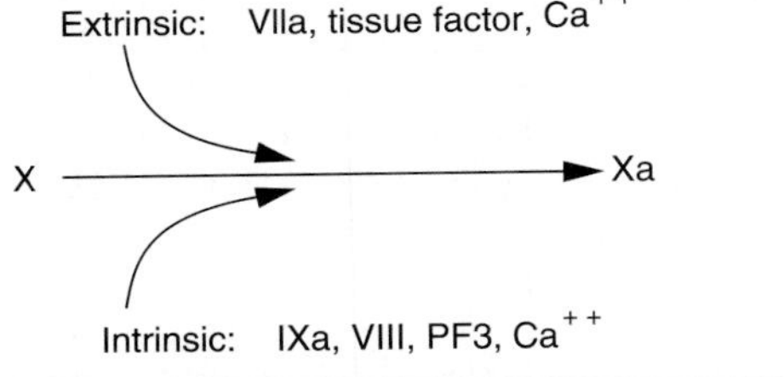


**figure 29.12**

Summary of the activation of factor X by the intrinsic and extrinsic pathways.

Activated factor X (factor Xa) can function only when it forms a complex with factor V and calcium (factor IV) (both cofactors) on a phospholipid surface (PF3). This complex of factor Xa, factor V, $Ca^{2+}$, and PF3 is called the **prothrombinase complex** because it activates prothrombin (factor II) to thrombin (factor IIa). This complex is similar to the factor IXa/factor VIII/$Ca^{2+}$/PF3 complex in that both factors V and VIII are markedly enhanced by activated thrombin (factor IIa), and factors Xa and IXa are structurally similar and belong to the prothrombin group of factors.

Factor V is a single polypeptide chain that binds to the platelet phospholipid surface so that activated factor X (factor Xa) can bind and interact with the substrate prothrombin (factor II) to produce factor IIa (figure 29.13).

### *Conversion of Prothrombin to Thrombin*

Prothrombin is also a single chain polypeptide that is cleaved by factor Xa. The factor V/$Ca^{2+}$/PF3 complex consists of two portions: a fragment commonly referred to as the **prethrombin 2 portion** and a **fragment 1,2 portion.** The prethrombin 2 portion is further divided into two smaller polypeptide chains joined by a disulfide bridge. This is the potent enzyme thrombin

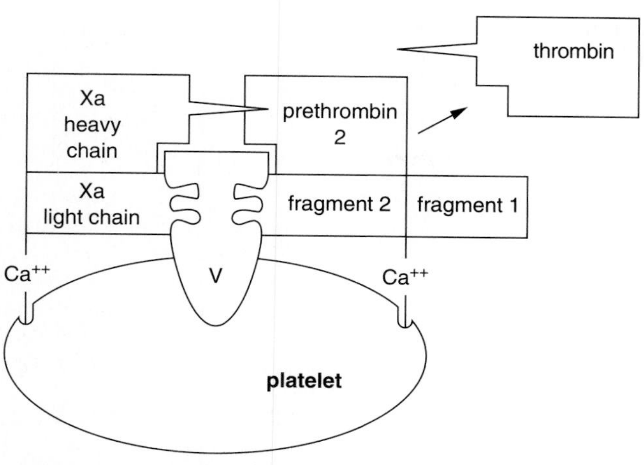


**figure 29.13**
The activation of thrombin from prothrombin involves four different factors: factor Xa, factor V, $Ca^{++}$, and PF3 which is provided by the platelet membrane.

(factor IIa), which in turn is able to cleave the fragment 1,2 portion into fragments 1 and 2, but the significance of this event is unknown (figure 29.14). The major function of activated thrombin is to cleave the protein fibrinogen into fibrin.

*Cleavage of Fibrinogen to Fibrin*

Fibrinogen is a complex protein composed of three pairs of polypeptide chains, making it possible to separate this large molecule into identical halves. Each half has one α-, one β-, and one γ-chain. The individual polypeptide chains are joined by several disulfide bridges. The six polypeptide chains form a molecular structure containing three wider areas or "nodes" (figure 29.15). The outer nodes are referred to as the **D domains,** and the central node is called the **E domain.** The D domains are composed of the carboxy terminals of the β- and γ-chains plus a short sequence of the α-chain. The remaining α-chain forms a long

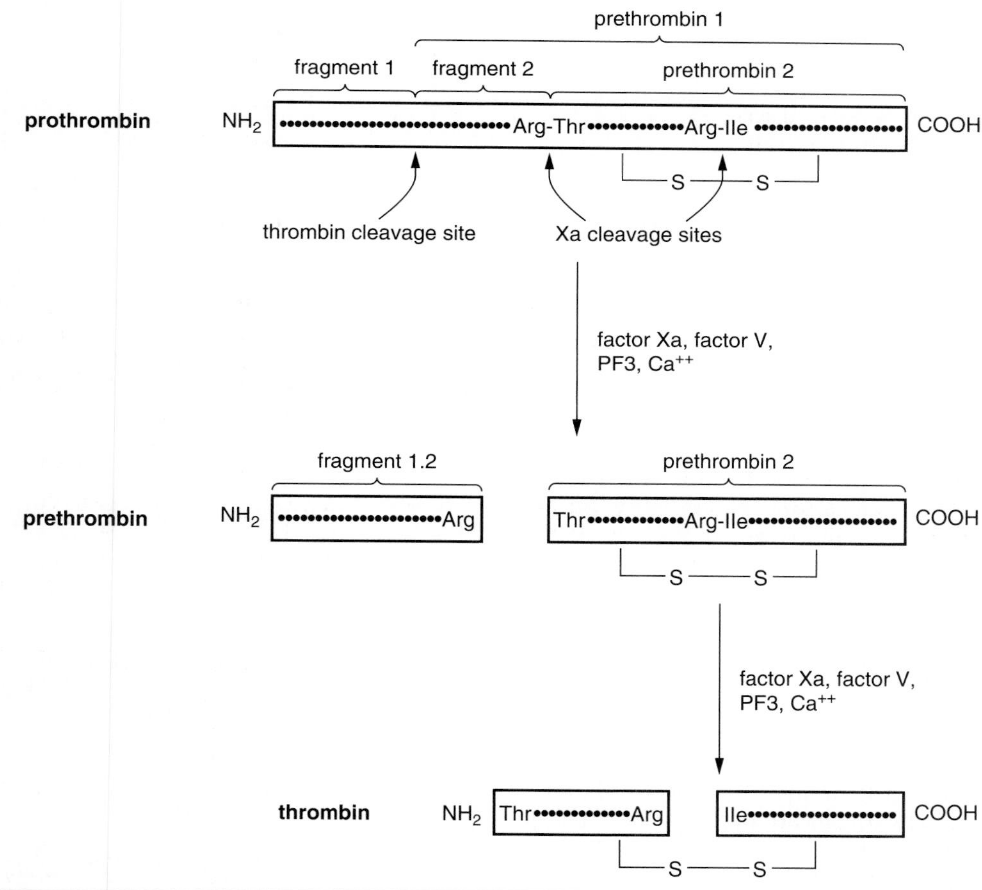


**figure 29.14**
The conversion of prothrombin to thrombin is a two-step process involving the intermediate production of prethrombin.

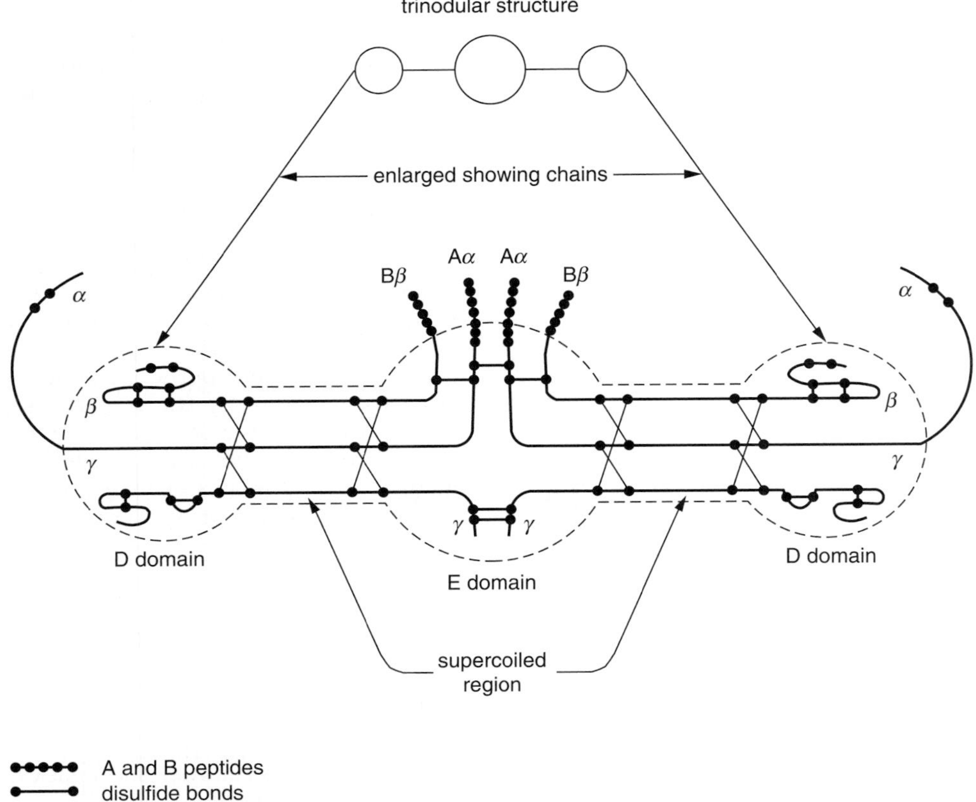


**figure 29.15**

A diagrammatic representation of the structure of fibrinogen showing its trinodal structure. Note that thrombin cleaves the A and B fragments from the α- and β-chains (near the E domain).

polar appendage at the C-terminal end. The E node is composed of the N-terminal sections of the α- and β-chains, plus a short N-terminal section of the γ-chains. The three nodes are connected by elongated supercoiled regions.

The function of activated thrombin is to cleave short portions of the N-terminal ends of the α- and β-chains. These cleaved sections are referred to as the A and B portions of the α- and β-chains, respectively. The A and B portions consist of approximately 3% of the total fibrinogen molecule. What remains of the fibrinogen molecule after they are cleaved is known as a **fibrin monomer.**

*Polymerization of Fibrin*

With the removal of the A and B portions of the fibrinogen molecule by thrombin, the negative charges around the E domain that normally cause fibrinogen molecules to repel each other, are eliminated. The net charge changes from negative to positive in the central node. The D domains, however, retain their negative charges. The D nodes of each fibrin monomer can then bind to the E domain of other fibrin monomers, thus resulting in longitudinal and lateral growth of fibrin. This process is known as **fibrin polymerization** (figure 29.16). Although the

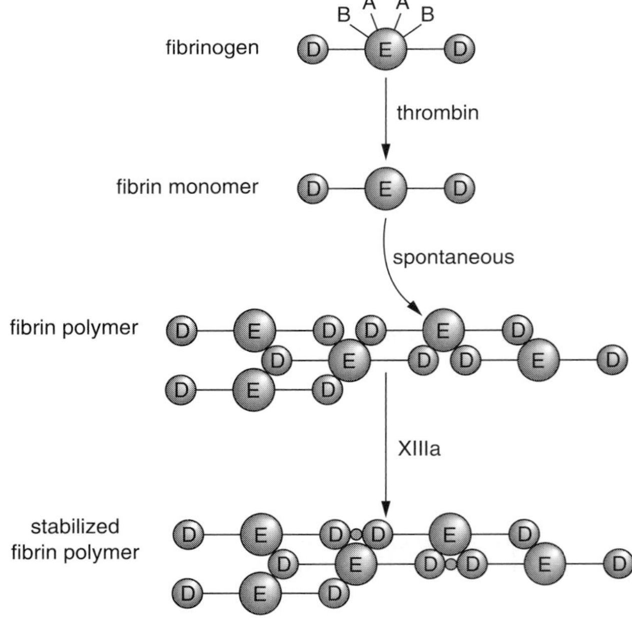


**figure 29.16**

The formation of fibrin polymers. Note that the positively charged E domains attach to the negatively charged D domains of other fibrin monomers.

charge attractions are sufficient for polymerization of fibrin, the structural configuration of the nodes and coils also contributes to the snug fit between the various fibrin monomers.

### *Action of Factor XIII on Fibrin Polymers*

The final reaction involved in fibrin strand formation is the stabilization of fibrin polymers by factor XIII. Initially the connections made by fibrin monomers are hydrogen bonds that readily dissociate in weak acids. Therefore to strengthen the fibrin connections, factor XIII is necessary. The activation of thrombin stimulates the conversion of factor XIII to factor XIIIa. This is a **calcium-dependent transamidase** responsible for forming covalent bonds between glutamine and lysine residues on the γ-chains of one monomer with the α-chains of neighboring monomers. Thus the various monomers become firmly cross-linked, resulting in fibrin strands of greater mechanical strength and increased resistance to the proteolytic action of plasmin. Hence factor XIII is known as the **fibrin stabilizing factor (FSF).**

Factor XIII is produced in both the liver and the megakaryocytes and is responsible for cross-linking another protein, called **fibronectin,** between the fibrin and the exposed collagen. Fibronectin is synthesized by endothelial cells and is found in both plasma and platelets. It adheres the fibrin clot to the exposed vessel surface. Fibronectin also assists in wound healing and tissue repair by supporting new cell growth.

## AMPLIFICATION OF HEMOSTASIS

Primary and secondary hemostasis resulting in the formation of a strong hemostatic plug is a rapid series of events. To form such a plug in a short period of time requires activation of clotting factors in a steplike fashion together with a millionfold amplification of the various steps to generate a maximum number of activated coagulation factors.

Many factors contribute to amplification of the coagulation cascade including kallikrein, which activates large quantities of factor XII; thrombin, which activates factors V and VIII (both cofactors); factor Xa, which activates factor VII, which in turn again activates factor X.

## CONTROL OF HEMOSTASIS

Except for the maintenance of vascular integrity by platelets, hemostatic mechanisms do not operate under normal physiological conditions. They remain in protective readiness and are activated as needed.

Further, as previously mentioned, when a vascular break occurs, initiation of coagulation must remain localized to the area of injury, as activated clotting factors that leak into the general circulation may cause severe clotting problems elsewhere in the body. Hence the activated coagulation factors must remain contained in a localized area. This containment is achieved in several ways. First, normal, healthy, noninjured endothelium does not promote coagulation. Second, many necessary clotting factors are locally concentrated by virtue of their selective adsorption to the platelet phospholipid surface and the selective accumulation of platelets at the site of injury. Third, circulating blood contains biochemical inhibitors that prevent amplification of the coagulation cascade in locations away from the hemostatic plug. The major plasma coagulation inhibitors are antithrombin, $\alpha_2$-macroglobulin, $\alpha_1$-antitrypsin, C1 inactivator, protein C, and protein S. Plasma antithrombin is especially important because it binds to activated thrombin that have escaped from the site of injury. Fourth, macrophages in the blood and tissues have an affinity for activated clotting factors and readily remove them from body fluids. Fifth, coagulation also automatically initiates fibrinolysis, ensuring that any clots formed away from the site of injury will be attacked and degraded almost immediately.

However, despite the presence of these inhibitory mechanisms, activated clotting factors occasionally leak from the site(s) of injury and form clots at other sites in the body. This hazard is present during major surgical operations and may cause clots in the brain **(strokes),** the lungs **(embolisms),** the heart **(infarcts),** and other organs **(thrombosis).**

## PROCESS OF FIBRINOLYSIS

Blood contains proteolytic enzymes known as the **fibrinolytic system,** which digest the fibrin clot and reestablishes circulation after the tear in the circulatory system has been repaired. Clot formation is immediately followed by repair of the blood vessel wall with production of new endothelial cells and formation of new basement membrane by activated fibroblasts. This is accomplished by **growth factors** produced by the activated platelets (platelet-derived growth factor) and by fibroblast growth factors produced by various cells, including macrophages and nerve cells.

The major protein of the fibrinolytic system is **plasminogen.** This compound is also known as fibrinolysin and is present in an inactive form in plasma in concentrations of 10 to 20 mg/dL. Plasminogen is present as an $\alpha_2$-globulin and has a molecular weight of about 90,000 daltons. Similar to other coagulation factors, plasminogen is also synthesized in the liver. Once formed, a clot can adsorb large amounts of plasminogen. This inactive compound can then be cleaved into an active enzyme called **plasmin.** Plasmin now interacts with fibrin and degrades large fibrin polymers into small ineffective fibrin fragments. These small fragments are known as **fibrin degradation products (FDP)** or **fibrin-split products (FSP)** (figure 29.17). FSPs are rapidly cleared from the circulation by the liver. Plasmin is also capable of digesting and fragmenting coagulation factor XII. These factor XII fragments can activate more plasminogen to plasmin, thus further amplifying the fibrinolytic

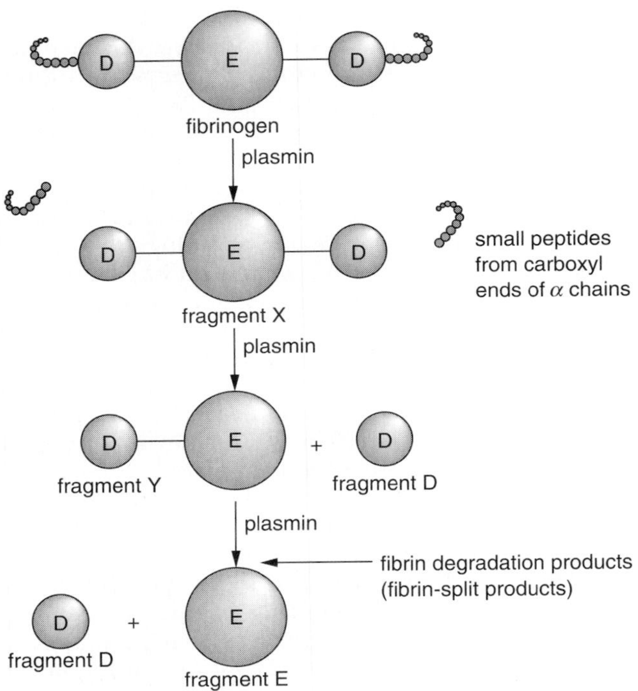


**figure 29.17**

The degradation of fibrinogen by plasmin. This process occurs in several steps. First, the small peptides from the carboxyl ends of the alpha chains are cleaved off. Then the D domains are cleaved off one at a time. This produces the final fibrin degradation products (FDPs).

system. They can also activate prekallikrein to kallikrein, which activates kininogen into kinins such as bradykinin, which is a powerful inflammatory agent. Factor XII fragments have less procoagulant activity than normal activated factor VII (i.e., factor XIIa). Further, plasmin also digests factors V, VIII, and XIII, all of which belong to the fibrinogen group of factors, which are consumed during the coagulation process. Finally, plasmin can activate complement factors. Activated complement factors increase inflammation through reactions associated with mast cells and basophils, which respond by releasing histamine.

Thus plasmin must be limited in its action to the area of the clot since free plasmin may affect the various coagulation and complement factors. However this potential danger is prevented by another $\alpha_2$-plasma globulin, known as **antiplasmin,** which readily complexes with plasmin and inactivates it.

### Plasminogen Activators

The factors that activate plasminogen into plasmin may be divided into the following three major categories:

1. **Intrinsic activators** are plasminogen activators present in the blood. The most important of these is activated factor XII (factor XIIa). Hence the activation of coagulation also initiates the fibrinolytic system. Another plasma activator is kallikrein, which is also produced by factor XIIa when it reacts with prekallikrein (figure 29.5).
2. **Extrinsic activators** are located in the body tissues. The most important extrinsic activator is the so-called tissue activator, which is produced primarily by endothelial cells and is especially prevalent in the walls of the blood vessels of the heart and kidneys. Extrinsic tissue activator reacts more effectively on adsorbed plasminogen than intrinsic plasminogen activators. Several coagulation factors enhance the activity of this tissue activator, such as factor Xa, thrombin, bradykinin, protein C, and platelet-activating factor (PAF).

   Release of tissue activators from endothelial cells is promoted not only by fibrin clots but also by exercise, hypotensive shock, and certain therapeutic drugs.
3. **Exogenous activators** are substances usually found outside the physiological milieu of the body that also can activate plasminogen. Urokinase, an enzyme frequently present in urine, can activate plasminogen. Streptokinase, another enzyme produced by certain streptococcal bacteria, is a powerful activator of plasminogen. Streptokinase is routinely used as a therapeutic agent to dissolve clots that may have formed inappropriately in the body.

From the foregoing discussion, we can readily see that the fibrinolytic and coagulation systems are intimately related. In rare circumstances in which the fibrinolytic system is activated without the simultaneous activation of the coagulation system, there will be an inappropriate breakdown of inactive fibrinogen, as well as other clotting factors. This may result in severe bleeding problems.

## Further Readings


Hirsch, J., and Brian, E. A. An Introduction to Normal Hemostatic Mechanisms. Chapter 1 in *Hemostasis and Thrombosis, A Conceptual Approach,* 2nd ed. Churchill Livingstone, New York, 1983.

McKenzie, S. B. Secondary Hemostasis. Chapter 21 in *Textbook of Hematology.* Lea & Febiger, Philadelphia, 1988.

Ogston, D. The Coagulation System. Chapter 3 in *The Physiology of Hemostasis.* Harvard University Press, Cambridge, Mass., 1983.

———. The Fibrinolytic Enzyme System. Chapter 4 in *The Physiology of Hemostasis.* Harvard University Press, Cambridge, Mass., 1983.

Thompson, A. R. Coagulation. Chapter 3 in *Manual of Hemostasis and Thrombosis,* 3rd ed. F. A. Davis, Co., Philadelphia, 1983.


## Review Questions

1. Distinguish between primary and secondary hemostasis.
2. Name the 12 original coagulation factors.
3. Why were factors III and IV later deleted from this list? What are the universal factors?
4. Which 2 factors were later added to bring the number of coagulation factors back to 12?
5. Divide the currently accepted coagulation factors into three major categories. What is the name of each category? Name the factors belonging to each category. Describe the characteristics of each category.
6. Distinguish between the extrinsic, intrinsic, and common pathways of coagulation. Which factors belong to each pathway?
7. What activates the intrinsic and extrinsic pathways?
8. Describe how factor XII is normally activated and how its activation is amplified.
9. Describe the coagulation cascade of the intrinsic pathway.
10. Distinguish between the structure of factors XII and XIIa, factors XI and XIa, and factors IX and IXa. How can these three compounds be classified from the functional point of view?
11. List the four factors needed to activate factor X in the intrinsic pathway.
12. Describe the coagulation cascade of the extrinsic pathway.
13. List the four factors involved in the activation of factor X in the extrinsic coagulation cascade.
14. Describe the two components of factor VIII. What are the names and functions of each?
15. Distinguish between the structure of factors X and Xa. What is the function of factor Xa?
16. List four factors needed to change factor II to factor IIa. What are other names for factors II and IIa?
17. Describe the function of factor IIa.
18. Describe the structure of fibrinogen. Explain the difference between fibrinogen and fibrin.
19. What is the normal level of fibrinogen in the blood?
20. Describe how fibrin monomers are polymerized into long fibrin strands.
21. Explain the function of factor XIII. Provide another name for factor XIII.
22. Explain how the coagulation process is amplified.
23. List five different factors that ensure that activated clotting factors are restricted to the area of injury.
24. Describe the function of the fibrinolytic system.
25. Distinguish between plasminogen and plasmin. What is the normal level of plasminogen in the blood?
26. What activates plasminogen?
27. List the major functions of plasmin.
28. List the three classes of plasminogen activators. Give examples of each.
29. List five different factors that enhance the activity of tissue activator.
30. What important therapeutic agent is used in clinical medicine to dissolve intravascular clots?

# Chapter Thirty

## *Disorders of Hemostasis*

Two major categories of disorders may occur as a result of abnormal hemostasis. Defects in the hemostatic mechanism may lead to prolonged bleeding conditions **(bleeding disorders)** or inappropriate clot formations **(intravascular thrombosis).** This chapter discusses the nature, causes, and treatment for various disorders of hemostasis.

## BLEEDING DISORDERS

The three major components of the hemostatic mechanism are the **blood vessels**, the **platelets**, and the **plasma protein factors** involved in coagulation and fibrinolysis. Thus this chapter first discusses vascular disorders that cause prolonged bleeding, followed by a description of platelet disorders that result in increased bleeding. Finally, bleeding disorders associated with abnormalities in the quality and quantity of the coagulation and fibrinolytic factors are discussed.

### Vascular Disorders

The **vascular disorders** are a heterogeneous group of syndromes characterized by easy bruising and spontaneous bleeding from the blood vessels. In most instances the bleeding is rather mild and is usually limited to bleeding into the skin and the mucous membranes in the form of petechiae and ecchymoses.

Vascular disorders resulting in abnormal hemostasis may be classified into two major categories: **inherited** and **acquired.**

#### *Inherited Vascular Disorders*

Inherited vascular bleeding disorders are rare; the most common is known as **hemorrhagic telangiectasia.** This disorder is transmitted as an autosomal dominant trait. Telangiectases are abnormally dilated blood vessels that begin to appear in the skin and mucous membranes early in adult life. These abnormal dilations develop because the subendothelial connective tissue (i.e., the basement membrane) of the small blood vessels is thin and fragile. Hence these blood vessels have difficulty maintaining their shape and dilate readily when blood pressure increases, which in turn leads to easy bleeding. Frequently such individuals with hemorrhagic telangiectasia also suffer from iron deficiency anemia due to chronic blood loss.

Another even rarer inherited vascular disorder is the **Ehlers-Danlos syndrome.** This disease is associated with abnormal collagen production, which also results in a defective basement membrane structure.

#### *Acquired Vascular Disorders*

Acquired vascular disorders are more common than inherited vascular disorders and include the following: (1) simple easy bruising, (2) senile purpura, (3) steroid purpura, (4) scurvy, (5) Henoch-Schönlein syndrome.

**Simple easy bruising** This common benign condition is observed frequently in otherwise healthy women who bruise easily as the result of even the smallest trauma. This phenomenon occurs predominantly on the legs and on the trunk and is due to the fragility of the blood vessels of the skin. The cause for this phenomenon is unknown. Simple bruising is not serious and does not require any special treatment.

**Senile purpura** A common vascular disorder in older people, senile purpura is characterized by patches of purplish discoloration on the arms and hands. The purpuric dark splotches are usually large and irregular. The skin on the arms and hands of these individuals is inelastic and thin, indicating a marked atrophy of collagen. Senile purpura is not a serious condition and does not require any special treatment.

**Steroid purpura** Bruising is common in people who have been treated for prolonged periods of time with large doses of steroid drugs. The same phenomenon also occurs in people suffering from **Cushing's disease,** which may be defined as a condition in which there is an overproduction of corticosteroids. Although the exact connection between large doses of steroids and the development of purpura is unknown, excess steroids may destroy the collagen fibers in the subcutaneous and subendothelial tissues, resulting in a weakness of the skin and the small blood vessels. Thus such persons are predisposed to bleed more readily, even with minimal trauma.

**Scurvy** Scurvy may be described as a disorder associated with easy bruising and the development of petechiae and ecchymoses in the skin, especially of the legs. Scurvy is caused by a deficiency of vitamin C, which is necessary for normal collagen formation. Lack of collagen will result in weak blood vessels, and this vascular fragility will be expressed in easy bruising. Treatment for this condition involves the ingestion of large doses of vitamin C.

**Henoch-Schönlein syndrome** This condition may be a hypersensitivity reaction—that is, Henoch-Schönlein syndrome is associated with acute inflammation of the small blood vessels of the microcirculation, resulting in increased vascular permeability and easy bleeding.

Inflamed blood vessels may occur anywhere in the body and may be associated with skin purpura, abdominal pain, joint pain, or hematuria. This condition is most common in children, although it may also occur in adults. Henoch-Schönlein syndrome is usually self-limiting. Although its cause is unknown, it occurs frequently after the ingestion of certain iatrogenic drugs or as the result of a group A streptococcal infection—hence this syndrome may also be an allergic reaction.

## Platelet Disorders

The two major groups of platelet disorders involve **quantitative abnormalities** or **qualitative abnormalities.** When platelet function is impaired because of a reduction in the number of platelets in the bloodstream, the condition is called **thrombocytopenia. Thrombocytopathia** occurs when the quality of the platelets is impaired. Both categories are discussed in the following material.

### *Quantitative Abnormalities (Thrombocytopenia)*

The many different causes for thrombocytopenia can be organized into five different categories: (1) decreased platelet production, (2) increased platelet destruction, (3) abnormal platelet distribution, (4) dilutional loss, (5) and idiopathic (cause(s) unknown).

**Decreased platelet production** In general, decreased platelet production results from a selective depression of the megakaryocyte mass or from generalized bone marrow failure. The latter is the most common cause of decreased platelet production. The major reasons for general bone marrow failure are the same that produce aplastic anemia and leukocytopenia and include (1) radiation; (2) iatrogenic drugs (e.g., those used in cancer treatment); (3) toxic chemicals (e.g., benzene); (4) myelophthisis (i.e., replacement of normal hematopoietic tissue by neoplastic cells); and (5) competition for nutrients by cancer cells.

Selective megakaryocyte depression interferes with production of megakaryocyte growth factors, especially production of thrombopoietin by drugs, chemicals, or infections.

**Increased platelet destruction** A number of causes may result in premature destruction of platelets in the bloodstream. However increased destruction of platelets does not necessarily lead to thrombocytopenia, as the megakaryocyte mass in the bone marrow may be able to compensate by increasing the number of platelets. Therefore, if the percentage of platelet destruction is rather mild, the platelet count will remain within the normal range. If large-scale platelet destruction occurs over a long period of time, the megakaryocyte mass will no longer be able to compensate, and thrombocytopenia will become apparent. The major causes for increased platelet destruction are:

**1. Autoimmune activity**
The major cause of increased platelet destruction is through autoimmune activity in which the body develops antibodies against its own platelets. These antibodies bind to the platelet surface and cause activation of complement, resulting in lysis of the affected platelets. The body can develop platelet antibodies in several ways, most commonly by creating neoantigens when drugs or chemicals bind to self-antigens on the platelet surface. A second method involves binding of immune complexes to the platelet surface. An immune complex is produced when the body develops an antibody against a plasma protein to which antibody then binds. The resulting antibody-antigen complex (= immune complex) attaches to the platelet surface, and also activates complement with a similar result—that is, lysis of the platelet (figure 30.1). Such autoimmune reactions produce acute thrombocytopenia.

**2. Idiopathic**
Another cause of thrombocytopenia is also associated with activation of the immune system and is known as **chronic immune thrombocytopenia**. This disorder is commonly found in young adults, particularly young females. These patients produce antiplatelet antibodies, but the low antibody level results in fewer antibodies binding to the platelets. This prevents complement activation on a large scale, and platelet destruction results from increased phagocytosis of antibody-covered platelets by mononuclear phagocytic cells of the liver and spleen. These phagocytic cells have special receptors (Fc receptors) for these antibodies. The reason for production of the low antibody level is unknown, thus the cause is termed **idiopathic.**

**3. Increased platelet aggregation**
Thrombocytopenia is frequently associated with disseminated intravascular coagulation (DIC). A major symptom of DIC is platelet aggregation, or clumping, which results from prothrombin activation. The small clumps of platelets become trapped in the microcirculation. Hence the number of platelets in the general circulation is reduced.

**4. Mechanical destruction of platelets**
Use of heart-lung machines, kidney dialysis machines, or prosthetic heart valves tends to produce lower than normal numbers of platelets, due to the accelerated platelet destruction that results when these fragile cytoplasmic particles come into contact with hard mechanical surfaces.

**5. Septicemia**
Septicemia, or blood poisoning, is associated with the presence of pathogenic microorganisms in the blood. The toxins produced by these pathogens may also cause platelet aggregation, resulting in the clogging of the microcirculation and lower than normal numbers of platelets.

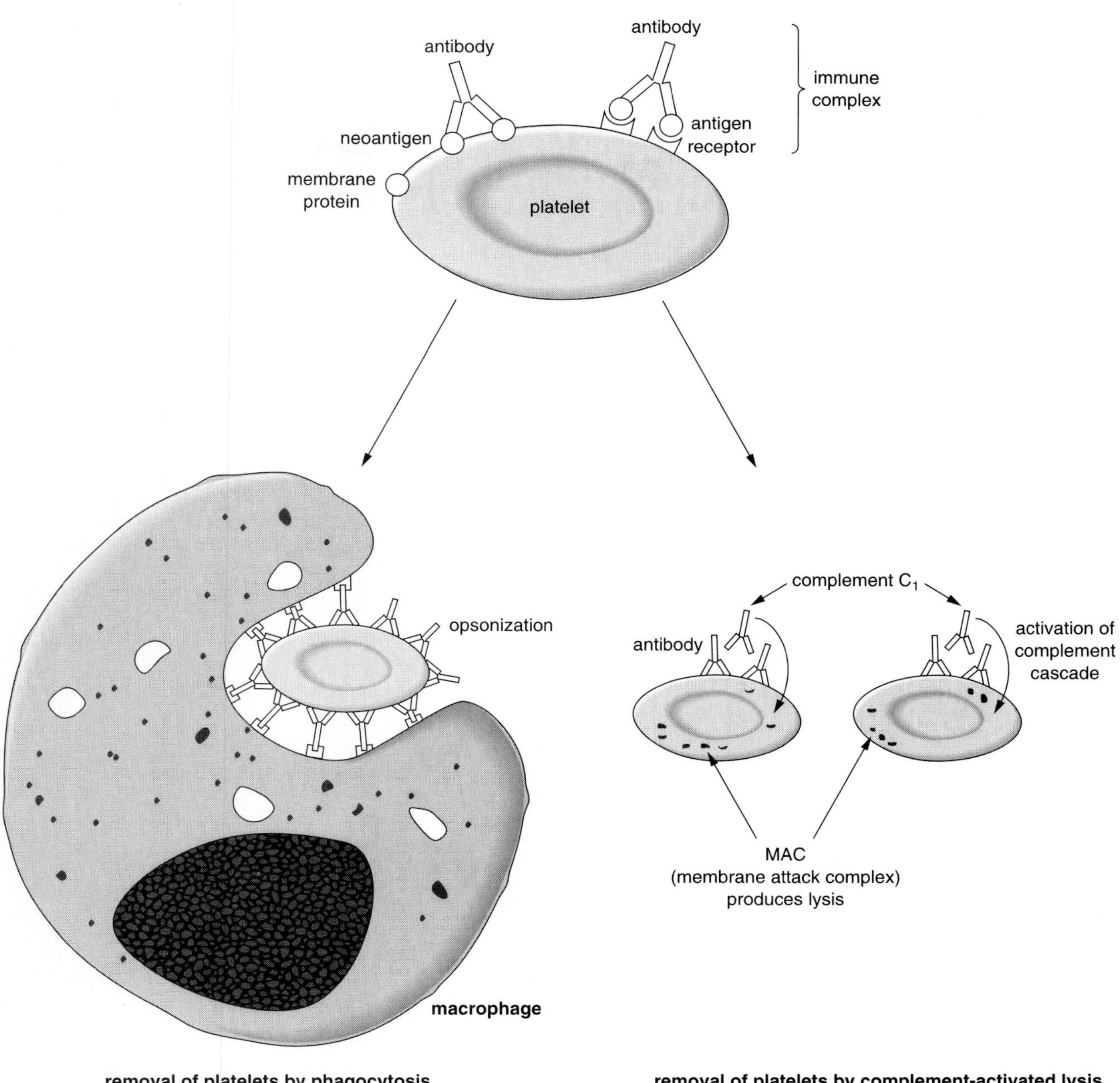


**figure 30.1**

Platelet opsonization may result in phagocytosis or complement-activated lysis of the platelets.

**Abnormal platelet distribution** Abnormal platelet distribution usually occurs as a result of splenomegaly. About one-third of all platelets in the peripheral circulation are found in the spleen. However when the spleen becomes enlarged, an increased proportion of the platelets will pool in this organ. If the bone marrow is unable to compensate for the increased splenic pool size, the percentage of platelets in the bloodstream will be reduced.

**Dilutional loss** Infused blood is usually composed of only packed red blood cells with few if any platelets. Hence the numbers of platelets in the blood becomes greatly reduced in persons who have undergone multiple blood transfusions. However this dilutional loss is usually a short-term condition, resolving when the bone marrow is able to compensate for the platelets lost to excess bleeding.

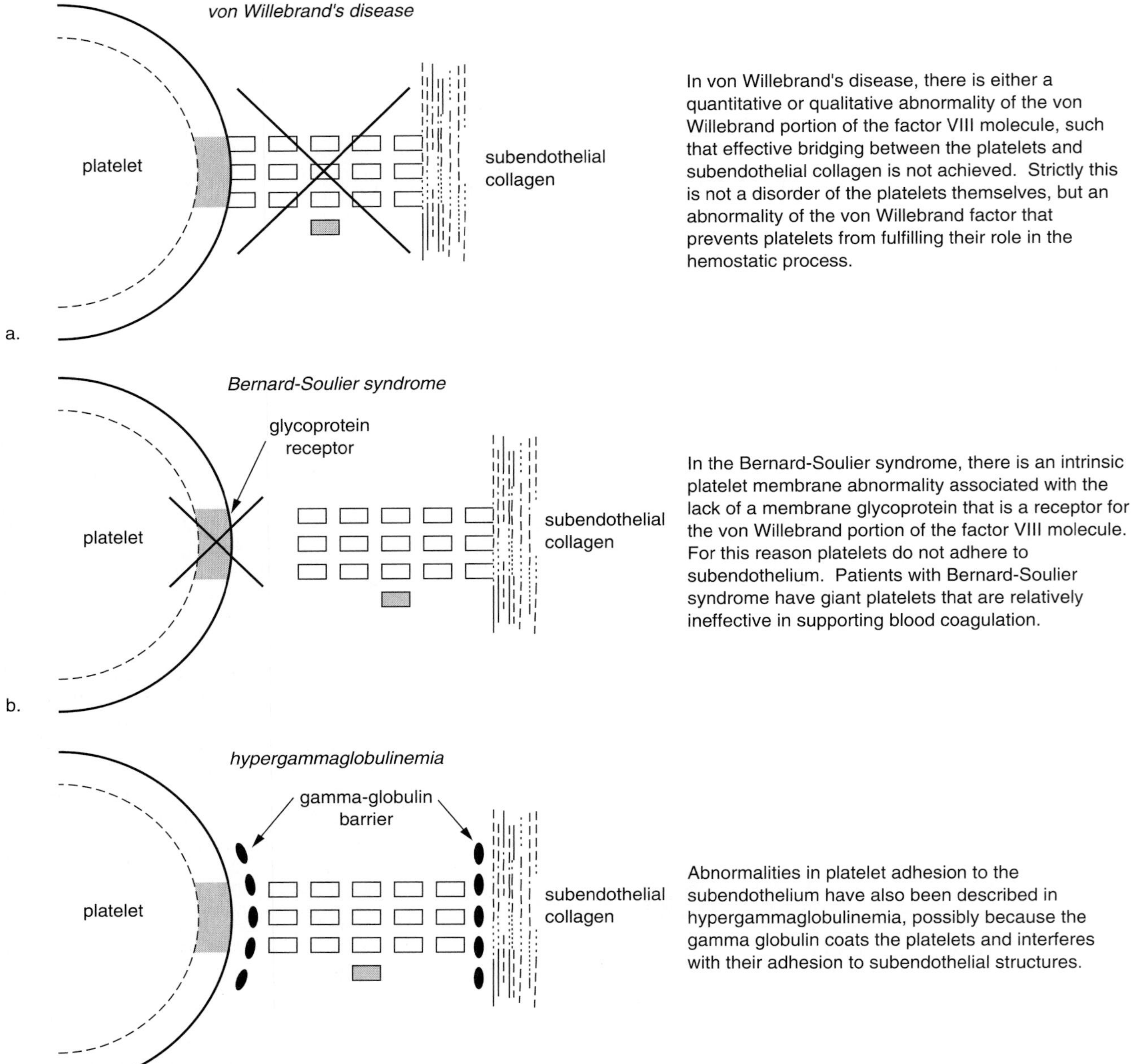


**figure 30.2**

Failure of the platelets to adhere to exposed basement membrane may be the result of (*a*) lack of von Willebrand factor (VIII:vWF); (*b*) lack of glycoprotein receptors on the platelet for the von Willebrand factor; and (*c*) excess gamma globulin coating the platelets and collagen, thus interfering with the process of adhesion.

*Qualitative Abnormalities (Thrombocytopathia)*

Qualitative abnormalities in platelets lead to platelet function disorders. The five major causes for defects in platelet function result from the failure of platelets to (1) adhere properly to the exposed basement membrane, (2) produce sufficient quantities of ADP, (3) respond to ADP, (4) produce thromboxane $A_2$ ($TXA_2$), and (5) bind to coagulation factors.

**Failure of platelets to adhere** Von Willebrand's disease, Bernard-Soulier syndrome, and hypergammaglobulinemia are three disorders that result in the inability of platelets to adhere.

**1. Von Willebrand's Disease (vWD)**
In vWD, quantitative or qualitative abnormalities of von Willebrand factor (factor VIII:vW) occur. Factor VIII:vW is necessary to bind platelets to basement membrane collagen. Although strictly speaking, vWD is not a disorder of the platelets, it is included here because it prevents platelets from binding to the vascular subendothelial layer (figure 30.2*a*).

**2. Bernard-Soulier Syndrome**
This disorder is an intrinsic platelet abnormality resulting from a lack of a platelet membrane glycoprotein that acts as a receptor for the von Willebrand factor. Patients with Bernard-Soulier syndrome have very large platelets that are nonfunctional in hemostatic plug formation (figure 30.2*b*).

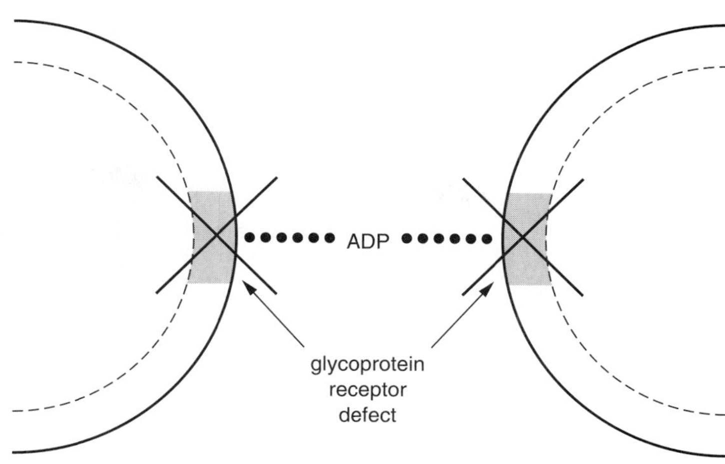


**figure 30.3**
Thrombasthenia is caused by a lack of specific glycoprotein receptor that is believed to be responsible for the interaction of ADP on the platelet surface. Patients with thrombasthenia may have a severe bleeding abnormally with bruising, petechiae, and purpura.

**3. Hypergammaglobulinemia**
In patients who produce excess immunoglobulins, antibodies coat the platelets and interfere with the adhesion of platelets to the basement membrane (figure 30.2*c*).

**Failure of platelets to produce and release ADP** Two different phenomena are involved in the failure of platelets to produce and release ADP. One condition involves platelets that are unable to produce sufficient amounts of ADP in their dense granules. This is an inherited condition. In the other condition, there may be sufficient production of ADP but an inability to release ADP from the platelets may exist. The platelets are unable to synthesize specific prostaglandins, which results in decreased concentrations of thromboxane $A_2$ ($TXA_2$), which is an important mediator molecule in the release of ADP from the platelets.

**Failure of platelets to respond to ADP** The inability of platelets to respond to ADP and thus their inability to aggregate is known as **thrombasthenia.** This rare disorder is associated with lack of a specific glycoprotein receptor on the platelet surface that is necessary to bind platelets to each other (figure 30.3). Thrombasthenia is an inherited disorder frequently associated with easy bruising and purpura.

**Failure of platelets to produce $TXA_2$** When platelets are exposed to collagen, they respond by activating the platelet surface enzyme phospholipase $A_2$, which cleaves arachidonic acid from membrane phospholipid. Arachidonic acid is then oxidized by the enzyme cyclooxygenase to the endoperoxides $PGG_2$ (prostaglandin $G_2$) and $PGH_2$ (prostaglandin $H_2$). These in turn are converted to $TXA_2$ by the enzyme thromboxane synthetase (figure 30.4*a*). $TXA_2$ stimulates the platelets to release ADP, which then acts as a chemotactic stimulus for platelet aggregation. The failure to produce $TXA_2$ may be associated with the absence of either cyclooxygenase (which is responsible for the production of endoperoxides), or thromboxane synthetase (which produces $TXA_2$). The inhibition of these enzymes is frequently related to iatrogenic drug use, especially aspirin therapy. The major effect of aspirin is its interference with cyclooxygenase activity (figure 30.4*b*).

**Failure of platelets to bind to coagulation factors** As explained in chapter 28, when platelets aggregate they change shape, exposing large numbers of receptor proteins on their surfaces. The ability of coagulation factors to bind to these receptors, however, may be impaired if (1) platelets fail to aggregate, (2) platelets fail to change shape or platelet receptors are abnormal (which occurs in Bernard-Soulier syndrome and in thrombasthenia), or (3) there is a shortage of coagulation factors.

## Disorders of Plasma Proteins Involved in Coagulation and Fibrinolysis

Two major categories of bleeding disorders caused by abnormalities associated with plasma proteins involve coagulation and fibrinolysis: inherited disorders and acquired disorders.

### *Inherited Disorders*

Inherited coagulation disorders are the result of inherited deficiencies of coagulation factors. These deficiencies may be expressed as either reduced production or abnormal synthesis of one or more coagulation factors. In the past most coagulation disorders were considered to result from the absence of one or more coagulation factors; however in most cases the body produces these coagulation factors but they are structurally abnormal and thus unable to function in the coagulation cascade.

All coagulation factors are synthesized in the liver, with the exception of factor VIII. The synthesis of factor VIII is complicated as it is made up of two active portions: factor VIII:vW and factor VIII:C. The

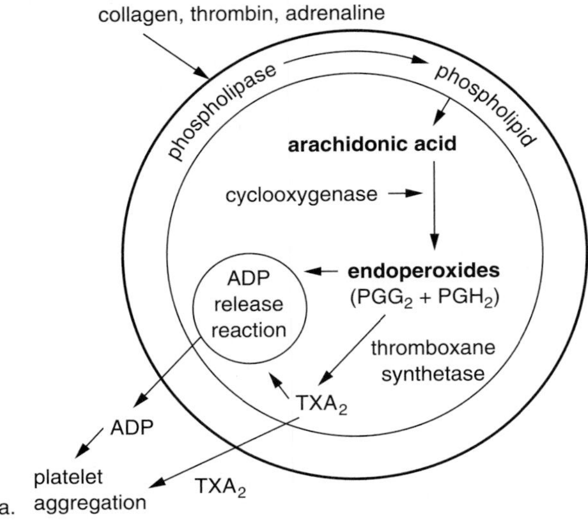


When platelets are exposed to various stimuli (including collagen, thrombin, and adrenaline), the platelet surface enzyme phospholipase $A_2$ is activated. This in turn cleaves arachidonic acid from membrane phospholipid. The arachidonic acid is then oxidized by cyclooxygenase to endoperoxides $PGG_2$ and $PGH_2$. $PGH_2$ is converted by thromboxane synthetase to thromboxane $A_2$, which stimulates platelets to release adenosine diphosphate and also to aggregate independently of ADP release.

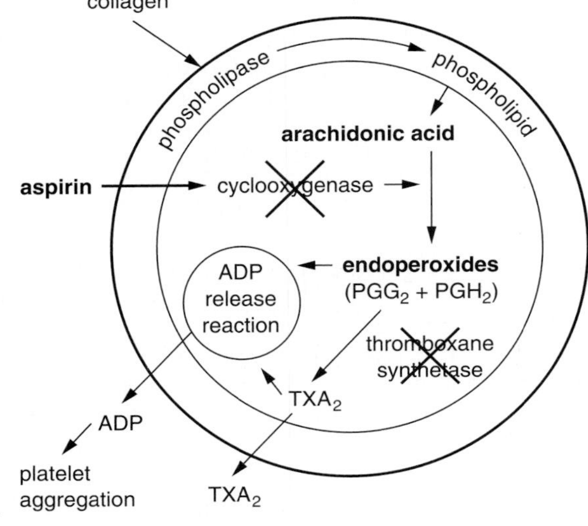


Abnormalities of the prostaglandin synthetic pathway can either be at the level of the enzyme cyclooxygenase (which converts arachidonic acid into the endoperoxides $PGG_2$ and $PGH_2$) or at the level of the enzyme thromboxane synthetase (which converts $PGH_2$ to thromboxane $A_2$). Since this is one of the mediators of ADP release, synthesis failure results in defective ADP release. This type of abnormality is most commonly seen in patients who have ingested aspirin but it is also seen with myeloproliferative disorders, following cardiac bypass surgery in which a pump oxygenator has been used, and as an inherited primary platelet function defect.

**figure 30.4**

Stimulated platelets will produce thromboxane ($TXA_2$). (*a*) $TXA_2$ induces platelet aggregation. (*b*) Aspirin and other substances that interfere with $TXA_2$ synthesis prevent platelets from aggregating.

site of factor VIII:C production is unknown, however we do know that its synthesis is under control of the X chromosome. Factor VIII:C is unstable in the plasma in the absence of factor VIII:vW, which functions as its carrier protein.

The factor VIII:vW portion of the molecule is produced by endothelial cells of the blood vessels and possibly also by megakaryocytes. Its synthesis is regulated by an autosomal chromosome. This large molecule is produced originally as a series of smaller subunits, which then polymerize to form the large polymer known as factor VIII:vW (figure 30.5). Factor VIII:C noncovalently binds to factor VIII:vW. Factor VIII:vW may play a role in regulating the synthesis and release of factor VIII:C.

Abnormal synthesis of coagulation factors will result in a bleeding disorder known as **hemophilia.** There are many different forms of hemophilia, each of which results from the lack of one or more clotting factors or the presence of abnormal coagulation factors. Hereditary deficiencies pertaining to 10 of the traditional coagulation factors have been described (not including factors III and IV). Many of these hemophilias are rare, with the exception of hemophilias resulting from deficiencies of factors VIII, IX, and XI (table 30.1). Since factor VIII consists of two portions, it is associated with two types of hemophilia. If factor VIII:vW is absent, von Willebrand's disease results. If factor VIII:C is missing, hemophilia A results. Lack of factor IX is known as hemophilia B (or Christmas disease). Lack of factor XI is known as hemophilia C (or Rosenthal syndrome).

**Von Willebrand's disease** Three different causes of von Willebrand's disease (vWD) have been described (figure 30.6): (1) factor VIII:vW is not produced; (2) production of factor VIII:vW is reduced; and (3) abnormal low-molecular-weight polymers of factor

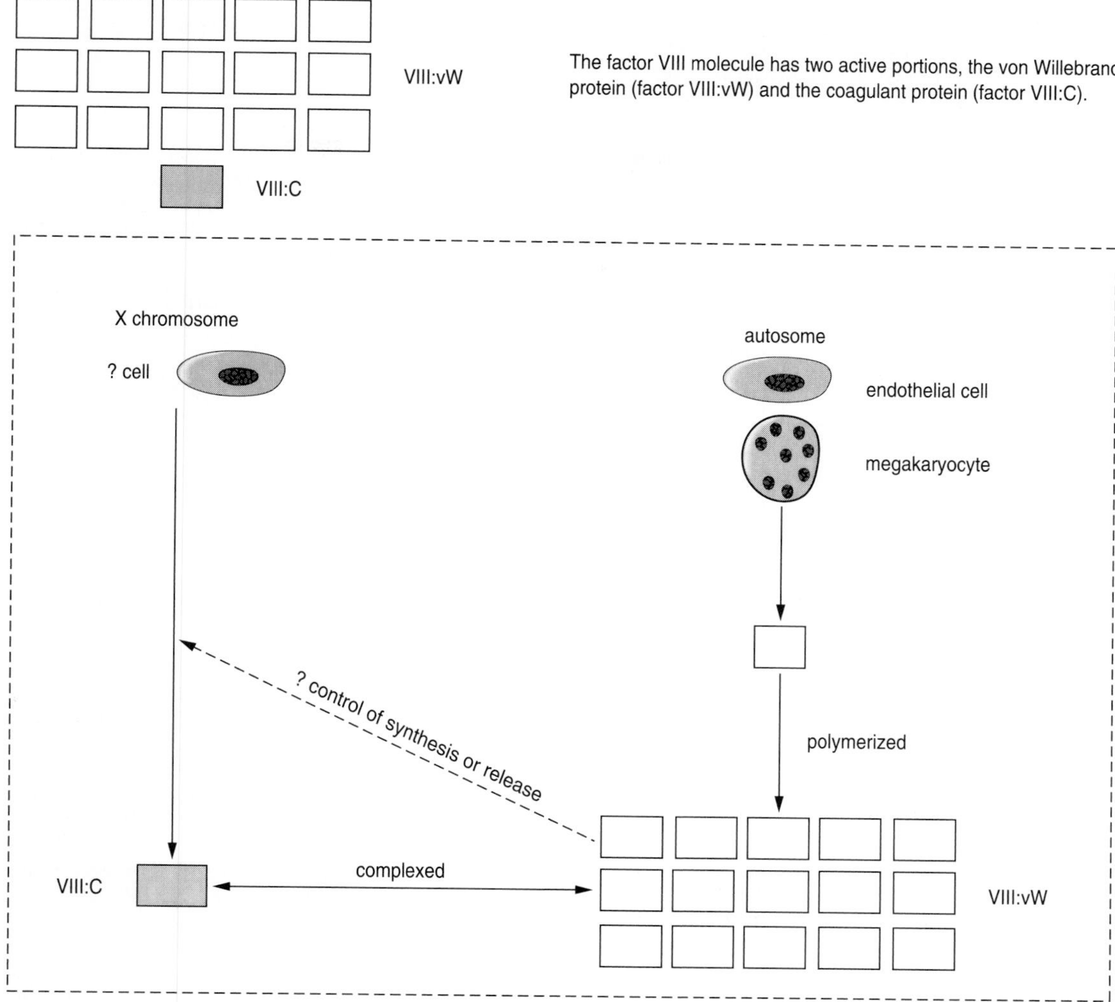


**figure 30.5**

The synthesis of factor VIII involves two different genes: one for factor VIII:C (this is found on the X chromosome) and one for factor VIII:vW, which is found on an autosome. Factor VIII:vW is synthesized as subunits that polymerize with each other and with factor VIII:C.

VIII:vW are present. Coagulation defects and prolonged bleeding occur in all cases; however, a different degree of clinical severity is associated with each form. A total lack of factor VIII:vW will have the most severe effect, because not only will platelet function be impaired, but lack of factor VIII:vW will also result in reduced levels of factor VIII:C, as it is denatured rapidly without its carrier molecule.

Because the factor VIII:vW defect is transmitted as an autosomal dominant trait, it can affect either parent, and hence can be transmitted to all their children irrespective of their sex (figure 30.7). Von Willebrand's disease is characterized by prolonged bleeding time, low factor VIII:C level, and defective platelet functioning. This platelet functioning defect can be tested in the laboratory by measuring platelet aggregation when

**Table 30.1** Diseases Associated with Defects in Various Coagulation Factors: their Mode of Inheritance and their Incidence

| Defects | Mode of Inheritance | Incidence |
|---|---|---|
| XII | Autosomal recessive | Very rare |
| XI (Hemophilic C) | Autosomal recessive | Rare |
| IX (Christmas disease) | Sex-linked recessive | Uncommon |
| VIII:C (Hemophilia A) | Sex-linked recessive | Uncommon |
| VIII:vW (von Willebrand's disease) | Autosomal dominant | Uncommon |
| VII | Autosomal recessive | Very rare |
| X | Autosomal recessive | Very rare |
| V | Autosomal recessive | Very rare |
| II | Autosomal recessive | Very rare |
| I | Autosomal recessive | Very rare |
| XIII | Autosomal recessive | Very rare |

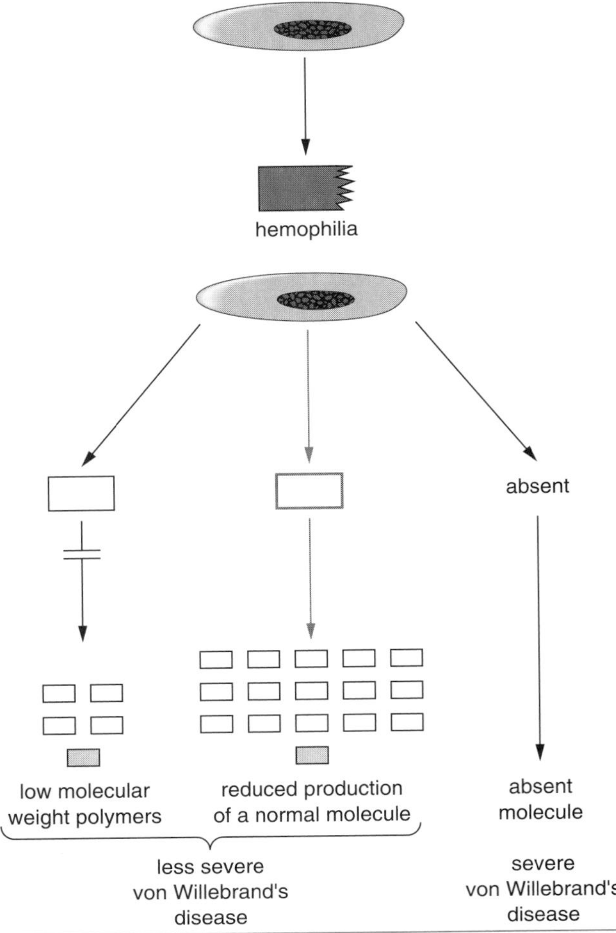


**figure 30.6**

The three different forms of von Willebrand's disease.

the antibiotic **ristocetin** is added. Ristocetin binds to factor VIII:vW which causes aggregation of platelets. Because of its effect on platelets, ristocetin cannot be used as an antibiotic internally (figure 30.8).

As previously mentioned, the severity of vWD varies considerably in different patients, even those who are closely related. Usually the bleeding is rather mild, but on occasion it can be severe. Treatment for this disease involves infusion of fresh plasma or cryoprecipitate. Shortly after the transfusion factors VIII:vW and VIII:C will rise sharply and reach a maximum peak within 6 to 12 hours but will return to pretreatment levels within 48 hours (figure 30.9).

**Hemophilia A** This condition may be defined as an X-linked inherited disease associated with a lack of or abnormal production of factor VIII:C. Since hemophilia A is X-linked, usually only males are affected, whereas females are normally the carriers (figure 30.10). Hemophilia A is the most common of all inherited coagulation disorders, as it comprises about 85% of all cases of hemophilia. The degree of bleeding varies greatly and is closely associated with the amount and type of factor VIII:C produced. Severe bleeding is common in males who produce less than 1% of normal levels of factor VIII:C. Individuals who produce 1 to 5% of normal levels of factor VIII:C usually exhibit moderate bleeding, and those with levels of 5 to 20% usually have mild bleeding problems.

The two major types of hemophiliac bleeding are spontaneous bleeding and posttraumatic hemorrhage.

**Spontaneous bleeding** occurs in severely affected hemophiliacs and is frequently associated with bleeding into the muscles, which may result in hematomas. Bleeding may also occur into the joints, which is known as **hemarthrosis,** and may result in permanent crippling unless treated promptly.

**Posttraumatic bleeding** produces large hematomas in the body. This can be dangerous as it may increase pressure on nerves and blood vessels. Vascular obstruction may develop into local gangrene.

Treatment of hemophilia A includes replacement therapy with factor VIII:C, especially when bleeding occurs as a result of vessel damage or from surgical procedures. Generally bleeding can be controlled when the level of factor VIII:C is above 20% of the normal level. In case of surgery or posttraumatic bleeding, a level of 60% or higher is necessary. Factor VIII:C replacement can be achieved by administering either fresh frozen plasma or a freeze-dried, purified concentrate of factor VIII. Several dangers are associated with transfusing fresh plasma or factor VIII concentrates. First is the possibility of developing anti-factor VIII antibodies. If the antibody concentration reaches high titers, future replacement therapy

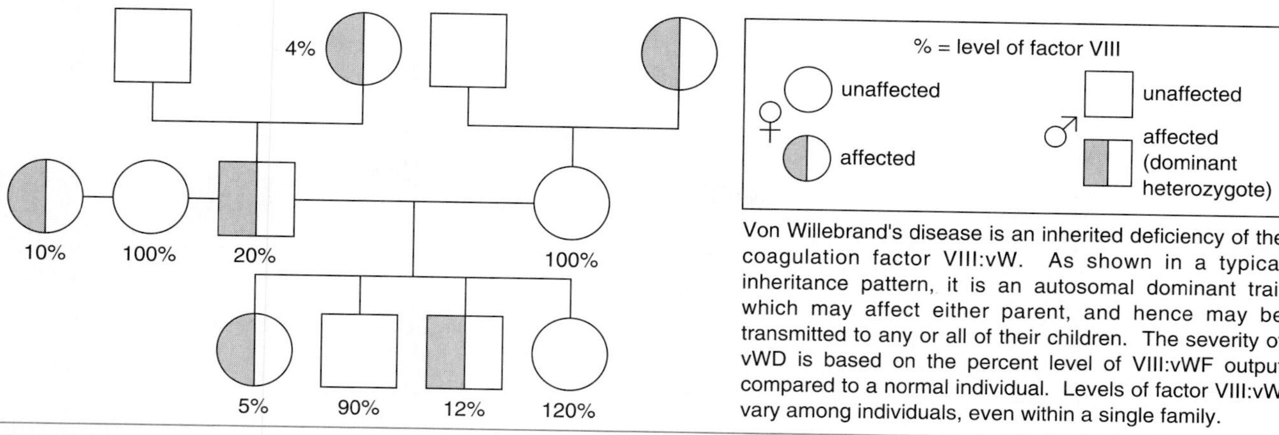


**figure 30.7**

Von Willebrand's disease is a dominant trait that is transmitted from parent to child irrespective of their sex.

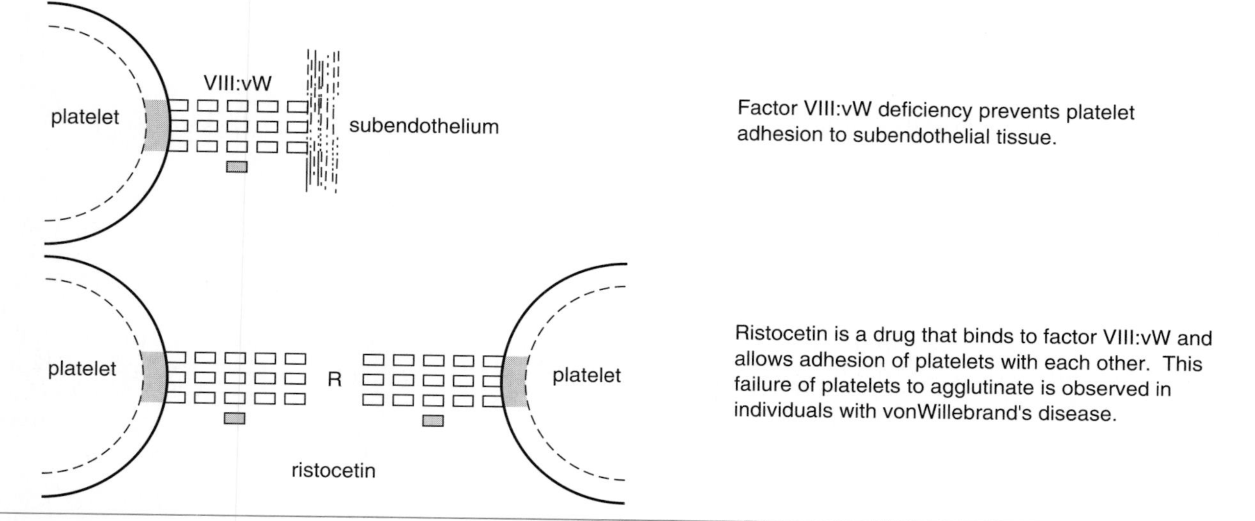


**figure 30.8**

Ristocetin-induced platelet agglutination test. This laboratory screening test is used to detect von Willebrand's disease.

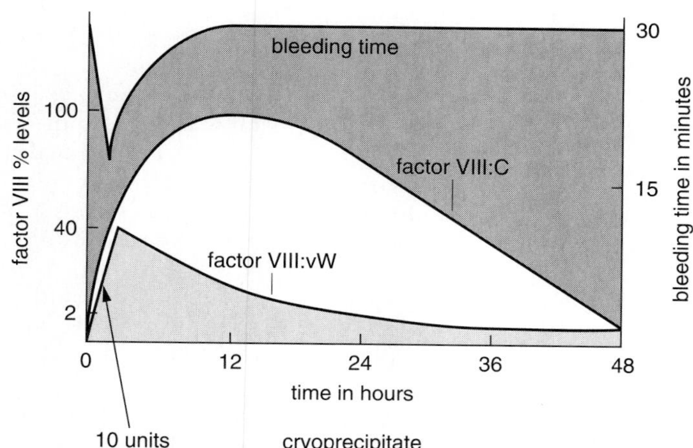


The percent of factor VIII:C and factor VIII:vW is plotted versus time after an initial infusion with plasma rich in antihemophilic factor VIII (i.e., cryoprecipitate). Immediately following a cryoprecipitate transfusion, the concentration of both factors increases. Within three hours, however, the level of VIII:vWF begins to decline while the level of factor VIII:C continues to rise, reaching a maximum in 6–12 hours. Bleeding time, which correlates with the level of VIII:vWF, declines immediately following the initial infusion with the effect lasting approximately three hours.

**figure 30.9**

Following a transfusion with cryoprecipitate there is an immediate rise in both factor VIII:C and factor VIII:vW. Within three hours the level of factor VIII:vW begins to drop, while the level of factor VIII:C lasts longer. It only begins to drop after about 12 hours.

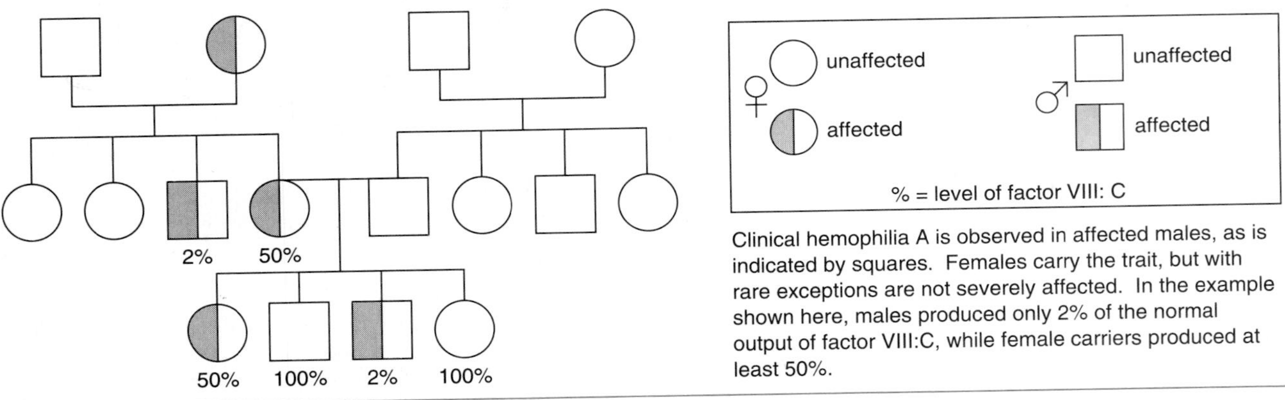


**figure 30.10**
An inheritance pattern of hemophilia A shows that females are the carriers, while males are the affected individuals.

may be difficult. Second, with every transfusion of blood or blood constituents there is the danger of transmitting infectious agents, especially hepatitis viruses and the human immunodeficiency virus (HIV).

Supportive care is important. Patients may experience episodes of bleeding, which leads to high levels of anxiety. Psychological support should prepare patients for the inevitable occurrence of bleeding and teach them how best to deal with the problem. Patients should also be taught how to avoid injury.

**Hemophilia B** Hemophilia B occurs as a result of a deficiency of factor IX. It is also known as **Christmas disease** and accounts for approximately 10% of all hemophilias. Like hemophilia A it is also an X-linked recessive trait. Although there is a direct correlation between the severity of the disease and the level of factor IX deficiency, bleeding is usually not as severe because factor IX is more stable than factor VIII:C.

Treatment for hemophilia B is similar to that of hemophilia A, which also includes both supportive and replacement therapies. Because of the stability of factor IX stored plasma can be used. Transfusions with fresh plasma or factor IX concentrates can be given less frequently as factor IX is more stable and has a longer half-life than factor VIII:C.

**Hemophilia C** The least common of the three major types of hemophilia, hemophilia C may be defined as a bleeding disease caused by a deficiency of factor XI. It is inherited as an autosomal recessive trait and represents less than 5% of all hemophilias. Hemophilia C is also known as **Rosenthal syndrome** and is not uncommon among Ashkenazi Jews. The symptoms and treatment are similar to that of the other hemophilias previously discussed.

**Excessive fibrinolysis** A rare bleeding disorder known as excessive fibrinolysis may be caused by an inherited deficiency of the fibrinolytic inhibitor substance called $\alpha_2$-antiplasmin. When this factor is decreased, the clots formed are almost immediately degraded by plasmin, so a stable hemostatic plug cannot form, resulting in prolonged bleeding.

Excessive fibrinolysis can be diagnosed by determining the level of $\alpha_2$-antiplasmin and may be treated by replacement therapy using plasma or plasma extracts.

*Acquired Disorders*

Acquired coagulation disorders may be more common than inherited ones. Unlike inherited coagulation defects, acquired coagulation disorders are caused by a deficiency or inhibition of more than one coagulation factor. Many different scenarios produce acquired coagulation deficiencies, but they can all be classified into the following five categories: (1) vitamin K deficiency, (2) liver disease, (3) massive transfusion syndrome, (4) diffuse intravascular thrombosis, and (5) primary pathological fibrinolysis.

**Vitamin K deficiency** As discussed in chapter 29, vitamin K is essential for synthesis and activation of the prothrombin group of coagulation factors (factors II, VII, IX, X). Vitamin K is required for carboxylation of glutamic acid residues on these clotting factors (figure 30.11). Only $\gamma$-carboxylated coagulation factors can bind to phospholipid receptor sites on platelet surfaces in the presence of calcium.

Vitamin K is present in many foods, especially in green vegetables. Since it is a fat-soluble vitamin it is dependent on bile, pancreatic lipase, and a healthy gut surface for absorption into the body. Vitamin K deficiency in adults usually occurs from impaired fat absorption in the gut, resulting from such conditions as biliary obstruction (usually by gallstones or tumors), pancreatic disease (as a result of pancreatitis or tumors), which will result in decreased lipase production, and from impaired intestinal absorption (e.g., caused by tropical sprue or tapeworms). Some antibiotics also interfere with absorption or utilization of vitamin K. Treatment of vitamin K deficiency involves replacement therapy with vitamin K. In cases of severe bleeding, blood transfusions may be necessary.

**Liver disease** Most coagulation factors are produced in the liver: thus liver impairment influences synthesis

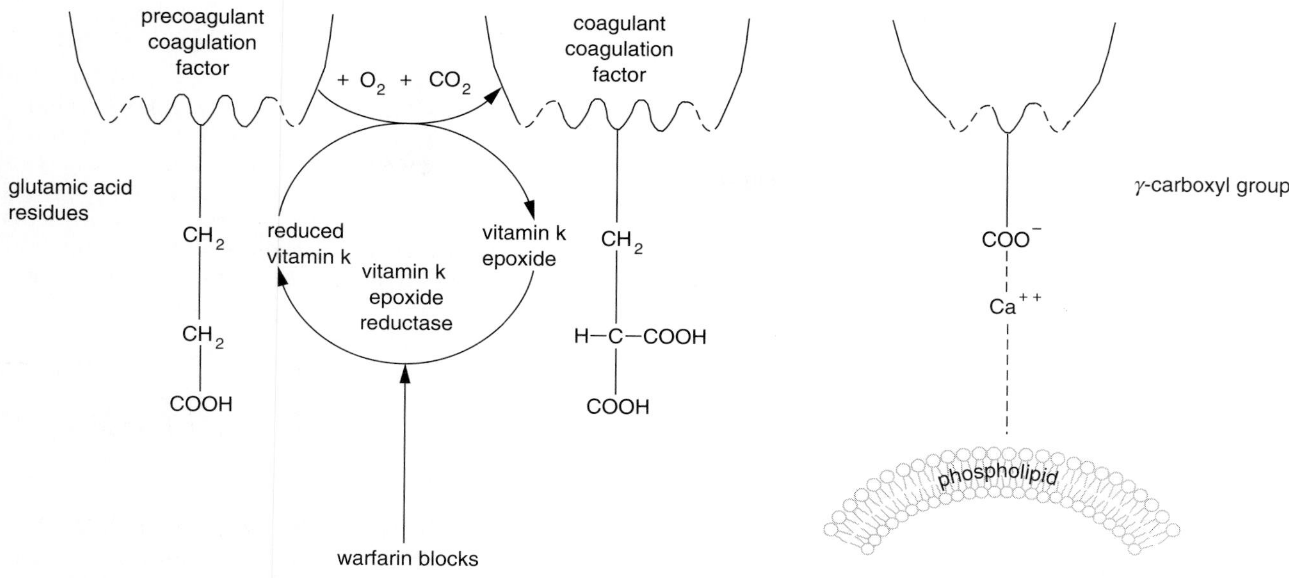


**figure 30.11**

The mechanisms of action of vitamin K: It promotes the binding of coagulation factors to the phospholipid receptor sites on platelets.

of the various clotting factors. Major causes of liver disease include cirrhosis due to drugs or alcohol, hepatitis due to infectious pathogens, liver cancer, and chronic liver disease due to chemicals or iatrogenic drugs.

**Massive transfusion syndrome** Patients may acquire this condition from transfusion of large amounts of stored blood. When blood is stored, certain labile coagulation factors—especially those belonging to the fibrinogen group—undergo chemical changes and become inactive. Platelets also tend to disappear from stored blood after 4 or 5 days. Hence if a patient receives large quantities of stored blood, coagulation deficiencies and thrombocytopenia may follow because of dilution of the patient's blood and from increased bleeding. The severity of the bleeding defects in such patients depends on the amount of blood transfused, the age of the stored blood, the rate of the transfusion, and the amount of bleeding.

**Diffuse intravascular thrombosis** This bleeding disorder is caused by a reduction of coagulation factors and platelets in the general circulation due to continuous low-level intravascular coagulation activity. As a result coagulation factors are consumed at a faster rate than they are being produced, which may cause an overall hemostatic defect. Diffuse intravascular thrombosis is also known as **consumption coagulopathy** and has three possible causes (figure 30.12). The first cause is the release of activated tissue factor. **(Factor III)** in the blood as a result of surgery or the birth process. Tissue factor may activate the extrinsic coagulation cascade and thus remove essential coagulation factors from the circulation.

The second cause of diffuse intravascular thrombosis is the production of activated thrombi by viruses

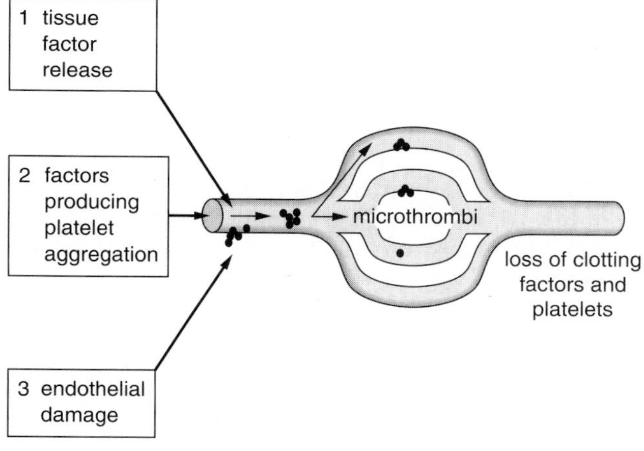


**figure 30.12**

Three major factors that cause diffuse intravascular thrombosis.

or other infectious agents. Activated thrombin causes platelets to release ADP, which acts as a chemotactic factor, bringing more platelets to a given area, thus promoting platelet aggregation. Platelet aggregation in turn activates the coagulation factors.

The third cause is extensive endothelial damage, which may result from vessel inflammation (=vasculitis) or from burns. Here again factors may be released that activate the coagulation cascade. Consumption coagulopathy may be diagnosed by (1) a low platelet count; (2) diminished levels of coagulation factors, especially fibrinogen; and (3) increased levels of fibrin-split products (as a result of increased fibrinolytic activity).

Treatment involves eliminating the causative agent (e.g., bacteria or viruses), replacement of depleted coagulation factors and perhaps even platelets, and inhibition of coagulation by the use of anticoagulants.

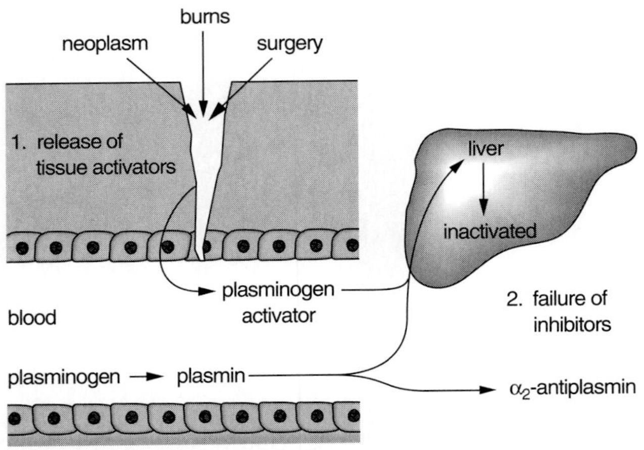


**figure 30.13**
The causes of primary pathological fibrinolysis.

**Primary pathological fibrinolysis** Primary pathological fibrinolysis is a very rare bleeding condition that occurs as a result of increased fibrinolytic activity. In inherited pathological fibrinolysis, the patient is unable to produce $\alpha_2$-antiplasmin, which is an inhibitor of plasmin activity (excess plasmin degrades fibrin clots prematurely).

Another cause for primary pathological fibrinolysis is the release of large quantities of plasminogen activator into the bloodstream. Plasminogen activator changes inactive plasminogen into active plasmin, which then attacks the fibrin polymers. Release of excessive plasminogen activator is especially common in people with cancer or impaired liver function, and those with severe burns or who undergo extensive surgery. The healthy liver usually inactivates any excess plasminogen activator as well as activated plasmin. If this is not possible, excess fibrinolysis will occur (figure 30.13).

## CLOTTING DISORDERS

The opposite of bleeding problems are clotting disorders. Inappropriate activation of clotting mechanisms in intact blood vessels results in fibrin clot formation, a phenomenon known as **intravascular thrombosis.** Production of intravascular clots is, however, a rather rare event as the body has many mechanisms to prevent such occurrences. The few situations in which thrombosis may occur involve either overwhelming thrombogenic stimulation or a breakdown in protective mechanisms.

### Overwhelming Thrombogenic Stimulation

When a thrombogenic stimulus is so great that it overwhelms the protective mechanisms of the body, a fibrin clot may form. Such thrombogenic clot formation may occur in the following situations: (1) large-scale vascular damage, (2) stimulation of platelet aggregation, and (3) activation of coagulation factors.

#### *Large-Scale Vascular Damage*

When vascular damage occurs on a large scale, it is frequently associated with widespread platelet aggregation and coagulation factor activation. As more clotting factors are activated, the danger increases that some of these activated clotting factors will leave the location of injury and start forming clots elsewhere in intact blood vessels. This is especially hazardous after major surgery and in severe traumatic injury.

#### *Stimulation of Platelet Aggregation*

Any factor that promotes platelet aggregation in intact blood vessels will promote clot formation.

#### *Activation of Coagulation Factors*

Any compound that stimulates activation of clotting factors in intact blood vessels will promote fibrin clot formation. However this rarely happens because any activated clotting factor that leaves the area of injury will be immediately neutralized by circulating inhibitor substances. Further, coagulation complexes are readily removed by macrophages and other mononuclear phagocytic cells. However if more fibrin is produced than can be eliminated, intravascular clotting may result.

### Failure of Protective Mechanisms

Lack of inhibitory factors, malfunctioning macrophages, and poor blood flow are among the reasons why protective mechanisms fail to function properly when confronted with activated clotting factors.

#### *Lack of Inhibitory Factors*

When an individual lacks sufficient inhibitory factors to neutralize activated clotting components, there is a greater chance that intravascular clots may develop.

#### *Malfunctioning Macrophages*

When macrophages are unable to rapidly remove activated clotting complexes from the circulation, an increased predisposition for intravascular thrombosis will result.

#### *Poor Blood Flow*

Individuals with impaired blood circulation may develop intravascular clots. Inadequate blood flow may be the result of heart failure, abnormally dilated blood vessels, or the compression of blood vessels due to prolonged cramped positions or the wearing of tight elastic bands on arms or legs. In these situations inhibitory substances and macrophages will have difficulty reaching activated clotting factors or aggregated platelets.

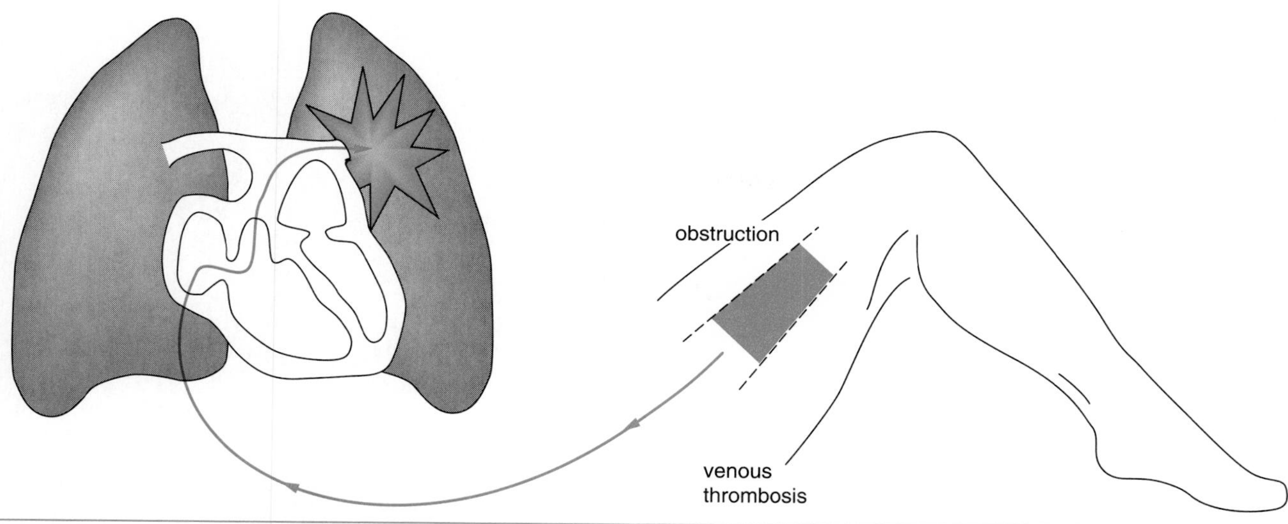


**figure 30.14**

The greatest danger in venous thrombosis is that an embolus may form and travel through the heart to the lungs, where it will produce a pulmonary embolism. Large-scale embolisms may be fatal.

Blood clots may form in any blood vessel but are more common in veins than in arteries because blood flows more slowly through veins. Most blood clots tend to occur in the veins of the legs, possibly interfering with venous return of blood to the heart. Blood clots also may result in inflammation, edema, discoloration, and ulceration of the legs, which is known as **thrombophlebitis.**

The most serious complication of venous thrombosis is **pulmonary embolism,** which occurs when a clot from the legs moves via the heart to the lungs (figure 30.14), where it obstructs the pulmonary circulation.

**Arterial thrombosis** is rather rare and is usually associated with large blood vessel damage. However the consequences of arterial clots are usually more severe than venous thrombi. They may lead to myocardial infarction (heart attack), cerebral infarction (stroke), or limb infarction, which may result in gangrene (figure 30.15).

Microcirculatory thrombosis leads to local ischemia and necrosis (=cell death). When this occurs on a large scale, it is known as **disseminated intravascular coagulation (DIC).**

### Treatment of Intravascular Thrombosis

The treatment for intravascular clotting depends on the blood vessel—that is, whether it is a vein, artery, or a tiny blood vessel (microcirculation).

#### *Venous Thrombosis*

Treatment should be three-pronged in venous thrombosis. First, the patient should receive **anticoagulants** to prevent further coagulation. Anticoagulation drugs interfere with the formation of fibrin clots. Heparin is the anticoagulant of choice because of its rapid action. Second, drugs that speed the degradation of existing clots should be administered. Usually these drugs are fibrinolytic enzymes that will quickly dissolve the thrombus or embolus. The major **thrombolytic agents** are streptokinase and urokinase, both of which rapidly activate plasminogen, converting it into the active enzyme plasmin, which then dissolves the clots. Dextran copolymerizes with the fibrin monomers, making the fibrin clot more susceptible to fibrinolysis. Further, rapid mobilization of the patient will reduce clot formation by increasing the flow of blood in the veins of the legs. This may be accomplished by mechanical devices or by massaging the leg. Blood flow may also be increased by reducing blood viscosity by volume expanders such as dextran. The preceding treatments are summarized in figure 30.16.

#### *Arterial Thrombosis*

Most arterial clots occur as a result of **atherosclerosis.** Arterial thrombosis frequently results in tissue damage because the thrombus blocks the arterial circulation, resulting in an arterial occlusion. Several techniques are used to eliminate these occlusions. The most important techniques are (1) bypass of the blockage with a venous graft **(bypass surgery)**; (2) dilation of the blocked vessel by **angioplasty,** a technique in which an empty balloon is inserted into the blocked vessel and filled with air; (3) dissolution of thrombi with fibrinolytic agents; (4) surgical resection of the blocked segment of the artery **(endarterectomy);** and (5) removal of the clot from the artery **(thrombectomy** or **embolectomy)** (figure 30.17).

#### *Clots in the Microcirculation*

Microclots in the small blood vessels may be caused by either platelet aggregation or fibrin clot formation or

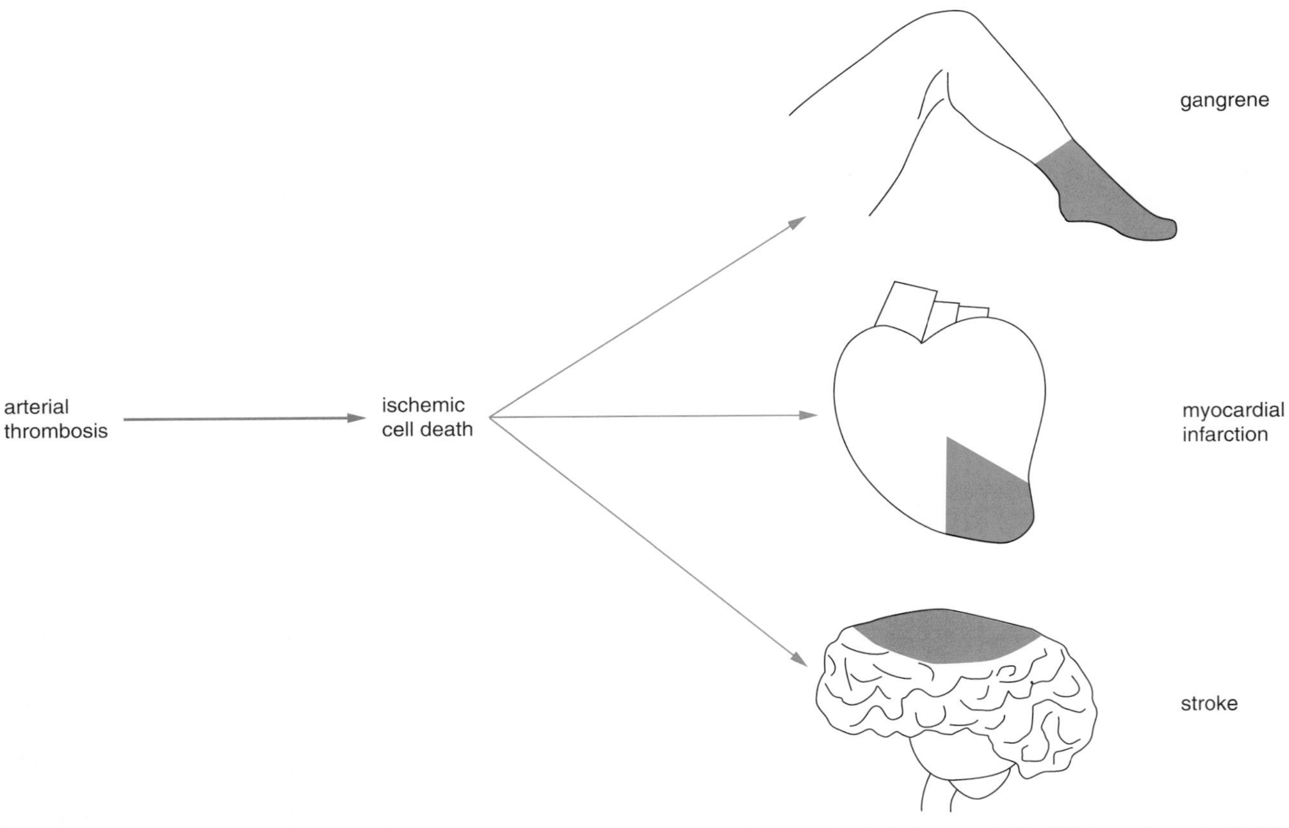


**figure 30.15**

The dangers of arterial thrombosis. Arterial thrombi lodge in the limbs, where they may cause gangrene; in the heart, where they may produce myocardial infarctions; or in the brain, where they may produce cerebral infarctions (strokes).

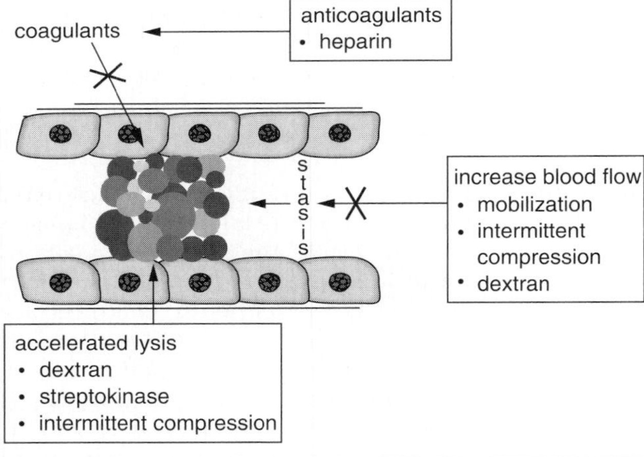


**figure 30.16**

Prevention of venous thrombosis.

both. Platelet aggregation may be inhibited by antiplatelet drugs such as aspirin. Aspirin inhibits the synthesis of cyclic endoperoxides and $TXA_2$, which are needed for platelet function including aggregation. Fibrin clots may be degraded by fibrinolytic enzymes and new fibrin clots may be prevented by anticoagulants.

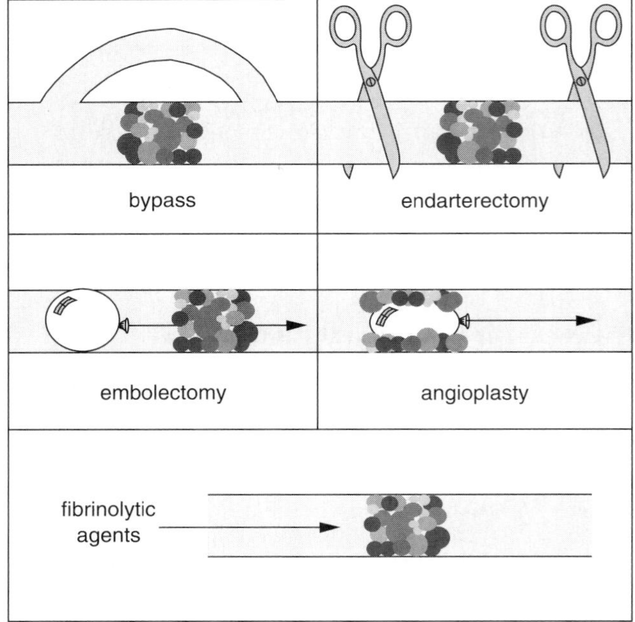


**figure 30.17**

Common treatments for arterial thrombosis.

# Further Readings

Handin, R. I. Hemorrhagic Disorders III: Disorder of Primary and Secondary Hemostasis. Lecture 28 in *Hematology*, 4th ed., William S. Beck, Ed. The MIT Press, Cambridge, Mass., 1985.

Hirsch, J. and Brian, E. A . *Hemostasis and Thrombosis, A Conceptual Approach,* 2nd ed. Churchill Livingstone, New York, 1983.

Larson, L. Disorders of Hemostasis. Chapter 22 in *Textbook of Hematology,* S. McKenzie, Ed. Lea & Febiger, Philadelphia, 1988.

# Review Questions

1. Name the two major categories of defects in hemostasis.
2. List the three major categories of bleeding disorders.
3. Name the two categories of vascular disorders.
4. Define hemorrhagic telangiectasia.
5. List five acquired vascular defects that result in increased bleeding.
6. Distinguish between senile and steroid purpura. What is the cause of each?
7. What causes scurvy? Why does it cause bleeding?
8. Describe the Henoch-Schönlein syndrome. When does it normally occur? What is its cause? List the major symptoms.
9. Distinguish between thrombocytopenia and thrombocytopathia.
10. List the five major causes of thrombocytopenia.
11. List the five major causes of deceased platelet production.
12. Describe the five major causes of increased platelet destruction.
13. What is DIC? What causes DIC?
14. List the five major causes of thrombocytopathia.
15. Give the three major reasons why platelets fail to adhere to collagen.
16. What is thrombasthenia?
17. What is the major reason for the failure of platelets to produce $TXA_2$? Explain why.
18. List the major inherited disorders of plasma proteins involved in coagulation.
19. Describe von Willebrand's disease. Describe the three forms of vWD.
20. Distinguish between hemophilia A, B, and C regarding (a) cause, (b) incidence, and (c) treatment.
21. What is (a) Christmas disease and (b) Rosenthal syndrome?
22. List the five major categories of acquired coagulation disorders.
23. Describe the mechanism of action of vitamin K. What foods contain vitamin K?
24. Explain why massive blood transfusions may produce prolonged bleeding.
25. Explain why liver disease may result in bleeding disorders.
26. List the three major causes for diffuse intravascular thrombosis.
27. What is consumption coagulopathy? Describe the treatment for this condition.
28. Describe the two major causes of clotting disorders.
29. List three important situations that cause overwhelming thrombogenic stimulation.
30. Explain how reduced blood flow may result in clotting problems.
31. Why are blood clots more common in veins than in arteries?
32. What is thrombophlebitis? What are some common symptoms?
33. List three treatments for venous thrombosis.
34. List five ways to treat arterial occlusions.
35. Describe the major methods to treat clots in the microcirculation.

# Chapter Thirty-One

## *Diagnosis of Defects in Hemostasis*

Abnormalities in the hemostatic mechanism may be detected in two ways: by clinical observation and by laboratory analysis. The initial diagnosis of defects in hemostasis usually is made when the patient exhibits overt clinical symptoms suspected to have occurred as a result of bleeding or clotting. These symptoms will prompt a physician to request laboratory tests to confirm the initial diagnosis and to pinpoint the cause of either the bleeding or clotting problem.

This chapter discusses and describes the clinical and laboratory findings associated with bleeding and clotting disorders.

## DIAGNOSIS OF BLEEDING DISORDERS

### Clinical Diagnosis of Bleeding Disorders

Many different overt clinical conditions may occur as a result of abnormal bleeding. The 10 most common are discussed in the following material.

*Petechiae*

Multiple, small red spots on the skin, each the size of the head of a pin are **petechiae,** which result from red blood cells leaving the small blood vessels at a given point. They usually occur in groups on certain areas of the body (e.g., the arm, the back) (figure 31.1). Petechiae are frequently observed in patients who lack the normal number of platelets (thrombocytopenia) or who have problems with platelet function (thrombocytopathia).

Normal platelets play an essential function in blood vessel integrity. A drastic reduction in platelets or abnormally functioning platelets will result in a loss of endothelial vessel wall integrity, allowing red blood cells to escape at certain points.

*Purpura*

Multiple red spots larger than petechiae are known as **purpura,** which usually occurs from the fusion of several petechiae. Purpura spots are formed by the same causes as petechiae—that is, thrombocytopenia and thrombocytopathia (figure 31.1).

*Ecchymosis*

Commonly known as a bruise, an **ecchymosis** *(pl. ecchymoses)* may be defined as a large area of blood under the skin due to simultaneous breaks of several blood vessels. Ecchymoses usually occur as a result of physical trauma. Bruises are particularly common in people with vascular or platelet disorders (figure 31.1).

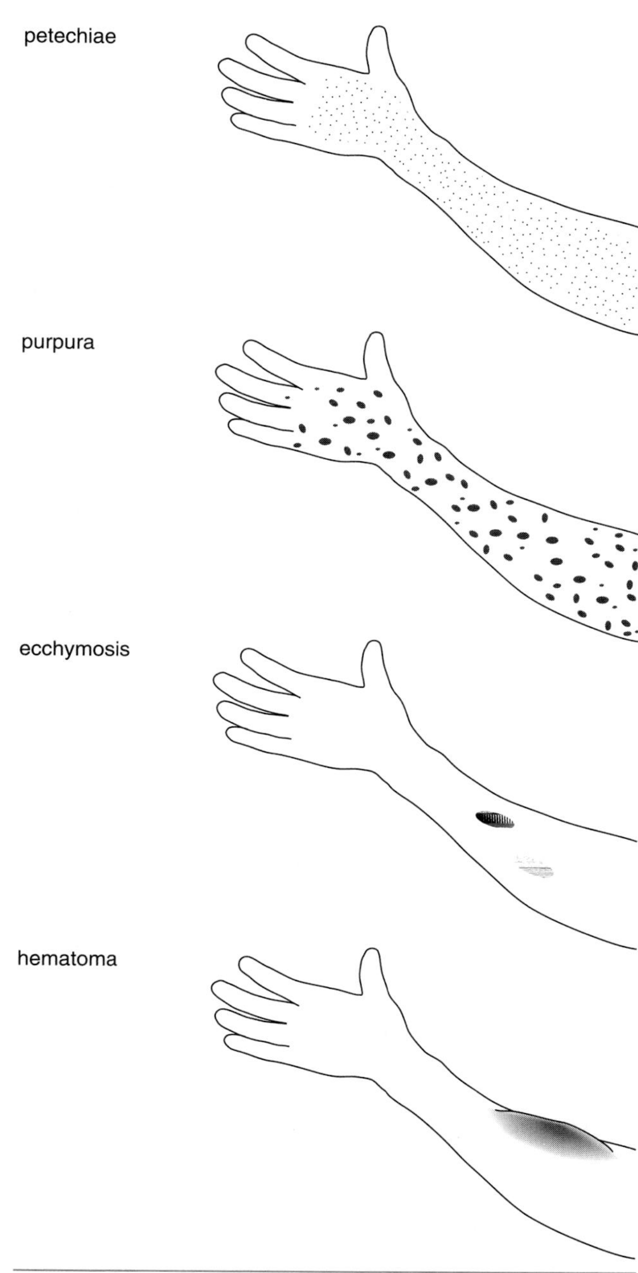


**figure 31.1**
Variations of bleeding in the skin.

*Hematoma*

A large bruise in which blood has penetrated subcutaneous tissue and produced swelling as well as discoloration is called a **hematoma** (figure 31.1). Hematomas are common in people with coagulation disorders as well as those with high levels of anticoagulants in their blood.

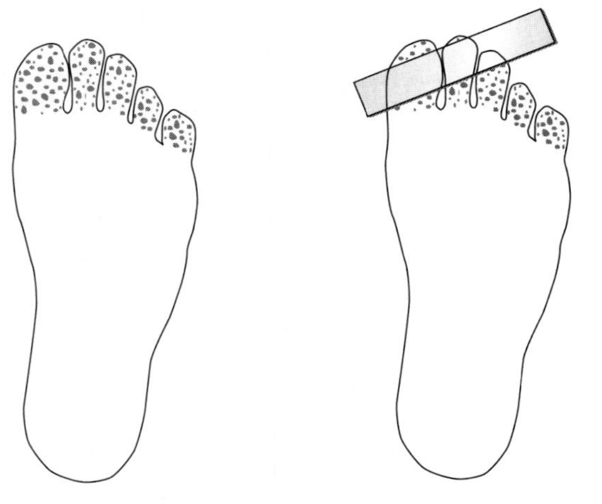

figure 31.2

Diagnosis of telangiectasia. This condition may be diagnosed by pressing a clear microscope slide against the skin over the lesion. If blanching occurs, telangiectasia is present.

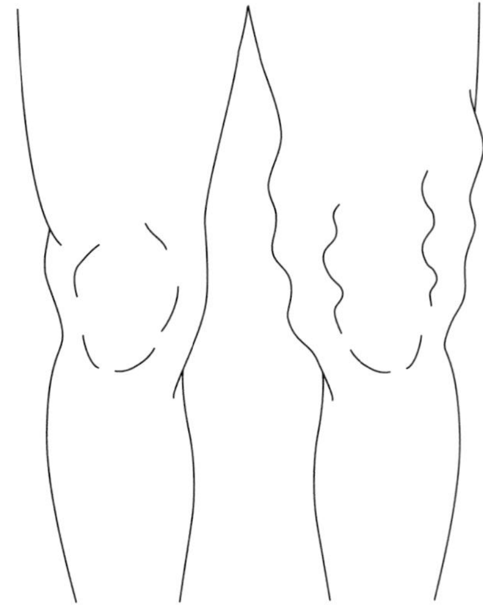

figure 31.3

The condition of hemarthrosis, i.e., bleeding into the joints. This condition is commonly seen in hemophiliacs.

*Telangiectasis*

Red spots or patches on the skin and mucous membranes resulting from abnormally dilated blood vessels are known as **telangiectases.** The difference between these spots and splotches caused by purpura is that telangiectases blanch under pressure. If a clear microscope slide is pressed against the skin, the patches disappear (figure 31.2).

*Hemarthrosis*

This term may be defined as a hemorrhage into a joint cavity (figure 31.3). **Hemarthrosis** is usually observed in patients with severe coagulation disorders, such as hemophilia, and may be associated with severe disfigurement.

*Hematuria*

Blood occurring in the urine is known as **hematuria.** The presence of discolored urine may have many causes, most commonly renal lesions, kidney cancer, hemophilia, and anticoagulant therapy.

*Blood in Stool*

This condition arises usually when there is gastric or intestinal bleeding.

*Hemorrhoids*

Bleeding from the veins of the anal canal, or **hemorrhoids,** results from passing hard stools. Hemorrhoids produce mechanical damage to the local blood vessels. Risk factors include obesity, pregnancy, and frequent constipation.

*Excess Vaginal Bleeding*

This may occur as a result of endometriosis, which is an inflammation of the inner lining of the uterus, or as a result of dysmenorrhea, which is excessive and often painful menstrual bleeding, frequently as a result of abnormal prostaglandin synthesis.

**Laboratory Diagnosis of Bleeding Disorders**

Although the nature of a bleeding disorder can frequently be suspected from the medical history and the physical features observed in the patient as a result of a thorough examination, laboratory tests are needed to confirm the diagnosis. Since the hemostatic mechanism consists of three main ingredients—blood vessels, platelets, and plasma proteins—tests are designed to identify the involvement of each of these components of hemostasis. Hence laboratory analysis may be divided into three categories to test for (1) involvement of vascular factors, (2) abnormalities in platelet structure and function, and (3) qualitative and quantitative presence of the soluble factors involved in coagulation.

*Tests for Vascular Factors*

Tests for vascular factors include the capillary fragility test (also known as the cuff test, the tourniquet test, or the capillary resistance test), and tests for bleeding time.

**Capillary fragility test** This test measures the ability of capillaries to withstand increased stress. A blood pressure cuff is placed above the patient's elbow and inflated to a maximum pressure of 100 mm Hg. Before the cuff is inflated, the arm is examined for

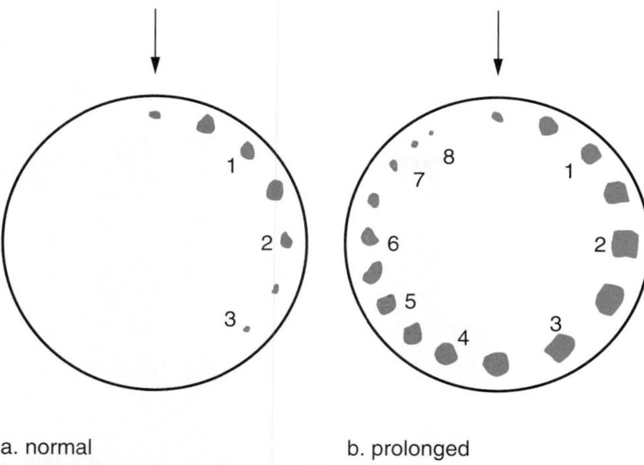


**figure 31.4**

The determination of bleeding time. *(a)* Normal blood spots on filter paper and *(b)* blood spots from a patient with a platelet defect.

petechiae. If any are present, they are noted by circling them with ink. After the cuff is inflated, it is kept in place at 100 mm Hg for 5 minutes after which it is deflated and removed. The number of new petechiae that have appeared are then counted in a circle 5 cm in diameter drawn below the bend of the elbow. If more than 10 new petechiae have appeared during that period, the test is considered to be positive, indicating the existence of capillary weakness or thrombocytopenia or both. To determine if thrombocytopenia is involved, a platelet count is performed.

**Bleeding time test** The bleeding time measures the time required for the cessation of bleeding after a standardized puncture through the skin. The time required depends on capillary integrity, number of platelets, and platelet function. For a normal bleeding time, the individual must possess an adequate number of platelets in the bloodstream as well as a normally functioning microcirculation. The bleeding time is almost always prolonged when the platelet count drops below 50,000/μL. A normal platelet count and prolonged bleeding time usually indicate diminished vascular integrity. Many different bleeding time tests were developed over the years, including the Duke test and the Ivy test. The **Duke test** involves puncturing the earlobe with a disposable lancet. Drops of blood are blotted onto a blotting paper every 30 seconds and the time at which bleeding stops is noted to the nearest second. Normal times for the Duke test are 1 to 3 minutes; borderline times are 3 to 6 minutes (figure 31.4).

The **Ivy test** is performed in a similar manner, except in this case a blood pressure cuff is used. After the blood pressure cuff is placed on the upper arm and inflated to 40 mm Hg, the skin is pierced with a lancet in two areas on the lower forearm. Again the blood is blotted every 30 seconds until the bleeding ceases. Normal times for the Ivy test are between 2 and 6 minutes.

The problem with both the Duke and the Ivy tests is the depth of the puncture wound. Although the puncture should be 3 mm deep, this is difficult to control. Wounds deeper than 3 mm are likely to involve blood vessels larger than capillary size, which might prolong the bleeding. Wounds less than 3 mm deep might not adequately test the hemostatic actions on the capillaries. For that reason, a new more standardized bleeding time test was developed to ensure a more standard cut. This test is the **template bleeding time test** and is a revised version of the Ivy test. The template bleeding time test uses a special template to ensure that a cut is exactly 9 mm long and penetrates 1 mm into the skin. With such shallow cuts, it is almost impossible to cut venules. As with the Ivy test, a blood pressure cuff is placed on the upper arm and inflated to 40 mm Hg. After the cuff is inflated, two shallow cuts (9 mm long and 1 mm deep) are made. Timing starts as soon as bleeding begins. The blood is blotted periodically until bleeding ceases. Normal bleeding times are 3 to 10 minutes. Several types of templates are available, including the Simplate II lancet template for cuts that are 5 mm long and 1 mm deep. These template methods are considered the best bleeding time test devices because the skin puncture is uniformly deep.

*Tests for Platelet Factors*

Tests for platelet factors include the quantitative platelet count, the qualitative platelet aggregation test, and the platelet adhesiveness assay, as well as the bleeding time and the clot retraction time test.

**Platelet count** Platelets can be counted automatically with an electronic cell counter but can also be counted manually. A commonly used method is the so-called **Rees and Ecker method,** in which the platelets are stained with brilliant cresyl blue, a dye that stains platelets a bluish color. The platelets are then counted visually with a **hemocytometer.** Details of this technique are given in the Appendix.

**Platelet aggregation test** Measurement of platelet aggregation is important in investigating suspected platelet function problems. In this test an aggregating agent (e.g., activated thrombin, ristocetin, or epinephrine) is added to a suspension of platelet-rich plasma (PRP), and the response is measured in a spectrophotometer. Significant aggregation will result in turbidity, and less light will be transmitted—that is, the more light transmitted, the less aggregation. Special machines called **aggregometers** are available to measure platelet aggregation.

**Platelet adhesiveness test** This test measures the ability of platelets to adhere to glass surfaces. When anticoagulated blood is passed through a tube containing glass beads, platelets will adhere to the beads. Adhesiveness can be determined by counting the number of platelets in the blood before they are passed through the column with glass beads, and by counting them again after they have passed through the column. In normal individuals 75 to 95% of the platelets will be removed from the blood that has percolated through the beads. If more than 50% of the platelets remain in the blood sample, after passing through the glass bead column, adhesion problems may exist.

**Clot retraction test** Normal blood will clot completely in a glass tube, and clot retraction will start within an hour. After 24 hours the clot should have retracted completely, and most of the serum should be expressed from the clot. Clot retraction primarily depends on normal platelet function. However fibrinogen levels may also influence clot retraction. If lower than normal levels of fibrinogen are present, the clot will not retract as well. However if normal levels of fibrinogen are present and clot retraction is still incomplete, platelet numbers or abnormal function may be involved. When the platelet count falls below 100,000/μL, poor clot retraction is usually seen (figure 31.5).

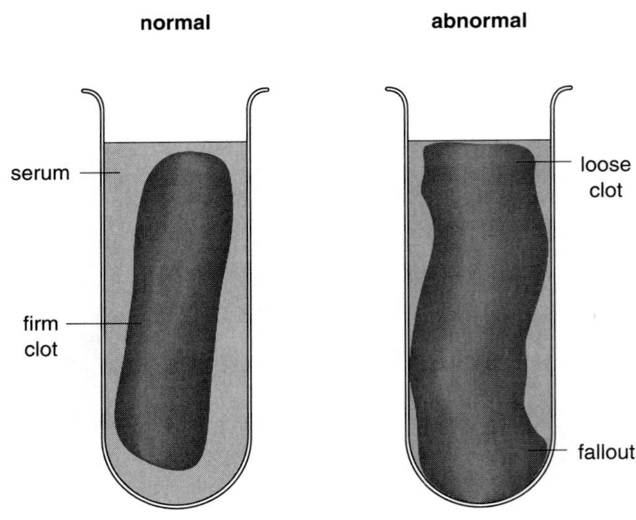


**figure 31.5**

Normal and abnormal clot retraction after 24 hours of incubation.

*Tests for Plasma Factors Involved in Coagulation*

Many tests are available to qualitatively and quantitatively analyze the presence of various coagulation factors. The three most common tests are the **fibrinogen determination test,** the **prothrombin time (PT) test,** and the **partial thromboplastin time (PTT) test.** In the normal PTT test the glass walls of the test tube activate factor XII. Because of the different kinds of glass and the varying degrees of cleanliness of test tubes, the reproducibility of the PTT is low. For that reason, the nonactivated PTT has been replaced by a variation of the test that now uses a chemical activator of factor XII. This newer test is known as the **activated partial thromboplastin time (APTT) test.**

**Fibrinogen determination test** Fibrinogen is the most important coagulation factor that must be present in large quantities to be effective; therefore a decrease or defect in fibrinogen will have an effect on the coagulation process. The fibrinogen determination test, which is also known as the **fibrinogen deficiency test** or **thrombin time (TT) test,** is a simple test by which activated thrombin is added to blood plasma. There is a direct correlation between the time it takes to form a clot and the amount of fibrinogen present. Test kits for fibrinogen determination will graphically show the correlation between the onset of the fibrin clot formation and the amount of fibrinogen present in the plasma. The normal fibrinogen levels in the blood are 200 to 400 mg of fibrinogen per deciliter of plasma.

If the fibrinogen level is normal, an increased coagulation time may be caused by either the absence of or inhibition of other clotting factors. Since two different pathways can induce coagulation, two different types of tests are available to determine the absence or inhibition of certain clotting factors. For the extrinsic pathway, the screening test used is the PT test, whereas the APTT is the test most commonly used to determine problems with the intrinsic coagulation pathway.

**Prothrombin time test** The PT test is performed by adding tissue extract (i.e., factor III = tissue factor) and calcium to the plasma. This initiates the activation of factor VII. Activated factor VII in turn activates factor X, which in the presence of factor V will activate factor II (=prothrombin). Activated thrombin will then convert fibrinogen into fibrin.

If the prothrombin time is prolonged, the cause could be a deficiency or inhibition of one or more of the following coagulation factors: factors VII, X, V, or II. If the prothrombin time is normal, then prolonged bleeding could be due to a deficiency or inhibition of one or more of the coagulation factors involved in the intrinsic pathway. This is investigated by the APTT test.

**Activated partial thromboplastin time test** This is a screening test for factor deficiency of the intrinsic pathway—that is, whether one or more of the following factors is deficient: factors XII, XI, IX, or VIII. The APTT test is performed by first adding kaolin and phospholipid to the plasma. Kaolin activates factors XII and XI, and phospholipid substitutes for platelets in the activation of factor VIII by factor IXa. Activated factor IX (factor IXa) and factor VIII and phospholipid and calcium then activate factor X and the remaining common pathway.

**Determination of factor deficiency or factor inhibition** A prolonged PT or APTT may be due to factor deficiency or factor inhibition, which can be determined by a simple screening test. If factor deficiency is the cause, then the prolonged

**Table 31.1** Common Laboratory Tests used to Determine Abnormalities in Hemostasis and their Interpretations

| Tests | Conclusion | |
|---|---|---|
| Prolonged bleeding time and normal platelet count | Vascular or platelet function disorder<br>or von Willebrand's disease | |
| Prolonged bleeding time and platelets ↓ | Thrombocytopenia +/- platelet function defect | |
| Prolonged bleeding time and abnormal PTT | Von Willebrand's disease | |
| Prolonged PTT (normal PT) | Bleeding | Factor VIII↓<br>(hemophilia A or von Willebrand's disease)<br>Factor IX↓ (Christmas disease)<br>Factor XI↓<br>Heparin* |
| | No Bleeding | "Lupuslike" inhibitor<br>Factor XII↓<br>Prekallikrein↓<br>Kininogen↓ |
| Prolonged PT (normal PTT) | Factor VII↓ (unusual) | |
| Prolonged PTT and PT | Factor II↓ (prothrombin)<br>Factor V↓<br>Factor X↓<br>Factor I↓ (fibrinogen)<br>Vitamin K deficiency<br>Warfarin therapy<br>DIC<br>Therapeutic fibrinolysis<br>Liver disease | |
| Prolonged TT | Fibrinogen↓<br>DIC<br>Therapeutic fibrinolysis<br>Liver disease<br>Heparin | } Usually in combination with coagulation defects |
| No abnormalities on routine screening tests | Bleeding | Factor XIII↓<br>Mild coagulation defects that do not give abnormal screening tests. $\alpha_2$-antiplasmin deficiency |

*PT sometimes slightly prolonged.

coagulation time should be corrected when **standard normal plasma (SNP)** is added to the patient's plasma. SNP contains all clotting factors in quantities sufficient to substitute for any factor deficiency in the patient's plasma. Therefore adding SNP to the patient's plasma and repeating either the PT or the APTT should return the clotting time back to normal. If the PT or the APTT are still prolonged, then the problem is caused by factor inhibition rather than factor deficiency.

**Determination of missing factors** When the PT or APTT is corrected by the SNP, then we could conclude that prolonged bleeding was caused by factor deficiency. The next step is to determine which of the factors is deficient. This is done by repeating either the PT and/or the APTT with special plasmas containing all the coagulation factors except one (e.g., SNP minus factor XII or SNP minus factor VII). In this way we can determine which of the factors is missing, as the PT or PTT will be prolonged with the special plasma that lacks the deficient factor(s).

**Interpretation of laboratory tests** If the factors belonging to the prothrombin group are absent, it may indicate **vitamin K deficiency** or **liver disease.** Von Willebrand's disease is usually indicated by a prolonged bleeding time and abnormal APTT. If the PT is normal and the APTT is abnormal, it usually indicates a lack of factor VIII or perhaps factors IX or XI. For a more complete interpretation of these laboratory tests refer to table 31.1.

*Risks of Excessive Bleeding*

Many serious consequences are associated with excessive and abnormal bleeding. The most important of these are as follows:

1. Excessive blood loss may result in either acute or chronic anemia, as described in earlier chapters.
2. Acute blood loss may result in circulatory failure, hypovolemic shock, and cardiac arrest.
3. Chronic blood loss may easily cause iron deficiency anemia.
4. Abnormal bleeding in the tissues may produce excessive pressure on vital organs. In case of bleeding in the brian, it may lead to neurological problems. Pressure on certain important blood vessels may lead to gangrene.
5. Bleeding into the joints and muscles may cause deformities, because fibrous tissue is deposited in these areas. These fibrous masses are produced by the body in an effort to contain these blood masses.

## DIAGNOSIS OF CLOTTING DISORDERS

The opposite of bleeding complications are the clotting disorders. Inappropriate activation of coagulation mechanisms in intact blood vessels results in thrombosis—that is, intravascular fibrin clot formation. Production of intravascular clots, however, is a rare event as the body has many protective mechanisms to prevent such occurrences. As explained in chapter 30, in the few situations in which thrombosis occurs, it involves either overwhelming thrombogenic stimulation or a breakdown in the protective mechanisms.

### Clinical Diagnosis of Clotting Disorders

Intravascular thrombi may occur in the veins, arteries, or microcirculation. Clinical symptoms and risks associated with such clots depend on where they are found in the body.

*Manifestations of Venous Thrombosis*

The majority of all thromboses are venous thromboses, which occur most commonly in the veins of the legs. Blood moves more slowly through the veins of the legs as it must work against gravity. This causes stress on the walls of these blood vessels, and as a result leg veins may become damaged more easily. Increased stress on the blood vessels and slow-moving blood may also contribute to the legs becoming more prone to inflammation and subsequent venous obstruction. All these factors predispose the leg veins to the formation of clots, especially in older people whose circulation has already slowed due to less efficient functioning of the heart and the lack of exercise.

The common clinical symptoms of venous thrombosis are as follows:

1. Pain and tenderness in the leg caused by local inflammation of vein walls.
2. Swelling of the leg due to local edema, which is usually due to obstruction in a large vein caused by the presence of a clot and the resulting inflammation.
3. Leg discoloration may be pale, bluish, or reddish. Paleness of the leg indicates that insufficient blood is circulating through the veins. This is relatively uncommon in thrombosis, and if it occurs is present only in the early stages of blockage by a fibrin clot. More common is a bluish color due to cyanosis caused by impaired venous return and a stagnant blood flow. Rarely the leg may appear red and hot due to marked perivascular inflammation.

In many cases venous thromboses are asymptomatic. Although most venous clots occur in the legs as a result of local inflammation—a condition known as **phlebitis**—they do not always remain there. The danger of venous thrombosis **(thrombophlebitis)** occurs when clots dislodge from the leg veins and move to other parts of the body, most commonly the lungs. The most common complication of venous thrombosis is the formation of a pulmonary embolus **(embolism)**. If the clot is small the effects will be minor, but if the clot is large it may produce pulmonary infarction, which may block the oxygen supply to part of the lung. This has two effects: it destroys part of the lung tissue, and as a consequence it impairs the ability of the lung to absorb oxygen. Lung infarction may result in local inflammation, which may be manifested by chest pain.

If the clot lodges in the pulmonary artery, it may interfere with the pumping action of the heart. This may affect the normal functioning of the heart, resulting in chest pain and abnormal heart rhythms.

*Manifestations of Arterial Thrombosis*

Intravascular clots in the arterial side of the circulation always affect the organ supplied by the blocked artery or arteries. The most common organs affected are the heart, resulting in myocardial infarctions (=heart attack), or the brain, producing stokes. Arterial embolisms may also affect the legs, which may result in the development of gangrene; the kidneys; and the mesenteric arteries—that is, the arteries supplying the intestines.

Arterial clots usually develop from deterioration of the arterial walls and are usually the result of either atherosclerosis or arteriosclerosis. **Atherosclerosis** results from the deposition of plaques rich in cholesterol on the wall of the blood vessels, thus reducing the diameter of the vessels. This fatty material may later be replaced with dense connective tissue and calcium deposits. **Arteriosclerosis** is a degenerative condition of the large arteries, making them less elastic. This condition is often referred to as "hardening of the arteries."

The development of atherosclerosis is influenced by several factors. Lack of exercise, smoking, obesity, diabetes, hypertension, and a diet high in low-density lipoprotein (LDL) cholesterol all seem to increase the risk of developing atherosclerosis. A genetic component also seems to be involved as this condition is more prevalent in certain families. In contrast, arteriosclerosis appears to be more a function of old age.

However all the factors producing arteriosclerosis or atherosclerosis cause endothelial damage, resulting in exposure of the underlying basement membrane. This produces platelet aggregation, which is often the first step in clot formation. The major arterial clotting disorders are discussed in the following material.

**Coronary arterial occlusion** Patients with partially blocked arteries frequently suffer from chest pain, a condition known as angina pectoris, or simply **angina.** Acute myocardial infarction is caused by the development of a clot in an artery of reduced diameter, resulting in a complete blockage of that artery. This condition is usually associated with severe chest pain that may radiate up into the jaws or down into the arms. Frequently myocardial infarction results in severe general malaise as well as congestive heart failure and arrhythmia.

A strong relationship exists between diabetes mellitus and the development of coronary artery disease. Diabetic men have twice the frequency and diabetic women three times the frequency of coronary heart disease compared to nondiabetic people.

**Stroke syndrome** A stroke may be defined as a complex symptom caused by a disorder of the blood vessels serving the brain, with impaired vascular supply to part of the brain. The three major causes of strokes are cerebral thrombosis, cerebral embolism, and cerebral hemorrhage.

**Cerebral thrombosis** is caused by a clot in a blood vessel of the brain and is the most common cause of stroke. Most often the thrombosis, or clot, is caused by atherosclerosis, where there is a narrowing of the vessel. The thrombosis produces ischemia, edema, and congestion of the brain tissue surrounding the area.

**Cerebral embolism** is caused by a small mass of material circulating in the blood vessels, usually a detached portion of a thrombus that settles in a cerebral vessel. Damage from embolisms is often less severe and recovery more rapid than in strokes from thrombosis and cerebral hemorrhage.

A rupturing of a blood vessel, usually an artery within the brain, is the cause of **cerebral hemorrhage.** The hemorrhage is associated with preexisting hypertension and weakening of blood vessel walls.

The symptoms of stroke vary depending on its cause, location in the brain, and extent of damage to the brain cells. Patients usually experience neurological symptoms such as loss of sensation, reflex changes, and hemiplegia (paralysis of one side of the body).

The major neurological events associated with the stroke syndrome are categorized according to whether the event is temporary or permanent and reversible or irreversible. **Transient ischemic attack (TIA)** is usually temporary and reversible, whereas **completed stroke (CS)** exhibits permanent symptoms of severe cerebral ischemia, which is an interrupted or deficient supply of blood to the brain. A TIA is an event characterized by temporary loss of function (e.g., speech or movement) usually lasting less than 24 hours and appearing to have no aftereffects.

**Gangrene and claudication** Peripheral arterial occlusions in the leg frequently lead to gangrene. Complete blockage of the artery will cause local ulceration, which may develop into gas gangrene. Clots in the arteries of the lower limbs often produce sudden pain due to **ischemia,** a deficiency of blood in part of a blood vessel. In other cases occlusive arterial disease is associated with numbness of the affected leg as well as **claudication,** which may be defined as reproducible leg pain that occurs as a result of exercise and can be relieved with rest.

**Visceral arterial embolism** Visceral arterial embolism may develop into infarction of the arteries associated with the kidneys or the intestines. Renal infarction is usually associated with sudden flank pain, hematuria, and the development of hypertension.

Acute mesenteric infarction produces severe abdominal pain, usually associated with vomiting and bowel evacuation.

#### *Manifestations of Microcirculatory Thrombosis*

Thrombosis of the microcirculation is a complication of disseminated intravascular coagulation (DIC) or disseminated platelet aggregation. It produces local ischemia and tissue necrosis, which may lead to local pain. However the most important consequence paradoxically seems to be development of a bleeding condition associated with the administration of platelets and clotting factors.

### Laboratory Diagnosis of Thrombosis

The laboratory diagnosis of thrombosis includes tests for venous and arterial occlusions. The presence of microcirculatory clots may be deduced by using a combination of tests of hemostasis previously described.

#### *Laboratory Tests for Venous Occlusions*

Many procedures are used to investigate clot formation in the venous circulation—for example, **venography.** In this test dye is injected into the veins, and an X ray is taken to follow the flow of the dye. A venous blockage will inhibit the flow of the dye. Venography is especially useful in cases of phlebitis.

Another test for venous occlusions uses ultrasound. This technique employs a Doppler ultrasound flow meter to identify changes in the blood flow velocity in the veins.

A third method uses radioactive fibrinogen. In this test $^{125}I$ fibrinogen is injected into a vein in the arm (after first administering potassium iodide to prevent uptake of the radioactive iodine into the thyroid gland). If a thrombus is present in the leg, the radioactive fibrinogen will be incorporated into the clot, and a higher level of radioactivity will be detected at that site.

*Laboratory Tests for Arterial Occlusions*

Several diagnostic techniques are commonly used to detect arterial occlusions. The first is **arteriography,** which is similar to venography. Radioactive scanning may involve any organ of the body.

The second method involves ultrasound, which is especially valuable in locating clots in the leg.

*Laboratory Tests for Microcirculatory Microthrombi*

Test results diagnosing disseminated intravascular thrombosis of the microcirculation include a prolonged bleeding time, a reduced platelet count, as well as a low fibrinogen determination—all of which can indicate small clots that may be present in the microcirculation.

## Further Readings


Bick, R. L. Congenital and Acquired Coagulation Protein Defects Associated with Hemorrhage and Thrombosis. Chapter 24 in *Diagnostic Hematology, Clinical and Technical Principles,* L. W. Powers, Ed. The C. V. Mosby Company, St. Louis, 1989.

Hirsch, J., and Brian, E. A. *Hemostasis and Thrombosis, A Conceptual Approach,* 2nd ed. Churchill Livingstone, New York, 1983.

Linné, J. J., and Ringsrud, K. M. *Basic Techniques in Clinical Laboratory Science,* 3rd ed. Mosby-Yearbook, St. Louis, 1992.

Powers, L. W. Platelet Disorders. Chapter 23 in *Diagnostic Hematology, Clinical and Technical Principles.* The C. V. Mosby Company, St. Louis, 1989.


## Review Questions

1. Describe two ways in which abnormalities in the hemostatic mechanism can be detected.
2. What are petechiae? What causes their development?
3. Define purpura. What causes the development of this condition?
4. Define ecchymosis. Why do ecchymoses develop?
5. Define hematoma. Distinguish between an ecchymosis and hematoma. Why do hematomas develop?
6. Define telangiectasia. Distinguish between splotches made by telangiectasia and those made by purpura conditions.
7. Define hemarthrosis. With what condition is hemarthrosis usually associated?
8. Define hematuria. What are some common causes of hematuria?
9. Blood in the stool is usually an indication of what condition?
10. Define the term hemorrhoids. What factors predispose people toward hemorrhoids?
11. List the major causes of excess vaginal bleeding.
12. List the three categories of tests designed to discover the causes of bleeding disorders.
13. List the major tests that determine if vascular factors are the cause of excessive bleeding.
14. What is the aim of the capillary fragility test? How is it performed?
15. What is measured in the bleeding time test? How is it performed?
16. What is the usual bleeding time? What is the usual reason for an increase in bleeding time?
17. Distinguish between the Duke and Ivy tests for bleeding time. What is the problem with these tests? How can this problem be overcome?
18. How is the template bleeding time test performed?
19. List the major tests used to identify the involvement of platelet factors in bleeding disorders.
20. How is a platelet count performed? Describe the Rees and Ecker method.
21. What is the platelet aggregation test? How is it performed?
22. What is the platelet adhesiveness test? How is it performed?
23. List the factors that influence the clot retraction time.
24. List the major tests used to investigate the involvement of plasma factors in bleeding disorders.
25. What does the fibrinogen determination test measure? What are the normal levels of fibrinogen in the blood?
26. How is the fibrinogen determination test performed?
27. Compare the use of the PT and the APTT tests.
28. When the PT test is prolonged, it is an indication of either factor deficiency or factor inhibition. How would you determine which of the two is involved?

29. Which coagulation factors can be tested with the PT test? How would you determine which of the coagulation factors is deficient?
30. Which coagulation factors are investigated in the APTT test?
31. Name some of the major dangers associated with excessive or prolonged bleeding.
32. List the three major groups of clotting disorders.
33. Why are most venous thromboses found in the legs?
34. Why are venous thromboses more common in the elderly than in the young?
35. List the major clinical symptoms associated with venous thrombosis. What is thrombophlebitis?
36. What is the greatest danger associated with venous thrombosis in the legs?
37. Why are arterial occlusions generally more dangerous than venous thromboses?
38. List the major clinical conditions associated with arterial occlusions.
39. Distinguish between atherosclerosis and arteriosclerosis. Which factors predispose an individual to the development of these conditions?
40. Define the following terms: (a) acute myocardial infarction, (b) stroke, (c) CS, (d) TIA, (e) gangrene, (f) claudication, and (g) mesenteric infarction.
41. Define DIC. What major clinical symptoms are associated with DIC?
42. Explain why DIC causes increased bleeding.
43. List the major laboratory tests for venous thromboses.
44. Name the major laboratory tests for arterial occlusions.
45. List the major laboratory tests for the presence of microcirculatory thrombi.
46. Define venography and arteriography.

# Appendix

## *Hematology Laboratory Manual*

## GENERAL RULES AND REGULATIONS

The first and most important rule in the hematology laboratory is to **wear gloves at all times** when handling or testing blood specimens. All blood samples must be considered potentially infectious, and great care should be taken in handling blood. Blood should never come in direct contact with the skin of the laboratory worker.

**No food or drink** should be allowed in the laboratory. All workbenches should be considered contaminated with infectious agents.

**Keep the laboratory neat.** You may be dealing with multiple blood specimens from many different patients, so it is imperative that the laboratory is orderly at all times. Untidy laboratory space can lead to mixing specimens and inaccurate recording.

**Take good care of the instruments.** Many instruments used in the hematology laboratory are very expensive pieces of equipment that require careful handling according to the manufacturer's instructions. Do not use unfamiliar instruments without first reading the instruction manual or without a demonstration by an instructor.

**Always follow the exact start-up and shut-down procedures** recommended by the manufacturer. Do not use shortcuts. Follow the quality control procedures described by the manufacturer.

**After analyses blood samples should be placed in special receptacles** and disposed of according to the Occupational Safety and Health Administration (OSHA) procedures. Do not pour blood down the sink.

## HEMOCYTOMETRY

Hemocytometry is the process of counting the formed elements of the blood. The elements counted in routine practice are red blood cells (RBCs), white blood cells (WBCs), and platelets. Counting is done either by an electronic device (electronic cell counter) or manually with the aid of a specially designed glass slide known as a hemocytometer. There are several advantages and disadvantages of electronic cell counting over the manual technique. First, electronic cell counting avoids human error, provided the machine is properly calibrated and subjected to constant quality control. Second, the electronic device is not only more accurate but also much faster in counting large numbers of blood samples. Manual counts are inherently labor intensive and thus cost intensive. On the other hand, manual cell counting does not require the expensive equipment and reagents associated with electronic cell counting. It also does not require constant calibration and quality control. In other words, manual cell counting is much less expensive than electronic cell counting. Further, manual cell counting may be performed where there is a microscope present. Electronic cell counting usually requires a special laboratory setting.

### General Cell-Counting Procedures

Since many different formed elements are in the blood in high concentrations, blood must be diluted before counting can be done. Diluting fluid is selected for its ability to (1) dilute the blood, (2) to lyse cell types not wanted in the count, or (3) to stain a particular cell type so that it stands out and can be counted more readily.

Both electronic and the manual cell counts require only minute samples of blood, usually 20 microliters (μL) or less. Since such small quantities are used, errors in counting can occur if exact procedures are not followed.

Blood cell counts are usually reported in number of cells per cubic millimeter ($mm^3$), or per microliter (μL); 1 μL is equivalent to 1 $mm^3$. The International Committee for Standardization in Hematology (ICSH) has recommended a new unit of measurement. This new SI unit measures the number of blood cells per liter of whole blood. Thus the SI unit of measurement equivalent to 1 μL is $\mathbf{10^{-6}}$L. However, presently most cell counts are reported as the number of cells per microliter or per cubic millimeter.

### Electronic Cell Counting

Presently a wide array of electronic cell-counting devices are available commercially. Some are completely automated, others are semiautomated, requiring an extra step (i.e., the dilution of the blood) before the instrument is able to count the cells. Most electronic cell counters use a **voltage pulse counting principle.** Since this technique was originally developed by Coulter in the 1950s, it is also known as the **Coulter principle.** The basic principle behind this method of counting is that cells are poor electrical conductors compared with the saline diluting fluid in which they are suspended. Each time a cell passes between two charged electrodes, there will be a drop in voltage. The major components of the instrument are a vacuum pump, a glass tube with a small aperture and two electrodes (one inside and one outside the tube), a pulse amplifier, and a recorder.

Counting involves immersing the glass tube with the electrodes into the diluted cell suspension to be counted (figure A.1). The vacuum pump is switched on, and an exact amount of the diluted blood is drawn into the glass tube through the aperture. Usually a volume of between 0.5 and 1.0 mL is drawn into the tube, depending on the instrument. Each cell that passes through the aperture increases the electrical resistance between the electrodes and thus produces a small voltage change. This drop in voltage is then amplified and counted by another part of the machine. After all the voltage pulses (each pulse

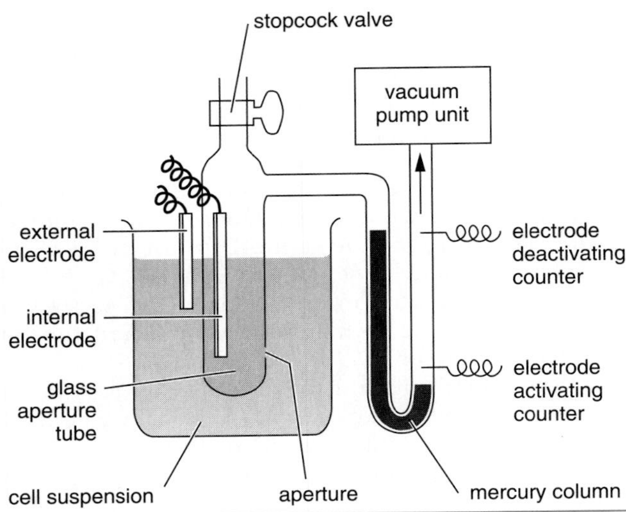


**figure A.1**

Schematic diagram of a cell counter based on voltage pulse counting (Coulter principle). Cells flow through an aperture that separates two compartments. The electrical potential between the electrodes changes as the cells pass. The number of impulses translates to the cell count, and the amplitude of the pulse depends on the cell volume that displaces the conductive fluid.

representing one cell) are counted, the exact number of cells per microliter is displayed on a digital screen or printed on a recording device, or both. Quality control specimens containing known numbers of cells are run periodically to determine the accuracy of the instrument.

The exact procedures for the automated counting of RBCs, WBCs, and platelets differs from instrument to instrument, and the user should follow the manufacturer's directions carefully.

### Manual White Blood Cell Count

The following reagents and instruments are needed for manual white blood cell counting:

1. **Anticoagulated blood.** For blood counts the anticoagulant of choice is **EDTA (ethylenediamine tetraacetic acid)** as this anticoagulant prevents blood coagulation and preserves the morphology of the cellular elements. When refrigerated properly, EDTA blood specimens can be preserved for up to 24 hours and still give relatively accurate readings. It is, of course, still advisable to perform the blood cell counts as soon as possible after the blood is withdrawn from the patient.
2. **White blood cell diluting pipette with rubber aspiration tubes and mouthpiece.** The aspiration tubing and mouthpiece are necessary to allow blood and diluting fluid to be drawn into the WBC pipette. The WBC pipette is known as a **Thoma pipette.** This is not a brand name but type of pipette that allows a 1:20 dilution of the blood. It does not measure blood in microliters but provides a dilution ratio. The WBC pipette consists of a stem, a bulb, and a top part that attaches to the rubber tubing. The stem is actually a graduated capillary tube divided into 10 parts and marked 0.5 at the fifth mark and 1.0 at the tenth mark. The bulb contains a white bead that facilitates the proper mixing of the blood with the diluent, and the piece above the bulb contains a graduation mark labeled 11.0 (figure A.2).
3. **White blood cell diluting fluid.** The WBC diluting fluid most frequently used is 2 mL/100 mL of **acetic acid.** This diluent does two things. First, it hemolyzes the red blood cells so that the white blood cells can be counted more readily. Second, it darkens the nuclei of the white blood cells somewhat, so that they are easier to see microscopically. **Türk's solution,** which adds gentian violet to the glacial acetic acid solution, can also be used. Gentian violet stains the white blood cell nuclei thus making them more readily identifiable.
4. **Hemocytometer and coverslip.** A hemocytometer is a specially designed counting chamber for WBCs, RBCs, and platelets. The most commonly used hemocytometer is the **improved Neubauer.** It is a thick glass slide with an H-shaped trough forming two ruled counting areas and two raised ridges (figure A.3). Each counting area is precision-ruled and is 3 mm wide and 3 mm long. The two raised ridges are 0.1 mm (1/10 mm) higher than the central counting areas which allows for the placement of a coverslip (or coverglass) (figure A.4). This creates a space between the counting area and the coverslip that is 0.1 mm deep. This space will be filled by the diluted blood from the Thoma pipette when the hemocytometer is "charged."

   When the ruled area of the counting chamber is viewed under the microscope, note that it is divided into 9 squares each 1 $mm^2$ (figure A.5). Each of the four corner squares (labeled A, B, C, and D) are further divided by single lines into 16 smaller squares. The white blood cells are counted in these four corner squares.
5. **Microscope with a low-power (10×) objective lens.** The WBCs are counted under low magnification (10× objective).

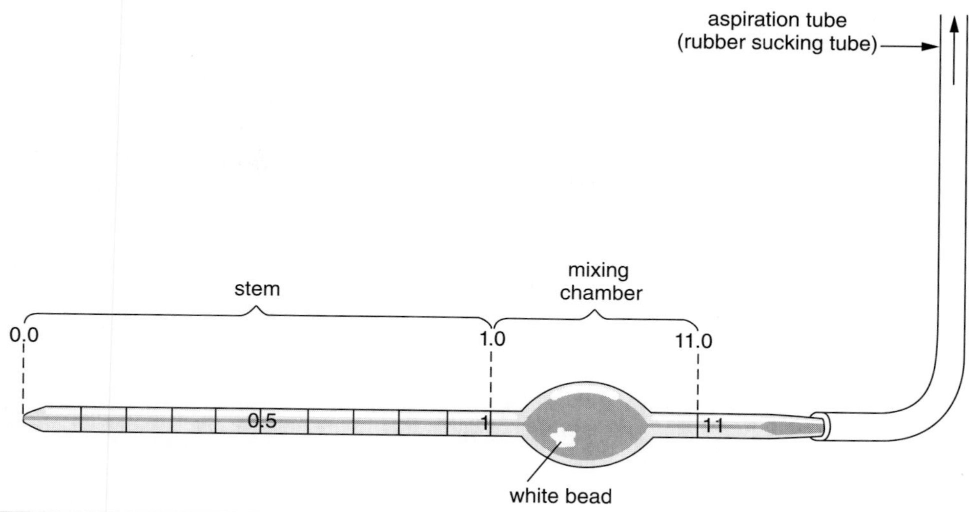


**figure A.2**

White cell Thoma pipette attached to rubber aspiration tube (or rubber sucking tube).

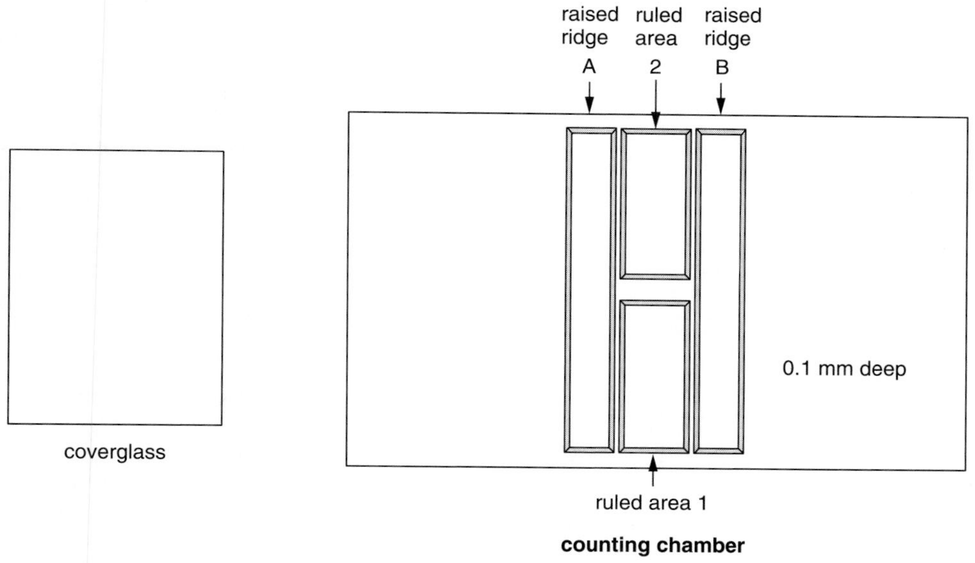


**figure A.3**

Hemocytometer counting chamber and coverglass.

6. **Aspirator connected to a faucet with running water.** An aspirator is a suction device used to clean the Thoma pipette after the cell count is completed. The aspirator works by sucking air out of the pipette once it is connected to a tap with running water (figure A.6). If the stem of the pipette is dipped in a beaker of distilled water, the water will be sucked into the pipette, allowing the inside of the pipette to be cleaned. After the pipette is cleaned, it can be dried by placing it either into a drying oven for a few hours or by dipping the pipette for a few seconds into ethanol and acetone. If the pipette becomes clogged by coagulated blood, it can be cleaned by placing it in a sonifier (a machine that produces high-frequency sound waves that will break up clots) or in a solution of 10% nitric acid. (When using nitric acid, exercise caution to avoid burns.)

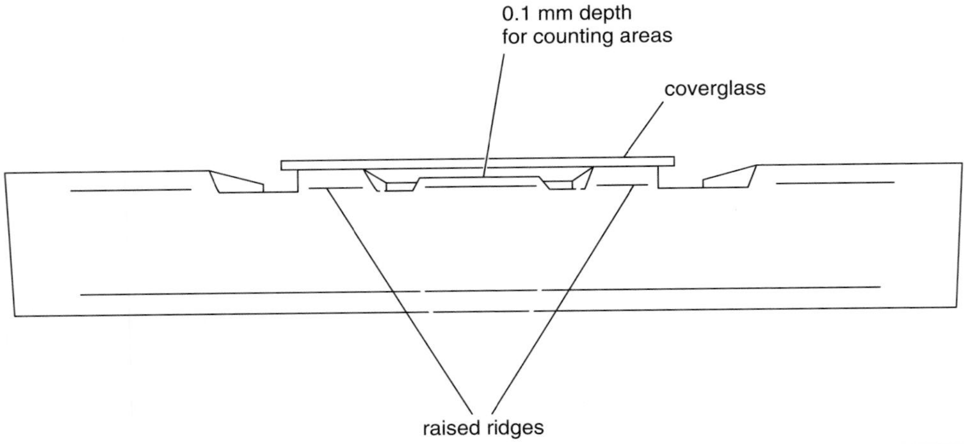


**figure A.4**

Side view of hemocytometer counting chamber.

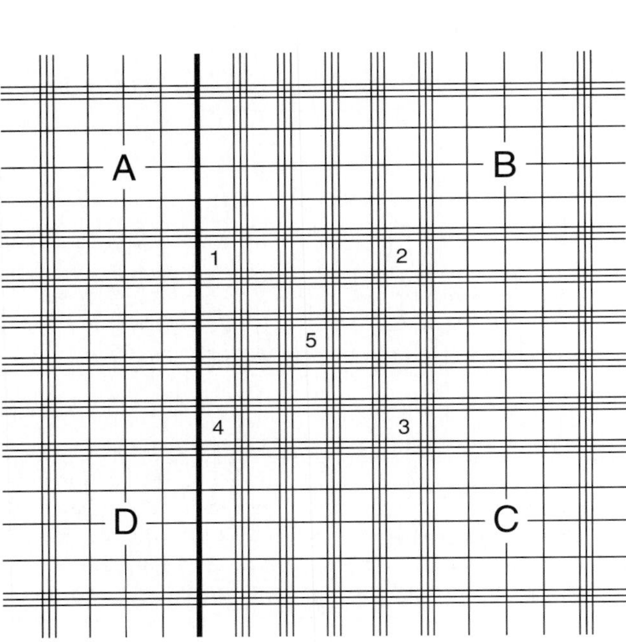


**figure A.5**

The ruled Neubauer hemocytometer. White blood cells are counted in areas A, B, C, and D (4 $mm^2$); red blood cells in areas 1, 2, 3, 4, and 5 (0.2 $mm^2$); and platelets in the central (25 smaller) squares (1 $mm^2$).

*Procedure for Manual WBC Count*

**Diluting the blood**

1. Assemble all equipment and reagents previously outlined and check that the WBC pipette is clean and in proper working order. Attach the rubber tube to the WBC Thoma pipette and the aspiration tube to the WBC Thoma pipette and the mouthpiece to the other end of the tube.
2. Pour a small amount of the diluting fluid (2% acetic acid or Türk's solution) from the stock bottle into a small jar or beaker.
3. Gently invert the fresh blood sample to ensure proper mixing; remove the stopper from the blood collection tube.
4. Insert the stem of the WBC Thoma pipette into the blood sample and withdraw the blood into the stem to slightly above the 0.5 mark.
5. Remove the pipette from the blood tube and cap the blood sample. Wipe the outside of the pipette stem with lint-free tissue and adjust the blood level to the 0.5 mark exactly by lightly tapping the tip of the pipette to the tissue. Be careful as it is easy to remove too much blood, and you will have to begin again.
6. Dip the tip of the pipette into the beaker with diluting fluid, making sure that the tip is well below the surface of the liquid.

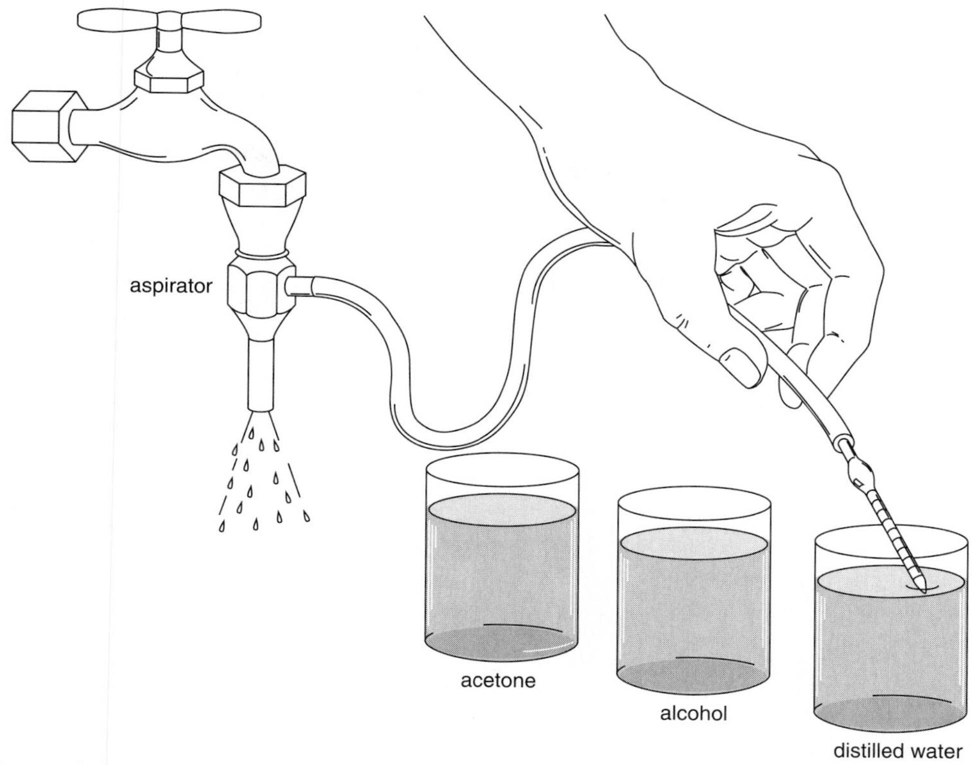


**figure A.6**
Cleaning a Thoma pipette by an aspirator connected to a faucet with running water.

7. Using constant suction, draw the diluent into the pipette until the bulb is filled and the fluid reaches the 11 mark slightly above the bulb of the pipette. While doing this, tap the bulb lightly with a finger so that the bead keeps moving and the blood is mixed with the diluent. Make sure that bubbles do not appear in the pipette bulb. If you observe bubbles, clean the pipette with the aspirator as previously described and repeat the procedure.
8. The fluid must remain close to the 11 mark and should never deviate more than 1 mm from it. This can be achieved by dropping the mouthpiece from the mouth as soon as the 11 mark is reached and simultaneously removing the pipette from the beaker and holding it in a horizontal position. Then gently remove the rubber tubing from the top of the pipette.
9. Place the pipette in a pipette shaker to mix the blood. If a pipette shaker is unavailable, place the pipette between the thumb and third finger as shown in figure A.7 and shake gently for 15 to 30 seconds.

### Filling the counting chamber

10. The blood is now ready to be placed in the hemocytometer counting chamber. Before filling or "charging" the hemocytometer, discard the first 5 drops (figure A.8) onto clean tissue paper or gauze to eliminate the cell-free diluent present in the pipette stem. Charge the chamber by touching the tip of the pipette gently to the edge of the coverglass (figure A.9). Blood will be drawn into the counting chamber by capillary action. Control the flow by placing the tip of your finger over the open end of

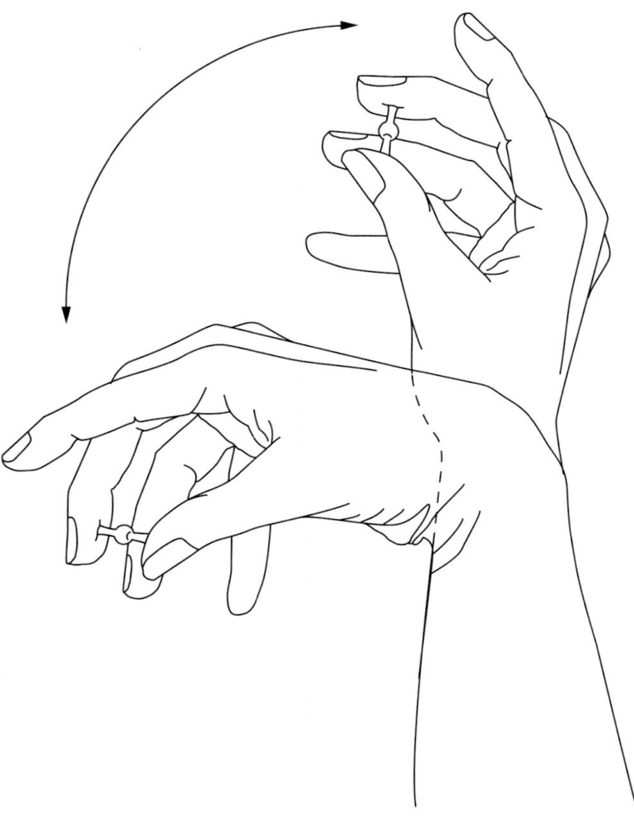

**figure A.7**
The proper method of manually mixing blood in a Thoma pipette. The hand is moved in a circular motion.

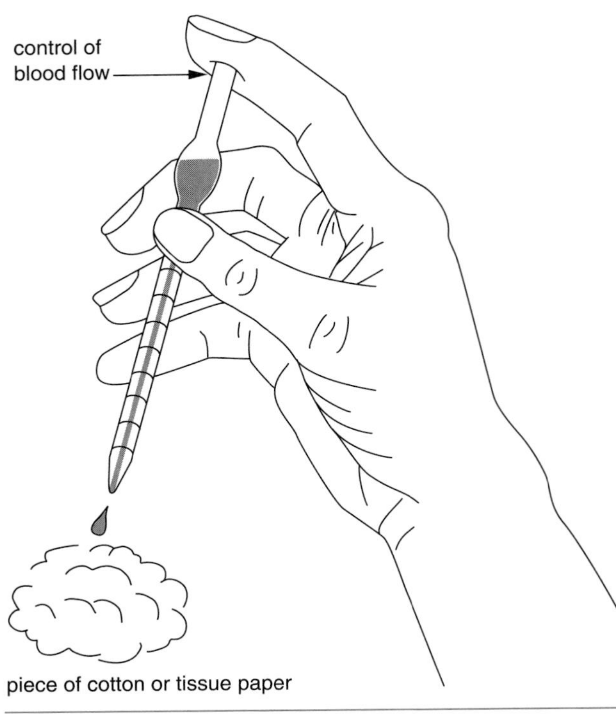


**figure A.8**
Discarding the first 5 drops from a Thoma pipette.

the pipette. Make sure the coverslip is not moved and that the chamber is uniformly charged. There should be no bubbles in the counting chamber and the fluid should not overflow the counting chamber into the grooves (figure A.10). If either happens, clean the hemocytometer and charge the chamber again.

## Counting the WBCs

11. Place the hemocytometer on the microscope stage and clamp in place. The center of the counting chamber should be directly under the light in the center of the stage.
12. Find the lined squares with the low-power objective and focus properly. The WBCs in each of the four corner squares will be counted. Notice that each corner square is divided into 16 smaller squares. Count all the white blood cells present in each of the 16 small squares of corner square A. Then repeat the same with corner squares B, C, and D. Cells touching the lines on the left side and the top of each small square are counted; those touching the right side and bottom of each small square are not counted (figure A.11). In this way cells are not counted twice. The total number of WBCs in each of the 4 corner squares should be relatively uniform and the values should agree within 10 cells. When the values do not agree within 10 cells, there is obviously an irregular distribution of WBCs in the counting chamber, and the procedure should be repeated—that is, the counting chamber should be cleaned and charged again.

## Calculating the WBC count

13. The final calculation should include (a) the total number of WBCs counted in all of the corner squares, (b) the dilution of the blood sample, (c) the total area of the 4 squares counted, and (d) the depth of the counting chamber. If the square areas and the depth are considered as volume (volume = area × depth = 4 $mm^2$ × 0.1 mm = 0.4 $mm^3$ or 0.4 μL) then the general formula is:

$$\text{WBCs}/\mu\text{L} = \frac{\text{Cells counted in 4 squares} \times \text{Dilution factor of the blood}}{\text{Volume}}$$

For instance, if the total WBC count in the 4 areas is 140, then the total WBC count would be:

$$\frac{140 \times 20}{0.4} = 7{,}000 \text{ WBCs}/\mu\text{L, or } 7 \times 10^3/\mu\text{L}$$

If SI units are used (i.e., number of WBCs per liter of blood) then the WBCs per microliter should be multiplied by $10^6$:

$$7 \times 10^3/\mu\text{L} \times 10^6 = 7 \times 10^9/\text{L}$$

Or more simply, the calculation can be determined by multiplying the total number of WBCs in the 4 squares by 50 (= 20 ÷ 0.4):

$$140 \times 50 = 7{,}000 \text{ WBCs}/\mu\text{L}$$

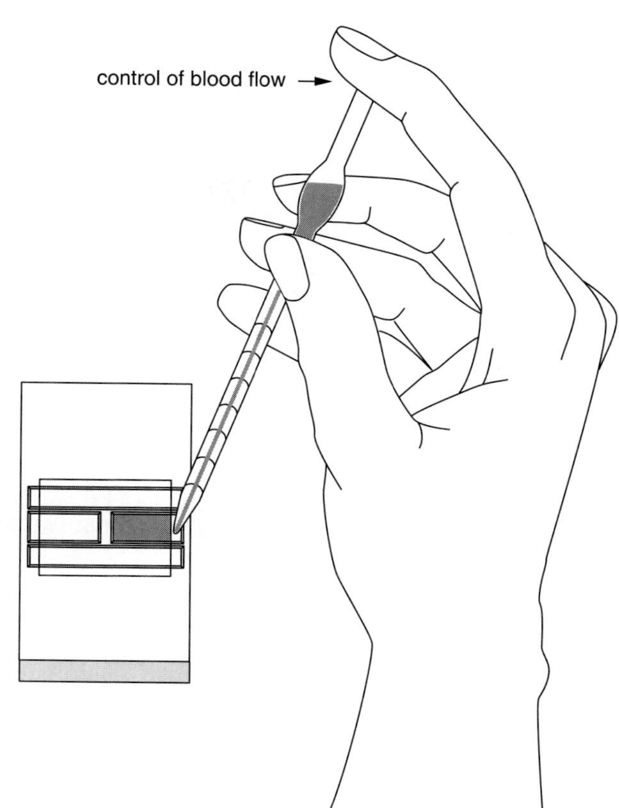


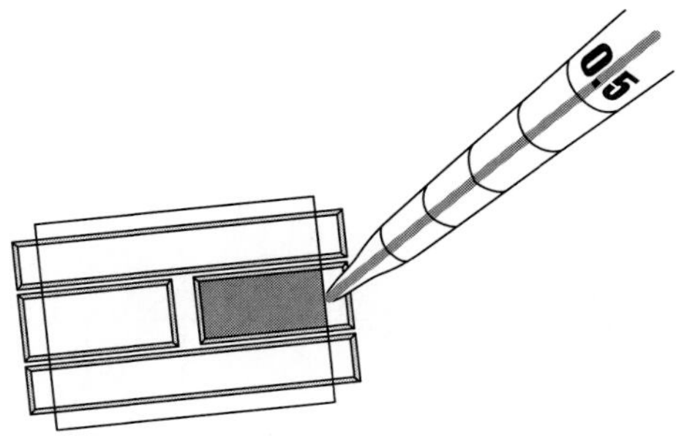


**figure A.9**
Filling the counting chamber of a hemocytometer.

Repeat counts should be within 10% of the original count. Thus if the original count was 7,000/µL, then the second count should be between 7,700 and 6,300 WBCs/µL. The normal range for total WBC counts is between 5,000 and 10,000 WBCs/µL.

## Manual Red Blood Cell Count

The following reagents and instruments are needed for a red blood cell count:

1. **Anticoagulated blood.** The same anticoagulated blood used for the WBC count is also used for the RBC count. Here again EDTA is the anticoagulant of choice.

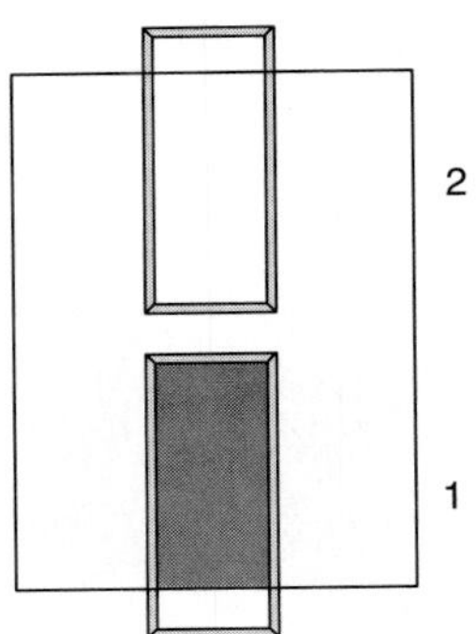


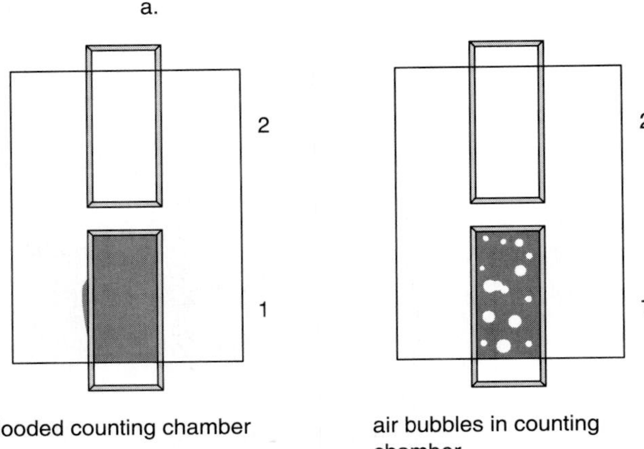


figure A.10

(*a*) Properly filled counting chamber. (*b*) Improperly filled counting chambers. Applying excess blood sample may cause overflow (flooding) of counting chamber. Air bubbles may be caused by an unclean chamber or coverglass or air bubbles in the pipette.

2. **Red blood cell pipette with rubber aspiration tube and mouthpiece.** The pipette used is also a Thoma pipette, but the bulb is much larger than the one used for the white blood cell pipette and contains a red bead. Like the white blood cell pipette, the stem of the red blood cell pipette is divided into 10 parts with an 0.5 sign at the fifth mark and a 1.0 sign at the tenth mark. There is a final mark above the bulb with a 101 sign. If blood is drawn into the stem to the 0.5 mark and this is mixed with the diluting fluid to the 101 mark, the blood will be diluted 200 times (0.5 in 100) (figure A.12).
3. **Red blood cell diluting fluid.** The two most common RBC diluting fluids are Gower's solution and Hayem's solution. An important characteristic of RBC diluting fluids is that they are isotonic with red blood cells; otherwise the osmotic balance is disturbed and the shapes of the red blood cells will become distorted and in extreme cases the cells may be destroyed. The diluting fluids should prevent red blood cells from clumping and allow them to be suspended evenly in the dilution fluid. Both Hayem's and Gower's solutions fulfill these characteristics. Gower's solution is composed of sodium sulfate, glacial acetic acid, and water, and Hayem's solution is composed of mercuric chloride, sodium chloride, and sodium sulfate in water; hence, Hayem's solution is slightly more toxic.
4. **Hemocytometer and coverslip.** The hemocytometer described for the WBC count is also used for the RBC count. The same precision-ruled 3 mm area is used to count the number of red blood cells per microliter. However the 1 mm$^2$ squares are not used. Instead, for the RBC count, the central square is used (figure A.13). This central 1 mm$^2$ square is divided into 16 smaller squares, not by single lines (as was the case with the corner squares used in the WBC count) but by triple lines. Further observation will show that each of these 16 smaller squares is divided into 16

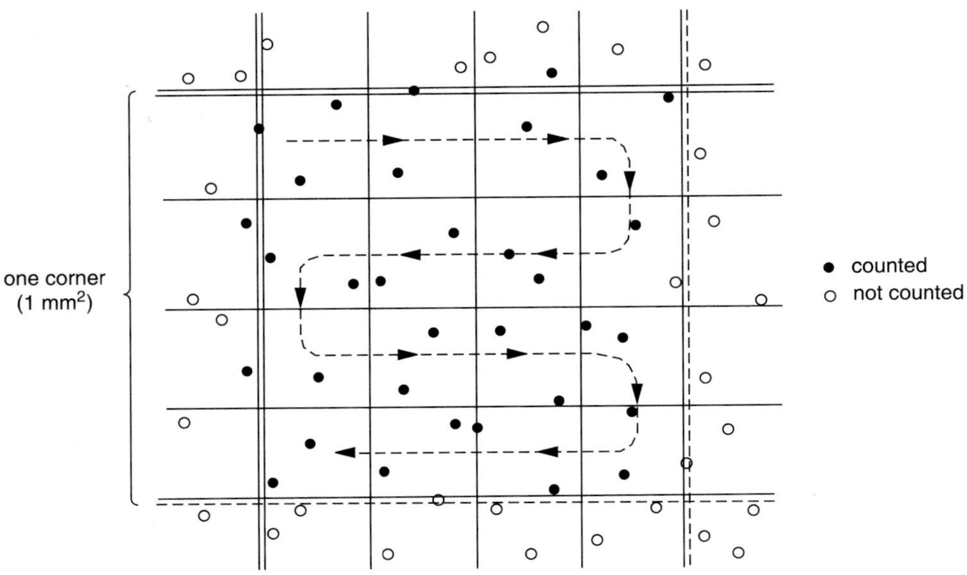


figure A.11

Examples of white blood cells counted (and not counted) in a representative square.

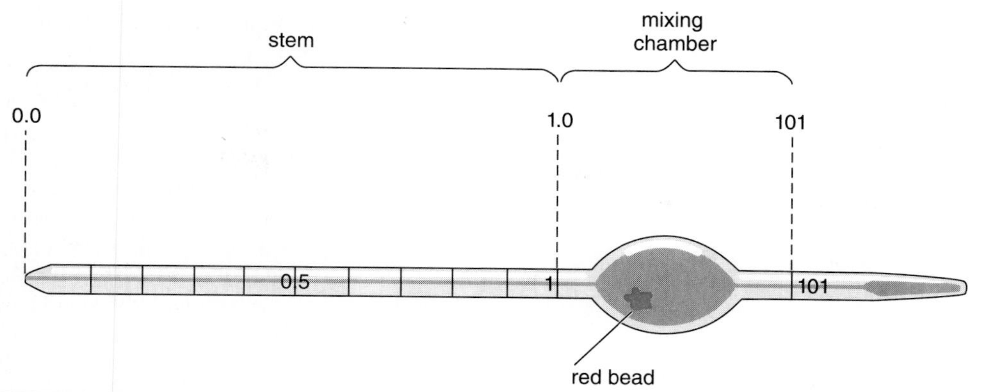


**figure A.12**

Red cell Thoma pipette (1:200 dilution).

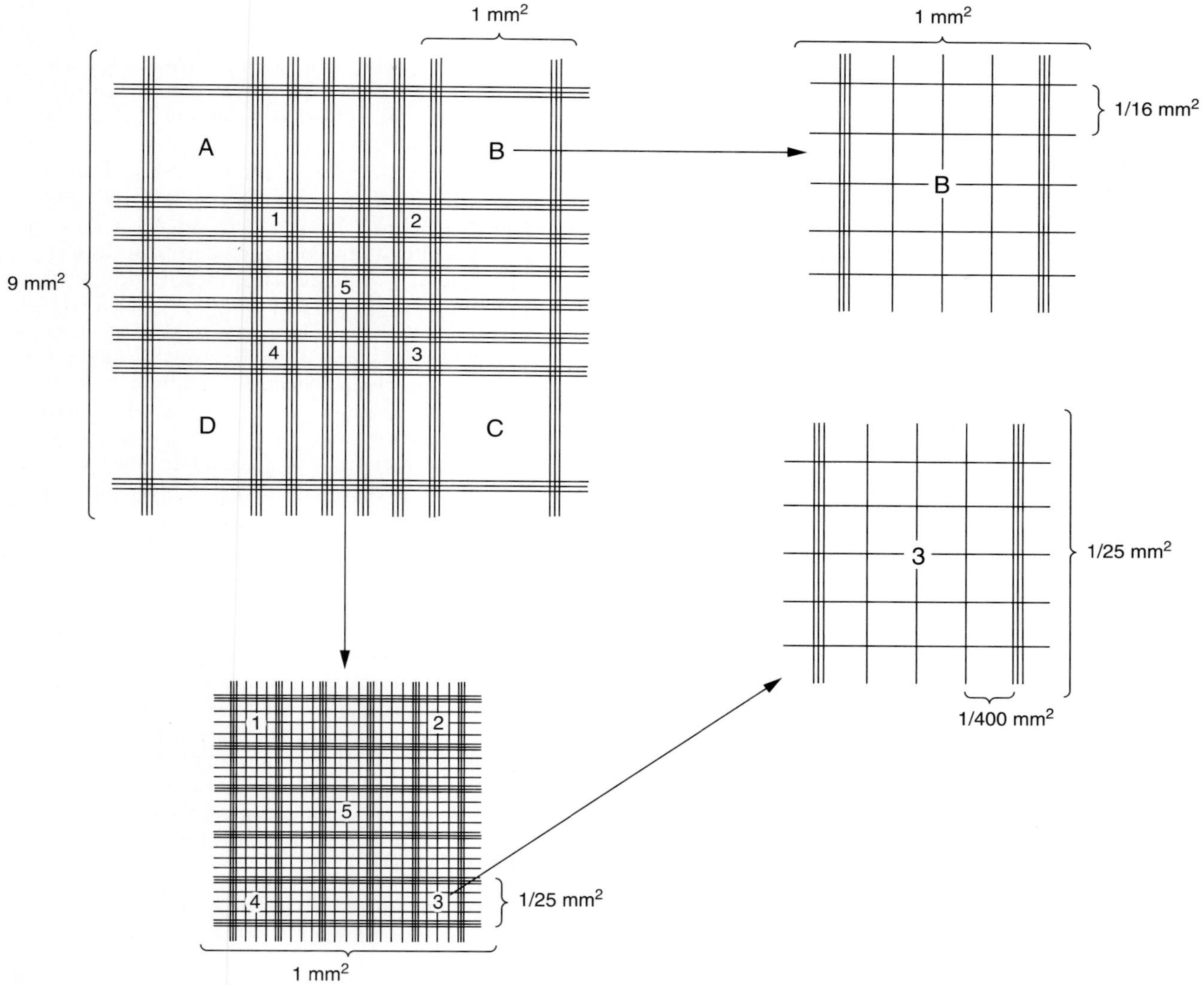

**figure A.13**

Red blood cells are counted in areas 1, 2, 3, 4, and 5 of the hemocytometer.

tiny squares made by single lines. Five of these 16 smaller squares are used for the counting of red blood cells—that is, the 4 corner squares and the central square.

5. **Microscope with both a low-power (10×) and a high-power (40×) objective.** The low-power objective is used to find the precision-ruled area and to focus on that area. Then the 40× objective is used to magnify specific areas (i.e., each of the 5 RBC counting squares).
6. **Aspirator connected to a faucet with running water.** The same procedure previously described for the WBC pipettes is also used to clean the RBC pipettes (see figure A.6).

*Procedure for Manual RBC Count*

**Diluting the blood**

1. Assemble all the instruments and reagents previously listed and check that the RBC pipette is clean and in proper working order. Attach rubber tubing to the RBC pipette and the red mouthpiece to the other side of the tubing.
2. Pour a small amount of diluting fluid (either Gower's or Hayem's solution) from the stock bottle into a small beaker.
3. Gently invert the tube of fresh blood to ensure proper mixing. Remove the stopper from the tube of blood.
4. Insert the stem of the RBC Thoma pipette into the blood sample and withdraw the blood into the stem to slightly above the 0.5 mark.
5. Remove the pipette from the blood, cap the tube, and replace in rack. Wipe the outside of the pipette stem with lint-free tissue and adjust the level of the blood in the stem to the 0.5 mark exactly by lightly tapping the tip of the pipette to the tissue. Be careful as it is easy to remove too much blood, and you will have to begin again.
6. Dip the pipette into the diluting fluid, making sure that the tip is well below the surface of the fluid. Using constant suction, draw the diluent into the pipette until the bulb is filled and the fluid reaches the 101 mark. While doing this, tap the bulb lightly with a finger so that the bead keeps moving and the blood is mixed with the diluent. Make sure that bubbles do not appear in the pipette bulb. If you observe bubbles, clean the pipette with the aspirator as previously described and repeat the procedure.
7. The fluid must remain close to the 101 mark and should never deviate more than 1 to 2 mm from it. If you draw in too much fluid, clean the pipette with the aspirator and repeat the procedure. After the dilution is complete, remove the rubber tubing from the pipette.
8. Place the pipette in a pipette shaker to mix the blood. If a pipette shaker is unavailable, place the pipette between the thumb and the third finger as shown in figure A.7 and shake gently for 30 seconds.

**Filling the counting chamber**

9. The blood is now ready to be added to the hemocytometer counting chamber. Before charging the instrument, discard the first 5 drops onto some clean tissue paper to eliminate the cell-free diluent present in the pipette stem.
10. Charging the chamber is performed the same way as was described for the WBC count. Very gently touch the tip of the pipette to the edge of the coverslip (see figure A.9). The diluted blood will be drawn into the counting chamber by capillary action. The flow of the fluid can be controlled by placing a finger over the end of the pipette. Make sure that the chamber is uniformly charged. Any movement of the coverglass, bubbles in the counting chamber, or an overflow into the grooves would invalidate the count. If these mistakes occur, clean and charge the hemocytometer again.

**Counting the RBCs**

11. Place the hemocytometer on the microscope stage and clamp in place. The center of the counting chamber should be directly under the light in the center of the stage.
12. First find the lined area with the low-power (10×) objective and focus properly. Now switch to high power (40×) and find one of the small corner squares in the center of the lined area. They can easily be recognized by the triple lines (figure A.13). Notice that each of these smaller squares is divided into 16 tiny squares. Count all the red blood cells in each of these 16 tiny squares. Cells touching the left side and the top of each tiny square are counted; those touching the right side and the bottom line are not counted. In this way cells are not counted twice. Count the number of RBCs in each of the tiny squares. Now repeat the other 3 corner squares and the central square.
13. The total number of RBCs in each of the squares should be relatively uniform. If there are great differences between the numbers of each of the 5 squares, an unequal distribution must be presumed and the count should be repeated. Because the manual RBC count (using a Thoma pipette) is a relatively inaccurate procedure, each blood sample should be tested at least twice (i.e., each sample should also be pipetted twice). The counts from the two pipette samples should be within 10% of each other. If this is not the case, the test should be repeated.

**Calculating the RBC count**

14. Calculating the red blood cell count is based on the same principles as those used for the white blood cell count. It includes the same four parameters: (a) the total number of RBCs counted in all 5 squares, (b) the dilution factor of the blood sample, (c) the total area of the 5 squares counted, and (d) the depth of the counting chamber.

One tiny square occupies 1/400 $mm^2$. The total number of tiny squares counted is 80 ($5 \times 16$), so the total area counted is $80/400\ mm^2 = \frac{1}{5}\ mm^2$(0.2 $mm^2$). As with the WBC count, volume is needed, and this is determined by multiplying the area times the depth, which is 0.1 mm. So the volume is $0.1 \times 0.2 = 0.02\ mm^3 = 0.02\ \mu L$. We can calculate the total RBC count according to the following formula:

$$\text{RBCs}/\mu\text{L} = \frac{\text{Cells counted in } 0.2\ \text{mm}^2 \times \text{Dilution factor}}{\text{Volume}}$$

For instance, if the total number of RBCs counted in the 5 squares was 500, then the total RBC count would be:

$$\frac{500 \times 200}{0.02} = 5{,}000{,}000\ \text{RBCs}/\mu\text{L}$$

To report this result in SI units (i.e., cells counted per liter of blood) multiply the number of cells per microliter times $10^6$, that is

$$5.0 \times 10^6/\mu\text{L} \times 10^6 = 5.0 \times 10^{12}/\text{L}$$

The total RBC count can also be calculated by multiplying the total number of RBCs in 5 squares times 10,000 (= $200 \div 0.02$):

$$500 \times 10{,}000 = 5{,}000{,}000\ \text{RBCs}/\mu\text{L}$$

The normal range for total RBC counts is 4.0 to $5.5 \times 10^6/\mu L$ for women and 4.5 to $6.0 \times 10^6/\mu L$ for men.

## Manual Platelet Count

The following reagents and instruments are necessary for a manual platelet count:

1. **Fresh blood.** This can be obtained directly from a patient via a finger puncture or from a freshly drawn sample preserved with the anticoagulant, EDTA.
2. **Red blood cell pipette with rubber tubing and mouthpiece.** The Thoma pipette used for RBC counting is also used for the manual platelet count. This pipette must be perfectly clean; anything that allows the platelets to adhere would reduce the actual platelet count.
3. **Rees-Ecker diluting fluid.** A good platelet diluting fluid must have certain important properties. First, it should prevent platelets from adhering to the glass surface of the pipette. It should also prevent the platelets from aggregating. Further, the diluting fluid should not promote premature platelet hemolysis. Finally, it should make platelets readily visible under a microscope. The Rees-Ecker solution fulfills these requirements. It contains sodium citrate, which prevents coagulation. It contains formalin, which fixes the platelets and thus prevents their premature hemolysis, and it contains the dye brilliant cresyl blue, which allows the platelets to be observed more clearly under a microscope.
4. **Hemocytometer and coverslip.** The hemocytometer used for the RBC and WBC counts can also be used for the platelet count. Platelets found in the central square millimeter (made up of 25 small squares) are counted.
5. **Microscope with low-power (10×) and a high-power (40×) objective lenses.** The low-power objective is used to find the ruled area and to focus on the central square. The high-power objective is used for counting.
6. **Aspirator connected to a faucet with running water.** The aspirator and its use were described in the WBC count section (see figure A.6).

*Procedure for Manual Platelet Count*

### Diluting the blood

1. The capillary bore of the RBC pipette is rinsed with Rees-Ecker diluting fluid. All excess fluid should be removed from the pipette before blood is drawn into the pipette.
2. Draw blood into the bore of the pipette to slightly above the 0.5 mark. Clean the outside of the pipette if an anticoagulated blood sample is used. Adjust the blood level to the exact 0.5 mark with a tissue paper. Do not linger over this process.
3. The blood should be diluted quickly with Rees-Ecker solution. It should be drawn up to the 101 mark of the RBC pipette. Two pipettes should be used simultaneously, and each should be used to charge one side of the counting chamber.
4. After dilution the pipettes should be shaken immediately for at least 1 minute.

### Charging the counting chamber

5. After mixing, the first few drops from each pipette should be discarded and the subsequent drops used to charge the counting chamber. Each side of the hemocytometer should be charged with a different pipette.
6. After the chambers are charged, the platelets should be allowed to settle properly for about 10 to 15 minutes. To prevent evaporation, the hemocytometers are kept in a covered container (usually a Petri dish) containing moistened tissue paper.

### Counting the platelets

7. The charged hemocytometer is carefully placed on the microscope stage so as not to disturb the platelets. The platelets are counted in the central square millimeter that is divided into 25 smaller squares by double or triple lines (see figure A.5). It is the same square millimeter used in the RBC count. However in the RBC count only 5 of the smaller squares are counted. The platelets appear as small, roundish, unevenly shaped structures.
8. After the platelets in the central square of the first chamber are counted, the procedure is repeated for the central square of the second chamber. The platelet counts should be within 10% of each other; otherwise the procedure should be repeated.

### Calculating the manual platelet count

As with the RBC and WBC counts, the important parameters for calculating the platelet counts include (1) the average number of platelets per square millimeter; (2) the dilution factor (which is 100, just like the RBC count); (3) the volume of the diluted blood counted, which is area times depth (1 $mm^2 \times 0.1$ mm $= 0.1$ μL).

The following formula is usually used:

$$\text{Platelets} = \frac{\text{Average number of platelets per mm}^2 \times 200}{0.1\ \mu\text{L}}$$

Thus if an average of 150 platelets were counted in each central square, then the platelet count should be:

$$\frac{150 \times 200}{0.1} = 300{,}000/\mu\text{L}\ (= 300 \times 10^9/\text{L})$$

Or simply multiply the platelets counted by 2,000 (= 200 ÷ 0.1) which is calculated as 150 × 2,000 = 300,000/μL. The normal range for the platelet count is between 150,000 and 400,000/μL or 150 to 400 × $10^9$/μL.

### Numerical estimate of platelets from a blood smear

Since so many variables may interfere with an accurate manual platelet count, a blood smear is also frequently used to verify the manual platelet count. The blood smear is prepared at the same time that the blood is drawn into the Thoma pipette.

For an estimate of platelet concentration from a blood smear to be valid, the platelets should not be clumped to any degree. The platelets should be observed in the same area in which differential white blood cell counts are made. In a normal individual each oil-immersion field should yield between 8 and 20 platelets. An estimate can also be calculated from the formula:

$$\text{Platelets}/\mu\text{L} = \text{Number of platelets per 100 RBCs} \times \text{Hematocrit (in \%)} \times 100$$

## Unopette System for Blood Dilution

The accurate manual count of RBCs, WBCs, and platelets depends to a large extent on the correct use of the Thoma pipettes. Since only small amounts of blood are used, even a slight variation in the amount of blood pulled up in the bore of the Thoma pipette may result in significant variations in counting. Further, small variations in the amount of diluting fluid may also result in large fluctuations in numbers of the same amount of blood. To reduce the errors in sampling and diluting, several different self-filling, self-measuring dilution micropipettes have been developed. The most well-known of these micropipette systems is the Unopette System. The Unopette System consists of a series of self-filling micropipettes available in different sizes, depending on the procedure to be performed. Special Unopettes are available for RBC counts, WBC counts, platelet counts, reticulocyte counts, eosinophil counts, hemoglobin measurement, and erythrocyte fragility tests. All Unopette pipettes have a similar shape but can be distinguished from each other by specific color codes. Each Unopette micropipette consists of a self-filling precalibrated glass capillary pipette attached to a plastic overflow chamber. The glass pipette tip is protected by a plastic shield (figure A.14). The other component of the Unopette System is the reservoir containing a premeasured volume of diluting fluid. The container is sealed with a plastic cover and is punctured with the tip of the shield just prior to use.

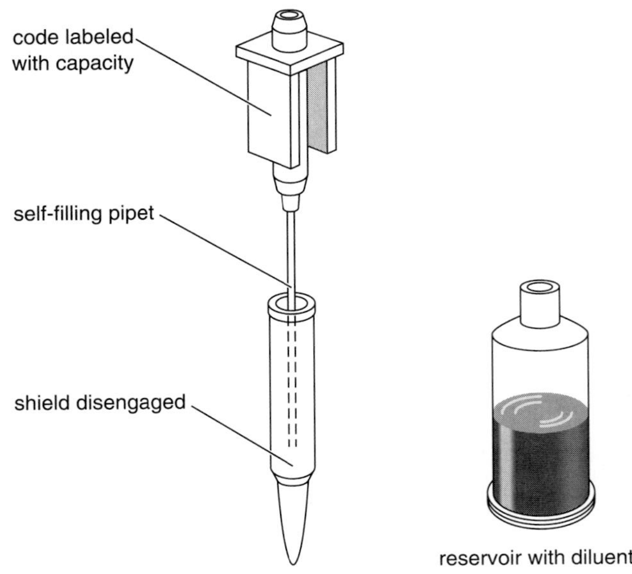


**figure A.14**
Self-filling disposable pipette, shield, and diluent reservoir.

These precalibrated pipettes and premeasured fluid reservoirs have become very popular in recent years because there is less dependence on human judgment regarding drawing exact quantities of blood or precise volumes of diluent, and thus less chance of error. Even more important is the fact that blood no longer needs to be pulled into a tube by mouth aspiration. Because of the danger of HIV and hepatitis infections, any device that relies on mouth aspiration is inherently more dangerous than one that does not depend on such technique. For that reason Thoma pipettes have been replaced in most laboratories by automated analyzers and the Unopette System.

For white blood cell counts a 25 μL capillary Unopette pipette fills readily with blood by capillary action. This blood is then released into a reservoir that contains 0.475 mL of acetic acid. The micropipette is rinsed, and the blood is mixed thoroughly with the diluent. The original Unopette pipette and reservoir is readily converted into a dropper assembly that may be used for charging the counting chamber.

For red blood cell counts a 10 μL capillary Unopette pipette is used. The blood is released in a reservoir containing 1.99 mL of physiological

(0.85%) saline. The blood is mixed and the pipette and reservoir used as a dropper system to charge the counting chamber.

For platelet counts a 20 μL capillary Unopette pipette is used. The blood is released into a reservoir containing 1.98 mL of 1% ammonium oxalate. Again the blood is well mixed and the pipette and reservoir used as a dropper system to charge the counting chamber.

## HEMOGLOBINOMETRY

Hemoglobin levels are determined as grams of hemoglobin per deciliter of blood. The test for hemoglobin is performed on free-flowing capillary blood from a finger puncture or from venous blood preserved with an anticoagulant. The anticoagulant of choice is EDTA. The hemoglobin concentrations remain fairly constant in blood for several days, provided it is properly refrigerated.

Hemoglobin levels are usually measured with a spectrophotometer. First, the red blood cells must be hemolyzed and a stable form of dissolved hemoglobin produced. When hemoglobin is mixed with **Drabkin's solution,** it is readily converted to cyanmethemoglobin, which is a stable form of hemoglobin that can easily be measured quantitatively in a spectrophotometer. This stable cyanmethemoglobin can be measured by various automated and semiautomated techniques.

### Automated Hemoglobinometry

Most automatic blood cell counters can also measure hemoglobin levels. Usually the blood sample used to count white blood cells is also used to measure hemoglobin levels. When blood is diluted for WBC counts, the diluent will hemolyze the red blood cells and develop a stable form of the resulting free hemoglobin, which is then automatically measured by a built-in spectrophotometer.

The results can be displayed on a digital screen and/or printed out separately.

### Nonautomated Hemoglobinometry

Before any unknown solution can be measured in a spectrophotometer, the instrument must be standardized—that is, a standard curve must be prepared against which the unknown samples can be read. To do this, samples of a known concentration of the solution (known as standard solutions) must be prepared and read first in the spectrophotometer using a specific wavelength.

For hemoglobin determinations, cyanmethemoglobin standard solutions are needed as well as a diluent, which is usually a modified Drabkin's solution. The cyanmethemoglobin standard solution can be obtained commercially, and Drabkin's solution (the diluent) can be prepared in the laboratory.

*Development of the Standard Cyanmethemoglobin Curve*

1. Usually a blank and five different dilutions are made. To guard against error, each standard is repeated twice, hence a total of 11 cuvettes or high-quality glass tubes are needed. They are labeled blank; 1,1 duplicate; 2,2 duplicate; 3,3 duplicate; 4,4 duplicate; and 5,5 duplicate. Each dilution made is equivalent to a certain amount of hemoglobin, measured in grams per deciliter.
2. Using volumetric pipettes, pipette the following amounts into the cuvettes:

| Tube Labeling | Cyanmet–hemoglobin Standard (mL) | Drabkin's Solution (diluent) (mL) | Hemoglobin Concentration (g/dL) |
|---|---|---|---|
| Blank | 0 | 5 | 0 |
| 1<br>1 duplicate | 1 | 4 | 4 |
| 2<br>2 duplicate | 2 | 3 | 8 |
| 3<br>3 duplicate | 3 | 2 | 12 |
| 4<br>4 duplicate | 4 | 1 | 16 |
| 5<br>5 duplicate | 5 | 0 | 20 |

3. Mix the contents of each tube well and read them in the spectrophotometer at a wavelength setting of 540 nm. First place the cuvette labeled blank in the cuvette well and zero the instrument so that it reads 0 absorbance (or 100% transmission).
4. Place each of the cuvettes in the well and read the amount of absorbance (or transmission) and record the result. The duplicates should accurately reflect the numbers of the original tubes. If not, the procedure should be repeated.
5. Plot the results on linear graph paper for absorbance readings (or semilogarithmic graph paper for transmittance readings). On one axis the various dilutions should be marked, on the other axis the various hemoglobin concentrations should be given, from 0 to 20 g/dL. A straight line is drawn through the points to give the best fit for the graph.
6. Retain this graph as subsequent hemoglobin concentrations can be read from this straight line.

*Manual Method for Hemoglobin Determination*

To read the hemoglobin concentration of a fresh blood sample, 3 clean cuvettes should be used: one is labeled as the blank, the second as sample, the third as duplicate. Fill each of the 3 cuvettes with 5 mL of modified Drabkin's solution. Add 0.02 mL (=20 μL) of blood to the sample and to the duplicate tubes. Attach a rubber sucking tube with a mouthpiece to the end of the **Sahli capillary pipette** and draw up blood to the 20 mm line. This exact amount of blood is then deposited into each of the 2 tubes. To make sure that all the blood is released into the cuvette, the Sahli pipette should be rinsed thoroughly by sucking up the solution into the tube and blowing it out again several times. Extreme caution should be taken that no blood (which must be considered to be a biohazard) or Drabkin's solution (which is extremely poisonous) gets into the mouth. The red blood cells will lyse. The cuvettes should be mixed thoroughly until the cyanmethemoglobin is evenly distributed. The tubes should then set for a few minutes for the correct color development to take place. They are now ready to be read.

Because of the dangers associated with mouth aspiration, the use of Sahli pipettes has been drastically reduced in recent years and has been largely replaced by the Unopette System.

The Unopette System consists of a glass capillary pipette that holds 0.02 mL (or 20 μL) attached to a plastic holder. When not in use the glass pipette is protected by a shield. The second component of the Unopette System is a reservoir that contains 5 mL of modified Drabkin's solution. The procedure for using the Unopette System is as follows:

1. Puncture the neck of the reservoir with the shield of the pipette, thus making an opening through which the blood sample can be deposited into the reservoir.
2. Remove the shield and hold the pipette in an almost horizontal position when the glass tip comes in contact with either capillary blood or anticoagulated venous blood. In this way the blood will be drawn automatically into the capillary pipette until it is filled. When filled, wipe the outside of the pipette, making sure not to remove any blood from the inside of the pipette.
3. Insert the capillary pipette into the reservoir and squeeze the reservoir. At the same time cover the overflow chamber of the pipette with the index finger. Next place the pipette firmly into the neck of the reservoir, then release the finger from the tip of the pipette, and stop squeezing the reservoir. The blood will be drawn into the diluent by suction.
4. Squeeze the reservoir gently 2 to 3 times to rinse the capillary bore.
5. Replace the index finger over the opening of the pipette overflow chamber and gently invert the whole assembly a few times to ensure a thorough mixing of the blood with the diluent.
6. Wait for 3 to 5 minutes for the correct color development to take place.

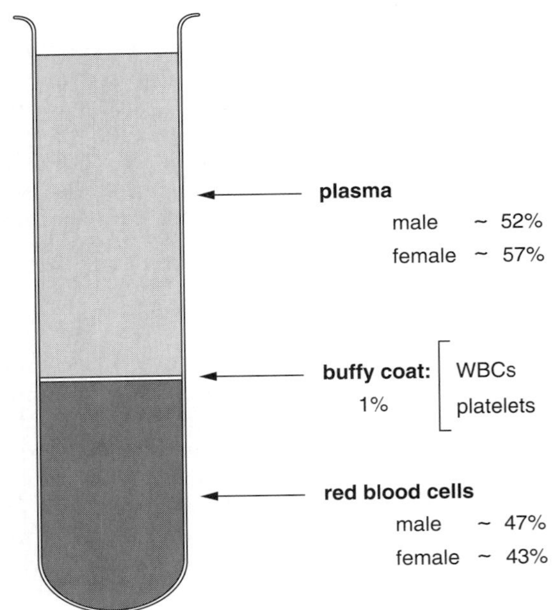


**figure A.15**
Layers of normal blood after centrifugation.

7. Remove the pipette assembly and invert the reservoir tube over a cuvette and squeeze the entire contents of the reservoir into the cuvette. The tube is now ready for reading.

The reading of the cuvette(s) is done using a spectrophotometer as follows: First read the blank cuvette filled with diluent at 540 nm and adjust the dial to 0 absorbance or 100% transmission. Then read the unknown sample(s) using the same wavelength. Record the reading and compare the amount of absorbance (or % transmission) with that of the standardized curve prepared earlier. This will give a direct reading of the amount of hemoglobin present in grams per deciliter. Report the hemoglobin reading to the first decimal place.

If a duplicate determination is made, the results should agree within ± 0.4 dL. If the discrepancy is greater, the procedure should be repeated.

The normal reference values for hemoglobin concentration are as follows: adult male, 14 to 18 g/dL; adult female, 12 to 16 g/dL; young child, 11 to 15 g/dL; and older child, 12 to 16 g/dL.

## HEMATOCRIT

Centrifuged whole blood samples can be separated into plasma, packed red blood cells, and a buffy coat layer (consisting of WBCs and platelets) (figure A.15).

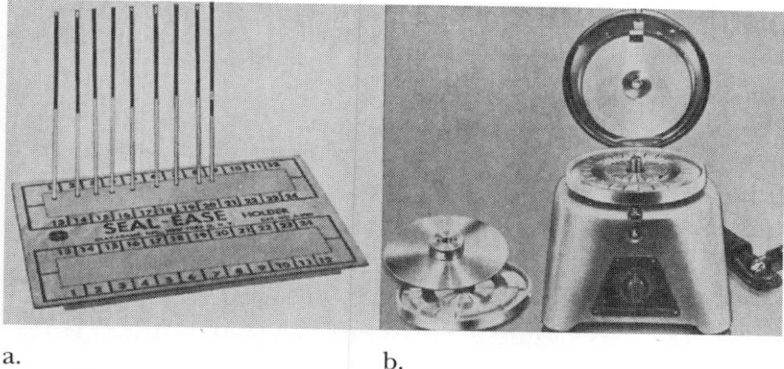

a. b.

**figure A.16**

Equipment for the microhematocrit method: (*a*) clay for sealing capillary tubes; and (*b*) high-speed microhematocrit centrifuge for capillary blood tubes.

Thus the hematocrit (hct), also known as **packaged cell volume (PCV),** may be defined as the percentage of the packed cell volume to the total amount of blood. This is a relatively simple test that relies on a visual observation of a small amount of centrifuged blood. It provides some useful information about the size and the content of red blood cells when it is correlated with other information such as the total RBC count and the hemoglobin concentration. The numbers generated by these tests enable us to calculate the so-called red blood cell indices. The red blood cell indices provide important data that help physicians to pinpoint and/or exclude certain types of anemia.

Two methods commonly used in determining the hematocrit value are: the macrohematocrit method, known as the **Wintrobe method,** and the **microhematocrit method.** The advantage of the latter method is that it requires less blood and also less time to determine a hematocrit. Hence the microhematocrit is the method of choice.

### Microhematocrit Method

To perform the microhematocrit method either capillary blood or anticoagulated venous blood can be used. For this technique special glass capillary tubes are used. These tubes are 1 mm in diameter and 7 cm long. If capillary blood is used, then the inside of the glass capillary tube should be coated with either heparin or EDTA anticoagulant. If anticoagulated blood is used, the capillary tubes used should not be coated on the inside with anticoagulant, as excess anticoagulant may cause cell shrinkage, and thus produce false low values. Both anticoagulated and nonanticoagulated tubes are readily available from commercial sources.

Blood is drawn into the tubes by capillary action. By holding the tubes in an almost horizontal manner the blood flows in more readily. An exact amount does not need to be drawn into the tube, but it should be at least half full, or ideally $2/3$ to $3/4$ full. For each blood sample at least two microhematocrit capillary tubes are filled, one functioning as the control.

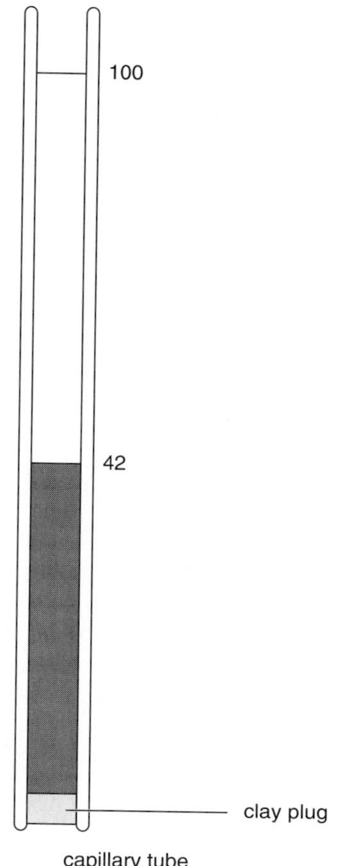


**figure A.17**

The microhematocrit method. The centrifuged capillary tube is read as a percentage of the total blood volume.

After the tubes are filled, the dry ends of each tube are sealed with a specially manufactured sealing clay that can be bought commercially (figure A.16). The sealed tubes are then placed in a special microhematocrit centrifuge with the sealed ends toward the outside of a special tube holder. The holder is then secured and the lid of the centrifuge is closed. The centrifuge is switched on and allowed to spin for 5 minutes. Most special microhematocrit centrifuges create centrifugal fields of up to 10,000 *g* with speeds of 10,000 rpm or greater.

After centrifugation the microhematocrit result is read with a graphic reading device that allows the hematocrit to be read directly as a percentage of the total blood volume (figure A.17). The duplicate should be within 1% of the original sample tube.

> Hematocrit results are reported as a percentage or a decimal fraction. Thus a 42% hematocrit is reported as 0.42. For normal adults the reference values are as follows: females, $0.40 \pm 0.05$ (35 to 45%) and males, $0.46 \pm 0.05$ (41 to 51%).

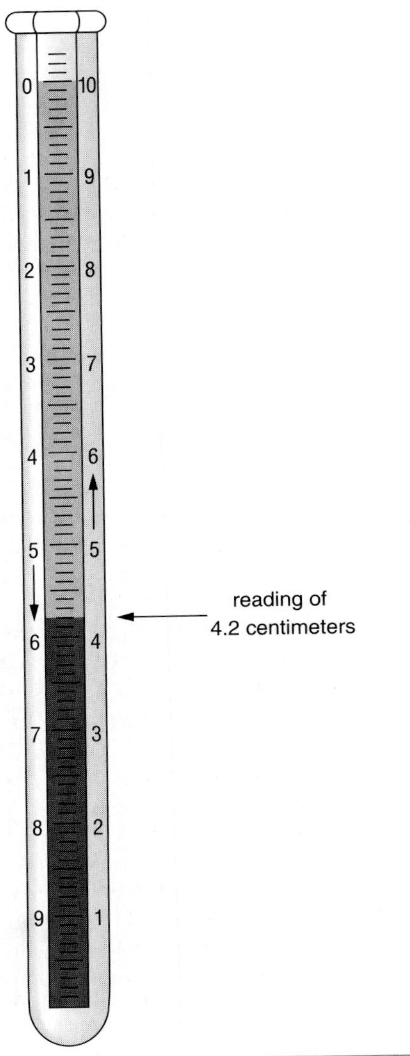


**figure A.18**

The Wintrobe tube used in the macrohematocrit method. The upper lighter portion of the tube is plasma. The lower darker portion is the packed red cells.

### Macrohematocrit Method (Wintrobe Method)

For the macrohematocrit method a special thick-walled hematocrit tube known as the Wintrobe tube is used (figure A.18). This glass tube has a fairly narrow inner diameter, and the outside of the tube is graduated from 1 to 105. It has a rubber cap, which prevents evaporation during the long centrifugation process. The Wintrobe tube is filled by using a long-tipped Pasteur pipette. Since Wintrobe tubes are difficult to clean after use, disposable tubes are now available.

Wintrobe tubes have a narrow bore and are difficult to fill. To avoid air bubbles, the tip of the narrow stem of the Pasteur pipette (filled with blood) should go to the bottom of the tube. While the blood is being expelled the Pasteur pipette should be gradually withdrawn. The tube is filled when the blood reaches the 100 mark.

The tube is sealed with a cap and placed in a standard bench centrifuge. The centrifuge should be balanced—that is, each tube container filled with a Wintrobe tube should be matched with another Wintrobe tube at the opposite side. Centrifugation is at full speed which is normally 2,300 to 2,500 *g*. To pack the cells, it is necessary to centrifuge the tubes for at least 30 minutes.

After centrifugation, three layers can be observed in the tube. At the bottom are the red cells, above these is a whitish "buffy coat," which consists mainly of white blood cells and platelets, and on top is the liquid layer of plasma. If the tube was filled exactly to the 100 mark, then the packed red blood cell volume can be read directly as a percentage at the interface of the top of the red blood cell mass and the buffy coat.

When the plasma level is not exactly at the 100 mark, then a simple calculation can be made to correct for this according to the following formula:

$$\text{Hct (\%)} = \frac{\text{Packed red blood cell height}}{\text{Total blood height}} \times 100$$

Because the macrohematocrit method is so time-consuming and difficult to perform, it is seldom used.

### Automated Hematocrit Method

Electronic cell counters are used to calculate the hematocrit value from a blood sample by computing individual cell volumes with the total amount of red blood cells in a given volume of blood. Since this method does not rely on centrifugation and problems with plasma that may have been trapped in a column of red blood cells, the hematocrit values obtained by electronic methods are usually slightly lower than those obtained by centrifugation methods.

## RED BLOOD CELL INDICES

The morphological classification of the anemias is based on the average size of the red blood cells, the hemoglobin concentration of the red blood cells, as well as the hemoglobin content of these cells. These parameters are known as (1) the MCV = mean corpuscular volume, (2) the MCH = mean corpuscular hemoglobin, and (3) the MCHC = mean corpuscular hemoglobin concentration. All three parameters can be calculated once the total RBC count, hemoglobin determination, and hematocrit value are known.

### MCV

The **MCV** determines the volume or size of the average RBC in femtoliters (fL), fL = $10^{-15}$/L. It is calculated by dividing the volume of RBCs per liter by the number of RBCs per liter, using the formula:

$$\text{MCV (fL)} = \frac{\text{Hct (\%)} \times 10}{\text{RBC count } (\times 10^{12}/\text{L})}$$

where the factor 10 is introduced to convert the hematocrit reading (in %) from volume of packed cells per deciliter to volume per liter (=1,000 mL).

For example if the hematocrit reading is 40% and the RBC count is 5,000,000 ($=5.0 \times 10^{12}$/L) per liter:

$$\text{MCV} = \frac{40 \times 10}{5.0} = 80 \text{ fL}$$

The MCV in normal adults is between 80 and 98 fL.

### MCH

The **MCH** is the average weight of the hemoglobin content in a red blood cell in picograms (pg = $10^{-12}$ g). This parameter is obtained by dividing the hemoglobin content of 1 L of blood (in g/L) by the number of RBCs (in $10^{12}$/L), according to the formula:

$$\text{MCH (pg)} = \frac{\text{Hemoglobin (g/dL)} \times 10}{\text{RBC count } (\times 10^{12}/\text{L})}$$

For example, if the hemoglobin content was 15 g/dL and the RBC count was $5.0 \times 10^{12}$/L, then

$$\text{MCH} = \frac{15 \times 10}{5.0} = 30 \text{ pg}$$

The normal range for the MCH is between 27 and 33 pg.

### MCHC

The **MCHC** is an expression of the average hemoglobin concentration per unit volume of packed red blood cells. It is expressed in grams per deciliter. It may be calculated from the MCV and the MCH or from hemoglobin and hematocrit values by using either of the following two formulas:

$$\text{MCHC (g/dL)} = \frac{\text{MCH}}{\text{MCV}} \times 100$$

or

$$\text{MCHC (g/dL)} = \frac{\text{Hemoglobin (g/dL)}}{\text{Hematocrit (as a decimal)}}$$

For instance, if the hemoglobin concentration was 15 g/dL and the hematocrit value was 0.40 (40%), then

$$\text{MCHC} = \frac{15}{0.40} = 37.5 \text{ g/dL}$$

The normal values of the MCHC range from 33 to 36 g/dL.

## DIFFERENTIAL LEUKOCYTE COUNT

The differential WBC count is also part of the complete blood cell count (CBC). This battery of tests is performed almost routinely as part of a physical checkup. The CBC includes five different tests: the RBC count, the WBC count, the hemoglobin determination, the hematocrit, and the differential leukocyte count. The differential WBC count encompasses the preparation, staining, and examination of a thin film of blood smear on a glass slide. Of the five tests of the CBC, the differential leukocyte count is the most important. This test not only allows the observer to classify and count the various white blood cells, but it also provides the viewer the opportunity to verify the hematocrit and hemoglobin values obtained.

Further, it also permits the observer to verify any variations in the numbers of WBCs, RBCs, and platelets, as well as note any abnormalities in their structures.

Finally, apart from the numbers generated, the blood smear is the only permanent record of a routine CBC that may be kept for a prolonged period of time. It allows the clinician to go back to previous times and to compare current with previous blood smears to see if the patient has made any improvements.

The sources of blood used for the routine blood smear are either capillary blood drawn from a finger or toe puncture, or venous blood preserved with EDTA as anticoagulant. Most routine blood smears are from venous blood which should be prepared within a few hours after being drawn.

### Preparing the Blood Smear

The glass slide used for making a blood smear should be exceptionally clean. Any dirt on the slide or an oily smudge from the fingers will interfere with the production of a proper smear. Therefore only new slides that have not been touched, or slides that have been cleaned with alcohol and wiped dry, should be used. Place the "specimen slide" on a flat surface and add a small drop of blood (about 2 to 3 mm in diameter) approximately 1 mm from the end of the slide (figure A.19).

While holding the other end of the specimen slide in place with the fingers of one hand, a "spreader slide" is placed in front of the drop of blood with the other hand. Make sure that the edges of the spreader slide are clean and free of chips, otherwise streaking will occur. The spreader slide should be held as follows: the middle finger should be below the spreader slide while the slide is stabilized by the edges of the adjoining fingers. The spreader slide should be held at a 45-degree angle and slowly drawn backward into the drop of blood. Once the edge of the spreader slide is in contact with the blood, the angle of the slide is lowered to about 25 degrees to

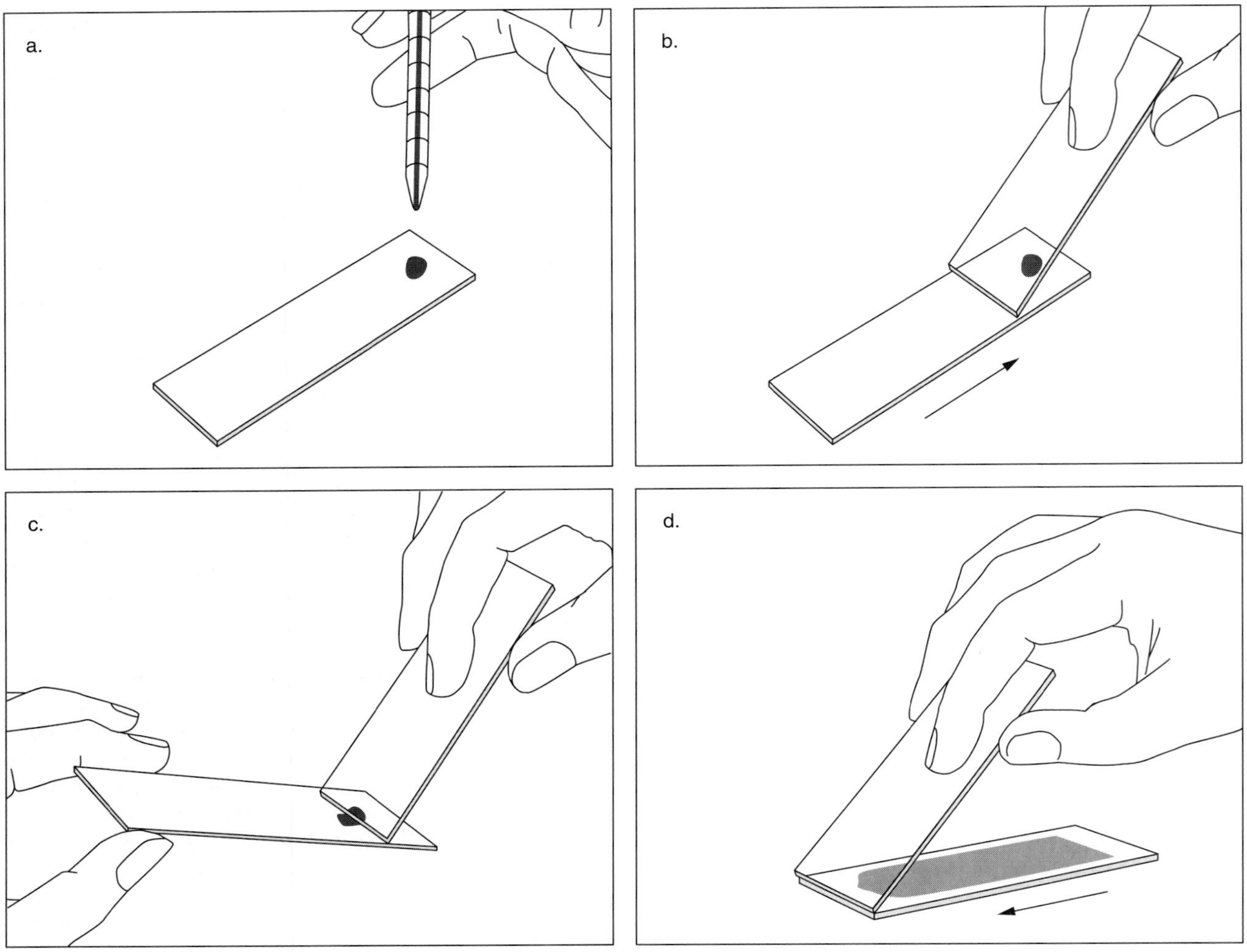

**figure A.19**
Preparing the blood smear.

allow the blood to flow evenly across the edge of the spreader slide. Once the blood has spread evenly across the edge of the spreader slide, quickly push the slide across the entire length of the slide containing the blood. As the spreader moves, a thin film of blood will be left on the specimen slide. Ideally the blood smear should cover ⅔ to ¾ of the slide when correctly prepared. The goal is to achieve a wedge-shaped smear with a thin edge (figure A.20).

Do not use the thumb and index finger to push the spreader slide over the specimen slide, as this results in too much pressure on the blood smear. This will cause an accumulation of white blood cells and platelets at the end of the smear, and thus a slide with an unequal distribution of blood cells.

Further, the blood smear should be prepared immediately after the drop of blood is placed on the

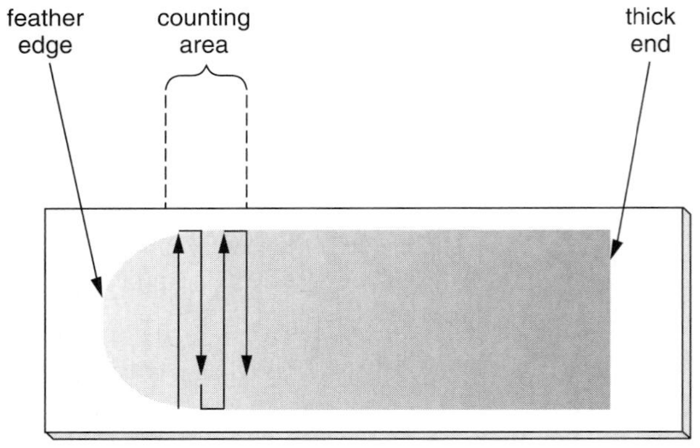


**figure A.20**
Blood smear showing a thin edge and the pathway for the differential cell count.

slide. If the drop of blood is allowed to dry, an uneven distribution of cells, clumping of platelets, and rouleau formation will result.

Only a small drop of blood should be used. A large drop of blood will make it impossible to produce a thin wedge at the end. Moreover a thick blood smear will make counting difficult and obscure the morphology of the cells.

From every blood specimen, two blood smears should be prepared: one to be stained, the other to be kept in reserve should the first film appear less than ideal for observation. The second blood smear is also analyzed when the first one shows many abnormal forms to verify if the abnormalities are artifacts (e.g., because of staining) or real changes.

The spreader slide is used first on the one slide, then turned over and used to make the second blood smear.

The blood smears should be air-dried. They should also be labeled immediately with the patient's name and the date (with a special pencil) at the end of the slide.

### Staining the Blood Smear

After the slide has dried, the blood smear should be stained as soon as possible, certainly within 1 or 2 hours. Conversely, the blood smear should be thoroughly dried before staining is performed. When wet films are stained they may show artificial abnormalities. The stain most often used for the examination of blood smears is Wright's stain, a type of Romanowsky stain that uses a mixture of acidic and basic analine dyes in methyl alcohol. Wright's stain contains the basic dye eosin and the acidic dye methylene blue. These reagents can be obtained commercially. After the stain is placed on the prepared blood smear, buffered distilled water is added, which is used for color differentiation. Each manufacturer has its own instructions, which should be carefully followed. After the final stain, the slides are washed in distilled water to remove the excess dye. The backs of the slides are dried with gauze or lint-free paper and air-dried. Do not use a slide for examination until it is thoroughly dry.

### Examining the Blood Smear

For the examination of the blood smear, a microscope with a low-power objective (10×) and an oil-immersion objective (100×) is necessary. The blood smear is first viewed under low power to find an area for viewing—that is, an area where the cells are not overlapping but are in only one layer close to one another. This area is usually located near the wedge-shaped end of the blood smear. A drop of immersion oil is then placed on the selected site and the 100× lens is moved into the oil while observing from the side. There must be contact between the lens and the oil on the slide. Once the lens and slide are in place, the microscope should be focused with the fine-adjustment knob. More light may be needed for viewing a slide under oil; this can be achieved by opening the iris diaphragm under the stage or by turning up the rheostat. The smear should be examined following a zigzag path as shown in figure A.20, moving toward the thicker part of the smear.

In a normal blood smear viewed under oil immersion, the red blood cells appear as round reddish bodies that are darker at the edges than in the center. The white blood cells appear as larger cells with deeply stained nuclei of various forms, and many of the cells have distinctly stained granules in their cytoplasm. The platelets appear as small, ovoid, bluish, slightly irregular bodies about ⅓ to ¼ the diameter of the red blood cells.

The differential leukocyte count consists of identifying and counting the first 100 white blood cells encountered. This gives the percentages of the cells present. For instance, if 25 of the 100 white blood cells were lymphocytes, then the percentage of lymphocytes is 25%.

To correctly identify each white blood cell, a color guide with pictures of the white blood cells should be used (e.g., *The Morphology of the Human Blood Cells* by Diggs, Sturm, and Bell, published by Abbott Laboratories). Such a guide is also helpful when abnormal cells are encountered, such as nucleated red blood cells in certain cases of hemolytic anemia.

The blood smear is also used to identify abnormal red blood cells. The presence of macrocytes, microcytes, poikilocytes and various inclusions frequently provide the clinician an indication of the type and cause of anemia. A guide showing the various sizes and shapes of abnormal red blood cells will help to identify certain disorders.

## RETICULOCYTE COUNT

Reticulocytes are the fifth stage in the developmental series of the erythrocytic cell line. They are young red blood cells that have lost their nucleus but still contain a certain amount of cytoplasmic RNA. This is the stage that leaves the bone marrow and enters the bloodstream to replace the red blood cells that have reached the end of their life span. Since the normal life span of erythrocytes is approximately 120 days, approximately 1% of the red blood cells need to be replaced each day. Hence the reticulocyte level should be around 1%. However, since people also lose blood via wounds, ulcers, menstruation, and so on, the normal level of reticulocytes is usually between 1 and 2%.

Reticulocytes are not easily detected in the normal blood smear, as the Wright's stain does not reveal the presence of the cytoplasmic RNA. Some immature

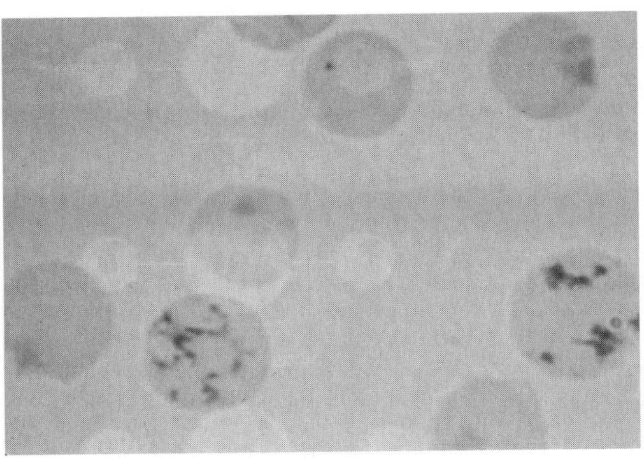

**figure A.21**

Reticulocytes.

From The American Society of Hematology Slide Bank, 3rd edition 1990, used with permission.

red blood cells may show some polychromatophilia—that is, the cytoplasm does not stain uniformly pink but may have a bluish hue (the remnants of the basophilic cytoplasm of the early stages of the erythroblast). To detect the presence of cytoplasmic RNA strands in reticulocytes, special supervital dyes need to be used. The supravital stains that are most commonly used for the reticulocyte count are brilliant cresyl blue and new methylene blue. Both dyes exhibit similar results; they stain the red blood cells a pale grayish-blue, and the RNA in the cells can be seen as dark blue strands and dots. Frequently the RNA will form a dark mesh or network, known as a reticulum—hence the term reticulocytes (figure A.21).

For the reticulocyte count both fresh capillary blood from a finger or toe puncture, or anticoagulated (EDTA) venous blood may be used. However, if venous blood is used, the test should be performed within 1 or 2 hours after the blood has been drawn.

### Manual Reticulocyte Count

Three drops of supravital dye are placed in a small test tube to which 2 drops of blood are added. The blood and the dye are mixed thoroughly and allowed to stand for 15 minutes.

After 15 minutes the cells are resuspended by gently shaking the test tube and then at least two blood smears are prepared by the method described for the differential WBC count. A small drop of the cell-dye mixture is placed on one end of two clean microscope slides and each drop is spread out into a thin film by a spreader slide. The slides are air-dried before microscopic observation.

The low-power objective is used to focus the slide and to find an area in which the red blood cells are well distributed—that is, close together but not overlapping. A drop of oil is placed on the selected area and viewed with the oil-immersion (100×) lens. A total of 1,000 erythrocytes is counted (500 on each of the two slides). The number of reticulocytes encountered should be recorded. The two slides should agree within 5 cells of each other. If not, the procedure should be repeated.

The reticulocyte count is recorded as the percentage of reticulocytes in the total of red blood cells counted, according to the following formula:

$$\% \text{ Reticulocytes} = \frac{\text{Reticulocytes}}{\text{RBCs}} \times 100$$

Thus if the total number of reticulocytes counted was 18, then

$$\% \text{ Reticulocytes} = \frac{18}{1{,}000} \times 100 = 1.8\%$$

The normal reticulocyte counts are as follows: adult male, 1.1 to 2.1%, and adult female, 0.9 to 1.9%.

## RBC OSMOTIC FRAGILITY TEST

The osmotic fragility test is a commonly used laboratory procedure to test the strength of red blood cells. As a result of changes in the content, membrane composition, or shape, red blood cells become weaker and are readily hemolyzed. Hence they are associated with certain anemias. The osmotic fragility test is designed to test the ability of red blood cells to withstand hemolysis in various hypotonic solutions of sodium chloride.

When red blood cells are placed in hypotonic solutions of sodium chloride, they tend to take in water and begin to swell. When a cell has reached beyond its maximum capacity of water absorption, it will burst (i.e., hemolyze). Normal red blood cells are able to withstand dilutions of about 0.5% NaCl before they begin to hemolyze, whereas more fragile cells begin hemolyzing at 0.75% NaCl. When RBCs are placed in an isotonic solution of 0.85% NaCl (physiological saline), no lysis should occur. The normal plasma environment is the osmotic equivalent of 0.85% NaCl. Increased RBC fragility is associated with spherocytic or ovalocytic cells—that is, associated with hereditary spherocytic anemia and hereditary ovalocytic anemia and certain other forms of hemolytic anemia. Decreased osmotic fragility is associated with iron deficiency anemia, sickle cell anemia, and thalassemia, especially when many target cells are present.

For the osmotic fragility test freshly drawn venous blood that is anticoagulated with heparin is preferred, although blood with other forms of anticoagulant can also be used.

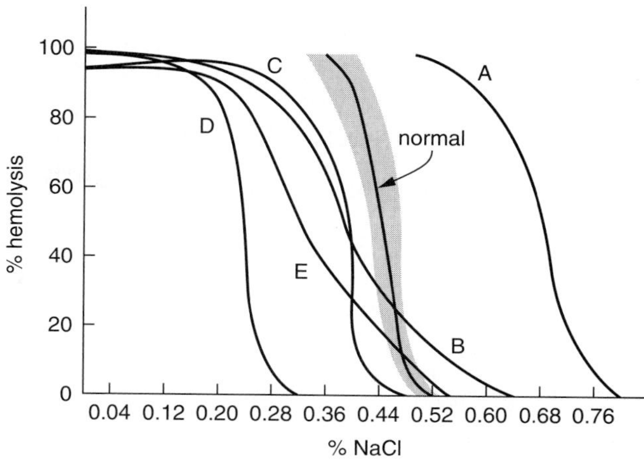


**figure A.22**

Osmotic fragility curves. *(A)* Hereditary spherocytosis. *(B)* Thalassemia major. *(C)* Thalassemia minor. *(D)* HbE disease. *(E)* Iron deficiency anemia.

### Manual Osmotic Fragility Test

Although both manual and automated methods are available to determine the osmotic fragility of red blood cells, the manual Unopette System is most commonly used. The Unopette System is a kit that consists of a series of 10 reservoirs of varying NaCl dilutions, (the dilutions are 0.85, 0.65, 0.60, 0.55, 0.50, 0.45, 0.40, 0.35, 0.30, and 0.00% NaCl) and 20 μL Unopette pipettes with a shield. The reservoir is punctured and 20 μL of blood is placed in each reservoir, using a new pipette for each reservoir. Carefully rinse the pipette with the solution in the reservoir and mix the sample in the reservoir.

Incubate the cell suspensions undisturbed at room temperature for 20 minutes.

After 20 minutes mix the contents of the reservoir again and carefully transfer the contents of each reservoir to a labeled centrifuge tube. Each of the tubes is then centrifuged at 2,000 rpm for 5 minutes.

Carefully remove the supernatant of each centrifuge tube into a spectrophotometer cuvette and read in the spectrophotometer at 540 nm. The percent transmittance should be read, using distilled water as the blank (i.e., has 100% transmittance).

Record the percentage transmittance for each cuvette and convert to absorbance.

Calculate the percent hemolysis as follows:

$$\% \text{ Hemolysis} = \frac{\text{Absorbance of supernatant}}{\text{Absorbance of 100\% hemolysis tube}} \times 100$$

Using linear graph paper, plot the percent hemolysis on the vertical axis and the NaCl concentration on the horizontal axis.

For normal red blood cells hemolysis begins at 0.45 to 0.50% and is complete at 0.20 to 0.30% (figure A.22).

# GLOSSARY

**abscess** A localized collection of pus in tissues.

**acanthocyte** A "thorny cell"; an abnormally shaped erythrocyte with spiny projections on the outer surface; seen in alcohol liver disease and postsplenectomy.

**achlorhydria** The absence of hydrochloric acid in gastric juice; common in persons with pernicious anemia.

**acid hydrolases** A variety of degradative enzymes found in primary granules of neutrophils.

**acquired disorders** The disorders caused by extrinsic factors or agents; not inherited.

**actin dysfunction syndrome** A rare inherited disorder of polymorphonuclear neutrophils (PMNs) affecting their locomotion and ingestion. Actin molecules are unable to polymerize properly into filaments.

**acute inflammation** Immediate-type hypersensitivity reactions, which occur rapidly as a result of exposure to antigens.

**acute leukemias** Leukemias that develop rapidly and are generally more severe; blasts and other immature cell forms predominate.

**adrenergic drugs** A group of compounds that mimic the sympathetic nervous system and counteract the effects of histamine. In this group epinephrine counteracts the effects of anaphylaxis.

**agammaglobulinemia** The absence or severe deficiency of gamma-globulin (antibody) production.

**agglutination** An antibody-antigen complex formed when an immunoglobulin binds to two antigenic particles; agglutination enhances engulfment of particles by phagocytic cells.

**aggregometer** An instrument used to measure platelet aggregation.

**AGL** Acute granulocytic leukemia; also known as acute myelocytic leukemia (AML).

**agranulocyte** A nongranular leukocyte; a white blood cell containing no specific-staining granules.

**AIDS** Acquired immunodeficiency syndrome; associated with pronounced lack of specific immunity. AIDS is caused by the retrovirus HIV (human immunodeficiency virus), which primarily infects the helper T lymphocyte.

**AIHA** Autoimmune hemolytic anemia.

**albumin** A plasma protein that functions in maintaining osmotic pressure of the blood. The osmotic pressure of albumin maintains balance between the intravascular and the extravascular fluid compartments.

**Alder-Reilly anomaly** An inherited disorder that results from abnormal polysaccharide metabolism in leukocytes, which prevents formation of secondary specific-staining granules.

**alkaline phosphatase** An enzyme that occurs in the secondary granules of neutrophils. The enzyme hydrolyzes phosphate esters, liberates inorganic phosphate, and has an optimum pH of about 10.

**ALL** Acute lymphocytic leukemia.

**allergic rhinitis** Hay fever.

**allostery** A change in shape that alters the activity of an enzyme.

**aminopeptidase** An enzyme that occurs in secondary granules of neutrophils. The enzyme hydrolyzes amino acids from polypeptides.

**AML** Acute myelocytic leukemia; also known as acute granulocytic leukemia (AGL).

**anamnestic response** A secondary response to an antigen enhanced by the presence of memory B and T lymphocytes. The serum level of IgG is 10 to 100 times that of the initial primary response.

**anaphylactic hypersensitivity (type I hypersensitivity)** The production of local or systemic inflammation often associated with IgE antibodies. These reactions are often hereditary and include disorders such as allergic rhinitis, urticaria, bronchial asthma, and food allergies.

**anaphylaxis** Exaggerated reaction to a foreign protein or other substance to which an individual has previously become sensitized.

**anaphylatoxins** The small-molecular-weight compounds that stimulate mast cells and basophils to release vasoactive amines.

**anemia** A condition in which there is a reduction below normal of the number of erythrocytes, quantity of hemoglobin, or volume of packed red cells in the blood.

**angular stomatitis** Sores in the corner of the mouth often associated with iron deficiency anemia.

**anisocytosis** An abnormal variation in the size of red blood cells, commonly found in persons suffering from anemia.

**antibody** *See* immunoglobulin.

**antibody-dependent cell-mediated cytotoxicity (ADCC)** The binding of certain white blood cells via the Fc receptors to infected and neoplastic cells, followed by release of their cytotoxic chemicals.

**antigen** Any substance capable of inducing a specific immune response and able to react with the products of that response such as specific antibody and leukocytes. Antigens may be soluble (toxins and foreign protein) or particulate (bacteria and tissue cells).

**antigen-presenting cells (APCs)** White blood cells, such as monocytes and macrophages, which present the processed antigen to the appropriate lymphocytic cell, thereby activating the immune system.

**antihistamine** A drug that counteracts the effect of histamine by binding to either the H1 or the H2 receptors on mast cells and basophils.

**aplastic anemia** Bone marrow damage by some physical (radiation), chemical (aromatic hydrocarbons), or infectious (viral) agent, all of which may kill the stem cells and stop further cell proliferation.

**APTT test** The activated partial thromboplastin time; a test for factor deficiency of the intrinsic coagulation pathway: factors XIII, IX, XI, and XII.

**arachidonic acid** A naturally occurring fatty acid that is a precursor of prostaglandin. It is principally involved in platelet metabolism.

**arthrosclerosis** A degenerative condition of large arteries making them less elastic; also referred to as "hardening of the arteries," which is more a function of old age.

**ataxia-telangiectasia** An inherited combined immunodeficiency disease characterized by improper balance and movement and dilations of small vessels.

**atherosclerosis** The deposition of plaques on the wall of arteries, thus reducing the diameter of the vessel. This condition is influenced by such factors as smoking, obesity, diabetes, and a diet high in LDL cholesterol.

**atrophic glossitis** The red, swollen, beefy appearance of the tongue commonly seen in megaloblastic anemia.

**atropine** An antihistamine drug that inhibits the action of acetylcholine, thus preventing production of cyclicguanosine monophosphate (cGMP) and degranulation by basophils and mast cells.

**Auer rods** The azurophilic rods observed in the cytoplasm of myeloblastic cells of patients with acute myelocytic leukemia (AML).

**autoantibodies** An antibody produced in response to, and reacting against, an individual's own cells and tissues.

**autosome** Any of the 22 pairs of chromosomes not associated with determination of sex.

**B lymphocyte** A lymphocyte that plays a major role in humoral immunity. When stimulated by a specific antigen, they mature into plasma cells capable of synthesizing antibody against that antigen.

**band cell** A "stab cell"; the last juvenile form of the neutrophilic series.

**basophil** A granulocyte characterized by deep blue-black cytoplasmic granules containing vasoconstricting amines. Basophils make up less than 1% of the total white blood cell population.

**Bence-Jones protein** An abnormal protein found in urine of patients with multiple myeloma. The protein, a monomer or dimer of Ig light chains, is characterized by unusual solubility properties. It precipitates when heated to 40–60° C but redissolves on further heating to 80–100° C.

**Bernard-Soulier syndrome** An intrinsic platelet abnormality associated with a lack of one of the platelet membrane glycoproteins. Platelets become nonfunctional in plug formation.

**bilirubin** The compound formed from biliverdin upon degradation of the porphyrin moiety of heme.

**bilirubin test** A procedure that measures the level of unconjugated bilirubin (indirect bilirubin) and the conjugated level of bilirubin (direct bilirubin) in blood or urine.

**biliverdin** A tetrapyrrole produced from the porphyrin moiety of heme. Biliverdin is the first bile pigment formed from the breakdown of heme. Biliverdin is further reduced to bilirubin.

**bleeding time test** A procedure that measures the time required for the cessation of bleeding after a standardized puncture through the skin.

**blood coagulation** The process by which platelets and soluble plasma proteins interact in a series of complex enzymatic reactions to finally convert fibrinogen into long strands of fibrin.

**bone marrow** The soft spongelike material in the cavities of bones. Its principal function is the manufacture of erythrocytes, leukocytes, and platelets.

**bradykinin** The precursor for the active kinins, lysylbradykinin and methionyl-lysyl-bradykinin, which function in blood coagulation. Bradykinin is a powerful vasodilator that increases capillary permeability.

**Bruton's disease** An inherited X-linked agammaglobulinemia disorder characterized by a total lack of immunoglobulin (antibody) production.

**buffy coat** The small whitish band of white blood cells formed between packed red blood cells and the plasma after centrifugation of whole blood that has not been allowed to coagulate.

**Burkitt's lymphoma** A malignant lymphoma of B cells characterized by chromosomal abnormalities and infection with Epstein-Barr virus (EBV).

**burr cell** An abnormal red blood cell characterized by the presence of short spines extending from the outer edge of the plasma membrane. These cells are seen in cases of anemia and neonatal liver disease.

**Cabot rings** The purple or reddish staining ringlike structures found in red blood cells; often seen in cases of severe anemia.

**CALLA** Common acute lymphocytic leukemia antigen; a common lymphocytic cell surface marker associated with a type of lymphocytic leukemia that cannot be classified as either B or T lymphocytes.

**capillary fragility test** A procedure that measures the ability of capillaries to withstand increased pressure. A positive test (formation of petechiae) indicates capillary weakness and/or thrombocytopenia.

**cell-mediated immunity (CMI)** The type of immunity involved in defending the body against intracellular pathogens, especially viruses and neoplastic cells. CMI consists primarily of cytotoxic and $T_{DTH}$ cells.

**Charcot-Leyden crystal (CLC) protein** A compound, also known as lipophospholysine, found in the plasma membrane of eosinophils. CLC breaks down the phospholipid membrane in cells of parasitic helminths.

**Chédiak-Higashi syndrome** A disorder associated with the inability of neutrophilic secondary granules to fuse with the phagosomes containing ingested microorganisms.

**chemotaxis** The movement or migration of certain white blood cells (e.g., macrophages, neutrophils) from the circulation or the tissues.

**Christmas disease** An inheritied deficiency of the plasma coagulation factor, factor IX.

**chronic granulomatous disease (CGD)** A rare congenital disorder in which the phagocytic cells are unable to produce the respiratory burst necessary for the generation of hydrogen peroxide and superoxide needed for microbial killing.

**chronic leukemia** Leukemia that develops slowly.

**cIg** Cytoplasmic immunoglobulin; a specific marker used to characterize pre-B lymphocytes, the first cell in B-lymphocyte differentiation.

**classical pathway** The activation of complement proteins, via an antibody-antigen complex, which ends with the formation of the membrane attack complex (e.g., cell lysis).

**CLL** Chronic lymphocytic leukemia

**clonal anergy** The natural suppression of B-cell cloning against self-antigens.

**clonal deletion** The destruction of T-cell clones against self-antigens.

**CML** Chronic myelocytic leukemia; also known as chronic granulocytic leukemia (CGL).

**cobalamins** Vitamin $B_{12}$ group. A cobalt-containing complex common to all members of the compound.

**collagen** A protein found in bone, cartilage, skin, and connective tissue.

**colony-stimulating factors (CSFs)** Growth factors that stimulate hematopoiesis.

**colony-forming unit (CFU)** Colonizing capacity of bone marrow stem cells and differentiation into specific cell lines.

**common pathway** The final steps in coagulation. The convergence of the intrinsic and extrinsic pathways, which includes activation of factor X, conversion of prothrombin to thrombin, cleavage of fibrinogen to fibrin, and the action of factor XIII on fibrin polymers.

**complement** A group of serum proteins, activated sequentially, involved in the immune reaction, which have cytolytic and chemotactic properties.

**contact coagulation group** The plasma protein factors (factor XI, factor XII, HMWK, and prekallikrein) involved in the initial phase of the coagulation process. These factors require contact with a negatively charged surface for their activation.

**Coombs' test** An agglutination test in which red blood cells agglutinate after Coombs' serum (anti-Ig antiserum) is added; occurs only if the red blood cells are covered with IgG molecules.

**crenation** The shrinkage of red blood cells in a hypertonic media or in conditions of increased pH.

**cytokinetics** The study of methods by which cell populations proliferate. Cytokinetics is concerned with size, maturation time, life span, and turnover rate of cells.

**cytotoxic T lymphocyte (CTL)** The cell type mainly responsible for cell-mediated immunity. CTLs can attack and destroy cells that have been invaded by viruses. Also called cytotoxic T cell ($T_c$).

**cytotoxic hypersensitivity** Type II hypersensitivity, involving IgM and IgG, resulting in lysis of cells (e.g., AIHA, Rh disease, and transfusion reactions).

**Dalton's law** A law stating that the total pressure exerted by a mixture of gases is equal to the sum of the partial pressures of each of the components of the mixture.

**delayed-type hypersensitivity** Type IV hypersensitivity or T-cell-dependent hypersensitivity. Inflammation caused by $T_{DTH}$ production of lymphokines 24–48 hours after exposure to certain antigens (e.g., eczema and contact dermatitis).

**demarcation membrane system (DMS)** The stage of thrombocyte production (platelet shedding) in which small areas of the megakaryocytic cytoplasm become separated from the rest of the cytoplasm.

**diapedesis** The passage of blood cells (WBCs) from the lumen of a blood vessel to extravascular tissue spaces without damage to the vessel itself.

**DIC** Disseminated intravascular coagulation; a disturbance in the hemostatic balance by a procoagulant stimulus. DIC results in the release of tissue factors into the circulation.

**diffuse intravascular thrombosis** A bleeding disorder caused by a reduced concentration of coagulation factors and platelets in the general circulation due to a continuous low level of intravascular coagulation activities. Also known as "consumption coagulopathy."

**DiGeorge's syndrome** A congenital aplasia of the thymus and parathyroid glands associated with lymphocytopenia in the peripheral bloodstream and absence of cell-mediated immunity (CMI).

**DiGuglielmo disease** A form of acute myelocytic leukemia (AML). Also referred to as erythroleukemia; characterized by abnormal normoblasts in the bone marrow.

**Döhle's bodies** Gray-blue cytoplasmic inclusions usually seen along the outer edges of mature neutrophils, thought to be ribosome-containing remnants of the promyelocyte cytoplasm. Döhle's bodies are commonly associated with severe infections and burns.

**Duke test** A procedure for measuring bleeding time; the time required for the cessation of bleeding after a standardized puncture of the earlobe.

**dysphagia** Not eating; a symptom of iron deficiency anemia.

**dyspnea** Difficulty in breathing.

**EBV** Epstein-Barr virus; the causative agent of Burkitt's lymphoma and infectious mononucleosis.

**eccentric** The location of the nucleus, moved to one side of the cell.

**ecchymosis** A bruise or skin hemorrhage, usually larger than 3 mm in diameter.

**eosinophilic chemotactic factor of anaphylaxis (ECF-A)** An eosinophilic chemotactic factor released from mast cells and basophils.

**edema** Tissue swelling due to the accumulation of excess fluid.

**Ehlers-Danlos syndrome** An inherited vascular disorder associated with abnormal collagen production, resulting in defective basement membrane structure.

**eicosanoids** A family of compounds produced within the plasma membranes of cells (e.g., prostaglandins).

**electrolytes** The chemical compounds that play an important role in maintaining the acid-base balance of the body (e.g., calcium ions, sodium ions, potassium ions).

**electrophoresis, protein** A technique that separates plasma proteins into alpha-, beta-, and gamma-globulin fractions based on molecular weight and charge.

**Embden-Meyerhof (EM) pathway** Glycolytic pathway.

**emperipoiesis** The movement of certain lymphocytes through the cytoplasm of endothelial cells.

**eosinophil** A granulated white blood cell characterized by large reddish staining cytoplasmic granules. Eosinophilia is associated with helminth infection and allergy.

**eosinophil cationic proteins (ECPs)** A series of seven cationic proteins, mainly produced by eosinophils, which promote clotting and inhibition of plasminogen activity.

**eosinophil-derived neurotoxin (EDN)** A toxin produced by eosinophils that causes damage to myelinated sheaths of neurons. EDN is thought to be effective against helminth parasite infections.

**eosinophil peroxidase** An enzyme found in eosinophilic granules that function in degradation proteins and assist in killing helminth parasites.

**erythrocyte** Red blood cell.

**erythron** The circulating red cells in the blood, their precursors, and the elements involved in their production.

**erythropoiesis** The proliferation and differentiation of erythrocytes.

**erythropoietin (EPO)** A growth hormone that stimulates proliferation and differentiation of erythrocytes in the bone marrow. EPO is produced by a renal erythropoietin factor in response to hypoxia.

**esophageal webbing** Lesions of the esophagus, a clinical symptom of iron deficiency anemia.

**extramedullary hematopoiesis** Blood cell production outside the bone marrow, such as lymphoid organs.

**extrinsic factor** The blood-clotting process activated as a result of damage to the endothelial cells lining the blood vessel walls. This pathway requires a tissue factor not normally present in the blood (factor III).

**FAB classification** The French-American-British system for classifying acute leukemias. The FAB group publishes and updates the classification based on morphological and cytochemical characteristics of these leukemias.

**Fabry's disease** A sex-linked syndrome characterized by a deficiency of alpha-galactosidase, which results in accumulation of trihexosyl ceramide in macrophages.

**Fanconi's anemia** An inherited hyperplasia of primary marrow; a hypoproliferation or pancytopenia of bone marrow cells.

**Fc receptor** The Fc fraction of an immunoglobulin molecule usually bound to white blood cell membranes.

**ferriheme** The trivalent iron ($Fe^{3+}$), which binds to the porphyrin ring of heme.

**ferritin** The major storage form of iron in tissues and developing erythrocytes.

**ferroheme** Ferrous heme, the bivalent iron ($Fe^{2+}$), which binds to the porphyrin ring.

**fibrin** The end product of blood coagulation.

**fibrin degradation products (FDPs)** Fragments of proteins split enzymatically from fibrinogen to form fibrin. Increased levels of FDP are seen in disseminated intravascular coagulation (DIC).

**fibrinogen** The beta-globulin fraction of plasma protein, which is the substrate for thrombin to form fibrin.

**fibrinolysis** The degradation of the fibrin plug.

**fibronectin** A plasma protein involved in the adhesion of blood cells to surfaces. Fibronectin binds to exposed fibroblasts and damaged endothelial cells.

**folate** or **folic acid** A common name for pteroylmonoglutamate, present in foods as polyglutamate. A deficiency of folic acid results in impairment of nucleic acid synthesis and consequently affects formation of erythrocytes and leukocytes.

**FSF** Fibrin stabilizing factor (factor XIII). FSF stabilizes the fibrin polymer.

**gamma globulin** Immunoglobulin or antibody.

**Gaucher's disease** A hereditary deficiency of the enzyme, beta-glucosidase, resulting in the accumulation of glucocerebrosides in macrophages and histiocytes.

**gelsolin** A compound, including profilin, which controls the disassembly and rescaffolding of actin polymers.

**GEMM-CFU** Granulocyte-erythrocyte-megakaryocyte-monocyte colony-forming unit.

**germinal follicles** The area of the spleen consisting mainly of B lymphocytes.

**glossitis** Inflammation of the tongue, often associated with sores. A clinical symptom of iron deficiency anemia.

**glucose-6-phosphate dehydrogenase** An enzyme needed in the hexose monophosphate shunt. Deficiencies of this enzyme are inherited and result in hemolytic anemia.

**glycocalyx** A glycoprotein complex that covers cell surfaces (e.g., the outer surface of platelets).

**glycosylated hemoglobin (HbG)** A variation of $HbA_1$. HbG contains a molecule of glucose attached to the N-terminal valine of the beta-chains of hemoglobin.

**GM-CFU** Granulocyte-monocyte colony-forming unit.

**gout** A disorder caused by excessive production and deposition of microcrystalline uric acid in tissues and joints.

**Gower hemoglobin** Embryonic hemoglobin (Gower 1 and Gower 2).

**granulocyte** White blood cells: neutrophils, eosinophils, and basophils. Grouped according to the staining characteristics of their cytoplasmic granules.

**granulocyte colony-stimulating factor (G-CSF)** A growth factor associated with the development and differentiation of neutrophils. G-CSF is produced by monocytes, macrophages, and fibroblasts.

**granulocyte-monocyte colony-stimulating factor (GM-CSF)** A growth factor associated with the development and differentiation of granulocytes and monocytes.

**granuloma** A nodular aggregate of macrophages and lymphocytes involved in immune activities.

**hairy cell leukemia (HCL)** A form of chronic lymphocytic leukemia characterized by fine cytoplasmic extensions surrounding the nucleus of T lymphocytes.

**Ham test** A procedure used to diagnose paroxysmal nocturnal hemoglobinuria (PNH). Increased red blood cell lysis occurs at a pH of 6.4 in patients with PNH.

**haptoglobin** A plasma glycoprotein produced by liver cells. Haptoglobin functions primarily against loss of iron by binding free hemoglobin during intravascular hemolysis of red blood cells.

**$HbA_1$** Adult hemoglobin, composed of two alpha- and two beta-chains; accounts for 97% of total HbA.

**$HbA_2$** Adult hemoglobin, composed of two alpha- and two delta-chains; accounts for 3% of total HbA.

**Hb Barts** An abnormal hemoglobin, composed of four gamma-chains.

**HbF** Fetal hemoglobin, composed of two alpha- and two gamma-chains.

**HbH** An abnormal hemoglobin, consisting of four beta-chains; commonly found in $\alpha$-thalassemia.

**Hb Portland** Embryonic hemoglobin with two zeta- and two gamma-chains; occurs early in gestation.

**heavy chain disease** A form of plasma cell leukemia characterized by excess production of either alpha-, gamma-, or mu-chains of the immunoglobulin molecule. Also known as Franklin's disease.

**Heinz bodies** Particles of precipitated or denatured hemoglobin in red blood cells. Their presence indicates oxidative injury to red blood cells and is found in persons with hemolytic anemia.

**helper T lymphocyte (HTL)** The cell type responsible for activating B lymphocytes and cytotoxic T lymphocytes. Also called helper T cell ($T_H$).

**hemarthrosis** A hemorrhage into joint cavities, commonly seen in coagulation disorders such as hemophilia.

**hematocrit** A test used to determine the ratio of blood cells to plasma.

**hematoma** A large bruise that is a collection of extravascular blood localized within a tissue or cavity; the blood has penetrated the subcutaneous tissue producing swelling and discoloration.

**hematopoiesis** The process of the development and maturation of blood cells.

**hematuria** Blood in the urine.

**heme** The nonprotein iron protoporphyrin constituent of hemoglobin, composed of a porphyrin ring with an iron atom inserted into the center and attached to four nitrogen atoms.

**hemochromatosis** A qualitative disorder of monocytes and macrophages due to excess deposits of iron. This syndrome is either inherited or acquired by multiple blood transfusions, alcoholic cirrhosis, etc.

**hemocytoblast** The pluripotent mesenchymal stem cell having the capacity to give rise to all blood cell lines.

**hemoglobin (Hb)** The pigment of red blood cells that carries oxygen and carbon dioxide. Hb contains four different polypeptide chains, each linked to a heme group.

**hemoglobin H disease** Alpha-thalassemia; reduced synthesis of the alpha-chains of hemoglobin: three out of four globin chains are absent.

**hemolytic anemia** Anemias caused by accelerated destruction of peripheral red blood cells, resulting in shortened life span of red blood cells.

**hemolytic disease of the newborn (HDN)** A hemolytic anemia of a fetus due to maternal antibodies, which results from previous exposure (Rh factor). Also called erythroblastosis fetalis or Rh disease.

**hemophilia** Hereditary deficiencies of plasma coagulation factors, resulting in bleeding disorders.

**hemophilia A** A sex-linked deficiency of the plasma coagulation factor, factor XIII:C.

**hemophilia B** Christmas disease; an inherited deficiency of the plasma coagulation factor, factor IX.

**hemophilia C** Parahemophilia or Rosenthal syndrome, a deficiency of plasma coagulation factor, factor XI.

**hemorrhage** Escape of blood from a ruptured vessel.

**hemorrhagic telangiectasia** An inherited vascular bleeding disorder.

**hemorrhoid** Enlarged veins in the mucous membrane inside or just outside the rectum. Leads to bleeding from the veins of the anal canal.

**hemosiderin** Aggregates of ferritin that are indicative of iron overload.

**hemostasis** Halting of bleeding.

**Henoch-Schönlein syndrome** A condition associated with acute inflammation of small blood vessels, causing increased vascular permeability and easy bleeding.

**Henry's law** The solubility of a gas in a liquid is directly proportional to the partial pressure of this gas.

**heparin** An anticoagulant that prevents blood clotting by binding with antithrombin III. Heparin, an acid mucopolysaccharide, is found in granules of basophils.

**hepatomegaly** An enlarged liver, seen in many hematological disorders.

**hereditary elliptocytosis** An inherited disorder that causes a defect in the actin-spectrin-ankyrin cytoskeleton of red blood cells.

**hereditary spherocytic anemia (HSA)** Hereditary hemolytic anemia associated with splenomegaly and the presence of spherocytic red blood cells. Red blood cell membranes lack spectrin, an important structural protein.

**heterophile antibody** An antibody that reacts with an antigen of an apparent unrelated species. For instance, sheep and horse red blood cells agglutinate by serum of persons infected with mononucleosis.

**HIM** Hematopoietic inductive microenvironment.

**histamine** An amine found in secondary specific-staining granules of mast cells and basophils. Histamine plays a major role in the production of acute inflammation and is produced by the decarboxylation of the amino acid, histidine.

**histiocyte** An immature tissue macrophage.

**HIV** Human immunodeficiency virus.

**HLA** Human leukocyte antigen; inherited antigenic determinants found on most nucleated cells of the body. Also known as the major histocompatibility complex (MHC).

**HMWK** High-molecular-weight kininogen, also called Fitzgerald's factor, is an important blood coagulation factor.

**Hodgkin's disease** A malignant lymphoma characterized by the presence of Reed-Sternberg cells in lymph nodes.

**Howell-Jolly bodies** Small, well-defined particles in erythrocytes that stain deep blue, red, or purple depending on the staining technique. These particles are thought to be nuclear fragments occurring in cases of severe anemia and postsplenectomy.

**humoral immunity** A component of the immune system that is due to the release of antibodies and complement, defending the body against extracellular antigens. "Humor" refers to the body fluids.

**hydrops fetalis** A hereditary deficiency of the alpha-chain of hemoglobin. This condition is incompatible with life and leads to death either *in utero* or soon after birth.

**hypersegmented neutrophils** The abnormal presence of neutrophils with nuclei having more than five lobes.

**hypersensitivity** Allergy.

**hypochromia** Red blood cells that show increased central pallor associated with reduced MCHC; commonly seen in cases of anemia.

**hypovolemic** A large decrease in the volume of blood, causing increased heart rate, rapid breathing, and intense thirst.

**hypoxia** Low oxygen tension in respiring tissues.

**iatrogenic factors** Factors that may interfere with the normal development and function of red blood cells, resulting from the side-effects of medical treatment.

**idiopathic** A disease occurring without known cause.

**immune complex hypersensitivity** Type II hypersensitivity; reactions involving IgM and IgG molecules forming complexes with antigens that locate in joints (rheumatoid arthritis), small blood vessels (vasculitis), kidney glomeruli (glomerulonephritis), or lymph nodes (lymphadenitis).

**immunocytes** Leukocytes that play a major role in specific immune responses, such as lymphocytes and macrophages.

**immunoglobulin (Ig)** The gamma-globulin (antibody) fraction of plasma; important in the body's defense against invading pathogens.

**ineffective erythropoiesis** The death of megaloblasts in the bone marrow, leading to failure to generate erythrocytes for release into the general circulation.

**infectious mononucleosis** A lymphoproliferative disorder or reactive lymphocytosis associated with viral infection (Epstein-Barr virus). This syndrome is also referred to as glandular fever, Pfeiffer's disease, and kissing disease.

**inflammation** A localized response produced in response to injury, which dilutes, walls off, or destroys the injuring agent.

**insidious diseases** The symptoms of disease are expressed gradually.

**interleukins (ILs)** The growth factors produced by a variety of white blood cells. IL influences the development of the early stages of hematopoiesis.

**intravascular thrombosis** Clotting disorders.

**intrinsic factor (IF)** A glycoprotein that binds and allows absorption of vitamin $B_{12}$ from the gastrointestinal tract. IF is produced by parietal cells in the wall of the stomach.

**intrinsic pathway of coagulation** The clotting process that occurs in the absence of a damaged blood vessel.

**iron deficiency anemia (IDA)** An anemia associated with a decreased level of iron available for hemoglobin synthesis.

**ischemia** A deficiency of blood in part of a blood vessel caused by functional constriction or obstruction.

**Ivy test** A bleeding time test used to measure cessation of bleeding.

**jaundice** A condition caused by accumulation of bilirubin, characterized by a yellow discoloration of the skin, mucous membranes, and sclerae. Jaundice is symptomatic of liver disease, hemolytic disease, and other abnormalities.

**kernicterus** The deposition of unconjugated bilirubin in the central nervous system of the newborn, resulting in irreversible brain damage (e.g., hemolytic disease of the newborn).

**koilonychia** Flat, spoon-shaped nails; a clinical symptom of iron deficiency anemia.

**lactoferrin** An iron-binding protein that functions in phagocytic cells, nonoxidatively, to kill microbes.

**large granular lymphocyte (LGL)** Also referred to as natural killer (NK) cells.

**leptocyte** A red blood cell that appears very thin and flat with only a thin rim of hemoglobin coloring the edges.

**leukemia** An acute or chronic progressive, malignant proliferation of one or more white blood cell lines.

**leukemogens** Leukemia-causing agents (e.g., benzene).

**leukin** An antibacterial compound derived from nucleohistone in phagocytic cells.

**leukocyte** White blood cell.

**leukopenia** Low or reduced number of leukocytes in the peripheral blood.

**leukopoiesis** The proliferation and differentiation of leukocytes.

**leukotrienes** A group of compounds produced by leukocytes; potent vasodilators that greatly increase vascular permeability and local edema.

**LM-CFU** Lymphoid-myeloid colony-forming unit.

**Löffler's syndrome** A clinical condition characterized by an infiltration of eosinophils in the pulmonary system. Results from helminth migration through the lungs as part of their life cycle.

**lymphadenopathy** A term used to denote any disease of the lymph nodes.

**lymphocyte** The group of white blood cells, B lymphocytes, T lymphocytes, and natural killer cells, which comprise 20–35% of the total leukocyte population.

**lymphokines** Substances produced by leukocytes that play a role in macrophage activation, lymphocyte transformation, and cell-mediated immunity.

**lymphoma** A neoplastic disorder of the lymphoid tissue.

**lysosome** A cytoplasmic organelle consisting of a membrane-bound sac of hydrolytic enzymes that functions in digestion within the cell.

**MAC** Membrane attack complex. An activated complement complex that forms a transmembranal channel; causes lysis of a cell membrane.

**macrocytic anemia** Anemia that usually produces larger than normal erythrocytes.

**macrocytosis** Increase in the size of red blood cells exceeding 9 μm in diameter.

**macrogranulocyte** A larger than normal granulocyte commonly found in megaloblastic anemia.

**macrophage** The large phagocytic cell involved in specific and nonspecific immune responses. Macrophages differentiate from circulating monocytes.

**major basic protein (MBP)** A protein found in the electron-dense core of secondary specific-staining granules of eosinophils. MBP plays a role in helminth destruction.

**mast cell** A connective tissue cell that produces granules containing histamine, heparin, and, in some cases, serotonin.

**MCHC** Mean corpuscular hemoglobin concentration.

**MCV** Mean corpuscular volume.

**medullary hematopoiesis** Blood cell production in the bone marrow.

**megakaryocyte** Precursor cell type of thrombocytes (platelets).

**megakaryoblast** Precursor cell type of megakaryocytes.

**megaloblast** A larger than normal blast cell in bone marrow associated with deficiencies of vitamin $B_{12}$ and folic acid, exhibiting delayed nuclear and cytoplasmic development.

**megaloblastic anemia** Synonymous with macrocytic anemia.

**Meg-CFU** Colony-forming units-megakaryocyte; the myeloid stem cells that differentiate under the influence of hormonal growth factors (e.g., GM-CSF, IF-3, EPO, thrombopoietin) into megakaryoblasts and megakaryocytes.

**memory cells** The immunocompetent B and T lymphocytes that develop into activated cell clones upon stimulation by a specific antigen.

**meningeal leukemia** The infiltration of malignant lymphoblasts (acute lymphocytic leukemia) into the brain, causing headache and nausea, followed by convulsion and palsylike symptoms.

**mesenchyme** Embryonic connective tissue derived from the mesoderm that later becomes the connective tissue and vessels of the body.

**metaplasia** The conversion of adult cells within a tissue (e.g., bone marrow, liver, or spleen) to a form that is abnormal for that tissue.

**methemalbuminemia** Methemoglobin bound to albumin; seen in hemolytic anemias.

**methemoglobinemia** The form of hemoglobin unable to transport oxygen, containing ferric iron rather than ferrous iron. One percent of the total circulating hemoglobin is normally converted to the ferric state each day by the methemoglobin reductive system.

**methylxanthines** The group of antihistamine compounds, including caffeine and theophylline, that inhibits the action of phosphodiesterase, thus preventing degranulation of basophils and mast cells.

**MHC** Major histocompatibility complex. Epitopes that function as antigenic determinants (human leukocyte antigens) on most nucleated cells of the body.

**microcytic anemia** Anemia that produces smaller than normal erythrocytes.

**microcytosis** Decrease in the size of red blood cells to less than 6 μm in diameter.

**microphage** Also called a neutrophil.

**mIg** Membrane immunoglobulin. The IgM antibody located on the membrane of the second series of B cells (virgin B cells) in differentiation.

**monocyte** The largest of the white blood cells in the bloodstream. Monocytes are phagocytic and differentiate into tissue macrophages.

**monocyte colony-stimulating factor (M-CSF)** A chemical stimulant (monokine) secreted by monocytes, fibroblasts, and endothelial cells that promotes growth and differentiation of monocytes and macrophages.

**mononuclear phagocytic system** The different forms that macrophages take on, depending on where they are found in the body: bone, osteoclasts; liver, Kupffer cells; brain, microglial cells; and lungs, alveolar macrophages.

**multi-colony-stimulating factor (multi-CSF)** A granulocytic growth factor also known as interleukin 3 (IL-3).

**multiple myeloma** A neoplastic disease of plasma cells in the bone marrow characterized by the presence of high levels of monoclonal antibody in blood and urine.

**myeloid metaplasia** The development of bone marrow tissue in sites where this does not normally occur; this tissue becomes hematopoietic.

**myeloma** A malignant disease developing from a clone of plasma cells, leading to overproduction of specific antibodies.

**myeloperoxidase deficiency syndrome** An inherited disease in which neutrophils are unable to produce the enzyme myeloperoxidase, which is needed to oxidize hydrogen peroxide, thereby reducing microbicidal activity.

**myelofibrosis** A condition where normal hematopoietic bone marrow tissue is replaced by noncancerous fibrous tissue.

**myelophthisis** The crowding out of healthy blood stem cells by tumor or cancer cells.

**myelophthisic anemia** Defective bone marrow caused by a decreased proliferation of new erythrocytes; normal marrow may be crowded out by a tumor in the bone or other tissue.

**myeloproliferative disorders** Conditions associated with nonreactive neutrophilias, usually the result of unregulated proliferation of white blood cells in the bone marrow.

**myxedema** A condition resulting from advanced hypothroidism, or deficiency of thyroxine.

**natural killer cells (NK cells)** Large granular lymphocytes (LGLs) which do not possess any of the markers characteristic of B and T lymphocytes. NK cells have Fc membrane receptors for IgG that trigger the release of cytotoxins.

**necrosis** Tissue destruction.

**neoantigen** The ability of certain chemicals (e.g., drugs) to bind and alter proteins on red blood cell surfaces, resulting in the formation of a nonself complex.

**neoplasm** The formation of an abnormal mass of cells often exhibiting uncontrolled and progressive growth.

**neutralization** The binding and inactivation of toxic compounds or viruses by antibodies.

**neutrophil** A phagocytic granulocyte, characterized by numerous small cytoplasmic granules; comprises 55–75% of the total leukocyte population. Neutrophils are the first line of defense against invading microorganisms.

**Niemann-Pick disease** An inherited disease, commonly seen in Ashkenazi Jews; caused by a deficiency of the enzyme sphingomyelinase, resulting in the accumulation of sphingomyelin in macrophages.

**non-Hodgkin's lymphoma** A malignant lymphoma characterized on the basis of the absence of the Reed-Sternberg cell, which is indicative of Hodgkin's disease.

**normocyte** An erythrocyte that is normal in size, shape, and color.

**nuclear cytoplasmic asynchronism (dissociation)** The nucleus of the megaloblastic erythroid cell retains its primitive appearance, and the cytoplasm matures normally; seen in macrocytic anemias.

**null cells** Lymphocytes that do not possess any of the markers of B or T lymphocytes.

**ontogeny** Embryonic development of blood cells.

**opsonin** A substance, such as an antibody, that enhances phagocytosis by binding to a receptor on an antigenic cell or molecule.

**osmotic fragility test** A procedure used to diagnose macrocytic anemias, such as spherocytosis. Red blood cells are suspended in a series of saline solutions of decreased concentration. Compared with normal red blood cells, macrocytic cells show increased osmotic fragility.

**ovalocyte** An oval-shaped erythrocyte.

**pancytopenia** Low numbers of white blood cells, red blood cells, and platelets in the bone marrow and in the peripheral blood system.

**Pappenheimer bodies** Iron-containing granules found at the edge of red blood cells; seen in cases of sideroblastic anemia, thalassemia, and other anemias.

**paroxysmal nocturnal hemoglobinuria (PNH)** An acquired hemolytic anemia that occurs as a result of complement action and lysis of red blood cell membranes.

**Pelger-Huët anomaly** An inherited abnormality characterized by a failure of the normal nuclear segmentation of neutrophils.

**pernicious anemia** The inability of the body to absorb vitamin $B_{12}$.

**petechiae** Small multiple hemorrhages in the skin, usually less than 3 mm in diameter.

**phagocytes** The white blood cells, such as macrophages and neutrophils, which engulf and destroy invading pathogens and neoplastic cells.

**phagolysosome** The structure formed when the phagosome fuses with a number of lysosomes in the phagocytic process.

**Philadelphia chromosome** An unique chromosomal abnormality associated with chronic myelocytic leukemia, which results from the translocation of a portion of the long arm of chromosome 22 to chromosome 9.

**phylogeny** Evolution of blood cells in the animal kingdom.

**pica** A craving for unusual, nonnutritional foods.

**plasma cell leukemia** Multiple myeloma.

**plasma** The liquid phase of blood composed of about 91% water.

**plasmacyte (plasma cell)** The mature antibody-secreting B cell found mainly in the cords of lymph nodes and activated lymphoid follicles of lymphoid tissues.

**plasmin** The activated form of plasminogen capable of degrading strands of fibrin into small nonreactive fragments.

**plasminogen** A protein activated by serine proteins to form plasmin.

**platelet** Thrombocyte.

**pluripotent stem cell (PPSC)** A common ancestral cell in bone marrow from which all blood cells are ultimately derived.

**PMN** Polymorphonuclear neutrophil. Refers to white blood cells with a segmented nucleus (e.g., neutrophil).

**poikilocytosis** The presence or variation in the shape of red blood cells observed on a blood smear.

**polychromatophilia** Erythrocytes that stain various shades of blue or gray with tinges of pink, which indicates the presence of immature red blood cells.

**polycythemia vera** A myeloproliferation disorder characterized by a neoplastic production of erythrocytic, granulocytic, megakaryocytic, and fibrocytic cell lines. The most obvious symptom is overproliferation of erythrocytes.

**porphyrias** Disorders associated with inborn errors of porphyrin metabolism.

**porphyrin** The ring structure present in hemoglobin, myoglobin, cytochrome, chlorophyll, etc.

**precipitin** Antibody that causes precipitation of its soluble antigen.

**prekallikrein** Fletcher factor; a plasma coagulation factor.

**prostaglandins** A group of fatty acids having numerous biological functions; produced by cells in response to a variety of external stimuli, such as tissue damage. Prostaglandins function in vasodilation and inflammation responses.

**protein A** A protein produced by the bacterium, *Staphylococcus aureus,* which interferes with the Fc fraction of the antibody molecule, preventing binding and phagocytosis by neutrophils.

**prothrombin** A glycoprotein in plasma that is converted to thrombin during blood clotting.

**PT test** The prothrombin test, which measures coagulation time. A prolonged PT time indicates deficiency or inhibition of one or more of the coagulation factors: factors VII, X, V, or II.

**purpura** Presence of purple patches in the skin.

**radiation therapy** The treatment of cancer and other diseases by ionizing radiation, which includes high-energy X rays, gamma rays, or implantation of isotopes.

**radiotherapy** *See* radiation therapy.

**RBC** Red blood cell; erythrocyte.

**reactive lymphocyte proliferation** A proliferation of atypical lymphocytes; seen frequently in infectious mononucleosis and other viral infections.

**red pulp** The area of the spleen, consisting of cords and sinuses, which gives the spleen its purplish appearance.

**red marrow** Active bone marrow; hematopoietic tissue.

**Reed-Sternberg cell** A cell type present in the lymph nodes of persons with Hodgkin's disease.

**Rees and Ecker method** A procedure used to enumerate platelets. Platelets are stained and counted visually with a hemocytometer.

**rouleau formation** An aggregation of red blood cells stacked like a pile of coins. This is often encountered as an artifact of smear preparation; however it may indicate the presence of increased plasma globulin and multiple myeloma.

**Schilling test** A procedure used to determine the serum level of cobalamin. Low levels indicate lack of intrinsic factor and vitamin $B_{12}$ absorption.

**schistocyte** Fragmented red blood cells commonly seen in hemolytic anemias, severe burns, and DIC.

**SCID** Severe combined immunodeficiency disease, an inherited disorder that involves the inability of the early stem cells to develop along the lymphocytic pathways; the absence, decrease, or malfunctioning of B and T lymphocytes.

**scurvy** Disorder caused by deficiency of vitamin C associated with the development of petechiae and ecchymosis.

**secretory immunoglobulin (sIg)** A type of IgA with antiviral properties; present in nonvascular fluids such as saliva, bile, synovial fluid, and intestinal and respiratory tract secretions.

**serine esterase** A proteolytic enzyme that has a specificity for the amino acid, serine.

**serotonin** A vasoconstrictor found in serum and in platelets, released during aggregation of platelets.

**shift neutropenia** A shift of neutrophilic cells from the circulating pool to the marginating pool. This condition may be caused by cirrhosis of the liver, macroglobulinemia, multiple myeloma, idiopathic hemolytic anemia, etc.

**shift platelets** The release of large premature platelets caused by increased need in cases of bleeding.

**sickle cell anemia** An inherited homozygous hemolytic anemia found predominantly in central African origin. If both parents are heterozygous (carriers), there will be a 25% chance of producing a child with sickle cell anemia.

**sickle cells** Erythrocytes shaped like a crescent or a sickle, resulting from deoxygenation and semisolid gelation of hemoglobin.

**sideroblastic anemias** A group of heterogeneous disorders, inherited or acquired, associated with hypochromia as a result of defective heme synthesis.

**siderocyte** A red blood cell containing ferritin (nonhemoglobin iron).

**specific immunity** Resistance specifically directed against a particular microorganism or its products.

**spherocyte** A spherical-shaped red blood cell lacking central pallor; may be seen in patients with hereditary spherocytic anemia.

**splenomegaly** A condition characterized by an enlarged spleen.

**stagnant loop syndrome** A cobalamin deficiency caused by bacterial overgrowth taking up vitamin $B_{12}$, thus preventing intestinal absorption of vitamin $B_{12}$.

**suppressor T lymphocyte** A cell type that inhibits immune responses, especially B-lymphocyte activity. Also called suppressor T cell ($T_s$ cell).

**systemic lupus erythematosus (SLE)** An autoimmune disease.

**$T_c$ cell** *See* cytotoxic T lymphocyte (CTL).

**$T_{DTH}$ cells** Delayed-type hypersensitivity (DTH) lymphocyte. A type of cytotoxic T lymphocyte (CTL) that secretes a series of lymphokines in chemotaxis, stimulation, development, and differentiation of monocytes and neutrophils.

**$T_H$ cell** *See* helper T lymphocyte (HTL).

**T lymphocytes** Cell types that originate in the bone marrow, differentiate in the thymus, and finally migrate to the lymph nodes, at which point they are available for cell-mediated immunity. T lymphocytes are classified as T-helper, T-suppressor, T-cytotoxic, and natural killer (NK) cells. Also known as T cells.

**Tangier disease** An inherited disorder in which the patient is unable to produce high-density lipoprotein (HDL), which results in an accumulation of cholesterol esters in macrophages.

**target cells** Abnormal form of erythrocytes in which there is a spot or disc of hemoglobin in center of the cell surrounded by a clear area. Target cells are commonly found in thalassemia and sickle cell anemia.

**T-cell receptors (TCRs)** The membrane receptors, such as CD4 and CD8, which differentiate the T-lymphocyte types: T-helper, $CD^{4+8+}$; T-cytotoxic, $CD^{4-8+}$; and natural killer, $CD^{4-8-}$ cells.

**thalassemia** Inherited group of anemias resulting from abnormal hemoglobin synthesis. Clinical manifestations of thalassemia include reduced MCV and MCHC, ineffective hematopoiesis, and accelerated hemolysis of red blood cells. Thalassemias are grouped into two categories, alpha- and beta-thalassemia, according to the globin chain that is deficient.

**thrombasthenia** An inherited disorder characterized by defective clot retraction and impaired ADP-induced platelet aggregation.

**thrombocyte** Also referred to as a platelet; a fragile, non-nucleated cell that functions in hemostasis.

**thrombocytopathia** Any qualitative problem associated with platelet malfunction.

**thrombophlebitis** The formation of blood clots in a blood vessel, most of which occur in leg veins. The blood clots cause inflammation, edema, discoloration, and ulceration.

**thrombopoietin** The growth factor that affects the rate of megakaryocyte cytoplasmic maturation and platelet release.

**thromboxane** A product from activated platelets that acts to constrict smooth muscles of the arteries.

**TIBC** Total iron-binding capacity.

**tissue factor** Tissue thromboplastin (factor III), released upon cell injury; activates the extrinsic pathway of blood coagulation.

**tissue necrosis factor (TNF)** A cytokine produced by macrophages that has a toxic effect on tumor cells but not on normal cells.

**transferrin** A beta-globulin that binds and transports iron from the intestinal tract to sites of hemoglobin synthesis.

**$T_s$ cell** *See* suppressor T lymphocyte.

**urobilinogen** A colorless product of bilirubin formed by the action of intestinal microorganisms. Increased levels occur in hepatitis and hemolytic anemias.

**urticaria** Hives.

**vasculitis** Inflammation of blood vessels.

**venography** A laboratory test for venous occlusions (blood clots).

**von Willebrand's disease** An inherited disorder caused by the absence of the coagulation factor, factor VIII:vW.

**Waldenström's macroglobulinemia** A form of plasma cell leukemia characterized by high levels of monoclonal IgM, the high-molecular-weight immunoglobulin.

**WBC** White blood cell or leukocyte.

**white pulp** The area of the spleen consisting of mainly lymphoid tissue made up of follicles and germinal centers.

**Wiskott-Aldrich syndrome** An inherited combined immunodeficiency disease characterized by lack of platelets, eczema, and a progressive deteriorization of the immune system.

**Wolman disease** An inherited disorder associated with a deficiency of acid esterase and an accumulation of triglycerides and cholesterol in monocytes and macrophages.

**yellow marrow** Inactive bone marrow composed of fatty tissue; seen in long bones of adults.

**yolk sac** A sac of embryonic tissue; site of primitive blood cells giving rise to immature erythrocytes.

# CREDITS

## PHOTOGRAPHS

### Chapter One

**Figure 1.1, 1.4, 1.5:** The Bettmann Archives; **1.6:** North Wind Picture Archives; **1.7:** The Bettmann Archives; **1.8:** North Wind Picture Archives; **1.10, 1.12, 1.13, 1.14, 1.15:** The Bettmann Archives; **1.16:** North Wind Picture Archives; **1.17, 1.18:** The Bettmann Archives

### Chapter Four

**Figure 4.6:** © David M. Phillips/Visuals Unlimited

### Chapter Ten

**Figure 10.2:** © SPL/Custom Medical Stock Photos

### Chapter Eleven

**Figure 11.5:** From A. V. Hoffbrand and J. E. Pettit, *Essential Haematology,* 2nd ed. © Blackwell Scientific Publications Limited

### Chapter Twelve

**Figure 12.8:** From A. V. Hoffbrand and J. E. Pettit, *Essential Haematology,* 2nd ed. © Blackwell Scientific Publications Limited

### Chapter Fourteen

**Figure 14.11:** © Omikron/Photo Researchers, Inc.

### Chapter Fifteen

**Figure 15.10:** From The American Society of Hematology Slide Bank, 3rd edition 1990, used with permission

### Chapter Eighteen

**Figure 18.6:** From Martin Cline, *The White Cell,* 1st ed. Harvard University Press

### Chapter Nineteen

**Figure 19.5:** ©Don Fawcett/Visuals Unlimited

### Chapter Twenty-One

**Figure 21.10A,B:** Courtesy of Dr. J. Tranum-Jensen, University of Copenhagen

### Chapter Twenty-Two

**Figure 22.2:** From J. B. Miale, *Laboratory Medicine: Hematology,* 6th ed. © 1982 Mosby-Year Book, Inc.; **22.3:** From The American Society of Hematology Slide Bank, 3rd edition 1990, used with permission

### Chapter Twenty-Five

**Figure 25.1:** From S. B. McKenzie, *Textbook of Hematology,* 1st ed. © 1988 Lea & Febiger; **25.3(left):** From J. B. Miale, *Laboratory Medicine: Hematology,* 6 ed. © 1982 Mosby-Year Book, Inc.; **25.3(middle):** From J. D. Bauer, *Clinical Laboratory Methods,* 9th ed. The Mosby-Year Book, Inc.; **25.3(right):** From J. D. Bauer, *Clinical Laboratory Methods,* 9th ed. The Mosby-Year Book, Inc.; **25.4A,B, 25.5A,B:** From J. B. Miale, *Laboratory Medicine: Hematology,* 6th ed. © 1982 Mosby-Year Book, Inc.; **25.6:** From J. D. Bauer, *Clinical Laboratory Methods,* 9th ed. The Mosby-Year Book, Inc.

### Chapter Twenty-Six

**Figure 26.2, 26.4A,B:** From The American Society of Hematology Slide Bank, 3rd edition 1990, used with permission; **26.5:** Armed Forces Institute of Pathology; **26.6A,B:** From R. L. Edelson, et al "Morphologic and Functional Properties of the Atypical Lymphocytes of the Sezary Syndrome," *Mayo Clinic Proceedings* 49:558–566, Aug. 1974; **26.8, 26.9:** From The American Society of Hematology Slide Bank, 3rd edition 1990, used with permission

### Chapter Twenty-Seven

**Figure 27.6A:** © CDC/Science Source/Photo Researchers, Inc.

### Appendix A

**A.17(both):** From Charles Seiverd, *Hematology for Medical Technologists,* 5th ed. © 1983 Lea & Febiger; **A.21:** From The American Society of Hematology Slide Bank, 3rd edition 1990, used with permission

## TEXT/LINE ART

### Chapter Two

**2.3:** From John W. Hole, Jr., *Human Anatomy and Physiology,* 6th ed. Copyright © 1993 Wm. C. Brown Communications, Inc., Dubuque, Iowa. All Rights Reserved. Reprinted by permission; **2.4:** From Stuart Ira Fox, *Human Physiology,* 4th ed. Copyright © 1993 Wm. C. Brown Communications, Inc., Dubuque, Iowa. All Rights Reserved. Reprinted by permission.

### Chapter Three

**3.1:** From Bong Hak Hyun, et al., *Practical Hematology.* Copyright © 1975 W. B. Saunders Publishing Company, Philadelphia, PA. Reprinted by permission; **3.2:** From Bong Hak Hyun, et al., *Practical Hematology.* Copyright © 1975 W. B. Saunders Publishing Company, Philadelphia, PA. Reprinted by permission; **3.7:** Reprinted by permission of the publishers from *The White Cell* by M. J. Cline, Cambridge, Mass.: Harvard University Press, Copyright © 1975 by the President and Fellows of Harvard College.

### Chapter Four

**4.1:** From Bong Hak Hyun, et al., *Practical Hematology.* Copyright © 1975 W. B. Saunders Publishing Company, Philadelphia, PA. Reprinted by permission; **4.10:** From Joan Creager, *Human Anatomy and Physiology,* 2d ed. Copyright © 1992 Joan Creager. Reprinted by permission of Wm. C. Brown Communications, Inc., Dubuque, Iowa. All Rights Reserved; **4.13:** From S. Lux and S. B. Shohet, "The Erythrocyte Membrane Skeleton: Biochemistry" in *Hospital Practice,* Volume 21, issue 10, page 82. Illustration by Alan D. Ielin.

### Chapter Five

**5.1:** From Joan Creager, *Human Anatomy and Physiology,* 2d ed. Copyright © 1992 Joan Creager. Reprinted by permission of Wm. C. Brown Communications, Inc., Dubuque, Iowa. All Rights Reserved; **5.6:** From Joan Creager, *Human Anatomy and Physiology,* 2d ed. Copyright © 1992 Joan Creager. Reprinted by permission of Wm. C. Brown Communications, Inc., Dubuque, Iowa. All Rights Reserved; **5.13:** Reprinted with permission from *Molecular Aspects of Medicine,* 1(2):156, J. M. White, "Haemoglobin Structure and Function," 1977, Elsevier Science Ltd., Pergamon Imprint, Oxford, England; **5.14:** Reprinted with permission from *Molecular Aspects of Medicine,* 1(2):156, J. M. White, "Haemoglobin Structure and Function," 1977, Elsevier Science Ltd., Pergamon Imprint, Oxford, England.

### Chapter Six

**6.1:** From David Goodsell, "A Look Inside the Living Cell," in *American Scientist,* September/October 1992, page 462. Reprinted by permission of the author; **6.2:** From Stuart Ira Fox, *Human Physiology,* 4th ed. Copyright © 1993 Wm. C. Brown Communications, Inc., Dubuque, Iowa. All Rights Reserved. Reprinted by permission; **6.3:** From Stuart Ira Fox, *Human Physiology,* 4th ed. Copyright © 1993 Wm. C. Brown Communications, Inc., Dubuque, Iowa. All Rights Reserved. Reprinted by permission; **6.4:** From Dr. M. F. Perutz, Cambridge, England; **6.6:** From Stuart Ira Fox, *Human Physiology,* 4th ed. Copyright © 1993 Wm. C. Brown Communications, Inc., Dubuque, Iowa. All Rights Reserved. Reprinted by permission; **6.7:** From Stuart Ira Fox, *Human Physiology,* 3d ed. Copyright © 1990 Wm. C. Brown Communications, Inc., Dubuque, Iowa. All Rights Reserved. Reprinted by permission; **6.8:** From Joan Creager, *Human Anatomy and Physiology,* 2d ed. Copyright © 1992 Joan Creager. Reprinted by permission of Wm. C. Brown Communications, Inc., Dubuque, Iowa. All Rights Reserved; **6.9:** From Stuart Ira Fox, *Human Physiology,* 4th ed. Copyright © 1993 Wm. C. Brown Communications, Inc., Dubuque, Iowa. All Rights Reserved. Reprinted by permission; **6.10:** From Stuart Ira Fox, *Human Physiology,* 4th ed. Copyright © 1993 Wm. C. Brown Communications, Inc., Dubuque, Iowa. All Rights Reserved. Reprinted by permission.

### Chapter Seven

**7.2:** From Richard A. Rifkind, et al., *Fundamentals of Hematology,* 2d ed. Copyright © 1980 C. V. Mosby Company Publishers, St. Louis, MO. Reprinted by permission.

### Chapter Ten

**10.1:** From Bong Hak Hyun, et al., *Practical Hematology.* Copyright © 1975 W. B. Saunders Publishing Company, Philadelphia, PA. Reprinted by permission; **10.9:** From Richard A. Rifkind, et al., *Fundamentals of Hematology,* 2d ed. Copyright © 1980 C. V. Mosby Company Publishers, St. Louis, MO. Reprinted by permission.

### Chapter Eleven

**11.1:** Reprinted by permission of *The New England Journal of Medicine,* R. R. Crichton, 284:1413, 1971. Copyright © 1971 Massachusetts Medical Society, Waltham, MA.; **11.4:** Figure from *Human Design: Molecular, Cellular, and Systematic Physiology* by William S. Beck, copyright © 1971 Harcourt Brace & Company, reproduced by permission of the publisher.

### Chapter Twelve

**12.7:** Reprinted with permission. From Arthur Bank, "Genetic Disorders of Hemoglobin Synthesis," in *Hospital Practice,* September 15, 1985, page 115. Copyright © 1985 The Maclean Hunter Medical Communications, Group, Inc., New York, NY.

### Chapter Fourteen

**14.8:** Reprinted with permission from *Molecular Aspects of Medicine,* 1(2):156, J. M. White, "Haemoglobin Structure and Function," 1977, Elsevier Science Ltd., Pergamon Imprint, Oxford, England.

### Chapter Fifteen

**15.2:** From John W. Hole, Jr., *Human Anatomy and Physiology,* 6th ed. Copyright © 1993 Wm. C. Brown Communications, Inc., Dubuque, Iowa. All Rights Reserved. Reprinted by permission; **15.4:** From Roitt, et al., *Immunology,* 2d ed. Copyright © Mosby-Wolfe, London, England. Reprinted by permission; **15.5:** From Roitt, et al., *Immunology,* 2d ed. Copyright © Mosby-Wolfe, London, England. Reprinted by permission.

### Chapter Sixteen

**16.1:** From Bong Hak Hyun, et al., *Practical Hematology.* Copyright © 1975 W. B. Saunders Publishing Company, Philadelphia, PA. Reprinted by permission; **16.6:** Reprinted by permission of the publishers from *The White Cell* by M. J. Cline, Cambridge, Mass.: Harvard University Press, Copyright © 1975 by the President and Fellows of Harvard College; **16.7:** Reprinted by permission of the publishers from *The White Cell* by M. J. Cline, Cambridge, Mass.: Harvard University Press, Copyright © 1975 by the President and Fellows of Harvard College.

### Chapter Seventeen

**17.1:** From Stuart Ira Fox, *Human Physiology,* 4th ed. Copyright © 1993 Wm. C. Brown Communications, Inc., Dubuque, Iowa. All Rights Reserved. Reprinted by permission; **17.5:** From Stuart Ira

Fox, *Human Physiology,* 4th ed. Copyright © 1993 Wm. C. Brown Communications, Inc., Dubuque, Iowa. All Rights Reserved. Reprinted by permission.

## Chapter Eighteen

**18.3:** Reprinted by permission of the publishers from *The White Cell* by M. J. Cline, Cambridge, Mass.: Harvard University Press, Copyright © 1975 by the President and Fellows of Harvard College; **18.7:** From M. Bessis, 1977, Blood Smears Reinterpreted Springer International and Kristic, R. V. 1978, Die Gewebe des Manschen und der Saugetiere, Springer Verlag. Reprinted by permission of Springer-Verlag, Heidelberg, Germany.

## Chapter Nineteen

**19.6:** From J. A. Bellanti, *Immunology II.* Copyright © 1978 W. B. Saunders Publishing Company, Philadelphia, PA; **19.7:** From Abul K. Abbas, et al., *Cellular and Molecular Immunology.* Copyright © 1991 W. B. Saunders Publishing Company, Philadelphia, PA. Reprinted by permission. **19.8:** From Abul K. Abbas, et al., *Cellular and Molecular Immunology.* Copyright © 1991 W. B. Saunders Publishing Company, Philadelphia, PA. Reprinted by permission. **19.13:** From T. P. Stossel, "How Cells Crawl" in *American Scientist,* 78, 1990. Reprinted by permission of *American Scientist,* journal of Sigma Xi, The Scientific Research Society; **19.14:** From T. P. Stossel, "How Cells Crawl" in *American Scientist,* 78, 1990. Reprinted by permission of *American Scientist,* journal of Sigma Xi, The Scientific Research Society; **19.15:** From T. P. Stossel, "How Cells Crawl" in *American Scientist,* 78, 1990. Reprinted by permission of *American Scientist,* journal of Sigma Xi, The Scientific Research Society; **19.16:** From T. P. Stossel, "How Cells Crawl" in *American Scientist,* 78, 1990. Reprinted by permission of *American Scientist,* journal of Sigma Xi, The Scientific Research Society.

## Chapter Twenty

**20.3:** From J. A. Bellanti, *Immunology II.* Copyright © 1978 W. B. Saunders Publishing Company, Philadelphia, PA.; **20.7:** From *Immunology* by Janis Kuby. Copyright © 1992 by W. H. Freeman and Company. Used with permission.; **20.9:** From J. A. Bellanti, *Immunology II.* Copyright © 1978 W. B. Saunders Publishing Company, Philadelphia, PA; **20.17:** From J. A. Bellanti, *Immunology II.* Copyright © 1978 W. B. Saunders Publishing Company, Philadelphia, PA.; **20.20:** From *Immunology* by Janis Kuby. Copyright © 1992 by W. H. Freeman and Company. Used with Permission.

## Chapter Twenty-One

**21.1:** From Roitt, et al., *Immunology,* 2d ed. Copyright © Mosby-Wolfe, London, England. Reprinted by permission; **21.7:** From Robert M. Coleman, et al., *Fundamental Immunology,* 2d ed. Copyright © 1992 Wm. C. Brown Communications, Inc., Dubuque, Iowa. All Rights Reserved. Reprinted by permission; **21.8:** From Robert M. Coleman, et al., *Fundamental Immunology,* 2d ed. Copyright © 1992 Wm. C. Brown Communications, Inc., Dubuque, Iowa. All Rights Reserved. Reprinted by permission; **21.9:** From Robert M. Coleman, et al., *Fundamental Immunology,* 2d ed. Copyright © 1992 Wm. C. Brown Communications, Inc., Dubuque, Iowa. All Rights Reserved. Reprinted by permission; **21.17:** From Robert M. Coleman, et al., *Fundamental Immunology,* 2d ed. Copyright © 1992 Wm. C. Brown Communications, Inc., Dubuque, Iowa. All Rights Reserved. Reprinted by permission; **21.18:** From Robert M. Coleman, et al., *Fundamental Immunology,* 2d ed. Copyright © 1992 Wm. C. Brown Communications, Inc., Dubuque, Iowa. All Rights Reserved. Reprinted by permission.

## Chapter Twenty-Two

**22.1:** From Roitt, et al., *Immunology,* 2d ed. Copyright © Mosby-Wolfe, London, England. Reprinted by permission.

## Chapter Twenty-Seven

**27.1:** From Abul K. Abbas, et al., *Cellular and Molecular Immunology.* Copyright © 1991 W. B. Saunders Publishing Company, Philadelphia, PA. Reprinted by permission.; **27.10:** From Abul K. Abbas, et al., *Cellular and Molecular Immunology.* Copyright © 1991 W. B. Saunders Publishing Company, Philadelphia, PA. Reprinted by permission.

## Chapter Twenty-Eight

**28.5:** From A. R. Thompson and L. A. Harker, *Manual of Emostasis and Thrombosis,* 3d ed. Copyright © 1982 F. A. Davis Company, Philadelphia, PA. Reprinted by permission.

## Chapter Twenty-Nine

**29.1:** Jack Hirsch and Elizabeth Brain, *Hemostasis and Thrombosis,* 2d ed., Churchill Livingstone, New York, NY, 1983. Reprinted by permission; **29.3:** From Shirlyn B. McKenzie, *Textbook of Hematology.* Copyright © 1988 Lea & Febiger Publishing Company, Malvern, PA. Reprinted by permission of the publisher and author; **29.6:** Adapted from Schiffman, S.: Factor XII. In *CRC Handbook series in Clinical Factor Science.* Edited by D. Seligson. *Section I: Hematology Vol. III.* Edited by R. M. Schmidt, Boca Raton, CRC Press, Inc. 1980. Reprinted by permission of CRC Press, Boca Raton, Florida; **29.7:** Adapted from Griffin, J. W., and Bouma, B. N. *Blood coagulation Factor XI. In CRC Handbook Series in Clinical Laboratory Science.* Edited by D. Seligson. *Section I: Hematology Vol. III.* Edited by R. M. Schmidt, Boca Raton, CRC Press, 1980. Reprinted by permission of CRC Press, Boca Raton, Florida; **29.8:** Adapted from Marlar, R. A. *Factor IX: Activation and function. In Prolthrombin and Other Vitamin K Proteins.* Edited by W. H. Seegers and D. A. Walsy, Boca Raton, CRC Press, 1986. Reprinted by permission of CRC Press, Boca Raton, Florida; **29.11:** From Shirlyn B. McKenzie, *Textbook of Hematology.* Copyright © 1988 Lea & Febiger Publishing Company, Malvern, PA. Reprinted by permission of the publisher and author; **29.14:** From Shirlyn B. McKenzie, *Textbook of Hematology.* Copyright © 1988 Lea & Febiger Publishing Company, Malvern, PA. Reprinted by permission of the publisher and author; **29.15:** From Shirlyn B. McKenzie, *Textbook of Hematology.* Copyright © 1988 Lea & Febiger Publishing Company, Malvern, PA. Reprinted by permission of the publisher and author.

## Chapter Thirty

**30.2:** Jack Hirsch and Elizabeth Brain, *Hemostasis and Thrombosis,* 2d ed., Churchill Livingstone, New York, NY, 1983. Reprinted by permission; **30.3:** Jack Hirsch and Elizabeth Brain, *Hemostasis and Thrombosis,* 2d ed., Churchill Livingstone, New York, NY, 1983. Reprinted by permission; **30.4:** Jack Hirsch and Elizabeth Brain, *Hemostasis and Thrombosis,* 2d ed., Churchill Livingstone, New York, NY, 1983. Reprinted by permission; **30.5:** Jack Hirsch and Elizabeth Brain, *Hemostasis and Thrombosis,* 2d ed., Churchill Livingstone, New York, NY, 1983. Reprinted by permission.

## Appendix A

**A.5:** From Jean Jorgenson Linne and Karen Munson Ringsrud, *Basic Techniques in Clinical Laboratory Science,* 3d ed. Copyright © 1992 C. V. Mosby Company Publishers, St. Louis, MO. Reprinted by permission; **A.11:** From Jean Jorgenson Linne and Karen Munson Ringsrud, *Basic Techniques in Clinical Laboratory Science,* 3d ed. Copyright © 1992 C. V. Mosby Company Publishers, St. Louis, MO. Reprinted by permission; **A.13:** From Jean Jorgenson Linne and Karen Munson Ringsrud, *Basic Techniques in Clinical Laboratory Science,* 3d ed. Copyright © 1992 C. V. Mosby Company Publishers, St. Louis, MO. Reprinted by permission.

## Illustrators

### Publications Services, Inc.

1.2, 1.3, 1.9, 1.11, 2.1, 2.2, 2.5, 3.1, 3.2, 3.3, 3.4, 3.5, 3.6, 3.7, 3.8, 4.1, 4.2, 4.3, 4.4, 4.7, 4.8, 4.11, 4.12, 4.13, 4.14, 5.2, 5.3, 5.7, 5.8, 5.9, 5.10, 5.11, 5.12, 5.13, 5.14, 6.1, 6.3, 6.4, 6.5, 7.1, 7.2, 7.3, 7.4, 7.5, 7.6, 8.2, 10.1, 10.3, 10.4, 10.5, 10.6, 10.7, 10.8, 10.9, 11.1, 11.2, 11.3, 11.4, 12.1, 12.2, 12.3, 12.4, 12.5, 12.6, 12.7, 13.1, 13.2, 14.1, 14.2, 14.4, 14.5, 14.6, 14.7, 14.8, 14.9A. 14.9B, 14.10, 14.12, 15.1, 15.3, 15.4, 15.5, 15.6, 15.7, 15.8, 15.9, 16.1, 16.2, 16.3, 16.4, 16.5, 16.6, 16.7, 16.8, 17.1, 17.2, 17.3, 17.4, 17.6, 17.8, 17.9, 18.1, 18.2, 18.3, 18.4, 18.5, 18.7, 18.8, 18.9, 18.10, 19.1, 19.2, 19.3, 19.6, 19.7, 19.8, 19.9, 19.10, 19.11, 19.12, 19.13, 19.14, 19.15, 19.16, 20.3, 20.4, 20.6, 20.7, 20.9, 20.10, 20.12, 20.13, 20.14, 20.15, 20.16, 20.17, 20.18, 20.19, 20.20, 20.21, 20.22, 20.23, 20.24, 20.25, 20.26, 21.1, 21.2, 21.3, 21.4, 21.5, 21.6, 21.7, 21.8, 21.9, 21.16, 21.17, 21.18, 22.1, 24.1, 24.2, 24.3, 24.4, 24.5, 24.6, 25.2, 26.1, 26.7, 27.1, 27.2, 27.3, 27.4, 27.5, 27.6B, 27.9, 27.10, 28.1, 28.2, 28.3, 28.4, 28.5, 28.6, 28.7, 28.8, 29.1, 29.2, 29.3, 29.4, 29.5, 29.6, 29.7, 29.8, 29.9, 29.10, 29.11, 29.12, 29.13, 29.14, 29.15, 29.16, 29.17, 30.1, 30.2, 30.3, 30.4, 30.5, 30.6, 30.7, 30.8, 30.9, 30.10, 30.11, 30.12, 30.13, 30.14, 30.15, 30.16, 30.17, 31.1, 31.2, 31.3, 31.4, 31.5, A.1, A.2, A.3, A.4, A.5, A.6, A.7, A.8, A.9, A.10, A.11, A.12, A.13, A.14, A.15, A.16, A.18, A.19, A.20, A.22.

### Sam Collins

17.5

### Carlyn Iverson

4.10

### Fineline

6.6, 6.7, 6.8, 15.21, 15.22

### Illustrious, Inc.

5.6, 6.2

### Nancy Marshburn

2.3

### Rob Gordon

2.4

# INDEX

## A

# D

# H

# I

## J



## K



## L

# M

# R



# S

# T

# U



# V



# W

## Y



## Z